Fluid Mechanics

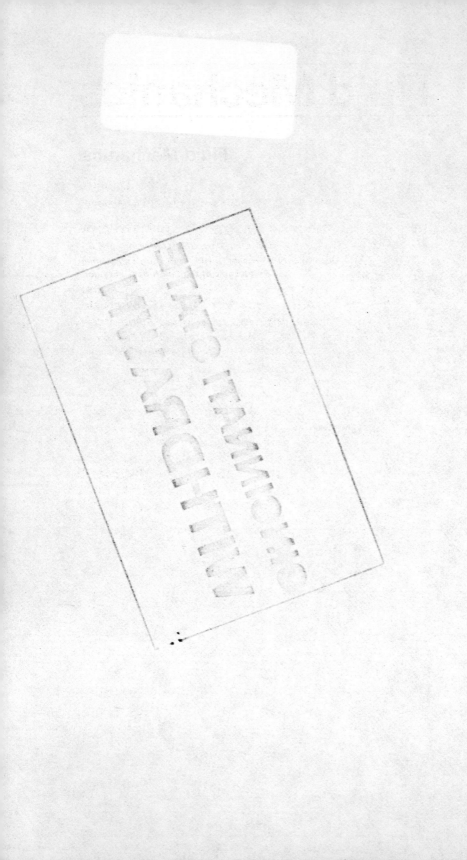

Fluid Mechanics

J F Douglas
MSc (Eng), PhD, ACGI, DIC, CEng, MICE, MIMechE

J M Gasiorek
BSc (Eng), PhD, CEng, MIMechE, MCIBS
Department of Mechanical and Production Engineering
Polytechnic of the South Bank, London

and

J A Swaffield
BSc, MPhil, PhD, CEng, MRAeS, FIPHE, MCIBS
Department of Mechanical Engineering
Brunel University, Middlesex

Second edition

Pitman

PITMAN PUBLISHING LIMITED
128 Long Acre, London, WC2E 9AN
PITMAN PUBLISHING INC
1020 Plain Street, Marshfield, Massachusetts 02050

Associated Companies
Pitman Publishing Pty Ltd, Melbourne
Pitman Publishing New Zealand Ltd, Wellington
Copp Clark Pitman, Toronto

©J. F. Douglas, J. M. Gasiorek and J. A. Swaffield, 1979, 1985

First edition 1979
Second edition 1985

Library of Congress Cataloging in Publication Data
Douglas, John F.
Fluid mechanics.

Includes bibliographies and index.
1. Fluid mechanics. I. Gasiorek, J. M. (Janusz Maria),
1927–   II. Swaffield, J. A., 1943–   III. Title.
TA357.D68 1985b   620.1'06   84-25493
ISBN 0-273-02134-6

British Library Cataloguing in Publication Data
Douglas, J. F.
    Fluid mechanics. – 2nd ed.
    1. Fluid mechanics
    I. Title   II. Gasiorek, J. M.   III. Swaffield, J. A.
    532   QC145.Z
    ISBN 0-273-02134-6

Printed in Great Britain at The Bath Press, Avon

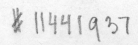

Contents

Preface to Second Edition

In the preparation of this second edition we have retained the aims of the
original text, namely to provide a broad-based treatment of the essentials of
Fluid Mechanics, while at the same time demonstrating the application of the
subject, particularly to the study and solution of higher level problems in
selected areas. In retaining this 'applications' approach we are both aware and
pleased that this technique currently features in the UK Engineering Council
statements on the training, education and 'formation' of engineers,
strengthening our view that this is one of the most efficient and relevant
methods of helping students in general to understand our subject. We believe
that such an approach should also include the use of improved computer-
based numerical solutions as these will become part of the engineer's everyday activities.

In the five years since the first edition was published there has been a
significant change in the availability of and access to micro and other computers for both the student and the practising engineer. Computers and
programs are of course not ends in themselves but rather they are powerful
tools that we can utilize to dispense with many tedious and repetitive calculations, thereby allowing the study, within an educational framework, of
problems of greater complexity and relevance, including time-dependent
phenomena that were previously beyond our capability without recourse to
simplifying assumptions. This second edition therefore includes a series of
computer programs chosen to illustrate these aspects of computer application
and to be of direct use to both student and practising engineer alike. While
the programs have been written in BBC Basic they may be transferred with
little difficulty to Apple, Commodore or Sinclair machines. A program
cassette tape will also be available to support the text.

None of this of course removes the necessity to provide a thorough basis
for the subject and this remains one of the text's main objectives. We have
included new material in areas that have been found particularly interesting
by our readers, as well as updating and refining the existing text. The treatment of incompressible flows around a body has been extended to include

the study of wakes, while the coverage of fluid machinery has been strengthened by the inclusion of a major new chapter on positive displacement machines. The existing treatment of unsteady flow has been extended to allow the application of numerical modelling techniques to unsteady open channel or partially filled pipe flows. Taken together with the introduction of computing methods we view these additions as supporting and reaffirming the aims and objectives of the original text.

Once again we would like to thank all our colleagues in many universities and polytechnics in the UK and overseas who have encouraged us by their positive response to and constructive comments on the first edition. All have helped us to formulate this new edition which we hope will fulfil a useful role for both the student and the practising engineer.

J.F. Douglas
J.M. Gasiorek
J.A. Swaffield

London, May 1984

Preface to First Edition

This is a textbook for all manner of engineers. Whether the reader is concerned with Civil, Mechanical or Chemical Engineering, Building Services or Environmental Engineering, the principles of fluid mechanics remain the same. Drawing on our joint experience of teaching students in all these disciplines, we have tried to set out these principles simply and clearly and to illustrate their application by examples drawn from the various branches of engineering.

In the planning of this book we are indebted to our colleagues in other Colleges, Polytechnics and Universities for the opportunity to study their syllabuses and examination papers which has enabled us to cover the general requirements of the Honours Degree and Professional examinations. We have also deliberately dealt with the elementary aspects of the subject very fully and so the book will meet the requirements of those studying for the Higher National Diploma or for the Higher Diploma or Higher Certificate of the Business and Technician Education Council (B.T.E.C.).

For ease of reference the contents has been divided into Parts which are substantially self-contained and we hope that they will provide a convenient source of information for the practising engineer in his day to day activities.

J. F. Douglas
J. M. Gasiorek
J. A. Swaffield

List of symbols

a	acceleration, area	i	hydraulic gradient
A	area, constant	I	moment of inertia
b	width, breadth	k	constant, radius of gyration
B	width, breadth, constant	K	bulk modulus
c	chord length, velocity of sound	l	length
		L	lift
c_p	specific heat at constant pressure	m	mass, area ratio, doublet strength, hydraulic mean depth
c_v	specific heat at constant volume		
		M	molecular weight
C	constant	n	number of, polytropic index
C_c	coefficient of contraction	N	rotational speed
C_d	coefficient of discharge	p	pressure
C_D	coefficient of drag	P	force, power, wetted perimeter
C_f	coefficient of friction		
C_L	coefficient of lift	q	flow rate per unit width or unit depth
C_v	coefficient of velocity		
d	diameter	Q	volumetric flow rate
D	drag, diameter, depth	r	radius, radial distance
e	base of natural logarithms	R	radius, reaction force
e	error, internal energy per unit mass	R	gas constant
		s	slope, distance, arbitrary coordinate within Cartesian system, slip
E	modulus of elasticity, energy		
f	friction factor	S	surface, entropy
$f(\ )$	reflected pressure wave	t	time
F	force, stress	T	temperature, torque surface width
$F(\ )$	pressure wave		
g	gravitational acceleration	u	velocity, peripheral blade velocity
h	vertical height, depth		
h	head loss	U	internal energy, velocity
H	head, enthalpy	v	velocity

V	volume	ϵ	absolute roughness, eddy viscosity
v_f	velocity of flow		
v_r	relative velocity	ζ	vorticity
v_x	velocity component in x direction	η	efficiency
		θ	angle
v_y	velocity component in y direction	μ	coefficient of dynamic viscosity
v_z	velocity component in z direction	ν	coefficient of kinematic viscosity, Poisson's ratio
v_r	radial velocity	ρ	mass density
v_θ	tangential velocity	σ	relative density (specific gravity), surface tension
w	specific weight		
W	weight, work	τ	shear stress
x, y, z	orthogonal coordinates	ϕ	shear strain, angle
y	gas content (%)	Φ	velocity potential
Z	potential head, depth	Ψ	stream function
		ω	angular (rotational) velocity, stage variable
a	angle, angular acceleration		
β	angle		
γ	adiabatic index (c_p/c_v)	Fr	Froude number
Γ	circulation	Ma	Mach number
δ	difference, increment	Re	Reynolds number
Δ	change in	Str	Strouhal number
		We	Weber number

List of Computer Programs

PART I
Elements of fluid mechanics

Fluid mechanics, as the name indicates, is that branch of applied mechanics which is concerned with the statics and dynamics of liquids and gases. The analysis of the behaviour of fluids is based upon the fundamental laws of applied mechanics which relate to the conservation of mass-energy and the force-momentum equation, together with other concepts and equations with which the student who has already studied solid-body mechanics will be familiar. There are, however, two major aspects of fluid mechanics which differ from solid-body mechanics. The first is the nature and properties of the fluid itself, which are very different from those of a solid. The second is that, instead of dealing with individual bodies or elements of known mass, we are frequently concerned with the behaviour of a continuous stream of fluid, without beginning or end.

A further problem is that it can be extremely difficult to specify either the precise movement of a stream of fluid or that of individual particles within it. It is, therefore, often necessary – for the purpose of theoretical analysis – to assume ideal, simplified conditions and patterns of flow. The results so obtained may then be modified by introducing appropriate coefficients and factors, determined experimentally, to provide a basis for the design of fluid systems. This approach has proved to be reasonably satisfactory – in so far as the theoretical analysis usually establishes the form of the relationship between the variables; the experimental investigation corrects for the factors omitted from the theoretical model and establishes a quantitative relationship.

1 Fluids and their properties

1.1. Fluids

In everyday life, we recognize three states of matter: solid, liquid and gas. Although different in many respects, liquids and gases have a common characteristic in which they differ from solids: they are fluids, lacking the ability of solids to offer permanent resistance to a deforming force. Fluids *flow* under the action of such forces, deforming continuously for as long as the force is applied. A fluid is unable to retain any unsupported shape; it flows under its own weight and takes the shape of any solid body with which it comes into contact.

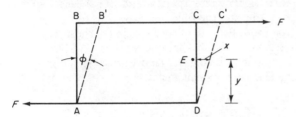

Figure 1.1 Deformation caused by shearing forces

Deformation is caused by *shearing* forces, i.e. forces such as F (Fig. 1.1), which act tangentially to the surfaces to which they are applied and cause the material originally occupying the space ABCD to deform to AB'C'D. This leads to the definition:

A fluid is a substance which deforms continuously under the action of shearing forces, however small they may be.

Conversely, it follows that

If a fluid is at rest, there can be no shearing forces acting and, therefore, all forces in the fluid must be perpendicular to the planes upon which they act.

1.2. Shear stress in a moving fluid

Although there can be no shear stresses in a fluid at rest, shear stresses *are* developed when the fluid is in motion, if the particles of the fluid move relative to each other so that they have different velocities, causing the original shape of the fluid to become distorted. If, on the other hand, the velocity of the fluid is the same at every point, no shear stresses will be produced, since the fluid particles are at rest relative to each other.

Usually, we are concerned with flow past a solid boundary. The fluid in contact with the boundary adheres to it and will, therefore, have the same velocity as the boundary. Considering successive layers parallel to the boundary (Fig. 1.2), the velocity of the fluid will vary from layer to layer as y increases.

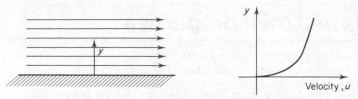

Figure 1.2   Variation of velocity with distance from a solid boundary

If ABCD (Fig. 1.1) represents an element in a fluid with thickness s perpendicular to the diagram, then the force F will act over an area A equal to BC x s. The force per unit area F/A is the *shear stress* τ and the deformation, measured by the angle ϕ (the *shear strain*), will be proportional to the shear stress. In a solid, ϕ will be a fixed quantity for a given value of τ, since a solid can resist shear stress permanently. In a fluid, the shear strain ϕ will continue to increase with time and the fluid will flow. It is found experimentally that, in a true fluid, the rate of shear strain (or shear strain per unit time) is directly proportional to the shear stress.

Suppose that in time t a particle at E (Fig. 1.1) moves through a distance x. If E is a distance y from AD then, for small angles,

Shear strain, $\phi = x/y$,

Rate of shear strain $= x/yt = (x/t)/y = u/y$,

where $u = x/t$ is the velocity of the particle at E. Assuming the experimental result that shear stress $\propto$ shear strain, then

$$\tau = \text{constant} \times u/y. \tag{1.1}$$

The term u/y is the change of velocity with y and may be written in the differential form $\mathrm{d}u/\mathrm{d}y$. The constant of proportionality is known as the *dynamic viscosity* μ of the fluid. Substituting into equation (1.1),

$$\tau = \mu \, \frac{\mathrm{d}u}{\mathrm{d}y}, \tag{1.2}$$

which is Newton's law of viscosity. The value of μ depends upon the fluid under consideration.

1.3. Differences between solids and fluids

To summarize, the differences between the behaviours of solids and fluids under an applied force are as follows.

(i) For a solid, the strain is a function of the applied stress, providing that the elastic limit is not exceeded. For a fluid, the rate of strain is proportional to the applied stress.

(ii) The strain in a solid is independent of the time over which the force is applied and, if the elastic limit is not exceeded, the deformation disappears when the force is removed. A fluid continues to flow for as long as the force is applied and will not recover its original form when the force is removed.

In most cases, substances can be classified easily as either solids or fluids. However, certain cases (e.g. pitch, glass) appear to be solids because their rate of deformation under their own weight is very small. Pitch is actually a fluid which will flow and spread out over a surface under its own weight −but it will take days to do so rather than milliseconds! Similarly, solids will flow and become plastic when subjected to forces sufficiently large to produce a stress in the material which exceeds the elastic limit. They will also 'creep' under sustained loading, so that the deformation increases with time. A plastic substance does *not* meet the definition of a true fluid, since the shear stress must exceed a certain minimum value before flow commences.

1.4. Newtonian and non-Newtonian fluids

Even among substances commonly accepted as fluids, there is wide variation in behaviour under stress. Fluids obeying Newton's law of viscosity (equation (1.2)) and for which μ has a constant value are known as *Newtonian fluids*. Most common fluids fall into this category, for which shear stress is linearly related to velocity gradient (Fig. 1.3). Fluids which do not obey Newton's

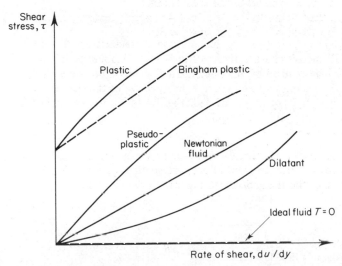

Figure 1.3 Variation of shear stress with velocity gradient

law of viscosity are known as non-Newtonian and fall into one of the following groups.

(i) *Plastic*, for which the shear stress must reach a certain minimum value before flow commences. Thereafter, shear stress increases with the rate of shear according to the relationship

$$\tau = A + B\left(\frac{du}{dy}\right)^{n},$$

where A, B and n are constants. If $n = 1$, the material is known as a Bingham plastic (e.g. sewage sludge).

(ii) *Pseudo-plastic*, for which dynamic viscosity decreases as the rate of shear increases (e.g. colloidal solutions, clay, milk, cement).

(iii) *Dilatant substances*, in which dynamic viscosity increases as the rate of shear increases (e.g. quicksand).

(iv) *Thixotropic substances*, for which the dynamic viscosity decreases with the time for which shearing forces are applied (e.g. thixotropic jelly paints).

(v) *Rheopectic materials*, for which the dynamic viscosity increases with the time for which shearing forces are applied.

(vi) *Viscoelastic materials*, which behave in a manner similar to Newtonian fluids under time-invariant conditions but, if the shear stress changes suddenly, behave as if plastic.

The above is a classification of actual fluids. In analysing some of the problems arising in fluid mechanics we shall have cause to consider the behaviour of an *ideal fluid*, which is assumed to have no viscosity. Theoretical solutions obtained for such a fluid often give valuable insight into the problems involved, and can, where necessary, be related to real conditions by experimental investigation.

1.5. Liquids and gases

Although liquids and gases both share the common characteristics of fluids, they have many distinctive characteristics of their own. A liquid is difficult to compress and, for many purposes, may be regarded as incompressible. A given mass of liquid occupies a fixed volume, irrespective of the size or shape of its container, and a free surface is formed (Fig. 1.4 (a)) if the volume of the container is greater than that of the liquid.

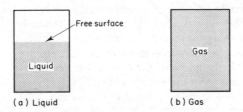

(a) Liquid   (b) Gas

Figure 1.4   Behaviour of a fluid in a container

A gas is comparatively easy to compress. Changes of volume with pressure are large, cannot normally be neglected and are related to changes of temperature. A given mass of gas has no fixed volume and will expand continuously unless restrained by a containing vessel. It will completely fill any vessel in which it is placed and, therefore, does not form a free surface (Fig. 1.4 (b)).

1.6. Molecular structure of materials

Solids, liquids and gases are all composed of molecules in continuous motion. However, the arrangement of these molecules, and the spaces between them, differ, giving rise to the characteristic properties of the three different states of matter. In solids, the molecules are densely and regularly packed and movement is slight, each molecule being restrained by its neighbours. In liquids, the structure is looser; individual molecules have greater freedom of movement and, although restrained to some degree by the surrounding molecules, can break away from this restraint, causing a change of structure. In gases, there is no formal structure, the spaces between molecules are large and the molecules can move freely.

The molecules of a substance exert forces on each other which vary with their intermolecular distance. Consider, for simplicity, a monatomic substance in which each molecule consists of a single atom. An idea of the nature of the forces acting may be formed from observing the behaviour of such a substance on a macroscopic scale.

(i) If two pieces of the same material are far apart, there is no detectable force exerted between them. Thus, the forces between molecules are negligible when widely separated and tend to zero as the separation tends towards infinity.

(ii) Two pieces of the same material can be made to weld together if they are forced into very close contact. Under these conditions, the forces between the molecules are attractive when the separation is very small.

(iii) Very large forces are required to compress solids or liquids, indicating that a repulsive force between the molecules must be overcome to reduce the spacing between them.

It appears from these observations that interatomic forces vary with the distance of separation (Fig. 1.5 (a)) and that there are two types of force, one attractive and the other repulsive. At small separations, the repulsive force is dominant; at larger separations, it becomes insignificant by comparison with the attractive force.

These conclusions can also be expressed in terms of the potential energy, defined as the energy required to bring one atom from infinity to a distance r from the second atom. The potential energy is zero if the atoms are infinitely far apart and is positive if external energy is required to move the first atom towards the second. Since Fig. 1.5(a) is the graph of the force F between the atoms vs. the distance of separation, the potential energy curve (Fig. 1.5(b)) will be the integral of this curve from ∞ to r, which is the shaded area in Fig. 1.5(a).

At r_0, there is a condition of minimum energy, corresponding to $F = 0$ and representing a position of stable equilibrium, accounting for the inherent stability of solids and liquids in which the molecules are sufficiently densely packed for this condition to exist. Figure 1.5(b) also indicates that a pair of atoms can be separated completely, so that $r = \infty$, by the application of a finite amount of energy ΔE, which is called the dissociation or binding energy.

Considering a large number of particles of a substance, each particle will have kinetic energy $\frac{1}{2}mu^2$, where m is the mass of the particle and u its velocity. If a particle collides with a pair of particles, it will only cause them to separate if it can transfer to the pair energy in excess of ΔE. Thus, the possibility of stable pairs forming will depend on the average value of $\frac{1}{2}mu^2$ in relation to ΔE.

(i) If the average value of $\frac{1}{2}mu^2 \gg \Delta E$, no stable pairs can form. The system will behave as a gas, consisting of individual particles moving rapidly with no apparent tendency to aggregate or occupy a fixed space.

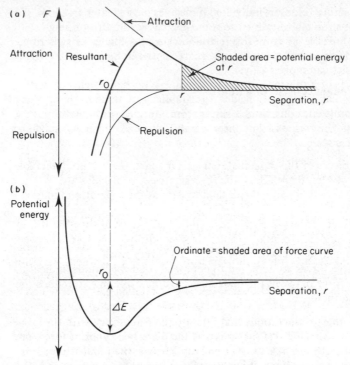

Figure 1.5   (a) Variation of force with separation. (b) Variation of potential energy
with separation.

(ii) If the average value of $\frac{1}{2}mu^2 \ll \Delta E$, no dissociation of pairs is possible
and the colliding particle may be captured by the pair. The system has the
properties of a solid, forming a stable conglomeration of particles which can
only be dissociated by supplying energy from outside (e.g. by heating to
produce melting and, subsequently, boiling).

(iii) If the average value of $\frac{1}{2}mu^2 \simeq \Delta E$, we have a system intermediate
between (i) and (ii), corresponding to the liquid state, since some particles
will have values of $\frac{1}{2}mu^2 > \Delta E$, causing dissociation, while others will have
values of $\frac{1}{2}mu^2 < \Delta E$ and will aggregate.

Summing up, in a solid, the individual molecules are closely packed and their
movement is restricted to vibrations of small amplitude. The kinetic energy is
small compared to the dissociation energy, so that the molecules do not
become separated but retain the same relative conditions.

In a liquid, the molecules are still closely packed, but their movement is
greater. Certain of the molecules will have sufficient kinetic energy to break
through the surrounding molecules, so that the relative positions of the
molecules can change from time to time. The material will cease to be rigid
and can flow under the action of applied forces. However, the attraction
between molecules is still sufficient to ensure that a given mass of liquid has a
fixed volume and that a free surface will be formed.

In a gas, the spacing between molecules is some ten times as great as in a
liquid. The kinetic energy is far greater than the dissociation energy. The

attractive forces between molecules are very weak and intermolecular effects are negligible, so that molecules are free to travel until stopped by a solid·or a liquid boundary. A gas will, therefore, expand to fill a container completely, irrespective of volume.

1.7. The continuum concept of a fluid

Although the properties of a fluid arise from its molecular structure, engineering problems are usually concerned with the bulk behaviour of fluids. The number of molecules involved is immense, and the separation between them is normally negligible by comparison with the distances involved in the practical situation being studied. Under these conditions, it is usual to consider a fluid as a continuum — a hypothetical continuous substance — and the conditions at a point as the average of a very large number of molecules surrounding that point within a distance which is large compared with the mean intermolecular distance (although very small in absolute terms). Quantities such as velocity and pressure can then be considered to be constant at any point, and changes due to molecular motion may be ignored. Variations in such quantities can also be assumed to take place smoothly, from point to point. This assumption breaks down in the case of rarefied gases, for which the ratio of the mean free path of the molecules to the physical dimensions of the problem is very much larger.

In this book, fluids will be assumed to be continuous substances and, when the behaviour of a small element or particle of fluid is studied, it will be assumed that it contains so many molecules that it can be treated as part of this continuum.

PROPERTIES OF FLUIDS

The following properties of fluids are of general importance to the study of fluid mechanics. For convenience, a fuller list of the values of these properties for common fluids is given in the Appendix, but typical values, SI units and dimensions in the MLT system (*see* Chapter 25) are given here.

1.8. Density

The density of a substance is that quantity of matter contained in unit volume of the substance. It can be expressed in three different ways, which must be clearly distinguished.

1.8.1. Mass density

Mass density ρ is defined as the mass of the substance per unit volume. As mentioned above, we are concerned, in considering this and other properties, with the substance as a continuum and not with the properties of individual molecules. The mass density at a point is determined by considering the mass δm of a very small volume δV surrounding the point. In order to preserve the concept of the continuum, δV cannot be made smaller than x^3, where x is a linear dimension which is large compared with the mean distance between molecules. The density at a point is the limiting value as δV tends to x^3:

$$\rho = \lim_{\delta V \to x^3} \frac{\delta m}{\delta V}.$$

Units: kilogrammes per cubic metre (kg m^{-3}).
Dimensions: $M L^{-3}$.
Typical values at $p = 1 \cdot 013 \times 10^5$ N m^{-2}, $T = 288 \cdot 15$ K:
water, 1000 kg m^{-3}; air, $1 \cdot 23$ kg m^{-3}.

1.8.2. Specific weight

Specific weight w is defined as the weight per unit volume. Since weight is dependent on gravitational attraction, the specific weight will vary from point to point, according to the local value of gravitational acceleration g. The relationship between w and ρ can be deduced from Newton's second law, since

Weight per unit volume = Mass per unit volume x g

$$w = \rho g.$$

Units: newtons per cubic metre (N m^{-3}).
Dimensions: $M L^{-2} T^{-2}$.
Typical values: water, $9 \cdot 81 \times 10^3$ N m^{-3}; air, $12 \cdot 07$ N m^{-3}.

1.8.3. Relative density

Relative density (or specific gravity) σ is defined as the ratio of the mass density of a substance to some standard mass density. For solids and liquids, the standard mass density chosen is the maximum density of water (which occurs at $4 \,^\circ\text{C}$ at atmospheric pressure):

$$\sigma = \rho_{\text{substance}}/\rho_{H_2 O} \text{ at } 4 \,^\circ\text{C}.$$

For gases, the standard density may be that of air or of hydrogen at a specified temperature and pressure, but the term is not used frequently.
    Units: since relative density is a ratio of two quantities of the same kind, it is a pure number having no units.
    Dimensions: as a pure number, its dimensions are $M^0 L^0 T^0 = 1$.
    Typical values: water, $1 \cdot 0$; oil, $0 \cdot 9$.

1.8.4. Specific volume

In addition to these measures of density, the quantity specific volume is sometimes used, being defined as the reciprocal of mass density, i.e. it is used to mean volume per unit mass.

1.9. Viscosity

A fluid at rest cannot resist shearing forces, and, if such forces act on a fluid which is in contact with a solid boundary (Fig. 1.2), the fluid will flow over the boundary in such a way that the particles immediately in contact with the boundary have the same velocity as the boundary, while successive layers of fluid parallel to the boundary move with increasing velocities. Shear stresses opposing the relative motion of these layers are set up, their magnitude depending on the velocity gradient from layer to layer. For fluids obeying Newton's law of viscosity, taking the direction of motion as the x direction and v_x as the velocity of the fluid in the x direction at a distance y from the boundary, the shear stress in the x direction is given by

$$\tau_x = \mu \frac{dv_x}{dy}. \tag{1.3}$$

More generally, for three-dimensional flow, in which the velocity v has components v_x, v_y and v_z along the x, y and z axes,

$$\tau_x = \mu \frac{\partial v_x}{\partial y}.$$

1.9.1. Coefficient of dynamic viscosity

The coefficient of dynamic viscosity μ can be defined as the shear force per unit area (or shear stress τ) required to drag one layer of fluid with unit velocity past another layer unit distance away from it in the fluid. Rearranging equation (1.3),

$$\mu = \tau \Big/ \frac{dv}{dy} = \frac{\text{Force}}{\text{Area}} \Big/ \frac{\text{Velocity}}{\text{Distance}} = \frac{\text{Force} \times \text{Time}}{\text{Area}} \text{ or } \frac{\text{Mass}}{\text{Length} \times \text{Time}}$$

Units: newton-seconds per square metre (N s m^{-2}) or kilogrammes per metre per second ($\text{kg m}^{-1} \text{ s}^{-1}$). (But note that the coefficient of viscosity is often measured in poise (P); $10 \text{ P} = 1 \text{ kg m}^{-1} \text{ s}^{-1}$.)
Dimensions: $M L^{-1} T^{-1}$.
Typical values: water, $1.14 \times 10^{-3} \text{ kg m}^{-1} \text{ s}^{-1}$; air, $1.78 \times 10^{-5} \text{ kg m}^{-1} \text{ s}^{-1}$.

1.9.2. Kinematic viscosity

The kinematic viscosity ν is defined as the ratio of dynamic viscosity to mass density:

$$\nu = \mu/\rho$$

Units: square metres per second ($\text{m}^2 \text{ s}^{-1}$). (But note that kinematic viscosity is often measured in stokes (St); $10^4 \text{ St} = 1 \text{ m}^2 \text{ s}^{-1}$.)
Dimensions: $L^2 T^{-1}$.
Typical values: water, $1.14 \times 10^{-6} \text{ m}^2 \text{ s}^{-1}$; air, $1.46 \times 10^{-5} \text{ m}^2 \text{ s}^{-1}$.

1.10. Causes of viscosity in gases

When a gas flows over a solid boundary, the velocity of flow in the x direction, parallel to the boundary, will change with the distance y, measured perpendicular to the boundary. In Fig. 1.6, the velocity in the x direction is v_x at a distance y from the boundary and $v_x + \delta v_x$ at a distance $y + \delta y$. As the molecules of gas are not rigidly constrained, and cohesive forces are small, there will be a continuous interchange of molecules between adjacent layers which are travelling at different velocities. Molecules moving from the slower layer will exert a drag on the faster, while those moving from the faster layer will exert an accelerating force on the slower.

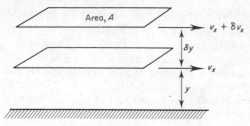

Figure 1.6

Assuming that the mass interchange per unit time is proportional to the area A under consideration, and inversely proportional to the distance δy between them,

Mass interchange per unit time $= kA/\delta y$,

where k is a constant of proportionality;

Change of velocity $= \delta v_x$;

Force exerted by one layer on the other

$= $ Rate of change of momentum

$= $ Mass interchange per unit time × Change of velocity

$$F = kA \, \frac{\delta v_x}{\delta y} \, ;$$

Viscous shear stress, $\tau = F/A = k \, \dfrac{\delta v_x}{\delta y}.$

Thus, from consideration of molecular mass interchange occurring in a gas, we arrive at Newton's law of viscosity.

If the temperature of a gas increases, the molecular interchange will increase. The viscosity of a gas will, therefore, increase as the temperature increases. According to the kinetic theory of gases, viscosity should be proportional to the square root of the absolute temperature; in practice, it increases more rapidly. Over the normal range of pressures, the viscosity of a gas is found to be independent of pressure, but it is affected by very high pressures.

1.11. Causes of viscosity in a liquid

While there will be shear stresses due to molecular interchange similar to those developed in a gas, there are substantial attractive, cohesive forces between the molecules of a liquid (which are very much closer together than those of a gas). Both molecular interchange and cohesion contribute to viscous shear stress in liquids.

The effect of increasing the temperature of a fluid is to reduce the cohesive forces while simultaneously increasing the rate of molecular interchange. The former effect tends to cause a decrease of shear stress, while the latter causes it to increase. The net result is that liquids show a reduction in viscosity with increasing temperature which is of the form

$$\mu_T = \mu_0/(1 + A_1 T + B_1 T^2) \tag{1.4}$$

where μ_T is the viscosity at T °C, μ_0 is the viscosity at 0 °C and A_1 and B_1 are constants depending upon the liquid. For water, $\mu_0 = 0.0179$ P, $A_1 = 0.033\ 68$ and $B_1 = 0.000\ 221$. When plotted, equation (1.4) gives a hyperbola, viscosity tending to zero as temperature tends to infinity. An alternative relationship is

$$\mu/\mu_0 = A_2 \exp\left[B_2(1/T' - 1/T_0)\right] \tag{1.5}$$

where A_2 and B_2 are constants and T' is the *absolute* temperature.

High pressures also affect the viscosity of a liquid. The energy required for the relative movement of the molecules is increased and, therefore, the viscosity increases with increasing pressure. The relationship depends on the nature of the liquid and is exponential, having the form

$$\mu_p = \mu_0 \exp\left\{C(p - p_0)\right\}, \tag{1.6}$$

where C is a constant for the liquid and μ_p is the viscosity at pressure p. For oils of the type used in oil hydraulic machinery, the increase in viscosity is of the order of 10 to 15 per cent for a pressure increase of 70 atm. Water, however, behaves rather differently from other fluids, since its viscosity only doubles for an increase in pressure from 1 to 1000 atm.

1.12. Surface tension

Although all molecules are in constant motion, a molecule within the body of the liquid is, on average, attracted equally in all directions by the other molecules surrounding it, but, at the surface between liquid and air, or the interface between one substance and another, the upward and downward attractions are unbalanced, the surface molecules being pulled inward towards the bulk of the liquid. This effect causes the liquid surface to behave as if it were an elastic membrane under tension. The surface tension σ is measured as the force acting across unit length of a line drawn in the surface. It acts in the plane of the surface, normal to any line in the surface, and is the same at all points. Surface tension is constant at any given temperature for the surface of separation of two particular substances, but it decreases with increasing temperature.

The effect of surface tension is to reduce the surface of a free body of liquid to a minimum, since to expand the surface area molecules have to be brought to the surface from the bulk of the liquid against the unbalanced attraction pulling the surface molecules inwards. For this reason, drops of liquid tend to take a spherical shape in order to minimize surface area. For such a small droplet, surface tension will cause an increase of internal pressure p in order to balance the surface force.

Considering the forces acting on a diametral plane through a spherical drop of radius r,

Force due to internal pressure $= p \times \pi r^2$,

Force due to surface tension round the perimeter

$$= 2\pi r \times \sigma.$$

For equilibrium,

$$p\pi r^2 = 2\pi r\sigma \quad \text{or} \quad p = 2\sigma/r.$$

Surface tension will also increase the internal pressure in a cylindrical jet of fluid, for which $p = \sigma/r$. In either case, if r is very small, the value of p becomes very large. For small bubbles in a liquid, if this pressure is greater than the pressure of vapour or gas in a bubble, the bubble will collapse.

In many of the problems with which engineers are concerned, the magnitude of surface tension forces are very small compared with the other forces acting on the fluid and may, therefore, be neglected. However, they can cause serious errors in hydraulic scale models and through capillary effects. Surface tension forces can be reduced by the addition of detergents.

Example 1.1

Air is introduced through a nozzle into a tank of water to form a stream of bubbles. If the bubbles are intended to have a diameter of 2 mm, calculate by how much the pressure of the air at the nozzle must exceed that of the surrounding water. Assume that $\sigma = 72{\cdot}7 \times 10^{-3}$ N m^{-1}.

Solution
Excess pressure,

$$p = 2\sigma/r.$$

Putting $r = 1$ mm $= 10^{-3}$ m, $\sigma = 72{\cdot}7 \times 10^{-3}$ N m^{-1}:

Excess pressure,

$$p = 2 \times 72{\cdot}7 \times 10^{-3}/1 \times 10^{-3}$$
$$= 143{\cdot}4 \text{ N m}^{-1}.$$

1.13. Capillarity

If a fine tube, open at both ends, is lowered vertically into a liquid which wets the tube, the level of the liquid will rise in the tube (Fig. 1.7(a)). If the liquid does not wet the tube, the level of liquid in the tube will be depressed below the level of the free surface outside (Fig. 1.7(b)). If θ is the angle of contact

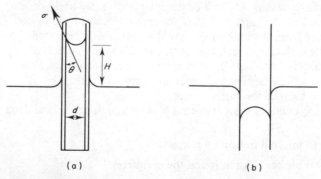

(a)   (b)

Figure 1.7   Capillarity

between liquid and solid and d is the tube diameter (Fig. 1.7(a)),

Upward pull due to surface tension =

Component of surface tension acting upwards x Perimeter of tube

$$= \sigma \cos \theta \times \pi d. \qquad (1.7)$$

The atmospheric pressure is the same inside and outside the tube, and, therefore, the only force opposing this upward pull is the weight of the vertical-sided column of liquid of height H, since, by definition, there are no shear stresses in a liquid at rest. Therefore, in Fig. 1.7, there will be no shear stress on the vertical sides of the column of liquid under consideration.

$$\text{Weight of column raised} = \rho g (\pi/4) d^2 H, \qquad (1.8)$$

where ρ is the mass density of the liquid.

Equating the upward pull to the weight of the column, from equations (1.7) and (1.8),

$$\sigma \cos \theta \times \pi d = \rho g (\pi/4) d^2 H,$$

Capillary rise, $H = 4\sigma \cos \theta / \rho g d$.

Capillary action is a serious source of error in reading liquid levels in fine gauge tubes, particularly as the degree of wetting and, therefore, the contact angle θ are affected by the cleanness of the surfaces in contact. For water in a tube of 5 mm diameter, the capillary rise will be approximately 4·5 mm, while for mercury the corresponding figure would be −1·4 mm (Fig. 1.8). Gauge glasses for reading the level of liquids should have as large a diameter as is conveniently possible, to minimize errors due to capillarity.

1.14. Vapour pressure

Since the molecules of a liquid are in constant agitation, some of the molecules in the surface layer will have sufficient energy to escape from the attraction of the surrounding molecules into the space above the free surface. Some of these molecules will return and condense, but others will take their place. If the space above the liquid is confined, an equilibrium will be reached so that the number of molecules of liquid in the space above the free surface is constant. These molecules produce a partial pressure known as the vapour pressure in the space.

The degree of molecular activity increases with increasing temperature, and, therefore, the vapour pressure will also increase. Boiling will occur when the vapour pressure is equal to the pressure above the liquid. By reducing the pressure, boiling can be made to occur at temperatures well below the boiling point at atmospheric pressure; for example, if the pressure is reduced to 0·2 bar (0·2 atm), water will boil at a temperature of 60 °C.

1.15. Cavitation

Under certain conditions, areas of low pressure can occur locally in a flowing fluid. If the pressure in such areas falls below the vapour pressure, there will be local boiling and a cloud of vapour bubbles will form. This phenomenon is

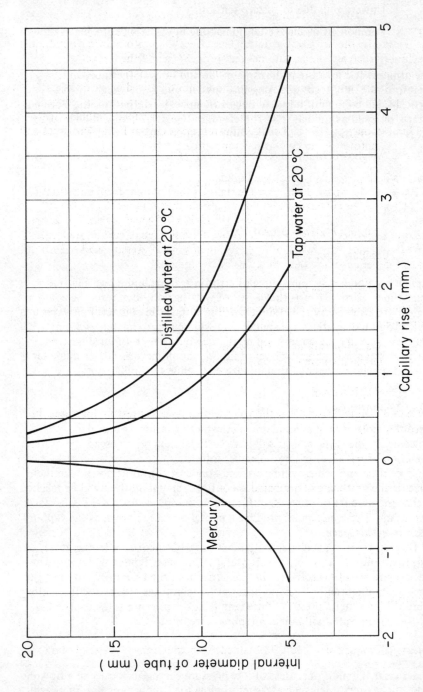

Figure 1.8   Capillary rise in glass tubes of circular cross-section

known as cavitation and can cause serious problems, since the flow of liquid can sweep this cloud of bubbles on into an area of higher pressure where the bubbles will collapse suddenly. If this should occur in contact with a solid surface, very serious damage can result due to the very large force with which the liquid hits the surface. Cavitation can affect the performance of hydraulic machinery such as pumps, turbines and propellers, and the impact of collapsing bubbles can cause local erosion of metal surfaces.

Cavitation can also occur, if a liquid contains dissolved air or other gases, since the solubility of gases in a liquid decreases as the pressure is reduced. Gas or air bubbles will be released in the same way as vapour bubbles, with the same damaging effects. Usually, this release occurs at higher pressures and, therefore, before vapour cavitation commences.

1.16. Compressibility and the bulk modulus

All materials, whether solids, liquids or gases, are compressible, i.e. the volume V of a given mass will be reduced to $V - \delta V$ when a force is exerted uniformly all over its surface. If the force per unit area of surface increases from p to $p + \delta p$, the relationship between change of pressure and change of volume depends on the bulk modulus of the material:

Bulk modulus = Change in pressure/Volumetric strain.

Volumetric strain is the change in volume divided by the original volume, therefore,

$$\frac{\text{Change in volume}}{\text{Original volume}} = \frac{\text{Change in pressure}}{\text{Bulk modulus}},$$

$$-\delta V/V = \delta p/K,$$

the minus sign indicating that the volume decreases as pressure increases. In the limit, as $\delta p \to 0$,

$$K = -V\frac{dp}{dV}. \qquad (1.9)$$

Considering unit mass of a substance,

$$V = 1/\rho. \qquad (1.10)$$

Differentiating,

$$V \, d\rho + \rho \, dV = 0$$

$$dV = -(V/\rho) \, d\rho.$$

Substituting for V from equation (1.10),

$$dV = -(1/\rho^2) \, d\rho. \qquad (1.11)$$

Putting the values of V and dV obtained from equations (1.10) and (1.11) in equation (1.9),

$$K = \rho\frac{dp}{d\rho}. \qquad (1.12)$$

The value of K is shown by equation (1.12) to be dependent on the relation-ship between pressure and density and, since density is also affected by temperature, it will depend on how the temperature changes during compression. If the temperature is constant, conditions are said to be isothermal, while, if no heat is allowed to enter or leave during compression, conditions are adiabatic. The ratio of the adiabatic bulk modulus to the isothermal bulk modulus is equal to γ, the ratio of the specific heat of a fluid at constant pressure to that at constant volume. For liquids, γ is approxi-mately unity and the two conditions need not be distinguished; for gases, the difference is substantial (for air, $\gamma = 1\cdot4$).

The concept of the bulk modulus is mainly applied to liquids, since for gases the compressibility is so great that the value of K is not a constant, but proportional to pressure and changes very rapidly. The relationship between pressure and mass density is more conveniently found from the characteristic equation of a gas (1.13). For liquids, the value of K is high and changes of density with pressure are small, but increasing pressure does bring the molecules of the liquid closer together, increasing the value of K. For water, the value of K will double if the pressure is increased from 1 to 3500 atm. An increase of temperature will cause the value of K to fall.

For liquids, the changes in pressure occurring in many fluid mechanics problems are not sufficiently great to cause appreciable changes in density. It is, therefore, usual to ignore such changes and to treat liquids as incompress-ible. Where, however, sudden changes of velocity generate large inertial forces, high pressures can occur and compressibility effects cannot be disregarded in liquids (*see* Chapter 19). Gases may also be treated as incompressible, if the pressure changes are very small, but, usually, compressibility cannot be ignored. In general, compressibility becomes important when the velocity of the fluid exceeds about one-fifth of the velocity of a pressure wave (e.g. the velocity of sound) in the fluid.

Units: since volumetric strain is the ratio of two volumes, the units of bulk modulus will be the same as those of pressure, newtons per square metre $(N\ m^{-2})$.

Dimensions: $M\ L^{-1}\ T^{-2}$.

Typical values: water, $2\cdot05 \times 10^9\ N\ m^{-2}$; oil, $1\cdot62 \times 10^9\ N\ m^{-2}$.

1.17. Equation of state of a perfect gas

The mass density of a gas varies with its absolute pressure p and absolute temperature T. For a perfect gas,

$$p = \rho RT, \tag{1.13}$$

where R is the gas constant for the gas concerned. Most gases at pressures and temperatures well removed from liquefaction follow this characteristic equation closely, but it does not apply to vapours.

Units: the gas constant is measured in joules per kilogramme per kelvin $(J\ kg^{-1}\ K^{-1})$.

Dimensions: $L^2\ T^{-2}\ \Theta^{-1}$.

Typical values: air, $287\ J\ kg^{-1}\ K^{-1}$; hydrogen, $4110\ J\ kg^{-1}\ K^{-1}$.

1.18. The universal gas constant

From equation (1.13) ρR is constant for a given value of pressure p and temperature T. By Avogadro's hypothesis, all pure gases have the same number of molecules per unit volume at the same temperature and pressure, so that ρ is proportional to the molar mass M (kg kmol^{-1}). Therefore, the quantity MR will be constant for all perfect gases, and is known as the universal gas constant R_0.

$$R_0 = MR = 8 \cdot 314 \text{ kJ kmol}^{-1} \text{K}^{-1}.$$

1.19. Specific heats of a gas

Since pressure, temperature and density of a gas are interrelated, the amount of heat energy H required to raise the temperature of a gas from T_1 to T_2 will depend upon whether the gas is allowed to expand during the process, so that some of the energy supplied is used in doing work instead of raising the temperature of the gas. Two different specific heats are, therefore, given for a gas, corresponding to the two extreme conditions of constant volume and constant pressure.

(i) *Specific heat at constant volume* c_v. For a temperature change from T_1 to T_2 at constant volume,

Heat supplied per unit mass, $H = c_v(T_2 - T_1)$.

Since there is no change in volume, no external work is done, so that the increase of internal energy per unit mass of gas is $c_v(T_2 - T_1)$ heat units.

(ii) *Specific heat at constant pressure* c_p. If the pressure is kept constant, the gas will expand as the temperature changes from T_1 to T_2:

Heat supplied per unit mass = $c_p(T_2 - T_1)$ heat units.

Only part of this energy is used to raise the temperature of the gas, the rest goes to external work.

Thus, $c_p > c_v$:

$$c_p(T_2 - T_1) = c_v(T_2 - T_1) + \text{External work (in heat units)}.$$

It can be shown that $R = (c_p - c_v)$, where R, c_p and c_v have the same units.
  Units: specific heat is measured in joules per kilogramme per kelvin, as is R.
  Dimensions: $L^2 \, T^{-2} \, \Theta^{-1}$.
  Typical values: air, $c_p = 1 \cdot 005 \text{ kJ kg}^{-1} \text{ K}^{-1}$, $c_v = 0 \cdot 718 \text{ kJ kg}^{-1} \text{K}^{-1}$.

1.20. Expansion of a gas

When a gas expands, the amount of work done will depend upon the relationship between pressure and volume, which, in turn, depends upon whether the gas receives or loses heat during the process.

If unit mass of a gas has a volume V_1 at pressure p_1 and volume V_2 at pressure p_2, as shown in Fig. 1.9, then,

Work done during expansion

= Area under p-V curve between V_1 and V_2 (shaded)

$$= \int_{V_1}^{V_2} p \, \mathrm{d}V. \tag{1.14}$$

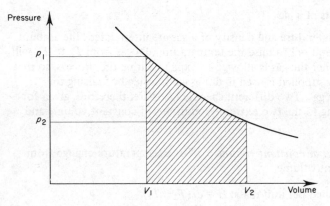

Figure 1.9   Expansion of a gas

(i) If the expansion is isothermal, the absolute temperature T (in kelvins) of the gas remains unchanged and the characteristic equation $p = \rho RT$ becomes $p/\rho = $ constant; or, putting $V = $ volume of unit mass $= 1/\rho$,

$$pV = \text{constant} = p_1 V_1 = RT,$$

$$p = p_1 V_1 (1/V).$$

From equation (1.14),

$$\text{Work done per unit mass} = p_1 V_1 \int_{V_1}^{V_2} \frac{\mathrm{d}V}{V}$$

$$= p_1 V_1 \log_e (V_2/V_1)$$

$$= RT \log_e (V_2/V_1). \tag{1.15}$$

(ii) For any known relationship between pressure and mass density of the form $p/\rho^n = $ constant, putting $V = 1/\rho$,

$$pV^n = p_1 V_1^n = \text{constant}.$$

Therefore,

$$p = p_1 V_1^n V^{-n}.$$

Work done by gas per unit mass $= \displaystyle\int_{V_1}^{V_2} p \; \mathrm{d}V$

$$= p_1 V_1^n \int_{V_1}^{V_2} V^{-n} \; \mathrm{d}V$$

$$= \{p_1 V_1^n / (1-n)\} \{V_2^{(1-n)} - V_1^{(1-n)}\}$$

$$= (1-n)^{-1} (p_1 V_1^n V_2^{(1-n)} - p_1 V_1)$$

or, since $p_1 V_1^n = p_2 V_2^n$,

Work done by gas per unit mass $= (p_2 V_2 - p_1 V_1)/(1-n)$

$$= (p_1 V_1 - p_2 V_2)/(n-1)$$

$$= R(T_1 - T_2)/(n-1). \tag{1.16}$$

(iii) If the compression is carried out adiabatically, no heat enters or leaves the system. Now, for any mode of compression, considering unit mass,

Heat supplied = Change of internal energy + Work done (in heat units).

Change of internal energy $= c_v(T_2 - T_1)$,

Mechanical work done $= (p_2 V_2 - p_1 V_1)/(1-n)$.

Thus, in general, if H is the heat supplied,

$$H = c_v(T_2 - T_1) + (p_2 V_2 - p_1 V_1)/(1-n).$$

Now, $R = (c_p - c_v)$ or $c_v = R/(c_p/c_v - 1)$.

Also $R(T_2 - T_1) = (p_2 V_2 - p_1 V_1)$.

Thus, $H = (p_2 V_2 - p_1 V_1)/(c_p/c_v - 1) + (p_2 V_2 - p_1 V_1)/(1-n)$

For an adiabatic change, $H = 0$, so that

$$(p_2 V_2 - p_1 V_1)/(c_p/c_v - 1) = -(p_2 V_2 - p_1 V_1)/(1-n)$$

$$= (p_2 V_2 - p_1 V_1)/(n-1),$$

and, therefore,

$$n = c_p/c_v = \gamma.$$

Thus, for an adiabatic change, the relationship between pressure and density is given by

$$pV^{\gamma} = p/\rho^{\gamma} = \text{constant}, \tag{1.17}$$

and, from (ii),

Work done by gas per unit mass $= (p_1 V_1 - p_2 V_2)/(\gamma - 1)$

$$= R(T_1 - T_2)/(\gamma - 1). \tag{1.18}$$

2 Pressure and head

2.1. Statics of fluid systems

The general rules of statics apply to fluids at rest, but, from the definition of a fluid (Section 1.1), there will be no shearing forces acting and, therefore, all forces (such as F in Fig. 2.1(a)) exerted between the fluid and a solid boundary must act at right angles to the boundary. If the boundary is curved (Fig. 2.1(b)), it can be considered to be composed of a series of chords on each of which a force $F_1, F_2, \ldots, F_n$ acts perpendicular to the surface at the section concerned. Similarly, considering any plane drawn through a body of fluid at rest (Fig. 2.1(c)), the force exerted by one portion of the fluid on the other acts at right angles to this plane.

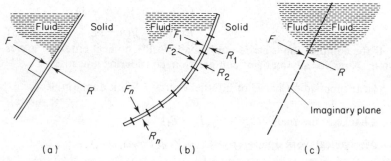

Figure 2.1 Forces in a fluid at rest

Shear stresses due to viscosity are only generated when there is relative motion between elements of the fluid. The principles of fluid statics can, therefore, be extended to cases in which the fluid is moving as a whole but all parts are stationary relative to each other.

In the analysis of a problem it is usual to consider an element of the fluid defined by solid boundaries or imaginary planes. A free body diagram can be drawn for this element, showing the forces acting on it due to the solid boundaries or surrounding fluid. Since the fluid is at rest, the element will be in equilibrium, and the sum of the component forces acting in any direction must be zero. Similarly, the sum of the moments of the forces about any point must be zero. It is usual to test equilibrium by resolving along three mutually perpendicular axes and, also, by taking moments in three mutually perpendicular planes.

Although a body or element may be in equilibrium, it can also be of interest to know what will happen if it is displaced from its equilibrium position. For example, in the case of a ship it is of the utmost importance to know whether it will overturn when it pitches or rolls or whether it will tend to right itself and return to its original position. There are three possible conditions of equilibrium.

(i) *Stable equilibrium*. A small displacement from the equilibrium position generates a force producing a righting moment tending to restore the body to its equilibrium position.

(ii) *Unstable equilibrium.* A small displacement produces an overturning moment tending to displace the body further from its equilibrium position.

(iii) *Neutral equilibrium.* The body remains at rest in any position to which it is displaced.

These conditions are typified by the three positions of a cone on a horizontal surface shown in Fig. 2.2.

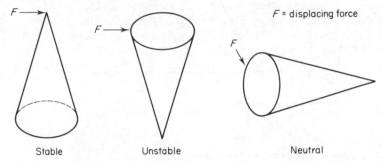

Figure 2.2   Types of equilibrium

2.2. Pressure

A fluid will exert a force normal to a solid boundary or any plane drawn through the fluid. Since problems may involve bodies of fluid of indefinite extent and, in many cases, the magnitude of the force exerted on a small area of the boundary or plane may vary from place to place, it is convenient to work in terms of the *pressure p* of the fluid, defined as the force exerted per unit area. If the force exerted on each unit area of a boundary is the same, the pressure is said to be uniform:

$$\text{Pressure} = \frac{\text{Force exerted}}{\text{Area of boundary}} \quad \text{or} \quad p = \frac{F}{A}.$$

If, as is more commonly the case, the pressure changes from point to point, we consider the element of force δF normal to a small area δA surrounding the point under consideration:

$$\text{Mean pressure,} \, p = \frac{\delta F}{\delta A}.$$

In the limit, as $\delta A \to 0$ (but remains large enough to preserve the concept of the fluid as a continuum),

$$\text{Pressure at a point,} \, p = \lim_{\delta A \to 0} \frac{\delta F}{\delta A} = \frac{dF}{dA}$$

Units: newtons per square metre (N m^{-2}). (Note that an alternative metric unit is the bar; 1 bar = $10^5 \, \text{N m}^{-2}$.)

Dimensions: $M \, L^{-1} \, T^{-2}$.

2.3. Pascal's law for pressure at a point

By considering the equilibrium of a small fluid element in the form of a triangular prism surrounding a point in the fluid (Fig. 2.3), a relationship can be established between the pressures p_x in the x direction, p_y in the y direction and p_s normal to any plane inclined at any angle θ to the horizontal at this point.

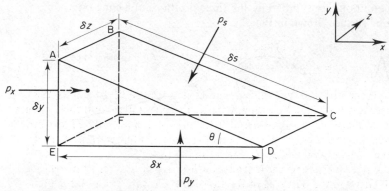

Figure 2.3 Equality of pressure in all directions at a point

If the fluid is at rest, p_x will act at right angles to the plane ABFE, p_y at right angles to CDEF and p_s at right angles to ABCD. Since the fluid is at rest, there will be no shearing forces on the faces of the element and the element will not be accelerating. The sum of the forces in any direction must, therefore, be zero.

Considering the x direction:

Force due to $p_x = p_x \times$ Area ABFE $= p_x \delta y \delta z$;

Component of force due to $p_s = -(p_s \times$ Area ABCD$) \sin \theta$

$$= -p_s \delta s \delta z \frac{\delta y}{\delta s}$$

$$= -p_s \delta y \delta z$$

(since $\sin \theta = \delta y / \delta s$). As p_y has no compound in the x direction, the element will be in equilibrium if

$$p_x \delta y \delta z + (-p_s \delta y \delta z) = 0,$$

$$p_x = p_s \tag{2.1}$$

Similarly, in the y direction,

Force due to $p_y = p_y \times$ Area CDEF $= p_y \delta x \delta z$;

Component of force due to $p_s = -(p_s \times$ Area ABCD$) \cos \theta$

$$= -p_s \delta s \delta z \frac{\delta x}{\delta s}$$

$$= -p_s \delta x \delta z$$

(since $\cos \theta = \delta x / \delta s$).

Weight of element $= -$ Specific weight $\times$ Volume

$$= -\rho g \times \tfrac{1}{2} \delta x \delta y \delta z.$$

As p_x has no component in the y direction, the element will be in equilibrium if

$$p_y \delta x \delta z + (-p_s \delta x \delta z) + (-\rho g \times \tfrac{1}{2} \delta x \delta y \delta z) = 0.$$

Since δx, δy and δz are all very small quantities, $\delta x \delta y \delta z$ is negligible in comparison with the other two terms, and the equation reduces to

$$p_y = p_s. \tag{2.2}$$

Thus, from equations (2.1) and (2.2),

$$p_s = p_x = p_y. \tag{2.3}$$

Now p_s is the pressure on a plane inclined at *any* angle θ; the x, y, and z axes have not been chosen with any particular orientation, and the element is so small that it can be considered to be a point. Equation (2.3), therefore, indicates that the pressure at a point is the same in all directions. This is known as *Pascal's law* and applies to a fluid at rest.

If the fluid is flowing, shear stresses will be set up as a result of relative motion between the particles of the fluid. The pressure at a point is then considered to be the mean of the normal forces per unit area (stresses) on three mutually perpendicular planes. Since these normal stresses are usually large compared with shear stresses it is generally assumed that Pascal's law still applies.

2.4. Variation of pressure vertically in a fluid under gravity

Figure 2.4 shows an element of fluid consisting of a vertical column of constant cross-sectional area A and totally surrounded by the same fluid of mass density ρ. Suppose that the pressure is p_1 on the underside at level z_1 and p_2 on the top at level z_2. Since the fluid is at rest the element must be in equilibrium and the sum of all the vertical forces must be zero. The forces acting are:

Force due to p_1 on area A acting up $= p_1 A$,

Force due to p_2 on area A acting down $= p_2 A$,

Force due to the weight of the element $= mg$

$= $ Mass density $\times g \times$ Volume $= \rho g A (z_2 - z_1)$.

Since the fluid is at rest, there can be no shear forces and, therefore, no vertical forces act on the side of the element due to the surrounding fluid. Taking upward forces as positive and equating the algebraic sum of the forces acting to zero,

$$p_1 A - p_2 A - \rho g A (z_2 - z_1) = 0,$$

$$p_2 - p_1 = -\rho g (z_2 - z_1). \tag{2.4}$$

Thus, in any fluid under gravitational attraction, pressure decreases with increase of height z.

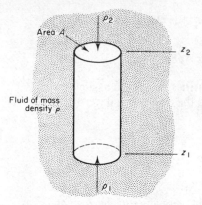

Figure 2.4   Vertical variation of pressure

Example 2.1

A diver descends from the surface of the sea to a depth of 30 m. What would be the pressure under which the diver would be working above that at the surface assuming that the density of sea water is 1025 kg m^{-3} and remains constant?

Solution
In equation (2.4), taking sea level as datum, $z_1 = 0$. Since z_2 is lower than z_1 the value of z_2 is −30 m. Substituting these values and putting $\rho = 1025$ kg m^{-3}:

$$\text{Increase of pressure} = p_2 - p_1$$

$$= -1025 \times 9{\cdot}81 \ (-30 - 0)$$

$$= 301{\cdot}7 \times 10^3 \ \text{N m}^{-2}.$$

2.5. Equality of pressure at the same level in a static fluid

If P and Q are two points at the same level in a fluid at rest (Fig. 2.5), a horizontal prism of fluid of constant cross-sectional area A will be in equilibrium. The forces acting on this element horizontally are p_1A at P and p_2A at Q. Since the fluid is at rest, there will be no horizontal shear

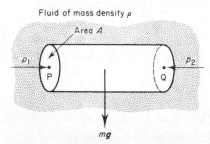

Figure 2.5   Equality of pressures at the same level

stresses on the sides of the element. For static equilibrium the sum of the horizontal forces must be zero:

$$p_1 A = p_2 A,$$

$$p_1 = p_2.$$

Thus, the pressure at any two points at the same level in a body of fluid at rest will be the same.

In mathematical terms, if (x, y) is the horizontal plane,

$$\frac{\partial p}{\partial x} = 0 \quad \text{and} \quad \frac{\partial p}{\partial y} = 0;$$

partial derivatives are used because pressure p could vary in three dimensions.

Pressures at the same level will be equal even though there is no direct horizontal path between P and Q, provided that P and Q are in the same continuous body of fluid. Thus, in Fig. 2.6, P and Q are connected by a

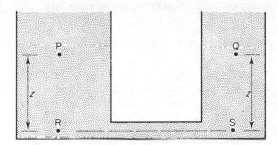

Figure 2.6   Equality of pressures in a continuous body of fluid

horizontal pipe, R and S being two points at the same level at the entrance and exit to the pipe. If the pressure is p_P at P, p_Q at Q, p_R at R and p_S at S, then, since R and S are at the same level,

$$p_R = p_S; \tag{2.5}$$

also $p_R = p_P + \rho g z$ and $p_S = p_Q + \rho g z.$

Substituting in equation (2.5),

$$p_P + \rho g z = p_Q + \rho g z,$$

$$p_P = p_Q.$$

2.6. General equation for the variation of pressure due to gravity from point to point in a static fluid

Let p be the pressure acting on the end P of a right prism of fluid of constant cross-sectional area A and $p + \delta p$ be the pressure at the other end Q (Fig. 2.7). The axis of the prism is inclined at an angle θ to the vertical, the height

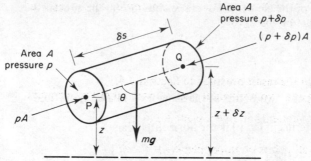

Figure 2.7   Variation of pressure in a stationary fluid

of P above a horizontal datum is z and that of Q is $z + \delta z$. The forces acting on the element are:

pA acting at right angles to the end face at P along the axis of the prism,

$(p + \delta p)A$ acting at Q along the axis in the opposite direction;

mg the weight of the element, due to gravity, acting vertically down

= Mass density × Volume × Gravitational acceleration

= $\rho \times A\delta s \times g$.

There are also forces due to the surrounding fluid acting normal to the sides of the element, since the fluid is at rest, and, therefore, perpendicular to its axis PQ.

For equilibrium of the element PQ, the resultant of these forces in any direction must be zero. Resolving along the axis PQ,

$$pA - (p + \delta p)A - \rho g A\delta s \cos \theta = 0,$$

$$\delta p = -\rho g \delta s \cos \theta,$$

or, in differential form,

$$\frac{dp}{ds} = -\rho g \cos \theta.$$

In the general three-dimensional case, s is a vector with components in the x, y and z directions. Taking the (x, y) plane as horizontal, if the axis of the element is also horizontal, $\theta = 90°$ and

$$\left(\frac{dp}{ds}\right)_{\theta = 90°} = \frac{\partial p}{\partial x} = \frac{\partial p}{\partial y} = 0, \tag{2.6}$$

confirming the results of Section 2.5 that, in a static fluid, pressure is constant everywhere in a horizontal plane. It is for this reason that the free surface of a liquid is horizontal.

If the axis of the element is in the vertical z direction, $\theta = 0°$ and

$$\left(\frac{dp}{ds}\right)_{\theta = 0°} = \frac{\partial p}{\partial z} = -\rho g,$$

and, since $\partial p/\partial x = \partial p/\partial y = 0$, the partial derivative $\partial p/\partial z$ can be replaced by the total differential dp/dz, giving

$$\frac{dp}{dz} = -\rho g \qquad (2.7)$$

which corresponds to the result obtained in Section 2.4.

Also, considering any two horizontal planes a vertical distance z apart,

Pressure at all points on lower plane $= p$,

Pressure at all points on upper plane $= p + z\dfrac{\partial p}{\partial z}$,

Difference of pressure $= z\dfrac{\partial p}{\partial z}$.

Since the planes are horizontal, the pressure must be constant over each plane, therefore, $\partial p/\partial z$ cannot vary horizontally. From equation (2.7), this implies that ρg shall be constant and, therefore, for equilibrium, the density ρ must be constant over any horizontal plane.

Thus, the conditions for equilibrium under gravity are:

(i) the pressure at all points on a horizontal plane must be the same;
(ii) the density at all points on a horizontal plane must be the same;
(iii) the change of pressure with elevation is given by $dp/dz = -\rho g$.

The actual pressure variation with elevation is found by integrating equation (2.6):

$$p = -\int \rho g\, dz \quad \text{or} \quad p_2 - p_1 = -\int_{z_1}^{z_2} \rho g\, dz, \qquad (2.8)$$

but this cannot be done unless the relationship between ρ and p is known.

2.7. Variation of pressure with altitude in a fluid of constant density

For most problems involving liquids it is usual to assume that the density ρ is constant and the same assumption can also be made for a gas if pressure differences are very small. Equation (2.8) can then be written

$$p = -\rho g \int dz = \rho g z + \text{constant}$$

or, for any two points at altitude z_1 and z_2 above datum,

$$p_2 - p_1 = -\rho g(z_2 - z_1).$$

2.8. Variation of pressure with altitude in a gas at constant temperature

The relation between pressure, density and temperature for a perfect gas is given by the equation $p/\rho = RT$. If conditions are assumed to be isothermal, so that temperature does not vary with altitude, ρ can be expressed in terms of p and the result substituted in equation (2.7),

$$\rho = \frac{p}{RT},$$

and, from equation (2.7),

$$\frac{dp}{dz} = -\rho g = -\frac{pg}{RT},$$

$$\frac{dp}{p} = -\frac{g}{RT}\, dz.$$

Integrating from $p = p_1$ when $z = z_1$, to $p = p_2$ when $z = z_2$,

$$\log_e (p_2/p_1) = -(g/RT)(z_2 - z_1),$$

$$p_2/p_1 = \exp \{-(g/RT)(z_2 - z_1)\}.$$

Example 2.2

At an altitude z, of 11 000 m, the atmospheric temperature T is $-56 \cdot 6\,°C$ and the pressure, p, is $22 \cdot 4$ kN m^{-2}. Assuming that the temperature remains the same at higher altitudes, calculate the density of the air at an altitude of z_2 of 15 000 m. Assume $R = 287$ J kg^{-1} K^{-1}.

Solution

Let p_2 be the absolute pressure at z_2. Since the temperature is constant,

$$p_2/p_1 = \exp \{-(g/RT)(z_2 - z_1)\}$$

Putting $p_1 = 22 \cdot 4$ kN m$^{-2} = 22 \cdot 4 \times 10^3$ N m^{-2}, $z_1 = 11\ 000$ m, $z_2 = 15\ 000$ m, $R = 287$ J kg^{-1} K^{-1}, $T = -56 \cdot 6\,°C = 216 \cdot 5$ K:

$$p_2 = 22 \cdot 4 \times 10^3 \times \exp \left\{ -\frac{9 \cdot 81 (15\ 000 - 11\ 000)}{287 \times 216 \cdot 5} \right\}$$

$$= 22 \cdot 4 \times 10^3 \times \exp (-0 \cdot 631) = 11 \cdot 91 \times 10^3\ \text{N m}^{-2}.$$

Also, from the equation of state for a perfect gas (see equation (1.13)), $p_2 = \rho_2 RT$ and so

$$\text{Density of air at 15 000 m, } \rho_2 = p_2/RT$$

$$= 11 \cdot 92 \times 10^3/287 \times 216 \cdot 5$$

$$= 0 \cdot 192\ \text{kg m}^{-3}.$$

2.9. Variation of pressure with altitude in a gas under adiabatic conditions

If conditions are adiabatic, the relationship between pressure and density is given by $p/\rho^\gamma = \text{constant} = p_1/\rho_1^\gamma$, so that

$$\rho = \rho_1 (p/p_1)^{1/\gamma}.$$

Substituting in equation (2.7),

$$\frac{dp}{dz} = -\frac{\rho_1 g}{p_1^{1/\gamma}}\, p^{1/\gamma},$$

$$dz = -\left(\frac{p_1^{1/\gamma}}{\rho_1 g} \right) p^{-1/\gamma}\, dp.$$

Integrating from $p = p_1$ when $z = z_1$, to $p = p_2$ when $z = z_2$,

$$z_2 - z_1 = -\frac{p_1^{1/\gamma}}{\rho_1 g}\left[\frac{p^{(\gamma-1)/\gamma}}{(\gamma-1)/\gamma}\right]_{p_1}^{p_2}$$

$$= -\left(\frac{\gamma}{\gamma-1}\right)\frac{p_1^{1/\gamma}}{\rho_1 g}\left(p_2^{(\gamma-1)/\gamma} - p_1^{(\gamma-1)/\gamma}\right)$$

$$= -\left(\frac{\gamma}{\gamma-1}\right)\frac{p_1}{\rho_1 g}\left\{\left(\frac{p_2}{p_1}\right)^{(\gamma-1)/\gamma} - 1\right\},$$

or, since $p_1/\rho_1 = RT$, for any gas,

$$z_2 - z_1 = -\left(\frac{\gamma}{\gamma-1}\right)\frac{RT_1}{g}\left\{\left(\frac{p_2}{p_1}\right)^{(\gamma-1)/\gamma} - 1\right\},$$

$$\left(\frac{p_2}{p_1}\right)^{(\gamma-1)/\gamma} = \frac{g(z_2 - z_1)}{RT_1}\left(\frac{\gamma-1}{\gamma}\right) + 1$$

$$\frac{p_2}{p_1} = \left\{1 - \frac{g(z_2 - z_1)}{RT_1}\left(\frac{\gamma-1}{\gamma}\right)\right\}^{\gamma/(\gamma-1)} \tag{2.9}$$

this can be extended to any isentropic process for which $p/\rho^n = $ constant, to give

$$\frac{p_2}{p_1} = \left\{1 - \frac{g(z_2 - z_1)}{RT_1}\left(\frac{n-1}{n}\right)\right\}^{n/(n-1)}. \tag{2.10}$$

The rate of change of temperature with altitude — the temperature lapse rate — can also be found for adiabatic conditions. From the characteristic equation, $\rho = p/RT$ and, since, from equation (2.7),

$$dz = -dp/\rho g,$$

substituting for ρ,

$$dz = -(RT/gp)\,dp. \tag{2.11}$$

For adiabatic conditions,

$$p/\rho^\gamma = p_1/\rho_1^\gamma,$$

or, since $p/\rho = RT$,

$$p = p_1(T_1/T)^{\gamma/(1-\gamma)}$$

and, differentiating,

$$dp = -\{\gamma/(1-\gamma)\}p_1 T_1^{\gamma/(1-\gamma)} T^{-1/(1-\gamma)}\,dT.$$

Substituting these values of p and dp in equation (2.11):

$$dz = -\frac{RT}{g}\frac{\{-[\gamma/(1-\gamma)]\,p_1 T_1^{\gamma/(1-\gamma)} T^{-1/(1-\gamma)}\,dT\}}{p_1 T_1^{\gamma/(1-\gamma)} T^{-\gamma/(1-\gamma)}}$$

$$= \{\gamma/(1-\gamma)\}(R/g)\,dT.$$

Temperature gradient,

$$\frac{dT}{dz} = -\{(\gamma-1)/\gamma\}(g/R). \tag{2.12}$$

Example 2.3

Calculate the pressure, temperature and density of the atmosphere at an altitude of 1200 m if at zero altitude the temperature is 15 °C and the pressure 101 kN m^{-2}. Assume that conditions are adiabatic ($\gamma = 1.4$) and $R = 287$ J kg^{-1} K^{-1}.

Solution
From equation (2.9),

$$p_2 = p_1 \left\{ 1 - \frac{g(z_2 - z_1)}{RT_1} \left(\frac{\gamma - 1}{\gamma} \right) \right\}^{\gamma/(\gamma-1)}.$$

Putting $p_1 = 101 \times 10^3$ N m^{-2}, $z_1 = 0$, $z_2 = 1200$ m, $T_1 = 15$ °C $= 288$ K, $\gamma = 1.4$, $R = 287$ J kg^{-1} K^{-1}:

$$p_2 = 101 \times 10^3 \left\{ 1 - \frac{9.81 \times 1200}{287 \times 288} \left(\frac{0.4}{1.4} \right) \right\}^{1.4/0.4} \text{N m}^{-2}$$

$$= 87.33 \times 10^3 \text{ N m}^{-2}.$$

From equation (2.12),

Temperature gradient,

$$\frac{dT}{dz} = -\{(\gamma - 1)/\gamma\}(g/R)$$

$$= -(0.4/1.4) \times (9.81/287)$$

$$= -9.76 \times 10^{-3} \text{ K m}^{-1}$$

$$T_2 = T_1 + \frac{dT}{dz}(z_2 - z_1)$$

$$= 288 - 9.76 \times 10^{-3} \times 1200$$

$$= 276.3 = \mathbf{3.3\ °C}.$$

From the equation of state,

Density at 1200 m,

$$\rho_2 = p_2/RT_2$$

$$= (87.33 \times 10^3)/(287 \times 276.3)$$

$$= 1.101 \text{ kg m}^{-3}.$$

2.10. Variation of pressure and density with altitude for a constant temperature gradient

Assuming that there is a constant temperature lapse rate (i.e. $dT/dz =$ constant) with elevation in a gas, so that its temperature falls by an amount δT for a unit change of elevation, then, if $T_1 =$ temperature at level z_1, $T =$ temperature at level z,

$$T = T_1 - \delta T(z - z_1), \tag{2.13}$$

From equation (2.7), $dp/dz = -\rho g$ and, since $p/\rho = RT$, putting $\rho = p/RT$,

$$\frac{dp}{dz} = -p\,\frac{g}{RT},$$

$$\frac{dp}{p} = -\frac{g}{RT}\,dz.$$

Substituting for T from equation (2.13),

$$dp/p = -\{g/R(T_1 - \delta T(z - z_1))\}\,dz.$$

Integrating between the limits p_1 and p_2 and z_1 and z_2,

$$\log_e(p_2/p_1) = (g/R\delta T)\log_e\{(T_1 - \delta T(z_2 - z_1))/T_1\},$$

$$p_2/p_1 = \{1 - (\delta T/T_1)(z_2 - z_1)\}^{g/R\delta T}. \qquad (2.14)$$

Comparing this with the result obtained in Section 2.9, and putting

$$\delta T = -\frac{dT}{dn} = \left(\frac{n-1}{n}\right)\left(\frac{g}{R}\right),$$

we have

$$g/R\delta T = n/(n-1).$$

Substituting in equation (2.14),

$$p_2/p_1 = \{1 - (g/RT_1)(z_2 - z_1)(n-1)/n\}^{n/(n-1)}$$

which agrees with equation (2.10).

To find the corresponding change of density ρ, since $p/\rho = RT$,

$$\frac{\rho_2}{\rho_1} = \frac{p_2}{p_1} \times \frac{T_1}{T_2} = \frac{p_2}{p_1} \times \frac{T_1}{T_1 - \delta T(z_2 - z_1)}$$

and, substituting from equation (2.14) for p_2/p_1,

$$\rho_2/\rho_1 = \{1 - (\delta T/T_1)(z_2 - z_1)\}^{g/R\delta T}\{1 - (\delta T/T_1)(z_2 - z_1)\}^{-1}$$

$$= \{1 - (\delta T/T_1)(z_2 - z_1)\}^{(g/R\delta T) - 1}. \qquad (2.15)$$

Example 2.4

Assuming that the temperature of the atmosphere diminishes with increasing altitude at the rate of $6 \cdot 5\,^\circ\text{C}$ per 1000 m, find the pressure and density at a height of 7000 m if the corresponding values at sea level are 101 kN m^{-2} and $1 \cdot 235$ kg m^{-3} when the temperature is 15 $^\circ$C. Take $R = 287$ J kg^{-1} K^{-1}.

Solution

From equation (2.14),

$$p_2 = p_1\{1 - (\delta T/T_1)(z_2 - z_1)\}^{g/R\delta T}.$$

Putting $p_1 = 101 \times 10^3$ N m^{-2}, $\delta T = 6 \cdot 5\,^\circ$C per 1000 m $= 0 \cdot 0065$ K m^{-1}, $T_1 = 15\,^\circ$C $= 288$ K, $(z_2 - z_1) = 7000$ m, $R = 287$ J kg^{-1} K^{-1}:

$$p_2 = 101 \times 10^3\{1 - (0 \cdot 0065/288) \times 7000\}^{9 \cdot 81/287 \times 0 \cdot 0065}$$

$$= \mathbf{40 \cdot 89 \times 10^3\ N\ m^{-2}}.$$

From the equation of state,

$$\text{Density}, \rho_2 = p_2/RT_2 = p_2/R\{T_1 - \delta T(z_2 - z_1)\}$$
$$= 40 \cdot 89 \times 10^3/287(288 - 0 \cdot 0065 \times 7000)$$
$$= 0 \cdot 588 \text{ kg m}^{-3}.$$

2.11. Variation of temperature and pressure in the atmosphere

A body of fluid which is of importance to the engineer is the atmosphere.
In practice, it is never in perfect equilibrium and is subject to large
incalculable disturbances. In order to provide a basis for the design of aircraft
an International Standard Atmosphere has been adopted which represents
the average conditions in Western Europe; the relations between altitude,
temperature and density have been tabulated (Table 2.1).

Altitude above sea level	Absolute pressure	Absolute temperature	Mass density	Kinematic viscosity
(m)	(bar)	(K)	(kg m^{-3})	(m^2 s^{-1})
0	1·013 25	288·15	1·2250	1·461 x 10^{-5}
1000	0·8988	281·7	1·1117	1·581
2000	0·7950	275·2	1·0066	1·715
4000	0·6166	262·2	0·8194	2·028
6000	0·4722	249·2	0·6602	2·416
8000	0·3565	236·2	0·5258	2·904
10 000	0·2650	223·3	0·4136	3·525
11 500	0·2098	216·7	0·3375	4·213
12 000	0·1940	216·7	0·3119	4·557
14 000	0·1417	216·7	0·2279	6·239
16 000	0·1035	216·7	0·1665	8·540
18 000	0·075 65	216·7	0·1216	11·686
20 000	0·055 29	216·7	0·088 91	15·989
22 000	0·040 47	218·6	0·064 51	22·201
24 000	0·029 72	220·6	0·046 94	30·743
26 000	0·021 88	222·5	0·034 26	42·439
28 000	0·016 16	224·5	0·025 08	58·405
30 000	0·011 97	226·5	0·018 41	80·134
32 000	0·008 89	228·5	0·013 56	109·620

Table 2.1. International Standard Atmosphere

Essentially, the standard atmosphere comprises the troposphere —
extending from sea level to 11 000 m — in which the temperature lapse rate
is constant at approximately 0·0065 K m^{-1} and the pressure–density
relationship is p/ρ^n = constant, where $n = 1·238$. Above 11 000 m lies the
stratosphere, in which conditions are assumed to be isothermal, with the
temperature constant at $-56\,^\circ$C. Figure 2.8 shows the variation of pressure
with altitude in the International Standard Atmosphere. The atmospheric
pressure at sea level is assumed to be equivalent to 760 mm of mercury, the
temperature 15 $^\circ$C and the density 1·225 kg m^{-3}.

In the real atmosphere, the troposphere extends to an average of 11 000 m,
but can vary from 7600 m at the poles to 18 000 m at the equator. While

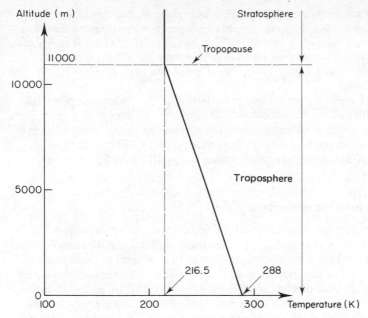

Figure 2.8   Variation of temperature with altitude in the Standard Atmosphere

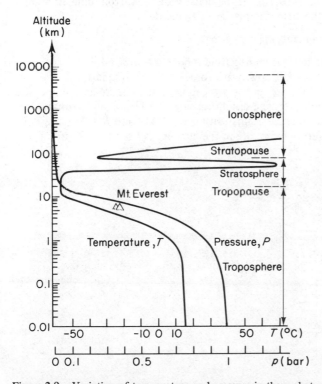

Figure 2.9   Variation of temperature and pressure in the real atmosphere

the temperature, in general, falls steadily with altitude, meteorological conditions can arise in the lower layers which produce temperature inversion — the temperature increasing with altitude. In the stratosphere, the temperature remains substantially constant up to approximately 32 000 m; it then rises to about 70 °C before falling again. Figure 2.9 shows typical values for pressure and temperature.

Because of vertical currents, the composition of the air remains practically constant, both in the troposphere and the stratosphere, except that there is a negligible amount of water vapour in the stratosphere and a slight reduction in the ratio of oxygen to nitrogen above an altitude of 20 km. Nine-tenths of the mass of the atmosphere is contained below 20 km and 99 per cent below 60 km.

2.12. Stability of the atmosphere

We have seen that there are variations of density from point to point in the atmosphere when it is at rest. In practice, there are local disturbances due to air currents. There are also changes of density as a result of local thermal effects, which cause the movement of elements of air into regions where they are surrounded by air of slightly different density and temperature. If the density of the surrounding air is less than that of the newly arrived element, there is a tendency for the element to return to its original position — since the net upward force exerted by the surrounding fluid is less than the weight of the element. In Fig. 2.10, if ρ_1 is the density of the surrounding air and ρ_2 is the density of the air in the displaced element,

$$\text{Weight of element, } mg = \rho_2 g A \delta z,$$

Upward force due to surrounding fluid = $\delta p \times A = \rho_1 g \delta z A$, and there is, therefore, a net downward force of $(\rho_2 - \rho_1) g \delta z A$.

As an element of fluid rises in the atmosphere, its pressure and temperature fall. Air is a poor conductor, and conditions are, therefore, approximately adiabatic. From equation (2.9), substituting $\gamma = 1.414$ and $R = 287 \text{ J kg}^{-1} \text{ K}^{-1}$, the adiabatic temperature lapse rate in the element is $\delta T = 0.01 \text{ K m}^{-1}$. The

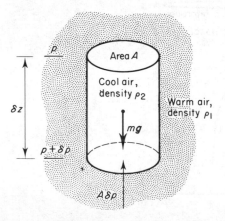

Figure 2.10   Stability of the atmosphere

natural temperature lapse rate $\delta T'$ occurring in the atmosphere is found to be of the order of $0{\cdot}0065$ K m^{-1}. Since the lapse rates differ for the ascending element of air and the surrounding atmosphere the changes of density with altitude will also differ and can be calculated from equation (2.15). For example, if ρ_1 = density of air at sea level, ρ_2 = density of air at an elevation of 1000 m, $R = 287$ J kg^{-1} K^{-1}, then:

(i) Assuming $\delta T = 0{\cdot}01$ K m^{-1}, for the air in the displaced element,

$$\frac{g}{R\delta T} - 1 = \frac{9{\cdot}81}{287 \times 0{\cdot}01} - 1 = 2{\cdot}418,$$

and from equation (2.15),

$$\rho_2 = \rho_1 \left\{1 - \frac{0{\cdot}01 \times 1000}{288}\right\}^{2{\cdot}48},$$

$$\frac{\rho_2}{\rho_1} = 0{\cdot}9181$$

for air in the element;

(ii) Assuming $\delta T = 0{\cdot}0065$ K m^{-1} for the surrounding atmosphere,

$$\frac{g}{R\delta T} - 1 = \frac{9{\cdot}81}{287 \times 0{\cdot}0065} - 1 = 4{\cdot}258,$$

$$\frac{\rho_2}{\rho_1} = \left\{1 - \frac{0{\cdot}0065 \times 1000}{288}\right\}^{4{\cdot}258} = 0{\cdot}9018$$

for the surrounding air.

Thus, the density of the ascending element expanding adiabatically decreases less rapidly than that of the surrounding air; the element eventually becomes denser than the surroundings and tends to fall back to its original level. The atmosphere, therefore, tends to be stable under normal conditons. If, however, the natural temperature lapse rate were to exceed the adiabatic lapse rate, equilibrium would be unstable and an element displaced upwards would continue to rise. Such conditions can arise in thundery weather.

2.13. Pressure and head

In a fluid of constant density, $dp/dz = -\rho g$ can be integrated immediately to give

$$p = -\rho g z + \text{constant}.$$

In a liquid, the pressure p at any depth z, measured downwards from the free surface so that $z = -h$ (Fig. 2.11), will be

$$p = \rho g h + \text{constant}$$

and, since the pressure at the free surface will normally be atmospheric pressure p_{atm},

$$p = \rho g h + p_{\text{atm}}. \tag{2.16}$$

It is often convenient to take atmospheric pressure as a datum. Pressures measured above atmospheric pressure are known as *gauge pressures*.

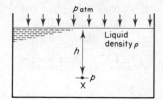

Figure 2.11   Pressure and head

Since atmospheric pressure varies with atmospheric conditions, a perfect vacuum is taken as the absolute standard of pressure. Pressures measured above perfect vacuum are called *absolute pressures*:

Absolute pressure = Gauge pressure + Atmospheric pressure.

Taking p_{atm} as zero, equation (2.16) becomes

$$p = pgh,\tag{2.17}$$

which indicates that, if g is assumed constant, the gauge pressure at a point X (Fig. 2.11) can be defined by stating the vertical height h, called the *head*, of a column of a given fluid of mass density ρ which would be necessary to produce this pressure. Note that when pressures are expressed as head, it is essential that the mass density ρ is given or the fluid named. For example, since from equation (2.17) $h = p/\rho g$ a pressure of 100 kN m^{-2} can be expressed in terms of water ($\rho_{H_2O} = 10^3$ kg m^{-3}) as a head of $(100 \times 10^3)/(10^3 \times 9\cdot81) = 10\cdot19$ m of water. Alternatively, in terms of mercury (relative density 13·6) a pressure of 100 kN m^{-2} will correspond to a head of $(100 \times 10^3)/(13\cdot6 \times 10^3 \times 9\cdot81) = 0\cdot75$ m of mercury.

Example 2.5

A cylinder contains a fluid at a gauge pressure of 350 kN m^{-2}. Express this pressure in terms of a head of (a) water ($\rho_{H_2O} = 1000$ kg m^{-3}), (b) mercury (relative density 13·6).

What would be the absolute pressure in the cylinder if the atmospheric pressure is 101·3 kN m^{-2}?

Solution
From equation (2.17), head, $h = p/\rho g$.

(a) Putting $p = 350 \times 10^3$ N m^{-2}, $\rho = \rho_{H_2O} = 1000$ kg m^{-3},

$$\text{Equivalent head of water} = \frac{350 \times 10^3}{10^3 \times 9\cdot81} = \textbf{35·68 m.}$$

(b) For mercury $\rho_{Hg} = \sigma\rho_{H_2O} = 13\cdot6 \times 1000$ kg m^{-3},

$$\text{Equivalent head of water} = \frac{350 \times 10^3}{13\cdot6 \times 10^3 \times 9\cdot81} = \textbf{2·62 m.}$$

Absolute pressure = Gauge pressure + Atmospheric pressure

$$= 350 + 101\cdot3 = \textbf{451·3 kN m}^{-2}.$$

2.14. The hydrostatic paradox

From equation (2.17) it can be seen that the pressure exerted by a fluid is dependent only on the vertical head of fluid and its mass density ρ; it is not affected by the weight of the fluid present. Thus, in Fig. 2.12 the four vessels

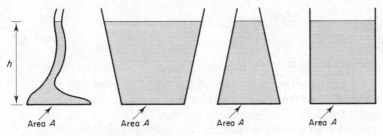

Figure 2.12   The hydrostatic paradox

all have the same base area A and are filled to the same height h with the same liquid of density ρ.

Pressure on bottom in each case, $p = \rho g h$,

Force on bottom = Pressure × Area = $pA = \rho g h A$.

Thus, although the weight of fluid is obviously different in the four cases, the force on the bases of the vessels is the same, depending on the depth h and the base area A.

2.15. Pressure measurement by manometer

The relationship between pressure and head is utilized for pressure measurement in the manometer or liquid gauge. The simplest form is the pressure tube or *piezometer* shown in Fig. 2.13, consisting of a single vertical tube,

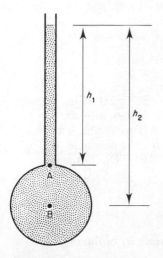

Figure 2.13   Pressure tube or piezometer

open at the top, inserted into a pipe or vessel containing liquid under pressure which rises in the tube to a height depending on the pressure. If the top of the tube is open to the atmosphere, the pressure measured is 'gauge' pressure:

Pressure at A = Pressure due to column of liquid of height h_1,

$$p_A = \rho g h_1.$$

Similarly,

Pressure at B = $p_B = \rho g h_2$.

If the liquid is moving in the pipe or vessel, the bottom of the tube must be flush with the inside of the vessel, otherwise the reading will be affected by the velocity of the fluid. This instrument can only be used with liquids, and the height of the tube which can conveniently be employed limits the maximum pressure that can be measured.

Example 2.6

What is the maximum gauge pressure of water that can be measured by means of a piezometer tube 2 m high? (Mass density of water $\rho_{H_2O} = 10^3$ kg m^{-3}.)

Solution

Since $p = \rho g h$ for maximum pressure, put $\rho = \rho_{H_2O} = 10^3$ and $h = 2$ m, giving

Maximum pressure, $p = 10^3 \times 9\cdot81 \times 2 = \mathbf{19\cdot62 \times 10^3\,N\,m^{-2}}$.

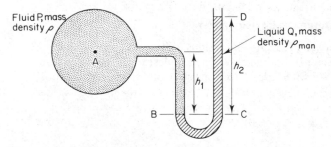

Figure 2.14   U-tube manometer

The U-tube gauge, shown in Fig. 2.14, can be used to measure the pressure of either liquids or gases. The bottom of the U-tube is filled with a mano-metric liquid Q which is of greater density ρ_{man} and is immiscible with the fluid P, liquid or gas, of density ρ, whose pressure is to be measured. If B is the level of the interface in the left-hand limb and C is a point at the same level in the right-hand limb,

Pressure p_B at B = Pressure p_C at C.

For the left-hand limb,

p_B = Pressure p_A at A + Pressure due to depth h_1 of fluid P

$= p_A + \rho g h_1$.

For the right-hand limb,

p_C = Pressure p_D at D + Pressure due to depth h_2 of liquid Q.

But p_D = Atmospheric pressure = Zero gauge pressure

and so $p_C = 0 + \rho_{man}gh_2$.

Since $p_B = p_C$,

$$p_A + \rho gh_1 = \rho_{man}gh_2,$$

$$p_A = \rho_{man}gh_2 - \rho gh_1 \qquad (2.18)$$

Example 2.7

A U-tube manometer similar to that shown in Fig. 2.14 is used to measure the gauge pressure of a fluid P of density $\rho = 800$ kg m^{-3}. If the density of the liquid Q is $13 \cdot 6 \times 10^3$ kg m^{-3}, what will be the gauge pressure at A if (a) $h_1 = 0 \cdot 5$ m and D is $0 \cdot 9$ m *above* BC, (b) $h_1 = 0 \cdot 1$ m and D is $0 \cdot 2$ m *below* BC?

Solution
(a) In equation (2.18), $\rho_{man} = 13 \cdot 6 \times 10^3$ kg m^{-3}, $\rho = 0 \cdot 8 \times 10^3$ kg m^{-3}, $h_1 = 0 \cdot 5$ m, $h_2 = 0 \cdot 9$ m, therefore:

$$p_A = 13 \cdot 6 \times 10^3 \times 9 \cdot 81 \times 0 \cdot 9 - 0 \cdot 8 \times 10^3 \times 9 \cdot 81 \times 0 \cdot 5$$

$$= \mathbf{116 \cdot 15 \times 10^3 \, N m^{-2}}.$$

(b) Putting $h_1 = 0 \cdot 1$ m and $h_2 = -0 \cdot 2$ m, since D is below BC:

$$p_A = 13 \cdot 6 \times 10^3 \times (-0 \cdot 2) - 0 \cdot 8 \times 10^3 \times 0 \cdot 1$$

$$= -\, \mathbf{27 \cdot 45 \times 10^3 \, N m^{-2}},$$

the negative sign indicating that p_A is below atmospheric pressure.

In Fig. 2.15, a U-tube gauge is arranged to measure the pressure difference between two points in a pipeline. As in the previous case, the principle involved in calculating the pressure difference is that the pressure at the same

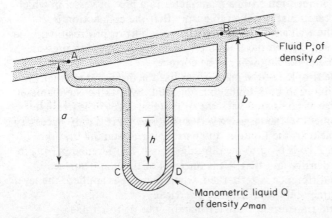

Figure 2.15   Measurement of pressure difference

level CD in the two limbs must be the same, since the fluid in the bottom of the U-tube is at rest. For the left-hand limb,

$$p_C = p_A + \rho g a$$

For the right-hand limb,

$$p_D = p_B + \rho g(b - h) + \rho_{man} g h.$$

Since $p_C = p_D$,

$$p_A + \rho g a = p_B + \rho g(b - h) + \rho_{man} g h,$$

Pressure difference $= p_A - p_B = \rho g(b - a) + hg\,(\rho_{man} - \rho).$ (2.19)

Example 2.8

A U-tube manometer is arranged, as shown in Fig. 2.15, to measure the pressure difference between two points A and B in a pipeline conveying water of density $\rho = \rho_{H_2O} = 10^3$ kg m^{-3}. The density of the manometric liquid Q is $13 \cdot 6 \times 10^3$ kg m^{-3}, and point B is $0 \cdot 3$ m higher than point A. Calculate the pressure difference when $h = 0 \cdot 7$ m.

Solution

In equation (2.19), $\rho = 10^3$ kg m^{-3}, $\rho_{man} = 13 \cdot 6 \times 10^3$ kg m^{-3}, $(b - a) = 0 \cdot 3$ m, and $h = 0 \cdot 7$ m.

Pressure difference $= p_A - p_B$

$$= 10^3 \times 9 \cdot 81 \times 0 \cdot 3 + 0 \cdot 7 \times 9 \cdot 81 (13 \cdot 6 - 1) \times 10^3 \, \text{N m}^{-2}$$

$$= 89 \cdot 467 \times 10^3 \, \text{N m}^{-2}.$$

In both the above cases, if the fluid P is a gas its density ρ can usually be treated as negligible compared to ρ_{man} and the equations (2.18) and (2.19) can be simplified.

In forming the connection from a manometer to a pipe or vessel in which a fluid is flowing, care must be taken to ensure that the connection is perpendicular to the wall and flush internally. Any burr or protrusion on the inside of the wall will disturb the flow and cause a local change in pressure so that the manometer reading will not be correct.

Industrially, the simple U-tube manometer has the disadvantage that the movement of the liquid in both limbs must be read. By making the diameter of one leg very large as compared with the other (Fig. 2.16), it is possible to make the movement in the large leg very small, so that it is only necessary to read the movement of the liquid in the narrow leg. Assuming that the manometer in Fig. 2.16 is used to measure the pressure difference $p_1 - p_2$ in a gas of negligible density and that XX is the level of the liquid surface when the pressure difference is zero, then, when pressure is applied, the level in the right-hand limb will rise a distance z vertically.

Volume of liquid transferred from left-hand leg to right-hand leg
$$= z \times (\pi/4)d^2;$$

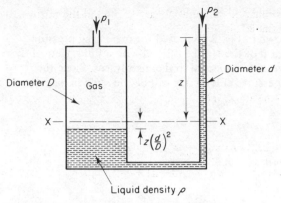

Figure 2.16   U-tube with one leg enlarged

Fall in level of the left-hand leg

$$= \frac{\text{Volume transferred}}{\text{Area of left-hand leg}}$$

$$= \frac{z \times (\pi/4) \, d^2}{(\pi/4) D^2} = z \left(\frac{d}{D}\right)^2.$$

The pressure difference, $p_1 - p_2$, is represented by the height of the manometric liquid corresponding to the new difference of level:

$$p_1 - p_2 = \rho g \{z + z \, (d/D)^2\} = \rho g z \{1 + (d/D)^2\},$$

or, if D is large compared with d,

$$p_1 - p_2 = \rho g z.$$

If the pressure difference to be measured is small, the leg of the U-tube may be inclined as shown in Fig. 2.17. The movement of the meniscus along

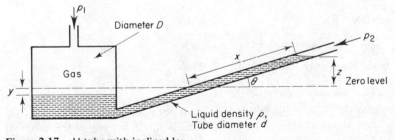

Figure 2.17   U-tube with inclined leg

the inclined leg, read off on the scale, is considerably greater than the change in level z:

$$\text{Pressure difference}, \, p_1 - p_2 = \rho g z = \rho g x \sin \theta.$$

The manometer can be made as sensitive as may be required by adjusting the angle of inclination of the leg and choosing a liquid with a suitable value of density ρ to give a scale reading x of the desired size for a given pressure difference.

The *inverted U-tube* shown in Fig. 2.18 is used for measuring pressure differences in liquids. The top of the U-tube is filled with a fluid, frequently air, which is less dense than that connected to the instrument. Since the fluid in the top is at rest, pressures at level XX will be the same in both limbs.

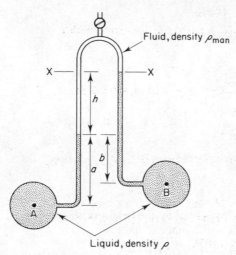

Figure 2.18   Inverted U-tube manometer

For the left-hand limb,

$$p_{XX} = p_A - \rho g a - \rho_{man} g h.$$

For the right-hand limb,

$$p_{XX} = p_B - \rho g (b + h).$$

Thus, $p_B - p_A = \rho g (b - a) + g h (\rho - \rho_{man}),$

or, if A and B are at the same level,

$$p_B - p_A = (\rho - \rho_{man}) g h.$$

If the top of the tube is filled with air ρ_{man} is negligible compared with ρ and $p_B - p_A = \rho g h$. On the other hand, if the liquid in the top of the tube is chosen so that ρ_{man} is very nearly equal to ρ, and provided that the liquids do not mix, the result will be a very sensitive manometer giving a large value of h for a small pressure difference.

Example 2.9

An inverted U-tube of the form shown in Fig. 2.18 is used to measure the pressure difference between two points A and B in an inclined pipeline through which water is flowing ($\rho_{H_2O} = 10^3$ kg m^{-3}). The difference of level

$h = 0.3$ m, $a = 0.25$ m and $b = 0.15$ m. Calculate the pressure difference $p_B - p_A$ if the top of the manometer is filled with (a) air, (b) oil of relative density 0.8.

Solution
In either case, the pressure at XX will be the same in both limbs, so that

$$p_{XX} = p_A - \rho g a - \rho_{man} g h = p_B - \rho g (b + h),$$

$$p_B - p_A = \rho g (b - a) + g h (\rho - \rho_{man}).$$

(a) If the top is filled with air ρ_{man} is negligible compared with ρ. Therefore,

$$p_B - p_A = \rho g (b - a) + \rho g h = \rho g (b - a + h)$$

Putting $\rho = \rho_{H_2O} = 10^3$ kg m^{-3}, $b = 0.15$ m, $a = 0.25$ m, $h = 0.3$ m:

$$p_B - p_A = 10^3 \times 9.81 \, (0.15 - 0.25 + 0.3)$$

$$= 1.962 \times 10^3 \, \text{N m}^{-2}.$$

(b) If the top is filled with oil of relative density 0.8, $\rho_{man} = 0.8 \rho_{H_2O}$ $= 0.8 \times 10^3$ kg m^{-3}.

$$p_B - p_A = \rho g (b - a) + g h (\rho - \rho_{man})$$

$$= 10^3 \times 9.81 \, (0.15 - 0.25) + 9.81 \times 0.3 \times 10^3 \, (1 - 0.8) \, \text{N m}^{-2}$$

$$= 10^3 \times 9.81 \, (-0.1 + 0.06)$$

$$= -392.4.$$

The manometer in its various forms is an extremely useful type of pressure gauge, but suffers from a number of limitations. While it can be adapted to measure very small pressure differences, it cannot be used conveniently for large pressure differences — although it is possible to connect a number of manometers in series and to use mercury as the manometric fluid to improve the range. A manometer does not have to be calibrated against any standard; the pressure difference can be calculated from first principles. However, for accurate work, the temperature should be known, since this will affect the density of the fluids. Some liquids are unsuitable for use because they do not form well-defined menisci. Surface tension can also cause errors due to capillary rise; this can be avoided if the diameters of the tubes are sufficiently large — preferably not less than 15 mm diameter. It is difficult to correct for surface tension, since its effect will depend upon whether the tubes are clean. A major disadvantage of the manometer is its slow response, which makes it unsuitable for measuring fluctuating pressures. Even under comparatively static conditions, slight fluctuations of pressure can make the liquid in the manometer oscillate, so that it is difficult to get a precise reading of the levels of the liquid in the gauge. These oscillations can be reduced by putting restrictions in the manometer connections. It is also essential that the pipes connecting the manometer to the pipe or vessel containing the liquid under pressure should be filled with this liquid and that there should be no air bubbles in the liquid.

FLUIDS IN RELATIVE EQUILIBRIUM

2.16. Relative equilibrium

If a fluid is contained in a vessel which is at rest, or moving with constant linear velocity, it is not affected by the motion of the vessel; but if the container is given a continuous acceleration, this will be transmitted to the fluid and affect the pressure distribution in it. Since the fluid remains at rest relative to the container, there is no relative motion of the particles of the fluid and, therefore, no shear stresses, fluid pressure being everywhere normal to the surface on which it acts. Under these conditions the fluid is said to be in relative equilibrium.

2.17. Pressure distribution in a liquid subject to horizontal acceleration

Figure 2.19 shows a liquid contained in a tank which has an acceleration a. A particle of mass m on the free surface at O will have the same acceleration

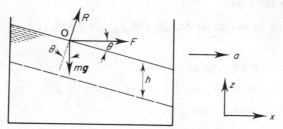

Figure 2.19   Effect of horizontal acceleration

a as the tank and so will be subjected to an accelerating force F. From Newton's law,

$$F = ma \qquad (2.20)$$

Also, F is the resultant of the fluid pressure force R, acting normally to the free surface at O, and the weight of the particle mg, acting vertically. Therefore,

$$F = mg \tan \theta \qquad (2.21)$$

Comparing equations (2.20) and (2.21),

$$\tan \theta = a/g \qquad (2.22)$$

and is constant for all points on the free surface. Thus, the free surface is a plane inclined at a constant angle θ to the horizontal.

Since the acceleration is horizontal, vertical forces are not changed and the pressure at any depth h below the surface will be ρgh. Planes of equal pressure lie parallel to the free surface.

2.18. Effect of vertical acceleration

If the acceleration is vertical, the free surface will remain horizontal. Considering a vertical prism of cross-sectional area A (Fig. 2.20), subject to

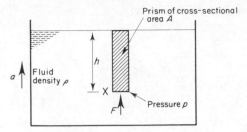

Figure 2.20   Effect of vertical acceleration

an upward acceleration a, then at depth h below the surface, where the pressure is p,

> Upward accelerating force, F = Force due to p − weight of prism
>
> $$= pA - \rho ghA.$$

By Newton's second law,

> F = Mass of prism × Acceleration
>
> $= \rho hA \times a.$

Therefore,

> $$pA - \rho ghA = \rho hAa,$$
>
> $$p = \rho gh(1 + a/g) \qquad\qquad (2.23)$$

2.19. General expression for the pressure in a fluid in relative equilibrium

If $\partial p/\partial x$, $\partial p/\partial y$ and $\partial p/\partial z$ are the rates of change of pressure p in the x, y and z directions (Fig. 2.21) and a_x, a_y and a_z the accelerations,

$$\text{Force in } x \text{ direction, } F_x = p\delta y\delta z - \left(p + \frac{\partial p}{\partial x}\delta x\right)\delta y\delta z$$

$$= -\frac{\partial p}{\partial x}\delta x\delta y\delta z$$

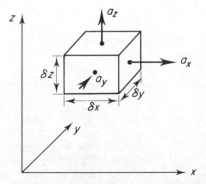

Figure 2.21   Relative equilibrium: the general case

By Newton's second law, $F_x = \rho\delta x\delta y\delta z \times a_x$, therefore,

$$-\frac{\partial p}{\partial x} = \rho a_x.$$

(2.24)

Similarly, in the y direction,

$$-\frac{\partial p}{\partial y} = \rho a_y.$$

(2.25)

In the vertical z direction, the weight of the element $\rho g\delta x\delta y\delta z$ must be considered:

$$F_z = p\delta x\delta y - \left(p + \frac{\partial p}{\partial z}\delta z\right)\delta x\delta y - \rho g\delta x\delta y\delta z$$

$$= -\frac{\partial p}{\partial z}\delta x\delta y\delta z - \rho g\delta x\delta y\delta z.$$

By Newton's second law,

$$F_z = \rho\delta x\delta y\delta z \times a_z,$$

therefore,

$$-\frac{\partial p}{\partial z} = \rho(g + a_z).$$

(2.26)

For an acceleration a_s in any direction in the x–z plane making an angle ϕ with the horizontal, the components of the acceleration are

$$a_x = a_s \cos\phi \quad \text{and} \quad a_z = a_s \sin\phi.$$

Now $\quad \dfrac{dp}{ds} = \dfrac{\partial p}{\partial x}\dfrac{dx}{ds} + \dfrac{\partial p}{\partial z}\dfrac{dz}{ds}.$

(2.27)

For the free surface and all other planes of constant pressure, $dp/ds = 0$. If θ is the inclination of the planes of constant pressure to the horizontal, $\tan\theta = dz/dx$. Putting $dp/ds = 0$ in equation (2.27),

$$\frac{\partial p}{\partial x}\frac{dx}{ds} + \frac{\partial p}{\partial z}\frac{dz}{ds} = 0$$

$$\frac{dz}{dx} = \tan\theta = -\frac{\partial p}{\partial x}\bigg/\frac{\partial p}{\partial z}.$$

Substituting from equations (2.24) and (2.26),

$$\tan\theta = -a_x/(g + a_z),$$

(2.28)

or, in terms of a_s,

$$\tan\theta = -\frac{a_s \cos\phi}{(g + a_s \sin\phi)}.$$

(2.29)

For the case of horizontal acceleration, $\phi = 0$ and equation (2.29) gives $\tan\theta = -a_s/g$, which agrees with equation (2.22). For vertical acceleration, $\phi = 90°$ giving $\tan\theta = 0$, indicating that the free surface remains horizontal.

Since, for the two-dimensional case,

$$dp = \frac{\partial p}{\partial x} dx + \frac{\partial p}{\partial z} dz,$$

the pressure at a particular point in the fluid can be found by integration:

$$p = \int dp = \int \frac{\partial p}{\partial x} dx + \int \frac{\partial p}{\partial z} dz.$$

Substituting from equations (2.24) and (2.26) and assuming that ρ is constant

$$p = \int (-\rho a_x) dx + \int \{-\rho (g + a_z)\} dz + \text{constant}$$

$$= -p (x a_s \cos \phi - gz - z a_s \sin \phi) + \text{constant},$$

or, since $x/z = \tan \theta$,

$$p = -z (a_s \tan \theta \cos \phi - g - a_s \sin \phi) + \text{constant}, \qquad (2.30)$$

where z is positive measured upwards from a horizontal datum fixed relative to the fluid.

Example 2.10

A rectangular tank $1\cdot2$ m deep and 2 m long is used to convey water up a ramp inclined at an angle ϕ of $30°$ to the horizontal (Fig. 2.22). Calculate

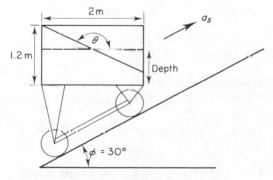

Figure 2.22   Acceleration up an inclined plane

the inclination of the water surface to the horizontal when (a) the acceleration parallel to the slope on starting from the bottom is 4 m s^{-2}, (b) the deceleration parallel to the slope on reaching the top is $4\cdot5$ m s^{-2}.

If no water is to be spilt during the journey what is the greatest depth of water permissible in the tank when it is at rest?

Solution

The slope of the water surface is given by equation (2.29).

During acceleration, $a_s = +4 \text{ m s}^{-2}$.

$$\tan \theta_A = \frac{-a_s \cos \phi}{g + a_s \sin \phi} = -\frac{4 \cos 30°}{9 \cdot 81 + 4 \sin 30°}$$

$$= -0 \cdot 2933$$

$$\theta_A = 163° 39'.$$

During retardation, $a_s = -4 \cdot 5 \text{ m s}^{-2}$.

$$\tan \theta_R = -\frac{(-4 \cdot 5) \cos 30°}{9 \cdot 81 - 4 \cdot 5 \sin 30°} = 0 \cdot 5154$$

$$\theta_R = 27° 16'.$$

Since $180° - \theta_R > \theta_A$, the worst case for spilling will be during retardation. When the water surface is inclined, the maximum depth at the tank wall will be

Depth + $\frac{1}{2}$ Length x tan θ,

which must not exceed 1·2 m if the water is not to be spilt. Putting length = 2 m, tan θ = tan θ_R = 0·5154,

$$\text{Depth} + (2 \cdot 0/2) \times 0 \cdot 5154 = 1 \cdot 2,$$

$$\text{Depth} = 1 \cdot 2 - 0 \cdot 5154$$

$$= 0 \cdot 6846 \text{ m}.$$

The equations derived in this section indicate:

(i) if there is no horizontal acceleration, $a_x = 0$ and tan $\theta = 0$ so that surfaces of constant pressure are horizontal;

(ii) in free space, g will be zero so that tan $\theta = -a_x/a_z$ (surfaces of constant pressure will therefore be perpendicular to the resultant acceleration);

(iii) since free surfaces of liquids are surfaces of constant pressure, their inclination will be determined by equation (2.29); thus, if a_x and a_y are zero, the free surface will be horizontal.

2.20. Forced vortex

A body of fluid, contained in a vessel which is rotating about a vertical axis with uniform angular velocity, will eventually reach relative equilibrium and rotate with the same angular velocity ω as the vessel, forming a forced vortex. The acceleration of any particle of fluid at radius r due to rotation will be $-\omega^2 r$ perpendicular to the axis of rotation, taking the direction of r as positive outward from the axis. Thus, from equation (2.24),

$$\frac{dp}{dr} = -\rho \omega^2 r.$$

Figure 2.23 shows a cylindrical vessel containing liquid rotating about its axis, which is vertical. At any point P on the free surface, the inclination θ of the

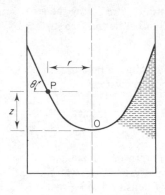

Figure 2.23   Forced vortex

free surface is given by equation (2.28),

$$\tan \theta = -\frac{a_x}{g + a_z} = \frac{\omega^2 r}{g} = \frac{dz}{dr}. \tag{2.31}$$

The inclination of the free surface varies with r and, if z is the height of P above O, the surface profile is given by integrating equation (2.31):

$$z = \int_0^x \frac{\omega^2 r}{g}\, dr = \frac{\omega^2 r^2}{2g} + \text{constant}. \tag{2.32}$$

Thus, the profile of the water surface is a paraboloid. Similarly, other surfaces of equal pressure will also be paraboloids.

If the container is closed and the fluid has no free surface, the paraboloid drawn to represent the imaginary free surface represents the variation of pressure head with radius. Thus, the pressure p at radius r is given by equation (2.32) as

$$z = p/\rho g = \omega^2 r^2/2g + \text{constant},$$

$$p = \rho \omega^2 r^2/2 + \text{constant}. \tag{2.33}$$

EXERCISES 2

2.1   Calculate the pressure in the ocean at a depth of 2000 m assuming that salt water is (a) incompressible with a constant density of 1002 kg/m^3, (b) compressible with a bulk modulus of 2·05 GN/m^2 and a density at the surface of 1002 kg/m^3.
[19.66 MN/m^2, 19.75 MN/m^2]
2.2   What will be (a) the gauge pressure, (b) the absolute pressure of water at a depth of 12 m below the free surface. Assume the density of water to be 1000 kg/m^3 and the atmospheric pressure 101 kN/m^2.
[117·72 kN/m^2, 218·72 kN/m^2]

2.3   Determine the pressure in N/m^2 at (a) a depth of 6 m below the free surface of a body of water and (b) at a depth of 9 m below the free surface of a body of oil of specific gravity 0·75.
[58·9 kN/m^2, 66·2 kN/m^2]

2.4   What depth of oil, specific gravity 0·8, will produce a pressure of 120 kN/m^2. What would be the corresponding depth of water.
[15·3 m, 12·2 m]

2.5   At what depth below the free surface of oil having a density of 600 kg/m^3 will the pressure be equal to 1 bar.
[17 m]

2.6   What would be the pressure in kN/m^2 if the equivalent head is measured as 400 mm of (a) mercury of specific gravity 13·6, (b) water, (c) oil of specific weight 7·9 kN/m^3, (d) a liquid of density 520 kg/m^3.
[53·4 kN/m^2, 3·92 kN/m^2, 3·16 kN/m^2, 2·04 kN/m^2]

2.7   A mass of 50 kg acts on a piston of area 100 cm^2. What is the intensity of pressure on the water in contact with the underside of the piston, if the piston is in equilibrium.
[4·905 x 10^4 N/m^2]

2.8   A vertical pipe 30 m long and 25 mm diameter has its lower end open and flush with the inner surface of the cover of a box which is 0·15 m high, square in plan view, each side being 0·6 m long. The bottom of the box is horizontal. If the box and pipe are completely filled with water calculate (a) the hydrostatic force exerted on the bottom of the box, (b) the force exerted on the floor upon which the box rests excluding the weight of the box and the pipe.
[106·5 kN, 0·674 kN]

2.9   The pressure head in a gas main at a point 120 m above sea level is equivalent to 180 mm of water. Assuming that the densities of air and gas remain constant and equal to 1·202 kg/m^3 and 0·561 kg/m^3 respectively, what will be the pressure head in mm of water at sea level.
[103 mm]

2.10   What is the gauge pressure and the absolute pressure at a point in water 10 m below the free surface. Assume atmospheric pressure to be 100 kN/m^2 and the density of water 1000 kg/m^3.
[98·1 kN/m^2, 198·1 kN/m^2]

2.11   What is the pressure in kN/m^2 absolute and gauge at a point 3 m below the free surface of a liquid having a mass density of 1·53 x 10^3 kg/m^3 if the atmospheric pressure is equivalent to 750 mm of mercury. Take the specific gravity of mercury as 13·6 and the density of water as 1000 kg/m^3.
[145 kN/m^2, 45 kN/m^2]

2.12   A manometer connected to a pipe in which a fluid is flowing indicates a negative gauge pressure head of 50 mm of mercury. What is the absolute pressure in the pipe in N/m^2 if the atmospheric pressure is 1 bar.
[93.3 kN/m^2]

2.13   What is the gauge pressure and the absolute pressure of the air in Fig. 2.24 if the barometric pressure is 780 mm of mercury and the liquid is (a) water of density 1000 kg/m^3, (b) oil of specific weight 7·5 x 10^3 N/m^3.
[4·9 kN/m^2, 99·1 kN/m^2; 3·7 kN/m^2, 100·3 kN/m^2]

2.14   An open tank contains oil of specific gravity 0·75 on top of water. If the depth of oil is 2 m and the depth of water 3 m, calculate the gauge and

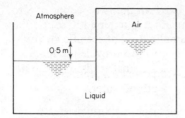

Figure 2.24

absolute pressures at the bottom of the tank when the atmospheric pressure is 1 bar.
[44·15 kN/m², 144·15 kN/m²]

2.15   A cylinder having a cross-sectional area of 0·2 m² has a vertical level gauge attached to its side and is filled with water to a height of 1 m above a horizontal datum. What will be the height of the water column in the level gauge above this datum if (*a*) 120 litres of oil of density 800 kg/m³ is poured on top of the water in the cylinder, (*b*) in addition a piston weighing 450 N is placed on top of the oil, there being no leakage of oil past the piston.
[1·48 m, 1·71 m]

2.16   A closed tank contains 0·5 m of mercury, 2 m of water, 3 m of oil of density 600 kg/m³ and there is an air space above the oil. If the gauge pressure at the bottom of the tank is 200 kN/m², what is the pressure of the air at the top of the tank.
[96 kN/m²]

2.17   An inverted cone 1 m high and open at the top contains water to half its height, the remainder being filled with oil of specific gravity 0·9. If half the volume of water is drained away find the pressure at the bottom (apex) of the inverted cone.
[9020 N/m²]

2.18   The gauge reading at A in Fig. 2.25 is 2 kN/m². Determine (*a*) the elevations from datum of the liquids in the open piezometer tubes E, F, and G, (*b*) the difference *h* between the mercury levels in the U-tube manometer.
[3·1 m, 4·896 m, 2·26 m; 213 mm]

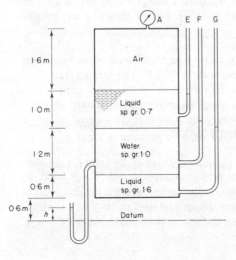

Figure 2.25

2.19 A hydraulic press has a diameter ratio between the two pistons of 8 to 1. The diameter of the larger piston is 600 mm and it is required to support a mass of 3500 kg. The press is filled with a hydraulic fluid of specific gravity 0·8. Calculate the force required on the smaller piston to provide the required force (a) when the two pistons are level, (b) when the smaller piston is 2·6 m below the larger piston.

[536·5 N, 626·6 N]

2.20 The diameter of the piston of an hydraulic jack is six times greater than the diameter of the plunger which is 20 mm. They are both of the same height and made of the same material. The piston weighs 100 N and supports a mass of 50 kg. The jack is filled with oil of density 900 kg/m³. Calculate the force required at the plunger to support the piston and the mass it carries at a level 10 cm above that of the plunger.

[13·9 N]

2.21 Show that the ratio of the pressures (p_2/p_1) and densities (ρ_2/ρ_1) for altitudes h_2 and h_1 in an isothermal atmosphere are given by

$$\frac{p_2}{p_1} = \frac{\rho_2}{\rho_1} = e^{-g(h_2 - h_1)/RT}$$

What increase in altitude is necessary in the stratosphere to halve the pressure. Assume a constant temperature of $-56·5\,°C$ and the gas constant $R = 287$ J/kg K.

[4390 m]

2.22 From observation it is found that at a certain altitude in the atmosphere the temperature is $-25\,°C$ and the pressure is 45·5 kN/m² while at sea level the corresponding values are $15\,°C$ and 101·5 kN/m². Assuming that the temperature decreases uniformly with increasing altitude, estimate the temperature lapse rate and the pressure and density of the air at an altitude of 3000 m.

[6·37°C/1000 m, 70·22 kN/m², 0·91 kg/m³]

2.23 Show that the ratio of the atmospheric pressure at an altitude h_1 to that at sea level may be expressed as $(p/p_0) = (T/T_0)^n$, a uniform temperature lapse rate being assumed. Find the ratio of the pressures and the densities at 10 700 m and at sea level taking the standard atmosphere as having a sea level temperature of $15\,°C$ and a lapse rate of $6·5\,°C$ per 1000 m to a minimum of $-56·5\,°C$.

[0·237, 0·312]

2.24 The barometric pressure of the atmosphere at sea level is equivalent to 760 mm of mercury and its temperature is 288 K. The temperature decreases with increasing altitude at the rate of 6·5 K per 1000 m until the stratosphere is reached in which the temperature remains constant at 216·5 K. Calculate the pressure in mm of mercury and the density in kg/m³ at an altitude of 14 500 m. Assume $R = 287$ J/kg K.

[97·52 mm, 0·209 kg/m³]

2.25 In the standard atmosphere the temperature at sea level is taken as $15\,°C$ and it is assumed that the lapse rate is constant at $6·5\,°C$ per 1000 m until the stratosphere is reached in which the temperature remains constant at $-56·5\,°C$. Calculate the relative pressure and relative density of the atmosphere at a height of 13 700 m taking the sea level values as unity and $R = 287$ J/kg K.

[0·146, 0·194]
2.26   A mercury U-tube manometer (Fig. 2.14) is used to measure the pressure above atmospheric of water in a pipe as shown. If the mercury, liquid Q, is 30 cm below A in the left hand limb and 20 cm above A in the right hand limb, what is the gauge pressure at A. Specific gravity of mercury = 13·6.

If the pressure at A is reduced by 40 kN/m^2 what will be the new difference in level of the mercury.
[63·8 kN/m^2, 18·8 cm]
2.27   In Fig. 2.26 fluid P is water and fluid Q is mercury. If the specific

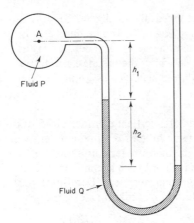

Figure 2.26

weight of mercury is 13·6 times that of water and the atmospheric pressure is 101·3 kN/m^2 what is the absolute pressure at A when $h_1 = 15$ cm and $h_2 = 30$ cm.
[59·8 kN/m^2]
2.28   A U-tube manometer (Fig. 2.27) measures the pressure difference between two points A and B in a liquid of density ρ_1. The U-tube contains

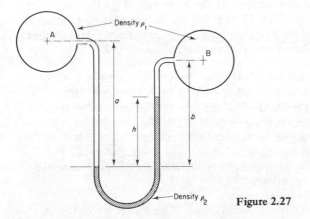

Figure 2.27

mercury of density ρ_2. Calculate the difference of pressure if $a = 1·5$ m,

$b = 0.75$ m and $h = 0.5$ m if the liquid at A and B is water and $\rho_2 = 13 \cdot 6\, \rho_1$.
[54·4 kN/m²]

2.29   The top of an inverted U-tube manometer is filled with oil of specific gravity 0·98 and the remainder of the tube with water of specific gravity 1·01. Find the pressure difference in N/m² between two points at the same level at the base of the legs when the difference of water level is 75 mm.
[22 N/m²]

2.30   An inverted U-tube manometer (Fig. 2.18) has air at the top of the tube. Find the difference of pressure of the water at A and B if $a = 60$ cm, $b = 180$ cm and $h = 45$ cm. Density of water = 1000 kg/m³.
[16·2 x 10³ N/m²]

2.31   An inclined manometer is required to measure an air pressure difference of about 3 mm of water with an accuracy of ±3%. The inclined arm is 8 mm diameter and the enlarged end is 24 mm diameter. The density of the manometer fluid is 740 kg/m³. Find the angle which the inclined arm must make with the horizontal to achieve the required accuracy assuming acceptable readability of 0·5 mm.
[7°35′]

2.32   An inclined tube manometer consists of a vertical cylinder of 35 mm diameter to the bottom of which is connected a tube of 5 mm diameter inclined upwards at 15° to the horizontal. The manometer contains oil of relative density 0·785. The open end of the inclined tube is connected to an air duct while the top of the cylinder is open to atmosphere. Determine the pressure in the air duct if the manometer fluid moves 50 mm along the inclined tube.

What is the error if the movement of the fluid in the cylinder is ignored.
[107·6 N/m², 7·87 N/m²]

2.33   An oil and water manometer consists of a U-tube 4 mm diameter with both limbs vertical. The right-hand limb is enlarged at its upper end to 20 mm diameter. The enlarged end contains oil with its free surface in the enlarged portion and the surface of separation between water and oil is below the enlarged end. The left-hand limb contains water only, its upper end being open to the atmosphere.

When the right-hand side is connected to a cylinder of gas the surface of separation is observed to fall by 25 mm, but the surface of the oil remains in the enlarged end. Calculate the gauge pressure in the cylinder. Assume that the specific gravity of the water is 1·0 and that of the oil 0·9.
[280 N/m²]

2.34   A manometer consists of a U-tube, 7mm internal diameter, with vertical limbs each with an enlarged upper end 44 mm diameter (Fig. 2.28). The left-hand limb and the bottom of the tube is filled with water and the top of the right-hand limb is filled with oil of specific gravity 0·83. The free surfaces of the liquids are in the enlarged ends and the interface between the oil and water is in the tube below the enlarged end. What would be the difference in pressures applied to the free surfaces which would cause the oil/water interface to move 1 cm.
[21 N/m²]

2.35   A vessel 1·4 m wide and 2·0 m long is filled to a depth of 0·8 m with a liquid of mass density 840 kg/m³. What will be the force in newtons on the bottom of the vessel (a) when being accelerated vertically upwards at 4 m/s,

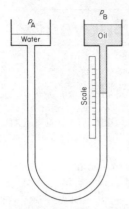

Figure 2.28

(*b*) when the acceleration ceases and the vessel continues to move at a
constant velocity of 7 m/s vertically upwards.
[25 985 N, 18 458 N]
2.36   A pipe 25 mm in diameter is connected to the centre of the top of a
drum 0·5 m in diameter, the cylindrical axis of the pipe and the drum being
vertical. Water is poured into the drum through the pipe until the water level
stands in the pipe 0·6 m above the top of the drum. If the drum and pipe are
now rotated about their vertical axis at 600 rev/min what will be the upward
force exerted on the top of the drum.
[13·26 kN]
2.37   A tube ABCD has the end A open to atmosphere and the end D
closed. The portion ABC is vertical while the portion CD is a quadrant of
radius 250 mm with its centre at B, the whole being arranged to rotate about
its vertical axis ABC. If the tube is completely filled with water to a height in
the vertical limb of 300 mm above C find (*a*) the speed of rotation which will
make the pressure head at D equal to the pressure head at C, (*b*) the value
and position of the maximum pressure head in the curved portion CD when
running at this speed.
[84·5 rev/min, 0·363 m of water]
2.38   A cylindrical vessel 0·6 m in diameter and 0·3 m high is open at the
top except for a lip 50 mm wide all round and normal to the side. It contains
water to a height of 0·2 m above the bottom and is rotated with its axis
vertical. Calculate the speed in rev/min at which the water reaches the inside
edge of the lip and the total pressure of water against the underside of the lip.
[87·5 rev/min, 49·5 N]
2.39   A closed air-tight tank 4 m high, and 1 m in diameter contains water
to a depth of 3·3 m. The air in the tank is at a pressure of 40 kN/m^2 gauge.
What are the absolute pressures at the centre and circumference of the base
of the tank when it is rotating about its vertical axis at a speed of 180 rev/
min. At this speed the water wets the top surface of the tank.
[15·94 m abs., 20·16 m abs]

3 Static forces on surfaces. Buoyancy

3.1. Action of fluid pressure on a surface

Since pressure is defined as force per unit area, when fluid pressure p acts on a solid boundary — or across any plane in the fluid — the force exerted on each small element of area δA will be $p\delta A$, and, since the fluid is at rest, this force will act at right angles to the boundary or plane at the point under consideration.

In a body of fluid, the pressure p may vary from point to point, and the forces on each element of area will also vary. If the fluid pressure acts on or across a plane surface, all the forces on the small elements will be parallel (Fig. 3.1) and can be represented by a single force, called the *resultant force*

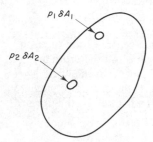

Figure 3.1   Forces on a plane surface

acting at right angles to the plane through a point called the *centre of pressure*.

Resultant force, R = Sum of forces on all elements of area

$$= p_1\delta A_1 + p_2\delta A_2 + \ldots + p_n\delta A_n = \Sigma\, p\,\delta A,$$

where Σ means 'the sum of'.

If the boundary is a curved surface, the elementary forces will act perpendicular to the surface at each point and will, therefore, not be parallel (Fig. 3.2). The resultant force can be found by resolution or by a polygon

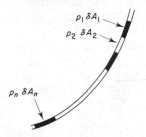

Figure 3.2   Forces on a curved surface

of forces, but will be less than $\Sigma\, p\,\delta A$. For example, in the extreme case of the curved surface of a bucket filled with water (Fig. 3.3), the elementary

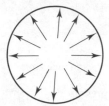

Figure 3.3   Forces on a cylindrical surface

forces acting radially on the vertical wall will balance and the resultant force
will be zero. If this were not so, there would be an unbalanced horizontal
force in some direction and the bucket would move of its own accord.

3.2. Resultant force and centre of pressure on a plane surface under uniform pressure

The pressure p on a plane horizontal surface in a fluid at rest will be the same
at all points, and will act vertically downwards at right angles to the surface.
If the area of the plane surface is A,

Resultant force $= pA$.

It will act vertically downwards and the centre of pressure will be the centroid
of the surface.

For gases, the variation of pressure with elevation is small and so it is
usually possible to assume that gas pressure on a surface is uniform, even
though the surface may not be horizontal. The resultant force is then
pA acting through the centroid of the plane surface.

3.3. Resultant force and centre of pressure on a plane surface immersed in a liquid

Figure 3.4 shows a plane surface PQ of any area A totally immersed in a
liquid of density ρ and inclined at an angle ϕ to the free surface. Considering

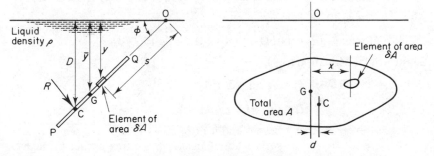

Figure 3.4   Resultant force on a plane surface immersed in a fluid

one side only, there will be a force due to fluid pressure p acting on each
element of area δA. The magnitude of p will depend on the vertical depth
y of the element below the free surface. Taking the pressure at the free

surface as zero, from equation (2.4), and measuring y downwards, $p = \rho g y$, therefore

Force on element of area, $\delta A = p\delta A = \rho g y \delta A$.

Summing the forces on all such elements over the whole surface, since these forces are all perpendicular to the plane PQ,

Resultant force, $R = \Sigma \, \rho g y \delta A$.

If we assume that ρ and g are constant,

$$R = \rho g \Sigma y \delta A. \tag{3.1}$$

The quantity $\Sigma y \delta A$ is the first moment of area of the surface PQ about the free surface of the liquid and is equal to $A\bar{y}$, where A = the area of the whole immersed surface PQ and $\bar{y}$ = the vertical depth to the centroid G of the immersed surface. Substituting in equation (3.1),

$$\text{Resultant force, } R = \rho g A\bar{y}. \tag{3.2}$$

This resultant force R will act perpendicular to the immersed surface at the centre of pressure C at some vertical depth D below the free surface, such that the moment of R about any point will be equal to the sum of the moments of the forces on all the elements δA about the same point. Thus, if the plane of the immersed surface cuts the free surface at O,

$$\text{Moment of } R \text{ about O} = \text{Sum of moments of forces on all elements} \\ \text{of area } \delta A \text{ about O}, \tag{3.3}$$

Force on any small element $= \rho g y \delta A = \rho g s \sin \phi \, . \, \delta A$,

since $h = s \sin \phi$.

Moment of force on element about O $= \rho g s \sin \phi \, . \, \delta A \times s$

$$= \rho g \sin \phi \, . \, \delta A \, . \, s^2.$$

Since ρ, g and ϕ are the same for all elements,

Sum of the moments of all such forces about O $= \rho g \sin \phi \, \Sigma s^2 \delta A$.

Also $R = \rho g A\bar{y}$, therefore,

Moment of R about O $= \rho g A\bar{y} \times \text{OC} = \rho g A\bar{y}(D/\sin \phi)$.

Substituting in equation (3.3),

$$\rho g A\bar{y}(D/\sin \phi) = \rho g \sin \phi \, \Sigma \, s^2 \delta A,$$

$$D = \sin^2 \phi \, (\Sigma \, s^2\delta A)/A\bar{y},$$

$\Sigma \, s^2 \delta A$ = Second moment of area of the immersed
surface about an axis in the free surface through O

$$= I_O = Ak_O^2,$$

where k_O = the radius of gyration of the immersed surface about O. Therefore,

$$D = \sin^2 \phi(I_O/A\bar{y}) = \sin^2 \phi(k_O^2/\bar{y}). \tag{3.4}$$

The values of I_O and k_O^2 can be found if the second moment of area of the immersed surface I_G about an axis through its centroid G parallel to the free surface is known by using the parallel axis rule,

or $Ak_O^2 = Ak_G^2 + A(\bar{y}/\sin \phi)^2$.

Thus $D = \sin^2 \phi \, [k_G^2 + (\bar{y}/\sin \phi)^2/\bar{y}] = \sin^2 \phi (k_G^2/\bar{y}) + \bar{y}$.   (3.5)

The geometrical properties of common figures are given in Table 3.1.

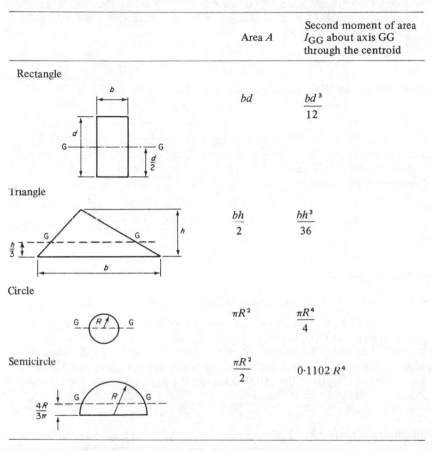

	Area A	Second moment of area I_{GG} about axis GG through the centroid
Rectangle	bd	$\dfrac{bd^3}{12}$
Triangle	$\dfrac{bh}{2}$	$\dfrac{bh^3}{36}$
Circle	πR^2	$\dfrac{\pi R^4}{4}$
Semicircle	$\dfrac{\pi R^2}{2}$	$0{\cdot}1102 \, R^4$

Table 3.1.   Geometrical properties of common figures

From equation (3.5) it can be seen that the centre of pressure will always be below the centroid G except when the surface is horizontal ($\phi = 0°$). As the depth of immersion increases, the centre of pressure will move nearer to the centroid, since for the given surface the change of pressure between the upper and lower edge becomes proportionately smaller in comparison with the mean pressure making the pressure distribution more uniform.

The lateral position of the centre of pressure can be found by taking moments about the line OG, which is the line of intersection of the immersed surface with a vertical plane through G:

$$R \times d = \text{Sum of moments of forces on small elements about OG}$$

$$= \Sigma \rho g \delta A y x$$

Putting $R = \rho g A \bar{y}$,

$$d = (\Sigma \delta A \pi x)/Ay.$$

If the area is symmetrical about a vertical plane through the centroid G, the moment of each small element on one side is balanced by that due to a similar element on the other side so that $\Sigma \delta A y = 0$. Therefore, $d = 0$ and the centre of pressure will be on the axis of symmetry.

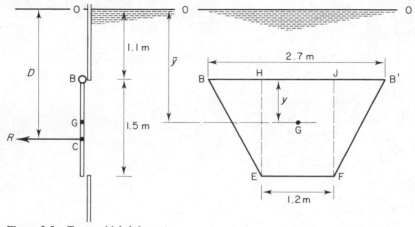

Figure 3.5   Trapezoidal sluice gate

Example 3.1

A trapezoidal opening in the vertical wall of a tank is closed by a flat plate which is hinged at its upper edge (Fig. 3.5). The plate is symmetrical about its centre line and is 1·5 m deep. Its upper edge is 2·7 m long and its lower edge is 1·2 m long. The free surface of the water in the tank stands 1·1 m above the upper edge of the plate. Calculate the moment about the hinge line required to keep the plate closed.

Solution

The moment required to keep the plate closed will be equal and opposite to the moment of the resultant force R due to the water acting at the centre of pressure C, i.e. $R \times CB$. From equation (3.2), $R = \rho g A \bar{y}$.

Area of plate, $A = \frac{1}{2}(2 \cdot 7 + 1 \cdot 2) \times 1 \cdot 5 = 2 \cdot 925$ m^2.

To find the position of the centroid G, take moments of area about BB′, putting the vertical distance GB = y:

$$A \times y = \text{Moment of areas BHE and FJB}' + \text{Moment of EFJH}$$

$$= 2 \times (\tfrac{1}{2} \times 1 \cdot 5 \times 0 \cdot 75) \times 0 \cdot 5 + (1 \cdot 2 \times 1 \cdot 5) \times 0 \cdot 75$$

$$2 \cdot 925 \, y = 0 \cdot 5625 + 1 \cdot 35 = 1 \cdot 9125,$$

$$y = 0 \cdot 654 \text{ m}.$$

Depth to the centre of pressure,

$$\bar{y} = y + \text{OB}$$

$$= 0.654 + 1.1 = 1.754 \text{ m.}$$

Substituting in equation (3.2),

$$\text{Resultant force, } R = 10^3 \times 9.81 \times 2.925 \times 1.754 \text{ N}$$

$$= 50.33 \text{ kN.}$$

From equation (3.4),

Depth to centre of pressure C, $D = \sin^2\phi(I_O/A\bar{y})$.

Using the parallel axis rule for second moments of area

I_O = Second moment of EFJH about O + Second moment of BEH
and B'FJ about O

$$= \left(\frac{1.2 \times 1.5^3}{12} + 1.2 \times 1.5 \times 1.85^2\right)$$

$$+ \left(\frac{1.5 \times 1.5^3}{36} + 1.5 \times 0.75 \times 1.6^2\right) \text{m}^4$$

$$= 9.5186 \text{ m}^4.$$

As the wall is vertical, $\sin\phi = 1$, therefore,

$$\text{Depth to centre of pressure, } D = \frac{9.5186}{2.925 \times 1.754}$$

$$= 1.8553 \text{ m}$$

$$\text{Moment about hinge} = R \times \text{BC} = 50.33 (1.8553 - 1.1)$$

$$= 38.01 \text{ kN m}$$

Example 3.2

The angle between a pair of lock gates (Fig. 3.6) is $140°$ and each gate is 6 m high and 1.8 m wide, supported on hinges 0.6 m from the top and bottom of the gate. If the depths of water on the upstream and downstream sides are 5 m and 1.5 m, respectively, estimate the reactions at the top and bottom hinges.

Solution

Figure 3.6(a) shows the plan view of the gates. F is the force exerted by one gate on the other and is assumed to act perpendicular to the axis of the lock if friction between the gates is neglected. P is the resultant of the water forces P_1 and P_2 (Fig. 3.6(b)) acting on the upstream and downstream faces of the gate, and R is the resultant of the forces R_T and R_B on the top and

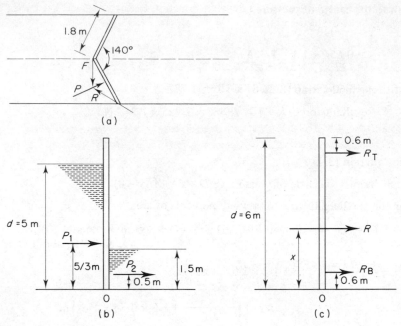

Figure 3.6   Lock gate

bottom hinges. Using equation (3.2)

Upstream water force,

$$P_1 = \rho g A_1 \bar{y}_1$$

$$= 10^3 \times 9 \cdot 81 \times (5 \times 1 \cdot 8) \times 2 \cdot 5$$

$$= 220 \cdot 725 \times 10^3 \text{ N},$$

Downstream water force,

$$P_2 = \rho g A_2 \bar{z}_2$$

$$= 10^3 \times 9 \cdot 81 \times (1 \cdot 5 \times 1 \cdot 8) \times 0 \cdot 75$$

$$= 19 \cdot 859 \times 10^3 \text{ N},$$

Resultant water force on one gate,

$$P = P_1 - P_2$$

$$= (220 \cdot 73 - 19 \cdot 86) \times 10^3 \text{ N}$$

$$= 200 \cdot 87 \times 10^3 \text{ N}.$$

The gates are rectangular, and so P_1 and P_2 will act at one-third of the depth of water (as shown in Fig. 3.6(b)), since, in equation (3.5), $\phi = 90°$, $\bar{y} = d/2$, $k_G^2 = d^2/12$, where d is the depth of the gate immersed (*see also* Section 3.4).

The height above the base at which the resultant force P acts can be found by taking moments. If P acts at a distance x from the bottom of the gate, then by taking moments about O,

$$Px = P_1 \times (5/3) - P_2 \times (1\cdot5/3)$$

$$= (220\cdot73 \times 5/3 - 19\cdot86 \times 0\cdot5) \times 10^3 = 357\cdot95 \times 10^3 \text{ N m},$$

$$x = 360\cdot1 \times 10^3/200\cdot87 \times 10^3 = 1\cdot782 \text{ m}.$$

Assuming that F, R and P are coplanar, they will meet at a point, and, since F is assumed to be perpendicular to the axis of the lock on plan, both F and R are inclined to the gate as shown at an angle of $20°$ so that $F = R$ and

$$P = F \sin 20° + R \sin 20° = 2 R \sin 20°,$$

$$R = \frac{P}{2 \sin 20°} = \frac{200\cdot87 \times 10^3}{2 \times 0\cdot342} = 293\cdot65 \times 10^3 \text{ N}.$$

If R is coplanar with P it acts at $1\cdot78$ m from the bottom of the gate. Taking moments about the bottom hinge,

$$4\cdot8\,R_T = 1\cdot18\,R$$

$$R_T = 1\cdot18/4\cdot8 \times 293\cdot65 \times 10^3 = 72\cdot2 \times 10^3 \text{ N} = 72\cdot2 \text{ kN},$$

$$R_B = R - R_T = 293\cdot65 - 72\cdot2 = \textbf{221}\cdot\textbf{45}$$

3.4. Pressure diagrams

The resultant force and centre of pressure can be found graphically for walls and other surfaces of constant vertical height for which it is convenient to calculate the horizontal force exerted per unit width. In Fig. 3.7, ABC is

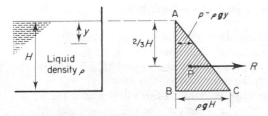

Figure 3.7 Pressure diagram for a vertical wall

the pressure diagram for the vertical wall of the tank containing a liquid, pressure being plotted horizontally against depth vertically. At the free surface A, the (gauge) pressure is zero. At depth y, $p = \rho g y$. The relationship between p and y is linear and can be represented by the triangle ABC. The area of this triangle will be the product of depth (in metres) and pressure (in newtons per square metre), and will represent, to scale, the resultant force R on unit width of the immersed surface perpendicular to the plane of the diagram (in newtons per metre).

Area of pressure diagram = $\frac{1}{2}$AB $\times$ BC = $\frac{1}{2}H \times \rho gH$.

Therefore,

$$\text{Resultant force, } R = \rho g H^2 / 2 \text{ for unit width}$$

and R will act through the centroid P of the pressure diagram, which is at a depth of $\frac{2}{3}H$ from A.

This result could also have been obtained from equations (3.2) and (3.5), since, for unit width,

$$R = \rho g A \bar{y} = \rho g (H \times 1) \times \tfrac{1}{2} H = \rho g H^2 / 2,$$

and, in equation (3.5), $\phi = 90°$, $\sin \phi = 1$, $\bar{y} = H/2$, $k_G^2 = H^2/12$, therefore,

$$D = \frac{H^2/12}{H/2} + \frac{H}{2} = \frac{2}{3} H$$

as before.

If the plane surface is inclined and submerged below the surface, the pressure diagram is drawn perpendicular to the immersed surface (Fig. 3.8)

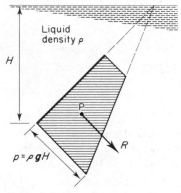

Figure 3.8   Pressure diagram for an inclined submerged surface

and will be a straight line extending from $p = 0$ at the free surface to $p = \rho g H$ at depth H. As the immersed surface does not extend to the free surface, the resultant force R is represented by the shaded area, instead of the whole triangle, and acts through the centroid P of this area.

It is also possible to draw pressure diagrams in three dimensions for immersed areas of various shapes as, for example, the triangular sluice gate in Fig. 3.9. However, such diagrams do little more than provide assistance in visualizing the situation.

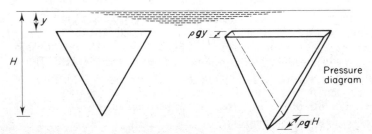

Figure 3.9   Pressure diagram for a triangular sluice gate

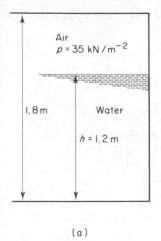

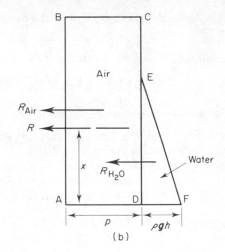

(a)   (b)

Figure 3.10 Pressure diagram

Example 3.3

A closed tank (Fig. 3.10), rectangular in plan with vertical sides, is 1·8 m deep and contains water to a depth of 1·2 m. Air is pumped into the space above the water until the air pressure is 35 kN m^{-2}. If the length of one wall of the tank is 3 m, determine the resultant force on this wall and the height of the centre of pressure above the base.

Solution

The air pressure will be transmitted uniformly over the whole of the vertical wall, and can be represented by the pressure diagram ABCD (Fig. 3.10(b)), the area of which represents the force exerted by the air per unit width of wall.

$$\text{Force due to air, } R_{Air} = (p \times AB) \times \text{Width}$$

$$= 35 \times 10^3 \times 1\cdot8 \times 3 = 189 \times 10^3 \text{ N}$$

and, since the wall is rectangular and the pressure uniform, R_{Air} will act at mid-height, which is 0·9 m above the base.

The pressure due to the water will start from zero at the free surface, corresponding to the point E, and reach a value DF equal to $\rho g h$ at the bottom. The area of the triangular pressure diagram EFD represents the force exerted by the water per unit width:

$$\text{Force due to water, } R_{H_2O} = \tfrac{1}{2} \times (\rho g h \times DE) \times \text{Width}$$

$$= \tfrac{1}{2} \times 10^3 \times 9\cdot81 \times 1\cdot2 \times 1\cdot2 \times 3$$

$$= 21\cdot19 \times 10^3 \text{ N},$$

and since the wall is rectangular, R_{H_2O} will act at $\tfrac{1}{3}h = 0\cdot4$ m from the base.

Total force due to both air and water,

$$R = R_{Air} + R_{H_2O}$$

$$= (189 + 21) \times 10^3$$

$$= 210\cdot19 \times 10^3 \text{ N}.$$

If x is the height above the base of the centre of pressure through which R acts,

$$R \times x = R_{air} \times 0\cdot9 + R_{H_2O} \times 0\cdot4,$$

$$x = (189 \times 0\cdot9 + 21 \times 0\cdot4)/210\cdot19 = \mathbf{0\cdot85\ m}.$$

3.5. Force on a curved surface due to hydrostatic pressure

If a surface is curved, the forces produced by fluid pressure on the small elements making up the area will not be parallel and, therefore, must be combined vectorially. It is convenient to calculate the horizontal and vertical components of the resultant force. This can be done in three dimensions, but the following analysis is for a surface curved in one plane only.

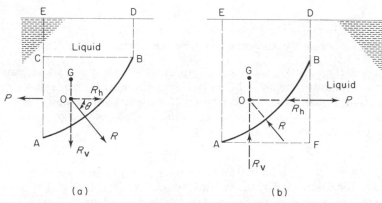

Figure 3.11   Hydrostatic force on a curved surface

In Fig. 3.11(a) and (b), AB is the immersed surface and R_h and R_v are the horizontal and vertical components of the resultant force R of the liquid on one side of the surface. In Fig. 3.11(a) the liquid lies above the immersed surface, while in Fig. 3.11(b) it acts below the surface.

In Fig. 3.11(a), if ACE is a vertical plane through A, and BC is a horizontal plane, then, since element ACB is in equilibrium, the resultant force P on AC must equal the horizontal component R_h of the force exerted by the fluid on AB because there are no other horizontal forces acting. But AC is the projection of AB on a vertical plane, therefore,

> Horizontal component R_h = Resultant force on the projection
> of AB on a vertical plane.

Also, for equilibrium, P and R_h must act in the same straight line; therefore, the horizontal component R_h acts through the centre of pressure of the projection of AB on a vertical plane.

Similarly, in Fig. 3.11(b), element ABF is in equilibrium, and so the horizontal component R_h is equal to the resultant force on the projection BF of the curved surface AB on a vertical plane, and acts through the centre of pressure of this projection.

In Fig. 3.11(a), the vertical component R_v will be entirely due to the weight of the fluid in the area ABDE lying vertically above AB. There are no other vertical forces, since there can be no shear forces on AE and BD because the fluid is at rest. Thus,

Vertical component, R_v = Weight of fluid vertically above AB,

and will act vertically downwards through the centre of gravity G of ABDE.

In Fig. 3.11(b), if the surface AB were removed and the space ABDE filled with the liquid, this liquid would be in equilibrium under its own weight and the vertical force on the boundary AB. Therefore,

Vertical component, R_v = Weight of the volume of the same fluid
which would lie vertically above **AB**,

and will act vertically upwards through the centre of gravity G of this imaginary volume of fluid.

In the case of closed vessels under pressure, a free surface does not exist, but an imaginary free surface can be substituted at a level $p/\rho g$ above a point at which the pressure p is known; ρ being the mass density of the actual fluid.

The resultant force R is found by combining the components vectorially. In the general case, the components in three directions may not meet at a point and, therefore, cannot be represented by a single force. However, in Fig. 3.11, if the surface is of uniform width perpendicular to the diagram, R_h and R_v will intersect at O. Thus,

$$\text{Resultant force, } R = \sqrt{(R_h^2 + R_v^2)}$$

and acts through O at an angle θ given by $\tan \theta = R_v/R_h$.

In the special case of a cylindrical surface, all the forces on each small element of area acting normal to the surface will be radial and will pass through the centre of curvature O (Fig. 3.12). The resultant force R must, therefore, also pass through the centre of curvature O.

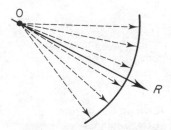

Figure 3.12   Resultant force on a cylindrical surface

Example 3.4

A sluice gate is in the form of a circular arc of radius 6 m as shown in Fig. 3.13. Calculate the magnitude and direction of the resultant force on the gate, and the location with respect to O of a point on its line of action.

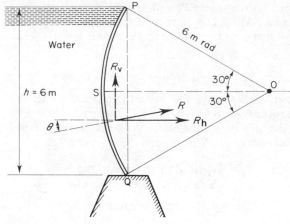

Figure 3.13   Sector gate

Solution
Since the water reaches the top of the gate,

> Depth of water, $h = 2 \times 6 \sin 30° = 6$ m,
>
> Horizontal component of force on gate $= R_h$ per unit length
>
> $\quad$ = Resultant force on PQ per unit length
>
> $\quad = \rho g \times h \times h/2 = \rho g h^2 /2$
>
> $\quad = (10^3 \times 9\cdot81 \times 36)/2$ N m^{-1} $= 176\cdot58$ kN m^{-1},
>
> Vertical component of force on gate $= R_v$ per unit length
>
> $\quad$ = Weight of water displaced by segment PSQ
>
> $\quad = (\text{Sector OPSQ} - \Delta\text{OPQ})\rho g$
>
> $\quad = ((60/360) \times \pi \times 6^2 - 6 \sin 30° \times 6 \cos 30°) \times 10^3 \times 9\cdot81$ N m^{-1}
>
> $\quad = 32\cdot00$ kN m^{-1},
>
> Resultant force on gate, $R = \sqrt{(R_h^2 + R_v^2)}$
>
> $\quad = \sqrt{(176\cdot58^2 + 32\cdot00^2)} = \mathbf{179\cdot46}$ **kN m^{-1}**.

If R is inclined at an angle θ to the horizontal,

> $\tan \theta = R_v/R_h = 32\cdot00/176\cdot58 = 0\cdot181$ 22
>
> $\theta = \mathbf{10\cdot27°}$ **to the horizontal.**

Since the surface of the gate is cylindrical, the resultant force R **must pass through O.**

3.6. Buoyancy

The method of calculating the forces on a curved surface applies to all shapes of surface and, therefore, to the surface of a totally submerged object

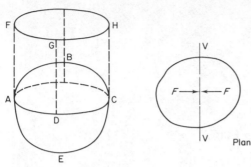

Figure 3.14   Buoyancy

(Fig. 3.14). Considering any vertical plane VV through the body, the projected area of each of the two sides on this plane will be equal and, as a result, the horizontal forces F will be equal and opposite. There is, therefore, no resultant horizontal force on the body due to the pressure of the surrounding fluid. The only force exerted by the fluid on an immersed body is vertical and is called the buoyancy or upthrust. It will be equal to the difference between the resultant forces on the upper and lower parts of the surface of the body. If ABCD is a horizontal plane,

Upthrust on body = Upward force on lower surface ADEC

$-$ Downward force on upper surface ABCD

= Weight of volume of fluid AECDGFH

$-$ Weight of volume of fluid ABCDGFH

= Weight of volume of fluid ABCDE

= Weight of fluid displaced by the body

and will act through the centroid of the volume of fluid displaced, which is known as the *centre of buoyancy*. This result is known as Archimedes' principle. As an alternative to the proof given above, it can be seen that, if the body were completely replaced by the fluid in which it is immersed, the forces exerted on the boundaries corresponding to the original body would exactly maintain the substituted fluid in equilibrium. Thus, the upward force on the boundary must be equal to the downward force corresponding to the weight of the fluid displaced by the body.

If a body is immersed so that part of its volume V_1 is immersed in a fluid of density ρ_1 and the rest of its volume V_2 in another immiscible fluid of mass density ρ_2 (Fig. 3.15),

Upthrust on upper part, $R_1 = \rho_1 g V_1$

acting through G_1, the centroid of V_1,

Upthrust on lower part, $R_2 = \rho_2 g V_2$

acting through G_2, the centroid of V_2,

Total upthrust $= \rho_1 g V_1 + \rho_2 g V_2$.

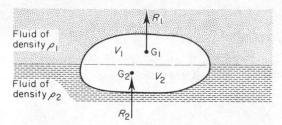

Figure 3.15   Body immersed in two fluids

The positions of G_1 and G_2 are not necessarily on the same vertical line, and the centre of buoyancy of the whole body is, therefore, not bound to pass through the centroid of the whole body.

Example 3.5

A rectangular pontoon has a width B of 6 m, a length l of 12 m, and a draught D of 1·5 m in fresh water (density 1000 kg m^{-3}). Calculate (a) the weight of the pontoon, (b) its draught in sea water (density 1025 kg m^{-3}) and (c) the load (in kilonewtons) that can be supported by the pontoon in fresh water if the maximum draught permissible is 2 m.

Solution

When the pontoon is floating in an unloaded condition,

Upthrust on immersed volume = Weight of pontoon.

Since the upthrust is equal to the weight of the fluid displaced,

Weight of pontoon = Weight of fluid displaced,

$$W = \rho g B l D.$$

(a) In fresh water, $\rho = 1000$ kg m^{-3} and $D = 1·5$ m, therefore,

Weight of pontoon, $W = 1000 \times 9·81 \times 6 \times 12 \times 1·5$ N

$$= \mathbf{1059·5 \ kN.}$$

(b) In sea water, $\rho = 1025$ kg m^{-3}, therefore,

Draught in sea water, $D = W/\rho g B l$

$$= \frac{1059·5 \times 10^3}{1025 \times 9·81 \times 6 \times 12} = \mathbf{1·46 \ m}$$

(c) For the maximum draught of 2 m in fresh water,

Total upthrust = Weight of water displaced = $g B l D$

$$= 1000 \times 9·81 \times 6 \times 12 \times 2 \ \text{N}$$

$$= 1412·6 \ \text{kN,}$$

Load which can be supported = Upthrust − Weight of pontoon

$$= 1412·6 - 1059·5 = \mathbf{353·1 \ kN.}$$

FLOATING BODIES

3.7. Equilibrium of floating bodies

When a body floats in vertical equilibrium in a liquid, the forces present are the upthrust R acting through the centre of buoyancy B (Fig. 3.16) and the

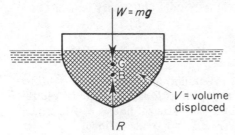

Figure 3.16   Body floating in equilibrium

weight of the body $W = mg$ acting through its centre of gravity. For equilibrium, R and W must be equal and act in the same straight line. Now, R will be equal to the weight of fluid displaced ρgV, where V is the volume of fluid displaced; therefore,

$$V = mg/\rho g = m/\rho.$$

As explained in Section 2.1, the equilibrium of a body may be stable, unstable or neutral, depending upon whether, when given a small displacement, it tends to return to the equilibrium position, move further from it or remain in the displaced position. For a floating body, such as a ship, stability is of major importance.

3.8. Stability of a submerged body

For a body totally immersed in a fluid, the weight $W = mg$ acts through the centre of gravity of the body, while the upthrust R acts through the centroid of the body B, which is the centre of buoyancy. Whatever the orientation of the body, these two points will remain in the same positions relative to the body. It can be seen from Fig. 3.17 that a small angular displacement θ from

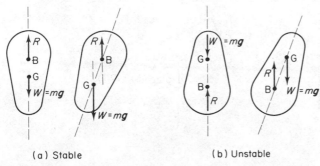

(a) Stable   (b) Unstable

Figure 3.17   Stability of submerged bodies

the equilibrium position will generate a moment $W \times BG \times \theta$. If the centre of gravity G is below the centre of buoyancy B (Fig. 3.13(a)), this will be a righting moment and the body will tend to return to its equilibrium position. However, if (as in Fig. 3.17(b)) the centre of gravity is above the centre of buoyancy, an overturning moment is produced and the body is unstable. Note that, as the body is totally immersed, the shape of the displaced fluid is not altered when the body is tilted and so the centre of buoyancy remains unchanged relative to the body.

3.9. Stability of floating bodies

Figure 3.18(a) shows a body floating in equilibrium. The weight $W = mg$ acts through the centre of gravity G and the upthrust R acts through the centre of buoyancy B of the displaced fluid in the same straight line as W. When the body is displaced through an angle θ (Fig. 3.18(b)), W continues to act through G; the volume of liquid remains unchanged since $R = W$, but the shape of this volume changes and its centre of gravity, which is the centre of buoyancy, moves relative to the body from B to B_1. Since R and W are no longer in the same straight line, a turning point proportional to $W \times \theta$ is produced, which in Fig. 3.18(b) is a righting moment and in Fig. 3.18(d) is an overturning moment. If M is the point at which the line of action of the upthrust R cuts the original vertical through the centre of gravity of the body G,

$$x = GM \times \theta,$$

provided that the angle of tilt θ is small, so that $\sin \theta = \tan \theta = \theta$ in radians.

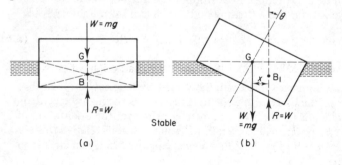

Stable

(a)   (b)

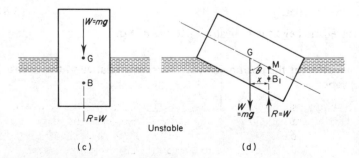

Unstable

(c)   (d)

Figure 3.18   Stable and unstable equilibrium

The point M is called the *metacentre* and the distance GM is the *metacentric height*. Comparing Fig. 3.18(b) and (d) it can be seen that

   (i) if M lies above G, a righting moment $W \times GM \times \theta$ is produced, equilibrium is stable and GM is regarded as positive;
   (ii) if M lies below G, an overturning moment $W \times GM \times \theta$ is produced, equilibrium is unstable and GM is regarded as negative;
   (iii) if M coincides with G, the body is in neutral equilibrium.

Since a floating body can tilt in any direction, it is usual, for a ship, to consider displacement about the longitudinal (rolling) and transverse (pitching) axes. The position of the metacentre and the value of the metacentric height will normally be different for rolling and pitching.

3.10. Determination of the metacentric height

The metacentric height of a vessel can be determined if the angle of tilt θ caused by moving a load P (Fig. 3.19) a known distance x across the deck is measured.

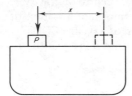

Figure 3.19   Determination of metacentric height

Overturning moment due to movement of load $P = Px$ (3.6)

If GM is the metacentric height and $W = mg$ is the total weight of the vessel, including P,

Righting moment $= W \times GM \times \theta$.   (3.7)

For equilibrium in the tilted position, the righting moment must equal the overturning moment so that, from equations (3.6) and (3.7),

$W \times GM \times \theta = Px$,

Metacentric height, $GM = Px/W\theta$ (3.8)

The true metacentric height is the value of GM as $\theta \rightarrow 0$.

3.11. Determination of the position of the metacentre relative to the centre of buoyancy

For a vessel of known shape and displacement, the position of the centre of buoyancy B is comparatively easily found and the position of the metacentre M relative to B can be calculated as follows. In Fig. 3.20, AC is the original waterline plane and B the centre of buoyancy in the equilibrium position. When the vessel is tilted through a small angle θ, the centre of buoyancy will

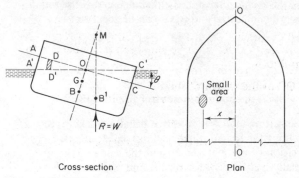

Cross-section   Plan

Figure 3.20   Height of metacentre above centre of buoyancy

move to B′ as a result of the alteration in the shape of the displaced fluid. A′C′ is the waterline plane in the displaced position. For small angles of tilt,

$$BM = BB'/\theta.$$

The movement of the centre of buoyancy, which is the centre of gravity of the displaced fluid, from B to B′ is the result of the removal of a volume of fluid corresponding to the wedge AOA′ and the addition of a wedge COC′. The total weight of fluid displaced remains unchanged, since it is equal to the weight of the vessel, therefore,

$$\text{Weight of wedge } AOA' = \text{Weight of wedge } COC'.$$

If a is a small area in the waterline plane at a distance x from the axis of rotation OO, it will generate a small volume, shown shaded, when the vessel is tilted.

$$\text{Volume swept out by } a = DD' \times a = ax\theta.$$

Summing all such volumes and multiplying by the specific weight ρg of the liquid,

$$\text{Weight of wedge } AOA' = \sum_{x=0}^{x=AO} \rho g a x\theta. \tag{3.9}$$

Similarly,

$$\text{Weight of wedge } COC' = \sum_{x=0}^{x=CO} \rho g a x\theta. \tag{3.10}$$

Since there is no change in displacement, we have, from equations (3.9) and (3.10),

$$\rho g\theta \sum_{x=0}^{x=AO} ax = \rho g\,\theta \sum_{x=0}^{x=CO} ax,$$

$$\Sigma ax = 0.$$

But Σax is the first moment of area of the water line plane about OO, therefore the axis OO must pass through the centroid of the water line plane.

The distance BB′ can now be calculated, since the couple produced by the movement of the wedge AOA′ to COC′ must be equal to the couple due to the movement of R from B to B′.

Moment about OO of the weight of fluid
swept out by area $a = \rho g a x \theta \times x$,

Total moment due to altered displacement $= \rho g \theta \Sigma a x^2$.

Putting $\Sigma a x^2 = I =$ Second moment of area of waterline plane about OO,

Total moment due to altered displacement $= \rho g \theta I$,   (3.11)

Moment due to movement of $R = R \times \text{BB}' = \rho g V \times \text{BB}'$,   (3.12)

where $V =$ volume of liquid displaced. Equating equations (3.11) and (3.12),

$$\rho g V \times \text{BB}' = \rho g \theta I,$$

$$\text{BB}' = \theta \, I/V,$$ (3.13)

giving $\text{BM} = \text{BB}'/\theta = I/V.$ (3.14)

The distance BM is known as the metacentric radius.

Example 3.6

A cylindrical buoy (Fig. 3.21) 1·8 m in diameter, 1·2 m high and weighing

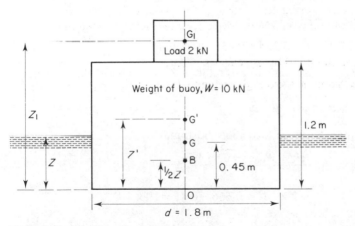

Weight of buoy, $W = 10$ kN

Load 2 kN

Z_1

Z

Z'

$\frac{1}{2}Z$

1.2 m

0.45 m

$d = 1.8$ m

Figure 3.21   Stability of a cylindrical buoy

10 kN floats in salt water of density 1025 kg m^{-3}. Its centre of gravity is 0·45 m from the bottom. If a load of 2 kN is placed on the top, find the maximum height of the centre of gravity of this load above the bottom if the buoy is to remain in stable equilibrium.

Solution

In Fig. 3.21, let G be the centre of gravity of the buoy, G_1 the centre of gravity of the load at a height Z_1 above the bottom, G' the combined centre of gravity of the load and the buoy at a height Z' above the bottom.

When the load is in position, let V be the volume of salt water displaced and Z the depth of immersion of the buoy.

Buoyancy force = Weight of salt water displaced

$$= \rho g V = \rho g (\pi/4) d^2 Z.$$

For equilibrium, buoyancy force must equal the combined weight of the buoy and the load $(W + W_1)$, therefore,

$$W + W_1 = \rho g (\pi/4) d^2 Z,$$

Depth of immersion,

$$Z = 4(W + W_1)/\rho g \pi d^2$$

$$= 4(10 + 2) \times 10^3/1025 \times 9{\cdot}81 \times 1{\cdot}8^2 \times \pi = 0{\cdot}47 \text{ m}.$$

The centre of buoyancy B will be at the centre of gravity of the displaced water, so that OB $= \frac{1}{2} Z = 0{\cdot}235$ m.

If the buoy and the load are just in stable equilibrium, the metacentre M must coincide with the centre of gravity G' of the buoy and load combined. The metacentric height G'M will then be zero and BG' = BM. From equation (3.14),

$$\text{BG}' = \text{BM} = \frac{I}{V} = \frac{\pi d^4/64}{\pi d^2 z/4} = \frac{1{\cdot}8^2}{16 \times 0{\cdot}47} = 0{\cdot}431 \text{ m}.$$

Thus, the position of G' is given by

$$Z' = \tfrac{1}{2} Z + \text{BG}' = 0{\cdot}235 + 0{\cdot}431 = 0{\cdot}666 \text{ m}.$$

The value of Z_1 corresponding to this value of Z' is found by taking moments about O:

$$W_1 Z_1 + 0{\cdot}45 \; W = (W + W_1) Z',$$

Maximum height of load above bottom,

$$Z_1 = \frac{(W + W_1)Z' - 0{\cdot}45 \; W}{W_1}$$

$$= \frac{12 \times 10^3 \times 0{\cdot}666 - 0{\cdot}45 \times 10 \times 10^3}{2 \times 10^3} \text{ m}$$

$$= 1{\cdot}746 \text{ m}.$$

3.12. Periodic time of oscillation

The displacement of a stable vessel through an angle θ from its equilibrium position produces a righting moment T which, from equation (3.7), is given by $T = W \times \text{GM} \times \theta$, where $W = mg$ is the weight of the vessel and GM is the metacentric height. This will produce an angular acceleration $d^2\theta/dt^2$, and, if I is the mass moment of inertia of the vessel about its axis of rotation,

$$\frac{d^2\theta}{dt^2} = \frac{T}{I} = -\frac{W \cdot GM \cdot \theta}{(W/g)k^2} = -\frac{GM \cdot \theta g}{k^2},$$

where k is the radius of gyration from its axis of rotation. The negative sign indicates that the acceleration is in the opposite direction to the displacement. Since this corresponds to simple harmonic motion,

$$\text{Periodic time, } t = 2\pi \sqrt{\left(\frac{\text{Displacement}}{\text{Acceleration}}\right)} = 2\pi \sqrt{\left(\frac{\theta}{GM \times \theta \times (g/k^2)}\right)}$$

$$= 2\pi \sqrt{(k^2 \, GM \times g)}, \qquad (3.15)$$

from which it can be seen that, although a large metacentric height will improve stability, it produces a short periodic time of oscillation, which results in discomfort and excessive stress on the structure of the vessel.

3.13. Stability of a vessel carrying liquid in tanks with a free surface

The stability of a vessel carrying liquid in tanks with a free surface (Fig. 3.22)

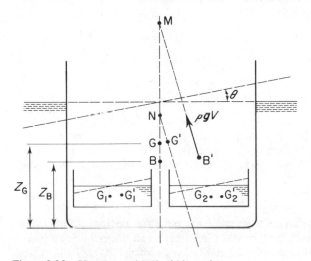

Figure 3.22   Vessel carrying liquid in tanks

is affected adversely by the movement of the centre of gravity of the liquid in the tanks as the vessel heels. Thus, G_1 will move to G_1' and G_2 to G_2'. The distance moved is calculated in the same way as the movement BB$'$ of the centre of buoyancy, given by equation (3.12):

$$G_1 G_1' = \theta \, I_1/V_1 \quad \text{and} \quad G_2 G_2' = \theta \, I_2/V_2,$$

where I_1 and I_2 are the second moments of area of the free surfaces, and V_1 and V_2 the volumes, of the liquid in the tanks. As a result of the movement of G_1 and G_2, the centre of gravity G of the whole vessel and contents will move to G'. If V is the volume of water displaced by the vessel and ρ is the mass density of water,

$$\text{Weight of vessel and contents} = \text{Weight of water displaced}$$

$$= \rho g V.$$

If the volume of liquid of density ρ_1 in the tanks is V_1 and V_2,

$$\text{Weight of contents of the first tank} \quad = \rho_1 g V_1,$$

$$\text{Weight of contents of the second tank} = \rho_1 g V_2.$$

Taking moments to find the change in the centre of gravity of the vessel and contents,

$$\rho g V \times GG' = \rho_1 g V_1 \times G_1 G_1' + \rho_1 g V_2 \times G_2 G_2'$$

$$= \rho_1 g V_1 \, \theta \, I_1/V_1 + \rho_1 g V_2 \, \theta \, I_2/V_2,$$

$$GG' = \frac{1}{V}(\rho_1/\rho) \, \theta \, (I_1 + I_2).$$

In the tilted position, the new vertical through B' intersects the original vertical through G at the metacentre M, but the weight W acts through G' instead of G and its line of action cuts the original vertical at N, reducing the metacentric height from GM to NM.

$$\text{Effective metacentric height, } NM = Z_B + BM - (Z_G + GM),$$

and, since $BM = I/V$ and $GN = GG'/\theta = \frac{1}{V}(\rho_1/\rho)(I_1 + I_2)$,

$$NM = Z_B - Z_G + \frac{1}{V}\{I - (\rho_1/\rho)(I_1 + I_2)\}. \tag{3.16}$$

Thus, the effect of the liquid in the tank is to reduce the effective metacentric height and impair stability, provided that the liquid in the tanks has a free surface so that its centre of gravity moves as the vessel tilts. Lateral subdivision of the tanks improves stability by reducing the sum of the second moments of area I_1, I_2, etc.

Example 3.7

A barge (Fig. 3.23) has vertical sides and ends and a flat bottom. In plan view it is rectangular, 20 m long by 6 m wide, but with an additional semicircular portion of 3 m radius at one end. The empty barge weighs 200 kN and floats upright in fresh water. The part of the vessel which is rectangular in plan is divided by a wall into two compartments 3 m wide by 20 m long. These compartments form open top tanks which are partly filled with liquid of relative density 0·8 to a depth of 0·8 m in one tank and 1·0 m in the other. The vessel rolls about a horizontal axis, but the flat end remains in a vertical plane. Ignoring the thickness of the material of the barge structure and assuming that the centre of gravity of the barge and contents is 0·45 m above the bottom, find the angle of roll.

Solution

In order to be able to determine the angle of roll, we must first find the effective metacentric height from equation (3.16).

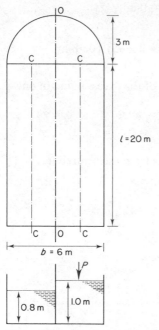

Figure 3.23 Barge containing liquids

For the whole vessel

$$I_{OO} = \frac{lb^3}{12} + \frac{\pi b^4}{128}$$

$$= 20 \times 6^3/12 + \pi \times 6^4/128 \text{ m}^4.$$

$$= 391.9 \text{ m}^4.$$

For each tank

$$I_{CC} = l \times (\tfrac{1}{2}b)^3/12 = 20 \times 3^3/12 = 45 \text{ m}^4.$$

Weight of barge   = 200 kN,

Weight of liquid load = $0.8 \times 10^3 \times 9.81 (20 \times 3 \times 1 + 20 \times 3 \times 0.8)$ N

$$= 846 \text{ kN},$$

Total weight of barge and contents = 1046 kN.

Area of water line plane of vessel = $20 \times 6 + \tfrac{1}{2}\pi \times 3^2 \text{ m}^2$

$$= 134.1 \text{ m}^2.$$

Volume of vessel submerged = Weight/(Density $\times g$)

$$= 1046 \times 10^3/10^3 \times 9.81 = 106.8 \text{ m}^3.$$

Depth submerged = $106.8/134.1 = 0.80$ m.

Height of centre of buoyancy B above bottom = $\tfrac{1}{2}$ Depth submerged

$$= 0.4 \text{ m}.$$

Putting these values in equation (3.16) with $\rho_1/\rho = 0.8$,

Effective metacentric height,

$$NM = 0.4 - 0.45 + (391.9 - 0.8 \times 2 \times 45)/106.8 \text{ m}$$
$$= 2.95 \text{ m}.$$

The overturning moment is caused by the weight of the excess liquid in one tank,

$$P = 0.8 \times 10^3 \times 9.81 \times 20 \times 3 \,(1.0 - 0.8)$$
$$= 94 \text{ kN}.$$

The centre of gravity of this excess liquid is 1.5 m from the centreline.

Overturning moment due to excess liquid $= P \times 1.5$,

Righting moment $= W \times NM \tan \theta$.

Thus, for equilibrium,

$$P \times 1.5 = W \times NM \tan \theta,$$
$$\tan \theta = 94 \times 1.5/1046 \times 2.95 = 0.0457,$$

Angle of roll, $\theta = 2° 37'$.

EXERCISES 3

3.1   A circular lamina 125 cm in diameter is immersed in water so that the distance of its edge measured vertically below the free surface varies from 60 cm to 150 cm. Find the total force due to the water acting on one side of the lamina, and the vertical distance of the centre of pressure below the surface.
[12 650 N, 1·1 m]

3.2   A rectangular plane area, immersed in water, is 1·5 m by 1·8 m with the 1·5 m side horizontal and the 1·8 m side vertical. Determine the magnitude of the force on one side and the depth of its centre of pressure if the top edge is (a) in the water surface, (b) 0·3 m below the water surface, (c) 30 m below the water surface.
[24 kN, 1·2 m; 32 kN, 1·42 m; 815 kN, 30·91 m]

3.3   One end of a rectangular tank is 1·5 m wide by 2 m deep. The tank is completely filled with oil of specific weight 9 kN/m³. Find the resultant pressure on this vertical end and the depth of the centre of pressure from the top.
[27 kN, 1·33 m]

3.4   What is the position of the centre of pressure of a vertical semicircular plane submerged in a homogeneous liquid with its diameter d at the free surface.
[Depth $3\pi d/32$]

3.5   A culvert draws off water from the base of a reservoir. The entrance to the culvert is closed by a circular gate 1·25 m in diameter which can be rotated about its horizontal diameter. Show that the turning moment on the gate is independent of the depth of water if the gate is completely immersed and find the value of this moment.
[1160 Nm]

3.6   A barge in the form of a closed rectangular tank 20 m long by 4 m wide floats in water. If the bottom is 1·5 m below the surface, what is the water force on one long side and at what level below the surface does it act.

If a uniform pressure of 50 kN/m² gauge is applied inside the barge what will be the new magnitude and point of action of the resultant force on the side. The deck is 0·2 m above water level.

[220·73 kN, 1·0 m; 1479·27 kN, 0·6 m below surface]

3.7   Figure 3.24 shows a gate in the side of a dam which opens auto-

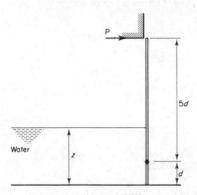

Figure 3.24

matically when the water level z exceeds a certain value. The gate is pivoted a distance d above its base and extends a distance $5d$ above the pivot. Determine the value of the ratio z/d for which the force P on the sill is a maximum and the ratio of z/d for which the gate just opens.

If $d = 1$ m and the gate is 3 m wide what is the maximum value of P.

[2, 3, 3·924 kN]

3.8   A rectangular sluice door (Fig. 3.25) is hinged at the top at A and kept closed by a weight fixed to the door. The door is 120 cm wide and

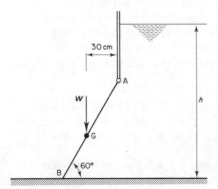

Figure 3.25

90 cm long and the centre of gravity of the complete door and weight is at G, the combined weight being 9810 N. Find the height of the water h on the inside of the door which will just cause the door to open.

[0·88 m]

3.9   A sluice gate closes a circular opening 0·30 m diameter and is hinged 1 m below the surface of the water which acts on its face. If the centre of the

opening lies at a depth of 1·25 m find the force on the gate due to the fluid pressure. Find also the minimum force that must be applied by a clamp which lies 0·5 m below the hinge, in order to keep the gate closed.
[866 N, 441 N]

3.10   A rectangular gate (Fig. 3.26) of negligible thickness, hinged at its top edge and of width b, separates two tanks in which there is the same liquid of density ρ. It is required that the gate shall open when the level in the left-hand tank falls below a distance H from the hinge. The level in the right-hand tank remains constant at a height y above the hinge. Derive an expression for the

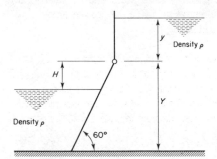

Figure 3.26

weight of the gate in terms of H, Y, y, b, and g. Assume that the weight of the gate acts at its centre of area.

$$\left[W = 0 \cdot 77\, gb\left(\frac{3Y^2(y+H) - H^3}{Y} \right) \right]$$

3.11   A tank with vertical sides contains water to a depth of 1·2 m and a layer of oil 800 mm deep which rests on top of the water. The relative density of the oil is 0·85 and above the oil there is air at atmospheric pressure. On one side of the tank an opening 500 mm wide is cut extending from top to bottom of the tank vertically and across this opening is welded a segment of a vertical circular cylinder of radius 700 mm. The chord length of this segment is 500 mm. Calculate the total thrust tending to separate this segment from the wall of the tank and the height of the line of action of this force above the base of the tank.
[8868 N, 651 mm]

3.12   The face of a dam is vertical to a depth of 7·5 m below the water surface and then slopes at 30° to the vertical. If the depth of water is 16·5 m specify completely the resultant force per metre run acting on the whole face.
[1475 kN at 24°36′ to horizontal through a point 5·5 m above the base and 2·93 m from the vertical face]

3.13   A masonry dam 6 m high has the water level with the top. Assuming that the dam is rectangular in section and 3 m wide, determine whether the dam is stable against overturning and whether tension will develop in the masonry joints. Density of masonry 1760 kg/m³.
[Stable, tension on the water face]

3.14   A square aperture in the vertical side of a tank has one diagonal vertical and is completely covered by a plane plate hinged along one of the

upper sides of the aperture. The diameters of the aperture are 2 m long and the tank contains a liquid of relative density 1·15. The centre of the aperture is 1·5 m below the free surface. Calculate the thrust exerted on the plate by the liquid, the moment of this thrust about the hinge and the position of the centre of pressure.
[33·84 kN, 29·24 kN m, 4 m above bottom]

3.15   A pair of lock gates, each 3 m wide, have their lower hinges at the bottom of the gates and their upper hinges 5 m from the bottom. The width of the lock is 5·5 m. Find the reaction between the gates when the water level is 4·5 m above the bottom on one side and 1·5 m on the other. Assuming that this force acts at the same height as the resultant force due to the water pressure find the reaction forces on the hinges.
[331 kN; 106·5 kN, 224·5 kN]

3.16   A spherical container is made up of two hemispheres, the joint between the two halves being horizontal. The sphere is completely filled with water through a small hole in the top. It is found that 50 kg of water are required for this purpose. If the two halves of the container are not secured together, what must be the mass of the upper hemisphere if it just fails to lift off the lower hemisphere.
[12·5 kg]

3.17   A sluice gate (Fig. 3.27) consists of a quadrant of a circle of radius 1·5 m pivoted at its centre O. Its centre of gravity is at G as shown. When the

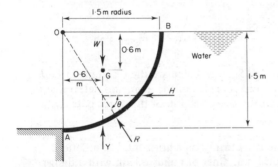

Figure 3.27

water is level with the pivot O, calculate the magnitude and direction of the resultant force on the gate due to the water and the turning moment required to open the gate. The width of the gate is 3 m and it has a mass of 6000 kg.
[61·6 kN, 57°28', 35·3 kN m]

3.18   A sector-shaped sluice gate having a radius of curvature of 5·4 m is as shown in Fig. 3.28. The centre of curvature C is 0·9 m vertically below the lower edge A of the gate and 0·6 m vertically above the horizontal axis passing through O about which the gate is constructed to turn. The mass of the gate is 3000 kg per metre run and its centre of gravity is 3·6 m horizontally from the centre O. If the water level is 2·4 m above the lower edge of the gate find per metre run (a) the resultant force acting on the axis at O, (b) the resultant moment about O.
[36·2 kN, 89 kN m]

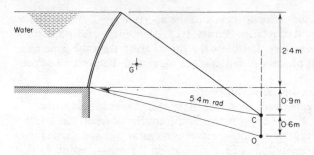

Figure 3.28

3.19   The face of a dam (Fig. 3.29) is curved according to the relation $y = x^2/2 \cdot 4$ where y and x are in metres. The height of the free surface above

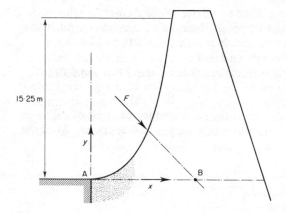

Figure 3.29

the horizontal plane through A is 15·25 m. Calculate the resultant force F due to the fresh water acting on unit breadth of the dam, and determine the position of the point B at which the line of action of this force cuts the horizontal plane through A.
[1290 kN/m, 11·84 m]

3.20   A flat circular baseplate rests firmly on a horizontal foundation. Placed above it is a hemispherical container of radius 60 cm with a flange bolted to the flat plate. The centre of the hemisphere is at the surface of the plate. The container is filled with water to a depth of 45 cm and in the air space there is a pressure of 35 kN/m² above atmosphere. If the container has a mass of 110 kg, calculate the total force tending to break the joint at the flange.
[39·48 kN]

3.21   A ship floating in sea water displaces 115 m³. Find (a) the weight of the ship if sea water has a density of 1025 kg/m³, and (b) the volume of fresh water of density 1000 kg/m³ which the ship would displace.
[118 000 kg, 118 m³]

3.22   A steel pipeline conveying gas has an internal diameter of 120 cm and an external diameter of 125 cm. It is laid across the bed of a river, completely immersed in water and is anchored at intervals of 3 m along its length. Calculate the buoyancy force in newtons/metre and the upward force in newtons on each anchorage. Density of steel = 7900 kg/m³, density of

water = 1000 kg/m^3.
[12 150 N/m, 13 950 N]
3.23   The ball-operated valve shown in Fig. 3.30 controls the flow from a
tank through a pipe to a lower tank, in which it is situated. The water level in

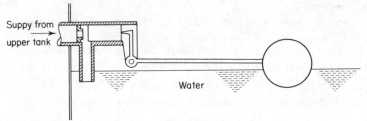

Suppy from
upper tank

Water

Figure 3.30

the upper tank is 7 m above the 10 mm diameter valve opening. Calculate the
volume of the ball which must be submerged to keep the valve closed.
[110 cm^3]
3.24   The shifting of a portion of cargo of mass 25 000 kg through a
distance of 6 m at right angles to the vertical plane containing the longitudinal
axis of a vessel causes it to heel through an angle of 5°. The displacement of
the vessel is 5000 metric tons and the value of I is 5840 m^4. The density of
sea water is 1025 kg/m^3. Find (a) the metacentric height and (b) the height
of the centre of gravity of the vessel above the centre of buoyancy.
[0·344 m, 0·851 m]
3.25   A buoy floating in sea water of density 1025 kg/m^3 is conical in
shape with a diameter across the top of 1·2 m and a vertex angle of 60°. Its
mass is 300 kg and its centre of gravity is 750 mm from the vertex. A flashing
beacon is to be fitted to the top of the buoy. If this unit has a mass of 55 kg
what is the maximum height of its centre of gravity above the top of the buoy
if the whole assembly is not to be unstable? (The centre of volume of a cone
of height h is at a distance $\frac{3}{4}h$ from the vertex.)
[2·31 m]
3.26   A uniform wooden cylinder has a relative density of 0·6. Determine
the ratio of diameter to length so that it will just float upright in water.
[1·386]
3.27   A rectangular pontoon 6 m by 3 m in plan floating in water has a
depth of immersion of 0·9 m and is subjected to a torque of 7600 N m about
the longitudinal axis. If the centre of gravity is 700 mm up from the bottom,
estimate the angle of heel.
[4°42']
3.28   A rectangular pontoon 10 m by 4 m in plan weighs 280 kN. A steel
tube weighing 34 kN is placed longitudinally on the deck. When the tube is in
a central position, the centre of gravity for the combined weight lies on the
vertical axis of symmetry 250 mm above the water surface. Find (a) the
metacentric height, (b) the maximum distance the tube may be rolled laterally
across the deck if the angle of heel is not to exceed 5°.
[1·67 m, 0·735 m]
3.29   A rectangular pontoon 12 m wide, 20 m long and 6 m deep weighs
5000 kN and floats in sea water of density 1025 kg/m^3. When a load of
2500 kN is added on the deck centreline the centre of gravity of the pontoon

88   Elements of fluid mechanics

and load is on the waterline. Determine the metacentric height if the movement of the 2500 kN load 1·0 m across the width of the deck results in an angle of roll of 5° for the pontoon.

Determine the maximum height of the load centre of gravity above the deck if the pontoon is to remain stable.

[3·79 m, 8·67 m]

3.30   A rectangular tank 90 cm long and 60 cm wide is mounted on bearings so that it is free to turn on a longitudinal axis. The tank has a mass of 68 kg and its centre of gravity is 15 cm above the bottom. When the tank is slowly filled with water it hangs in stable equilibrium until the depth of water is 45 cm after which it becomes unstable. How far is the axis of the bearings above the bottom of the tank.

[0·26 m]

3.31.   A rectangular pontoon 6 m wide, 15 m long and 2·1 m deep weighs 80 tonnes when loaded, but without ballast water. A vertical diaphragm divides the pontoon longitudinally into two compartments each 3 m wide and 15 m long. Twenty tonnes of water ballast are admitted to the bottom of each compartment, the water surface being free to move. The centre of gravity of the pontoon without ballast water is 1·5 m above the bottom and on the geometrical centre of the plan.

(a) Calculate the metacentric height for rolling.

(b) If two tonnes of the deck load is shifted 3 m laterally find the approximate angle of heel.

[1·28 m, 2° 14′]

3.32   A cylindrical buoy 1·35 m in diameter and 1·8 m high has a mass of 770 kg. Show that it will not float with its axis vertical in sea water of density 1025 kg/m^3.

If one end of a vertical chain is fastened to the base, find the pull required to keep the buoy vertical. The centre of gravity of the buoy is 0·9 m from its base.

[GM = −0·42 m, 4640 N]

3.33   A solid cylinder 1 m diameter and 0·8 m high is of uniform relative density 0·85. Calculate the periodic time of small oscillations when the cylinder floats with its axis vertical in still water.

[3·92 s]

3.34   A ship has a displacement of 5000 metric tonnes. The second moment of area of the waterline section about a fore and aft axis is 12 000 m^4 and the centre of buoyancy is 2 m below the centre of gravity. The radius of gyration is 3·7 m. Calculate the period of oscillation. Sea water has a density of 1025 kg/m^3.

[11·1 s]

PART II

Concepts of fluid flow

The study of fluid motion is complicated by the introduction of viscosity dependent shear forces that were absent in the preceding treatment of stationary fluids. In the majority of flow cases analysis relies upon a body of empirical work, supported by the concepts of dimensional analysis and similarity. In this part of the text we will establish the analytical techniques that will later be combined with the empirical representation of frictional forces to allow the study of 'real' fluid behaviour.

In order to deal effectively with flowing fluids it is first necessary to identify flow categories, defined in predominantly mathematical terms, that will allow the appropriate analysis to be undertaken by identifying suitable and acceptable simplifications. Examples of the categories to be introduced include variation of the flow parameters with time (steady or unsteady) or variations along the flow path (uniform or non-uniform). Similarly compressibility effects may be important in high-speed gas flows but may be ignored in many liquid flow situations.

In parallel to setting up these flow categories it is also necessary to develop a series of mathematically expressed principles that will allow the variations in flow parameters as a result of the motion of the fluid to be predicted.

The principles of continuity, energy and momentum are developed in this part of the text. The steady-flow energy equation is introduced and will be utilized later to describe the behaviour of real fluids by the inclusion of an empirically based friction term. The momentum equation will be introduced and its application illustrated for both fluid-to-solid boundary transfers, such as the calculation of forces acting on moving vanes or pipe nozzles, and for other flow situations, such as the formation of hydraulic jumps in open channel flows.

While the treatment of the behaviour of real fluid motion requires the introduction of viscous and, possibly, compressibility terms, the study of an ideal fluid freed from these constraints is useful and important, particularly in the consideration of flow patterns away from the influence of solid boundaries. Primarily a mathematical modelling tool, the study of ideal fluid flow has its roots in the work of eighteenth-century hydrodynamicists and has applications now in aerodynamics as it allows the introduction of a further flow classification, namely rotational or irrotational flow. The study of ideal flow allows flow patterns around aerofoil sections to be considered and therefore naturally leads to considerations of lift and vorticity.

Taken together with Part I this portion of the text provides the foundation upon which the study and application of the behaviour of real fluids may be based.

4 Motion of fluid particles and streams

4.1. Fluid flow

The motion of a fluid is usually extremely complex. The study of a fluid at rest, or in relative equilibrium, was simplified by the absence of shear forces, but when a fluid flows over a solid surface or other boundary, whether stationary or moving, the velocity of the fluid in contact with the boundary must be the same as that of the boundary, and a velocity gradient is created at right angles to the boundary (*see* Section 1.2). The resulting change of velocity from layer to layer of fluid flowing parallel to the boundary gives rise to shear stresses in the fluid. Individual particles of fluid move as a result of the action of forces set up by differences of pressure or elevation. Their motion is controlled by their inertia and the effect of the shear stresses exerted by the surrounding fluid. The resulting motion is not easily analysed mathematically, and it is often necessary to supplement theory by experiment.

If an individual particle of fluid is coloured, or otherwise rendered visible, it will describe a *pathline*, which is the trace showing the position at successive intervals of time of a particle which started from a given point. If, instead of colouring an individual particle, the flow pattern is made visible by injecting a stream of dye into a liquid, or smoke into a gas, the result will be a *streakline* or *filament line*, which gives an instantaneous picture of the positions of all the particles which have passed through the particular point at which the dye is being injected. Since the flow pattern may vary from moment to moment, a streakline will not necessarily be the same as a pathline. When using tracers or dyes it is essential to choose a material having a density and other physical properties as similar as possible to those of the fluid being studied.

In analysing fluid flow, we also make use of the idea of a *streamline*, which is an imaginary curve in the fluid across which, at a given instant, there is no flow. Thus, the velocity of every particle of fluid along the streamline is tangential to it at that moment. Since there can be no flow through solid boundaries, these can also be regarded as streamlines. For a continuous stream of fluid, streamlines will be continuous lines extending to infinity upstream and downstream, or will form closed curves as, for example, round the surface of a solid object immersed in the flow. If conditions are steady and the flow pattern does not change from moment to moment, pathlines and streamlines will be identical; if the flow is fluctuating this will not be the case. If a series of streamlines are drawn through every point on the perimeter of a small area of the stream cross-section, they will form a *streamtube* (Fig. 4.1). Since there is no flow across a streamline, the fluid inside the

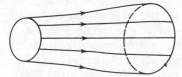

Figure 4.1   A streamtube

streamtube cannot escape through its walls, and behaves as if it were contained in an imaginary pipe. This is a useful concept in dealing with the flow of a large body of fluid, since it allows elements of the fluid to be isolated for analysis.

4.2. Uniform flow and steady flow

Conditions in a body of fluid can vary from point to point and, at any given point, can vary from one moment of time to the next. Flow is described as *uniform* if the velocity at a given instant is the same in magnitude and direction at every point in the fluid. If, at the given instant, the velocity changes from point to point, the flow is described as *non-uniform*. In practice, when a fluid flows past a solid boundary there will be variations of velocity in the region close to the boundary. However, if the size and shape of the cross-section of the stream of fluid is constant, the flow is considered to be uniform.

A *steady* flow is one in which the velocity, pressure and cross-section of the stream may vary from point to point but do not change with time. If, at a given point, conditions do change with time, the flow is described as *unsteady*. In practice, there will always be slight variations of velocity and pressure, but, if the average values are constant, the flow is considered to be steady.

There are, therefore, four possible types of flow.

(i) *Steady uniform flow*. Conditions do not change with position or time. The velocity and cross-sectional area of the stream of fluid are the same at each cross-section; e.g. flow of a liquid through a pipe of uniform bore running completely full at constant velocity.

(ii) *Steady non-uniform flow*. Conditions change from point to point but not with time. The velocity and cross-sectional area of the stream may vary from cross-section to cross-section, but, for each cross-section, they will not vary with time; e.g. flow of a liquid at a constant rate through a tapering pipe running completely full.

(iii) *Unsteady uniform flow*. At a given instant of time the velocity at every point is the same, but this velocity will change with time; e.g. accelerating flow of a liquid through a pipe of uniform bore running full, such as would occur when a pump is started up.

(iv) *Unsteady non-uniform flow*. The cross-sectional area and velocity vary from point to point and also change with time; e.g. a wave travelling along a channel.

4.3. Frames of reference

Whether a given flow is described as steady or unsteady will depend upon the situation of the observer, since motion is relative and can only be described in terms of some frame of reference which is determined by the observer. If a wave travels along a channel, then to an observer on the bank the flow in the channel will appear to vary with time, and, therefore, be unsteady. If, however, the observer were travelling on the crest of the wave, conditions would not appear to him to change with time, and the flow would be steady according to his frame of reference.

The frame of reference adopted for describing the motion of a fluid is usually a set of fixed coordinate axes, but the analysis of steady flow is usually simpler than that of unsteady flow and it is sometimes useful to use moving coordinate axes to convert an unsteady flow problem to a steady flow problem. The normal laws of mechanics will still apply, provided that the movement of the coordinate axes takes place with uniform velocity in a straight line.

4.4. Real and ideal fluids

When a real fluid flows past a boundary, the fluid immediately in contact with the boundary will have the same velocity as the boundary. As explained in Section 4.1, the velocity of successive layers of fluid will increase as we move away from the boundary. If the stream of fluid is imagined to be of infinite width perpendicular to the boundary, a point will be reached beyond which the velocity will approximate to the free stream velocity, and the drag exerted by the boundary will have no effect. The part of the flow adjoining the boundary in which this change of velocity occurs is known as the *boundary layer*. In this region, shear stresses are developed between layers of fluid moving with different velocities as a result of viscosity and the interchange of momentum due to turbulence causing particles of fluid to move from one layer to another. The thickness of the boundary layer is defined as the distance from the boundary at which the velocity becomes equal to 99 per cent of the free stream velocity. Outside this boundary layer, in a real fluid, the effect of the shear stresses due to the boundary can be ignored and the fluid can be treated as if it were an *ideal fluid*, which is assumed to have no viscosity and in which there are no shear stresses. If the fluid velocity is high and its viscosity low, the boundary layer is comparatively thin, and the assumption that a real fluid can be treated as an ideal fluid greatly simplifies the analysis of the flow and still leads to useful results.

Even in problems in which the effects of viscosity and turbulence cannot be neglected, it is often convenient to carry out the mathematical analysis assuming an ideal fluid. An experimental investigation can then be made to correct the theoretical analysis for the factors omitted and to bring the results obtained into agreement with the behaviour of a real fluid.

4.5. Compressible and incompressible flow

All fluids are compressible, so that their density will change with pressure, but, under steady flow conditions and provided that the changes of density are small, it is often possible to simplify the analysis of a problem by assuming that the fluid is incompressible and of constant density. Since liquids are relatively difficult to compress, it is usual to treat them as if they were incompressible for all cases of steady flow. However, in unsteady flow conditions, high pressure differences can develop (*see* Chapter 19) and the compressibility of liquids must be taken into account.

Gases are easily compressed and, except when changes of pressure and, therefore, density are very small, the effects of compressibility and changes of internal energy *must* be taken into account.

4.6. One-, two- and three-dimensional flow

Although, in general, all fluid flow occurs in three dimensions, so that velocity, pressure and other factors vary with reference to three orthogonal axes, in some problems the major changes occur in two directions or even in only one direction. Changes along the other axis or axes can, in such cases, be ignored without introducing major errors, thus simplifying the analysis.

Flow is described as *one-dimensional* if the factors, or parameters, such as velocity, pressure and elevation, describing the flow at a given instant, vary only along the direction of flow and not across the cross-section at any point. If the flow is unsteady, these parameters may vary with time. The one dimension is taken as the distance along the central streamline of the flow, even though this may be a curve in space, and the values of velocity, pressure and elevation at each point along this streamline will be the average values across a section normal to the streamline. A one-dimensional treatment can be applied, for example, to the flow through a pipe, but, since in a real fluid the velocity at any cross-section will vary from zero at the pipe wall (Fig. 4.2) to a maximum at the centre, some correction will be necessary

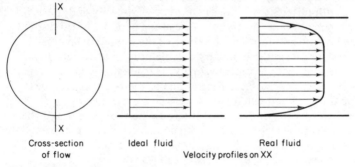

Cross-section   Ideal fluid   Real fluid
of flow   Velocity profiles on XX

Figure 4.2   Velocity profiles for one-dimensional flow

to compensate for this (*see* Chapter 8) if a high degree of accuracy is required.

In *two-dimensional* flow it is assumed that the flow parameters may vary in the direction of flow and in one direction at right angles, so that the streamlines are curves lying in a plane and are identical in all planes parallel to this plane. Thus, the flow over a weir of constant cross-section (Fig. 4.3)

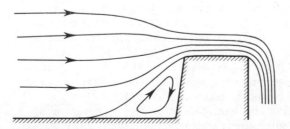

Figure 4.3   Two-dimensional flow

and infinite width perpendicular to the plane of the diagram can be treated as two-dimensional. A real weir has a limited width, but it can be treated as two-dimensional over its whole width and then an end correction can be introduced to modify the result to allow for the effect of the disturbance produced by the end walls (*see* Chapter 16).

A special case of two-dimensional flow occurs when the cross-section of the flow is circular and the flow parameters vary symmetrically about the axis. For example, ideally the velocity distribution in a circular pipe will be the same across any diameter, the velocity varying from zero at the wall to a maximum at the centre. Referred to orthogonal coordinate axes (x in the direction of motion, y and z in the plane of the cross-section) the flow is three-dimensional, but, since it is axisymmetric, it can be reduced to two-dimensional flow by using a system of *cylindrical* coordinates (x in the direction of flow and r the radius defining the position in the cross-section).

4.7. Analysing fluid flow

One difficulty encountered in deciding how to investigate the flow of a fluid is that, in the majority of problems, we are dealing with an endless stream of fluid. We have to decide what part of this stream shall constitute the element or system to be studied and what shall be regarded as the surroundings which act upon this system. There are two main alternatives.

(i) We can study the behaviour of a specific element of the fluid of fixed mass. Such an element constitutes a closed system. Its boundaries are a closed surface which may vary with time, but always contain the same mass of fluid. At any instant, a free body diagram can be drawn showing the forces exerted by the surrounding fluid and any solid boundaries on this element.

(ii) We can define the system to be studied as a fixed region in space, or in relation to some frame of reference, known as a *control volume*, through which the fluid flows, forming, in effect, an open system. The boundary of this system is its control surface and its shape does not change with time. The control volume for a particular problem is chosen arbitrarily for reasons of convenience of analysis. However, the control surface will usually follow solid boundaries where these are present, and where it cuts the flow direction it will do so at right angles. Where there are no solid boundaries the control volume may form a streamtube.

4.8. Motion of a fluid particle

Any particle or element of fluid will obey the normal laws of mechanics in the same way as a solid body. When a force is applied, its behaviour can be predicted from Newton's laws, which state:

(i) A body will remain at rest or in a state of uniform motion in a straight line until acted upon by an external force.

(ii) The rate of change of momentum of a body is proportional to the force applied and takes place in the direction of action of that force.

(iii) Action and reaction are equal and opposite.

Since momentum is the product of mass and velocity, for an element of fixed mass Newton's second law relates the change of velocity occurring in a given time (i.e. the acceleration) to the applied force. Working in a coherent system of units, such as SI, the proportionality becomes an equality and Newton's second law can be written

$$\text{Force} = \text{Mass} \times \frac{\text{Change of velocity}}{\text{Time}}$$

$$= \text{Mass} \times \text{Acceleration}.$$

The relationships between the acceleration a, initial velocity v_1, final velocity v_2 and the distance moved s in time t are given by the equations of motion:

$$v_2 = v_1 + at,$$

$$s = v_1 t + \tfrac{1}{2} a t^2,$$

$$v_2^2 = v_1^2 + 2as.$$

In any body of flowing fluid, the velocity at a given instant will generally vary from point to point over any specified region, and if the flow is unsteady the velocity at each point may vary with time. In this field of flow, at any given time, a particle at point A will have a different velocity from that of a particle at point B. The velocities at A and B may also change with time. Thus the change of velocity δv, which occurs when a particle moves from A to B through a distance δs in time δt, is given by

$$\begin{array}{c} \text{Total change} \\ \text{of velocity} \end{array} = \begin{array}{c} \text{Difference of} \\ \text{velocity between} \\ \text{A and B at the} \\ \text{given instant} \end{array} + \begin{array}{c} \text{Change of} \\ \text{velocity at} \\ \text{B occurring} \\ \text{in time } \delta t. \end{array} \qquad (4.1)$$

The velocity v depends on both distance s and time t. The rate of change of velocity with position at a given time is, therefore, expressed by the partial differential $\partial v / \partial s$ and the rate of change of velocity with time at a given point is expressed by the partial differential $\partial v / \partial t$. Since A and B are a distance δs apart,

Difference of velocity between A and B at the given instant

$$= \frac{\partial v}{\partial s} \cdot \delta s \ .$$

Also,

$$\text{Change of velocity at B in time } t = \frac{\partial v}{\partial t} \cdot \delta t.$$

Thus, in symbols, equation (4.1)

$$\mathrm{d}v = \frac{\partial v}{\partial s} \cdot \delta s + \frac{\partial v}{\partial t} \cdot \delta t. \qquad (4.2)$$

4.9. Acceleration of a fluid particle

The forces acting on a particle are related to the resultant acceleration

$\delta v/\delta t$ of the particle by Newton's second law. From equation (4.2) in the limit as $\delta t \to 0$,

Acceleration in the direction of flow, $a = \dfrac{dv}{dt} = \dfrac{\partial v}{\partial s}\dfrac{ds}{dt} + \dfrac{\partial v}{\partial t}$.

To denote that the derivative dv/dt is obtained by following the motion of a single particle, it is written Dv/Dt, and since $ds/dt = v$,

$$a = \frac{Dv}{Dt} = v\frac{\partial v}{\partial s} + \frac{\partial v}{\partial t}. \tag{4.3}$$

The derivative D/Dt is known as the *substantive derivative*. The total acceleration, known as the *substantive acceleration*, is composed of two parts, as shown in equation (4.3):

(i) the *convective acceleration* $v(\partial v/\partial s)$ due to the movement of the particle from one point to another point at which the velocity at the given instant is different;

(ii) the *local* or *temporal acceleration* $\partial v/\partial t$ due to the change of velocity at every point with time.

For steady flow, $\partial v/\partial t = 0$, while for uniform flow, $\partial v/\partial s = 0$.

We have so far assumed that the particle is accelerating in a straight line, but, if it is moving in a curved path, its velocity will be changing in direction and consequently there will be an acceleration perpendicular to its path, whether the velocity v is changing in magnitude or not. Figure 4.4 shows a

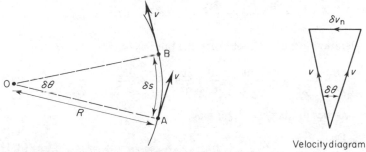

Velocity diagram

Figure 4.4   Change of velocity for a circular path

particle moving from A to B along a curved path of length δs subtending a small angle $\delta \theta$ at the centre of curvature. The change of velocity δv_n will be perpendicular to the direction of motion and, from the velocity diagram,

$$\delta v_n = v\delta\theta = v\delta s/R.$$

Dividing by δt, the time in which the change occurs, in the limit the acceleration perpendicular to the direction of motion is

$$a_n = \frac{dv_n}{dt} = \frac{v}{R}\frac{ds}{dt}$$

or, since

$$\frac{ds}{dt} = v,$$

$$a_n = v^2/R.$$

This is the convective term, and, if v has a component v_n towards the instantaneous centre of curvature, there will be a temporal term $\partial v_n / \partial t$ so that the substantial derivative is

$$a_n = \frac{v^2}{R} + \frac{\partial v_n}{\partial t}.$$

In general, the motion of a fluid particle is three-dimensional and its velocity and acceleration can be expressed in terms of three mutually perpendicular components. Thus, if v_x, v_y, and v_z are the components of the velocity in the x, y and z directions, respectively, and a_x, a_y and a_z the corresponding components of acceleration, the velocity field is described by

$$v_x = v_x(x, y, z, t),$$

$$v_y = v_y(x, y, z, t),$$

$$v_z = v_z(x, y, z, t),$$

and the velocity $\mathbf{v}$ at any point is given by

$$\mathbf{v} = v_x \mathbf{i} + v_y \mathbf{j} + v_z \mathbf{k},$$

where $\mathbf{i, j}$ and $\mathbf{k}$ are the unit vectors in the x, y and z directions.

The change of the component velocities in each direction as a particle moves in a fluid can now be calculated. Thus, in the x direction,

$$\delta v_x = \frac{\partial v_x}{\partial x} \cdot \delta x + \frac{\partial v_x}{\partial y} \cdot \delta y + \frac{\partial v_x}{\partial z} \cdot \delta z + \frac{\partial v_x}{\partial t} \cdot \delta t$$

and the acceleration in the x direction, in the limit as $\delta t \to 0$, will be

$$a_x = \frac{Dv_x}{Dt} = \frac{\partial v_x}{\partial x} \cdot \frac{dx}{dt} + \frac{\partial v_x}{\partial y} \cdot \frac{dy}{dt} + \frac{\partial v_x}{\partial z} \cdot \frac{dz}{dt} + \frac{\partial v_x}{\partial t}$$

or, since $dx/dt = v_x$, $dy/dt = v_y$, $dz/dt = v_z$,

$$a_x = \frac{Dv_x}{Dt} = v_x \frac{\partial v_x}{dx} + v_y \frac{\partial v_x}{\partial y} + v_z \frac{\partial v_x}{\partial z} + \frac{\partial v_x}{\partial t}. \qquad (4.4)$$

Similarly,

$$a_y = \frac{Dv_y}{Dt} = v_x \frac{\partial v_y}{\partial x} + v_y \frac{\partial v_y}{\partial y} + v_z \frac{\partial v_y}{\partial z} + \frac{\partial v_y}{\partial t}, \qquad (4.5)$$

$$a_z = \frac{Dv_z}{Dt} = v_x \frac{\partial v_z}{\partial x} + v_y \frac{\partial v_z}{\partial y} + v_z \frac{\partial v_z}{\partial z} + \frac{\partial v_z}{\partial t}. \qquad (4.6)$$

The first three terms in each of equations (4.4) to (4.6) represent the convective acceleration and the final term the local or temporal acceleration.

4.10. Laminar and turbulent flow

Observation shows that two entirely different types of fluid flow exist. This was demonstrated by Osborne Reynolds in 1883 through an experiment in

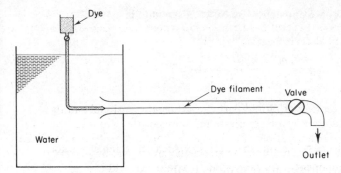

Figure 4.5   Reynolds' apparatus

which water was discharged from a tank through a glass tube (Fig. 4.5). The
rate of flow could be controlled by a valve at the outlet, and a fine filament
of dye injected at the entrance to the tube. At low velocities, it was found
that the dye filament remained intact throughout the length of the tube,
showing that the particles of water moved in parallel lines. This type of flow
is known as *laminar, viscous* or *streamline*, the particles of fluid moving in an
orderly manner and retaining the same relative positions in successive cross-
sections.

As the velocity in the tube was increased by opening the outlet valve, a
point was eventually reached at which the dye filament at first began to
oscillate and then broke up so that the colour was diffused over the whole
cross-section, showing that the particles of fluid no longer moved in an
orderly manner but occupied different relative positions in successive cross-
sections. This type of flow is known as *turbulent* and is characterized by
continuous small fluctuations in the magnitude and direction of the velocity
of the fluid particles, which are accompanied by corresponding small
fluctuations of pressure.

When the motion of a fluid particle in a stream is disturbed, its inertia
will tend to carry it on in the new direction, but the viscous forces due to
the surrounding fluid will tend to make it conform to the motion of the rest
of the stream. In viscous flow, the viscous shear stresses are sufficient
to eliminate the effects of any deviation, but in turbulent flow they are
inadequate. The criterion which determines whether flow will be viscous or
turbulent is therefore the ratio of the inertial force to the viscous force acting
on the particle.

Suppose l is a characteristic length in the system under consideration, e.g.
the diameter of a pipe or the chord of an aerofoil, and t is a typical time,
then lengths, areas, velocities and accelerations can all be expressed in terms
of l and t. For a small element of fluid of mass density ρ,

$$\text{Volume of element} = k_1 l^3,$$

$$\text{Mass of element} = k_1 \rho l^3,$$

$$\text{Velocity of element, } v = k_2 l/t,$$

$$\text{Acceleration of element} = k_3 l/t^2,$$

where k_1, k_2 and k_3 are constants. By Newton's second law,

$$\text{Inertial force} = \text{Mass} \times \text{Acceleration}$$
$$= k_1 \rho l^3 \times k_3 l/t^2$$
$$= k_1 k_3 \rho l^2 (l/t)^2$$
$$= (k_1 k_3 / k_2^2) \rho l^2 v^2 .$$

Similarly,

$$\text{Viscous force} = \text{Viscous shear stress} \times \text{Area on which stress acts}.$$

From Newton's law of viscosity (equation (1.2)),

$$\text{Viscous shear stress} = \mu \times \text{Velocity gradient} = \mu(v/k_4 l) ,$$

where μ = coefficient of dynamic viscosity.

$$\text{Area on which shear stress acts} = k_5 l^2.$$

Therefore,

$$\text{Viscous force} = \mu(v/k_4 l) \times k_5 l^2 = (k_5/k_4) \mu v l.$$

The ratio

$$\frac{\text{Inertial force}}{\text{Viscous force}} = \frac{k_1 k_3 k_4}{k_2^2 k_5} \frac{\rho l^2 v^2}{\mu v l} = \text{Constant} \times \frac{\rho v l}{\mu}.$$

Thus, the criterion which determines whether flow is viscous or turbulent is the quantity $\rho v l/\mu$, known as the Reynolds number. It is a ratio of forces and, therefore, a pure number and may also be written as $v l/\nu$, where ν is the kinematic viscosity ($\nu = \mu/\rho$).

Experiments carried out with a number of different fluids in straight pipes of different diameters have established that if the Reynolds number is calculated by making l equal to the pipe diameter and using the mean velocity $\bar{v}$ (Section 4.11), then, below a critical value of $\rho \bar{v} d/\mu = 2000$, flow will normally be laminar (viscous), any tendency to turbulence being damped out by viscous friction. This value of the Reynolds number applies only to flow in pipes, but critical values of the Reynolds number can be established for other types of flow, choosing a suitable characteristic length such as the chord of an aerofoil in place of the pipe diameter. For a given fluid flowing in a pipe of a given diameter, there will be a *critical velocity* of flow $\bar{v}_c$ corresponding to the critical value of the Reynolds number, below which flow will be viscous.

In pipes, at values of the Reynolds number >2000, flow will not necessarily be turbulent. Laminar flow has been maintained up to Re = 50 000, but conditions are unstable and any disturbance will cause reversion to normal turbulent flow. In straight pipes of constant diameter, flow can be assumed to be turbulent if the Reynolds number exceeds 4000.

4.11. Discharge and mean velocity

The total quantity of fluid flowing in unit time past any particular cross-section of a stream is called the *discharge* or flow at that section. It can be measured either in terms of mass, in which case it is referred to as the mass rate of flow $\dot{m}$ and measured in units such as kilogrammes per second, or it can be measured in terms of volume, when it is known as the volume rate of flow Q, measured in such units as cubic metres per second.

In an ideal fluid, in which there is no friction, the velocity u of the fluid would be the same at every point of the cross-section (Fig. 4.2). In unit time, a prism of fluid would pass the given cross-section and, if the cross-sectional area normal to the direction of flow is A, the volume passing would be Au. Thus

$$Q = Au.$$

In a real fluid, the velocity adjacent to a solid boundary will be zero. For a pipe, the velocity profile would be as shown in Fig. 4.6(a) for laminar flow and Fig. 4.6(b) for turbulent flow.

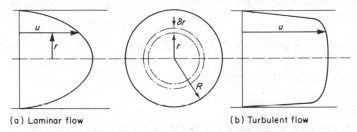

(a) Laminar flow   (b) Turbulent flow

Figure 4.6   Calculation of discharge for a circular section

If u is the velocity at any radius r, the flow δQ through an annular element of radius r and thickness δr will be

δQ = Area of element x Velocity

    $= 2\pi r\delta r \times u$,

and, hence,

$$Q = 2\pi \int_0^R ur \, dr \tag{4.7}$$

If the relation between u and r can be established, this integral can be evaluated.

In many problems, the variation of velocity over the cross-section can be ignored, the velocity being assumed to be constant and equal to the *mean velocity* $\bar{u}$, defined as volume rate of discharge Q divided by the area of cross-section A *normal* to the stream:

Mean velocity, $\bar{u} = Q/A$.

Example 4.1

Air flows between two parallel plates 80 mm apart. The following velocities were determined by direct measurement.

Distance from one plate (mm)	0	10	20	30	40	50	60	70	80
Velocity (m s^{-1})		0	23	28	31	32	29	22	14	0

Plot the velocity distribution curve and calculate the mean velocity.

Solution
Figure 4.7 shows the velocity distribution curve. The area enclosed by the

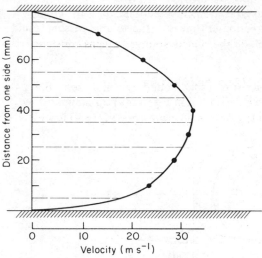

Figure 4.7   Velocity distribution curve

curve represents the product of velocity and distance, and since the two plates are parallel it will represent (to scale) the volume passing per second for unit width perpendicular to the diagram.

$$\text{Mean velocity}, \bar{u} = \frac{\text{Discharge per unit width}}{\text{Distance between plates}}.$$

The area under the graph can be determined in any convenient way and scaled to give the discharge per unit width. Using the mid-ordinate method, taking values from Fig. 4.7,

$$\bar{u} = (\Sigma \text{Mid-ordinates}/8)$$

$$= (17 \cdot 5 + 26 \cdot 0 + 29 \cdot 6 + 31 \cdot 9 + 30 \cdot 7 + 25 \cdot 4 + 18 \cdot 1 + 7 \cdot 7)/8$$

$$= 23 \cdot 36 \text{ m s}^{-1}.$$

4.12. Continuity of flow

Except in nuclear processes, matter is neither created nor destroyed. This principle of *conservation of mass* can be applied to a flowing fluid. Considering any fixed region in the flow (Fig. 4.8) constituting a control volume,

$$\begin{array}{c}
\text{Mass of fluid entering} \\
\text{per unit time}
\end{array} =
\begin{array}{c}
\text{Mass of fluid} \\
\text{leaving per unit} \\
\text{time}
\end{array} +
\begin{array}{c}
\text{Increase of mass} \\
\text{of fluid in the} \\
\text{control volume} \\
\text{per unit time.}
\end{array}$$

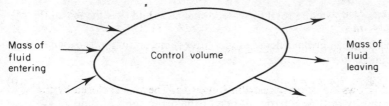

Figure 4.8   Continuity of flow

For steady flow, the mass of fluid in the control volume remains constant and the relation reduces to

Mass of fluid entering = Mass of fluid leaving
per unit time   per unit time.

Applying this principle to steady flow in a streamtube (Fig. 4.9) having

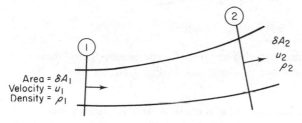

Figure 4.9   Continuous flow through a streamtube

a cross-sectional area small enough for the velocity to be considered as constant over any given cross-section, for the region between sections 1 and 2, since there can be no flow through the walls of a streamtube,

Mass entering per unit time = Mass leaving per unit time
at section 1   at section 2.

Suppose that at section 1 the area of the streamtube is δA_1, the velocity of the fluid u_1 and its density ρ_1, while at section 2 the corresponding values are $\delta A_2, u_2$ and ρ_2, then

Mass entering per unit time at $1 = \rho_1 \delta A_1 u_1$,

Mass leaving per unit time at $2 = \rho_2 \delta A_2 u_2$.

Then, for steady flow,

$$\rho_1 \delta A_1 u_1 = \rho_2 \delta A_2 u_2 = \text{Constant.} \tag{4.8}$$

This is the *equation of continuity* for the flow of a compressible fluid through a streamtube, u_1 and u_2 being the velocities measured at right angles to the cross-sectional areas δA_1 and δA_2.

For the flow of a real fluid through a pipe or other conduit, the velocity will vary from wall to wall. However, using the mean velocity $\bar{u}$, the equation of continuity for steady flow can be written as

$$\rho_1 A_1 \bar{u}_1 = \rho_2 A_2 \bar{u}_2 = \dot{m}, \tag{4.9}$$

where A_1 and A_2 are the total cross-sectional areas and $\dot{m}$ is the mass rate of flow.

If the fluid can be considered as incompressible, so that $\rho_1 = \rho_2$, equation (4.9) reduces to

$$A_1 \bar{u}_1 = A_2 \bar{u}_2 = Q. \tag{4.10}$$

The continuity of flow equation is one of the major tools of fluid mechanics, providing a means of calculating velocities at different points in a system.

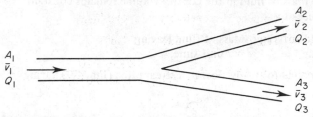

Figure 4.10   Applications of the continuity equation

The continuity equation can also be applied to determine the relation between the flows into and out of a junction. In Fig. 4.10, for steady conditions,

Total inflow to junction = Total outflow from junction,

$$\rho_1 Q_1 = \rho_2 Q_2 + \rho_3 Q_3.$$

For an incompressible fluid, $\rho_1 = \rho_2 = \rho_3$ so that

$$Q_1 = Q_2 + Q_3$$

or $\qquad A_1 \bar{v}_1 = A_2 \bar{v}_2 + A_3 \bar{v}_3.$

In general, if we consider flow towards the junction as positive and flow away from the junction as negative, then for steady flow at any junction the algebraic sum of all the mass flows must be zero:

$$\Sigma \rho Q = 0.$$

Example 4.2

Water flows through a pipe AB (Fig. 4.11) of diameter $d_1 = 50$ mm, which

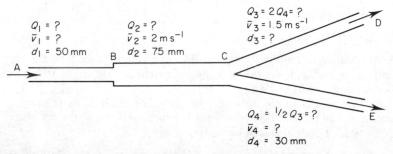

Figure 4.11   Relations between discharge, diameter and velocity

is in series with a pipe BC of diameter $d_2 = 75$ mm in which the mean velocity $\bar{v}_2 = 2$ m s^{-1}. At C the pipe forks and one branch CD is of diameter d_3 such that the mean velocity $\bar{v}_3$ is 1·5 m s^{-1}. The other branch CE is of diameter $d_4 = 30$ mm and conditions are such that the discharge Q_2 from BC divides so that $Q_4 = \tfrac{1}{2}Q_3$. Calculate the values of $Q_1, \bar{v}_1, Q_2, Q_3, d_3,$ Q_4 and $\bar{v}_4$.

Solution
Since pipes AB and BC are in series and water is 'incompressible', the volume rate of flow will be the same in each pipe, $Q_1 = Q_2$. But

$$Q_2 = \text{Area of pipe} \times \text{Mean velocity} = (\pi/4)\,d_2^2\bar{v}_2,$$

$$Q_1 = Q_2 = (\pi/4) \times (0\cdot075)^2 \times 2 = 8\cdot836 \times 10^{-3}\ \text{m}^3\ \text{s}^{-1}.$$

$$\text{Mean velocity in AB} = \bar{v}_1 = \frac{Q_1}{(\pi/4)\,d_1^2} = \frac{8\cdot836 \times 10^{-3}}{(\pi/4)(0\cdot050)^2},$$

$$\bar{v}_1 = \mathbf{4\cdot50\ m\ s^{-1}}.$$

Considering pipes BC, CD and DE, the discharge from BC must be equal to the sum of the discharges through CD and CE. Therefore $Q_2 = Q_3 + Q_4$, and since $Q_4 = \tfrac{1}{2}Q_3$ we have $Q_2 = 1\cdot5Q_3$, from which

$$Q_3 = Q_2/1\cdot5 = \mathbf{5\cdot891 \times 10^{-3}\,m^3\,s^{-1}}$$

and $Q_4 = \tfrac{1}{2}Q_3 = \mathbf{2\cdot945 \times 10^{-3}\ m^3\ s^{-1}}.$

Also, since $Q_3 = (\pi/4)d_3^2\bar{v}_3$, $d_3 = \sqrt{(4Q_3/\pi\bar{v}_3)}$,

$$d_3 = \sqrt{\left(\frac{4 \times 5\cdot891 \times 10^{-3}}{\pi \times 1\cdot5}\right)} = 0\cdot071\ \text{m} = \mathbf{71\ mm},$$

$$\bar{v}_4 = \frac{Q_4}{(\pi/4)d_4^2} = \frac{2\cdot945 \times 10^{-3}}{(\pi/4)(0\cdot030)^2} = \mathbf{4\cdot17\ m\ s^{-1}}.$$

4.13. Continuity equations for three-dimensional flow using Cartesian coordinates

The control volume ABCDEFGH in Fig. 4.12 is taken in the form of a small

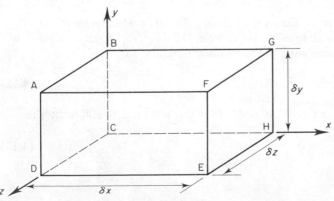

Figure 4.12   Continuity in three dimensions

rectangular prism with sides δx, δy and δz in the x, y and z directions, respectively. The mean values of the component velocities in these directions are v_x, v_y and v_z. Considering flow in the x direction,

Mass inflow through ABCD in unit time $= \rho v_x \delta y \delta z$.

In the general case, both mass density ρ and velocity v_x will change in the x direction and so

Mass outflow through EFGH in unit time $= \left\{ \rho v_x + \dfrac{\partial}{\partial x}(\rho v_x)\delta x \right\} \delta y \delta z$.

Thus,   Net outflow in unit time in x direction $= \dfrac{\partial}{\partial x}(\rho v_x)\delta x \delta y \delta z$.

Similarly,

Net outflow in unit time in y direction $= \dfrac{\partial}{\partial y}(\rho v_y)\delta x \delta y \delta z$,

Net outflow in unit time in z direction $= \dfrac{\partial}{\partial z}(\rho v_z)\delta x \delta y \delta z$.

Therefore,

Total net outflow in unit time $= \left\{ \dfrac{\partial}{\partial x}(\rho v_x) + \dfrac{\partial}{\partial y}(\rho v_y) + \dfrac{\partial}{\partial z}(\rho v_z) \right\} \delta x \delta y \delta z$.

Also, since $\partial \rho / \partial t$ is the change in mass density per unit time,

Change of mass in control volume in unit time $= -\dfrac{\partial \rho}{\partial t}\delta x \delta y \delta z$

(the negative sign indicating that a net outflow has been assumed). Then,

Total net outflow in unit time = change of mass in control
volume in unit time

$$\left\{ \frac{\partial}{\partial x}(\rho v_x) + \frac{\partial}{\partial y}(\rho v_y) + \frac{\partial}{\partial z}(\rho v_z) \right\} \delta x \delta y \delta z = -\frac{\partial \rho}{\partial t}\delta x \delta y \delta z$$

or $\dfrac{\partial}{\partial x}(\rho v_x) + \dfrac{\partial}{\partial y}(\rho v_y) + \dfrac{\partial}{\partial z}(\rho v_z) = -\dfrac{\partial \rho}{\partial t}.$ (4.11)

Equation (4.11) holds for every point in a fluid flow whether steady or unsteady, compressible or incompressible. However, for incompressible flow, the density ρ is constant and the equation simplifies to

$$\frac{\partial v_x}{\partial x} + \frac{\partial v_y}{\partial y} + \frac{\partial v_z}{\partial z} = 0. \tag{4.12}$$

For two-dimensional incompressible flow this will simplify still further to

$$\frac{\partial v_x}{\partial x} + \frac{\partial v_y}{\partial y} = 0. \tag{4.13}$$

Example 4.3

The velocity distribution for the flow of an incompressible fluid is given by

$v_x = 3 - x$, $v_y = 4 + 2y$, $v_z = 2 - z$. Show that this satisfies the requirements of the continuity equation.

Solution

For three-dimensional flow of an incompressible fluid, the continuity equation simplifes to equation (4.12);

$$\frac{\partial v_x}{\partial x} = -1, \quad \frac{\partial v_y}{\partial y} = +2, \quad \frac{\partial v_z}{\partial z} = -1,$$

and, hence,

$$\frac{\partial v_x}{\partial x} + \frac{\partial v_y}{\partial y} + \frac{\partial v_z}{\partial z} = -1 + 2 - 1 = 0,$$

which satisfies the requirement for continuity.

4.14. Continuity equation for cylindrical coordinates

The form of the continuity equation for a system of cylindrical coordinates r, θ and z, in which r and θ are measured in a plane corresponding to the x-y plane for Cartesian coordinates can be found by using the relations between polar and Cartesian coordinates:

$$x^2 + y^2 = r^2, \ (y/x) = \tan \theta,$$

$$v_x = v_r \cos \theta - v_\theta \sin \theta, \ v_y = v_r \sin \theta + v_\theta \cos \theta,$$

$$\frac{\partial}{\partial x} = \frac{\partial}{\partial r} \cdot \frac{\partial r}{\partial x} + \frac{\partial}{\partial \theta} \cdot \frac{\partial \theta}{\partial x}, \ \frac{\partial}{\partial y} = \frac{\partial}{\partial r} \cdot \frac{\partial r}{\partial y} + \frac{\partial}{\partial \theta} \cdot \frac{\partial \theta}{\partial y}.$$

This results in equation (4.12) becoming

$$\frac{1}{r} \cdot \frac{\partial}{\partial r} (r v_r) + \frac{1}{r} \frac{\partial v_\theta}{\partial \theta} + \frac{\partial v_z}{\partial z} = 0. \tag{4.14}$$

In the case of two-dimensional flow, this can be simplified further. Putting $\partial v_z / \partial z = 0$ and writing

$$\frac{\partial}{\partial r} (r v_r) = \left(r \frac{\partial v_r}{\partial r} + v_r \right),$$

equation (4.14) becomes

$$\frac{v_r}{r} + \frac{\partial v_r}{\partial r} + \frac{1}{r} \frac{\partial v_\theta}{\partial \theta} = 0.$$

EXERCISES 4

4.1   The velocity of a fluid varies with time t. Over the period from $t = 0$ to $t = 8$ s the velocity components are $u = 0$ m/s and $v = 2$ m/s; while from $t = 8$ s to $t = 16$ s the components are $u = 2$ m/s and $v = -2$ m/s. A dye streak is injected into the flow at a certain point commencing at time $t = 0$ and the path of a particle of fluid is also traced from that point starting at $t = 0$. Draw to scale the streakline, pathline of the particle and the streamlines at time $t = 12$ s.

4.2   The velocity distribution for a two-dimensional field of flow is given by

$$u = \frac{2}{3 + t} \text{ m/s and } v = 2 - \frac{t^2}{32} \text{ m/s.}$$

For the period of time from $t = 0$ to $t = 12$ s draw a streakline for an injection of dye through a certain point A and a pathline for a particle of fluid which was at A when $t = 0$. Draw also the streamlines for $t = 6$ s and $t = 12$ s.

4.3   A nozzle is formed so that its cross-sectional area converges linearly along its length. The inside diameters are 75 mm and 25 mm at inlet and exit and the length of the nozzle is 300 mm. What is the convective acceleration at a section halfway along the length of the nozzle if the discharge is constant at 0·014 m³/s.
[174·6 m/s²]

4.4   During a wind tunnel test on a sphere of radius $r = 150$ mm it is found that the velocity of flow u along the longitudinal axis of the tunnel passing through the centre of the sphere at a point upstream which is a distance x from the centre of the sphere is given by

$$u = U_0 \left(1 - \frac{r^3}{x^3}\right)$$

where U_0 is the mean velocity of the undisturbed airstream. If $U_0 = 60$ m/s what is the convective acceleration when the distance x is (a) 300 mm, (b) 150 mm.
[3937 m/s², 0]

4.5   Two parallel co-axial discs are 250 mm in radius and are brought together so that their faces approach each other with a velocity of 30 mm/s causing the air between them to be expelled radially at their periphery. What will be the convective acceleration at the periphery when the discs are 5 mm apart. Assume that the velocity distribution is uniform.
[4·5 m/s²]

4.6   The velocity along the centreline of a nozzle of length L is given by

$$u = 2 t \left(1 - 0 \cdot 5 \frac{x}{L}\right)^2$$

where u is the velocity in m/s, t is the time in seconds from the commencement of flow, x is the distance from the inlet to the nozzle. Find the convective acceleration and the local acceleration when $t = 3$ s, $x = \frac{1}{2}L$ and $L = 0 \cdot 8$ m.
[18·99 m/s², 1·125 m/s²]

4.7   Water flows through a pipe 25 mm in diameter at a velocity of 6 m/s. Determine whether the flow will be laminar or turbulent assuming that the dynamic viscosity of water is $1 \cdot 30 \times 10^{-3}$ kg/ms and its density 1000 kg/m³. If oil of specific gravity 0·9 and dynamic viscosity $9 \cdot 6 \times 10^{-2}$ kg/ms is pumped through the same pipe what type of flow will occur.
[Turbulent, Re = 115 385; laminar, Re = 1406]

4.8   If 70 litres of fuel oil are discharged from a tank in 50 s, find the rate of discharge in m³/s. If the discharge takes place through a pipe 50 mm in diameter find the mean velocity of the oil in the pipe.
[1·4 × 10⁻³ m³/s, 0·713 m/s]

4.9   An air duct is of rectangular cross-section 300 mm wide by 450 mm deep. Determine the mean velocity in the duct when the rate of flow is 0·42 m³/s. If the duct tapers to a cross-section 150 mm wide by 400 mm deep, what will be the mean velocity in the reduced section assuming that the density remains unchanged.
[3·11 m/s, 7·0 m/s]

4.10   A sphere of diameter 300 mm falls axially down a 305 mm diameter vertical cylinder which is closed at its lower end and contains water. If the sphere falls at a speed of 150 mm/s, what is the mean velocity relative to the cylinder wall of the water in the gap surrounding the midsection of the sphere.
[4·46 m/s]

4.11   The air entering a compressor has a density of 1·2 kg/m³ and a velocity of 5 m/s, the area of the intake being 20 cm². Calculate the mass flow rate. If air leaves the compressor through a 25 mm diameter pipe with a velocity of 4 m/s what will be its density.
[12×10^{-3} kg/s, 9·55 kg/m³]

4.12   Water flows through a pipe AB 1·2 m in diameter at 3 m/s and then passes through a pipe BC which is 1·5 m in diameter. At C the pipe forks. Branch CD is 0·8 m in diameter and carries one-third of the flow in AB. The velocity in branch CE is 2·5 m/s. Find (*a*) the volume rate of flow in AB, (*b*) the velocity in BC, (*c*) the velocity in CD, (*d*) the diameter of CE.
[3·393 m³/s, 1·92 m/s, 2·25 m/s, 1·073 m]

4.13   A closed tank of fixed volume is used for the continuous mixing of two liquids which enter at A and B and are discharged completely mixed at C. The diameter of the inlet pipe at A is 150 mm and the liquid flows in at the rate of 56 dm³/s and has a specific gravity of 0·93. At B the inlet pipe is 100 mm diameter, the flow rate is 30 dm³/s and the liquid has a specific gravity of 0·87. If the diameter of the outlet pipe at C is 175 mm, what will be the mass flow rate, velocity and specific gravity of the mixture discharged.
[78·18 kg/s, 3·58 m/s, 0·909]

4.14   In a 0·6 m diameter duct carrying air the velocity profile was found to obey the law $u = -5r^2 + 0.45$ m/s where u is the velocity at radius r. Calculate the volume rate of flow of the air and the mean velocity.
[0·0628 m³/s, 0·222 m/s]

4.15   A piston has a diameter of 50 mm and moves in a cylinder of 51 mm internal diameter. The cylinder is closed and contains water. If the piston is forced into the cylinder at 20 mm/s, what will be the velocity at which the water escapes between the piston and the cylinder wall.
[0·496 m/s]

4.16   During a test on a circular duct 2 m in diameter it was found that the fluid velocity was zero at the duct surface and 6 m/s on the axis of the duct when the flow rate was 9 m³/s. Assuming the velocity distribution to be given by

$$u = c_1 - c_2 r^n,$$

where u is the fluid velocity at any radius r, determine the values of the constants c_1, c_2, and n, specifying the units of c_1 and c_2.

Evaluate the mean velocity and determine the radial position at which a Pitot tube must be placed to measure this mean velocity.
[6 m/s, 6 m$^{-0.825}$ s^{-1}, 1·825; 2·86 m/s, 0·708 m]

4.17   Oil flows between two parallel flat plates which are a distance
$t = 12$ mm apart. The velocity of the oil has a maximum·value of 0.5 m/s on
a plane midway between the plates and at any distance y from this plane the
velocity is

$$u = C(\tfrac{1}{4}t^2 - y^2)$$

Calculate the volume rate of flow per unit width measured perpendicular to
the direction of flow and also the mean velocity. At what distance from the
surface of the plates would the velocity be equal to the mean velocity.
[4 dm^3/s, 0·333 m/s, 2·536 m]

4.18   Air flows through a rectangular duct which is 30 cm wide by 20 cm
deep in cross-section. To determine the volume rate of flow experimentally
the cross-section is divided into a number of imaginary rectangular elements
of equal area and the velocity measured at the centre of each element with
the following results:

Distance from bottom of duct (cm)	Distance from side of duct (cm)				
	3	9	15	21	27
18	1·6	2·0	2·2	2·0	1·7
14	1·9	3·4	6·9	3·7	2·0
10	2·1	6·8	10·0	7·0	2·3
6	2·0	3·5	7·0	3·8	2·1
2	1·8	2·0	2·3	2·1	1·9

Calculate the volume rate of flow and the mean velocity in the duct.
[0·202 m^3/s, 3·364 m/s]

4.19   If a two-dimensional flow field were to have velocity components

$$u = U(x^3 + xy^2) \text{ and } v = U(y^3 + yx^2)$$

would the continuity equation be satisfied?

4.20   Determine whether the following expressions satisfy the continuity
equation:

(a) $u = 10xt$, $v = -10yt$, $\rho = $ constant

(b) $u = U\left(\dfrac{y}{\delta}\right)^{\frac{1}{7}}$, $v = 0, 8$, $\rho = $ constants

5 The momentum equation and its applications

5.1. Momentum and fluid flow

In mechanics, the momentum of a particle or object is defined as the product of its mass m and its velocity v:

Momentum = mv.

The particles of a fluid stream will possess momentum, and, whenever the velocity of the stream is changed in magnitude or direction, there will be a corresponding change in the momentum of the fluid particles. In accordance with Newton's second law (page 95), a force is required to produce this change, which will be proportional to the rate at which the change of momentum occurs. The force may be provided by contact between the fluid and a solid boundary (e.g. the blade of a propeller or the wall of a bend in a pipe) or by one part of the fluid stream acting on another. By Newton's third law, the fluid will exert an equal and opposite force on the solid boundary or body of fluid producing the change of velocity. Such forces are known as dynamic forces, since they arise from the motion of the fluid and are additional to the static forces (*see* Chapter 3) due to pressure in a fluid which occur even when the fluid is at rest.

To determine the rate of change of momentum in a fluid stream, consider a control volume in the form of a straight section of a streamtube ABCD (Fig. 5.1). Assuming steady flow, so that the boundaries of the streamtube do

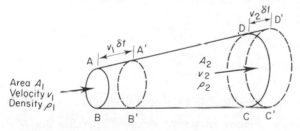

Figure 5.1   Momentum of a flowing fluid

not alter with time, and a uniform velocity at all points in a given cross-section, suppose that at AB, velocity = v_1, area = A_1, density = ρ_1, and at CD, velocity = v_2, area = A_2, density = ρ_2. After a small interval of time δt, the fluid originally contained in ABCD will have moved to $A'B'C'D'$:

Distance $AA' = v_1 \delta t$,

Distance $CC' = v_2 \delta t$.

Since the mass of fluid remains unchanged,

Mass now occupying $CC'D'D$ = Mass previously occupying $AA'B'B$,

$$\rho_2 A_2 v_2 \delta t = \rho_1 A_1 v_1 \delta t.$$

The momentum of the fluid ABCD will change in moving to A'B'C'D', because this movement is equivalent to adding the fluid CC'D'D, which has a velocity v_2, and losing the fluid AA'B'B, which has a velocity v_1. Thus,

$$
\begin{array}{llll}
\text{Change of momentum} & & \text{Increase due to} & \text{Momentum of} \\
\text{of fluid ABCD in} & = & \text{momentum of fluid} \; - & \text{fluid} \qquad (5.1) \\
\text{time } \delta t & & \text{CC'D'D} & \text{AA'B'B,}
\end{array}
$$

Momentum of fluid CC'D'D = Mass x Velocity

$$
= \rho_2 A_2 v_2 \delta t \times v_2
$$

$$
= \rho_2 A_2 v_2^2 \delta t.
$$

Similarly,

Momentum of fluid AA'B'B = $\rho_1 A_1 v_1^2 \delta t$.

Substituting in equation (5.1),

$$
\begin{array}{l}
\text{Change of momentum} \\
\text{of fluid between} \\
\text{AB and CD in unit} \\
\text{time}
\end{array}
= \rho_2 A_2 v_2^2 \delta t - \rho_1 A_1 v_1^2 \delta t.
$$

Dividing by δt,

$$
\begin{array}{l}
\text{Rate of change of} \\
\text{momentum of fluid} \\
\text{between AB and CD}
\end{array}
= \rho_2 A_2 v_2^2 - \rho_1 A_1 v_1^2. \qquad (5.2)
$$

For continuity of flow,

Mass passing AB per unit time = Mass passing CD per unit time,

$$
\rho_1 A_1 v_1 = \rho_2 A_2 v_2 = \dot{m}.
$$

Substituting in equation (5.2),

$$
\begin{array}{l}
\text{Rate of change of momentum} \\
\text{between AB and CD}
\end{array}
= \dot{m}(v_2 - v_1)
$$

$$
= \text{Mass per unit time flowing} \times \text{Change of velocity.} \qquad (5.3)
$$

Note that this is the *increase* of momentum per unit time in the direction of motion, and according to Newton's second law will be caused by a force F, such that

$$
F = \dot{m}(v_2 - v_1). \qquad (5.4)
$$

This is the resultant force acting on the fluid element ABCD in the direction of motion. By Newton's third law, the fluid will exert an equal and opposite reaction on its surroundings.

5.2. Momentum equation for two- and three-dimensional flow along a streamline

In Section 5.1, the momentum equation (5.4) was derived for one-dimensional flow in a straight line, assuming that the incoming and outgoing velocities v_1 and v_2 were in the same direction. Figure 5.2 shows a two-dimensional

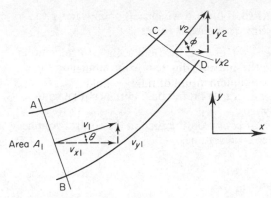

Figure 5.2   Momentum equation for two-dimensional flow

problem in which v_1 makes an angle θ with the x axis while v_2 makes a corresponding angle ϕ. Since both momentum and force are vector quantities, they can be resolved into components in the x and y directions and equation (5.4) applied. Thus, if F_x and F_y are the components of the resultant force on the element of fluid ABCD,

$$F_x = \text{Rate of change of momentum of fluid in } x \text{ direction}$$

$$= \text{Mass per unit time} \times \text{Change of velocity in } x \text{ direction}$$

$$= \dot{m}(v_2 \cos \phi - v_1 \cos \theta) = \dot{m}(v_{x2} - v_{x1}).$$

Similarly,

$$F_y = \dot{m}(v_2 \sin \phi - v_1 \sin \theta) = \dot{m}(v_{y2} - v_{y1}).$$

These components can be combined to give the resultant force,

$$F = \sqrt{(F_x^2 + F_y^2)}.$$

Again, the force exerted by the fluid on the surroundings will be equal and opposite.

For three-dimensional flow, the same method can be used, but the fluid will also have component velocities v_{z1} and v_{z2} in the z direction and the corresponding rate of change of momentum in this direction will require a force

$$F_z = \dot{m}(v_{z2} - v_{z1}).$$

To summarize the position, we can say, in general, that

Total force *exerted on*   =   Rate of change of momentum
the fluid in a control   in the given direction of
volume in a given   the fluid passing through
direction   the control volume,

$$F = \dot{m}(v_{\text{out}} - v_{\text{in}}).$$

The value of F is positive in the direction in which v is assumed to be positive.

For any control volume, the total force F which acts upon it in a given direction will be made up of three component forces:

F_1 = Force exerted *in the given direction* on the fluid in the control volume by any *solid body* within the control volume or coinciding with the boundaries of the control volume;

F_2 = force exerted *in the given direction* on the fluid in the control volume by *body forces such as gravity*;

F_3 = force exerted *in the given direction* on the fluid in the control volume by the fluid outside the control volume.

Thus,

$$F = F_1 + F_2 + F_3 = \dot{m}(v_{\text{out}} - v_{\text{in}}). \qquad (5.5)$$

The force R *exerted by* the fluid on the solid body inside or coinciding with the control volume in the given direction will be equal and opposite to F_1 so that $R = -F_1$.

5.3. Momentum correction factor

The momentum equation (5.5) is based on the assumption that the velocity is constant across any given cross-section. When a real fluid flows past a solid boundary, shear stresses are developed and the velocity is no longer uniform over the cross-section. In a pipe, for example, the velocity will vary from zero at the wall to a maximum at the centre. The momentum per unit time for the whole flow can be found by summing the momentum per unit time through each element of the cross-section, provided that these are sufficiently small for the velocity perpendicular to each element to be taken as uniform. Thus, if the velocity perpendicular to the element is u and the area of the element is δA,

Mass passing through element in unit time = $\rho \delta A \times u$,

$\begin{array}{l}\text{Momentum per unit time} \\ \text{passing through element}\end{array}$ = Mass per unit time × Velocity

$$= \rho \delta A u \times u = \rho u^2 \delta A,$$

$\begin{array}{l}\text{Total momentum per unit} \\ \text{time passing whole} \\ \text{cross-section}\end{array}$ $= \int \rho u^2 \, \mathrm{d}A. \qquad (5.6)$

To evaluate this integral, the velocity distribution must be known.

If we consider turbulent flow through a pipe of radius R (Fig. 5.3), the velocity u at any distance y from the pipewall is given approximately by Prandtl's one-seventh power law:

$$u = u_{\text{max}}(y/R)^{1/7},$$

the maximum velocity, u_{max}, occurring at the centre of the pipe. Since the velocity is constant at any radius $r = R - y$, it is convenient to take the

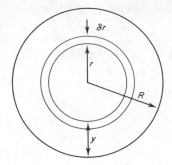

Figure 5.3   Calculation of momentum correction factor

element of area δA in equation (5.6) as an annulus of radius r and width δr,

$$\delta A = 2\pi r \delta r,$$

and, from equation (5.6), for the whole cross-section,

$$\text{Total momentum per unit time} = \int_0^R \rho u^2 \, dA$$

$$= \int_0^R \rho u_{\max}^2 (y/R)^{2/7} \, 2\pi r \, dr$$

$$= \left(\frac{2\pi\rho}{R^{2/7}}\right) u_{\max}^2 \int_0^R y^{2/7} r \, dr. \qquad (5.7)$$

Since $r = R - y$, $dr = -dy$, and so, substituting for r and dr in equation (5.7) and changing the limits (because $y = 0$ when $r = R$),

$$\frac{\text{Total momentum}}{\text{per unit time}} = \frac{2\pi\rho u_{\max}^2}{R^{2/7}} \int_R^0 y^{2/7}(R - y)(-dy)$$

$$= \frac{2\pi\rho u_{\max}^2}{R^{2/7}} \int_R^0 (y^{9/7} - Ry^{2/7}) \, dy$$

$$= \frac{2\pi\rho u_{\max}^2}{R^{2/7}} \left[\frac{7}{16} y^{16/7} - \frac{7}{9} Ry^{9/7}\right]_R^0$$

$$= \frac{2\pi\rho u_{\max}^2}{R^{2/7}} R^{16/7} \left(\frac{7}{9} - \frac{7}{16}\right)$$

$$= \frac{49}{72} \pi\rho R^2 u_{\max}^2. \qquad (5.8)$$

In practice, it is usually more convenient to use the mean velocity $\bar{u}$ instead of the maximum velocity u_{max}:

$$\text{Mean velocity, } \bar{u} = \frac{\text{Total volume per unit time passing section}}{\text{Total area of cross-section}}$$

$$= \frac{1}{\pi R^2} \int_0^R u\,\delta A.$$

Putting $u = u_{max}(y/R)^{1/7}$ and $\delta A = 2\pi r\,\delta r$

$$\bar{u} = \frac{1}{\pi R^2} \int_0^R u_{max}\left(\frac{y}{R}\right)^{1/7} 2\pi r\,dr = \frac{2u_{max}}{R^{15/7}} \int_0^R y^{1/7} r\,dr.$$

Putting $r = R - y$, $dr = -dy$, and changing the limits,

$$\bar{u} = \frac{2u_{max}}{R^{15/7}} \int_R^0 y^{1/7}(R-y)(-dy)$$

$$= \frac{2u_{max}}{R^{15/7}} \int_R^0 (y^{8/7} - Ry^{1/7})\,dy$$

$$= \frac{2u_{max}}{R^{15/7}} \left[\frac{7}{15}y^{15/7} - \frac{7}{8}Ry^{8/7}\right]_R^0$$

$$= \frac{49}{60} u_{max},$$

$$u_{max} = \frac{60}{49}\bar{u}. \tag{5.9}$$

Substituting from equation (5.9) in equation (5.8),

$$\text{Total momentum per unit time} = \frac{49}{72}\pi\rho R^2 \left(\frac{60}{49}\right)^2 \bar{u}^2$$

$$= 1{\cdot}02\rho\pi R^2\,\bar{u}^2, \tag{5.10}$$

or, since $\rho\pi R^2\,\bar{u} = $ mass per unit time,

Momentum per unit time = $1{\cdot}02 \times$ Mass per unit time $\times$ Mean velocity.

If the momentum per unit time of the stream had been calculated from the mean velocity without considering the velocity distribution, the value obtained would have been $\rho\pi R^2\,\bar{u}^2$. To take the velocity distribution into account, a momentum correction factor β must be introduced, so that, for the whole stream,

True momentum per unit time = $\beta \times$ Mass per unit time $\times$ Mean velocity.

The value of β depends upon the shape of the cross-section and the velocity distribution.

APPLICATIONS OF THE MOMENTUM EQUATION

5.4. Gradual acceleration of a fluid in a pipeline neglecting elasticity

It is frequently the case that the velocity of the fluid flowing in a pipeline has to be changed, thus causing the momentum of the whole mass of fluid in the pipeline to change. This will require the action of a force which can be calculated from the rate of change of momentum of the mass of fluid and is produced as a result of a change in the pressure difference between the ends of the pipeline. In the case of liquids flowing in rigid pipes, an approximate value of the change of pressure can be obtained by neglecting the effects of elasticity — providing that the acceleration or deceleration is small.

Example 5.1

Water flows through a pipeline 60 m long at a velocity of $1 \cdot 8$ m s^{-1} when the pressure between the inlet and outlet ends is 25 kN m^{-2}. What increase of pressure difference is required to accelerate the water in the pipe at the rate of $0 \cdot 02$ m s^{-2}? Neglect elasticity effects.

Solution
Let A = cross-sectional area of the pipe, l = length of pipe, ρ = mass density of water, a = acceleration of water, δp = increase in pressure at inlet required to produce acceleration a.

As this is not a steady flow problem, consider a control mass comprising the whole of the water in the pipe. By Newton's second law,

Force due to δp in direction of motion $=$ Rate of change of momentum of water in the whole pipe

$\qquad\qquad = $ Mass of water in pipe $\times$ Acceleration,   (I)

Force due to δp $\quad=$ Cross-sectional area $\times\ \delta p = A\delta p$,

Mass of water in pipe $\quad=$ Mass density $\times$ Volume $= \rho A l$.

Substituting in (I),

$$A\delta p = \rho A l a,$$

$$\delta p = \rho l a = 10^3 \times 60 \times 0 \cdot 02 \text{ N m}^{-2}$$

$$= 1 \cdot 2 \text{ kN m}^{-2}.$$

In this example, the change of pressure difference is small because the acceleration is small, but very large pressures can be developed by sudden accelerations or decelerations, such as may occur when valves are shut suddenly. The elasticity of the fluid and of the pipe must then be taken into account, as explained in Chapter 19.

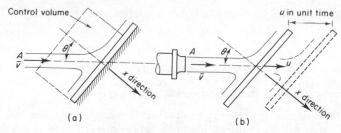

Figure 5.4   Force exerted on a flat plate

5.5. Force exerted by a jet striking a flat plate

When a jet of fluid strikes a stationary flat plate at an angle θ (Fig. 5.4), it does not rebound, but flows out over the plate in all directions. In a direction normal to the surface of the plate, the velocity of the stream will be reduced to zero and the momentum normal to the plate destroyed. There will, therefore, be a force exerted between the jet and the plate equal to the rate of change of momentum normal to the plate acting on the plate in the direction of motion, with an equal and opposite reaction by the plate on the jet.

   In a direction parallel to the plate, the force exerted will depend on the shear stress between the fluid and the surface of the plate. For an ideal fluid, there would be no shear stress and no force parallel to the plate. The fluid would flow out over the plates so that the total momentum per second parallel to the plate remained unchanged.

Example 5.2

A jet of water from a fixed nozzle has a diameter d of 25 mm and strikes a flat plate at an angle θ of 30° to the normal to the plate. The velocity of the jet $\overline{v}$ is 5 m s^{-1}, and the surface of the plate can be assumed to be frictionless.

   Calculate the force exerted normal to the plate (a) if the plate is stationary, (b) if the plate is moving with a velocity u of 2 m s^{-1} in the same direction as the jet.

Solution

For each case, the control volume taken is fixed relative to the plate (*see* Fig. 5.4) and, since we wish to find the force exerted normal to the plate, the x direction is chosen perpendicular to the surface of the plate, as shown. From equation (5.5),

Force exerted by fluid on plate in x direction,

$$R = -F_1 = F_2 + F_3 - \dot{m}(v_{\text{out}} - v_{\text{in}})_x.$$

The gravity force F_2 is negligible, and, if the fluid in the jet is assumed to be at atmospheric pressure throughout, F_3 is zero, thus,

$$R = -\dot{m}(v_{\text{out}} - v_{\text{in}})_x = \dot{m}(v_{\text{in}} - v_{\text{out}})_x,$$

where $\dot{m}$ is the mass per unit time of the fluid entering the control volume; v_{out} and v_{in} are measured relative to the control volume, which is fixed relative to the plate, so that

$$(v_{in} - v_{out})_x = \text{(Initial velocity } - \text{ Final velocity) relative to the plate in the } x \text{ direction.}$$

$$\begin{array}{l}\text{Force exerted by} \\ \text{fluid on plate in} \\ x \text{ direction, } R\end{array} = \begin{array}{l}\text{Mass per unit} \\ \text{time entering} \\ \text{control} \\ \text{volume}\end{array} \times \begin{array}{l}\text{(Initial velocity } - \text{ Final} \\ \text{velocity) relative to plate} \\ \text{in } x \text{ direction.}\end{array} \quad \text{(I)}$$

(a) If the plate is stationary (Fig. 5.4(a)),

$$\begin{array}{l}\text{Mass entering control} \\ \text{volume per unit time}\end{array} = \begin{array}{l}\text{Mass leaving nozzle} \\ \text{per unit time}\end{array}$$

$$= \rho A \bar{v},$$

where A = area of jet.

$$\begin{array}{l}\text{Initial component of velocity relative} \\ \text{to plate in } x \text{ direction}\end{array} = \bar{v} \cos \theta,$$

$$\begin{array}{l}\text{Final component of velocity relative} \\ \text{to plate in } x \text{ direction}\end{array} = 0,$$

$$\begin{array}{l}\text{(Initial velocity } - \text{ Final velocity)} \\ \text{relative to plate in } x \text{ direction}\end{array} = \bar{v} \cos \theta.$$

Substituting in (I),

$$\text{Force exerted on plate in } x \text{ direction} = \rho A \bar{v} (\bar{v} \cos \theta)$$

$$= \rho A \bar{v}^2 \cos \theta.$$

Putting $\rho = 1000$ kg m^{-3}, $A = \pi d^2/4 = \pi \times (0\cdot025)^2/4$ m^2, $\bar{v} = 5$ m s^{-1}, $\theta = 30°$,

$$\begin{array}{l}\text{Force exerted on plate} \\ \text{in } x \text{ direction}\end{array} = 1000 \times (\pi/4)(0\cdot025)^2 \times 5^2 \times 0\cdot866 \text{ N}$$

$$= 10\cdot63 \text{ N}.$$

(b) If the plate moves in the same direction as the jet with velocity u, the jet will be continually extending by a length u per unit time (Fig. 5.4(b)) and, taking the control volume as fixed relative to the plate,

$$\begin{array}{l}\text{Mass per unit time} \\ \text{entering control} \\ \text{volume}\end{array} = \begin{array}{l}\text{Mass per unit} \\ \text{time leaving} \\ \text{nozzle}\end{array} - \begin{array}{l}\text{Mass per unit} \\ \text{time required to} \\ \text{extend jet}\end{array}$$

$$= \rho A \bar{v} - \rho A u = \rho A (\bar{v} - u).$$

$$\begin{array}{l}\text{Initial component of velocity} \\ \text{relative to plate in } x \text{ direction}\end{array} = (\bar{v} - u) \cos \theta,$$

$$\begin{array}{l}\text{Final component of velocity} \\ \text{relative to plate in } x \text{ direction}\end{array} = 0,$$

(Initial velocity − Final
velocity) relative to plate in $= (\bar{v} - u)\cos\theta$.
x direction

Substituting in (I),

Force exerted on plate in x direction $= \rho A(\bar{v} - u)^2 \cos\theta$.   (II)

Inserting the numerical values in (II), above,

Force exerted on plate in x direction

$= 1000 \times (\pi/4)(0\cdot025)^2 \times (5 - 2)^2 \times 0\cdot866$ N

$= 3\cdot83$ N.

5.6. Force due to the deflection of a jet by a curved vane

Both velocity and momentum are vector quantities and, therefore, even if the
magnitude of the velocity remains unchanged, a change in direction of a
stream of fluid will give rise to a change of momentum. If the stream is
deflected by a curved vane (Fig. 5.5), entering and leaving tangentially without impact, a force will be exerted between the fluid and the surface of the

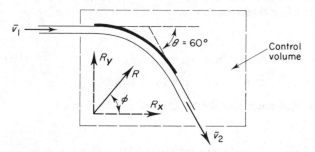

Figure 5.5   Force exerted on a curved vane

vane to cause this change of momentum. It is usually convenient to calculate
the components of this force parallel and perpendicular to the direction of
the incoming stream by calculating the rate of change of momentum in these
two directions. The components can then be combined to give the magnitude
and direction of the resultant force which the vane exerts on the fluid, and
the equal and opposite reaction of the fluid on the vane.

Example 5.3

A jet of water from a nozzle is deflected through an angle $\theta = 60°$ from its
original direction by a curved vane which it enters tangentially (*see* Fig. 5.5)
without shock with a mean velocity $\bar{v}_1$ of 30 ms^{-1} and leaves with a mean
velocity $\bar{v}_2$ of 25 m s^{-1}. If the discharge $\dot{m}$ from the nozzle is 0·8 kg s^{-1},
calculate the magnitude and direction of the resultant force on the vane if the
vane is stationary.

Solution

The control volume will be as shown in Fig. 5.5. The resultant force R *exerted by the fluid* on the vane is found by determining the component forces R_x and R_y in the x and y directions, as shown. Using equation (5.5),

$$R_x = -F_1 = F_2 + F_3 - \dot{m}(v_{out} - v_{in})_x.$$

Neglecting force F_2 due to gravity and assuming that for a free jet the pressure is constant everywhere, so that $F_3 = 0$,

$$R_x = \dot{m}(v_{in} - v_{out})_x, \tag{I}$$

and, similarly,

$$R_y = \dot{m}(v_{in} - v_{out})_y. \tag{II}$$

Since the nozzle and vane are fixed relative to each other,

$$\begin{matrix} \text{Mass per unit} \\ \text{time entering} \\ \text{control} \\ \text{volume} \end{matrix} = \dot{m} = \begin{matrix} \text{Mass per unit} \\ \text{time leaving} \\ \text{nozzle.} \end{matrix}$$

In the x direction,

$$v_{in} = \text{Component of } \bar{v}_1 \text{ in } x \text{ direction} = \bar{v}_1,$$

$$v_{out} = \text{Component of } \bar{v}_2 \text{ in } x \text{ direction} = \bar{v}_2 \cos \theta.$$

Substituting in (I),

$$R_x = \dot{m}(\bar{v}_1 - \bar{v}_2 \cos \theta). \tag{III}$$

Putting $\dot{m} = 0.8 \text{ kg s}^{-1}$, $\bar{v}_1 = 30 \text{ m s}^{-1}$, $\bar{v}_2 = 25 \text{ m s}^{-1}$, $\theta = 60°$,

$$R_x = 0.8 (30 - 25 \cos 60°) = 14 \text{ N}.$$

In the y direction,

$$v_{in} = \text{Component of } \bar{v}_1 \text{ in } y \text{ direction} = 0,$$

$$v_{out} = \text{Component of } \bar{v}_2 \text{ in } y \text{ direction} = \bar{v}_2 \sin \theta.$$

Thus, from (II),

$$R_y = \dot{m} \bar{v}_2 \sin \theta. \tag{IV}$$

Putting in the numerical values,

$$R_y = 0.8 \times 25 \sin 60° = 17.32 \text{ N}.$$

Combining the rectangular components R_x and R_y,

$$\begin{matrix} \text{Resultant force exerted} \\ \text{by fluid on vane, } R \end{matrix} = \sqrt{(R_x^2 - R_y^2)}$$

$$= \sqrt{(14^2 + 17.32^2)} = 22.27 \text{ N}.$$

This resultant force R will be inclined to the x direction at an angle $\phi = \tan^{-1} (R_y/R_x) = \tan^{-1} (17.32/14) = 51°3'.$

5.7. Force exerted when a jet is deflected by a moving curved vane

If a jet of fluid is to be deflected by a moving curved vane without impact at the inlet to the vane, the relation between the direction of the jet and the tangent to the curve of the vane at inlet must be such that the relative velocity of the fluid at inlet is tangential to the vane. The force in the direction of motion of the vane will be equal to the rate of change of momentum of the fluid in the direction of motion, i.e. the mass deflected per second multiplied by the change of velocity in that direction. The force at right angles to the direction of motion will be equal to the mass deflected per second times the change of velocity at right angles to the direction of motion.

Example 5.4

A jet of water 100 mm in diameter leaves a nozzle with a mean velocity $\bar{v}_1$ of 36 m s^{-1} (Fig. 5.6) and is deflected by a series of vanes moving with a velocity u of 15 m s^{-1} in a direction at $30°$ to the direction of the jet, so that it leaves the vane with an absolute mean velocity $\bar{v}_2$ which is at right angles to the direction of motion of the vane. Due to friction, the velocity of the fluid relative to the vane at outlet $\bar{v}_{r2}$ is equal to 0·85 of the relative velocity $\bar{v}_{r1}$ at inlet.

Calculate (a) the inlet angle α and outlet angle β of the vane which will permit the fluid to enter and leave the moving vane tangentially without shock, (b) the force exerted on the series of vanes in the direction of motion u.

Solution

If the absolute velocity $\bar{v}_2$ is to be at right angles to the direction of motion, the vane must turn the fluid so that it leaves with a relative velocity $\bar{v}_{r2}$ which has a component velocity equal and opposite to u as shown in the outlet velocity triangle (Fig. 5.6).

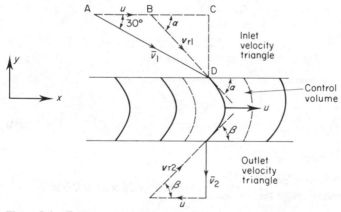

Figure 5.6   Force exerted on a series of moving vanes

(a) To determine the inlet angle α, consider the inlet velocity triangle. The velocity of the fluid relative to the vane at inlet: $\bar{v}_{r1}$, must be tangential to the vane and make an angle α with the direction of motion,

$$\tan \alpha = CD/BC = \bar{v}_1 \sin 30° / (\bar{v}_1 \cos 30° - u).$$

Putting $\bar{v}_1 = 36 \text{ m s}^{-1}$ and $u = 15 \text{ m s}^{-1}$,

$$\tan \alpha = 36 \times 0{\cdot}5/(36 \times 0{\cdot}866 - 15) = 1{\cdot}113,$$

$$\alpha = 48° 3'.$$

To determine the outlet angle β, if $\bar{v}_2$ has no component in the direction of motion, the outlet velocity triangle is right-angled, $\cos \beta = u/\bar{v}_{r2}$, but $\bar{v}_{r2}$ = $0{\cdot}85\ \bar{v}_{r_1}$ and, from the inlet triangle,

$$\bar{v}_{r_1} = CD/\sin \alpha = \bar{v}_1 \sin 30°/\sin \alpha.$$

Therefore

$$\cos \beta = \frac{u \sin \alpha}{0{\cdot}85\ v_1 \sin 30°} = \frac{15 \times 0{\cdot}744}{0{\cdot}85 \times 36 \times 0{\cdot}5} = 0{\cdot}729,$$

$$\beta = 43° 11'.$$

(b) Since the jet strikes a series of vanes, perhaps mounted on the periphery of a wheel, so that as each vane moves on its place is taken by the next in the series, the average length of the jet does not alter and the whole flow from the nozzle of diameter d is deflected by the vanes.

Neglecting the force due to gravity and assuming a free jet that does not fill the space between the vanes completely, so that the pressure is constant everywhere, the component forces in the x and y directions (Fig. 5.6) can be found from equation (5.5) putting $R = -F_1$ and $F_2 = F_3 = 0$. In the direction of motion, which is the x direction,

$$R_x = \dot{m}(v_{\text{in}} - v_{\text{out}})_x \qquad\qquad (I)$$

Mass per unit time entering control volume = $\dot{m}$ = Mass per unit time leaving nozzle

$$= \rho(\pi/4)d^2\bar{v}_1,$$

v_{in} = Component of $\bar{v}_1$ in x direction = $\bar{v}_1 \cos 30°$,

v_{out} = Component of $\bar{v}_2$ in x direction = $\bar{v}_2 \cos 90° = 0$.

Substituting in (I),

Force on vanes in direction of motion = $R_x = \rho(\pi/4)d^2\bar{v}_1 \times \bar{v}_1 \cos 30°$.

Putting in the numerical values,

Force on vanes in direction of motion = $1000 \times (\pi/4)(0{\cdot}1)^2 \times 36 \times 36 \times 0{\cdot}866$ N

$$= 8816 \text{ N}.$$

5.8. Force exerted on pipe bends and closed conduits

Figure 5.7 shows a bend in a pipeline containing fluid. When the fluid is at rest, it will exert a static force on the bend because the lines of action of the

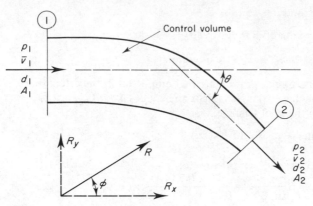

Figure 5.7   Force on a tapering bend

forces due to pressures p_1 and p_2 do not coincide. If the bend tapers, the magnitude of the static forces will also be affected.

When the fluid is in motion, its momentum will change as it passes round the bend due to the change in its direction and, if the pipe tapers, any consequent change in magnitude of its velocity; there must,. therefore, be an additional force acting between the fluid and the pipe.

Example 5.5

A pipe bend tapers from a diameter of d_1 of 500 mm at inlet (*see* Fig. 5.7) to a diameter of d_2 of 250 mm at outlet and turns the flow through an angle θ of 45°. Measurements of pressure at inlet and outlet show that the pressure p_1 at inlet is 40 kN m^{-2} and the pressure p_2 at outlet is 23 kN m^{-2}. If the pipe is conveying oil which has a density ρ of 850 kg m^{-3}, calculate the magnitude and direction of the resultant force on the bend when the oil is flowing at the rate of 0·45 m^3 s^{-1}. The bend is in a horizontal plane.

Solution

Referring to Fig. 5.7, take the x direction parallel to the incoming velocity $\bar{v}_1$ and the y direction as shown. The control volume is bounded by the inside wall of the bend and the inlet and outlet sections 1 and 2.

$$\frac{\text{Mass per unit time entering}}{\text{control volume}} = \rho Q$$

The forces *acting on the fluid* will be F_1 exerted by the walls of the pipe, F_2 due to gravity (which will be zero), and F_3 due to the pressures p_1 and p_2 of the fluid outside the control volume acting on areas A_1 and A_2 at sections 1 and 2. The force exerted by the fluid on the bend will be $R = -F_1$. Using equation (5.5), putting $F_2 = 0$ and resolving in the x direction:

$$(F_1 + F_3)_x = \dot{m}(v_{\text{out}} - v_{\text{in}})_x$$

and, since $R_x = -(F_1)_x$,

$$R_x = (F_3)_x - \dot{m}(v_{\text{out}} - v_{\text{in}})_x. \tag{I}$$

Now $(F_3)_x = p_1 A_1 - p_2 A_2 \cos \theta,$

v_{out} = Component of $\bar{v}_2$ in x direction = $\bar{v}_2 \cos \theta,$

v_{in} = Component of $\bar{v}_1$ in x direction = $\bar{v}_1.$

Substituting in (I),

$$R_x = p_1 A_1 - p_2 A_2 \cos \theta - \rho Q(\bar{v}_2 \cos \theta - \bar{v}_1). \tag{II}$$

Resolving in the y direction,

$$(F_1 + F_3)_y = \dot{m}(v_{out} - v_{in})_y$$

and, since $R_y = -(F_1)_y,$

$$R_y = (F_3)_y - \dot{m}(v_{out} - v_{in})_y. \tag{III}$$

Now,

$\cdot (F_3)_y = 0 + p_2 A_2 \sin \theta,$

v_{out} = Component of $\bar{v}_2$ in y direction = $-\bar{v}_2 \sin \theta,$

v_{in} = Component of $\bar{v}_1$ in y direction = 0.

Substituting in (III),

$$R_y = p_2 A_2 \sin \theta + \rho Q \bar{v}_2 \sin \theta.$$

For the given problem,

$A_1 = (\pi/4) d_1^2 = (\pi/4)(0.5)^2 = 0.196\ 38\ \text{m}^2,$

$A_2 = (\pi/4) d_2^2 = (\pi/4)(0.25)^2 = 0.049\ 09\ \text{m}^2,$

$Q = 0.45\ \text{m}^3\ \text{s}^{-1},$

$\bar{v}_1 = Q/A_1 = 0.45/0.196\ 38 = 2.292\ \text{m s}^{-1},$

$\bar{v}_2 = Q/A_2 = 0.45/0.049\ 09 = 9.167\ \text{m s}^{-1}.$

Putting $\rho = 850\ \text{kg m}^{-3}$, $\theta = 45^\circ$, $p_1 = 40\ \text{kN m}^{-2}$, $p_2 = 23\ \text{kN m}^{-2}$, and substituting in equation (II)

$R_x = 40 \times 10^3 \times 0.196\ 35 - 23 \times 10^3 \times 0.049\ 09 \cos 45^\circ$

$\qquad\qquad - 850 \times 0.45\ (9.167 \cos 45^\circ - 2.292)\ \text{N}$

$\quad = 10^3\ (7.855 - 0.798 - 1.603)\ \text{N}$

$\quad = \textbf{5.454} \times \textbf{10}^3\ \textbf{N}.$

Substituting in equation (III)

$R_y = 23 \times 10^3 \times 0.049\ 09 \sin 45^\circ + 850 \times 0.45 \times 9.167 \sin 45^\circ\ \text{N}$

$\quad = 10^3\ (0.798 + 2.479)\ \text{N}$

$\quad = \textbf{3.277} \times \textbf{10}^3\ \textbf{N}.$

Combining the x and y components,

$$\text{Resultant force on bend, } R = \sqrt{(R_x^2 + R_y^2)}$$

$$= \sqrt{(5 \cdot 454^2 + 3 \cdot 277^2)} \text{ kN}$$

$$= 6 \cdot 362 \text{ kN}.$$

The inclination of R to the x direction is given by

$$\phi = \tan^{-1}(R_y/R_x) = \tan^{-1} 3 \cdot 277/5 \cdot 454 = 31°.$$

5.9. Reaction of a jet

Whenever the momentum of a stream of fluid is increased in a given direction in passing from one section to another, there must be a net force acting on the fluid in that direction, and, by Newton's third law, there will be an equal and opposite force exerted by the fluid on the system which is producing the change of momentum. A typical example is the reaction force exerted when a fluid is discharged in the form of a high velocity jet, and which is applied to the propulsion of ships and aircraft through the use of propellers, pure jet engines, and rocket motors. The propulsive force can be determined from the application of the linear momentum equation (5.5) to flow through a suitable control volume.

Example 5.6

A jet of water of diameter d = 50 mm issues with velocity $\bar{v}$ = 4·9 m s^{-1} from a hole in the vertical side of an open tank which is kept filled with water to a height of 1·5 m above the centre of the hole (Fig. 5.8). Calculate the reaction of the jet on the tank and its contents (a) when it is stationary, (b) when it is moving with a velocity u = 1·2 m s^{-1} in the opposite direction to the jet while the velocity of the jet relative to the tank remains unchanged. In the latter case, what would be the work done per second?

Solution

Take the control volume shown in Fig. 5.8. In equation (5.5), the direction under consideration will be that of the issuing jet, which will be considered as positive in the direction of motion of the jet; therefore, $F_2 = 0$, and, if the jet is assumed to be at the same pressure as the outside of the tank, $F_3 = 0$.

Force exerted by
fluid system in $R = -F_1 = -\dot{m}(v_{\text{out}} - v_{\text{in}})$,
direction of motion,

or in words,

Reaction force
in direction   Mass discharged   Increase of velocity   (I)
opposite to that   =   per unit time   × in direction of jet.
of the jet

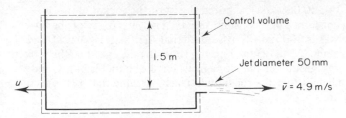

Figure 5.8   Reaction of a jet

In the present problem,

Mass discharged per unit time $\dot{m} = \rho(\pi/4)\,d^2\bar{v} = 1000 \times (\pi/4)\,(0.05)^2 \times 4.9 \text{ kg s}^{-1}$

$$= 9.62 \text{ kg s}^{-1}.$$

(a) If the tank is stationary,

$$v_{\text{out}} = \bar{v} = 4.9 \text{ m s}^{-1},$$

v_{in} = Component of velocity of the free surface in the direction of the jet = 0.

Substituting in equation (I),

Reaction of jet on tank = $9.62 \times (4.9 - 0) \text{ N}$

$$= 47.14 \text{ N}$$

in the direction opposite to that of the jet.

(b) If the tank is moving with a velocity u in the opposite direction to that of the jet, the effect is to superimpose a velocity of $-u$ on the whole system:

$$v_{\text{out}} = \bar{v} - u,$$

$$v_{\text{in}} = -u$$

$$v_{\text{out}} - v_{\text{in}} = \bar{v}.$$

Thus, the reaction of the jet R remains unaltered at **47.14 N**.

Work done per second = Reaction x Velocity of tank

$$= R \times u = 47.14 \times 1.2 = 56.57 \text{ W}.$$

A rocket motor is, in principle, a simple form of engine in which the thrust is developed as the result of the discharge of a high velocity jet of gas produced by the combustion of the fuel and oxidizing agent. Both the fuel and the oxidant are carried in the rocket and so it can operate even in outer space. It does not require atmospheric air either for combustion or for the jet to push against; the thrust is entirely due to the reaction developed from the momentum per second discharged in the jet.

Example 5.7

The mass of a rocket m_r is 150 000 kg and, when ready to launch, it carries a mass of fuel m_{fo} of 300 000 kg. The initial thrust of the rocket motor is 5 MN and fuel is consumed at a constant rate $\dot{m}$. The velocity $\bar{v}_r$ of the jet relative to the rocket is 3000 m s^{-1}. Assuming that the flight is vertical, and neglecting air resistance, find (a) the burning time, (b) the speed of the rocket and the height above ground at the moment when all the fuel is burnt, and (c) the maximum height that the rocket will reach. Assume that g is constant and equal to 9·81 m s^{-2}.

Solution

(a) From equation (I), Example 5.6,

$$\text{Initial thrust, } T = \dot{m}\,\bar{v}_r,$$

$$\text{Rate of fuel consumption, } \dot{m} = T/\bar{v}_r$$

$$= 5 \times 10^6/3000 = 1667 \text{ kg s}^{-1},$$

$$\text{Initial mass of fuel, } m_{fo} = 300\,000 \text{ kg,}$$

$$\text{Burning time} = m_{fo}/\dot{m} = 300\,000/1667 = \textbf{180 s.}$$

(b) If there is no air resistance, the forces acting on the rocket and the fuel which it contains during vertical flight are the thrust T acting upwards and the weight $(m_r + m_{ft})g$ acting downwards, where m_{ft} is the mass of the fuel in the rocket at time t.

From Newton's second law,

$$T - (m_r + m_{ft})g = (m_r + m_{ft})\,\frac{dv_t}{dt},$$

where v_t is the velocity of the rocket at time t.

$$\frac{dv_t}{dt} = \frac{T - (m_r + m_{ft})g}{m_r + m_f}.$$

Since the fuel is being consumed at a rate $\dot{m}$,

$$\text{Mass of fuel at time } t,\ m_{ft} = m_{fo} - \dot{m}t.$$

Also, $T = \dot{m}\bar{v}_r$ and so

$$\frac{dv_t}{dt} = \frac{\dot{m}\bar{v}_r - (m_r + m_{fo} - \dot{m}t)g}{m_r + m_{fo} - \dot{m}t}.$$

Substituting numerical values,

$$\frac{dv_t}{dt} = \frac{1667 \times 3000 - (150\,000 + 300\,000 - 1667\,t) \times 9\cdot81}{150\,000 + 300\,000 - 1667\,t}$$

$$= -9\cdot81 + \frac{3000}{269\cdot95 - t} \text{ m s}^{-2}.$$

Integrating,

$$v_t = -9\cdot81\,t - 3000\log_e(269\cdot95 - t) + \text{constant.}$$

Putting $v_t = 0$ when $t = 0$, the value of the constant is $3000\log_e 269\cdot95$, giving

$$v_t = -9\cdot81\,t - 3000\log_e(1 - t/269\cdot95). \tag{I}$$

From (a), all the fuel will be burnt out when $t = 180$ s. Substituting in equation (I),

$$v_t = -9 \cdot 81 \times 180 - 3000 \log_e (1 - 180/269 \cdot 95) \text{ m s}^{-1}$$

$$= -1765 \cdot 8 + 1531 \cdot 2 = \mathbf{3296 \cdot 9 \text{ m s}^{-1}}.$$

The height at time $t = 180$ s is given by

$$Z_1 = \int_0^{180} v_t \, dt = -9 \cdot 81 \int_0^{180} t \, dt - 3000 \int_0^{180} \log_e (1 - t/269 \cdot 95) \, dt$$

$$= -[4 \cdot 9 \, t^2]_0^{180} - 3000 \, [269 \cdot 95 \, (1 - t/269 \cdot 95) \, \{\log_e (1 - t/269 \cdot 95) - 1\}]_0^{180}$$

$$= -158 \, 760 + 243 \, 451 \cdot 9 = 84 \, 691 \cdot 9 \text{ m}$$

$$= \mathbf{84 \cdot 692 \text{ km}.}$$

(c) When the fuel is exhausted, the rocket will have reached an altitude of 84 692 m and will have kinetic energy $m_r \, v_t^2/2g$. It will, therefore, continue to rise a further distance Z_2 until this kinetic energy has been converted into an increase of potential energy.

$$Z_2 = v_t^2/2g = 1531 \cdot 2^2 /2 \times 9 \cdot 81 = 119 \, 468 \text{ m},$$

$$\text{Maximum height reached} = Z_1 + Z_2 = 84 \, 692 + 119 \, 468 \text{ m}$$

$$= \mathbf{204 \cdot 2 \text{ km}.}$$

For aircraft or missiles propelled in the atmosphere it is not necessary to employ a self-contained system, the propulsive force being obtained from the reaction of a jet of atmospheric air which is taken in and accelerated by means of a propeller, turboprop or jet engine and expelled at the rear of the craft.

In the case of the jet engine, air is taken in at the front of the engine and mixed with a small amount of fuel which, on burning, produces a stream of hot gas to be discharged at a much higher velocity at the rear. Figure 5.9(a) shows a jet engine moving through still air. It is convenient to take a control volume which is fixed relative to the engine and to reduce the system to a steady state by imposing a rearward velocity v upon it (Fig. 5.9(b)). Relative to the control volume,

Intake velocity, $\bar{v}_1 = v$,   Jet velocity $= \bar{v}_r$,

Total force exerted on fluid in the control volume in the direction of the jet $=$ Increase of momentum in direction of the jet,

$$F = \dot{m}_2 \bar{v}_2 - \dot{m}_1 \bar{v}_1.$$

Since the mass per unit time of the hot gases discharged will be greater than

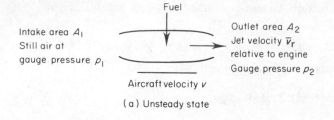

Intake area A_1
Still air at
gauge pressure p_1

Fuel

Outlet area A_2
Jet velocity $\bar{v}_r$
relative to engine
Gauge pressure p_2

Aircraft velocity v

(a) Unsteady state

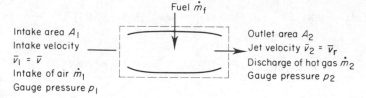

Intake area A_1
Intake velocity
$\bar{v}_1 = \bar{v}$
Intake of air $\dot{m}_1$
Gauge pressure p_1

Fuel $\dot{m}_f$

Outlet area A_2
Jet velocity $\bar{v}_2 = \bar{v}_r$
Discharge of hot gas $\dot{m}_2$
Gauge pressure p_2

(b) Steady state - control volume moves with engine

Figure 5.9   Jet engine

that of the air entering the control volume, owing to the addition of fuel,

$$\dot{m}_2 = \dot{m}_1 + \dot{m}_f.$$

Putting $\bar{v}_2 = \bar{v}_r$ and $\bar{v}_1 = v$,

$$F = (\dot{m}_1 + \dot{m}_f)\,\bar{v}_r - \dot{m}_1 v$$

$$= \dot{m}_1(\bar{v}_r - v) + \dot{m}_f\bar{v}_r.$$

If r is the ratio of the mass of fuel burned to the mass of air taken in,

$$F = \dot{m}_1\{(1+r)\,\bar{v}_r - v\}.$$

If T is the thrust exerted on the engine by the fluid, taken as positive in the direction of the jet, the force F_1 exerted by the engine on the fluid is equal to $-T$. There will be no gravity forces acting on the fluid in horizontal flight, but there will be a force $(p_1 A_1 - p_2 A_2)$ exerted on the fluid due to the fluid outside the control volume, so that

$$F = -T + (p_1 A_1 - p_2 A_2).$$

Substituting for F in the previous equation

$$T = (p_1 A_1 - p_2 A_2) - \dot{m}_1\{(1+r)\,\bar{v}_r - v\}$$

Force on engine
in forward $= -T = \dot{m}_1\{(1+r)\,\bar{v}_r - v\} - (p_1 A_1 - p_2 A_2).$ (5.11)
direction

Example 5.8

A jet engine consumes 1 kg of fuel for each 40 kg of air passing through the engine. The fuel consumption is $1 \cdot 1$ kg s^{-1} when the aircraft is travelling in still air at a speed of 200 m s^{-1}. The velocity of the gases which are discharged at atmospheric pressure from the tailpipe is 700 m s^{-1} relative to the engine.

Calculate (a) the thrust of the engine, (b) the work done per second, and (c) the efficiency.

Solution

(a) From equation (5.11), putting $r = 1/40$,

$$\dot{m}_1 = \dot{m}_f/r = 40 \times 1\cdot1 = 44 \text{ kg s}^{-1}$$

$v = 200 \text{ m s}^{-1}$, $\bar{v}_r = 700 \text{ m s}^{-1}$, $p_1 = p_2 = 0$, therefore,

$$\text{Thrust} = 44 \left\{ \left(1 + \frac{1}{40}\right) 700 - 200 \right\} = 22\,770 \text{ N}$$

$$= 22\cdot77 \text{ kN.}$$

(b) Work done per second = Thrust × Forward velocity

$$= T \times v = 22\cdot77 \times 200 = 4554 \text{ kW.}$$

(c) In addition to the useful work done on the aeroplane, work is also done in giving the exhaust gases discharge from the tailpipe kinetic energy. Relative to the earth, the velocity of the air at outlet is $(\bar{v}_r - v)$, while at intake it is zero for still air. Since the mass discharge is $\dot{m}_1(1 + r)$,

$$\begin{array}{l} \text{Loss of kinetic energy} \\ \text{per second} \end{array} = \tfrac{1}{2}\dot{m}_1(1 + r)(\bar{v}_r - v)^2$$

$$= \tfrac{1}{2} \times 44 \left(1 + \frac{1}{40}\right)(700 - 200)^2 \text{ W}$$

$$= 5638 \text{ kW.}$$

$$\text{Efficiency} = \frac{\text{Work done per second}}{\text{Work done per second} + \text{Loss}}$$

$$= \frac{4554}{4554 + 5638} = 0\cdot447 = 44\cdot7 \text{ per cent.}$$

Jet propulsion can also be applied to boats. Water is taken in through an opening either in the bows of the vessel (Fig. 5.10(a)) or on either side (Fig. 5.10(b)) and pumped out of a jet pipe at the stern at high velocity. In both cases, the control volume taken for analysis is fixed relative to the vessel. The two cases differ in that the water entering at the bows has a velocity relative to the vessel in the direction of the jet equal to the absolute velocity of the vessel u, while in Fig. 5.10(b), for side intake, the water entering has no component velocity in the direction of the jet.

Example 5.9

Derive a formula for the propulsion efficiency of a jet propelled vessel in still water if u is the absolute velocity of the vessel, v_r the velocity of the jet relative to the vessel when the intake is (a) at the bows facing the direction of motion, (b) amidships at right angles to the direction of motion.

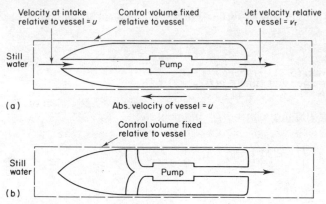

Figure 5.10 Jet propulsion of vessels. (a) Intake in direction of motion. (b) Intake in side of vessel

Solution

(a) For intake in the direction of motion,

Mass of fluid entering control volume in unit time $= \rho Q$,

v_{in} = Mean velocity of water at inlet in direction of motion relative to control volume $= u$,

v_{out} = Mean velocity of water at outlet in direction of motion relative to control volume $= v_r$.

From equation (5.26), assuming that the pressure in the water is the same at outlet and inlet,

$$\text{Propelling force} = \rho Q(v_{out} - v_{in}) = \rho Q(v_r - u),$$

Work done per unit time = Propelling force × Speed of boat

$$= \rho Q(v_r - u)u$$

In unit time, a mass of water ρQ enters the pump intake with a velocity u and leaves with a velocity v_r.

Kinetic energy per unit time at inlet $= \frac{1}{2}\rho Q u^2$,

Kinetic energy per unit time at outlet $= \frac{1}{2}\rho Q v_r^2$,

Kinetic energy per unit time supplied by pump $= \frac{1}{2}\rho Q(v_r^2 - u^2)$,

$$\text{Hydraulic efficiency} = \frac{\text{Work done per unit time}}{\text{Energy supplied per unit time}}$$

$$= \rho Q(v_r - u)u / \tfrac{1}{2}\rho Q(v_r^2 - u^2)$$

$$= 2u/(v_r + u).$$

(b) For intake at right angles to the direction of motion (Fig. 5.10(b)), the control volume used will be the same as in (a), as will the rate of change of momentum through the control volume, and therefore the propelling force. Hence,

$$\text{Work done per unit time} = \rho Q(v_r - u)u.$$

As, however, the intake to the pumps is at right angles to the direction of motion, the forward velocity of the vessel will not assist the intake of water to the pumps and, therefore, the whole of the energy of the outgoing jet must be provided by the pumps.

$$\text{Energy supplied per unit time} = \tfrac{1}{2}\rho Q v_r^2,$$

$$\text{Hydraulic efficiency} = \frac{\text{Work done per unit time}}{\text{Energy supplied per unit time}}$$

$$= \rho Q(v_r - u)u / \tfrac{1}{2}\rho Q v_r^2$$

$$= 2(v_r - u)u / v_r^2.$$

5.10. Drag exerted when a fluid flows over a flat plate

When a fluid flows over a stationary flat surface, such as the upper surface of the smooth flat plate shown in Fig. 5.11, there will be a shear stress τ_0

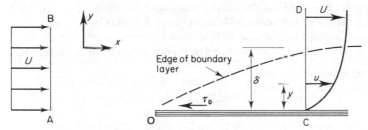

Figure 5.11   Drag on a flat plate

between the surface of the plate and the fluid, acting to retard the fluid. At a section AB of the flow well upstream of the tip of the plate O, the velocity will be undisturbed and equal to U. The fluid in contact with the surface of the plate will be at rest, and, at a cross-section such as CD, the velocity u of the adjacent fluid will increase gradually with the distance y away from the plate until it approximates to the free stream velocity at the outside of the *boundary layer* when $y = \delta$. The limit of this boundary layer, in which the drag of the stationary boundary affects the velocity of the fluid, is defined as the distance δ at which $u/U = 0.99$. The value of δ will increase from zero at the leading edge O, since the drag force D exerted on the fluid due to the shear stress τ_0 will increase as x increases. The value of D can be found by applying the momentum equation.

Consider a control volume PQSR (Fig. 5.12) consisting of a section of the boundary layer of length Δx at a distance x from the upstream edge of the plate. Fluid enters the control volume through section PQ and through the

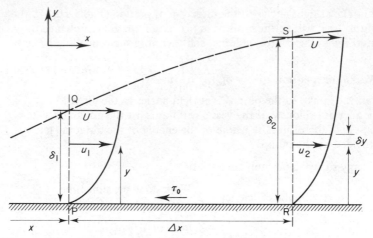

Figure 5.12   Momentum equation applied to a boundary layer

upper edge of the boundary layer QS, leaving through section RS. Applying the momentum equation,

> Force acting on fluid   Rate of increase of momentum in
> in control volume in   = x direction of fluid passing
> x direction   through control volume.

Since the velocity u in the boundary layer varies with the distance y from the surface of the plate, the momentum efflux through RS must be determined by integration. Consider an element of thickness δy, through which the velocity in the x direction is u_2. For a width B perpendicular to the diagram,

> Momentum per second
> passing through element   = Mass per second x Velocity

$$= \rho B \delta y u_2 \times u_2,$$

> Total momentum per second
> passing through RS $= \rho B \displaystyle\int_0^{\delta_2} u_2^2 \, \mathrm{d}y.$ (5.12)

Similarly, for the control surface PQ,

> Total momentum per second
> passing through PQ $= \rho B \displaystyle\int_0^{\delta_1} u_1^2 \, \mathrm{d}y,$ (5.13)

where u_1 is the velocity through PQ at a distance y from the surface. For the control surface QS, for continuity of flow,

> Rate of flow into   Rate of flow   Rate of flow
> the control volume, Q $=$ through RS $-$ through PQ

$$Q = B \int_0^{\delta_2} u_2 \, \mathrm{d}y - B \int_0^{\delta_1} u_1 \, \mathrm{d}y,$$

Momentum in x direction
entering through QS $= \rho Q U$

$$= \rho B \left\{ \int_0^{\delta_2} u_2 \, dy - \int_0^{\delta_1} u_1 \, dy \right\} U \quad (5.14)$$

Force exerted on the fluid by
the boundary in the x direction $= -\tau_0 B \Delta x. \qquad (5.15)$

Equating the force given by equation (5.15) with the sum of the x momenta from equations (5.12), (5.13) and (5.14),

$$-\tau_0 B \Delta x = \rho B \left\{ \int_0^{\delta_2} u_2^2 \, dy - \int_0^{\delta_1} u_1^2 \, dy - U \left(\int_0^{\delta_2} u_2 \, dy - \int_0^{\delta_1} u_1 \, dy \right) \right\}$$

$$= \rho B \left\{ \int_0^{\delta_2} (u_2^2 - U u_2) \, dy - \int_0^{\delta_1} (u_1^2 - U u_1) \, dy \right\}.$$

The term in brackets is the difference between $\displaystyle\int_0^{\delta} (u^2 - Uu) \, dy$ at sections

RS and PQ, which can be written as $\displaystyle\Delta \left[\int_0^{\delta} u \, (u - U) \, dy \right]$ so that

$$-\tau_0 B \Delta x = \rho B \Delta \left[\int_0^{\delta} u (u - U) \, dy \right].$$

In the limit, as Δx tends to zero,

$$\tau_0 = \rho U^2 \frac{d}{dx} \int_0^{\delta} \frac{u}{U} \left(1 - \frac{u}{U} \right) dy. \qquad (5.16)$$

The drag D on one surface of the plate will be given by

$$D = B \int_0^x \tau_0 \, dx.$$

If the fluid acts on both the upper and the lower surface of the plate, this force will of course be doubled.

5.11. Angular motion

In Section 4.8, we set out the equations of motion for a particle or element of fluid moving in a straight line. If the particle or element is rotating about a fixed point, similar equations can be written to describe its angular motion. Angular displacement will be measured as the angle θ in radians through which the particle or element has moved about the centre measured from a reference direction. Angular velocity ω will be the rate of change of displacement θ with time, i.e.

$$\omega = \dot{\theta} = \frac{\mathrm{d}\theta}{\mathrm{d}t},$$

and the angular acceleration α will be the rate of change of ω with time, so that

$$\alpha = \ddot{\theta} = \frac{\mathrm{d}^2\theta}{\mathrm{d}t^2}.$$

The laws of angular motion will be similar to those for linear motion (*see* Section 4.8):

$$\omega_2 = \omega_1 + \alpha t,$$
$$\theta = \omega_1 t + \tfrac{1}{2}\alpha t^2,$$
$$\omega_2^2 = \omega_1^2 + 2\alpha\theta.$$

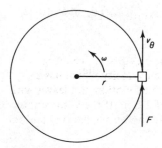

Figure 5.13   Angular motion

For a particle (Fig. 5.13) which at a given instant is rotating about a fixed point with angular velocity ω at a radius r,

Tangential linear velocity, $v_\theta = \omega r$,

Momentum of particle, $m v_\theta = m \omega r$.

If the angular velocity changes from ω to zero in time t under the influence of a force F acting at radius r,

Rate of change of momentum of particle $= m r \omega / t$.

By Newton's second law,

$$F = m r \omega / t.$$

This force produces a turning moment or torque T about the centre of rotation,

$$T = Fr = mr^2 \omega/t = \text{Angular momentum/Time.}$$

Now consider a particle moving in a curved path, so that in time t it moves from a position at which it has an angular velocity ω_1 at radius r_1 to a position in which the corresponding values are ω_2 and r_2. The effect will be equivalent to first applying a torque to reduce the particle's original angular momentum to zero, and then applying a torque in the opposite direction to produce the angular momentum required in the second position:

$$\text{Torque required to eliminate original angular momentum} = mr_1^2 \omega_1/t,$$

$$\text{Torque required to produce new angular momentum} = mr_2^2 \omega_2/t,$$

$$\text{Torque required to produce change of angular momentum} = (m/t)(\omega_2 r_2^2 - \omega_1 r_1^2)$$

$$= (m/t)(v_{\theta 2} r_2 - v_{\theta 1} r_1),$$

where v_θ = tangential velocity = ωr.

This analysis applies equally to a stream of fluid moving in a curved path, since m/t is the mass flowing per unit time, $\dot{m} = \rho Q$. The torque which must be acting on the fluid will be

$$T = \rho Q(v_{\theta 2} r_2 - v_{\theta 1} r_1), \tag{5.17}$$

and, of course, the fluid will exert an equal and opposite reaction.

Example 5.10

A water turbine rotates at 240 rev min^{-1}. The water enters the rotating impeller at a radius of 1·2 m with an absolute mean velocity which has a tangential component of 2·3 m s^{-1} in the direction of motion and leaves with a tangential component of 0·2 m s^{-1} at a radius of 1·6 m. If the volume rate of flow through the turbine is 10 m^3 s^{-1}, calculate the torque exerted and the theoretical power output.

Solution

In equation (5.17), $\rho = 1000$ kg m^{-3}, $Q = 10$ m^3 s^{-1}, $\bar{v}_{\theta 2} = 0 \cdot 2$ m s^{-1}, $\bar{v}_{\theta 1} = 2 \cdot 3$ m s^{-1}, $r_2 = 1 \cdot 6$ m, $r_1 = 1 \cdot 2$ m. Hence,

$$\text{Torque acting on fluid} = 1000 \times 10 \, (0 \cdot 2 \times 1 \cdot 6 - 2 \cdot 3 \times 1 \cdot 2) \text{ N m}$$

$$= 10\,000 \, (0 \cdot 32 - 2 \cdot 76) = -24\,400 \text{ N m.}$$

The torque exerted by the fluid on the rotor will be equal and opposite:

$$\text{Torque exerted by fluid} = 24\,400 \text{ N m.}$$

If n is the rotational speed in revolutions per second,

$$n = 240/60 = 4,$$

$$\text{Power output} = 2\pi n T = 2\pi \times 4 \times 24\,400 \text{ W}$$

$$= 613\,318 \text{ W} = \textbf{613·32 kW.}$$

5.12. Euler's equation of motion along a streamline

From consideration of the rate of change of momentum from point to point along a streamline and the forces acting due to the effects of the surrounding pressures and changes of elevation, it is possible to derive a relationship between velocity, pressure, elevation and density along a streamline.

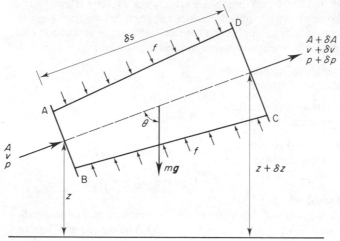

Figure 5.14   Euler's equation

Figure 5.14 shows a short section of a streamtube surrounding the streamline and having a cross-sectional area small enough for the velocity to be considered constant over the cross-section. AB and CD are two cross-sections separated by a short distance δs. At AB the area is A, velocity v, pressure p and elevation z, while at CD the corresponding values are $A + \delta A$, $v + \delta v$, $p + \delta p$ and $z + \delta z$. The surrounding fluid will exert a pressure f on the sides of the element and, if the fluid is assumed to be inviscid, there will be no shear stresses on the sides of the streamtube and f will act normally. The weight of the element mg will act vertically downward at an angle θ to the centreline.

$$\text{Mass per unit time flowing} = \rho A v,$$

$$\begin{matrix} \text{Rate of increase of momentum} \\ \text{from AB to CD} \end{matrix} = \rho A v\{(v + \delta v) - v\}$$

$$= \rho A v \delta v. \qquad (5.18)$$

The forces acting to produce this increase of momentum in the direction of motion are

Force due to p in direction of motion $= pA$,

Force due to $p + \delta p$ opposing motion $= (p + \delta p)(A + \delta A)$,

Force due to f producing a component in the direction of motion $= f \delta A$,

Force due to mg producing a component opposing motion $= mg \cos \theta$,

Resultant force in the direction of motion $= pA - (p + \delta p)(A + \delta A) + f \delta A - mg \cos \theta$.

The value of f will vary from p at AB to $p + \delta p$ at CD and can be taken as $p + k\delta p$, where k is a fraction,

$$\text{Weight of element, } mg = \rho g \times \text{Volume} = \rho g \left(A + \tfrac{1}{2} \delta A \right) \delta s,$$

$$\cos \theta = \delta z / \delta s,$$

Resultant force in the direction of motion $= -p\delta A - A\delta p - \delta p \delta A + p\delta A + k\delta p \cdot \delta A$

$$- \rho g \left(A + \tfrac{1}{2} \delta A \right) \delta s \cdot (\delta z / \delta s).$$

Neglecting products of small quantities,

$$\text{Resultant force in the direction of motion} = -A\delta p - \rho g A \delta z. \quad (5.19)$$

Applying Newton's second law from equations (5.18) and (5.19),

$$\rho A v \delta v = -A\delta p - \rho g A \delta z.$$

Dividing by $\rho A \delta s$,

$$\frac{1}{\rho} \frac{\delta p}{\delta s} + v \frac{\delta v}{\delta s} + g \frac{\delta z}{\delta s} = 0, \quad (5.20)$$

or, in the limit as $\delta s \to 0$,

$$\frac{1}{\rho} \frac{dp}{ds} + v \frac{dv}{ds} + g \frac{dz}{ds} = 0. \quad (5.21)$$

This is known as *Euler's equation*, giving, in differential form, the relationship between pressure p, velocity v, density ρ_1 and elevation z along a streamline for steady flow. It cannot be integrated until the relationship between density ρ and pressure p is known.

For an incompressible fluid, for which ρ is constant, integration of equation (5.21) along the streamline, with respect to s, gives

$$p/\rho + v^2/2 + gz = \text{constant}. \quad (5.22)$$

The terms represent energy per unit mass. Dividing by g

$$p/\rho g + v^2/2g + z = \text{constant} = H, \quad (5.23)$$

in which the terms represent the energy per unit weight. Equation (5.23) is known as *Bernoulli's equation* and states the relationship between pressure, velocity and elevation for steady flow of a frictionless fluid of constant density. An alternative form is

$$p + \tfrac{1}{2} \rho v^2 + \rho g z = \text{constant}, \quad (5.24)$$

in which the terms represent the energy per unit volume.

These equations apply to a single streamline. The sum of the three terms is constant along any streamline, but the value of the constant may be different for different streamlines in a given stream.

If equation (5.21) is integrated along the streamline between any two points indicated by suffixes 1 and 2,

$$p_1/\rho g + v_1^2/2g + z_1 = p_2/\rho g + v_2^2/2g + z_2. \quad (5.25)$$

For a compressible fluid, the integration of equation (5.21) can only be partially completed, to give

$$\int \frac{\mathrm{d}p}{\rho g} + \frac{v^2}{2g} + z = H.$$

The relationship between ρ and p must then be inserted for the given case. For gases, this will be of the form $p\rho^n$ = constant, varying from adiabatic to isothermal conditions, while, for a liquid, $\rho(\mathrm{d}p/\mathrm{d}\rho) = K$, the bulk modulus.

5.13. Pressure waves and the velocity of sound in a fluid

In a real fluid, any change of pressure at a point or any cross-section will be associated with a change in density of the fluid, so that the particles of fluid will change their positions, moving closer together or further apart. Adjacent particles will, in turn, change their positions, and so the change of pressure and density will spread very rapidly through the fluid. Clearly, if the fluid were incompressible, every particle would have to change its position simultaneously and the speed of propagation of the disturbance or pressure wave would, theoretically, be infinite. However, the elasticity of a compressible fluid allows the particles to adjust their positions one after the other, so that the disturbance spreads with a finite velocity. The speed of propagation of a pressure change is very rapid and, in some problems, it is sufficient to assume that pressure changes are propagated instantaneously throughout the fluid. However, when studying abrupt changes of pressure, such as those occurring when a valve on a pipeline is closed suddenly, or when fluid velocities are high relative to a solid body (as in the case of aircraft in flight), the speed of propagation of pressure changes in the fluid can be a factor of major importance from the practical point of view.

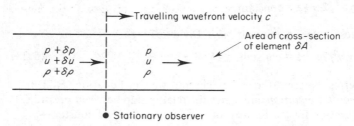

Figure 5.15   Pressure wave. Unsteady flow relative to a stationary observer

In Fig. 5.15, a pressure wave is moving through a fluid from left to right with a velocity c relative to a stationary observer. The fluid to the right of the wavefront will not have been affected by the pressure wave and will have its original pressure p, velocity u relative to the observer and density ρ as indicated. To the left, the fluid behind the wavefront will be at the new pressure $p + \delta p$, velocity $u + \delta u$, and density $\rho + \delta \rho$. From a terrestrial frame of reference, conditions in the fluid are not steady, since at a point fixed with reference to a stationary observer conditions will change with time. The usual equations for steady flow cannot, therefore, be applied. However, to an observer moving with the wave at velocity c the wave will appear stationary; conditions will not change with time and the flow is steady and can be analysed as such.

As shown in Fig. 5.16, the effect is equivalent to imposing a backward

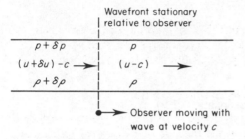

$$p + \delta p \quad | \quad p$$
$$(u + \delta u) - c \rightarrow | \quad (u - c) \quad \rightarrow$$
$$\rho + \delta \rho \quad | \quad \rho$$

Figure 5.16   Pressure wave. Steady state relative to a moving observer

velocity c on the system from right to left. Considering an element of cross-sectional area δA perpendicular to the direction of flow,

$$\begin{array}{c}\text{Mass per unit time flowing}\\\text{on the left of wavefront}\end{array} = \begin{array}{c}\text{Mass per unit time flowing}\\\text{on the right of wavefront}\end{array}$$

$$(\rho + \delta\rho)(u + \delta u - c)\,\delta A = \rho(u - c)\,\delta A,$$

$$\rho\delta u + u\delta\rho + \delta u\delta\rho - c\delta\rho = 0,$$

$$(c - u)\,\delta\rho = (\rho + \delta\rho)\,\delta u. \tag{5.26}$$

Due to the pressure difference δp across the wavefront, there will be a force acting to the right, in the direction of flow, which will cause an increase in momentum per unit time in this direction.

$$\text{Force due to } \delta p = \text{Increase of momentum per unit time to the right}$$

$$= \text{Mass per unit time} \times \text{Increase of velocity,}$$

$$\delta p \times \delta A = \rho(u - c)\,\delta A \times \{u - (u + \delta u)\},$$

$$\delta p = \rho(u - c)(-\delta u),$$

$$\delta u = \delta p / \rho(c - u). \tag{5.27}$$

Substituting from equation (5.27) for δu in equation (5.26),

$$(c - u)\,\delta\rho = (\rho + \delta\rho)\,\delta p / \rho(c - u),$$

$$(c - u)^2 = \left(1 + \frac{\delta\rho}{\rho}\right)\frac{\delta p}{\delta\rho}. \tag{5.28}$$

If the change of pressure and density across the wavefront is small, the pressure wave is said to be *weak* and, in the limit as δp and $\delta \rho$ tend to zero, equation (5.28) gives

$$(c - u) = \sqrt{\left(\frac{\mathrm{d}p}{\mathrm{d}\rho}\right)}.$$

Now $(c - u)$ is the velocity of the wavefront relative to the fluid, so that

$$\text{Velocity of propagation of a weak pressure wave,} \quad c - u = \sqrt{\left(\frac{\mathrm{d}p}{\mathrm{d}\rho}\right)}. \quad (5.29)$$

For a mass m of fluid of volume V and density ρ,

$$\rho V = m.$$

Differentiating,

$$\rho\,\mathrm{d}V + V\,\mathrm{d}\rho = 0,$$

$$\mathrm{d}\rho = -(\rho/V)\,\mathrm{d}V.$$

If K is the bulk modulus, then from equation (1.9),

$$K = -V\frac{\mathrm{d}p}{\mathrm{d}V} = \rho\frac{\mathrm{d}p}{\mathrm{d}\rho},$$

$$\frac{\mathrm{d}p}{\mathrm{d}\rho} = \frac{K}{\rho}.$$

Therefore,

$$\text{Velocity of propagation of a weak pressure wave,} \quad c - u = \sqrt{(K/\rho)}. \quad (5.30)$$

This equation applies to solids, liquids and gases. Note, however, that when c represents the velocity of sound in still air, $u = 0$.

Since sound is propagated in the form of very weak pressure waves, equation (5.30) gives the velocity of sound or *sonic velocity*, with $u = 0$. In a gas, the pressure and temperature changes occurring due to the passage of a sound wave are so small and so rapid that the process can be considered as reversible and adiabatic, so that p/ρ^γ = constant. Differentiating,

$$\frac{\mathrm{d}p}{\mathrm{d}\rho} = \frac{\gamma p}{\rho}$$

or, since, $p/\rho = RT$,

$$\frac{\mathrm{d}p}{\mathrm{d}\rho} = \gamma R T.$$

Substituting in equation (5.29), with $u = 0$,

$$\text{Sonic velocity, } c = \sqrt{(\gamma p/\rho)} = \sqrt{(\gamma R T)}. \quad (5.31)$$

The above equations apply only to weak pressure waves in which the change of pressure is very small compared with the pressure of the fluid. The pressure change involved in the passage of a sound wave in atmospheric air, for example, varies from about 3×10^{-5} N m^{-2} for a barely audible sound to 100 N m^{-2} for a sound so loud that it verges on the painful. These are small in comparison with atmospheric pressure of 10^5 N m^{-2}. For relatively large pressure changes, the velocity of propagation of the pressure wave would be greater.

The sonic velocity is important in fluid mechanics, because when the velocity of the fluid exceeds the sonic velocity, i.e. becomes *supersonic*, small pressure waves cannot be propagated upstream. At *subsonic* velocities, lower than the sonic velocity, small pressure waves can be propagated both upstream and downstream. This results in the flow pattern around an obstacle, for example, differing for supersonic and subsonic flow with consequent differences in the forces exerted. The ratio of the fluid velocity u to the sonic velocity $c - u$ is known as the Mach number $\text{Ma} = u/(c - u)$. If $\text{Ma} > 1$, flow is supersonic; if $\text{Ma} < 1$, flow is subsonic.

5.14. Velocity of propagation of a small surface wave

For flow with a free surface, as, for example, in open channels, the pressure cannot vary from point to point along the free surface. A disturbance in the fluid will be propagated as a surface wave rather than as a pressure wave. Using the approach adopted in Section 5.13, assume that a surface wave of

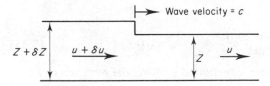

(a) Unsteady flow as seen by stationary observer

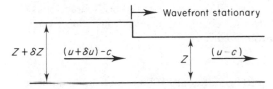

(b) Steady flow as seen by moving observer

Figure 5.17   Velocity of propagation of a small surface wave. (a) Unsteady flow as seen by a stationary observer. (b) Steady flow as seen by a moving observer

height δZ (Fig. 5.17(a)) is being propagated from left to right in the view of a stationary observer. If this wave is brought to rest relative to the observer by imposing a velocity c equal to the wave velocity on the observer, conditions will now appear steady, as shown in Fig. 5.17(b). Considering a width B of the flow, perpendicular to the plane of the diagram,

Mass per unit time
flowing on the left   =   Mass per unit time
flowing on the right
of the wavefront   of the wavefront,

$$\rho \times B(Z + \delta Z) \times (u + \delta u - c) = \rho \times B \times Z \times (u - c),$$

the mass density ρ being the same on both sides of the wavefront since the pressure is unchanged. Simplifying,

$$Z\delta u + u\delta Z + \delta Z\delta u - c\delta Z = 0,$$

$$(c - u)\,\delta Z = (Z + \delta Z)\,\delta u. \tag{5.32}$$

The change of momentum occurring as a result of the change of velocity across the wavefront is produced as a result of the hydrostatic force due to the difference of level δZ acting on the cross-sectional area BZ. By Newton's second law,

$$\text{Hydrostatic force due to } \delta Z = \text{Mass per unit time} \times \text{Change of velocity,}$$

$$\rho g \, \delta Z \, BZ = BZ(u - c) \times (-\delta u),$$

$$\delta u = g \, \delta Z / (c - u).$$

Substituting from equation (5.32) for δu,

$$(c - u) \, \delta Z / (Z + \delta Z) = g \, \delta Z / (c - u),$$

$$(c - u)^2 = (Z + \delta Z) g$$

$$= gZ$$

if the wave height δZ is small.

$$\text{Velocity of propagation of the wave relative to the fluid} = c - u = \sqrt{(gZ)}. \qquad (5.33)$$

Taking velocities in a downstream direction as positive, the wave velocity c relative to the bed of the channel is given by $\sqrt{(gZ)} + u$ if it is travelling downstream, and $\sqrt{(gZ)} - u$ if it is travelling upstream. Thus, if the stream velocity $u > \sqrt{(gZ)}$, the wave cannot travel upstream relative to the bed, while if $u < \sqrt{(gZ)}$ a surface wave will be propagated in both directions. The ratio of the stream velocity u to the velocity of propagation $c - u$ of the wave in the fluid is known as the *Froude number* Fr:

$$\text{Fr} = u/(c - u) = u/\sqrt{(gZ)}. \qquad (5.34)$$

Thus, the condition for a wave to be stationary is that the Froude number Fr = 1. The Froude number is also a criterion of the type of flow. If Fr > 1 flow is fast and shallow and is said to be *rapid* or *shooting*, while if Fr < 1 flow is slow and deep and is said to be *tranquil* or *streaming*. Analogies can usefully be drawn between compressible flow and flow with a free surface, and the latter can be used for the experimental investigation of the former.

EXERCISES 5

5.1   Oil flows through a pipeline 0·4 m in diameter. The flow is laminar and the velocity at any radius r is given by $u = (0·6 - 15r^2)$ m/s. Calculate (a) the volume rate of flow, (b) the mean velocity, (c) the momentum correction factor.
[0·0377 m³/s, 0·30 m/s, 1·333]

5.2   A liquid flows through a circular pipe 0·6 m in diameter. Measurements of velocity taken at intervals along a diameter are:

Distance from wall m	0	0·05	0·1	0·2	0·3	0·4	0·5	0·55	0·6
Velocity m/s		0	2·0	3·8	4·6	5·0	4·5	3·7	1·6	0

(a) Draw the velocity profile, (b) calculate the mean velocity, (c) calculate the momentum correction factor.
[2·88 m/s, 1·26]

5.3   Calculate the mean velocity and the momentum correction factor for a velocity distribution in a circular pipe given by $(v/v_0) = (y/R)^{1/n}$ where v is the velocity at a distance y from the wall of the pipe, v_0 is the centreline velocity, R the radius of the pipe and n an unspecified power.

$$\left[\frac{2v_0 n^2}{(n+1)(2n+1)}, \frac{(n+1)(2n+1)^2}{4n^2(n+2)} \right]$$

5.4   A pipeline is 120 m long and 250 mm in diameter. At the outlet there is a nozzle 25 mm in diameter controlled by a shut-off valve. When the valve is fully open water issues as a jet with a velocity of 30 m/s. Calculate the reaction of the jet.

If the valve can be closed in 0·2 s what will be the resulting rise in pressure at the valve required to bring the water in the pipe to rest in this time. Assume no change in density of the water and no expansion of the pipe.
[438 N, 180 kN/m²]

5.5   Water flows along a pipeline 60 m long with a mean velocity of 3 m/s. A valve at the outlet end of the pipe is turned off so that the water in the pipe is brought to rest in half a second with uniform retardation. Calculate the pressure rise at the valve neglecting the effect of the elasticity of the pipe and the water.
[360 kN/m²]

5.6   A uniform pipe 75 m long containing water is fitted with a plunger. The water is initially at rest. If the plunger is forced into the pipe in such a way that the water is accelerated uniformly to a velocity of 1·7 m/s in 1·4 s what will be the increase of pressure on the face of the plunger assuming that the water and the pipe are not elastic.

If instead of being uniformly accelerated the plunger is driven by a crank 0·25 m long and making 120 rev/min so that the plunger moves with simple harmonic motion, what would be the maximum pressure on the face of the piston.
[91 kN/m², 2961 kN/m²]

5.7   A flat plate is struck normally by a jet of water 50 mm in diameter with a velocity of 18 m/s. Calculate (a) the force on the plate when it is stationary, (b) the force on the plate when it moves in the same direction as the jet with a velocity of 6 m/s, (c) the work done per second and the efficiency in case (b).
[636·2 N, 282·7 N, 1696·5 W, 29·6%]

5.8   A jet of water 50 mm diameter with a velocity of 18 m/s strikes a flat plate inclined at an angle of 25° to the axis of the jet. Determine the normal force exerted on the plate (a) when the plate is stationary, (b) when the plate is moving at 4·5 m/s in the direction of the jet, and (c) determine the work done and the efficiency for case (b).
[269 N, 151·2 N, 5%]

5.9   In an undershot waterwheel the cross-sectional area of the stream striking the series of radial flat vanes of the wheel is 0·1 m² and the velocity of the stream is 6 m/s. The velocity of the vanes is 3 m/s. Calculate the force exerted on the series of vanes by the stream, the work done and the hydraulic efficiency.

[1·8 kN, 5400 W, 50%]

5.10 A square plate of mass 12·7 kg, uniform thickness and 300 mm length of edge, is hung so that it can swing freely about its upper horizontal edge. A horizontal jet, 20 mm in diameter, strikes the plate with a velocity of 15 m/s. The centreline of the jet is 150 mm below the upper edge of the plate so that when the plate is vertical the jet strikes the plate normally at the centre. Find (*a*) what force must be applied at the lower edge of the plate to keep it vertical, (*b*) what inclination to the vertical the plate will assume under the action of the jet if allowed to swing freely.

[35·34 N, 34·57°]

5.11 A jet of water delivers 85 dm^3/s at 36 m/s onto a series of vanes moving in the same direction as the jet at 18 m/s. If stationary the water which enters tangentially would be diverted through an angle of 135°. Friction reduces the relative velocity at exit from the vanes to 0·80 of that at entrance. Determine the magnitude of the resultant force on the vanes and the efficiency of the arrangement. Assume no shock at entry.

[2546 N, 0·783]

5.12 A stream of water issues from a nozzle of area 2580 mm^2 with a velocity of 30 m/s. The water is deflected by a curved vane through an angle of 2·62 rad when the vane is stationary. If the vane moves with a velocity of 12 m/s in the initial direction of the jet, find (*a*) the force on the vane in the direction of motion of the vane, (*b*) the force perpendicular to that direction, (*c*) the work done per second on the vane.

[1580 N, 424 N, 19·3 kW]

5.13 A jet of water strikes a stationary curved vane without shock and is deflected 150° from its original direction. The discharge from the jet is 0·68 kg/s and the jet velocity is 24 m/s. Assume that there is no reduction of relative velocity due to friction and determine the magnitude and direction of the reaction on the vane.

If the vane is now allowed to move with a velocity of 8 m/s in the direction of the jet, calculate the power developed.

[31·6 N at 15° to jet direction, 108 W]

5.14 A 5 cm diameter jet delivering 56 litres of water per second impinges without shock on a series of vanes moving at 12 m/s in the same direction as the jet. The vanes are curved so that they would, if stationary, deflect the jet through an angle of 135°. Fluid resistance has the effect of reducing the relative velocity by 10 per cent as the water traverses the vanes. Determine (*a*) the magnitude and direction of the resultant force on the vanes, (*b*) the work done per second by the vanes and (*c*) the efficiency of the arrangements.

[1390 N at 21°18', 18·16 kW, 79·6%]

5.15 A reducing bend is incorporated in a pipeline so that the direction of flow is turned through 60° in the horizontal plane and the pipe diameter is reduced from 0·25 m to 0·15 m. The velocity and pressure at the entry to the bend are 1·5 m/s and 300 kN/m^2 gauge respectively and at the exit the pressure is 287·2 kN/m^2 gauge. Find the magnitude and direction of the reaction force on the bend in the horizontal plane due to the flowing water.

[13 kN at 21·4°]

5.16 Figure 5.18 shows a cross-section of the end of a circular duct through which air (density $1 \cdot 2$ kg/m³) is discharged to atmosphere through a circumferential slot, the exit velocity being 30 m/s. Find the force exerted on the duct by the air if the gauge pressure at A is 2065 N/m² below the pressure at outlet.

[138·3 N]

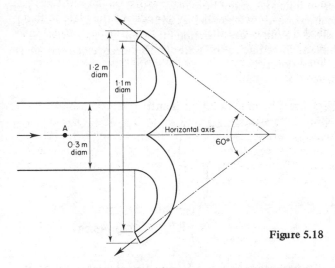

Figure 5.18

5.17 The cascade of vanes shown in Fig. 5.19 is placed at the corner of a rectangular air duct to accelerate the flow from $11 \cdot 55$ m/s to 20 m/s and to divert it through an angle of 90°. The depth of the duct is $1 \cdot 75$ m at all sections and the spacing of the vanes measured along the length of the cascade

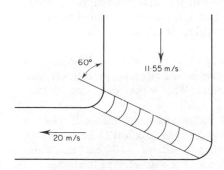

Figure 5.19

is 0·2 m. If the pressure difference across the cascade is 160 N/m² determine the magnitude and direction of the resultant force on each vane of the cascade. Assume that the density of the air is $1 \cdot 2$ kg/m³ and viscosity effects are negligible.

[112·7 N, at 29·8°]

5.18 A venturi meter with a throat diameter of 0·1 m is fitted in a 0·2 m diameter horizontal pipe carrying water at a rate of 0·04 m³/s. If the pressure in the pipe is 2·5 bar (gauge) and all losses are assumed negligible, determine the axial forces acting on the throat of the venturi meter and determine whether it is in tension or compression. Assume that the pipe does not provide any resisting force.

[5·83 kN compression]

5.19   Water flows through the pipe bend and nozzle arrangement shown in Fig. 5.20 which lies with its axis in the horizontal plane. The water issues from the nozzle into the atmosphere as a parallel jet with a velocity of 16 m/s and the pressure at A is 128 kN/m² gauge. Friction may be neglected. Find the moment of the resultant force due to the water on this arrangement about a vertical axis through the point X.

[60·36 N m clockwise]

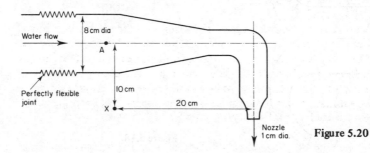

Figure 5.20

5.20   The nozzle of a firehose produces a 50 mm diameter jet of water. If the discharge is 85 dm³/s, calculate the reaction of the jet if the jet velocity is 10 times that of the water in the hose.

[3320 N]

5.21   Water flows through a pipeline 75 cm in diameter. At a certain point the pipe changes its direction from horizontal to an inclination of 45° downwards. Find the magnitude and direction of the force exerted by the water on the bend if the rate of flow is 2 m³/s and the pressure at the bend is equivalent to a head of 9 m of water. Neglect friction.

[31·7 kN at 67°30′]

5.22   A ram-jet engine consumes 20 kg of air per second and 0·6 kg of fuel per second. The exit velocity of the gases is 520 m/s relative to the engine and the flight velocity is 200 m/s absolute. What is the power developed.

[1340 kW]

5.23   A boat driven by reaction jets discharging astern is found to have a resistance to motion of 3·33 kN when moving at 25 knots (1 knot = 0·514 m/s). The cross-sectional area of the jets at discharge is 100 cm² and water enters the pumping system from the sides of the vessel. Pipe losses are assumed to be 5 per cent of the jet energy relative to the ship. Find the absolute jet velocity.

If the pumps are 73 per cent efficient, find the brake power of the driving engines. If the latter are 31 per cent efficient, determine the overall efficiency of the propelling system. Sea water weighs 10 kN/m³.

[24·5 knot, 93·2 kW, 14·3%].

5.24   The resistance of a ship is given by $5 \cdot 55\, u^6 + 978\, u^{1 \cdot 9}$ N at a speed of u m/s. It is driven by a jet propulsion system with intakes facing forward, the efficiency of the jet drive being 0·8 and the efficiency of the pumps 0·72. The vessel is to be driven at 3·4 m/s. Find (a) the mass of water to be pumped astern per second, (b) the power required to drive the pump.
[10 928 kg/s, 109·66 kW]

5.25   . A rocket is fired vertically starting from rest. Neglecting air resistance what velocity will it attain in 68 s if its initial mass is 13 000 kg and fuel is burnt at the rate of 124 kg/s the gases being ejected at a velocity of 1950 m/s relative to the rocket.

If the fuel is exhausted after 68 s what is the maximum height that the rocket will reach. Take $g = 9 \cdot 8$ m/s^2 .
[1372 m/s, 131·8 km]

5.26   A submarine cruising well below the surface of the sea leaves a wake in the form of a cylinder which is symmetrical about the longitudinal axis of the submarine. The wake velocity on the longitudinal axis is equal to the speed of the submarine through the water which is 5 m/s and decreases in direct proportion to the radius to zero at a radius of 6 m. Calculate the force acting on the submarine and the minimum power required to keep the submarine moving at this speed. Density of sea water 1025 kg/m^3.
[483 kN, 2415 kW]

5.27   If $u = ay + by^2$ represents the velocity of air in the boundary layer of a surface, a and b being constants and y the perpendicular distance from the surface, calculate the shear stress acting on the surface when the speed of the air relative to the surface is 75 m/s at a distance of 1·5 mm from the surface and 105 m/s when 3 mm from the surface. The viscosity of the air is 18×10^{-6} kg/ms.
[1·17 N/m^2]

5.28   A lawn sprinkler consists of a horizontal tube with nozzles at each end normal to the tube but inclined upward at 40° to the horizontal. A central bearing incorporates the inlet for the water supply. The nozzles are 3 mm diameter and are at a distance of 12 cm from the central bearing. If the speed of rotation of the tube is 120 rev/min when the velocity of the jets relative to the nozzles is 17 m/s calculate (a) the absolute velocity of the jets, (b) the torque required to overcome the frictional resistance of the tube and bearing.
[16·1 m/s, 10·6 N m/s]

5.29   Derive an expression for the velocity of transmission of a pressure wave through a fluid of bulk modulus K and mass density ρ. What will be the velocity of sound through water if $K = 2 \cdot 05 \times 10^9$ N/m^2 and $\rho = 1000$ kg/m^3 .
[1432 m/s]

5.30   Calculate the velocity of sound in air assuming an adiabatic process if the temperature is 20°C, $\gamma = 1 \cdot 41$ and $R = 287$ J/kgK.
[344·34 m/s]

5.31   Calculate the velocity of propagation relative to the fluid of a small surface wave along a very wide channel in which the water is 1·6 m deep. If the velocity of the stream is 2 m/s what will be the Froude number.
[3·96 m/s, 0·505]

6 The energy equation and its applications

6.1. Mechanical energy of a flowing fluid

An element of fluid, as shown in Fig. 6.1, will possess potential energy due

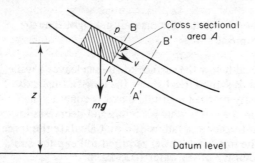

Figure 6.1 Energy of a flowing fluid

to its height z above datum and kinetic energy due to its velocity v, in the same way as any other object. For an element of weight mg.

$$\text{Potential energy} = mgz,$$

$$\text{Potential energy per unit weight} = z, \tag{6.1}$$

$$\text{Kinetic energy} = \tfrac{1}{2}mv^2,$$

$$\text{Kinetic energy per unit weight} = v^2/2g. \tag{6.2}$$

A steadily flowing stream of fluid can also do work because of its pressure. At any given cross-section, the pressure generates a force and, as the fluid flows, this cross-section will move forward and so work will be done. If the pressure at a section AB is p and the area of the cross-section is A,

$$\text{Force exerted on AB} = pA.$$

After a weight mg of fluid has flowed along the streamtube, section AB will have moved to A'B':

$$\text{Volume passing AB} = mg/\rho g = m/\rho.$$

Therefore,

$$\text{Distance AA}' = m/\rho A,$$

$$\text{Work done} = \text{Force} \times \text{Distance AA}'$$

$$= pA \times m/\rho A$$

$$\text{Work done per unit weight} = p/\rho g \tag{6.3}$$

The term $p/\rho g$ is known as the flow work or pressure energy. Note that the term pressure energy refers to the energy of a fluid when *flowing* under pressure as part of a continuously maintained stream. It must not be confused with the energy stored in a fluid due to its elasticity when it is compressed. The concept of pressure energy is sometimes found difficult to understand. In solid body mechanics, a body is free to change its velocity without restriction and potential energy can be freely converted to kinetic energy as its level falls. The velocity of a stream of fluid which has a steady volume rate of flow depends on the cross-sectional area of the stream. Thus, if the fluid flows, for example, in a uniform pipe and is incompressible, its velocity cannot change and so the conversion of potential energy to kinetic energy cannot take place as the fluid loses elevation. The surplus energy appears in the form of an increase in pressure. As a result, pressure energy can, in a sense, be regarded as potential energy in transit.

Comparing the results obtained in equations (6.1), (6.2) and (6.3) with equation (5.23) it can be seen that the three terms of Bernoulli's equation are the pressure energy per unit weight, the kinetic energy per unit weight and the potential energy per unit weight; the constant H is the total energy per unit weight. Thus, Bernoulli's equation states that, for steady flow of a frictionless fluid along a streamline, the total energy per unit weight remains constant from point to point although its division between the three forms of energy may vary:

Pressure	Kinetic	Potential	Total	
energy per +	energy per +	energy per =	energy per	= Constant,
unit weight	unit weight	unit weight	unit weight	

$$p/\rho g + v^2/2g + z = H. \tag{6.4}$$

Each of these terms has the dimension of a length, or head, and they are often referred to as the pressure head $p/\rho g$, the velocity head $v^2/2g$, the potential head z and the total head H. Between any two points, suffixes 1 and 2, on a streamline, equation (6.4) gives

$$\frac{p_1}{\rho_1 g} + \frac{v_1^2}{2g} + z_1 = \frac{p_2}{\rho_2 g} + \frac{v_2^2}{2g} + z_2 \tag{6.5}$$

or   Total energy per unit weight at 1 = Total energy per unit weight at 2,

which corresponds with equation (5.25).

In formulating equation (6.5), it has been assumed that no energy has been supplied to or taken from the fluid between points 1 and 2. Energy could have been supplied by introducing a pump; equally, energy could have been lost by doing work against friction or in a machine such as a turbine. Bernoulli's equation can be expanded to include these conditions, giving

Total energy		Total energy	Loss per	Work done	Energy
per unit	=	per unit	+ unit	+ per unit	− supplied
weight at 1		weight at 2	weight	weight	per unit
					weight

$$\frac{p_1}{\rho_1 g} + \frac{v_1^2}{2g} + z_1 = \frac{p_2}{\rho_2 g} + \frac{v_2^2}{2g} + z_2 + h + w - q. \tag{6.6}$$

Example 6.1

A fire engine pump develops a head of 50 m, i.e. it increases the energy per unit weight of the water passing through it by 50 N m N^{-1}. The pump draws

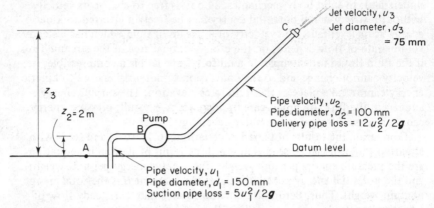

Figure 6.2

water from a sump at A (Fig. 6.2) through a 150 mm diameter pipe in which there is a loss of energy per unit weight due to friction $h_1 = 5u_1^2/2g$ varying with the mean velocity u_1 in the pipe, and discharges it through a 75 mm nozzle at C, 30 m above the pump, at the end of a 100 mm diameter delivery pipe in which there is a loss of energy per unit weight $h_2 = 12u_2^2/2g$. Calculate (a) the velocity of the jet issuing from the nozzle at C and (b) the pressure in the suction pipe at the inlet to the pump at B.

Solution

(a) We can apply Bernoulli's equation in the form of equation (6.6) between two points, one of which will be C, since we wish to determine the jet velocity u_3, and the other a point at which the conditions are known, such as a point A on the free surface of the sump where the pressure will be atmospheric, so that $p_A = 0$, the velocity v_A will be zero if the sump is large, and A can be taken as datum level so that $z_A = 0$. Then,

| Total energy per unit weight at A | = | Total energy per unit weight at C | + | Loss in inlet pipe | − | Energy per unit weight supplied by pump | + | Loss in discharge pipe, | (I) |

$$\text{Total energy per unit weight at A} = \frac{p_A}{\rho g} + \frac{v_A^2}{2g} + z_A = 0,$$

$$\text{Total energy per unit weight at C} = \frac{p_C}{\rho g} + \frac{u_3^2}{2g} + z_3,$$

$$p_C = \text{Atmospheric pressure} = 0,$$

$$z_3 = 30 + 2 = 32 \text{ m}.$$

Therefore,

Total energy
per unit weight $= 0 + u_3^2/2g + 32 = u_3^2/2g + 32$ m.
at C

Loss in inlet pipe, $h_1 = 5u_1^2/2g$,

Energy per unit weight supplied by pump $= 50$ m,

Loss in delivery pipe, $h_2 = 12u_2^2/2g$.

Substituting in (I),

$$0 = (u_3^2/2g + 32) + 5u_1^2/2g - 50 + 12\,u_2^2/2g,$$

$$u_3^2 + 5u_1^2 + 12u_2^2 = 2g \times 18. \qquad\qquad \text{(II)}$$

From the continuity of flow equation,

$$(\pi/4)\,d_1^2 u_1 = (\pi/4)\,d_2^2 u_2 = (\pi/4)\,d_3^2 u_3,$$

therefore,

$$u_1 = \left(\frac{d_3}{d_1}\right)^2 u_3 = \left(\frac{75}{150}\right)^2 u_3 = \frac{1}{4}u_3,$$

$$u_2 = \left(\frac{d_3}{d_2}\right)^2 u_3 = \left(\frac{75}{100}\right)^2 u_3 = \frac{9}{16}u_3.$$

Substituting in equation (II),

$$u_3^2 \{1 + 5 \times (\tfrac{1}{4})^2 + 12 \times (\tfrac{9}{16})^2\} = 2g \times 18,$$

$$5{\cdot}109\,u_3^2 = 2g \times 18$$

$$u_3 = 8{\cdot}314 \text{ m s}^{-1}.$$

(b) If p_B is the pressure in the suction pipe at the pump inlet, applying Bernoulli's equation to A and B,

Total energy   Total energy   Loss in
per unit $=$ per unit weight + inlet
weight at A   at B   pipe,

$$0 = (p_B/\rho g + u_1^2/2g + z_2) + 5u_1^2/2g,$$

$$p_B/\rho g = -z_2 - 6u_1^2/2g.$$

$z_2 = 2$ m, $u_1 = \tfrac{1}{4}u_3 = 8{\cdot}314/4 = 2{\cdot}079$ m s^{-1},

$p_B/\rho g = -(2 + 6 \times 2{\cdot}079^2/2g) = -(2 + 1{\cdot}32) = -3{\cdot}32$ m,

$p_B = -1000 \times 9{\cdot}81 \times 3{\cdot}32 = -32\,569$ N m^{-2}

$= 32{\cdot}569$ kN m^{-2} below atmospheric pressure.

6.2. Steady flow energy equation

Bernoulli's equation and its expanded form, as given in equation (6.6), were developed from Euler's equation (5.21) which, in turn, was derived from the momentum equation. It is possible to develop an energy equation for the steady flow of a fluid from the principle of conservation of energy which states:

> *For any mass system, the net energy supplied to the system equals the increase of energy of the system plus the energy leaving the system.*

Thus, if ΔE is the increase of energy of the system, ΔQ is the energy supplied to the system and ΔW the energy leaving the system, then, considering the energy balance for the system,

$$\Delta E = \Delta Q - \Delta W.$$

The energy of a mass of fluid will have the following forms:

(i) internal energy due to the activity of the molecules of the fluid forming the mass;

(ii) kinetic energy due to the velocity of the mass of fluid itself;

(iii) potential energy due to the mass of fluid being at a height above datum level and acted upon by gravity.

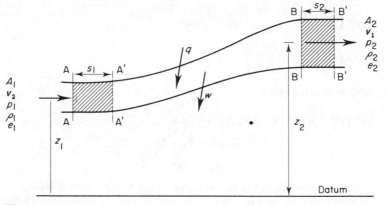

Figure 6.3   Steady flow energy equation

Suppose that at section AA (Fig. 6.3) through a streamtube the cross-sectional area is A_1, the pressure p_1, velocity v_1, density ρ_1, internal energy per unit mass e_1 and height above datum z_1, while the corresponding values at BB are $A_2, p_2, v_2, \rho_2, e_2$ and z_2. The fluid flows steadily with a mass flow rate $\dot{m}$ and between sections AA and BB the fluid receives energy at the rate of q per unit mass and loses energy at the rate of w per unit mass. For example, q may be in the form of heat energy while w might take the form of mechanical work.

$$\begin{array}{l}\text{Energy entering} \\ \text{at AA in unit} \\ \text{time, } E_1\end{array} = \begin{array}{c}\text{Kinetic} \\ \text{energy}\end{array} + \begin{array}{c}\text{Potential} \\ \text{energy}\end{array} + \begin{array}{c}\text{Internal} \\ \text{energy}\end{array}$$

$$= \dot{m}(\tfrac{1}{2}v_1^2 + gz_1 + e_1),$$

$$\begin{array}{l}\text{Energy leaving} \\ \text{at BB in unit} \\ \text{time, } E_2\end{array} = \dot{m}(\tfrac{1}{2}v_2^2 + gz_2 + e_2).$$

Therefore

$$\Delta E = E_2 - E_1 = \dot{m}\{\tfrac{1}{2}(v_2^2 - v_1^2) + g(z_2 - z_1) + (e_2 - e_1)\}. \tag{6.7}$$

This change of energy has occurred because energy has entered and left the fluid between AA and BB. Also, work is done on the fluid in the control volume between the two sections AA and BB by the fluid entering at AA and by the fluid in the control volume as it leaves at BB.

Energy entering per unit time between AA and BB $= \dot{m}q$,

Energy leaving per unit time between AA and BB $= \dot{m}w$

As the fluid flows, work will be done by the fluid entering at AA since a force $p_1 A_1$ is exerted on the cross-section by the pressure p_1 and, in unit time t, the fluid which was at AA will move a distance s_1 to $A'A'$:

Work done in unit time on the fluid at AA $= p_1 A_1 s_1/t$.

But,   Mass passing per unit time, $\dot{m} = \rho_1 A_1 s_1/t$,

therefore,

$A_1 s_1 = \dot{m}/\rho_1$,

Work done per unit time on the fluid at AA $= p_1 \dot{m}/\rho_1$.

Similarly,

Work done per unit time by the fluid at BB $= p_2 \dot{m}/\rho_2$.

Change of energy of $\quad=$ Work done on fluid at AA
the system, ΔE $\quad\quad$ $-$ Work done by fluid at BB
$\quad\quad\quad$ $+$ Energy entering between AA and BB
$\quad\quad\quad$ $-$ Energy leaving between AA and BB

$$= \dot{m}\,p_1/\rho_1 - \dot{m}\,p_2/\rho_2 + \dot{m}q - \dot{m}w$$

$$= \dot{m}(q - w + p_1/\rho_1 - p_2/\rho_2). \tag{6.8}$$

Comparing equations (6.7) and (6.8),

$$\tfrac{1}{2}(v_2^2 - v_1^2) + g(z_2 - z_1) + (e_2 - e_1) = p_1/\rho_1 - p_2/\rho_2 + q - w.$$

Thus, $$gz_1 + \tfrac{1}{2}v_1^2 + (p_1/\rho_1 + e_1) + q - w = gz_2 + \tfrac{1}{2}v_2^2 + (p_2/\rho_2 + e_2). \tag{6.9}$$

The terms $(p_1/\rho_1 + e_1)$ and $(p_2/\rho_2 + e_2)$ can be replaced by the enthalpies H_1 and H_2, giving

$$gz_1 + \tfrac{1}{2}v_1^2 + H_1 + q - w = gz_2 + \tfrac{1}{2}v_2^2 + H_2. \tag{6.10}$$

This steady flow energy equation can be applied to all fluids, real or ideal, whether liquids, vapours or gases, provided that flow is continuous and energy is transferred steadily to or from the fluid at constant rates q and w, conditions remaining constant with time and all quantities being constant across the inlet and outlet sections. In thermodynamics, it is usual to distinguish between heat and work and to treat q as the net inflow of heat and w as the net outflow of mechanical work per unit mass.

6.3. Kinetic energy correction factor

The derivation of Bernoulli's equation and the steady flow energy equation
has been carried out for a streamtube assuming a uniform velocity across the
inlet and outlet sections. In a real fluid flowing in a pipe or over a solid
surface, the velocity will be zero at the solid boundary and will increase as
the distance from the boundary increases. The kinetic energy per unit weight
of the fluid will increase in a similar manner. If the cross-section of the flow
is assumed to be composed of a series of small elements of area δA and the
velocity normal to each element is u, the total kinetic energy passing through
the whole cross-section can be found by determining the kinetic energy
passing through an element in unit time and then summing by integrating
over the whole area of the section.

Mass passing through element in unit time $= \rho \delta A \times u$,

Kinetic energy per unit time passing through element $= \frac{1}{2} \times$ Mass per unit time $\times$ (Velocity)^2

$$= \frac{1}{2} \rho \delta A u^3,$$

Total kinetic energy passing in unit time $= \int \frac{1}{2} \rho u^3 \delta A$,

Total weight passing in unit time $= \int \rho g u \delta A$.

Thus, taking into account the variation of velocity across the stream,

$$\text{True kinetic energy per unit weight} = \frac{1}{2g} \frac{\int \rho u^3 \delta A}{\int \rho u \delta A}, \tag{6.11}$$

which is not the same as $\bar{u}^2/2g$, where $\bar{u}$ is the mean velocity:

$$\bar{u} = \int (u/A) \, dA.$$

Thus,   True kinetic energy per unit weight $= \alpha \bar{u}^2/2g$,   (6.12)

where α is the kinetic energy correction factor, which has a value dependent
on the shape of the cross-section and the velocity distribution. For a circular
pipe, assuming Prandtl's one-seventh power law, $u = u_{max}(y/R)^{1/7}$, for the
velocity at a distance y from the wall of a pipe of radius R, the value of
$\alpha = 1 \cdot 058$.

6.4. Representation of energy changes in a fluid system

The changes of energy, and its transformation from one form to another
which occurs in a fluid system, can be represented graphically. In a real fluid
system, the total energy per unit weight will not remain constant. Unless
energy is supplied to the system at some point by means of a pump, it will
gradually decrease in the direction of motion due to losses resulting from
friction and from the disturbance of flow at changes of pipe section or as a
result of changes of direction. In Fig. 6.4, for example, the flow of water
from the reservoir at A to the reservoir at D is assisted by a pump which

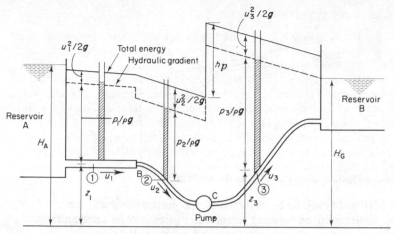

Figure 6.4   Energy changes in a fluid system

develops a head h_p, thus providing an addition to the energy per unit weight of h_p.

At the surface of reservoir A, the fluid has no velocity and is at atmospheric pressure (which is taken as zero gauge pressure), so that the total energy per unit weight is represented by the height H_A of the surface above datum.

As the fluid enters the pipe with velocity u_1, there will be a loss of energy due to disturbance of the flow at the pipe entrance and a continuous loss of energy due to friction as the fluid flows along the pipe, so that the total energy line will slope downwards. At B there is a change of section, with an accompanying loss of energy, resulting in a change of velocity to u_2. The total energy line will continue to slope downwards, but at a greater slope since u_2 is greater than u_1 and friction losses are related to velocity. At C, the pump will put energy into the system and the total energy line will rise by an amount h_p. The total energy line falls again due to friction losses and the loss due to disturbance at the entry to the reservoir, where the total energy per unit weight is represented by the height of the reservoir surface above datum (the velocity of the fluid being zero and the pressure atmospheric).

If a piezometer tube were to be inserted at point 1, the water would not rise to the level of the total energy line, but to a level $u_1^2/2g$ below it, since some of the total energy is in the form of kinetic energy. Thus, at point 1, the potential energy is represented by z_1, the pressure energy by $p_1/\rho g$ and the kinetic energy by $u_1^2/2g$, the three terms together adding up to the total energy per unit weight at that point.

Similarly, at points 2 and 3, the water would rise to levels $p_2/\rho g$ and $p_3/\rho g$ above the pipe, which are $u_2^2/2g$ and $u_3^2/2g$, respectively, below the total energy line. The line joining all the points to which the water would rise, if an open stand pipe was inserted, is known as the *hydraulic gradient*, and runs parallel to the total energy line at a distance below it equal to the velocity head.

If, as in Fig. 6.5, a pipeline rises above the hydraulic gradient, the pressure

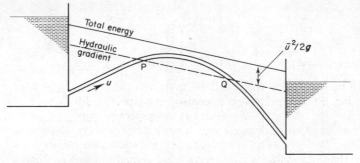

Figure 6.5   Pipeline rising above hydraulic gradient

in the portion PQ will be below atmospheric pressure and will form a
siphon. Under reduced pressure, air or other gases may be released from
solution or a vapour pocket may form and interrupt the flow.

6.5. The Pitot tube

The Pitot tube is used to measure the velocity of a stream and consists of
a simple L-shaped tube facing into the oncoming flow (Fig. 6.6(a)). If the
velocity of the stream at A is u, a particle moving from A to the mouth
of the tube B will be brought to rest so that u_0 at B is zero. By
Bernoulli's equation,

$$\text{Total energy per unit weight at A} = \text{Total energy per unit weight at B,}$$

$$u^2/2g + p/\rho g = u_0^2/2g + p_0/\rho g,$$

$$p_0/\rho g = u^2/2g + p/\rho g,$$

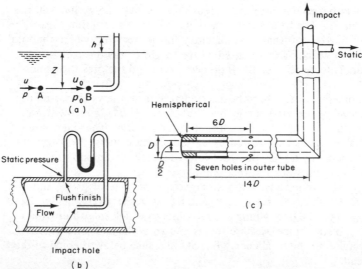

Figure 6.6   Pitot tube

since $u_0 = 0$. Thus, p_0 will be greater than p. Now $p/\rho g = z$ and $p_0/\rho g = h + z$. Therefore,

$$u^2/2g = (p_0 - p)/\rho g = h,$$

Velocity at $A = u = \sqrt{(2gh)}$.

When the Pitot tube is used in a channel, the value of h can be determined directly (as in Fig. 6.6(a)), but, if it is to be used in a pipe, the difference between the static pressure and the pressure at the impact hole must be measured with a differential pressure gauge, using a static pressure tapping in the pipe wall (as in Fig. 6.6(b)) or a combined Pitot-static tube (as in Fig. 6.6(c)). In the Pitot-static tube, the inner tube is used to measure the impact pressure while the outer sheath has holes in its surface to measure the static pressure.

While, theoretically, the measured velocity $u = \sqrt{(2gh)}$, Pitot tubes may require calibration. The true velocity is given by $u = C\sqrt{(2gh)}$, where C is the coefficient of the instrument and h is the difference of head measured in terms of the fluid flowing. For the Pitot-static tube shown in Fig. 6.6(c), the value of C is unity for values of Reynolds number $\rho u D/\mu > 3000$, where D is the diameter of the tip of the tube.

6.6. Changes of pressure in a tapering pipe

Changes of velocity in a tapering pipe were determined by using the continuity of flow equation (Section 4.12). Change of velocity will be accompanied by a change in the kinetic energy per unit weight and, consequently, by a change in pressure, modified by any change of elevation or energy loss, which can be determined by the use of Bernoulli's equation.

Example 6.2

A pipe inclined at $45°$ to the horizontal (Fig. 6.7) converges over a length l of 2 m from a diameter d_1 of 200 mm to a diameter d_2 of 100 mm at the upper end. Oil of relative density 0·9 flows through the pipe at a mean velocity $\bar{v}_1$ at the lower end of 2 m s^{-1}. Find the pressure difference across the 2 m length ignoring any loss of energy, and the difference in level that would be shown on a mercury manometer connected across this length. The relative density of mercury is 13·6 and the leads to the manometer are filled with the oil.

Solution

Let $A_1, \bar{v}_1, p_1, d_1, z_1$ and $A_2, \bar{v}_2, p_2, d_2, z_2$ be the area, mean velocity, pressure, diameter and elevation at the lower and upper sections, respectively. For continuity of flow, assuming the density of the oil to be constant,

$$A_1 \bar{v}_1 = A_2 \bar{v}_2$$

so that

$$\bar{v}_2 = (A_1/A_2)\bar{v}_1$$

$$A_1 = (\pi/4)d_1^2 \quad \text{and} \quad A_2 = (\pi/4)d_2^2,$$

thus, $\bar{v}_2 = (d_1/d_2)^2 \, \bar{v}_1 = (0·2/0·1)^2 \times 2 = 8 \text{ m s}^{-1}.$

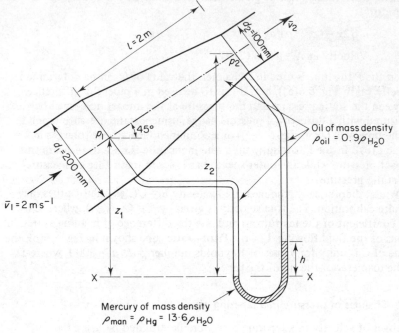

Figure 6.7   Pressure change in a tapering pipe

Applying Bernoulli's equation to the lower and upper sections, assuming no energy losses,

$$\begin{array}{ll} \text{Total energy per unit} \\ \text{weight at section 1} \end{array} = \begin{array}{ll} \text{Total energy per unit} \\ \text{weight at section 2,} \end{array}$$

$$p_1/\rho_0 g + \bar{v}_1^2/2g + z_1 = p_2/\rho_0 g + \bar{v}_2^2/2g + z_2,$$

$$p_1 - p_2 = \tfrac{1}{2}\rho_0(\bar{v}_2^2 - \bar{v}_1^2) + \rho_0 g(z_2 - z_1) \tag{I}$$

Now, $z_2 - z_1 = l \sin 45° = 2 \times 0.707 = 1.414$ m

and, since the relative density of the oil is 0.9, if ρ_{H_2O} = density of water, then $\rho_{oil} = 0.9\,\rho_{H_2O} = 0.9 \times 1000 = 900$ kg m^{-3}. Substituting in equation (I),

$$p_1 - p_2 = \tfrac{1}{2} \times 900(8^2 - 2^2) + 900 \times 9.81 \times 1.414 \text{ N m}^{-2}$$

$$= 8829(3.058 + 1.414) = 39\,484 \text{ N m}^{-2}.$$

For the manometer, pressure in each limb will be the same at level XX, therefore,

$$p_1 + \rho_{oil} g z_1 = p_2 + \rho_{oil} g(z_2 - h) + \rho_{man} g h,$$

$$(p_1 - p_2)/\rho_{oil} g + z_1 - z_2 = h(\rho_{man}/\rho_{oil} - 1),$$

$$h = \left(\frac{\rho_{oil}}{\rho_{man} - \rho_{oil}}\right)\left(\frac{p_1 - p_2}{\rho_{oil} g} + z_1 - z_2\right)$$

Putting $\rho_{oil} = 0.9 \rho_{H_2O} = 900$ kg m^{-3} and $\rho_{man} = 13.6 \rho_{H_2O}$,

$$h = \{0.9/(13.6 - 0.9)\}\{39\,484/900 \times 9.81 - 1.414\} \text{ m}$$

$$= 0.217 \text{ m}.$$

6.7. Principle of the venturi meter

As shown by equation (I) in the last example, the pressure difference between any two points on a tapering pipe through which a fluid is flowing depends on the difference of level $z_2 - z_1$, the velocities $\bar{v}_2$ and $\bar{v}_1$, and, therefore, on the volume rate of flow Q through the pipe. Hence, the pressure difference can be used to determine the volume rate of flow for any particular configuration. The venturi meter uses this effect for the measurement of flow in pipelines. As shown in Fig. 6.8, it consists of a short converging conical tube leading to a cylindrical portion, called the throat, of smaller diameter than that of the pipeline, which is followed by a diverging section in which the diameter increases again to that of the main pipeline. The pressure difference from which the volume rate of flow can be determined is measured between the entry section 1 and the throat section 2, often by means of a U-tube manometer (as shown). The axis of the meter may be inclined at any angle.

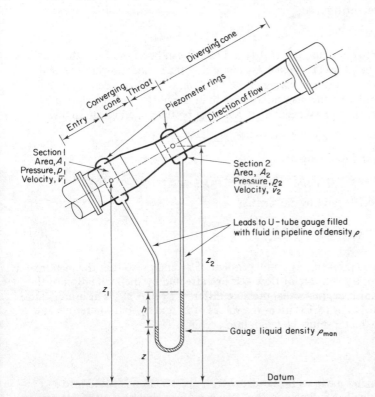

Figure 6.8   Inclined venturi meter and U-tube

Assuming that there is no loss of energy, and applying Bernoulli's equation to sections 1 and 2,

$$z_1 + p_1/\rho g + v_1^2/2g = z_2 + p_2/\rho g + v_2^2/2g,$$

$$v_2^2 - v_1^2 = 2g\{(p_1 - p_2)/\rho g + (z_1 - z_2)\}. \tag{6.13}$$

For continuous flow,

$$A_1 v_1 = A_2 v_2 \quad \text{or} \quad v_2 = (A_1/A_2)v_1.$$

Substituting in equation (6.13),

$$v_1^2 \{(A_1/A_2)^2 - 1\} = 2g\{(p_1 - p_2)/\rho g + (z_1 - z_2)\},$$

$$v_1 = \frac{A_2}{(A_1^2 - A_2^2)^{1/2}} \sqrt{\left\{2g\left(\frac{p_1 - p_2}{\rho g} + z_1 - z_2\right)\right\}}$$

Volume rate of flow, $Q = A_1 v_1 = \{A_1 A_2/(A_1^2 - A_2^2)^{1/2}\}\sqrt{(2gH)},$ (6.14)

where $H = (p_1 - p_2)/\rho g + (z_1 - z_2)$ or, if m = area ratio = A_1/A_2,

$$Q = \{A_1/(m^2 - 1)^{1/2}\}\sqrt{(2gH)}. \tag{6.15}$$

In practice, some loss of energy will occur between sections 1 and 2. The value of Q given by equations (6.14) and (6.15) is a theoretical value which will be slightly greater than the actual value. A coefficient of discharge C_d is, therefore, introduced:

Actual discharge, $Q_{\text{actual}} = C_d \times Q_{\text{theoretical}}$.

The value of H in equations (6.14) and (6.15) can be found from the reading of the U-tube gauge (Fig. 6.8). Assuming that the connections to the gauge are filled with the fluid flowing in the pipeline, which has a density ρ, and that the density of the manometric liquid in the bottom of the U-tube is ρ_{man}, then, since pressures at level XX must be the same in both limbs,

$$p_X = p_1 + \rho g(z_1 - z) = p_2 + \rho g(z_2 - z - h) + \rho_{\text{man}} h.$$

Expanding and re-arranging,

$$H = (p_1 - p_2)/\rho g + (z_1 - z_2) = h(\rho_{\text{man}}/\rho - 1).$$

Equation (6.19) can now be written

$$Q = \{a_1/(m^2 - 1)^{1/2}\}\sqrt{\left\{2gh\left(\frac{\rho_{\text{man}}}{\rho} - 1\right)\right\}}. \tag{6.16}$$

Note that equation (6.16) is independent of z_1 and z_2, so that the manometer reading h for a given rate of flow Q is not affected by the inclination of the meter. If, however, the actual pressure difference $(p_1 - p_2)$ is measured and equation (6.14) or (6.15) used, the values of z_1 and z_2, and, therefore, the slope of the meter, must be taken into account.

Example 6.3

A venturi meter having a throat diameter d_2 of 100 mm is fitted into a pipeline which has a diameter d_1 of 250 mm through which oil of specific gravity 0·9 is flowing. The pressure difference between the entry and throat tappings is measured by a U-tube manometer, containing mercury of specific gravity 13·6, and the connections are filled with the oil flowing in the pipeline. If the difference of level indicated by the mercury in the U-tube is 0·63 m, calculate the theoretical volume rate of flow through the meter.

Solution
Using equation (6.16),

$$\text{Area at entry, } A_1 = (\pi/4)\, d_1^{\,2} = (\pi/4)(0{\cdot}25)^2 = 0{\cdot}0491 \text{ m}^2,$$

$$\text{Area ratio, } m = A_1/A_2 = (d_1/d_2)^2 = (0{\cdot}25/0{\cdot}10)^2 = 6{\cdot}25,$$

$$h = 0{\cdot}63 \text{ m}, \; \rho_{Hg} = \rho_{man} = 13{\cdot}6 \times \rho_{H_2O}, \; \rho_{oil} = 0{\cdot}9\rho_{H_2O}.$$

where ρ_{Hg} = density of mercury, ρ_{H_2O} = density of water and ρ_{oil} = density of oil. Substituting in equation (6.16),

$$Q = \{0{\cdot}0491/(6{\cdot}25^2 - 1)^{1/2}\}\sqrt{\{2 \times 9{\cdot}81 \times 0{\cdot}63(13{\cdot}6/0{\cdot}9 - 1)\}} \text{ m}^3 \text{ s}^{-1}$$

$$= \mathbf{0{\cdot}105 \text{ m}^3 \text{ s}^{-1}}.$$

6.8. Pipe orifices

The venturi meter described in Section 6.7 operates by changing the cross-section of the flow, so that the cross-sectional area is less at the downstream pressure tapping than at the upstream tapping. A similar effect can be achieved by inserting an orifice plate which has an opening in it smaller than the internal diameter of the pipeline (as shown in Fig. 6.9). The orifice

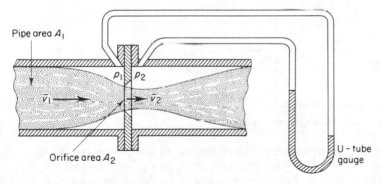

Figure 6.9   Pipe orifice meter

plate produces a constriction of the flow as shown, the cross-sectional area A_2 of the flow immediately downstream of the plate being approximately the same as that of the orifice. The arrangement is cheap compared to the cost of a venturi meter, but there are substantial energy losses. The theoretical discharge can be calculated from equation (6.14) or (6.15), but the actual discharge may be as little as two-thirds of this value. A coefficient of discharge must, therefore, be introduced in the same way as for the venturi meter, a typical value for a sharp-edged orifice being 0·65.

6.9. Limitation on the velocity of flow in a pipeline

Since Bernoulli's equation requires that the total energy per unit weight of a flowing fluid shall, if there are no losses, remain constant, any increase in velocity or elevation must be accompanied by a reduction in pressure. Furthermore, since the pressure can never fall below absolute zero, there will be a maximum velocity for a given configuration of a pipeline which cannot be exceeded. For a flowing liquid, the pressure will never fall to absolute zero since air or vapour will be released and form pockets in the flow well before this can occur.

6.10. Theory of small orifices discharging to atmosphere

An orifice is an opening, usually circular, in the side or base of a tank or reservoir, through which fluid is discharged in the form of a jet, usually into the atmosphere. The volume rate of flow discharged through an orifice will depend upon the head of the fluid above the level of the orifice and it can, therefore, be used as a means of flow measurement. The term 'small orifice' is applied to an orifice which has a diameter, or vertical dimension, which is small compared with the head producing flow, so that it can be assumed that this head does not vary appreciably from point to point across the orifice.

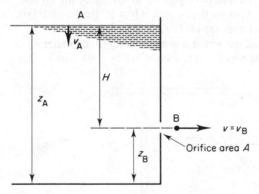

Figure 6.10   Flow through a small orifice

Figure 6.10 shows a small orifice in the side of a large tank containing liquid with a free surface open to atmosphere. At a point A on the free surface, the pressure p_A is atmospheric and, if the tank is large, the velocity v_A will be negligible. In the region of the orifice, conditions are rather uncertain, but at some point B in the jet, just outside the orifice, the pressure p_B will again be atmospheric and the velocity v_B will be that of the jet v. Taking the datum for potential energy at the centre of the orifice and applying Bernoulli's equation to A and B, assuming that there is no loss of energy,

$$\text{Total energy per unit weight at A} = \text{Total energy per unit weight at B,}$$

$$z_A + v_A^2/2g + p_A/\rho g = z_B + v_B^2/2g + p_B/\rho g.$$

Putting $z_A - z_B = H$, $v_A = 0$, $v_B = v$ and $p_A = p_B$,

Velocity of jet, $v = \sqrt{(2gH)}$.   (6.17)

This is a statement of *Torricelli's theorem*, that the velocity of the issuing jet is proportional to the square root of the head producing flow. Equation (6.17) applies to any fluid, H being expressed as a head of the fluid flowing through the orifice. For example, if an orifice is formed in the side of a vessel containing gas of density ρ at a uniform pressure p, the value of H would be $p/\rho g$. Theoretically, if A is the cross-sectional area of the orifice,

Discharge, $Q =$ Area × Velocity

$= A \sqrt{(2gH)}$.   (6.18)

In practice, the actual discharge is considerably less than the theoretical discharge given by equation (6.18), which must, therefore, be modified by introducing a *coefficient of discharge* C_d, so that

Actual discharge, $Q_{actual} = C_d \, Q_{theoretical}$

$= C_d A \sqrt{(2gH)}$.   (6.19)

There are two reasons for the difference between the theoretical and actual discharges. First, the velocity of the jet is less than that given by equation (6.17) because there is a loss of energy between A and B:

Actual velocity at B $= C_v \times v = C_v \sqrt{(2gH)}$,   (6.20)

where C_v is a *coefficient of velocity* which has to be determined experimentally and is of the order of 0·97.

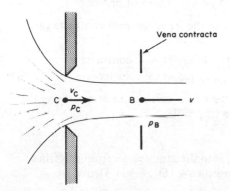

Figure 6.11   Contraction of issuing jet

Second, as shown in Fig. 6.11, the paths of the particles of the fluid converge on the orifice and the area of the issuing jet at B is less than the area of the orifice A at C. In the plane of the orifice, the particles have a component of velocity towards the centre and the pressure at C is greater than atmospheric pressure. It is only at B, a small distance outside the orifice, that the paths of the particles have become parallel. The section through B is called the *vena contracta*.

Actual area of jet at B $= C_c A$,   (6.21) where C_c is the *coefficient of contraction*, which can be determined experimentally and will depend on the profile of the orifice. For a sharp-edged orifice of the form shown in Fig. 6.11, it is of the order of 0·64.

We can now determine the actual discharge from equations (6.20) and (6.21)

$$\text{Actual discharge} = \text{Actual area at B} \times \text{Actual velocity at B}$$

$$= C_c A \times C_v \sqrt{(2gH)}$$

$$= C_c \times C_v A \sqrt{(2gH)}. \tag{6.22}$$

Comparing equation (6.22) with equation (6.19), we see that the relation between the coefficients is

$$C_d = C_c \times C_v.$$

The values of the coefficient of discharge, the coefficient of velocity and the coefficient of contraction are determined experimentally and values are available for standard configurations in British Standard Specifications.

To determine the coefficient of discharge, it is only necessary to collect, or otherwise measure, the actual volume discharge from the orifice in a given time and compare this with the theoretical discharge given by equation (6.18).

$$\text{Coefficient of discharge, } C_d = \frac{\text{Actual measured discharge}}{\text{Theoretical discharge}}.$$

Similarly, the actual area of the jet at the vena contracta can be measured,

$$\text{Coefficient of contraction, } C_c = \frac{\text{Area of jet at vena contracta}}{\text{Area of orifice}}.$$

In the same way, if the actual velocity of the jet at the vena contracta can be found,

$$\text{Coefficient of velocity, } C_v = \frac{\text{Velocity at vena contracta}}{\text{Theoretical velocity}}.$$

If the orifice is not in the bottom of the tank, one method of measuring the actual velocity of the jet is to measure its profile.

Example 6.4

A jet of water discharges horizontally into the atmosphere from an orifice in the vertical side of a large open-topped tank (Fig. 6.12). Derive an

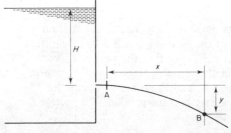

Figure 6.12   Determination of the coefficient of velocity

expression for the actual velocity v of the jet at the vena contracta if the jet falls a distance y vertically in a horizontal distance x, measured from the vena contracta. If the head of water above the orifice is H, calculate the coefficient of velocity.

If the orifice has an area of 650 mm^2 and the jet falls a distance y of 0·5 m in a horizontal distance x of 1·5 m from the vena contracta, calculate the values of the coefficients of velocity, discharge and contraction, given that the volume rate of flow is 0·117 m^3 and the head H above the orifice is 1·2 m.

Solution
Let t be the time taken for a particle of fluid to travel from the vena contracta A (Fig. 6.14) to the point B. Then

$$x = vt \quad \text{and} \quad y = \tfrac{1}{2}gt^2$$

or $\quad v = x/t \quad \text{and} \quad t = \sqrt{(2y/g)}.$

Eliminating t,

Velocity at the vena contracta, $v = \sqrt{(gx^2/2y)}.$

This is the *actual* velocity of the jet at the vena contracta.
From equation (6.17),

Theoretical velocity $= \sqrt{(2gH)},$

$$\text{Coefficient of velocity} = \frac{\text{Actual velocity}}{\text{Theoretical velocity}} = v/\sqrt{(2gH)}$$

$$= \sqrt{(x^2/4yH)}.$$

Putting $x = 1·5$ m, $y = 0·5$ m, $H = 1·2$ m and area, $A = 650 \times 10^{-6}$ m^2,

Coefficient of velocity, $C_v = \sqrt{(x^2/4yH)} = \sqrt{1·5^2/4 \times 0·5 \times 1·2}$

$$= 0·968,$$

Coefficient of discharge, $C_d = Q_{\text{actual}}/A \sqrt{(2gH)}$

$$= (0·117/60)/650 \times 10^{-6} \sqrt{(2 \times 9·81 \times 1·2)}$$

$$= 0·618,$$

Coefficient of contraction, $C_c = C_c/C_v = 0·618/0·968 = 0·639.$

6.11. Theory of large orifices

If the vertical height of an orifice is large, so that the head producing flow is substantially less at the top of the opening than at the bottom, the discharge calculated from the formula for a small orifice, using the head h measured to the centre of the orifice, will not be the true value, since the velocity will vary very substantially from top to bottom of the opening. The method adopted is to calculate the flow through a thin horizontal strip across the orifice (Fig. 6.13), and integrate from top to bottom of the opening to obtain the theoretical discharge, from which the actual discharge can be determined if the coefficient of discharge is known.

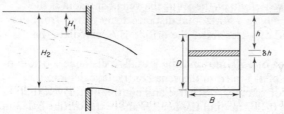

Figure 6.13   Flow through a large orifice

Example 6.5

A reservoir discharges through a rectangular sluice gate of width B and height D (Fig. 6.13). The top and bottom of the opening are at depths H_1 and H_2 below the free surface. Derive a formula for the theoretical discharge through the opening.

If the top of the opening is 0·4 m below the water level and the opening is 0·7 m wide and 1·5 m in height, calculate the theoretical discharge (in cubic metres per second), assuming that the bottom of the opening is above the downstream water level.

What would be the percentage error if the opening were to be treated as a small orifice?

Solution

Since the velocity of flow will be much greater at the bottom than at the top of the opening, consider a horizontal strip across the opening of height δh at a depth h below the free surface:

$$\text{Area of strip} = B\delta h,$$

$$\text{Velocity of flow through strip} = \sqrt{(2gh)},$$

$$\text{Discharge through strip}, \delta Q = \text{Area} \times \text{Velocity}$$

$$= B\sqrt{(2g)}h^{1/2}\delta h.$$

For the whole opening, integrating from $h = H_1$ to $h = H_2$,

$$\text{Discharge}, Q = B\sqrt{(2g)}\int_{H_1}^{H_2} h^{1/2}\,dh$$

$$= \tfrac{2}{3}B\sqrt{(2g)}(H_2^{3/2} - H_1^{3/2}).$$

Putting $B = 0.7$ m, $H_1 = 0.4$ m, $H_2 = 1.9$ m,

Theoretical discharge, $Q = \tfrac{2}{3} \times 0.7 \times \sqrt{(2 \times 9.81)}(1.9^{3/2} - 0.4^{3/2})$ m³ s⁻¹

$$= 2.067(2.619 - 0.253) = \mathbf{4.891\ m^3\ s^{-1}}.$$

For a small orifice, $Q = A\sqrt{(2gh)}$, where A is the area of the orifice and h is the head above the centreline. Putting

$$A = BD = 0.7 \times 1.5\ \text{m}^2,$$

$$h = \tfrac{1}{2}(H_1 + H_2) = \tfrac{1}{2}(0.4 + 1.9) = 1.15\ \text{m},$$

$$Q = 0.7 \times 1.5\ \sqrt{(2 \times 9.81 \times 1.15)} = \mathbf{4.988\ m^3\ s^{-1}}.$$

This result is greater than that obtained by the large orifice analysis.

Error = (4·988 − 4·891)/4·891 = 0·0198 = **1·98 per cent.**

6.12. Elementary theory of notches and weirs

A notch is an opening in the side of a measuring tank or reservoir extending above the free surface. It is, in effect, a large orifice which has no upper edge, so that it has a variable area depending upon the level of the free surface. A weir is a notch on a large scale, used, for example, to measure the flow of a river, and may be sharp-edged or have a substantial breadth in the direction of flow.

The method of determining the theoretical flow through a notch is the same as that adopted for the large orifice. For a notch of any shape (Fig. 6.14),

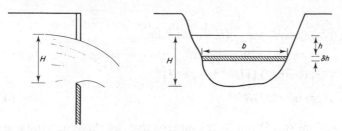

Figure 6.14   Discharge from a notch

consider a horizontal strip of width b at a depth h below the free surface and height δh.

Area of strip = $b\,\delta h$,

Velocity through strip = $\sqrt{(2gh)}$,

Discharge through strip, δQ = Area × Velocity

$$= b\,\delta h\sqrt{(2gh)}. \tag{6.23}$$

Integrating from $h = 0$ at the free surface to $h = H$ at the bottom of the notch,

$$\text{Total theoretical discharge, } Q = \sqrt{(2g)} \int_0^H bh^{1/2}\,dh. \tag{6.24}$$

Before the integration of equation (6.32) can be carried out, b must be expressed in terms of h.

For a *rectangular notch* (Fig. 6.15(a)), put b = constant = B in equation (6.24), giving

$$Q = B\sqrt{(2g)} \int_0^H h^{1/2}\,dh$$

$$= \tfrac{2}{3}B\sqrt{(2g)}H^{3/2}. \tag{6.25}$$

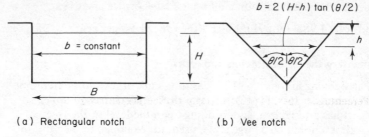

(a) Rectangular notch   (b) Vee notch

Figure 6.15   Rectangular and vee notches

For a *vee notch* with an included angle θ (Fig. 6.15(b)), put $b = 2(H-h)\tan(\theta/2)$ in equation (6.24), giving

$$Q = 2\sqrt{(2g)}\tan(\theta/2)\int_0^H (H-h)h^{1/2}\,dh$$

$$= 2\sqrt{(2g)}\tan(\theta/2)\,[\tfrac{2}{3}Hh^{3/2} - \tfrac{2}{5}h^{5/2}]_0^h$$

$$= \tfrac{8}{15}\sqrt{(2g)}\tan(\theta/2)H^{5/2}. \tag{6.26}$$

Inspection of equations (6.25) and (6.26) suggests that, by choosing a suitable shape for the sides of the notch, any desired relationship between Q and H could be achieved, but certain laws do lead to shapes which are not feasible in practice.

As in the case of orifices, the actual discharge through a notch or weir can be found by multiplying the theoretical discharge by a coefficient of discharge to allow for energy losses and the contraction of the cross-section of the stream at the bottom and sides.

Example 6.6

It is proposed to use a notch to measure the flow of water from a reservoir and it is estimated that the error in measuring the head above the bottom of the notch could be 1·5 mm. For a discharge of 0·28 m³ s⁻¹, determine the percentage error which may occur, using a right-angled triangular notch with a coefficient of discharge of 0·6.

Solution
For a vee notch, from equation (6.26),

$$Q = C_d\,\tfrac{8}{15}\sqrt{(2g)}\tan(\theta/2)H^{5/2}.$$

Putting $C_d = 0\cdot6$ and $\theta = 90°$,

$$Q = 0\cdot6 \times \tfrac{8}{15} \times \sqrt{(19\cdot62)} \times 1 \times H^{5/2}$$

$$= 1\cdot417H^{5/2}. \tag{I}$$

When $Q = 0\cdot28$ m³ s⁻¹, $H = (0\cdot28/1\cdot417)^{2/5} = 0\cdot5228$ m. The error δQ in the discharge, corresponding to an error δH in the measurement of H, can be found by differentiating equation (I):

$$\delta Q = 2 \cdot 5 \times 1 \cdot 417 H^{3/2} \delta H = 2 \cdot 5 Q \delta H / H,$$

$$\delta Q / Q = 2 \cdot 5 \delta H / H.$$

Putting $\delta H = 1 \cdot 5$ mm and $H = 0 \cdot 5228$ m,

$$\text{Percentage error} = (\delta Q / Q) \times 100 = (2 \cdot 5 \times 0 \cdot 0015 / 0 \cdot 5228) \times 100$$

$$= \textbf{0·72 per cent.}$$

In the foregoing theory, it has been assumed that the velocity of the liquid approaching the notch is very small so that its kinetic energy can be neglected; it can also be assumed that the velocity through any horizontal element across the notch will depend only on its depth below the free surface. This is a satisfactory assumption for flow over a notch or weir in the side of a large reservoir, but, if the notch or weir is placed at the end of a narrow channel, the *velocity of approach* to the weir will be substantial and the head h producing flow will be increased by the kinetic energy of the approaching liquid to a value

$$x = h + \alpha \bar{v}^2 / 2g, \tag{6.27}$$

where $\bar{v}$ is the mean velocity of the liquid in the approach channel and α is the kinetic energy correction factor to allow for the non-uniformity of velocity over the cross-section of the channel. Note that the value of $\bar{v}$ is obtained by dividing the discharge by the full cross-sectional area of the channel itself, not that of the notch. As a result, the discharge through the strip (shown in Fig. 6.14) will be

$$\delta Q = b \delta h \sqrt{(2gx)},$$

and, from equation (6.27), $\delta h = \delta x$, so that

$$\delta Q = b \sqrt{(2g)} x^{1/2} \, dx. \tag{6.28}$$

At the free surface, $h = 0$ and $x = \alpha \bar{v}^2 / 2g$, while, at the sill, $h = H$ and $x = H + \alpha \bar{v}^2 / 2g$. Integrating equation (6.28) between these limits,

$$Q = \sqrt{(2g)} \int_{\alpha \bar{v}^2 / 2g}^{(H + \alpha \bar{v}^2 / 2g)} b x^{1/2} \, dx.$$

For a rectangular notch, putting $b = B = \text{constant}$,

$$Q = \tfrac{2}{3} B \sqrt{(2g)} H^{3/2} \left[\left(1 + \frac{\alpha \bar{v}^2}{2gH} \right)^{3/2} - \left(\frac{\alpha \bar{v}^2}{2gH} \right)^{3/2} \right] \tag{6.29}$$

Example 6.7

A long rectangular channel $1 \cdot 2$ m wide leads from a reservoir to a rectangular notch $0 \cdot 9$ m wide with its sill $0 \cdot 2$ m above the bottom of the channel. Assuming that, if the velocity of approach is neglected, the discharge over the notch, in SI units, is given by $Q = 1 \cdot 84 BH^{3/2}$, calculate the discharge (in cubic metres per second) when the head over the bottom of the notch H is $0 \cdot 25$ m (a) neglecting the velocity of approach, (b) correcting for the velocity of approach assuming that the kinetic energy correction factor α is 1·1.

Solution

(a) Neglecting the velocity of approach,

$$Q_1 = 1\cdot84\, BH^{3/2}.$$

Putting $B = 0\cdot9$ m and $H = 0\cdot25$ m,

$$Q_1 = 1\cdot84 \times 0\cdot9 \times 0\cdot25^{3/2}$$

$$= 0\cdot207 \text{ m}^3\text{ s}^{-1}.$$

(b) Taking the velocity of approach into account, from equation (6.29) the correction factor k will be

$$k = \{(1 + \alpha\bar{v}^2/2gH)^{3/2} - (\alpha\bar{v}^2/2gH)^{3/2}\}$$

and the corrected value of Q will be $Q_2 = Q_1 \times k$, so that

$$Q_2 = 1\cdot84\, BH^{3/2}\left\{\left(1 + \frac{\alpha\bar{v}^2}{2gH}\right)^{3/2} - \left(\frac{\alpha\bar{v}^2}{2gH}\right)^{3/2}\right\}.$$

Putting $B = 0\cdot9$ m, $H = 0\cdot25$ m and $\alpha = 1\cdot1$,

$$Q_2 = 1\cdot84 \times 0\cdot9 \times 0\cdot25^{3/2}\left\{\left(1 + \frac{1\cdot1\bar{v}^2}{19\cdot62 \times 0\cdot25}\right)^{3/2} - \left(\frac{1\cdot1\bar{v}^2}{19\cdot62 \times 0\cdot25}\right)^{3/2}\right\}$$

$$= 0\cdot207\{(1 + 0\cdot224\,\bar{v}^2)^{3/2} - (0\cdot224\bar{v}^2)^{3/2}\}. \qquad (I)$$

Now,

$$V = \text{Velocity in approach channel} = \frac{\text{Discharge}}{\text{Area of channel}}$$

$$= Q_2/1\cdot2(H + 0\cdot2). \qquad (II)$$

Using (II), the solution to (I) can be found by successive approximation, taking $\bar{v} = 0$ for the first approximation – which gives $Q = 0\cdot207$ m^3 s^{-1}.
Inserting this value of Q in (II), with $H = 0\cdot25$ m,

$$\bar{v} = 0\cdot207/1\cdot2 \times 0\cdot45 = 0\cdot3833 \text{ m s}^{-1}.$$

Putting $\bar{v} = 0\cdot3833$ m s^{-1} in (I),

$$Q = 0\cdot207\{(1\cdot0329)^{3/2} - (0\cdot0329)^{3/2}\} = 0\cdot2161 \text{ m}^3\text{ s}^{-1}.$$

For the next approximation,

$$\bar{v} = 0\cdot2161/1\cdot2 \times 0\cdot45 = 0\cdot4002 \text{ m s}^{-1},$$

giving

$$Q = 0\cdot207\{(1\cdot0359)^{3/2} - (0\cdot0359)^{3/2}\} = 0\cdot2168 \text{ m}^3\text{ s}^{-1}.$$

A further approximation gives

$$\bar{v} = 0\cdot2168/1\cdot2 \times 0\cdot45 = 0\cdot4015 \text{ m s}^{-1}$$

and

$$Q = 0\cdot207\{(1\cdot0360)^{3/2} - (0\cdot0360)^{3/2}\} = 0\cdot2169 \text{ m}^3\text{ s}^{-1}.$$

6.13. The power of a stream of fluid

In Section 6.1, it was shown that a stream of fluid could do work as a result of its pressure p, velocity v and elevation z and that the total energy per unit weight H of the fluid is given by

$$H = p/\rho g + v^2/2g + z.$$

If the weight per unit time of fluid flowing is known, the power of the stream can be claculated, since

$$\text{Power} = \text{Energy per unit time} = \frac{\text{Weight}}{\text{Unit time}} \times \frac{\text{Energy}}{\text{Unit weight}}.$$

If Q is the volume rate of flow,

$$\text{Weight per unit time} = \rho g Q,$$

$$\text{Power} = \rho g Q H = \rho g Q \{p/\rho g + v^2/2g + z\}$$

$$= pQ + \tfrac{1}{2}\rho v^2 Q + \rho g Q z. \tag{6.30}$$

Example 6.8

Water is drawn from a reservoir, in which the water level is 240 m above datum, at the rate of $0 \cdot 13$ m³ s⁻¹. The outlet of the pipeline is at datum level and is fitted with a nozzle to produce a high speed jet to drive a turbine of the Pelton wheel type. If the velocity of the jet is 66 m s⁻¹, calculate (a) the power of the jet, (b) the power supplied from the reservoir, (c) the head used to overcome losses and (d) the efficiency of the pipeline and nozzle in transmitting power.

Solution

(a) The jet issuing from the nozzle will be at atmospheric pressure and at datum level so that, in equation (6.30), $p = 0$ and $z = 0$. Therefore,

$$\text{Power of jet} = \tfrac{1}{2}\rho v^2 Q.$$

Putting $\rho = 1000$ kg m⁻³, $v = 66$ m s⁻¹, $Q = 0 \cdot 13$ m³ s⁻¹,

$$\text{Power of jet} = \tfrac{1}{2} \times 1000 \times 66^2 \times 0 \cdot 13 = 283\ 410 \text{ W}$$

$$= 283 \cdot 14 \text{ kW}.$$

(b) At the reservoir, the pressure is atmospheric and the velocity of the free surface is zero so that, in equation (6.30), $p = 0$, $v = 0$. Therefore,

$$\text{Power supplied from reservoir} = \rho g Q z.$$

Putting $\rho = 1000$ kg m⁻³, $Q = 0 \cdot 13$ m³ s⁻¹, $z = 240$ m,

$$\text{Power supplied from reservoir} = 1000 \times 9 \cdot 81 \times 0 \cdot 13 \times 240 \text{ W}$$

$$= 306\ 072 \text{ W}$$

$$= 306 \cdot 07 \text{ kW}.$$

(c) If H_1 = total head at the reservoir, H_2 = total head at the jet, and h = head lost in transmission,

$$\text{Power supplied from reservoir} = \rho g Q H_1 = 306 \cdot 07 \text{ kW},$$

$$\text{Power of issuing jet} = \rho g Q H_2 = 283 \cdot 14 \text{ kW},$$

$$\text{Power lost in transmission} = \rho g Q h = 22 \cdot 93 \text{ kW},$$

$$\text{Head lost in pipeline} = h = \frac{\text{Power lost}}{\rho g Q}$$

$$= \frac{22 \cdot 93 \times 10^3}{1000 \times 9 \cdot 81 \times 0 \cdot 13} = \textbf{17} \cdot \textbf{98 m.}$$

(d)

$$\text{Efficiency of transmission} = \frac{\text{Power of jet}}{\text{Power supplied by reservoir}}$$

$$= 283 \cdot 14 / 306 \cdot 07 = 0 \cdot 925$$

$$= \textbf{92} \cdot \textbf{5 per cent.}$$

6.14. Radial flow

When a fluid flows radially inwards, or outwards from a centre, between two parallel planes as in Fig. 6.16, the streamlines will be radial straight lines and the streamtubes will be in the form of sectors. The area of flow will therefore increase as the radius increases, causing the velocity to decrease. Since the flow pattern is symmetrical, the total energy per unit weight H will be the same for all streamlines and for all points along each streamline if we assume that there is no loss of energy.

If v is the radial velocity and p the pressure at any radius r,

$$H = p/\rho g + v^2/2g = \text{constant}. \tag{6.31}$$

Applying the continuity of flow equation and assuming that the density of the fluid remains constant, as would be the case for a liquid,

$$\text{Volume rate of flow, } Q = \text{Area} \times \text{Velocity}$$

$$= 2\pi r b \times v,$$

where b is the distance between the planes. Thus,

$$v = Q/2\pi r b$$

and, substituting in equation (6.31),

$$p/\rho g + Q^2/8\pi^2 r^2 b^2 g = H,$$

$$p = \rho g \{ H - (Q^2/8\pi^2 b^2 g) \times (1/r^2) \}. \tag{6.32}$$

If the pressure p at any radius r is plotted as in Fig. 6.16(c), the curve will be parabolic and is sometimes referred to as Barlow's curve.

If the flow discharges to atmosphere at the periphery, the pressure at any point between the plates will be below atmospheric; there will be a force tending to bring the two plates together and so shut off flow. This phenomenon can

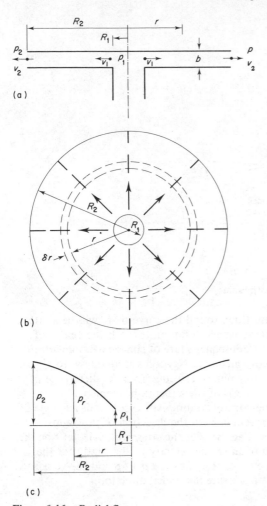

Figure 6.16   Radial flow

be observed in the case of a disc valve. Radial flow under the disc will cause the disc to be drawn down onto the valve seating. This will cause the flow to stop, the pressure between the plates will return to atmospheric and the static pressure of the fluid on the upstream side of the disc will push it off its seating again. The disc will tend to vibrate on the seating and the flow will be intermittent.

6.15. Flow in a curved path. Pressure gradient and change of total energy across the streamlines

Velocity is a vector quantity with both magnitude and direction. When a fluid flows in a curved path, the velocity of the fluid along any streamline will undergo a change due to its change of direction, irrespective of any alteration in magnitude which may also occur. Considering the streamtube

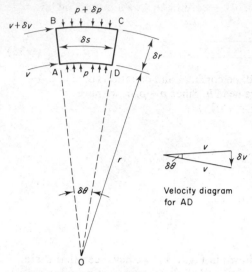

Figure 6.17   Change of pressure with radius

(shown in Fig. 6.17), as the fluid flows round the curve there will be a rate of change of velocity, that is to say an acceleration, towards the centre of curvature of the streamtube. The consequent rate of change of momentum of the fluid must be due, in accordance with Newton's second law, to a force acting radially across the streamlines resulting from the difference of pressure between the sides BC and AD of the streamtube element.

In Fig. 6.17, suppose that the control volume ABCD subtends an angle $\delta\theta$ at the centre of curvature O, has length δs in the direction of flow and thickness b perpendicular to the diagram. For the streamline AD, let r be the radius of curvature, p the pressure and v the velocity of the fluid. For the streamline BC, the radius will be $r + \delta r$, the pressure $p + \delta p$ and the velocity $v + \delta v$, where δp is the change of pressure in a radial direction.

From the velocity diagram,

Change of velocity in radial direction, $\delta v = v\delta\theta$

or, since $\delta\theta = \delta s/r$,

Radial change of velocity between AB and CD $= v\dfrac{\delta s}{r}$

Mass per unit time flowing through streamtube $=$ Mass density $\times$ Area $\times$ Velocity
$= \rho \times (b \times \delta r) \times v$

Change of momentum per unit time in radial direction $=$ Mass per unit time $\times$ Radial change of velocity

$$= \rho b \delta r v^2 \delta s/r. \tag{6.33}$$

This rate of change of momentum is produced by the force due to the pressure difference between faces BC and AD of the control volume:

$$\text{Force} = \{(p + \delta p) - p\}b\delta s. \tag{6.34}$$

Equating equations (6.33) and (6.34), according to Newton's second law,

$$\delta p b \delta s = \rho b \delta r v^2 \, \delta s / r,$$

$$\delta p / \delta r = \rho v^2 / r. \tag{6.35}$$

For an incompressible fluid, ρ will be constant and equation (6.35) can be expressed in terms of the pressure head h. Since $p = \rho g h$, we have $\delta p = \rho g \delta h$. Substituting in equation (6.35),

$$\rho g \, \delta h / \delta r = \rho v^2 / r,$$

$$\delta h / \delta r = v^2 / gr,$$

or, in the limit as δr tends to zero,

Rate of change of pressure head in radial direction

$$= \frac{dh}{dr} = \frac{v^2}{gr} \tag{6.36}$$

To produce the curved flow shown in Fig. 6.17, we have seen that there must be a change of pressure head in a radial direction. However, since the velocity v along streamline AD is different from the velocity $v + \delta v$ along BC, there will also be a change in the velocity head from one streamline to another:

Rate of change of velocity head radially

$$= \{(v + \delta v)^2 - v^2\}/2g\delta r$$

$$= \frac{v}{g} \frac{\delta v}{\delta r}, \text{ neglecting products of small quantities,}$$

$$= \frac{v}{g} \frac{dv}{dr}, \text{ as } \delta r \text{ tends to zero.} \tag{6.37}$$

If the streamlines are in a horizontal plane, so that changes in potential head do not occur, the change of total head H — i.e. the total energy per unit weight — in a radial direction, $\delta H / \delta r$, is given by

$$\delta H / \delta r = \text{Change of pressure head} + \text{Change of velocity head.}$$

Substituting from equations (6.36) and (6.37), in the limit,

Change of total energy with radius, $\dfrac{dH}{dr} = \dfrac{v^2}{gr} + \dfrac{v}{g}\dfrac{dv}{dr}$

$$= \frac{v}{g}\left(\frac{v}{r} + \frac{dv}{dr}\right). \tag{6.38}$$

The term $(v/r + dv/dr)$ is also known as the *vorticity* of the fluid (*see* Section 7.2).

In obtaining equation (6.38), it has been assumed that the streamlines are horizontal, but this equation also applies to cases where the streamlines are inclined to the horizontal, since the fluid in the control volume is in effect weightless, being supported vertically by the surrounding fluid.

If the streamlines are straight lines, $r = \infty$ and $dv/dr = 0$. From equation (6.38) for a stream of fluid in which the velocity is uniform across the cross-section, and neglecting friction, we have $dH/dr = 0$ and the total energy per unit weight H is constant for all points on all streamlines. This applies whether the streamlines are parallel or inclined, as in the case of radial flow (Section 6.14).

6.16. Vortex motion

In vortex motion, the streamlines form a set of concentric circles and the changes of total energy per unit weight will be governed by equation (6.38). Four types of vortex are recognized.

6.16.1. Forced vortex or flywheel vortex

The fluid rotates as a solid body with constant angular velocity ω, i.e. at any radius r,

$$v = \omega r \quad \text{so that} \quad \frac{dv}{dr} = \omega \quad \text{and} \quad \frac{v}{r} = \omega.$$

From equation (6.38),

$$\frac{dH}{dr} = \frac{\omega r}{g}(\omega + \omega) = \frac{2\omega^2 r}{g}.$$

Integrating,

$$H = \omega^2 r^2/g + C, \tag{6.39}$$

where C is a constant. But, for any point in the fluid,

$$H = p/\rho g + v^2/2g + z = p/\rho g + \omega^2 r^2/2g + z.$$

Substituting in equation (6.39),

$$p/\rho g + \omega^2 r^2/2g + z = \omega^2 r^2/g + C,$$

$$p/\rho g + z = \omega^2 r^2/2g + C. \tag{6.40}$$

If the rotating fluid has a free surface, the pressure at this surface will be atmospheric and therefore zero (gauge).

Putting $p/\rho g = 0$ in equation (6.40), the profile of the free surface will be given by

$$z = \omega^2 r^2/2g + C. \tag{6.41}$$

Therefore, the free surface will be in the form of a paraboloid (Fig. 6.18).

Similarly, for any horizontal plane, for which z will be constant, the pressure distribution will be given by

$$p/\rho g = \omega^2 r^2/2g + (C - z). \tag{6.42}$$

Example 6.9

A closed vertical cylinder 400 mm diameter and 500 mm high is filled with oil of relative density 0·9 to a depth of 340 mm, the remaining volume containing air at atmospheric pressure. The cylinder revolves about its

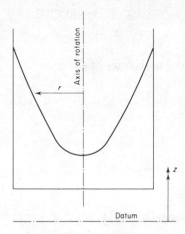

Figure 6.18   Forced vortex

vertical axis at such a speed that the oil just begins to uncover the base. Calculate (a) the speed of rotation for this condition and (b) the upward force on the cover.

Solution

   (a) When stationary, the free surface will be at AB (Fig. 6.19), a height Z_2 above the base.

$$\text{Volume of oil} = \pi r_1^2 Z_2.$$

When rotating at the required speed ω, a forced vortex is formed and the free surface will be the paraboloid CDE.

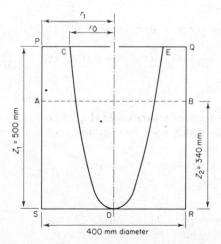

Figure 6.19   Forced vortex example

Volume of oil = Volume of cylinder PQRS − Volume of paraboloid CDE

$$= \pi r_1^2 Z_1 - \tfrac{1}{2}\pi r_0^2 Z_1,$$

since the volume of a paraboloid is equal to half the volume of the circumscribing cylinder.

No oil is lost from the container, therefore,

$$\pi r_1^2 Z_2 = \pi r_1^2 Z_1 - \tfrac{1}{2}\pi r_0^2 Z_1,$$

$$r_0^2 = 2r_1^2 (1 - Z_2/Z_1),$$

$$r_0 = r_1 \sqrt{\{2(1 - Z_2/Z_1)\}} = r_1 \sqrt{\{2(1 - 340/500)\}}$$

$$= 0 \cdot 8 r_1 = 0 \cdot 8 \times 200 = 160 \text{ mm.}$$

Also, for the free surface of the vortex from equation (6.41),

$$z = \omega^2 r^2 / 2g + \text{constant}$$

or, between points C and D, taking D as datum level,

$$Z_D = 0 \text{ when } r = 0 \quad \text{and} \quad Z_C = Z_1 \text{ when } r = r_0$$

giving

$$Z_1 - 0 = \omega^2 r_0^2 / 2g,$$

$$\omega = \sqrt{(2g Z_1 / r_0^2)}$$

$$= \sqrt{(2 \times 9 \cdot 81 \times 0 \cdot 5 / 0 \cdot 16^2)} = \textbf{19·6 rad s}^{-1}.$$

(b) The oil will be in contact with the top cover from radius $r = r_0$ to $r = r_1$. If p is the pressure at any radius r, the force on an annulus of radius r and width δr is given by

$$\delta F = p \times 2\pi r \delta r.$$

Integrating from $r = r_0$ to $r = r_1$,

$$\text{Force on top cover, } F = 2\pi \int_{r_0}^{r_1} p r \delta r \qquad\qquad \text{(I)}$$

From equation (6.42),

$$p/\rho g = \omega^2 r^2 / 2g + C.$$

Since the pressure at r_0 is atmospheric, $p = 0$ when $r = r_0$, so that

$$C = -\omega^2 r_0^2 / 2g$$

and

$$p = \rho g \left\{ \frac{\omega^2 r^2}{2g} - \frac{\omega^2 r_0^2}{2g} \right\} = \frac{\rho \omega^2}{2g} (r^2 - r_0^2).$$

Substituting in (I),

$$F = 2\pi \frac{\rho \omega^2}{2} \int_{r_0}^{r_1} (r^2 - r_0^2) r \, dr$$

$$= \rho \omega^2 \pi \int_{r_0}^{r_1} (r^3 - r_0^2 r) \, dr$$

$$= \rho \omega^2 \pi \left[\tfrac{1}{4} r^4 - \tfrac{1}{2} r_0^2 r^2 \right]_{r_0}^{r_1}$$

$$= \rho \omega^2 \pi \{ \tfrac{1}{4} r_1^4 - \tfrac{1}{4} r_0^4 - \tfrac{1}{2} r_0^2 r_1^2 + \tfrac{1}{2} r_0^4 \}$$

$$= \frac{\rho \omega^2 \pi}{4} (r_1^4 + r_0^4 - 2 r_0^2 r_1^2) = \frac{\pi}{4} \rho \omega^2 (r_1^2 - r_0^2)^2$$

$$= \frac{\pi}{4} \times (0 \cdot 9 \times 1000) \times 19 \cdot 6^2 (0 \cdot 2^2 - 0 \cdot 16^2)^2 \, \text{N}$$

$$= 56 \cdot 3 \, \text{N}.$$

6.16.2. Free vortex or potential vortex

In this case, the streamlines are concentric circles, but the variation of velocity with radius is such that there is no change of total energy per unit weight with radius, so that $dH/dr = 0$. Substituting in equation (6.38),

$$0 = \frac{v}{g} \left(\frac{v}{r} + \frac{dv}{dr} \right),$$

$$\frac{dv}{v} + \frac{dr}{r} = 0.$$

Integrating,

$$\log_e v + \log_e r = \text{constant}$$

or $vr = C$,

where C is a constant known as the *strength* of the vortex at any radius r;

$$v = C/r. \tag{6.43}$$

Since, at any point,

$$z + p/\rho g + v^2/2g = H = \text{constant},$$

substituting for v from equation (6.43)

$$z + p/\rho g + C^2/2gr^2 = H.$$

If the fluid has a free surface, $p/\rho g = 0$ and the profile of the free surface is given by

$$H - z = C^2/2gr^2. \tag{6.44}$$

which is a hyperbola asymptotic to the axis of rotation and to the horizontal plane through $z = H$, as shown in Fig. 6.20.

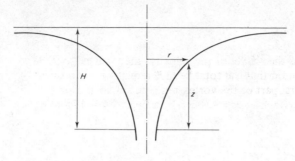

Figure 6.20   Free vortex

For any horizontal plane, z is constant and the pressure variation is given by

$$p/\rho g = (H - z) - C^2/2gr^2, \tag{6.45}$$

Thus, in the free vortex, pressure decreases and circumferential velocity increases as we move towards the centre.

Example 6.10

A point A on the free surface of a free vortex is at a radius $r_A = 200$ mm and a height $z_A = 125$ mm above datum. If the free surface at a distance from the axis of the vortex, which is sufficient for its effect to be negligible, is 180 mm above datum, what will be the height above datum of a point B on the free surface at a radius of 100 mm?

Solution
For point A, from equation (6.44),

$$H - z_A = C^2/2gr_A^2,$$

therefore,

$$C^2/2g = r^2(H - z_A).$$

Now H is the head above datum at an infinite distance from the axis of rotation, where the effect of the vortex is negligible, so that $H = 180$ mm $= 0.18$ m. Also, $z_A = 0.125$ m and $r_A = 0.2$ m. Substituting,

$$\frac{C^2}{2g} = 0.2^2(0.18 - 0.125) = 2.2 \times 10^{-3} \text{ m}^3.$$

For point B,

$$H - z_B = C^2/2gr_B^2$$
$$z_B = H - C^2/2gr_B^2$$
$$= 0.18 - (2.2 \times 10^{-3})/0.1^2 = -0.04 \text{ m}$$
$$= \textbf{40 mm below datum.}$$

6.16.3. Compound vortex

In the free vortex, $v = C/r$ and thus, theoretically, the velocity becomes infinite at the centre. The velocities near the axis would be very high and, since friction losses vary as the square of the velocity, they will cease to be negligible, and the assumption that the total head H remains constant will cease to be true. The central part of the vortex tends to rotate as a solid body, thus forming a forced vortex surrounded by a free vortex. Figure 6.21

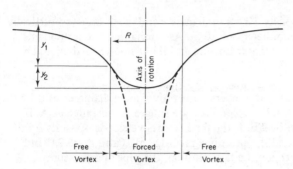

Figure 6.21   Compound vortex

shows the free surface profile of such a compound vortex, and also represents the variation of pressure with radius on any horizontal plane in the vortex. The velocity at the common radius R must be the same for the two vortices.

For the free vortex, if y_1 = depression of the surface at radius R below the level of the surface at infinity,

$$y_1 = C^2/2gR^2 = v^2/2g = \omega^2 R^2/2g.$$

For the forced vortex, if y_2 = height of the surface at radius R above the centre of the depression,

$$y_2 = R^2/2g = C^2/2gR^2.$$

Thus,   Total depression $= y_2 + y_1 = C^2/gR^2 = \omega^2 R^2/g.$ (6.46)

For the forced vortex, the velocity at radius R is ωR, while for the free vortex, from equation (6.43), the velocity at radius R is C/R. Therefore, the common radius, at which these two velocities will be the same, is given by

$$\omega R = C/R \quad \text{or} \quad R = \sqrt{(C/\omega)}.$$

In Section 7.9 it will be shown that $C = \Gamma/2\pi$, where Γ is the circulation, so that

$$\text{Common radius, } R = \sqrt{(C/\omega)} = \sqrt{(\Gamma/2\pi\omega)}.$$

EXERCISES 6

6.1   Water flowing in a pipeline 4 m above datum level has a velocity of 12 m/s and is at a gauge pressure of 36 kN/m². If the mass density of water is 1000 kg/m³ what is the total energy per unit weight of the water at this point, reckoned above datum level measuring pressure relative to that of the atmosphere?

[15·01 J/N]

6.2   The suction pipe of a pump rises at a slope of 1 vertical in 5 along the pipe and water passes through it at 1·8 m/s. If dissolved air is released when the pressure falls to more than 70 kN/m² below atmospheric pressure, find the greatest practicable length of pipe neglecting friction. Assume that the water in the sump is at rest.

[34·9 m]

6.3   A jet of water is initially 12 cm in diameter and when directed vertically upwards reaches a maximum height of 20 m. Assuming that the jet remains circular determine the rate of water flowing and the diameter of the jet at a height of 10 m.

[0·224 m³/s, 14·27 cm]

6.4   A pipe AB carries water and tapers uniformly from a diameter of 0·1 m at A to 0·2 m at B over a length of 2m. Pressure gauges are installed at A, B and also at C the midpoint of AB. If the pipe centreline slopes upwards from A to B at an angle of 30° and the pressures recorded at A and B are 2·0 and 2·3 bars respectively, determine the flow through the pipe and pressure recorded at C neglecting all losses.

[0·0723 m³/s, 2·17 bar]

6.5   The centreline of a tapered pipe AB slopes upwards from A to B at an angle of 30° to the horizontal. The distance AB is 5 m and the diameter increases uniformly from 10 cm at A to 15 cm at B. The pipe carries petrol (spec. gr. 0·74) and pressure gauges are installed at A and B. Find (a) the flow rate when the readings on the pressure gauges are equal, (b) the pressure difference across AB for the same rate of flow when the direction of taper is reversed.

[0·0613 m³/s, 54·4 kN/m²]

6.6   A pipe 300 m long tapers from 1·2 m diameter to 0·6 m diameter at its lower end and slopes downwards at 1 in 100. The pressure at the upper end is 69 kN/m². Neglecting friction losses find the pressure at the lower end when the rate of flow is 5·5 m³/min.

[98·5 kN/m²]

6.7   Air enters a compressor at the rate of 0·5 kg/s with a velocity of 6·4 m/s, specific volume 0·85 m³/kg and a pressure of 1 bar. It leaves the compressor at a pressure of 6·9 bar with a specific volume of 0·16 m³/kg and a velocity of 4·7 m/s. The internal energy of the air at exit is greater than that at entry by 85 kJ/kg. The compressor is fitted with a cooling system which removes heat at the rate of 60 kJ/s. Calculate the power required to drive the compressor and the cross-sectional areas of the inlet and outlet pipes.

[115·2 kW, 0·0664 m², 0·0170 m²]

6.8   A Pitot-static tube is used to measure air velocity. If a manometer connected to the instrument indicates a difference in pressure head between the tappings of 4 mm of water, calculate the air velocity assuming the coefficient of the Pitot tube to be unity. Density of air = 1·2 kg/m³.

[8·08 m/s]

6.9   A liquid flow through a circular pipe 0·6 m in diameter. Measurements of velocity taken at intervals along a diameter are:

Distance from the wall m	0	0.05	0·1	0·2	0·3	0·4	0·5	0·55	0·6
Velocity m/s	0	2·0	3·8	4·6	5·0	4·5	3·7	1·6	0

Draw the velocity profile, calculate the mean velocity and determine the kinetic energy correction factor.

[2·88 m/s, 1·735]

6.10   A venturi meter having a throat 100 mm in diameter is fitted in a pipeline 250 mm in diameter through which oil of specific gravity 0·9 is flowing at the rate of 0·1 m^3/s. The inlet and throat of the meter are connected differentially to a U-tube manometer containing mercury of specific gravity 13·6 with oil immediately above this. Starting from Bernoulli's equation find the coefficient of discharge for the meter if the difference in mercury levels is 0·63 m.

[0·95]

6.11   A venturi meter with a throat diameter of 100 mm is fitted in a vertical pipeline of 200 mm diameter with oil of specific gravity 0·88 flowing upwards at a rate of 0·06 m^3/s. The venturi meter coefficient is 0·96. Two pressure gauges calibrated in kN/m^2 are fitted at tapping points, one at the throat and the other in the inlet pipe 320 mm below the throat. The difference between the two gauge pressure readings is 28 kN/m^2.

Working from Bernoulli's equation determine (a) the volume flow rate of oil through the pipe, (b) the difference in level in the two limbs of a mercury manometer if it is connected to the tapping points and the connecting pipes are filled with the same oil.

[0·059 m^3/s, 202 mm]

6.12   The mean velocity of water in a 0·15 m diameter horizontal pipe is 5 m/s. Working from first principles calculate the pressure difference which would be shown by an inverted U-tube manometer across the 0·10 m diameter orifice meter in the pipe. For the meter C_d = 0·61.

[13·9 m]

6.13   An orifice plate is to be used to measure the rate of air flow through a 2 m diameter duct. The mean velocity in the duct will not exceed 15 m/s and a water tube manometer, having a maximum difference between water levels of 150 mm, is to be used. Assuming the coefficient of discharge to be 0·64, determine a suitable orifice diameter to make the full use of the manometer range. Take the density of air as 1·2 kg/m^3.

[1·31 m]

6.14   Air flows through a 20 cm diameter pipe and the maximum expected rate of flow is 3 m^3/min. If the available water manometer has a maximum reading of 200 mm, what would be a suitable diameter for an orifice plate to measure this flow? Assume the density of air to be 1·2 kg/m^3, C_v = 0·97, C_c = 0·65.

[4·2 cm]

6.15   A sharp-edged orifice, 5 cm in diameter, in the vertical side of a large tank discharges under a head of 5 m. If C_c = 0·62 and C_v = 0·98, determine (a) the diameter of the jet at the vena contracta, (b) the velocity of the jet at the vena contracta and (c) the discharge in m^3/s.

[3·94 cm, 9·71 m/s, 0·0119 m^3/s]

6.16   Find the diameter of a circular orifice to discharge 0·015 m^3/s under a head of 2·4 m using a coefficient of discharge of 0·6. If the orifice is in a

vertical plane and the jet falls 0·25 m in a horizontal distance of 1·3 m from the vena contracta, find the value of the coefficient of contraction.
[6·82 cm, 0·715]

6.17   Tank A has a cross-sectional area of 0·5 m^2, is open at the top to atmosphere and is allowed to empty through an orifice in its vertical side. During part of the period of emptying, the water is caught in a tank B, whose top edge is 0·2 m below the level of the orifice, and whose near side is 0·75 m and far side 1·25 m from the plane of the orifice. If the liquid level in tank A is initially 2·5 m above the orifice, find, from first principles, the minimum capacity of B to ensure that it does not overflow. Neglect friction.
[0·626 m^3]

6.18   A tank has a circular orifice 20 mm diameter in the vertical side near the bottom. The tank contains water to a depth of 1 m above the orifice with oil of relative density 0·8 for a depth of 1 m above the water. Acting on the upper surface of the oil is an air pressure of 20 kN/m^2 gauge. The jet of water issuing from the orifice travels a horizontal distance of 1·5 m from the orifice while falling a vertical distance of 0·156 m. If the coefficient of contraction of the orifice is 0·65, estimate the value of the coefficient of velocity and the actual discharge through the orifice.
[0·97, 1·72 dm^3/s]

6.19   Water flows from a reservoir through a rectangular opening 2 m high and 1·2 m wide in the vertical face of a dam. Calculate the discharge in m^3/s when the free surface in the reservoir is 0·5 m above the top of the opening assuming a coefficient of discharge of 0·64.
[8·16 m^3/s]

6.20   A vertical triangular orifice in the wall of a reservoir has a base 0·9 m long, 0·6 m below its vertex and 1·2 m below the water surface. Determine the theoretical discharge.
[1·19 m^3/s]

6.21   The discharge over a rectangular notch is to be 0·14 m^3/s when the water level is 23 cm above the sill. Assuming a coefficient of discharge of 0·6 and deducing the formula for theoretical discharge from first principles, calculate the width of notch required.
[0·72 m]

6.22   A rectangular channel 1·2 m wide has at its end a rectangular sharp-edged notch with an effective width after allowing for side contractions of 0·85 m and with its sill 0·2 m from the bottom of the channel. Assuming that the velocity head averaged over the channel is $\alpha V^2/2g$ where V is the mean velocity and $\alpha = 1·1$, calculate the discharge in m^3/s when the head is 250 mm above the sill allowing for the velocity of approach.
[0·204 m^3/s]

6.23   In an experiment on a 90° vee notch the flow is collected in a 0·9 m diameter vertical cylindrical tank. It is found that the depth of water increases by 0·685 m in 16·8 s when the head over the notch is 0·2 m. Determine the coefficient of discharge of the notch.
[0·615]

6.24   A sharp-edged notch is in the form of a symmetrical trapezium. The horizontal base is 10 cm wide, the top is 50 cm wide and the depth is 30 cm. Assuming the coefficient of discharge to be 0·6 and that the velocity of approach is negligible, calculate the height of the water level above the base of the notch if the discharge is $0.043 \text{ m}^3/\text{s}$.

[22.9 cm]

6.25 A turbine is supplied with water from a reservoir which is 200 m above the level of the discharge pipe. The volume rate of flow through the pipeline is $0.2 \text{ m}^3/\text{s}$. If the power output from the shaft of the turbine is 310 kW and it has a mechanical efficiency of 90 per cent, calculate (a) the power drawn from the reservoir, (b) the hydraulic power delivered to the turbine, (c) the loss of energy per unit weight in the pipeline.

[392.4 kW, 344.4 kW, 24.26 J/N]

6.26 A jet of water issues from a nozzle with a velocity of 40 m/s at the rate of $0.15 \text{ m}^3/\text{s}$. What is the power of the jet?

[120 kW]

6.27 A pump discharges $2 \text{ m}^3/\text{s}$ of water through a pipeline. If the pressure difference between the inlet and the outlet of the pump is equivalent to 10 m of water, what power is being transmitted to the water from the pump?

[196.2 kW]

6.28 Inward radial flow occurs between two horizontal discs 0.6 m in diameter and 75 mm apart, the water leaving through a central pipe 150 mm in diameter in the lower disc at the rate of $0.17 \text{ m}^3/\text{s}$. If the absolute pressure at the outer edge of the disc is 101 kN/m^2 calculate the pressure at the outlet. Find also the resultant force on the upper disc.

[90 kN/m^2, 567 N]

6.29 Two horizontal discs are 12.5 mm apart and 300 mm in diameter. Water flows radially outwards between the discs from a 50 mm diameter pipe at the centre of the lower disc. If the pressure at the outer edge of the disc is atmospheric, calculate the pressure in the supply pipe when the velocity of the water in the pipe is 6 m/s. Find also the resultant pressure on the upper disc, neglecting impact force.

[-17.5 kN/m^2, 126.7 N]

6.30 Describe briefly the causes and the nature of the secondary motion of a fluid flowing round a river bend and also a bend in a square section duct running full.

Air of density 1.21 kg/m^3 flows in a duct system of square cross-section $1.5 \text{ m} \times 1.5 \text{ m}$. At a certain section where the duct describes a horizontal circular arc of mean radius 3.75 m it is found that a pressure difference of 30 N/m^2 exists between the inner and outer walls of the duct. Assuming that the air flows round the bend under free vortex conditions so that the product of velocity and radius remains constant, calculate the volume flow rate in the duct.

[$17.23 \text{ m}^3/\text{s}$]

6.31 A hollow cylindrical drum with its axis vertical has an internal diameter of 600 mm and is full of water. A set of paddles 200 mm in diameter rotates concentrically with the axis of the drum at 120 rev/min and produces a compound vortex in the water. Assuming that all the water in the 200 cm core rotates as a forced vortex with the paddles and that the water outside this core moves as a free vortex, determine (a) the velocity of the water at 75 mm and 225 mm from the centre, (b) the pressure head at these radii above the pressure head at the centre.

[0.95 m/s, 0.57 m/s, 47 mm, 149 mm]

7 Two-dimensional ideal flow

In Chapter 4, a distinction was made between real and ideal fluids. The former exhibit the effects of viscosity and will be dealt with in the next part of the book, whereas the latter will be considered in this chapter.

An ideal fluid is a purely hypothetical fluid which is assumed to have no viscosity and no compressibility, also, in the case of liquids, no surface tension and no vaporization. The study of flow of such a fluid stems from the eighteenth century hydrodynamics developed by mathematicians, who, by making the above assumptions regarding the fluid, aimed at establishing mathematical models for fluid flow. Although the assumptions of ideal flow appear to be very far fetched, the introduction of the boundary layer concept by Prandtl in 1904 enabled the distinction to be made between two régimes of flow: that adjacent to the solid boundary, in which viscosity effects are predominant and, therefore, the ideal flow treatment would be erroneous, and that outside the boundary layer, in which viscosity has negligible effect so that the idealized flow conditions may be applied. This argument is developed further in Chapter 10 when dealing with external flow.

The ideal flow theory may also be extended to situations in which fluid viscosity is very small and velocities are high, since they correspond to very high values of Reynolds number, at which flows are independent of viscosity. Thus, it is possible to see ideal flow as that corresponding to an infinitely large Reynolds number and to zero viscosity. The applications of ideal flow theory are found in aerodynamics, in accelerating flow, tides and waves.

The study of ideal flow provides mathematical expressions for streamlines in elementary or basic flow patterns. By combining these basic flow patterns in various ways, it is possible to obtain complex flow patterns which, in many cases, resemble remarkably closely the real situations outside the boundary layer and any associated wakes.

7.1. Rotational and irrotational flow

Considerations of ideal flow lead to yet another flow classification, namely the distinction between rotational and irrotational flow.

Basically, there are two types of motion: translation and rotation. The two may exist independently or simultaneously, in which case they may be considered as one superimposed on the other. If a solid body is represented by a square, then pure translation or pure rotation may be represented as shown in Fig. 7.1(a) and (b), respectively.

If we now consider the square to represent a fluid element, it may be subjected to deformation. This can be either linear or angular, as shown in Fig. 7.2(a) and (b), respectively.

Now, consider a motion of a fluid in which rotation of fluid elements is superimposed on their translation. In time dt, then, point A on the fluid element aAb moves to A′ and the element assumes position a′A′b′, as shown in Fig. 7.3. The two angles of rotation α and β will not be the same if

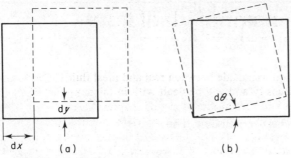

Figure 7.1   Translation and rotation

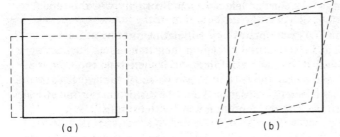

Figure 7.2   Linear and angular deformation

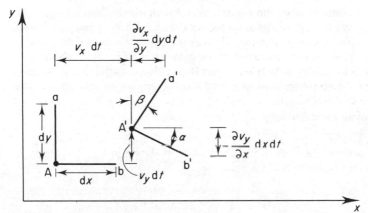

Figure 7.3   Rotation, translation and deformation

deformation takes place and, therefore, the average rate of rotation in time dt will be:

$$\omega = \frac{\alpha + \beta}{2} \times \frac{1}{dt} = \frac{1}{2} \frac{(\alpha + \beta)}{dt},$$

but, for small values and taking anticlockwise rotation as positive,

$$\alpha = \frac{\text{Arc}}{\text{Radius}} = \frac{\partial v_y}{\partial x} \, dx \, dt \, \frac{1}{dx} = \frac{\partial v_y}{\partial x} \, dt$$

and $\quad \beta = -\frac{\partial v_x}{\partial y} \, dy \, dt \, \frac{1}{dy} = \frac{\partial v_x}{\partial y} \, dt.$

The rate of rotation about the z axis is, therefore,

$$\omega = \frac{1}{2}\left(\frac{\partial v_y}{\partial x}\,dt - \frac{\partial v_x}{\partial y}\,dt\right)\frac{1}{dt} = \frac{1}{2}\left(\frac{\partial v_y}{\partial x} - \frac{\partial v_x}{\partial y}\right).$$

The expression in brackets,

$$\frac{\partial v_y}{\partial x} - \frac{\partial v_x}{\partial y} = \zeta, \tag{7.1}$$

is called the *vorticity* and is denoted by ζ. Thus,

$$\zeta = 2\omega_z, \tag{7.2}$$

where ω_z is the angular velocity of the fluid elements about their mass centre in the x-y plane. In three-dimensional flow, ω_z would represent only one of three components of the angular velocity ω and vorticity would be equal to 2ω.

The expression (7.1) was obtained by stipulating rotation of fluid elements to exist and to be superimposed on their translation. Such a flow is known as *rotational*. It follows, therefore, that, if there is no rotation, the expression (7.1) and, hence, the vorticity must be equal to zero. Thus, if the motion of particles is purely translational and the distortion is symmetrical, the flow is *irrotational* and the condition which it must satisfy is

$$\frac{\partial v_y}{\partial x} - \frac{\partial v_x}{\partial y} = 2\omega_z = 0. \tag{7.3}$$

The distinction between the rotational and irrotational flow is important because, for example, it will be shown later that Bernoulli's equation derived for a streamline applies to all streamlines in the flow field only if the flow is irrotational. It will also be shown that the generation of lift by such surfaces as aerofoils is associated with irrotational flow. Also, a useful and practical procedure of determining 'flow nets' can only be applied to irrotational flow.

7.2. Circulation and vorticity

Consider a fluid element ABCD in rotational motion. Let the velocity components along the sides of the element be as shown in Fig. 7.4. Since the

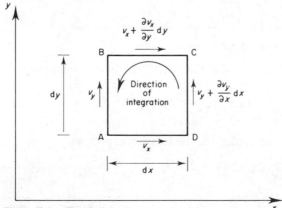

Figure 7.4   Circulation

element is rotating, being part of rotational flow, there must be a 'resultant' peripheral velocity. However, since the centre of rotation is not known, it is more convenient to relate rotation to the sum of products of velocity and distance around the contour of the element. Such a sum is, of course, the line integral of velocity around the element and it is called *circulation*, denoted by Γ.

Thus,   Circulation, $\Gamma = \oint v_s \, ds$. $\qquad\qquad$ (7.4)

Circulation is, by convention, regarded as positive for anticlockwise direction of integration. Thus, for the element ABCD, starting from side AD,

$$\Gamma_{ABCD} = v_x \, dx + \left(v_y + \frac{\partial v_y}{\partial x} \, dx\right) dy - \left(v_x + \frac{\partial v_x}{\partial y} \, dy\right) dx - v_y \, dy$$

$$= \frac{\partial v_y}{\partial x} \, dx \, dy - \frac{\partial v_x}{\partial y} \, dy \, dx$$

$$= \left(\frac{\partial v_y}{\partial x} - \frac{\partial v_x}{\partial y}\right) dx \, dy,$$

but $\qquad \left(\frac{\partial v_y}{\partial x} - \frac{\partial v_x}{\partial y}\right) = \zeta$

for the two-dimensional flow in the x–y plane and, therefore, is the vorticity of the element about the z axis, ζ_z. The product $dx \, dy$ is the area of the element dA. Thus,

$$\Gamma_{ABCD} = \left(\frac{\partial v_y}{\partial x} - \frac{\partial v_x}{\partial y}\right) dx \, dy = \zeta_z \, dA.$$

It is seen, therefore, that the circulation around a contour is equal to the sum of the vorticities within the area of the contour. This is known as Stokes' theorem and may be stated mathematically, for a general case of any contour C (Fig. 7.5) as

$$\Gamma_C = \oint v \cos \theta \, ds = \int_A \zeta \, dA. \qquad\qquad (7.5)$$

Figure 7.5   Circulation and vorticity

The concept of circulation is very important in the theory of lifting surfaces such as aerofoils, hydrofoils and blades of rotodynamic machines.

The above considerations indicate that, for irrotational flow, since vorticity is equal to zero, the circulation around a closed contour through which fluid is flowing, must be equal to zero.

7.3. Streamlines and the stream function

In Section 4.1, a distinction was made between streaklines, pathlines and streamlines. Of the three, the streamline is the one which is a purely theoretical line in space, defined as being tangential to instantaneous velocity vectors. From this definition of a streamline, it follows that there can be no flow across it, simply because a line cannot be tangential to a velocity vector which at the same time crosses it.

The concept of the streamline is very useful, especially in ideal flow, because it enables the fluid flow to be conceived as occurring in patterns of streamlines. These patterns may be described mathematically so that the whole system of analysis may be based on it. It requires, however, a mathematical definition of a streamline. Consider, in a two-dimensional case, the velocity and displacement vectors of a fluid at a point, together with their

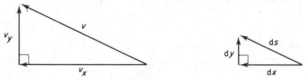

Figure 7.6   Velocity and displacement components

orthogonal components, as shown in Fig. 7.6. Since, by definition of a streamline, $ds \parallel v$, it follows that

$$dy \parallel v_y \quad \text{and} \quad dx \parallel v_x.$$

Thus, the velocity triangle and the displacement triangle are similar and, therefore,

$$\frac{dx}{v_x} = \frac{dy}{v_y}. \tag{7.6}$$

This constitutes the equation of a streamline.

Since the streamlines in a flow pattern describe it, it is useful to label them by some numerical system. Furthermore, it is possible to relate the numerical labels to the flowrate of the pattern which is being described. Thus, let aa and bb be two streamlines in a flow bounded by solid boundaries AA and BB in Fig. 7.7. If the streamline aa is denoted by Ψ_a, which will be labelled by a numerical value representing the flow rate per unit depth between AA and the streamline aa, then,

$$\Psi_a = Q_{0c}$$

and, similarly, if

$$\Psi_b = Q_{0e},$$

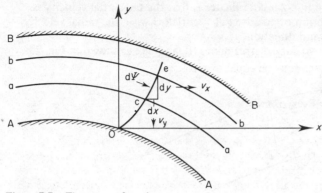

Figure 7.7   The stream function

it follows that

$$d\Psi = \Psi_b - \Psi_a = Q_{ce},$$

so that

$$d\Psi = v_x\, dy - v_y\, dx \tag{7.7}$$

and Ψ, which is called the *stream function*, is given by

$$\Psi = \int v_x\, dy - \int v_y\, dx. \tag{7.8}$$

Thus, the stream function depends upon position coordinates,

$$\Psi = f(x, y)$$

and, hence, the total derivative:

$$d\Psi = \frac{\partial \Psi}{\partial x}dx + \frac{\partial \Psi}{\partial y}dy. \tag{7.9}$$

Comparing equations (7.9) and (7.7), the relationships between the stream function and the velocity components are obtained:

$$v_x = \frac{\partial \Psi}{\partial y} \quad \text{and} \quad v_y = -\frac{\partial \Psi}{\partial x}. \tag{7.10}$$

(*Note:* the sign convention adopted here is for the flow to be positive from left to right.) Since the value of a stream function represents the flowrate between a given streamline described by the stream function and a reference boundary, it follows that it must be constant for the given streamline in order to satisfy the continuity equation combined with the requirement of no flow across a streamline.

In some curved flows, it is more convenient to use the polar coordinates in mathematical analysis. In these, since

$$\Psi = f(r, \theta),$$

by differentiation,

$$d\Psi = \frac{\partial \Psi}{\partial r}dr + \frac{\partial \Psi}{\partial \theta}d\theta. \tag{7.11}$$

The sign convention in polar coordinates is that the tangential velocity is positive in the direction of positive θ, i.e. anticlockwise; the radial velocity is positive in the outward direction.

Consider, now, two curved streamlines Ψ_a and Ψ_b, as shown in Fig. 7.8.

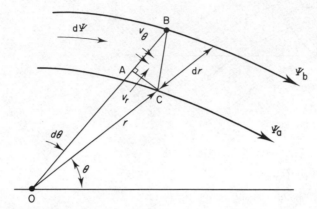

Figure 7.8   Stream function in polar coordinates

Assuming that $AB = dr$ when $d\theta = 0$ and applying the continuity equation, the following relationship is obtained:

$$v_r(r \, d\theta) - v_\theta \, dr = d\Psi. \tag{7.12}$$

Comparing equations (7.11) and (7.12), the following relationships are deduced:

$$v_\theta = -\frac{\partial \Psi}{\partial r} \quad \text{and} \quad v_r = \frac{1}{r}\frac{\partial \Psi}{\partial \theta}. \tag{7.13}$$

7.4. Velocity potential and potential flow

In connection with flow nets mentioned earlier, an important concept is that of the *velocity potential*. It is defined as

$$\Phi = \int_A^B v_s \, ds, \tag{7.14}$$

where A and B are two points in a potential field and v_s is the velocity tangential to the elementary path s.

To understand the above concept, it is necessary to appreciate the meaning of the term 'potential', so often used in mechanics as well as in other situations, which satisfy the same specific conditions. Consider, for example, points P and P′ in a gravitational field. If P′ has a greater potential than P, the difference in their potential δW is defined as the work required to move a particle from P to P′ against the gravitational force. If the distance between the points is δs and the force required is F then:

$$\delta W = F \, \delta s$$

or $$W = \int_{P}^{P'} F \, ds. \tag{7.15}$$

Clearly, the work done in such a case is independent of the path taken in doing the work and this property is a characteristic of potential fields only. In any other case, say against friction, the work done would depend upon the path taken. Therefore, not all fields are potential, but only those in which the path taken is immaterial.

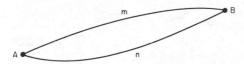

Figure 7.9   Different paths in the potential field

Returning to our velocity field, if points A and B in Fig. 7.9 belong to some potential field, then the integral

$$\int_{A}^{B} v \, ds$$

is independent of the path taken. Therefore,

$$\int_{A}^{B} v_{\mathrm{m}} \, ds = \int_{A}^{B} v_{\mathrm{n}} \, ds. \tag{7.16}$$

From the analogy with the gravitational field, it is apparent that the condition of equation (7.16) will only be satisfied if the field is potential. Fluid flow in such a field is known as *potential flow*.

Consider, now, the circulation around AnBm:

$$\Gamma_{\text{AnBm}} = \int_{A}^{B} v_{\mathrm{n}} \, ds + \int_{B}^{A} v_{\mathrm{m}} \, ds$$

$$= \int_{A}^{B} v_{\mathrm{n}} \, ds - \int_{A}^{B} v_{\mathrm{m}} \, ds. \tag{7.17}$$

But, for the flow to be potential, the two integrals in equation (7.17) must be equal and, therefore, $\Gamma_{\text{AnBm}} = 0$. If the circulation is equal to zero, it follows that vorticity must also be equal to zero and, therefore, the condition for potential flow is

$$\frac{\partial v_y}{\partial x} - \frac{\partial v_x}{\partial y} = 0, \tag{7.18}$$

which is identical with the condition for irrotational flow. Thus, potential flow is irrotational and vice versa.

In irrotational (potential) flow, therefore, the function (7.14)

$$\Phi = \int_A^B v_s \, ds$$

exists, from which it follows that, if v_x and v_y are the orthogonal components of v_s, then,

$$v_x = \frac{\partial \Phi}{\partial x} \quad \text{and} \quad v_y = \frac{\partial \Phi}{\partial y}, \tag{7.19}$$

so that

$$\Phi = \int v_x \, dx + \int v_y \, dy. \tag{7.20}$$

Similarly, in polar coordinates, if v_θ and v_r are tangential and radial components of v_s, then,

$$\Phi = \int v_r \, dr + \int v_\theta r \, d\theta, \tag{7.21}$$

from which

$$v_r = \frac{\partial \Phi}{\partial r} \quad \text{and} \quad v_\theta = \frac{\partial \phi}{r \partial \theta}. \tag{7.22}$$

It is now appropriate to consider the implications of potential flow to the applicability of Bernoulli's equation. It was originally derived in Section 5.12 to apply along a streamline, but not necessarily across streamlines, i.e. from one streamline to a neighbouring one. Furthermore, it was shown in Section 6.15 that in some curved flows the Bernoulli constant (or total head), defined as

$$H = p/\rho g + v^2/2g + z,$$

varies across streamlines, the variation in general being governed by equation (6.38), namely,

$$\frac{dH}{dr} = \frac{v}{g}\left(\frac{v}{r} + \frac{dv}{dr}\right).$$

However, in some flows the Bernoulli constant is the same for all streamlines, so that the Bernoulli equation may be applied to any points in the flow field. Clearly, for this to happen dH/dr must be equal to zero. One such obvious case is when $dr \to \infty$, i.e. in the case of straight line flows. The other possibility is for the expression in brackets to be equal to zero, i.e.

$$\frac{v}{r} + \frac{dv}{dr} = 0. \tag{7.23}$$

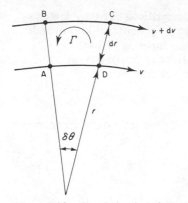

Figure 7.10   Circulation in polar coordinates

Let us examine such a case by considering an element of fluid in a curved flow, as shown in Fig. 7.10. The circulation around the element ABCD is

$$\Gamma_{ADCB} = vr\delta\theta - (v + dv)(r + dr)\delta\theta = -v\,dr\delta\theta - dv\,dr\delta\theta - r\,dr\delta\theta.$$

Neglecting infinitesimals of the third order, this reduces to

$$\Gamma_{ADCB} = -v\,dr\delta\theta - r\,dv\delta\theta.$$

But the area of the element is $r\,\delta\theta\,dr$, so that the vorticity is given by

$$\zeta = \frac{\Gamma}{\text{Area}} = \frac{-v\,dr\delta\theta - r\,dv\delta\theta}{r\delta\theta\,dr}$$

$$= -\left(\frac{v}{r} + \frac{dv}{dr}\right), \tag{7.24}$$

which is the same as the left-hand side of equation (7.23). Thus, if the vorticity is zero, there is no variation of Bernoulli's constant. This condition applies to irrotational (potential) flow.

In potential flow, then, Bernoulli's equation applies to the whole flow field and is not limited to individual streamlines.

7.5. Relationship between stream function and velocity potential; flow nets

Comparing equations (7.10) and (7.19), from (7.10),

$$v_x = \frac{\partial\Psi}{\partial y} \quad \text{and} \quad v_y = -\frac{\partial\Psi}{\partial x};$$

from (7.19),

$$v_x = \frac{\partial\Phi}{\partial x} \quad \text{and} \quad v_y = \frac{\partial\Phi}{\partial y}.$$

Thus, equating for v_x and v_y, we obtain

$$\frac{\partial\Psi}{\partial y} = \frac{\partial\Phi}{\partial x} \quad \text{and} \quad \frac{\partial\Phi}{\partial y} = -\frac{\partial\Psi}{\partial x}. \tag{7.25}$$

These equations are known as Cauchy-Rieman equations and they enable the stream function to be calculated if the velocity potential is known and vice versa in a potential flow.

It is now possible to return to the condition for potential flow and to restate it in terms of the stream function. The condition is

$$\frac{\partial v_y}{\partial x} - \frac{\partial v_x}{\partial y} = 0,$$

but $\quad v_y = -\dfrac{\partial \Psi}{\partial x} \quad$ and $\quad v_x = \dfrac{\partial \Psi}{\partial y},$

so that, by substitution,

$$\frac{\partial}{\partial x}\left(-\frac{\partial \Psi}{\partial x}\right) - \frac{\partial}{\partial y}\left(\frac{\partial \Psi}{\partial y}\right) = 0$$

and

$$\frac{\partial^2 \Psi}{\partial x^2} + \frac{\partial^2 \Psi}{\partial y^2} = 0. \tag{7.26}$$

This is the Laplace equation for stream function, which must be satisfied for the flow to be potential.

It is also interesting to note that Laplace's equation for the velocity potential must also be satisfied. This follows by substitution of equations (7.19) into the continuity equation for steady, incompressible, two-dimensional flow (equation (4.13)):

$$\frac{\partial v_x}{\partial x} + \frac{\partial v_y}{\partial y} = 0.$$

Substituting, now, for v_x and v_y from equations (7.19),

$$\frac{\partial}{\partial x}\left(\frac{\partial \Phi}{\partial x}\right) + \frac{\partial}{\partial y}\left(\frac{\partial \Phi}{\partial y}\right) = 0,$$

so that

$$\frac{\partial^2 \Phi}{\partial x^2} + \frac{\partial^2 \Phi}{\partial y^2} = 0. \tag{7.27}$$

Thus, the Laplace equation for the velocity potential must also be satisfied.

The fact that, for potential flow, both the stream function and the velocity potential satisfy the Laplace equation indicates that Ψ and Φ are interchangeable (Cauchy-Rieman equations) and that the lines of constant Ψ, i.e. streamlines, and the lines of constant Φ, called *equipotential* lines, are mutually perpendicular. This means that, if streamlines are plotted, points can be marked on them which have the same value of Φ and can be joined to form equipotential lines. Thus, a flow net of streamlines and equipotential lines is formed.

When the streamlines converge, the velocity increases and, therefore, for a given increment of $\delta\Psi$, the distance between the equipotential lines will also decrease.

The method of drawing a flow net consists of drawing by eye streamlines equispaced at $\delta\Psi$ at some section where the flow is rectilinear, such as AG or

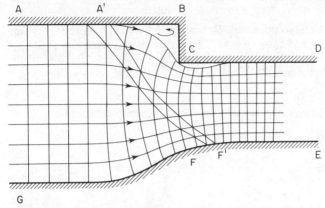

Figure 7.11   Example of a flow net

DE in Fig. 7.11, which shows an example of a network drawn for a rather
unusual converging section, of which the upper half constitutes a sudden
contraction but the lower half provides a smooth transition. The number of
streamlines drawn, or, rather, the size of intervals between them, depends
upon the accuracy required. The more streamlines one uses, the more
accurate will be the result, but, equally, the time spent in drawing the net
will be greater. The set of equipotential lines is drawn next at intervals
$\delta\Phi = \delta\Psi$ and in such a way that they cross each streamline at right angles.
Thus a set of 'squares' is obtained. The process is done by eye and requires
a series of successive adjustments to both streamlines and equipotential lines
until a satisfactory network of 'squares' is achieved. As a final check,
diagonals through the 'squares' may be drawn. They, too, should be smooth
lines and should form a net of squares. A pair of such diagonals from A' to
F' is shown in Fig. 7.11.

Where abrupt changes of the outer boundary occur, such as at points B
and C, it can be seen that the streamline $AA'D$ cannot follow the contour
and separates from the boundary. At B, where the streamline turns towards
the fluid, the velocity at the separation area will be zero and the fluid
trapped there will be stagnant. At point C, the streamline turns away from
the fluid, indicating high velocity in the separation bubble. This velocity is
spent in rotation of considerable vigour. Certainly, therefore, the assumption
of irrotational flow is not valid there. In general, then, wherever the
streamlines diverge or converge abruptly, separation may occur. Because
GFE is smooth and converging, no separation will occur there. Should,
however, the flow direction be reversed, although the flow net would remain
the same, separation might be expected downstream of F due to the
divergence of flow. Separation phenomena are discussed fully in Chapter 9 in
connection with boundary layer.

Constructing flow nets is a useful exercise which requires a lot of patience
and experience. The alternative is to use precise mathematical expressions for
stream function and velocity potential describing the flow from which a flow
net can be plotted exactly. The following sections of this chapter deal with
such mathematical expressions for some basic flows which may then be
combined to represent more complex flow patterns.

Example 7.1

In a two-dimensional, incompressible flow the fluid velocity components are given by: $v_x = x - 4y$ and $v_y = -y - 4x$. Show that the flow satisfies the continuity equation and obtain the expression for the stream function. If the flow is potential obtain also the expression for the velocity potential.

Solution

For incompressible, two-dimensional flow, the continuity equation is

$$\frac{\partial v_x}{\partial x} + \frac{\partial v_y}{\partial y} = 0,$$

but $v_x = x - 4y$ and $v_y = -y - 4x$

and $\dfrac{\partial v_x}{\partial x} = 1,$ $\dfrac{\partial v_y}{\partial y} = -1;$

therefore, $1 - 1 = 0$ and the flow satisfies the continuity equation.

To obtain the stream function, using equations (7.10),

$$v_x = \frac{\partial \Psi}{\partial y} = x - 4y, \tag{I}$$

$$v_y = \frac{\partial \Psi}{\partial x} = -(y + 4x). \tag{II}$$

Therefore, from (I),

$$\Psi = \int (x - 4y)\, dy + f(x) + C$$

$$= xy - 2y^2 + f(x) + C.$$

But, if $\Psi_0 = 0$ at $x = 0$ and $y = 0$, which means that the reference streamline passes through the origin, then $C = 0$ and

$$\Psi = xy - 2y^2 + f(x). \tag{III}$$

To determine $f(x)$, differentiate partially the above expression with respect to x and equate to $-v_y$, equation (II):

$$\frac{\partial \Psi}{\partial x} = y + \frac{\partial}{\partial x} f(x) = y + 4x,$$

$$f(x) = \int 4x\, dx = 2x^2$$

Substitute into (III),

$$\Psi = 2x^2 + xy - 2y^2.$$

To check whether the flow is potential, there are two possible approaches:

(a) Since

$$\frac{\partial v_y}{\partial x} - \frac{\partial v_x}{\partial y} = 0,$$

but

$$v_y = -(4x + y) \quad \text{and} \quad v_x = (x - 4y).$$

Therefore,

$$\frac{\partial v_y}{\partial x} = -4 \quad \text{and} \quad \frac{\partial v_x}{\partial y} = -4,$$

so that

$$\frac{\partial v_y}{\partial x} - \frac{\partial v_x}{\partial y} = -4 + 4 = 0$$

and flow is potential.

(b) Laplace's equation must be satisfied:

$$\frac{\partial^2 \Psi}{\partial x^2} + \frac{\partial^2 \Psi}{\partial y^2} = 0,$$

$$\Psi = 2x^2 + xy - 2y^2.$$

Therefore,

$$\frac{\partial \Psi}{\partial x} = 4x + y \quad \text{and} \quad \frac{\partial \Psi}{\partial y} = x - 4y,$$

$$\frac{\partial^2 \Psi}{\partial x^2} = 4 \quad \text{and} \quad \frac{\partial^2 \Psi}{\partial y^2} = -4.$$

Therefore $4 - 4 = 0$ and so flow is potential.

Now, to obtain the velocity potential,

$$\frac{\partial \Phi}{\partial x} = v_x = x - 4y,$$

therefore,

$$\Phi = \int (x - 4y)\, dx + f(y) + G.$$

But $\Phi_0 = 0$ at $x = 0$ and $y = 0$, so that $G = 0$. Therefore

$$\Phi = x^2/2 + 4yx + f(y).$$

Differentiating with respect to y and equating to v_y,

$$\frac{\partial \Phi}{\partial y} = 4x + \frac{d}{dy} f(y) = -(4x + y)$$

$$\frac{d}{dy} f(y) = -y \quad \text{and} \quad f(y) = -\frac{y^2}{2},$$

so that

$$\Phi = x^2/2 + 4yx - y^2/2.$$

7.6. Straight line flows and their combinations

The simplest flow patterns are those in which the streamlines are all straight lines parallel to each other (Fig. 7.12).

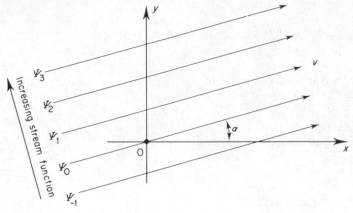

Figure 7.12   Rectilinear flow

The convention for numbering the streamlines is that the stream function is considered to increase to the left of an observer as he is looking down-stream, i.e. in the direction of flow along the streamlines, as indicated in Fig. 7.12.

If the velocity of the rectilinear flow v is inclined to the x axis at an angle α, then its components are:

$$v_x = v \cos \alpha \quad \text{and} \quad v_y = v \sin \alpha.$$

The stream function is obtained simply by substitution of the above expressions into

$$d\Psi = v_x \, dy - v_y \, dx,$$

whereupon

$$\Psi = \int v \cos \alpha \, dy - \int v \sin \alpha \, dx + \text{constant}.$$

Since in a uniform flow $v = $ constant and in a straight line flow α is also constant, the expression for the stream function becomes:

$$\Psi = vy \cos \alpha - vx \sin \alpha + \text{constant}.$$

The constant of integration may be made zero by choosing the reference streamline $\Psi_0 = 0$ to pass through the origin, so that when $x = 0$ and $y = 0$ the stream function $\Psi = \Psi_0 = 0$. Thus

$$\Psi = v(y \cos \alpha - x \sin \alpha). \tag{7.28}$$

Since v_x and v_y are constant, then $\partial v_x / \partial y$ and $\partial v_y / \partial x$ are both zero and, therefore, the flow is potential.

The velocity potential is obtained from

$$d\Phi = \frac{\partial \Phi}{\partial x} \, dx + \frac{\partial \Phi}{\partial y} \, dy = v_x \, dx + v_y \, dy.$$

Therefore, by substitution and integration,

$$\Phi = \int v \cos \alpha \, dx + \int v \sin \alpha \, dy + \text{constant},$$

but, if $\Phi = \Phi_0 = 0$ at $x = 0$ and $y = 0$, then

$$\Phi = v(x \cos \alpha + y \sin \alpha). \tag{7.29}$$

Some simple straight line flows may be illustrated as follows.

(i) Uniform, straight line flow in the direction Ox, velocity u, shown in Fig. 7.13.

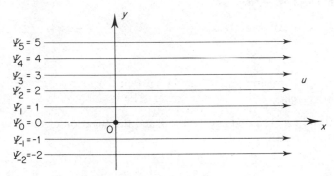

Figure 7.13   Straight line flow: $\Psi = uy$

Let streamline $\Psi_0 = 0$ be along the x axis. Now,

$$v_x = \frac{\partial \Psi}{\partial y} = u = \text{constant}.$$

Therefore,

$$\partial \Psi = u \partial y.$$

Integrating,

$$\Psi = uy + \text{constant},$$

but $\Psi_0 = 0$ at $x = 0$ and $y = 0$ so that constant $= 0$, and the equation of the stream function becomes

$$\Psi = uy.$$

Alternatively, the volume flowing between x-axis and any streamline, per unit depth, is $q = uy$ and, therefore,

$$\Psi = uy.$$

(ii) Uniform, straight line flow in the direction Oy, velocity v, shown in Fig. 7.14.

Let streamline $\Psi_0 = 0$ be along the y axis. Now,

$$v_y = -\frac{\partial \Psi}{\partial x} = v = \text{constant}.$$

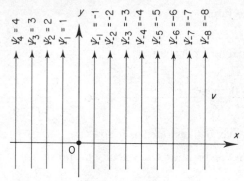

Figure 7.14   Straight line flow: $\Psi = vx$

Therefore,

$$\partial \Psi = -v \, dx.$$

Integrating,

$$\Psi = -vx + \text{constant},$$

but $\Psi_0 = 0$ at $x = 0$ and $y = 0$ so that constant $= 0$, and the equation of the stream function is

$$\Psi = -vx.$$

(iii) Combined flow consisting of a uniform flow $u = 10 \text{ m s}^{-1}$ along Ox and uniform flow $v = 20 \text{ m s}$ along Oy, shown in Fig. 7.15.

Choose a suitable scale for x and y, say 10 mm = 20 m. Draw horizontal streamlines $\Psi_0 = uy = 10y$ and label them. Draw vertical streamlines $\Psi_b = -vx = -20x$ and label them.

At point A the stream function due to v is $\Psi_b = 20$ and the stream function due to u is $\Psi_a = -20$. Therefore, the combined stream function (scalar quantity) is $\Psi = 20 - 20 = 0$. Similarly, it is zero at point B and the origin. Hence the stream function for the streamline passing through AOB is $\Psi = 0$.

By the same method, at point A$'$ the stream function due to v is $\psi_b = 0$ and the stream function for the streamline due to u is $\psi_a = -20$. Therefore, the combined stream function is $\Psi = 0 - 20 = -20$. Similarly, the combined stream function at B$'$ is also equal to -20. Thus, the straight line passing through A$'$B$'$ represents a streamline of the combined flow whose stream function is -20. By repeating the process of drawing lines through points at which the combined value of the stream function is the same, a new set of streamlines is obtained and it represents the combined flow pattern.

The same results may be obtained by the algebraic method. Since

$$\Psi = \Psi_a + \Psi_b,$$

it follows that

$$\Psi = uy - vx = 10y - 20x.$$

This equation represents a family of straight lines, each line being defined by the particular value of Ψ assigned to it.

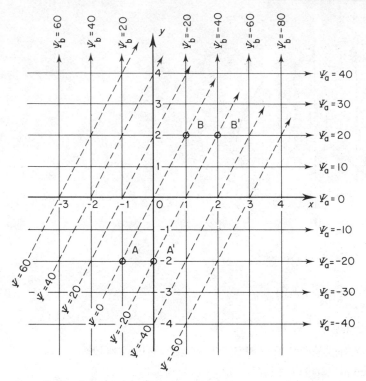

Figure 7.15   Combination of straight line flows

The other basic flow patterns in which the streamlines are straight lines are those in which the fluid flows radially either outwards from a point, in which case it is known as a *source*, or inwards into a point, in which case it is known as a *sink*. A sink flow is simply treated as a negative source flow and, thus, the mathematics of both may be explained by considering only the source flow, which is shown in Fig. 7.16.

Radial flows and their applications were already discussed in Section 6.14, but here we are concerned with the mathematical expressions for their stream function and velocity potential which lead to more complex and useful flow combinations.

In radial flows, it is seen that, since the velocity passes through the origin and is a function of θ only, the tangential component of velocity does not exist and

$$v = v_r.$$

Consider now a source of unit depth and let the steady rate of flow be q, known as the *strength* of the source. Then, at any radius r, the radial velocity is given by:

$$v_r = q/2\pi r. \tag{7.30}$$

The stream function and the velocity potential are obtained in a similar manner as for the rectilinear flow, but polar coordinates are used. Since,

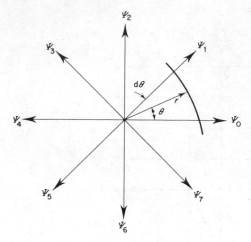

Figure 7.16   Radial flow: a source

from equation (7.12),

$$d\Psi = rv_r\, d\theta - v_\theta\, dr,$$

for radial flow $v_\theta = 0$, and for a source $v_r = q/2\pi r$, it follows that

$$d\Psi = r(q/2\pi r)\, d\theta = (q/2\pi)\, d\theta.$$

Integrating,

$$\Psi = q\theta/2\pi + \text{constant}.$$

If, however, $\Psi = \Psi_0 = 0$ when $\theta = 0$, the constant of integration becomes zero and

$$\Psi = q\theta/2\pi \text{ for a source,} \tag{7.31}$$

$$\Psi = -q\theta/2\pi \text{ for a sink.} \tag{7.32}$$

Similarly, it may be shown that

$$\Phi = (q/2\pi)\log_e r \text{ for a source,} \tag{7.33}$$

$$\Phi = -(q/2\pi)\log_e r \text{ for a sink.} \tag{7.34}$$

The simplest case of combining flow patterns is that in which a source is added to a uniform rectilinear flow. This is accomplished by the additions of the stream functions of the two types of flow. The stream function for a uniform rectilinear flow parallel to the x axis is

$$\Psi_R = v_0 y = v_0 r \sin\theta,$$

and that for a source is

$$\Psi_S = q\theta/2\pi.$$

Thus, the stream function for the combined flow is

$$\Psi = \Psi_R + \Psi_S = v_0 r \sin\theta + q\theta/2\pi. \tag{7.35}$$

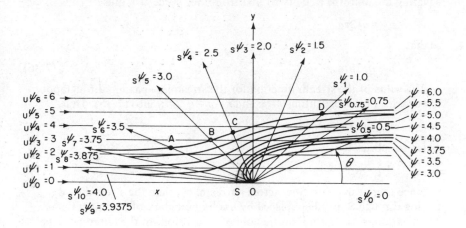

Figure 7.17   Combination of rectilinear flow and a source

Figure 7.17 shows graphically that this is the superposition of a system of radial streamlines onto a system of straight streamlines parallel to the x axis. By definition, a given streamline is associated with one particular value of the stream function and, therefore, if we join the points of intersection of the radial streamlines with the rectilinear streamlines where the sum of the stream functions is a given constant value, the resulting line will be one streamline of the combined flow pattern. If this procedure is repeated for a number of values of the combined stream function, the result will be a picture of the combined flow pattern. This is shown in Fig. 7.17, where numerical values were assigned to stream functions in order to illustrate the point. Observe, for example, the streamline of the combined flow $\Psi = 6$. It passes through points A, B, C and D such that:

at A, $\Psi_U = 2 \cdot 5$ and $\Psi_S = 3 \cdot 5$, therefore $\Psi = 2 \cdot 5 + 3 \cdot 5 = 6$;

at B, $\Psi_U = 3 \cdot 0$ and $\Psi_S = 3 \cdot 0$, therefore $\Psi = 3 \cdot 0 + 3 \cdot 0 = 6$;

at C, $\Psi_U = 4 \cdot 0$,   and $\Psi_S = 2 \cdot 0$, therefore $\Psi = 2 \cdot 0 + 4 \cdot 0 = 6$;

at D, $\Psi_U = 5 \cdot 0$ and $\Psi_S = 1$ therefore $\Psi = 5 \cdot 0 + 1 \cdot 0 = 6$.

All streamlines of the combined flow are obtained in this manner.

It is interesting to note, in this particular flow pattern, that the resulting streamlines are grouped into two distinct sets. In one set all the streamlines emerge from the origin ($\Psi = 3, 3 \cdot 5, 3 \cdot 75$) and in the other they approach the rectilinear flow asymptotically at some distance upstream ($\Psi = 4 \cdot 5, 5, 6$). The two sets are separated by the streamline $\Psi = 4$, which passes through point S. This point is a *stagnation point*, where the velocity from the source is equal to the uniform velocity of the parallel flow, so that the resultant velocity at S is zero. The distance OS = a may, therefore, be determined by equating the uniform velocity to that from the source at radius a. Thus,

$$v_0 = q/2\pi a,$$

so that

$$a = q/2\pi v_0. \tag{7.36}$$

The value of the stream function for the streamline passing through point S is obtained by substituting $\theta = \pi$ and $r = a = q/2\pi v_0$ into (7.35). Thus,

$$\Psi_S = v_0(q/2\pi v_0) \sin \pi + q\pi/2\pi,$$

which simplifies to

$$\Psi_S = \tfrac{1}{2}q. \tag{7.37}$$

Since there can be no flow across a streamline, then the streamline ψ_S passing through S may be replaced by a solid boundary of an object under investigation, such as a hill or the nose of an aerofoil. In the latter case, the

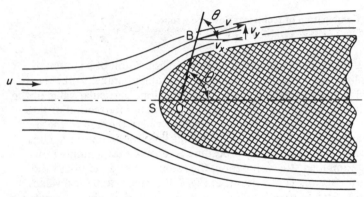

Figure 7.18   Rankine body

flow pattern below the x axis must also be used as shown in Fig. 7.18. It is then known as a *half-body* or *Rankine body*.

The general equation for the streamline through point S is

$$\Psi = \tfrac{1}{2}q = v_0 r \sin \theta + q\theta/2\pi, \tag{7.38}$$

from which the radial distance r_S to any point on this streamline is

$$r_S = q(\pi - \theta)/2\pi v_0 \sin \theta, \tag{7.39}$$

and it describes the contour of the Rankine body. It can be appreciated that as $x \to \infty$ this streamline becomes parallel to the x axis and, there, the perpendicular distance from the x axis to the streamline represents the maximum half-width of the Rankine body. The perpendicular distance is given by

$$y = r \sin \theta = q(\pi - \theta)/2\pi v_0.$$

But, as $x \to \infty$, the radius $r \to \infty$ and $\theta \to 0$, so that

$$y_{max} = q/2v_0. \tag{7.40}$$

Example 7.2

In the ideal flow around a half-body, the free stream velocity is 0·5 m s^{-1} and the strength of the source is 2·0 m^2 s^{-1}. Determine the fluid velocity and its direction at a point, $r \doteq 1\cdot0$ m and $\theta = 120°$.

Solution

The stream function for the flow around a half-body is given by

$$\Psi = v_0 r \sin \theta + q\theta/2\pi.$$

In this case, $v_0 = 0\cdot5$ m s^{-1}, $q = 2\cdot0$ m^2 s^{-1}. To determine the fluid velocity and its velocity vector at a point, it is first necessary to determine its tangential and radial components. These are

$$v_r = \frac{1}{r}\frac{\partial\Psi}{\partial\theta} \quad \text{and} \quad v_0 = -\frac{\partial\Psi}{\partial r}.$$

Therefore,

$$v_r = \frac{1}{r}\left(v_0 r \cos\theta + \frac{q}{2\pi}\right) = \frac{1}{1}\left(0\cdot5 \times 1 \times \cos 120° + \frac{2}{2\pi}\right)$$

$$= -0\cdot25 + 0\cdot318 = 0\cdot068 \text{ m s}^{-1},$$

$$v_\theta = -v_0 \sin\theta = -0\cdot5 \sin 120° = -0\cdot433 \text{ m s}^{-1},$$

which is in the clockwise direction. Therefore,

$$v = \sqrt{(v_r^2 + v_\theta^2)} = \sqrt{(0\cdot0047 + 0\cdot188)} = 0\cdot438 \text{ m s}^{-1}.$$

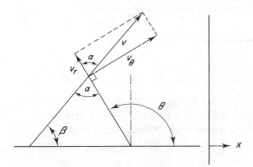

Figure 7.19

If β is the angle the velocity vector makes with the horizontal, as shown in Fig. 7.19, then

$$\beta = \theta - \alpha$$

and $\tan\alpha = v_\theta/v_r = 0\cdot438/0\cdot068 = 6\cdot44.$

Therefore

$$\alpha = 81\cdot2°, \text{ and } \beta = 120 - 81\cdot2 = 38\cdot8°.$$

7.7. Combined source and sink flows; doublet

Let us consider the flow pattern resulting from the combination of a source and a sink of equal strength, which means that the flow rate from the source is equal to the flow rate into the sink. Also, let them be placed symmetrically about the origin and on the x axis, as shown in Fig. 7.20.

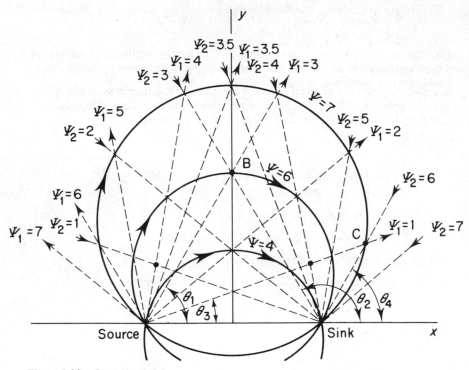

Figure 7.20   Source and sink

Let the stream function for the source be Ψ_1, for the sink be Ψ_2 and let the flow rate be q. Since the convention for stream functions is that they increase to the left while looking downstream, it follows that the stream functions for the source increase as the angle θ_1 increases and those for the sink decrease as angle θ_2 increases.

As discussed earlier, the value of a combined stream function is obtained by addition of the values of stream functions at their intersection. For example, if the combined stream function is $\Psi = 5$, it will pass through points of intersection of Ψ_1 and Ψ_2 such that their values add up to 5 (e.g. $\Psi_1 = 1$ and $\Psi_2 = 4$ or $\Psi_1 = 2$ and $\Psi_2 = 3$ and so on, as shown in Fig. 7.20). Figure 7.20 also shows that the combined streamlines of a source and a sink of equal strength are circles passing through the point source and the point sink. Mathematically,

$$\Psi = \Psi_1 + \Psi_2 = q\theta_1/2\pi - q\theta_2/2\pi,$$
$$= (q/2\pi)(\theta_1 - \theta_2), \tag{7.41}$$

which, since Ψ and q are constant for any given streamline, is a condition satisfied by a circle.

Figure 7.20 shows only half of the flow pattern, the other half, below the x axis, being the mirror image of that above it.

The velocity potential for such a combined flow is also obtained by addition of the velocity potentials for the source and the sink. Thus

$$\Phi = \Phi_{source} + \Phi_{sink} = (q/2\pi) \log_e r_1 - (q/2\pi) \log_e r_2 ,$$

$$\Phi = (q/2\pi)(\log_e r_1 - \log_e r_2). \tag{7.42}$$

7.7.1. Doublet

If a sink and a source of equal strength are brought together in such a way that the product of their strength and the distance between them remains constant, the resulting flow pattern is known as a *doublet*.

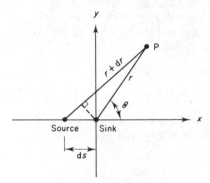

Figure 7.21   Source and sink

Consider point P (Fig. 7.21) on the velocity potential of a doublet. Let the velocity potential for the doublet be Φ_D, then,

$$\Phi_D = (q/2\pi) \log_e (r + dr) - (q/2\pi) \log_e r$$

$$= \frac{q}{2\pi} \log_e \frac{r + dr}{r} = \frac{q}{2\pi} \log_e \left(1 + \frac{dr}{r} \right).$$

Expanding,

$$\log_e \left(1 + \frac{dr}{r} \right) = \frac{dr}{r} - \tfrac{1}{2} \left(\frac{dr}{r} \right)^2 + \ldots.$$

Neglecting terms of second order and higher,

$$\Phi_D = \frac{q}{2\pi} \frac{dr}{r},$$

but, since by definition of a doublet $ds \to 0$, it follows that

$$dr \simeq ds \cos \theta$$

and $\Phi_D = (q/2\pi r) ds \cos \theta.$

Also for a doublet, by definition q ds = constant. Let this constant, known as the *strength of the doublet*, be denoted by m, then,

$$m = q \, ds$$

and $\Phi_D = (m/2\pi r) \cos \theta$ (7.43)

or, in rectangular coordinates,

$$\Phi_D = \frac{m}{2\pi}\left(\frac{x}{x^2 + y^2}\right)$$ (7.44)

From the above equations, the expressions for the stream function may be obtained, namely,

$$\Psi_D = -(m/2\pi r) \sin \theta = -(m/2\pi)\{y/(x^2 + y^2)\}.$$ (7.45)

Note that the above equations were derived for a doublet in which the source and the sink were placed on the x axis. Such a doublet is shown in Fig. 7.22(a). If, however, the source and the sink are placed on the y axis, the resulting doublet is oriented as in Fig. 7.22(b) and the expressions for the stream function and the velocity potential become

$$\Phi_{D(yy)} = (m/2\pi r) \sin \theta = (m/2\pi)\{y/(x^2 + y^2)\},$$ (7.46)

$$\Psi_{D(yy)} = -(m/2\pi r) \cos \theta = -m/2\pi\{x/(x^2 + y^2)\}.$$ (7.47)

The flow is always from the source to the sink, so that, if they are placed as shown in Fig. 7.21, the flow in the doublet is as shown in Fig. 7.22(a). If,

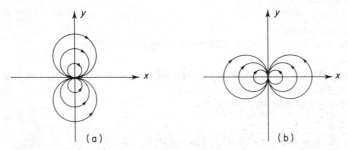

(a)   (b)

Figure 7.22   Doublets

however, the positions of the source and the sink are reversed, the flow directions are also reversed, which means that the expressions for the stream function and the velocity potential change signs.

Example 7.3

A source of strength 10 m^2 s^{-1} at $(1,0)$ and a sink of the same strength at $(-1,0)$ are combined with a uniform flow of 25 m s^{-1} in the $-x$ direction. Determine the size of Rankine body formed by the flow and the difference in pressure between a point far upstream in the uniform flow and the point $(1,1)$.

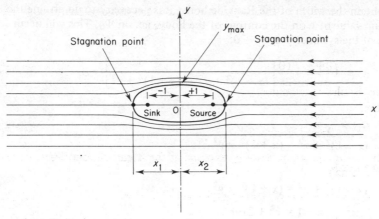

Figure 7.23

Solution
For the source (Fig. 7.23),

$$\Psi_{source} = \frac{q}{2\pi}\theta = \frac{10}{2\pi}\tan^{-1}\frac{y}{x-1};$$

for the sink,

$$\Psi_{sink} = -\frac{q}{2\pi}\theta = -\frac{10}{2\pi}\tan^{-1}\frac{y}{x+1};$$

for the uniform flow,

$$\psi_u = -25y.$$

Thus, the combined flow is represented by the stream function

$$\Psi = \frac{10}{2\pi}\left(\tan^{-1}\frac{y}{x-1} - \tan^{-1}\frac{y}{x+1}\right) - 25y.$$

To obtain stagnation points, $v_x = 0$. Thus,

$$v_x = \frac{\partial\Psi}{\partial y} = \frac{10}{2\pi}\left[\frac{x-1}{(x-1)^2+y^2} - \frac{x+1}{(x+1)^2+y^2}\right] - 25.$$

Now, $v_x = 0$ at $y = 0$, i.e. on the x axis, therefore,

$$\frac{10}{2\pi}\left[\frac{1}{x-1} - \frac{1}{x+1}\right] = 25,$$

$$(x+1)-(x-1) = 5\pi(x-1)(x+1),$$

$$2/5\pi = x^2 - 1,$$

$$x^2 = 2/5\pi + 1 = 1\cdot127,$$

$$x_{12} = \pm 1\cdot062 \text{ m},$$

and the length of the Rankine body is

$$l = x_1 + x_2 = 2\cdot124 \text{ m}.$$

To obtain the width of the Rankine body, it is necessary to determine the maximum value of y on the contour of the body, i.e. on Ψ_0. This will occur because of the symmetry at $v_y = 0$:

$$v_y = -\frac{\partial \Psi}{\partial x} = -\frac{10}{2\pi}\left[\frac{y}{(x-1)^2 + y^2} - \frac{y}{(x+1)^2 + y^2}\right] = 0,$$

Therefore,

$$\frac{y}{(x-1)^2 + y^2} = \frac{y}{(x+1)^2 + y^2},$$

but, since $y \neq 0$,

$$(x-1)^2 + y^2 = (x+1)^2 + y^2,$$

$$x^2 - 2x + 1 = x^2 + 2x + 1.$$

$$-4x = 0,$$

$$x = 0,$$

which is expected from the symmetry of the source and sink about the origin. To find the value of y_{max}, which will give the width of the body, substitute the above value of $x = 0$ into $\Psi = 0$:

$$0 = \frac{10}{2\pi}[\tan^{-1}(-y) - \tan^{-1} y] - 25y_{max},$$

but, since $|y| = |-y|$,

$$\frac{25 \times 2\pi}{10} y_{max} = 2\tan^{-1}y,$$

$$y_{max} = 0 \cdot 127 \tan^{-1}y$$

from which $y_{max} = 0 \cdot 047$ m and the width of the Rankine body is $2y_{max} = 0 \cdot 094$ m.

At point $(1,1)$,

$$v_x = \frac{10}{2\pi}(-\tfrac{2}{5}) - 25 = -25 \cdot 63 \text{ m s}^{-1},$$

$$v_y = -\frac{10}{2\pi}(1 - \tfrac{1}{5}) = -1 \cdot 27 \text{ m s}^{-1}.$$

Therefore,

$$v = (v_x^2 + v_y^2)^{\frac{1}{2}} = 25 \cdot 66 \text{ m s}^{-1}.$$

Since the flow is potential, Bernoulli's equation may be applied to any two points, such as one in the free stream ($v_\infty = 25$ m s^{-1}) and point $(1,1)$:

$$p_\infty/\rho g + v_\infty^2/2g = p_{(1,1)}/\rho g + v^2/2g,$$

and, hence,

$$p_{(1,1)} - p_\infty = (\rho/2)(v^2 - v_\infty^2) = (\rho/2) \, 33 \cdot 43 \text{ N m}^{-2}.$$

7.8. Flow past a cylinder

A flow pattern equivalent to that of an ideal fluid passing a stationary cylinder, with its axis perpendicular to the direction of flow, is obtained by combining a doublet with rectilinear flow. Figure 7.24 shows the resulting

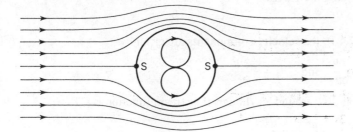

Figure 7.24   Flow around a cylinder

streamlines and the stagnation points S which are formed. The combined stream function and the velocity potential are

$$\Psi_c = \Psi_D + \Psi_R = -(m/2\pi r) \sin \theta + v_0 r \sin \theta,$$

$$\Psi_c = (v_0 r - m/2\pi r) \sin \theta \tag{7.48}$$

and $\quad \Phi_c = (v_0 r + m/2\pi r) \cos \theta. \tag{7.49}$

Since the flow pattern corresponds to that around a cylinder, it is of interest to obtain an expression for the radius of this cylinder. Since the distance between the two stagnation points is the diameter of this cylinder, say $2a$, and the flow at a stagnation point is zero, it follows that the streamline passing through S is $\Psi_0 = 0$. Thus

$$\Psi_0 = (v_0 a - m/2\pi a) \sin \theta = 0,$$

so that $\quad v_0 a = m/2\pi a$

and $\quad a = \sqrt{(m/2\pi v_0)}. \tag{7.50}$

For a given velocity of the uniform flow and a given strength of the doublet, the radius a is constant, which proves that the body so derived is circular. It is also possible to plot the flow pattern around a cylinder of radius a with uniform velocity v_0. From equation (7.50), the strength of the doublet is

$$m = 2\pi v_0 a^2$$

and the combined stream function becomes

$$\Psi_c = (v_0 r - 2\pi v_0 a^2 / 2\pi r) \sin \theta$$

$$= v_0 (r - a^2/r) \sin \theta. \tag{7.51}$$

Similarly, the velocity potential,

$$\Phi_c = v_0 (r + a^2/r) \cos \theta. \tag{7.52}$$

Example 7.4

If a 40 mm diameter cylinder is immersed in a stream having velocity of $1 \cdot 0$ m s^{-1}, determine the radial and normal components of velocity at a point on a streamline where $r = 50$ mm and $\theta = 135°$, measured from the positive x axis. Assume flow to be ideal. Also determine the pressure distribution with radial distance along the y axis.

Solution
By equations (7.13), the velocity components are given by

$$v_r = \frac{1}{r}\frac{\partial \Psi}{\partial \theta} \quad \text{and} \quad v_\theta = -\frac{\partial \Psi}{\partial r},$$

but, for the ideal flow around a cylinder, the stream function is

$$\Psi_c = v_0\left(r - \frac{a^2}{r}\right)\sin\theta.$$

Therefore,

$$v_r = \frac{1}{r}\frac{\partial}{\partial\theta}\left[v_0\left(r - \frac{a^2}{r}\right)\sin\theta\right] = \frac{v_0}{r}\left(r - \frac{a^2}{r}\right)\cos\theta$$

$$= \left(1 - \frac{a^2}{r^2}\right)v_0\cos\theta$$

and

$$v_\theta = -\frac{1}{\partial r}\left[v_0\left(r - \frac{a^2}{r}\right)\sin\theta\right] = -v_0\left(1 + \frac{a^2}{r^2}\right)\sin\theta.$$

Substituting the numerical values $a = 2$ cm, $r = 5$ cm, $v_0 = 1 \cdot 0$ m s^{-1}, $\theta = 135°$, the following values for velocity components are obtained:

$$v_r = \left(1 - \frac{4}{25}\right)\cos 135° = -\frac{21}{25}\frac{1}{\sqrt{2}} = -0 \cdot 592 \text{ m s}^{-1},$$

$$v_\theta = -\left(1 + \frac{4}{25}\right)\sin 135° = -\frac{29 \cdot}{25}\frac{1}{\sqrt{2}} = -0 \cdot 783 \text{ m s}^{-1}.$$

Remembering the sign convention for cylindrical coordinates, these components are as shown in Fig. 7.25.

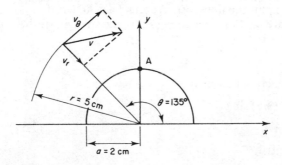

Figure 7.25

To obtain the pressure distribution along Ay it is first necessary to determine the velocity variation so that it may be used in applying the Bernoulli equation. Since for Ay, $\theta = 90°$, it follows that $v_r = 0$, and, hence,

$$v_\theta = -(1 + \theta^2/r^2)v_0,$$

which is the required velocity distribution for Ay. Now, applying the Bernoulli equation to a point far upstream where the velocity is v_0, the pressure is p_0, and to the section in the equation

$$p_0/\rho + v_0^2/2 = p/\rho + v_\theta^2/2,$$

so that

$$
\begin{aligned}
p - p_0 &= (\rho/2)(v_0^2 - v_\theta^2) = (\rho/2)\,[v_0^2 - (1 + a^2/r^2)^2 v_0^2] \\
&= (\rho/2)\,v_0^2[1 - (1 + 2\,a^2/r^2 + a^4/r^4)], \\
&= -(\rho v_0^2/2)(2a^2/r^2 + a^4/r^4).
\end{aligned}
$$

This equation shows that when $r \to \infty$, the pressure p approaches p_0, but at the surface of the cylinder (at point A), where $r = a$, the pressure is lower than that upstream by an amount equal to $\frac{3}{2}\rho v_0^2$.

7.9. Curved flows and their combinations

The previous sections of this chapter dealt with flows whose basic components were straight line flows, either rectilinear or radial. The third basic type of flow is such that the streamlines are concentric circles, as shown in Fig. 7.26. Such flows are known as vortex flows. Their characteristic is

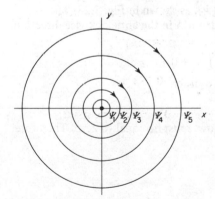

Figure 7.26   Vortex flow

that the radial component of velocity $v_r = 0$. This is so because, of course, there cannot be any flow across streamlines and, since in vortex flow they are circular, the flow must be confined to purely circular paths. Thus in any vortex flow

$$v_r = 0 \quad \text{and} \quad v = v_\theta. \tag{7.53}$$

There are two fundamental types of vortex flow distinguished by the nature of flow, namely rotational and irrotational. From these two basic types, various combinations of flows are possible.

Vortex flows and their applications were already discussed in Section 6.16. They were not, however, defined there with respect to rotational or irrotational flow, neither were the mathematics of their stream function and velocity potential where appropriate discussed.

Let us first consider the irrotational vortex flow which is known as the *free vortex*. Because it is irrotational, the vorticity and circulation across the stream must be equal to zero. Consider, in a free vortex flow, an element of

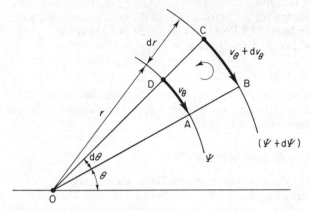

Figure 7.27   An element of vortex flow

fluid between streamlines Ψ and $(\Psi + d\Psi)$, as shown in Fig. 7.27. The circulation round the element, starting from A in the anticlockwise direction, is

$$\Gamma_{ABCD} = 0 - (v_\theta + dv_\theta)(r + dr)\, d\theta + 0 + v_\theta r\, d\theta,$$

and, neglecting infinitesimals of the third order,

$$\Gamma_{ABCD} = -(v_\theta\, dr + r\, dv_\theta)\, d\theta.$$

This, by the definition of irrotational flow, must be equal to zero. Therefore,

$$-(v_\theta\, dr + r\, dv_\theta)\, d\theta = 0,$$

so that

$$v_\theta\, dr + r\, dv_\theta = 0,$$

but this is a differential of a product,

$$d(rv_\theta) = 0,$$

which, when integrated, gives

$$rv_\theta = \text{constant.} \tag{7.54}$$

This equation defines the relationship between the velocity and radius for a free vortex. It shows that the velocity increases towards the centre of the vortex and tends to infinity when the radius tends to zero. The velocity decreases as the radius increases and tends to zero as the radius tends to infinity. One practical example of this type of vortex flow is the emptying of a container through a central hole.

The constant in equation (7.54) may be established by making use of the singularity which exists in the free vortex flow, namely the infinite velocity at the centre of the vortex which we mentioned above. At this point, the vorticity which is given (equation (7.24)) by

$$- \left(\frac{\partial v_\theta}{\partial r} + \frac{v_\theta}{r} \right)$$

becomes indeterminate on substitution of $r \to 0$ and $v_\theta \to \infty$. It can, however, be determined by evaluating the circulation around the centre, i.e. along any of the concentric streamlines. This does not violate the condition for irrotational flow, by which the free vortex is defined, because the condition states that vorticity (and circulation) must be zero for any closed loop across the flow (see Section 7.1). The circulation around a circular streamline:

$$\Gamma_C = \oint v \, ds = \text{Circumference} \times \text{Tangential velocity} = 2\pi r v_\theta$$

but, since $v_\theta r = \text{constant}$, it follows that this particular circulation is constant for any streamline and, therefore, for the whole vortex field. It may, therefore, be used to measure the intensity of the vortex and is known as the *vortex strength*. Thus, vortex strength

$\Gamma_C = 2\pi r v_\theta$ for the anticlockwise vortex.

The free vortex equation (7.54) may now be rewritten as

$$v_\theta r = \Gamma_C / 2\pi. \tag{7.55}$$

The stream function may be obtained from equation (7.12):

$$d\Psi = v_r r \, d\theta - v_\theta \, dr.$$

Since $v_r = 0$,

$$\Psi = - \int v_\theta \, dr,$$

and, substituting for v_θ from equation (7.55),

$$\Psi = - \int \frac{\Gamma_C}{2\pi r} \, dr = - \frac{\Gamma_C}{2\pi} \log_e r + \text{constant}.$$

The constant of integration is made zero by taking $\Psi = 0$ at $r = 1$, so that, finally,

$$\Psi = -(\Gamma_C / 2\pi) \log_e r \tag{7.56}$$

for anticlockwise rotation. The sign of the above expression becomes positive for clockwise rotation.

The velocity potential follows (equation (7.21)) from

$$d\Phi = v_r \, dr + r v_\theta \, d\theta,$$

whereas, upon substitution and making $\Phi_0 = 0$ at $\theta = 0$,

$$\Phi = \int \frac{\Gamma_C}{2\pi} d\theta = \frac{\Gamma_C}{2\pi} \theta. \tag{7.57}$$

Since the free vortex is irrotational, Bernoulli's constant remains the same for all streamlines.

Example 7.5

A two-dimensional fluid motion takes the form of concentric horizontal circular streamlines. Show that the radial pressure gradient is given by

$$\frac{dp}{dr} = \rho \frac{v^2}{r},$$

where ρ = density, v = tangential velocity, r = radius. Hence, evaluate the pressure gradient for such a flow defined by $\Psi = 2 \log_e r$, where Ψ = stream function, at a radius of 2 m and fluid density of 10^3 kg m^{-3}.

Solution

For two concentric streamlines the variation of total head or Bernoulli's constant is, in general, given by:

$$\frac{dH}{dr} = \frac{v_\theta}{g} \left(\frac{dv_\theta}{dr} + \frac{v_\theta}{r} \right),$$

but, for horizontal flow, $z = 0$ and, for vortex flow, $v = v_\theta$, so that

$$H = p/\rho g + v_\theta^2/2g,$$

therefore, differentiating,

$$\frac{dH}{dr} = \frac{1}{\rho g} \frac{dp}{dr} + \frac{v_\theta}{g} \frac{dv_\theta}{dr}.$$

Equating the two equations,

$$\frac{v_\theta}{g} \left(\frac{dv_\theta}{dr} + \frac{v_\theta}{r} \right) = \frac{1}{\rho g} \frac{dp}{dr} + \frac{v_\theta}{g} \frac{dv_\theta}{dr}.$$

from which

$$\frac{1}{\rho} \frac{dp}{dr} = v_\theta \frac{dv_\theta}{dr} + \frac{v_\theta^2}{r} - v_\theta \frac{dv_\theta}{dr} = \frac{v_\theta^2}{r}.$$

Therefore,

$$\frac{dp}{dr} = \rho \frac{v_\theta^2}{r}.$$

The stream function $\Psi = 2 \log_e r$ represents a free vortex, for which

$$\Psi = (\Gamma_C/2\pi) \log_e r,$$

and, hence,

$$\Gamma_C/2\pi = 2,$$

but, for a free vortex,

$$\Gamma_C/2\pi = v_\theta r,$$

so that

$$v_\theta r = 2 \quad \text{and} \quad v_\theta^2 = \frac{4}{r^2}$$

and, therefore,

$$\frac{dp}{dr} = \rho \frac{4}{r^3} = 10^3 \times \frac{4}{2^3} = 500 \text{ N m}^{-3}.$$

The most common example of a rotational vortex, which is considered of fundamental importance, is a *forced vortex*. It will be shown in what follows that in a forced vortex the fluid rotates as a solid body with a constant rotational velocity.

Consider circulation around a segmental element (such as in Fig. 7.27) of a forced vortex, remembering that $v_r = 0$:

$$\Gamma_{ABCD} = -(v_\theta \, dr + r \, dv_\theta) \, d\theta.$$

The area of the element is

$$A = r \, d\theta \, dr,$$

so that vorticity (as already shown in Section 7.2, equation (7.5)) is given by

$$\zeta = \frac{\Gamma_{ABCD}}{dA} = -\left(\frac{v_\theta}{r} + \frac{dv_\theta}{dr}\right),$$

but, if, for a solid body, rotation ω is the angular velocity which at any radius r is related to the tangential velocity v_θ by

$$\omega = v_\theta/r, \tag{7.58}$$

it follows, therefore, that for a forced vortex the vorticity,

$$\zeta = -2\omega \tag{7.59}$$

and is constant for a given vortex.

The flow is rotational and there is, therefore, variation of Bernoulli's constant with radius. Using equation (6.38),

$$\frac{dH}{dr} = \frac{v_\theta}{g}\left(\frac{v_\theta}{r} + \frac{dv_\theta}{dr}\right) = v_\theta \times \frac{2\omega}{g}$$

and, since $v_\theta = \omega r$,

$$\frac{dH}{dr} = \frac{2\omega^2 r}{g}. \tag{7.60}$$

In order to determine the pressure distribution or the surface gradient in a forced vortex, the above expression must be used in conjunction with Bernoulli's equation, as was shown in Section 6.16.

The stream function for a forced vortex is obtained in the same manner as for the free vortex, but using the appropriate relationship (equation (7.58)), namely that

$$v_\theta = \omega r,$$

which yields

$$\Psi = -\int \omega r \, dr = -\tfrac{1}{2}\omega r^2 + \text{constant}.$$

But, for $\Psi = 0$ at $r = 0$,

$$\Psi = -\tfrac{1}{2}\omega r^2 \tag{7.61}$$

for anticlockwise rotation.

Since the forced vortex is rotational, there is no velocity potential corresponding to it and the Laplace equations are not satisfied.

A free spiral vortex, mentioned in Section 6.16, is, in contrast to a forced vortex, irrotational and represents a potential flow. The free spiral vortex is the combination of a free vortex and radial flow. It is, therefore, obtained by superposition of the stream functions of a free vortex with either a sink or a source flow depending upon the direction of the radial flow.

For outward flow using a source and for a clockwise vortex,

$$\Psi_{sv} = \Psi_{\text{source}} + \Psi_{\text{free vortex}}$$

$$= q\theta/2\pi + (\Gamma_C/2\pi)\log_e r,$$

$$= (1/2\pi)(q\theta + \Gamma_C \log_e r) \tag{7.62}$$

and

$$\Phi_{sv} = \Phi_{\text{source}} + \Phi_{\text{free vortex}}$$

$$= (q/2\pi)\log_e r + (\Gamma_C/2\pi)\,\theta$$

$$= (1/2\pi)(q\log_e r + \Gamma_C\theta). \tag{7.63}$$

The resulting flow is shown in Fig. 7.28.

7.10. Flow past a cylinder with circulation; Kutta–Joukowski's law

Flow past a stationary cylinder may be obtained by superposition of a parallel flow and a doublet. This was discussed in Section 7.8. However, in the first half of the 19th century the German physicist H. G. Magnus observed experimentally that if the cylinder in a parallel flow stream is rotated about its axis a transverse force, which tends to move the cylinder across the parallel flow stream, is generated. This is known as *Magnus effect* or aerodynamic lift.

The hydrodynamic equivalent of rotating a cylinder in a flow stream is to add circulation by means of a free vortex to the doublet in a parallel flow. Such a flow pattern can be obtained directly from the results of Sections

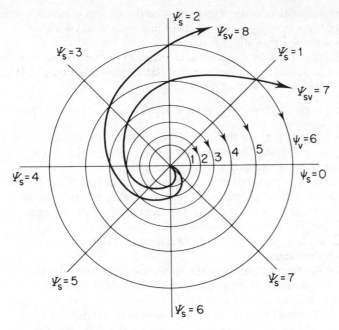

Figure 7.28   Free spiral vortex

7.8 and 7.9 by adding the stream function of a free vortex (equation (7.56)) for clockwise rotation to the stream function of flow past a cylinder, given by equation (7.51). Thus, the combined stream function is

$$\Psi = v_0 (r - a^2/r) \sin \theta + (\Gamma_C/2\pi) \log_e r. \qquad (7.64)$$

The addition of circulation to the ideal flow past a cylinder gives rise to an asymmetric flow pattern, as shown in Fig. 7.29. There is an increase of velocity on one side of the cylinder and a decrease on the other. In consequence of this, the stagnation points move from the axis of the parallel flow. Their positions depend upon the magnitude of the circulation and can be determined using equation (7.64), remembering that at stagnation points

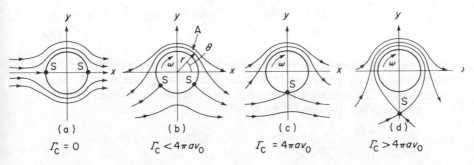

(a)   (b)   (c)   (d)

$\Gamma_C = 0$ $\Gamma_C < 4\pi a v_0$ $\Gamma_C = 4\pi a v_0$ $\Gamma_C > 4\pi a v_0$

Figure 7.29

$v_\theta = 0$. Thus, the tangential velocity is given by

$$v_\theta = -\frac{\partial \Psi}{\partial r} = -v_0 \left(1 + \frac{a^2}{r^2}\right) \sin \theta - \frac{\Gamma_C}{2\pi r} = 0.$$

Also, on the contour of the cylinder, $r = a$, therefore,

$$-2v_0 \sin \theta - \Gamma_C/2\pi a = 0 \quad \text{or} \quad \sin \theta = -\Gamma_C/4\pi a v_0. \tag{7.65}$$

The negative sign indicates that for the positive parallel flow (from left to right) and clockwise circulation, the stagnation points lie below the x axis. Furthermore, if the value of circulation $\Gamma_C < 4\pi a v_0$, then $\sin \theta < -1$ and the stagnation points will lie in positions such as those shown in Fig. 7.26(b). For $\Gamma_C = 4\pi a v_0$, the stagnation points merge on the negative y axis, as shown in Fig. 7.26(c), and, finally, for $\Gamma_C > 4\pi a v_0$, the stagnation points will be as shown in Fig. 7.26(d). Since $\sin \theta$ cannot be greater than one, this is only possible if a is increased, which means that the stagnation point moves away from the surface of the cylinder.

In order to establish the magnitude of the transverse force acting on the cylinder and mentioned earlier, it is necessary to obtain first the pressure distribution around the cylinder and then the forces arising from it. Since the flow is irrotational, Bernoulli's equation may be applied to a point some distance upstream in the parallel flow and to a point on the surface of the cylinder. Thus,

$$p/\rho + v_\theta^2/2 = p_0/\rho + v_0^2/2,$$

where p is the pressure on the cylinder and it varies with θ, p_0 is the pressure in the parallel flow some distance upstream, where the velocity is v_0. Rearranging and solving for pressure difference:

$$p - p_0 = (\rho/2)(v_0^2 - v_\theta^2) = (\rho v_0^2/2)(1 - v_\theta^2/v_0^2),$$

but $v_\theta = -2v_0 \sin \theta - \Gamma_C/2\pi a,$

so that

$$p - p_0 = (\rho v_0^2/2)[1 - (-2 \sin \theta - \Gamma_C/2\pi a v_0)^2]$$

$$= (\rho v_0^2/2)(1 - 4 \sin^2 \theta - 2\Gamma_C \sin \theta/\pi a v_0 - \Gamma_C^2/4\pi^2 a^2 v_0^2). \tag{7.66}$$

Consider, now, an element of the cylinder's surface. The force due to pressure acting on it is $(p - p_0)a \, d\theta$, and it may be resolved into vertical and horizontal components. The transverse force will, in our case, be the sum of the vertical components. Thus, the upward force

$$L = -\int_0^{2\pi} (p - p_0)a \sin \theta \, d\theta.$$

In equation (7.66), let

$$1 - \Gamma_C^2/4\pi^2 a^2 v_0^2 = A,$$

then, $p - p_0 = (\rho v_0^2/2)(A - 4\sin^2\theta - 2\Gamma_C\sin\theta/\pi a v_0)$

and $$L = -\int_0^{2\pi} (\rho v_0^2/2)(A - 4\sin^2\theta - 2\Gamma_C\sin\theta/\pi a v_0)a\sin\theta\,d\theta$$

$$= -\int_0^{2\pi} (\rho v_0^2 a/2)(A\sin\theta - 4\sin^3\theta - 2\Gamma_C\sin^2\theta/\pi a v_0)\,d\theta.$$

But, $$\int_0^{2\pi}\sin\theta\,d\theta = 0 \quad \text{and} \quad \int_0^{2\pi}\sin^3\theta\,d\theta = 0,$$

so that,

$$L = \frac{\rho a v_0^2}{2}\int_0^{2\pi}\frac{2\Gamma_C\sin^2\theta}{\pi a v_0}\,d\theta = \frac{\rho v_0\Gamma_C}{\pi}\left[\frac{1}{2}\theta - \frac{\sin 2\theta}{4}\right]_0^{2\pi}$$

$$L = \rho v_0\Gamma_C. \tag{7.67}$$

Thus, the force perpendicular to the direction of the parallel flow, or a main free stream, which in general is known as a *lift*, is, for a rotating cylinder of infinite length, independent of the diameter of the cylinder and equal to the product of fluid density, free stream velocity and circulation. This statement is known as *Kutta–Joukowski's law* and gives theoretical justification for the experimentally observed Magnus effect.

It is interesting to note that the horizontal component of the force on the cylinder due to pressure, which in general is called the *drag*, and for our case is given by

$$D = \int_0^{2\pi} (p - p_0)a\cos\theta\,d\theta,$$

is equal to zero. This result, obtained on the assumption of the ideal flow, is not supported by experiments. This is so because in real fluids viscous friction provides resistance to flow and separation may occur.

7.11. Computer Program "ROTCYL"

(1) The program calculates and displays the angular positions of stagnation points on a cylinder rotating in a uniform fluid stream, the value of the lift coefficient and the values of pressure coefficient along the x axis upstream of the cylinder. If there is only one stagnation point on the y axis the program calculates its distance from the cylinder surface.

(2) The required input is: cylinder diameter, D (mm); its rotational speed, N (rev min^{-1}); and the free-stream fluid velocity, U (m s)$^{-1}$.

(3) Calculations are based on potential flow theory. Use is made of equations (7.65), (7.66) and (7.67) together with the definitions of the coefficient of lift, equation (10.5), and the pressure coefficient, equation (26.1), remembering that the circulation on the surface of the cylinder is $\Gamma = \frac{1}{2} \Pi D^2 \omega$ so that

$$E = - \sin \Theta = - \frac{\Pi N D}{120 U_0}$$

in line 80.

(4) Input example: cylinder diameter D, 60mm; rotational speed N, 1245 rev min^{-1}; free-stream fluid velocity U, 20 ms^{-1}.

(5) Output:

D (MM)	N (R/M)	U (M/S)
60	1245	20

THE ANGULAR POSITIONS OF
STAGNATION POINTS ARE: -5·61   185·61

THE LIFT COEFFICIENT IS: 1.23

THE VALUES OF PRESSURE COEFFICIENT
ALONG THE NEGATIVE X-AXIS ARE:

X	CP
30	−0.962
150	−7.7E-2
270	−2.4E-2
390	−1.2E-2
510	−7E-3
630	−4E-3

WISH TO CHANGE INITIAL DATA? (Y/N) ?Y

List:

```
10 MODE 7
20 PRINT
30 PRINT "PROGRAM: ROTCYL"
40 GOSUB 520: PRINT "WHEN ? APPEARS TYPE IN NUMERICAL VALUES OF
DATA ASKED FOR."
50 GOSUB 530: GOSUB 510: INPUT "CYLINDER DIA. (MM)= ",D
60 GOSUB 510: INPUT "ITS ROTATIONAL SPEED (REV/MIN)= ",N
70 GOSUB 510: INPUT "UPSTREAM FLUID VELOCITY (M/S)= ",U
80 E=(22*N*D)/(120*U*7000): REM: 22/7 IS PI
90 C=4*22*E/7
100 CL=INT(C*100+.5)/100
110 CLS:PRINT: PRINT "D (MM)","N (R/M)","U (M/S)"
120 PRINT: PRINT: D; TAB(10)N; TAB(20)U
130 GOSUB 530: GOSUB 520
140 IF E>1 THEN 220
150 T=-57.3*ATN(1/(SQR(1/E^2-1)))
160 T1=INT(T*100+.5)/100
170 T2=180-T1
180 PRINT "THE ANGULAR POSITIONS OF"
190 PRINT "STAGNATION POINTS ARE:"; TAB(24)T1; TAB(32)T2
200 GOSUB 520
210 GOTO 300
220 S=(14*U*60000)/(22*N)-D/2
230 S1=INT(S*100+.5)/100
240 PRINT "THERE IS ONLY ONE STAGNATION"
250 PRINT "POINT AT 90 DEGREES FROM X-AXIS."
260 PRINT "ITS DISTANCE FROM THE SURFACE OF"
```

```
270 PRINT "THE CYLINDER IS:"
280 PRINT TAB(20) S1;"(MM)"
290 GOSUB 510: GOSUB 530
300 PRINT "THE LIFT COEFFICIENT IS:",CL
310 GOSUB 510: GOSUB 530
320 PRINT "THE VALUES OF PRESSURE COEFFICIENT"
330 PRINT "ALONG THE NEGATIVE X-AXIS ARE:"
340 GOSUB 510: PRINT TAB(5) "X", TAB(15) "CP"
350 GOSUB 510
360 FOR I=D/2 TO 11*D STEP 2*D
370 V=((22*N*(D)^2)/(7000*120*U*I))^2+(1-(D/2/I)^2)^2
380 P=V-1
390 CP=INT(P*1000+.5)/1000
400 X=INT(I*100+.5)/100
410 PRINT; TAB(3)X; TAB(13)CP: NEXT I
420 GOSUB 510: GOSUB 530: GOSUB 530
430 INPUT "WISH TO CHANGE INITIAL DATA? (Y/N)";A$
440 IF A$="N" THEN 470
450 CLS
460 GOTO 50
470 GOSUB 520: CLS
480 GOSUB 520: GOSUB 520: GOSUB 520
490 PRINT: PRINT "THANK YOU"
500 END
510 PRINT: RETURN
520 PRINT: PRINT: RETURN
530 FOR X=1 TO 4000: NEXT: RETURN
```

EXERCISES 7

7.1   The x- and y- components of fluid velocity in a two-dimensional flow field are $u = x$ and $v = -y$ respectively.

(*a*) Determine the stream function and plot the streamlines $\psi = 1, 2, 3$.

(*b*) If a uniform flow defined by $\psi = y$ is superimposed on the above flow, plot the resulting streamlines and label them with ψ-values.

(*c*) Determine the stream function and the velocity potential for the above combined flow.

$[\psi = xy, \phi = x^2/2 + x - y^2/2]$

7.2   The stream function for the two-dimensional flow of a liquid is given by $\psi = 2xy$. In the range of values of x and y between 0 and 5 plot the stream-lines and equipotential lines passing through coordinates $(1, 1)$, $(1, 2)$, $(2, 2)$. Also determine the velocity in magnitude and direction at the point $(1, 2)$.

$[3{\cdot}16, 63{\cdot}4°]$

7.3   A flow has a potential function ϕ given by

$$\phi = V(x^3 - 3xy^2)$$

Derive the corresponding stream function ψ and show that some of the stream lines are straight lines passing through the origin of coordinates. Find the inclinations of these lines. Evaluate also the magnitude and direction of the velocity at an arbitrary point x, y.

$[60°]$

7.4   The formula $\phi = 0{\cdot}04x^3 + axy^2 + by^3$ represents a two-dimensional potential flow where x and y are Cartesian coordinates measured in metres and ϕ is the potential function in m²/s. Evaluate the constants a and b and calculate the pressure difference between the points $(0, 0)$ and $(3\text{ m}, 4\text{ m})$ if the fluid has a density of 1300 kg/m³.

$[-0{\cdot}12, 0, 58{\cdot}5 \text{ kN/m}^2]$

7.5   A source of strength 30 m²/s is located at the origin, and another source of strength 20 m²/s is located at $(1, 0)$. Find the velocity components u and v at $(-1, 0)$ and $(1, 1)$. Also, if the dynamic pressure at infinity is zero

for $\rho = 2 \cdot 0$ kg/m^3 calculate the dynamic pressure at the above points.
[40·5 N/m^2, 39·6 N/m^2]

7.6   A source of strength m at the origin and a uniform flow of 15 m/s are combined in two-dimensional flow so that a stagnation point occurs at $(1,0)$. Obtain the velocity potential and stream function for this case.
$[\psi = -15 \tan^{-1}(y/x) + 15y, \phi = -7 \cdot 5 \ln(x^2 + y^2) + 15x]$

7.7   A source discharging 20m^3/s is located at $(-1, 0)$ and a sink of twice the strength is located at $(2, 0)$. For the pressure at the origin of 100 N/m^2 and density of 1·8 kg/m^3, find the velocity and pressure at points $(0, 1)$ and $(1, 1)$.

7.8   Show that the potential function for the flow generated by a source in a two-dimensional system is $a \ln(x^2 + y^2)$ where a is a constant. Hence derive an expression for potential function for a doublet and show that the stream-lines in a flow generated by a doublet are circular. Sketch these streamlines.

7.9   Show that the potential function $\phi = ax/(x^2 + y^2)$ represents the flow generated by a doublet. In which direction is the doublet oriented? A cylinder of radius 4 cm is held with its centre at the point $(0,0)$ in a fluid stream. At large distances from the cylinder the fluid velocity is constant at 30 m/s parallel to the x axis and in the direction of x increasing. Calculate the components of the fluid velocity at the point $x = -4$ cm, $y = 1$ cm.
$[-3 \cdot 217$ m/s, $15 \cdot 64$ m/s$]$

7.10   Under what circumstances does potential flow analysis give an accurate prediction of the flow of real fluids?
Show that the potential function

$$\phi = U\left(x + \frac{a^2 x}{x^2 + y^2}\right)$$

gives the potential flow round a cylinder of radius a. A small particle whose velocity is at all times equal to that of the fluid immediately surrounding it passes through the point $(-3a, 0)$ at time $t = 0$. At what time will it pass through the point $(-2a, 0)$?
$[1 \cdot 203 \, a/U_0]$

7.11   Sketch the pressure distributions which occur in a free vortex and in a forced vortex. Write down, or derive, fundamental relationships connecting pressure and velocity in both cases.
Two radii r_1 and r_2 $(r_2 > r_1)$, in the same horizontal plane, have the same values in a free vortex and in a forced vortex. The velocity of whirl at radius r_1 is the same in both cases. Determine, in terms of r_1, the radius r_2 at which the pressure difference between r_1 and r_2 in the forced vortex is twice that in the free vortex.
$[r_2 = r_1 \sqrt{2}]$

7.12   Show that a free vortex is an example of irrotational motion. A hollow cylinder 1 m diameter, open at the top, spins about its axis which is vertical, thus producing a forced vortex motion of the liquid contained in it. Calculate the height of the vessel so that the liquid just reaches the top when the minimum depth is 15 cm at 150 rev/min.
[3·29 m]

7.13   Prove that, in the forced vortex motion of a liquid, the rate of increase of pressure p with respect to the radius r at a point in the liquid is given by

$$\mathrm{d}p/\mathrm{d}r = \rho\omega^2 r$$

in which ω is the angular velocity of liquid and ρ its density. What will be the thrust on the top of a closed vertical cylinder of 15 cm diameter if it rotates about its axis at 400 rev/min and is completely filled with water.
[43·5 N]

7.14   An enclosed horizontal duct 2 m square has vertical sides. It is running full of water and at one point there is a curved right angled bend, the radius of curvature to the centre line of the duct being 6 m. If the flow in the bend is assumed to be frictionless and to have a free vortex distribution, calculate the rate of flow if the difference of pressure head between the inner and outer sides is 23 cm of water.
[8·07 m^3/s]

7.15   A compound vortex in a large tank of water comprises a forced vortex core surrounded by a free vortex. Determine the depth of water at the centre of the core below the free vortex level if the velocity is 2·5 m/s at the common radius of 18 cm.
[0·636 m]

7.16   A closed cylindrical container of radius R is full of water which is rotated by paddles of radius a. The axis of rotation of the paddles is vertical and their angular velocity is ω. The water within radius a thus has a forced vortex motion while the water between the radii a and R is assumed to have a free vortex motion. If the pressure at the centre of the top cover is atmospheric, develop an expression for the force exerted by the water on the top cover in terms of a, R, ω and the density ρ of the fluid.
[$F = \rho\pi\omega^2 a^4 (R^2/a^2 - \ln R/a - 3/4)$]

7.17   Define vorticity and discuss the significance of irrotational motion. Give the vorticity at 1 m and 3 m radius in a vortex whose speed is 1 m/s throughout, and calculate the difference in pressure between these two places, if the axis of rotation is (i) vertical, and (ii) horizontal.
[20·78 kN/m^2]

PART III
Behaviour of real fluids

In earlier chapters, the basic equations of continuity, energy and momentum were introduced and applied to fluid flow cases where the assumption of frictionless flow was made. The analysis presented in the following chapters will introduce concepts necessary to extend the previous work to real fluids in which viscosity is accepted, and hence leads to situations where frictional effects cannot be ignored. The concept of Reynolds number as an indication of flow type will be used extensively and the fluid boundary layer, already introduced in Chapter 5, which lies between the free stream and the surface passed by the fluid and in which all the flow resistance is concentrated, will be expanded.

It will be necessary to distinguish between two different situations: namely, that in which the fluid moves inside a pipe or duct or in a channel so that it is guided by a boundary surrounding the fluid and that in which the fluid flows around a solid body. In the first case, the flow is sometimes referred to as bounded flow and in the second case as external flow. The examples of the latter are fluid flow around a bridge pier or flow of wind around a house. Also to this category belong all the cases of solid objects moving through a stationary fluid, because it is the relative velocity between the fluid and the object that really matters. Thus an aeroplane in flight or a sailing ship are examples of such situations.

The bounded flow and the external flow round a body are both governed by the same basic principles. In all cases the fluid velocity at the boundary, i.e. where the fluid meets the solid surface, is equal to zero. This condition is sometimes referred to as the 'no slip' condition. The velocity then increases with distance perpendicular to the boundary, the rate of increase being governed by the particular law applicable to the type of flow, which may be either laminar or turbulent.

In the external flow, the fluid velocity at some distance away from the boundary reaches a free stream velocity, which is the velocity of undisturbed (by the solid object) fluid, usually taken some distance upstream of the object. Thus for a bridge pier the fluid velocity at its surface will be zero and will increase away from it until it reaches the velocity of the undisturbed river. For a ship the velocity of the fluid at its surface will be equal to that of the ship and will diminish down to zero at some distance away from the ship as the water of the sea may be taken as stationary.

For bounded flow, such as in a pipe, the velocity of the fluid is zero at the wall and increases to a maximum at the centre of the pipe where the boundary layers, starting from the diametrically opposite points on the wall, meet.

In all the above cases, there is a velocity gradient and, thus, shear stresses in the fluid. In order to maintain flow this shear stress must also be maintained and this can only be achieved by additional forces doing work on the fluid. In other words, there must be a continuous supply of energy for the flow to exist. This energy, supplied solely to maintain flow in a bounded system, is usually expressed per unit weight of the fluid flowing and thus is in units of fluid head.

$$\frac{\text{Energy supplied per unit time}}{\text{Weight of fluid flowing}} = \frac{\text{Force} \times \text{Distance/Time}}{\text{Specific weight} \times \text{Discharge}}$$

$$= \frac{pa \times s/t}{\rho g Q} = \frac{pav}{\rho g Q} = \frac{pQ}{\rho g Q} = \frac{p}{\rho g} = h$$

This head (or energy) is considered as lost because it cannot be used for any other purpose than to maintain flow and hence it is called *head loss*. Such losses will be discussed in detail in Chapters 8 and 9.

In external flows, the forces required to maintain the velocity gradient in the boundary layer and energy dissipation in separation wakes are called the *drag* and will be discussed fully in Chapters 10 and 11.

8 Laminar and turbulent flow in bounded systems

8.1. Incompressible, steady and uniform laminar flow between parallel plates

Consider first the case of steady laminar flow between inclined parallel plates, one of which is moving at a velocity U (Fig. 8.1) in the flow direction. It is required to calculate the velocity profile between the plates and hence the flow through the system.

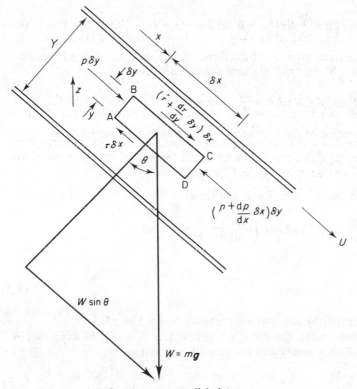

Figure 8.1   Laminar flow between parallel plates

This flow condition may be analysed by application of the momentum equation to an element of the flow — ABCD in Fig. 8.1 — and by consideration of the constraints imposed on the flow by limiting the analysis to steady, uniform, laminar flow.

The momentum equation may be stated as

Resultant force in flow direction = Rate of change of momentum
in flow direction.

However, as the flow is restricted to the steady, uniform case, then the acceleration is zero. (If the acceleration of the flow is described by the equation

$$\frac{d\bar{v}}{dt} = \frac{\partial \bar{v}}{\partial t} + \frac{\partial \bar{v}}{\partial x} \cdot \frac{\partial x}{\partial t},$$

then for steady flow $\partial \bar{v}/\partial t$ is zero and for uniform flow $\partial \bar{v}/\partial x$ is zero, hence the zero value of $d\bar{v}/dt$.) Thus, the resultant force acting on the fluid element ABCD is zero and the flow is in a state of equilibrium under the action of the forces illustrated.

If it is assumed that the plates are sufficiently wide to make edge effects negligible, then the resultant force, in the flow direction, on the fluid element may be expressed, for unit width of plate, as

$$p\delta y - \left(p + \frac{dp}{dx}\delta x\right)\delta y + W \sin\theta - \tau\delta x + \left(\tau + \frac{d\tau}{dy}\delta y\right)\delta x = 0, \qquad (8.1)$$

where p is the static pressure of the flow, τ is the shear stress, θ is the plate inclination and $W = \rho g \delta x \delta y$ per unit width. Therefore,

$$-\frac{dp}{dx}\delta x \delta y + W \sin\theta + \frac{d\tau}{dy}\delta y \delta x = 0.$$

If z is the elevation of the system above some horizontal datum, then

$$\sin\theta = -\frac{dz}{dx}$$

and, hence, by substitution for W and $\sin\theta$,

$$-\frac{dp}{dx}\delta x \delta y + \rho g \delta x \delta y \left(-\frac{dz}{dx}\right) + \frac{d\tau}{dy}\delta y \delta z = 0,$$

so that

$$\frac{d\tau}{dy} = \frac{d}{dx}(p + \rho gz), \qquad (8.2)$$

where $(p + \rho gz)$ is the *piezometric pressure*, denoted by p^*.

As previously stated, the shear stress in laminar flow may be expressed in terms of the fluid viscosity and the velocity gradient as

$$\tau = \mu \frac{du}{dy}, \qquad (8.3)$$

hence, integrating equation (8.2) and substituting for τ yields an expression in terms of the velocity gradient:

$$\tau = \mu \frac{du}{dy} = y \left[\frac{d}{dx}(p + \rho gz)\right] + C_1. \qquad (8.4)$$

This integration was possible as $(p + \rho gz)$ is assumed to vary only in the x direction. Integration of equation (8.4) with respect to y will yield an equation for the velocity distribution between the plates in the form $u = f(y)$, in terms of fluid viscosity, piezometric head and two constants of integration that may be evaluated by consideration of the system boundary conditions at $y = 0$ and $y = Y$:

$$u = \frac{1}{\mu}\frac{d}{dx}(p + \rho gz)\frac{y^2}{2} + y\frac{C_1}{\mu} + C_2. \tag{8.5}$$

At the interface between the fluid and the plates at $y = 0$ and $y = Y$, the relative velocity of the fluid to the plate is zero, i.e. the condition of no slip. Thus, at $y = 0$ it follows that the fluid velocity $u = 0$ as this plate is itself stationary. At $y = Y$ the fluid velocity relative to the plate is zero; however, as the plate is moving at a velocity U in the flow direction, the value of u at $y = Y$ must similarly be $u = U$. Substituting these two boundary conditions in turn into equation (8.5) yields:

$$y = 0, u = 0, \text{ therefore } C_2 = 0,$$

$$y = Y, u = U, \text{ therefore } C_1 = \mu\frac{U}{Y} - \frac{Y}{2}\frac{d}{dx}(p + \rho gz)$$

or $$u = \frac{y}{Y}U - \frac{1}{2\mu}\frac{d}{dx}(p + \rho gz)(Yy - y^2). \tag{8.6}$$

Equation (8.6) represents the velocity profile across the gap between the two plates, and is a general equation from which a number of restricted cases may be considered. For example,

(i) Horizontal plates with no movement of the upper plates, i.e. $U = 0$, $\sin \theta = 0$, hence $dz/dx = 0$ and

$$u = -\frac{1}{2\mu}\frac{dp}{dx}(Yy - y^2). \tag{8.7}$$

Note equation (8.7) represents a parabolic velocity profile, and the negative sign recognizes that dp/dx itself will be negative as the pressure drops in the flow direction.

(ii) Horizontal plates with upper plate motion:

$$u = \frac{y}{Y}U - \frac{1}{2\mu}\frac{dp}{dx}(Yy - y^2). \tag{8.8}$$

Equation (8.8) indicates that fluid flow may occur even if there is no pressure gradient in the x direction, provided the plates are in motion. In this case $u = (y/Y)U$, a straight line velocity distribution. This phenomenon is known as Couette flow.

The volume flow rate Q may be calculated for any of the above cases by integrating the expression for δQ, the flow through an element δy of the plate separation and of unit width, between the system boundary at $y = 0$ and $y = Y$.

Generally, $\delta Q = u\delta y$ per unit width, hence

$$Q = \int_{y=0}^{Y} u \, dy.$$

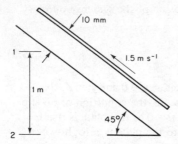

Figure 8.2

For the general case, illustrated in Fig. 8.1, Q per unit width becomes

$$Q = \int_{y=0}^{Y} \left\{ \frac{y}{Y} U - \frac{1}{2\mu} \frac{d}{dx} (p + \rho gz)(Yy - y^2) \right\} dy$$

$$= \left[\frac{U}{Y} \frac{y^2}{2} \right]_0^Y - \frac{1}{2\mu} \frac{d}{dx} (p + \rho gz) \left[Y \frac{y^2}{2} - \frac{y^3}{3} \right]_0^Y$$

Therefore,

$$Q = \frac{UY}{2} - \frac{1}{2\mu} \frac{d}{dx} (p + \rho gz) \frac{Y^3}{6}. \tag{8.9}$$

For flow between stationary horizontal plates this reduces to

$$Q = \frac{1}{12\mu} \frac{dp}{dx} Y^3 \text{ per unit width.} \tag{8.10}$$

Example 8.1

Laminar flow of a fluid of viscosity $\mu = 0.9$ N s m^{-2} and density $\rho = 1260$ kg m^{-3} occurs between a pair of parallel plates of extensive width, inclined at 45° to the horizontal, the plates being 10 mm apart. The upper plate moves with a velocity 1.5 m s^{-1} relative to the lower plate and in a direction opposite to the fluid flow. Pressure gauges, mounted at two points 1 m vertically apart on the upper plate, record pressures of 250 kN m^{-2} and 80 kN m^{-2}, respectively. Determine the velocity and shear stress distribution between the plates, the maximum flow velocity and the shear stress on the upper plate (Fig. 8.2).

Solution

Flow direction from direction of pressure gradient. At (1),

$$p_1 + \rho gz_1 = 250 + 9.81 \times 1.0 \times \frac{1260}{1000}$$

$$= 262.36 \text{ kN m}^{-2}.$$

At (2),

$$p_2 + \rho gz_2 = 80 \text{ kN m}^{-2}$$

as $z = 0$ if datum taken at (2). Flow is down slope, upper plate moves 'up' slope. Pressure gradient

$$\frac{dp^*}{dx} = -\frac{(262 \cdot 36 - 80)}{1 \cdot \sqrt{2}} = -182 \cdot 36/\sqrt{2}$$

$p^* = (p + \rho g z) = -128 \cdot 95 \text{ kN m}^{-2}$ per metre, i.e. $z = 1$.

From equation (8.6),

$$u = y \frac{U}{Y} - \frac{1}{2\mu} \frac{dp^*}{dx} (Yy - y^2),$$

where $U = -1 \cdot 5 \text{ m s}^{-1}$, $Y = 0 \cdot 01$ m and u is the local velocity at a point y above the lower plate. Thus the velocity profile is

$$u = \frac{-1 \cdot 5}{0 \cdot 01} y + \frac{128 \cdot 95 \times 10^3}{2 \times 0 \cdot 9} (0 \cdot 01 y - y^2)$$

$$= -150 y + 716 \cdot 4 y - 71 \cdot 64 \times 10^3 y^2,$$

$$= 566 \cdot 4 y - 71 \cdot 64 \times 10^3 y^2.$$

Shear stress distribution is given by

$$\tau_y = \mu \left(\frac{du}{dy} \right)_y,$$

$$\frac{du}{dy} = 566 \cdot 4 - 143 \cdot 28 \times 10^3 y,$$

$$\tau_y = 509 \cdot 76 - 128 \cdot 95 \times 10^3 y,$$

u_{max} occurs where $du/dy = 0$, $y = 566 \cdot 4 \times 10^{-3}/143 \cdot 28 = 0 \cdot 395 \times 10^{-2}$.

Hence,

$$u_{max} = 566 \cdot 4 \times 0 \cdot 003\ 95 - 71 \cdot 64 \times 10^3 \times 0 \cdot 003\ 95^2$$

$$= 2 \cdot 24 + 1 \cdot 117 = 3 \cdot 36 \text{ m s}^{-1}.$$

Shear stress on upper plate is given by

$$\tau_Y = \mu \left(\frac{du}{dy} \right)_{y=Y} = 509 \cdot 76 - 128 \cdot 95 \times 10^3 \times 0 \cdot 01$$

$$= 0 \cdot 78 \text{ kN m}^{-2}.$$

This is the fluid shear at the plate; hence, the shear force on the plate is $0 \cdot 78$ kN per unit area resisting plate motion.

Equation (8.10) may be applied to laminar flow between concentric cylinders, provided that the annulus is of small dimensions compared to the cylinder diameter. An example of this case would involve the leakage past a piston within a cylinder, as shown in Fig. 8.3. Hence, the leakage flow becomes

$$Q = \frac{1}{12\mu} \frac{p_1 - p_2}{l} (\Delta R)^3 \, 2\pi R_1, \tag{8.11}$$

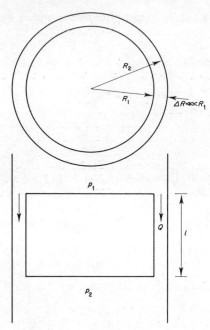

Figure 8.3   Leakage flow past a piston within a cylinder

where $\Delta R = R_1 - R_2$ and is the piston/cylinder separation and the total width of the 'parallel plates' is given by the piston circumference. In this case, it will be seen that plate width edge effects can be ignored as the 'parallel plates' are effectively continuous.

8.2. Incompressible, steady and uniform laminar flow in circular cross-section pipes

Steady, uniform, laminar flow in a circular cross-section pipe or annulus may be treated in the same manner as described for laminar flow between parallel plates. The analysis rests on the same basic principles, namely the application of the momentum equation to an element of flow within the conduit; the application of shear stress–velocity gradient relationship (8.3); and the knowledge of the flow condition at the pipe wall, which allows the constants of integration to be evaluated, namely the no slip condition.

Consider an annular element in the flow of internal radius r and radial thickness δr, as shown in Fig. 8.4, in an inclined tube, of radius R, carrying a fluid under laminar flow conditions. Applying the momentum equation to the situation illustrated in Fig. 8.4 yields an expression

$$p\,2\pi r\delta r - \left(p + \frac{\mathrm{d}p}{\mathrm{d}x}\,\delta x\right) 2\pi r\delta r + \tau\,2\pi r\,\mathrm{d}x$$

$$- \left[2\pi r\tau\delta x + \frac{\mathrm{d}}{\mathrm{d}r}\,(2\pi r\tau\,\mathrm{d}x)\,\delta r\right] + W\sin\theta = 0,$$

$$(8.12)$$

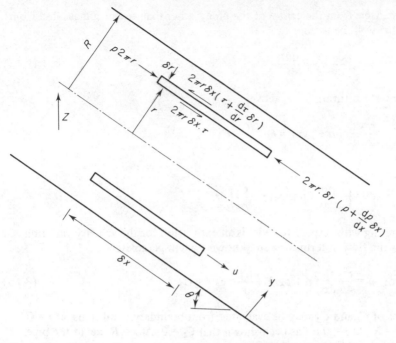

Figure 8.4   Forces acting on an annular element in a laminar pipe flow situation

where p is the flow static pressure, $W = mg$ is the element weight and τ is the shear stress at radius r. Due to the assumption of steady, uniform conditions the flow acceleration is zero and, hence, the resultant force on the element is zero. Putting $W = 2\pi r \delta r \delta x \rho g$ and $\sin \theta = -dz/dx$, where z is the elevation of the pipe above some horizontal datum, reduces the expression (8.12) to

$$-\frac{dp}{dx} - \frac{1}{r}\frac{d}{dr}(r\tau) - \rho g \frac{dz}{dx} = 0$$

by dividing by $2\pi r \delta r \delta x$. Rearranging,

$$\frac{d}{dx}(p + \rho g z) + \frac{1}{r}\frac{d}{dr}(r\tau) = 0. \tag{8.13}$$

The term $(p + \rho g z)$ is the flow piezometric pressure and is independent of r, enabling equation (8.13) to be integrated with respect to r. Hence,

$$\frac{r^2}{2}\frac{d}{dx}(p + \rho g h) + r\tau + C_1 = 0.$$

If conditions at the pipe centre line are substituted into the above expression, then $C_1 = 0$ as $r = 0$.

The shear stress–velocity gradient expression of equation (8.3) may be employed in a modified form to take note of the direction of measurement of distance r from the centre of the pipe, rather than use of y measured from the pipe wall, hence

$$\tau = \mu \frac{du}{dy} = -\mu \frac{du}{dr} \tag{8.14}$$

and, by substituting for τ, above,

$$\frac{r^2}{2} \frac{d}{dx} (p + \rho gz) = r\mu \frac{du}{dr} - C_1$$

and

$$du = \left(\frac{r}{2\mu} \frac{d}{dx} (p + \rho gz) + \frac{C_1}{r\mu} \right) dr.$$

Integrating with respect to r yields an expression for the velocity variation across the flow in terms of r and known system parameters.

$$u = \frac{r^2}{4\mu} \frac{d}{dx} (p + \rho gz) + \frac{C_1}{\mu} \log_e r + C_2. \tag{8.15}$$

Values of C_1 and C_2 may be evaluated from boundary conditions at $r = 0$ and $r = R$. At $r = 0$ it has been shown that $C_1 = 0$. At $r = R$, i.e. at the pipe wall, the local flow velocity u is zero; hence,

$$C_2 = -\frac{R^2}{4\mu} \frac{d}{dx} (p + \rho gz)$$

and

$$u = -\frac{(R^2 - r^2)}{4\mu} \frac{d}{dx} (p + \rho gz). \tag{8.16}$$

Equation (8.16) describes the variation of local fluid velocity u across the pipe and, from the form of the equation, this velocity profile may be seen to be parabolic. The negative sign is again present due to the fact that the pressure gradient will be negative in the flow direction.

The maximum velocity will occur on the pipe centre line, i.e. $r = 0$, hence

$$u_{max} = -\frac{R^2}{4\mu} \frac{d}{dx} (p + \rho gz). \tag{8.17}$$

The volume flow rate through the pipe under these flow conditions may be calculated by integrating the incremental flow δQ through an annulus of radial width δr at radius r across the flow from $r = 0$ to $r = R$ (see Fig. 8.4):

$$\delta Q = u \, 2\pi r \delta r,$$

$$Q = \int_0^R u \, 2\pi r \, dr. \tag{8.18}$$

Substituting for u at general radius r, yields an expression

$$Q = -\frac{\pi}{2\mu}\frac{d}{dx}(p + \rho gz)\int_0^R (R^2 r - r^3)\,dr$$

$$= -\frac{\pi}{2\mu}\frac{d}{dx}(p + \rho gz)\left[R^2\frac{r^2}{2} - \frac{r^4}{4}\right]_0^R$$

$$= -\frac{\pi}{8\mu}\frac{d}{dx}(p + \rho gz)R^4$$

or, in terms of a pressure drop Δp over a length l of pipe of diameter d,

$$Q = \Delta p \pi d^4/128\,\mu l. \tag{8.19}$$

The mean flow velocity is given by Q/A, where A is the pipe cross-sectional area $\pi d^2/4$. Hence,

$$\bar{u} = -\frac{\pi}{8\mu}\frac{d}{dx}(p + \rho gz)R^2 = \tfrac{1}{2}u_{max}, \tag{8.20}$$

as shown in Fig. 8.5.

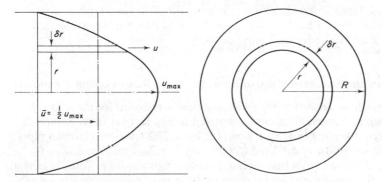

Figure 8.5   Velocity distribution in laminar flow in a circular pipe

Equation (8.19) may be rearranged for the pressure loss giving the well-known Hagen–Poiseuille equation:

$$\Delta p = 128\,\mu l Q/\pi d^4. \tag{8.21}$$

Alternatively, substituting for $Q = (\pi d^2/4)\bar{u}$,

$$\Delta p = 32\,\mu l \bar{u}/d^2. \tag{8.22}$$

Example 8.2

Glycerine of viscosity 0.9 N s m^{-2} and density 1260 kg m^{-3} is pumped along a horizontal pipe 65 m long of diameter $d = 0.01$ m at a flow rate of $Q = 180$ litre min^{-1}. Determine the flow Reynolds number and verify whether the flow is laminar or turbulent. Calculate the pressure loss in the pipe due to frictional effects and calculate the maximum flow rate for laminar flow conditions to prevail.

Solution

$$\text{Mean velocity}, \bar{u} = Q/A = \left[\left(\frac{180}{60} \right) \Big/ \frac{\pi d^2}{4} \right] \times 10^{-3} \text{ m s}^{-1}$$

$$= 38 \cdot 2 \text{ m s}^{-1}.$$

$$\text{Re} = \rho \bar{u} d / \mu = 1260 \times 38 \cdot 2 \times 0 \cdot 01 / 0 \cdot 9 = 535$$

Therefore, flow is laminar as Re = 535 < 2000 (*see* Section 4.10).

Frictional losses may be calculated from the Hagen–Poiseuille equation (8.21):

$$\Delta p = 128 \, \mu l Q / \pi d^4$$

$$= 128 \times 0 \cdot 9 \times 65 \times 3 \times 10^{-3} / \pi \times 0 \cdot 01^4$$

$$= 715 \times 10^6 \text{ N m}^{-2}.$$

Upper limit of laminar flow conditions is reached when

$$\text{Re} / \text{Re}_{crit} = Q / Q_{crit},$$

$$Q_{crit} = (Q / \text{Re}) \, \text{Re}_{crit}$$

$$= (180 / 535) \times 2000 \text{ litre min}^{-1},$$

therefore,

$$Q_{crit} = 6721 \text{ min}^{-1} = 0 \cdot 0112 \text{ m}^3 \text{s}^{-1}.$$

8.3. Incompressible, steady and uniform turbulent flow in bounded conduits

In the preceding sections expressions have been developed for the velocity distribution and pressure losses encountered during laminar flow. Reference to the Reynolds number of such flow, i.e. Re < 2000 in closed circular pipes, shows that such flow is restricted to relatively low flow rate conditions for all gases and those liquids that do not possess a high viscosity. Thus, in general, turbulent flow conditions are far more likely in most engineering situations. In the following section, expressions will be developed for the losses incurred in turbulent flow in both closed and open conduits. However, it will be seen that completely analytical solutions are not available and that empirical relationships are needed in order to produce the necessary expressions.

Consider a small element of fluid within a conduit, as shown in Fig. 8.6. The flow is assumed to be uniform and steady so that the fluid acceleration in the flow direction is zero. Applying the momentum equation to the fluid element in the flow direction yields

$$p_1 A - p_2 A - \tau_0 l P + W \sin \theta = 0, \tag{8.23}$$

where P is the wetted perimeter of the element defined as that part of the conduit's circumference which is in contact with the fluid. It will be seen that including the area over which the shear stress τ_0 acts in the form of lP, as above, effectively renders the derivation applicable to both open or closed conduits. Putting $W = \rho g A l$ and $\sin \theta = -\Delta z / l$ yields

$$A (p_1 - p_2) - \tau_0 l P - \rho g A \, \Delta z = 0,$$

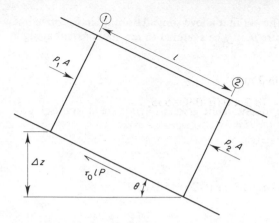

Figure 8.6   Turbulent flow in a bounded conduit

where p_1, p_2 are the static pressures in the flow at sections 1 and 2 (Fig. 8.6). Hence

$$\frac{1}{l}\{(p_1 - p_2) - \rho g \Delta z\} - \tau_0 \frac{P}{A} = 0$$

where the first term represents a drop in piezometric head over a length l of the conduit and the ratio A/P is known as the hydraulic mean depth, normally denoted by m, thus,

$$\tau_0 = m \frac{\mathrm{d}p^*}{\mathrm{d}x} \tag{8.24}$$

where $\mathrm{d}p^*/\mathrm{d}x$ is the rate of loss of piezometric head along the conduit and τ_0 is the wall or boundary shear stress.

In order to express τ_0 in equation (8.24), the concept of a flow friction factor f is introduced, which is a non-dimensional, experimentally measured factor normally introduced in the form

$$\tau_0 = f \rho \bar{v}^2 / 2. \tag{8.25}$$

where $\bar{v}$ is the mean flow velocity, Hence,

$$\frac{\mathrm{d}p^*}{\mathrm{d}x} = f \rho \bar{v}^2 / 2m. \tag{8.26}$$

If the frictional head loss down a length l of the conduit is denoted by h_f, then the rate of loss of piezometric pressure may be expressed as

$$\frac{\mathrm{d}p^*}{\mathrm{d}x} = f \rho v^2 / 2m = \rho g h_f / l$$

and $h_f = f l \bar{v}^2 / 2gm.$ \tag{8.27}

Now, as

$$\frac{\mathrm{d}p^*}{\mathrm{d}x} = \frac{\mathrm{d}}{\mathrm{d}x}(p + \rho g z),$$

where z is the elevation of the conduit above some datum, then for *open channels*, as the static pressure p may be assumed to remain constant along the channel, it follows that

$$\frac{dp^*}{dx} = \rho g \frac{dz}{dx} = \rho g \sin \theta$$

and, since for uniform flow the hydraulic gradient h_f/l is equal to the slope of the channel,

$$\frac{h_f}{l} = \sin \theta = i.$$

Then it follows, by equating equations (8.26) with (8.27), that

$$f \rho \bar{v}^2 / 2m = \rho g i,$$

so that

$$\bar{v} = \sqrt{(2g/f)} \times \sqrt{(mi)}.$$

If, now,

$$\sqrt{(2g/f)} = C \qquad\qquad (8.28)$$

is substituted, the expression known as the Chezy formula is obtained:

$$\bar{v} = C\sqrt{(mi)}. \qquad\qquad (8.29)$$

yielding the flow rate through a given channel of a given slope and roughness. Various values of C are employed in open channel design.

For pipes running full of fluid, the wetted perimeter becomes the internal circumference of the pipeline, hence $A/P = m = \pi D^2/4\pi D = D/4$, so that the equation (8.27) becomes for circular cross-sections

$$h_f = \frac{4fl}{d} \cdot \frac{\bar{v}^2}{2g}. \qquad\qquad (8.30)$$

This expression is in every way equivalent to the Chezy equation above and follows directly from the study of the general condition illustrated in Fig. 8.5. It is known as the Darcy-Weisbach equation for head loss in circular pipes. The Darcy equation is also the equivalent of the Poiseuille equation derived for laminar flow, with one important exception, namely the inclusion of an empirical factor f to describe the friction loss in turbulent flow, which was not necessary in the case of laminar flow. This fundamental difference arises from the complexity of turbulent flow, resulting in the fact that the relationship $\tau = \mu(du/dy)$ cannot be used and, therefore, an analytical solution is not possible.

8.4. Incompressible, steady and uniform turbulent flow in circular cross-section pipes

The head loss in turbulent flow in a closed section pipe is given by the Darcy equation:

$$h_f = \frac{4fL}{d} \cdot \frac{\bar{v}^2}{2g}.$$

It will be seen from the above expression that all the parameters, with the exception of the friction factor f, are measurable. Results of extensive experimentation in this area led to the establishment of the following proportional relationships:

(1) $h_f \propto l$;
(2) $h_f \propto \bar{v}^2$;
(3) $h_f \propto 1/d$;
(4) h_f depends on the surface roughness of the pipe walls;
(5) h_f depends on fluid density and viscosity;
(6) h_f is independent of pressure.

The value of f must be selected so that the correct value of h_f will always be given by the Darcy equation and so cannot be a single-value constant. The value of f must depend on all the parameters listed above. Expressed in a form suitable for dimensional analysis this implies that

$$f = \phi(\bar{v}, d, \rho, \mu, k, k', \alpha), \tag{8.31}$$

where k is a measure of the size of the wall roughness, k' is a measure of the spacing of the roughness particles, both having dimensions of length, and α is a form factor, a dimensionless parameter whose value depends on the shape of the roughness particles. In the general rough pipe case, dimensional analysis yields an expression

$$f = \phi_2 \left(\rho \bar{v} d / \mu, k/d, k'/d, \alpha\right)$$

or, in terms of Reynolds number,

$$f = \phi_2 (\text{Re}, k/d, k'/d, \alpha). \tag{8.32}$$

Dimensional analysis can only indicate the best combination of parameters for an empirical solution; the actual algebraic format of the relation for friction factor in terms of the variables listed must be determined by experimentation.

Blasius, in 1913, was the first to propose an accurate empirical relation for friction factor in turbulent flow in smooth pipes, namely

$$f = 0.079/\text{Re}^{1/4}. \tag{8.33}$$

This expression yields results for head loss to ±5 per cent for smooth pipes at Reynolds numbers up to 100 000.

At this point it may be useful to note that the value of friction factor quoted in many American texts is $4f$ in the notation employed in this text, and so the value of the constant in Blasius' equation will be changed. In this text the UK-recognized value of friction factor as defined by equation (8.30) will be used exclusively.

For rough pipes, Nikuradse, in 1933, proved the validity of the f dependence on the relative roughness ratio k/d by investigating the head loss in a number of pipes which had been treated internally with a coating of sand particles whose size could be varied. These tests in no way investigated the effect of particle spacing k'/d, or of particle shape factor α, on the friction factor, but did show that, for one type of roughness,

$$f = \phi_3 \, (\text{Re}, \, k/d). \tag{8.34}$$

It may well be argued that experimental problems would make it virtually impossible to hold k'/d and α constant so that the effect of roughness size k/d might be investigated in isolation. However, the accuracy of the results obtained by basing the value of f simply on Reynolds number and k/d does suggest that the effects of particle spacing and shape are negligible compared to that of the relative roughness based solely on k/d.

Thus, the calculation of losses in turbulent pipe flow is dependent on the use of empirical results and the most common reference source is the Moody chart, which is a logarithmic plot of f vs. Re for a range of k/d values. This type of data presentation is commonly referred to as a Stanton diagram. A typical *Moody chart* is presented as Fig. 8.7 and a number of distinct regions may be identified and commented on.

(i) The straight line labelled 'laminar flow', representing $f = 16/\text{Re}$, is a graphical representation of the Poiseuille equation (8.19), i.e.

$$Q = \Delta p \pi d^4 / 128 \, \mu l,$$

$$h_f = \Delta p / \rho g = 128 \, \mu l Q / \rho g \pi d^4.$$

Now, $Q = \pi (d^2/4) \bar{v}.$

Hence, from Darcy's equation,

$$h_f = 128 \mu l \pi d^2 \bar{v} / 4 \rho g \pi d^4 = 4 f l \bar{v}^2 / 2 g d$$

or $$f = \frac{128}{8} \, \mu / \rho \bar{v} d$$

$$= 16/\text{Re}. \tag{8.35}$$

Equation (8.35) plots as a straight line of slope -1 on a log–log plot and is independent of pipe surface roughness. This relation also shows that the Darcy equation may be applied to the laminar flow régime provided that the correct f value is employed.

(ii) For values of $k/d < 0.001$ the rough pipe curves of Fig. 8.7 approach the Blasius smooth pipe curve due to the presence of the laminar sub-layer (discussed in Chapter 9), which develops in turbulent flow close to the pipe wall and whose thickness decreases with increasing Reynolds number. Thus, for certain combinations of surface roughness and Reynolds number, the thickness of the laminar sub-layer is sufficient to cover the wall roughness and the flow behaves as if the pipe wall were smooth. For higher Reynolds numbers the roughness particles project above the now decreased thickness laminar sub-layer and contribute to an increased head loss.

(iii) At high Reynolds numbers, or for pipes having a high k/d value, all the roughness particles are exposed to the flow above the laminar sub-layer. In this condition, the head loss is totally due to the generation of a wake of eddies by each particle making up the pipe roughness. This form of head loss is known as 'form drag' and is directly proportional to the square of the mean flow velocity, thus $h_f \propto \bar{v}^2$ and, hence, from Darcy's equation, f is a constant, depending only on the roughness particle size. This condition

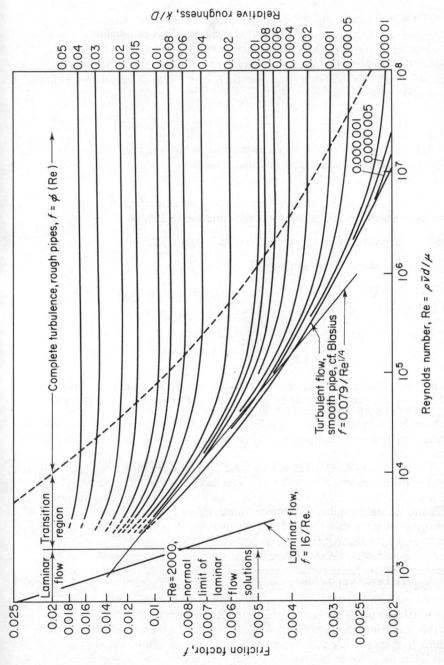

Figure 8.7   Variation of friction factor f with Reynolds number and pipe wall roughness for ducts of circular cross-section

is represented on the Moody chart by portions of the f vs. Re curves which are parallel to the Re axis and which occur at high values of Re and k/d.

Example 8.3

Calculate the loss of head due to friction and the power required to maintain flow in a horizontal circular pipe of 40 mm diameter and 750 m long when water (coefficient of dynamic viscosity 1.14×10^{-3} N s m^{-2}) flows at a rate: (a) 4.01 min^{-1}, (b) 301 min^{-1}. Assume that for the pipe the absolute roughness is $0.000\ 08$ m.

Solution

(a) In order to establish whether the flow is turbulent or laminar it is first necessary to calculate the Reynolds number:

$$\text{Re} = \rho \bar{v} d/\mu,$$

but $Q = 4.0 \times 10^{-3}/6.0 = 66.7 \times 10^{-6}\ \text{m}^3\ \text{s}^{-1}$

and   Pipe area, $A = \pi d^2/4 = \pi(0.04)^2/4 = 1.26 \times 10^{-3}\ \text{m}^2$

so that the mean velocity in the pipe is given by

$$\bar{v} = Q/A = 66.7 \times 10^{-6}/1.26 \times 10^{-3} = 52.9 \times 10^{-3}\ \text{m s}^{-1}.$$

Hence,

$$\text{Re} = \frac{10^3 \times 52.9 \times 10^{-3} \times 0.04}{1.14 \times 10^{-3}} = 1856.$$

Therefore, flow is laminar, since Re < 2000. So, the loss due to friction may therefore be calculated either by using Poiseuille's equation (i) or Darcy's equation and $f = 16/\text{Re}$ (ii):

(i) Poiseuille's equation:

$$\Delta p = \frac{128\,\mu l Q}{\pi d^4} = \frac{128 \times 1.14 \times 10^{-3} \times 750 \times 66.7 \times 10^{-6}}{\pi(0.04)^4} = 907.6\ \text{N m}^{-2}.$$

Therefore, head lost due to friction is given by

$$h_f = \frac{\Delta p}{\rho g} = \frac{907.6}{10^3 \times 9.81} = 92.4 \times 10^{-3}\ \textbf{m of water.}$$

(ii) Darcy's equation:

$$h_f = \frac{4 f l}{d}\frac{\bar{v}^2}{2g}, \quad \text{but} \quad f = \frac{16}{\text{Re}} = \frac{16}{1856} = 0.008\ 62.$$

Hence,

$$h_f = \frac{4 \times 0.008\ 62 \times 750}{0.04} \times \frac{(52.9 \times 10^{-3})^2}{2 \times 9.81} = 92.4 \times 10^{-3}\ \textbf{m of water.}$$

Power required to maintain flow, $P = \rho g h_f Q$

$$= 10^3 \times 9.81 \times 92.4 \times 10^{-3} \times 66.7$$
$$\times 10^{-6}$$

$$= 0.0605 \text{ W.}$$

(b) $Q = \dfrac{30 \times 10^{-3}}{60} = 0.5 \times 10^{-3} \text{ m}^3 \text{ s}^{-1}, \; \bar{v} = \dfrac{0.5 \times 10^{-3}}{1.26 \times 10^{-3}} = 0.4 \text{ m s}^{-1},$

therefore,

$$\text{Re} = \frac{10^3 \times 0.4 \times 0.04}{1.14 \times 10^{-3}} = 14\,035$$

and the flow is turbulent, so the Darcy equation must be used. To determine the value of friction factor:

Relative roughness $= k/d = 0.000\,08/0.04 = 0.002$.

From Moody's chart, for Re $= 1.4 \times 10^4$ and relative roughness of 0.002, $f = 0.008$. Therefore,

$$h_f = \frac{4fl}{d} \frac{\bar{v}^2}{2g} = \frac{4 \times 0.008 \times 750}{0.04} \times \frac{(0.4)^2}{2 \times 9.81} = \textbf{4.89 m of water.}$$

Power required, $P = \rho g h_f Q = 10^3 \times 9.81 \times 4.89 \times 0.5 \times 10^{-3}$

$$= \textbf{24.0 W.}$$

8.5. Steady and uniform turbulent flow in open channels

It was shown in Section 8.3 that the general equation for head losses in turbulent flow could be derived concurrently for both open and closed section conduits. The general equation (8.27)

$$h_f = f l \bar{v}^2 / 2 g m,$$

reduces to the Chezy equation (8.29),

$$\bar{v} = C \sqrt{(mi)},$$

when it is realized that, for open channels, provided the flow is steady and uniform, h_f/l is equal to the slope of the channel. This, however, is not the case in non-uniform flow discussed in Chapter 16.

Since $\bar{v}$ is the mean velocity in the channel of area A, it follows that the flow rate Q is given by:

$$Q = A\bar{v} = AC\sqrt{(mi)}. \tag{8.36}$$

Although C is referred to as the Chezy coefficient, implying a dimensionless constant, this is not so; since $C = 2g/f$, it has dimensions of $\text{L}^{1/2}\text{T}^{-1}$.

It has been shown that, for pipe flow, the value of f depends both on the flow Reynolds number and on the surface roughness of the pipe material, so it would be reasonable to expect C to vary with Re and k/m, where m, the mean hydraulic depth, is employed as the characteristic length for the system. Generally, the dependence of C on Reynolds number is small and k/m is the predominant factor. For almost all open channel work the flow may be assumed to be fully turbulent, with a high value of Reynolds number and, therefore, k/m may be taken as the sole factor affecting C values, provided the channel section shape remains simple.

Finally, it will be seen that, as values of C depend only on Re and k/m and not on the Froude number of the open channel flow, the Chezy equation applies equally to rapid or tranquil flow, defined in Chapter 15. However, in cases of non-uniform flow, an important distinction between the slope of the channel and the hydraulic gradient has to be made. This will be discussed in Chapter 16.

Example 8.4

A rectangular open channel has a width of 4·5 m and a slope of 1 vertical to 800 horizontal. Find the mean velocity of flow and the discharge when the depth of water is 1·2 m and if C in the Chezy formula is 49.

Solution
The mean velocity may be obtained using Chezy formula:

$$\bar{v} = C \sqrt{(mi)}.$$

In it, $i = 1/800$ and $m = A/P$. Now,

$$A = 4 \cdot 5 \times 1 \cdot 2 = 5 \cdot 4 \text{ m}^2,$$

$$P = 2 \times 1 \cdot 2 + 4 \cdot 5 = 6 \cdot 9 \text{ m},$$

so that

$$m = 5 \cdot 4/6 \cdot 9 = 0 \cdot 783 \text{ m}.$$

Substituting into the Chezy formula,

$$\bar{v} = 49 \sqrt{(0 \cdot 783/800)} = \mathbf{1 \cdot 53 \text{ m s}^{-1}}.$$

The discharge is given by

$$Q = \bar{v}A = 1 \cdot 53 \times 5 \cdot 4 = \mathbf{8 \cdot 27 \text{ m}^3 \text{ s}^{-1}}.$$

8.6. Velocity distribution in turbulent, fully developed pipe flow

Due to the nature of turbulent flow, there are difficulties in the derivation of expressions defining the distribution of velocity in pipe flow. The use of dimensional analysis, together with a series of assumptions based on the relative importance of the fluid viscosity and eddy viscosity terms in the laminar sub-layer which is present in a fully developed turbulent boundary layer, do, however, allow the prediction of the form of the velocity distribution expressions. The algebraic format of these equations has been developed from experimental investigations and, although now well established and accepted, are empirical in nature.

For fully developed turbulent pipe flow in a circular cross-section pipe, it would be reasonable to suppose that the local velocity u at a distance y from the pipe wall would be given by a general function of the form

$$u = \phi(\rho, \mu, \tau_0, R, y, k), \tag{8.37}$$

where ρ, μ are the fluid density and viscosity, R is the pipe radius, k is the roughness particle size and τ_0 is the wall shear stress.

Dimensional analysis suggests an expression of the form

$$\frac{\bar{u}}{\sqrt{(\tau_0/\rho)}} = \phi_1 \left\{ \left(\frac{\rho R}{\mu}\right)_{\sqrt{}} \left(\frac{\tau_0}{\rho}\right), \frac{y}{R}, \frac{k}{R} \right\}.$$

The term $\sqrt{(\tau_0/\rho)}$ has the dimensions of velocity and is referred to as the *shear stress velocity u**.

$$\bar{u}/u^* = \phi_1 \{\rho u^* R/\mu, y/R, k/R\}, \tag{8.38}$$

where $\rho u^* R/\mu$ is a form of the Reynolds number.

To proceed beyond equation (8.38), it is necessary to make some assumptions about the importance of the various groups.

Surface roughness, represented by k/R, will affect the value of u^*, but will only be a significant factor in the flow zone close to the wall. Similarly, the fluid viscosity will only be of major importance in the laminar sub-layer close to the pipe wall. Thus, the velocity in the central, turbulent core of the flow will be assumed to depend only on the positional group y/R. It is customary to express this relationship in terms of the velocity defect, or the difference between the local velocity u at position y from the wall and the flow maximum velocity on the pipe centre line $u_{\max}$. Hence,

$$(u_{\max} - u)/u^* = \phi_2 (y/R). \tag{8.39}$$

This expression is referred to as the *velocity defect distribution* and is well supported by experimental work which shows that, for a wide range of flow Reynolds numbers, the velocity profiles only differ in the region close to the pipe wall. It is apparent from experimental results that, as the Reynolds number increases, the friction factor f and the shear stress τ_0 terms become smaller, and the velocity profile across the central core of the flow becomes progressively more uniform.

Prandtl proposed an empirical velocity distribution for this turbulent central core of the form

$$u/u_{\max} = (y/R)^n, \tag{8.40}$$

where the value of $n = \frac{1}{7}$ for Re $> 10^5$ and decreases above this Reynolds number. This is well supported experimentally, but does break down at $y = R$ as symmetry here demands that $du/dy = 0$, which cannot be satisfied by the expression.

If the case of the smooth pipe is considered, the k/R group becomes unimportant and so, close to the pipe wall the effect of pipe radius R is negligible, so long as $y \ll R$; let $y = y_2$ be the limit for this assumption so that

$$u/u^* = \phi_3 (\text{Re}^*),\ 0 < y < y_2, \tag{8.41}$$

where Re$^* = \rho y u^*/\mu$, a group independent of pipe radius R.

If equation (8.39) is applicable from $y = y_1$ to R, and if experimental results which indicate that $y_2 > y_1$ are accepted, then it becomes apparent that there is a region in the flow, close to the pipe wall, where both equations (8.39) and (8.41) apply simultaneously.

For a smooth pipe, the relation

$$u/u^* = \phi_1 \, (\text{Re}, y/R) \tag{8.42}$$

applies for $y_1 < y < R$, where $\text{Re} = \rho R u^*/\mu$ and, for $y = R$,

$$u_{max}/u^* = \phi_4 (\text{Re}). \tag{8.43}$$

Adding equations (8.39) and (8.42) yields

$$u_{max}/u^* = \phi_4 (\text{Re}) = \phi_1 \, (\text{Re}, y/R) + \phi_2 \, (y/R) \tag{8.44}$$

for $y_1 < y < R$. Both equations (8.39) and (8.41) apply in the zone $y_1 < y < y_2$ and may be added to give

$$u_{max}/u^* = \phi_2 (y^*) + \phi_3 (\text{Re}^*) = \phi_4 (\text{Re}), \tag{8.45}$$

where $y^* = y/R$. Differentiating (8.45) with respect to Re yields

$$y^* \phi_3' (\text{Re}^*) = \phi_4' (\text{Re}) \tag{8.46}$$

as $\text{Re}^* = \text{Re} y^*$. Since $\phi_4' (\text{Re})$ is independent of y^*, so then is $y^* \phi_3'(\text{Re}^*)$, so that $\phi_3'(\text{Re}^*)$ has the form

$$(1/y^*)\phi(\text{Re}). \tag{8.47}$$

Similarly, differentiating equation (8.45) with respect to y^* yields

$$\phi_2(y^*) + \text{Re} \, \phi_3'(\text{Re}^*) = 0 \tag{8.48}$$

and so $\phi_3'(\text{Re}^*)$ has the form

$$(1/\text{Re})\phi(y^*) \tag{8.49}$$

since $\phi_2'(y^*)$ is independent of Re.

In order for equations (8.49) and (8.47) to be satisfied simultaneously in the zone $y_1 < y < y_2$, it is necessary for $\phi(\text{Re}) = \text{constant}/\text{Re}$ and $\phi(y^*) = \text{constant}/y^*$, therefore, $\phi_3' (\text{Re}^*) = A/\text{Re}^*$, where A is a constant.

From (8.46),

$$\phi_4' (\text{Re}) = y^* \phi_3' (\text{Re}) = y^* A/\text{Re}^* = A/\text{Re}$$

or $\phi_4 (\text{Re}) = A \log_e \text{Re} + \text{constant}.$ \hfill (8.50)

Similarly, from (8.48),

$$\phi_2' (y^*) = -\text{Re} \, \phi_3' (\text{Re}^*) = -\text{Re} \, A/\text{Re}^*$$
$$= -A/y^*$$

or, by integration,

$$\phi_2(y^*) = -A \log_e y^* + \text{constant}. \tag{8.51}$$

However, from equations (8.39), (8.45),

$$u/u^* = \phi_4 (\text{Re}) - \phi_2 (y^*).$$

Thus,

$$u/u^* = (A \log_e Re + constant) + (A \log_e y^* + constant),$$

$$u/u^* = A \log_e Re^* + A_1, \qquad (8.52)$$

where A and A_1 are constants to be determined experimentally.

Equation (8.52) is known as the *universal velocity distribution*. However, it is to be noted that, due to the restriction placed on equations (8.39) and (8.45), the expression only applies in the central turbulent core of the pipeline, as shown in Fig. 8.8.

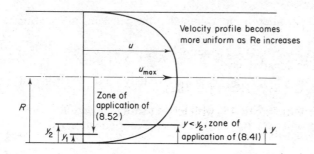

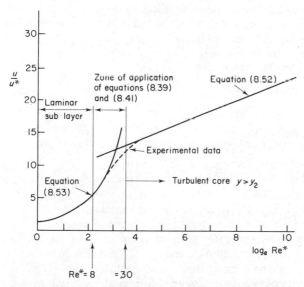

Figure 8.8   Zones of application of empirical relations defining velocity distribution in turbulent pipe flow

Close to the pipe wall, within the laminar sub-layer, the effect of fluid viscosity is predominant and so the expression $\tau = \mu(du/dy)$ may be integrated to describe the velocity distribution in this zone:

$$u = \tau(y/\mu) + E,$$

where $E = 0$ when $u = 0$ at $y = 0$, Hence,

$$u = \tau y / \mu$$

or $\quad u / \sqrt{(\tau/\rho)} = y \rho \tau^{1/2} / \mu \rho^{1/2},$

$$u/u^* = \text{Re}^*. \qquad (8.53)$$

Equation (8.53) may be plotted, as shown in Fig. 8.8, and intersects the straight line relation of equation (8.52) at a point which, theoretically, defines the upper limit of the laminar sub-layer. In practice, the upper limit of the laminar sub-layer is ill-defined, and experimental results tend to smooth the intersection (Fig. 8.8).

Nikuradse's results for smooth pipes show that the constants in equation (8.52) may be taken as

$$u/u^* = 2 \cdot 5 \log_e \text{Re}^* + 5 \cdot 5 = 5 \cdot 75 \log_{10} \text{Re}^* + 5 \cdot 5. \qquad (8.54)$$

As shown in Fig 8.8, equation (8.53) applies accurately up to $\text{Re}^* \simeq 8$ and (8.52) applies from $\text{Re}^* \simeq 30$.

By employing equation (8.52), it is possible to relate the friction factor to mean flow velocity and flow Reynolds number. From equation (8.25),

$$f = 2\tau / \rho \bar{u}^2,$$

where $\bar{u}$ is the mean flow velocity. Mean velocity may be calculated by integration of equation (8.52) across the pipe, assuming the thickness of the laminar sub-layer to be negligible, and dividing the result by the pipe cross-sectional area, as was done in the case of laminar flow (equation (8.20)). Substitution of friction factor for mean velocity through the relation above (equation (8.25)) yields an expression of the form

$$1/\sqrt{f} = F + G \log_e (\text{Re} \sqrt{f}), \qquad (8.55)$$

where $\text{Re} = \rho d\bar{u}/\mu$ and F, G are constants.

The constants in equation (8.55) may be obtained from experimental investigations, and Nikuradse's results for smooth pipes suggest an expression

$$1/\sqrt{f} = 4 \cdot 07 \log_{10} (\text{Re} \sqrt{f}) - 0 \cdot 6, \qquad (8.56)$$

although values of $4 \cdot 0$ and $-0 \cdot 4$ for G and F do yield improved results. This expression has been verified for Reynolds numbers in the range $5000 < \text{Re} < 3 \times 10^6$. However, the Blasius expression for smooth pipes (equation (8.33)) gives reasonable accuracy for Re values up to 10^5.

Similar relations for velocity distribution and friction factor may be determined for rough pipes, the form of the expressions being deduced by dimensional analysis techniques and the algebraic format of the relations being obtained empirically by extensive testing.

The Moody chart (Fig. 8.7) showed that rough pipe turbulent flow falls into two régimes, as far as friction factor calculation is concerned. First, a régime where friction factor is independent of Reynolds number and depends on surface roughness; second, a transitionary régime where the friction factor increases with Reynolds number for any particular pipe surface roughness, from the value appropriate for a smooth pipe up to the Reynolds number independent value mentioned. The mechanism responsible for this transition has already been explained in Section 8.4 in terms of the relation between roughness particle size and laminar sub-layer thickness.

Nikuradse showed that, by describing the flow Reynolds number in terms of the roughness particle size, then the three identifiable flow régimes could be described as follows:

(i) smooth pipe f results apply for $Re = \rho u^* k/\mu < 4$;
(ii) transition occurs for $4 < Re < 70$;
(iii) f is independent of Re for $70 < Re$.

For the f independent zone, the velocity profile may be expressed as

$$u/u^* = \phi(y/R, k/R) \qquad (8.57)$$

and experimental results verify an equation of the form

$$u/u^* = 5 \cdot 75 \log_{10}(y/k) + 8 \cdot 48. \qquad (8.58)$$

Integration to give mean velocity and introduction of friction factor via equation (8.25) yields an expression similar to (8.55):

$$1/\sqrt{f} = 4 \log_{10}(d/k) + 2 \cdot 28, \qquad (8.59)$$

where d is the pipe diameter.

For the transition régime, $4 < \rho u^* k/\mu < 70$, an expression

$$1/\sqrt{f} = -4 \log_{10}\{k/3 \cdot 71 d + 1 \cdot 26/Re\sqrt{f}\}, \qquad (8.60)$$

where $Re = \rho \bar{u} d/\mu$, has been shown to be applicable and may be seen to converge to equation (8.59) or (8.56) for fully rough pipes or smooth pipes characterized by $Re \rightarrow \infty$ or $k \rightarrow 0$. Equation (8.60) is known as the *Colebrook–White equation* and was employed by Moody in the preparation of the friction factor chart of Fig. 8.7 (*see* program 8.9).

As mentioned in the derivation of the laminar flow equations, the results only apply to fully developed pipe flow and so do not cover the entry length of a pipeline. However, as this is normally short in comparison to the pipe length, no appreciable error arises.

Example 8.5

Assuming the following velocity distribution in a circular pipe:

$$u = u_{max}(1 - r/R)^{1/7},$$

where u_{max} is the maximum velocity, calculate (a) the ratio between the mean velocity and the maximum velocity, (b) the radius at which the actual velocity is equal to the mean velocity.

Solution

(a) The elementary discharge through an annulus dr is given by

$$dQ = 2\pi r u \, dr$$

$$= 2\pi u_{max}(1 - r/R)^{1/7} \, dr$$

and discharge through the pipe by

$$Q = 2\pi u_{max} \int_0^R r(1 - r/R)^{1/7} \, dr.$$

Let $1 - r/R = x$, then

$$\frac{dx}{dr} = -\frac{1}{R} \quad \text{and} \quad dr = -R \, dx$$

so that

$$R - r = xR, \qquad \qquad \text{when } r = 0, x = 1,$$

$$r = R - xR = R(1 - x), \quad \text{when } r = R, x = 0.$$

Therefore, substituting,

$$Q = 2\pi u_{max} \int_1^0 R(1-x)x^{1/7}(-R \, dx) = 2\pi R^2 u_{max} \int_0^1 (1-x)x^{1/7} \, dx$$

$$= 2\pi R^2 u_{max} \left[\frac{7}{8} x^{8/7} - \frac{7}{15} x^{15/7} \right]_0^1$$

$$= 2\pi R^2 u_{max} \left(\frac{7}{8} - \frac{7}{15} \right) = 2\pi R^2 u_{max} \left(\frac{105 - 56}{120} \right) = \pi R^2 u_{max} \frac{49}{60}.$$

and

$$\bar{u} = Q/\pi R^2 = \pi R^2 u_{max} \tfrac{49}{60}/\pi R^2 = \tfrac{49}{60} u_{max}.$$

With the result that

$$\bar{u}/u_{max} = \mathbf{49/60}.$$

(b)

$$u = \bar{u} = 49u_{max}/60 = u_{max} (1 - r/R)^{1/7}.$$

Therefore,

$$(49/60)^7 = 1 - R/r$$

and $r/R = 1 - (49/60)^7 = 1 - 0.242 = 0.758.$

Hence,

$$r = 0.758R.$$

Example 8.6

Assuming the universal velocity distribution for the turbulent flow in a pipe,

$$\frac{u}{u*} = 5.5 + 5.75 \log_{10} Re*,$$

determine the radius at which the point velocity is equal to the mean velocity and the ratio of mean velocity to maximum velocity.

Solution

The point velocity is given by

$$u = u* \, (5\cdot5 + 5\cdot75 \log_{10} \mathrm{Re}*),$$

but

$$u* = \sqrt{(\tau_0/\rho)} = \text{constant and } \mathrm{Re}* = \rho y u*/\mu.$$

Let $\rho u*/\mu = a$ and, for a pipe, let $y = r$. Then,

$$u = u* \, (5\cdot5 + 5\cdot75 \log_{10} ar).$$

To obtain the mean velocity, it is necessary first to calculate the discharge which, when divided by the cross-sectional area of the pipe, will give the mean velocity. Thus, the elementary discharge through an annulus dr is given by

$$dQ = 2\pi r u \, dr = 2\pi u* \, (5\cdot5r + 5\cdot75r \log ar) \, dr$$

and discharge through the pipe by

$$Q = 2\pi u* \left[5\cdot5 \int_0^R r \, dr + 5\cdot75 \int_0^R r \log_{10} ar \, dr \right].$$

Now,

$$\int_0^R r \, dr = \left[\frac{r^2}{2} \right]_0^R = \frac{R^2}{2},$$

and to obtain

$$\int_0^R r \log_{10} ar \, dr = \frac{1}{\log_e 10} \int_0^R r \log_e ar \, dr,$$

put $\log_e ar = y$. Hence,

$$dy = \frac{1}{ar} a \, dr = \frac{dr}{r}$$

and $r \, dr = dx.$

Therefore,

$$x = r^2/2.$$

Integrating now by parts,

$$\int y \, dr = yx - \int r \, dy.$$

Substituting

$$\int r \log_{10} ar \, dr = \frac{1}{2} r^2 \log_e ar - \frac{1}{2} \int r^2 \frac{1}{r} \, dr$$

$$= \tfrac{1}{2} r^2 \log_e ar - r^2/4$$

$$= (r^2/2) \, (\log_e ar - \tfrac{1}{2}).$$

Therefore,

$$\int_0^R r \log_{10} ar \, dr = \frac{1}{\log_e 10} \int_0^R r \log_e ar \, dr$$

$$= \frac{1}{\log_e 10} \left[\frac{r^2}{2} \left(\log_e ar - \frac{1}{2} \right) \right]_0^R$$

$$= \left[\frac{r^2}{2} \left(\log_{10} ar - \frac{1}{2 \log_e 10} \right) \right]_0^R$$

$$= \frac{R^2}{2} \left(\log_{10} aR - \frac{1}{2 \log_e 10} \right)$$

$$= \frac{R^2}{2} \left(\log_{10} aR - 0 \cdot 217 \right).$$

Substituting into the equation for Q,

$$Q = 2\pi u^* \left\{ 5 \cdot 5 \frac{R^2}{2} + 5 \cdot 75 \frac{R^2}{2} \left(\log_{10} aR - 0 \cdot 217 \right) \right\}$$

$$= \pi u^* R^2 \left(5 \cdot 5 + 5 \cdot 75 \log_{10} aR - 1 \cdot 298 \right)$$

$$= \pi u^* R^2 \left(4 \cdot 252 + 5 \cdot 75 \log_{10} aR \right).$$

Therefore, mean velocity in the pipe,

$$\bar{u} = Q / \pi R^2 = u^* (4 \cdot 252 + 5 \cdot 75 \log_{10} aR).$$

The radius at which the point velocity is the same as the mean velocity is now obtained by equating the two expressions:

$$u^* (5 \cdot 5 + 5 \cdot 75 \log_{10} ar) = u^* (4 \cdot 252 + 5 \cdot 75 \log_{10} aR),$$

$$1 \cdot 248 = 5 \cdot 75 (\log_{10} aR - \log_{10} ar).$$

$$0 \cdot 217 = \log_{10} (R/r),$$

$$10^{0 \cdot 217} = R/r,$$

so that, finally,

$$r = \mathbf{0 \cdot 607} \, R.$$

The maximum velocity occurs at the centre of the pipe, where $r = 0$. Therefore,

$$u_{max} = 5 \cdot 5 u^*$$

and $\quad \bar{u}/u_{max} = u^* (4 \cdot 252 + 5 \cdot 75 \log aR)/5 \cdot 5 \, u^*$

$$= 0 \cdot 773 + 1 \cdot 045 \log_{10} aR$$

$$= \mathbf{0 \cdot 773 + 1 \cdot 045 \log_{10} (\rho u^* / \mu) R}.$$

8.7. Velocity distribution in fully developed, turbulent flow in open channels

In the use of the Chezy and related equations for open channel flow, the assumption is made that the flow is uniform across the channel. This is, in practice, never achieved, and, further, due to the lack of symmetry in open channel flow, the accepted central positon of maximum flow velocity for pipe flow is also not reproduced. The actual velocity profiles across the flow are influenced by the presence of the channel solid boundaries and the free surface. Irregularities in the solid boundaries are generally so large and random that each channel has its own individual velocity distribution and there is no direct equivalent to the velocity distribution expressions derived for pipe flow. Figure 8.9 illustrates a typical velocity distribution, the

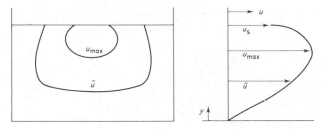

Figure 8.9   Velocity distribution in a simple open channel under fully developed turbulent flow conditions

maximum velocity occurring at some depth below the free surface, usually between 5 and 25 per cent of flow depth, and the mean velocity, which is usually some 80 to 85 per cent of the free surface velocity, occurs at about 60 per cent of the flow depth below the free surface.

Normally, in open channel calculations, the uncertainties involved in the flow parameters are so large as to render any variations of flow velocity away from the mean to be negligible and these are neglected.

8.8. Separation losses in pipe flow

Whenever the uniform cross-section of a pipeline is interrupted by the inclusion of a pipe fitting, such as a valve, bend, junction or flow measurement device, then a pressure loss will be incurred. The value of these losses, which are sometimes misleadingly referred to as 'minor losses', have to be included in a pipeline's total resistance, if errors in pump and system matching or flow calculations for a given pressure differential are to be avoided. In this treatment, the term 'separation loss' has been chosen to define pressure losses across such fittings, as it is felt that this term well describes the physical phenomena which occur at such obstructions in the pipeline. Generally, the flow separates from the pipe walls as it passes through the obstructing pipe fitting, resulting in the generation of eddies in the flow, with consequent pressure loss, as shown in Fig. 8.10 for the case of a sudden enlargement. For small complex pipe networks such as those found in some chemical process plants, aircraft fuel and hydraulic systems and in ventilation systems, the total effect of separation losses may be the predominant factor in the system

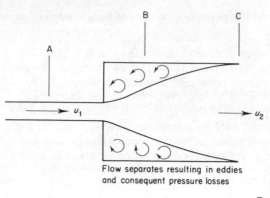

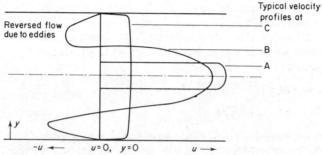

Figure 8.10   Separation loss in a sudden enlargement

pressure loss calculation, exceeding the contribution of pipe friction at the design flow rate. Conversely, in large pipe systems, such as water distribution networks or overland oil pipelines, the losses due to pipe fittings may be negligible compared to the friction loss and may often be ignored.

8.8.1. Losses in sudden expansions and contractions

Generally, the losses due to pipe or duct fittings are determined experimentally. However, the case of a sudden expansion in a pipe or duct may be determined analytically. Figure 8.11 illustrates a sudden enlargement; consider a control

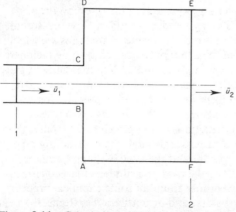

Figure 8.11   Calculation of the loss coefficient for a sudden enlargement

volume ABCDEF as shown and let p_1 and p_2 be the pressures at section 1 and 2, respectively, where the mean flow velocities are related by the continuity equation

$$A_1 \bar{u}_1 = A_2 \bar{u}_2 \tag{8.61}$$

and represent the duct cross-sectional areas A_1, A_2.

By application of the momentum equation between sections 1 and 2 the following relation may be derived:

$$\begin{array}{l} \text{Resultant force in} \\ \text{flow direction} \end{array} = \begin{array}{l} \text{Rate of change of momentum} \\ \text{in flow direction,} \end{array}$$

$$p_1 A_1 + p'(A_2 - A_1) - p_2 A_2 = \rho Q(\bar{u}_2 - \bar{u}_1), \tag{8.62}$$

where $Q = A_2 \bar{u}_2$ and p' is the pressure acting on the annulus represented by AB and CD, of cross-sectional area $(A_2 - A_1)$. It may be assumed that $p_1 = p'$, due to the small radial acceleration at entry to the larger diameter duct at section ABCD, a result which is well supported experimentally. Hence, equation (8.62) reduces to

$$(p_1 - p_2)A_2 = \rho Q(\bar{u}_2 - \bar{u}_1) = \rho \bar{u}_2 A_2 (\bar{u}_2 - \bar{u}_1),$$

$$p_1 - p_2 = \rho \bar{u}_2 (\bar{u}_2 - \bar{u}_1). \tag{8.63}$$

If Bernoulli's equation is now applied between sections 1 and 2, with a term h included to represent the separation loss, then an expression for the pressure differential $p_1 - p_2$ may be derived:

$$p_1/\rho g + \bar{u}_1^2/2g + Z_1 = p_2/\rho g + \bar{u}_2^2/2g + Z_2 + h,$$

where $Z_1 = Z_2$ if the enlargement is situated in a horizontal pipe or duct, thus,

$$h = (p_1 - p_2)/\rho g + (\bar{u}_1^2 - \bar{u}_2^2)/2g$$

and, substituting for $p_1 - p_2$ from equation (8.62),

$$h = \frac{\rho \bar{u}_2 (\bar{u}_2 - \bar{u}_1)}{\rho g} + \frac{\bar{u}_1^2 - \bar{u}_2^2}{2g} = \frac{1}{2g}(2\bar{u}_2^2 - 2\bar{u}_2\bar{u}_1 + \bar{u}_1^2 - \bar{u}_2^2)$$

$$= \frac{1}{2g}(\bar{u}_1^2 - 2\bar{u}_1\bar{u}_2 + \bar{u}_2^2) = \frac{(\bar{u}_1 - \bar{u}_2)^2}{2g}.$$

Thus, the loss due to sudden enlargement is given by

$$h = (\bar{u}_1 - \bar{u}_2)^2/2g \tag{8.64}$$

Alternatively, since, from equation (8.61),

$$\bar{u}_2 = \bar{u}_1 (A_1/A_2),$$

$$h = (\bar{u}_1^2/2g)(1 - A_1/A_2)^2 = \frac{\bar{u}_2^2}{2g}\left(\frac{A_2}{A_1} - 1\right)^2. \tag{8.65}$$

This expression is sometimes referred to as the Borda-Carnot relationship and is usually within a few per cent of the experimental result for the separation loss incurred by sudden enlargement in coaxial pipelines.

The loss at exit from a pipe into a reservoir may be obtained by considering equation (8.65). It will be seen that as $A_2 \to \infty$, so $\bar{u}_2 \to 0$ and $h \to \bar{u}_1^2/2g$, i.e. the kinetic energy of the approaching flow. This case is obviously representative of a pipe discharging into a large tank or a duct discharging to atmosphere and is the accepted expression for conduit exit loss.

Sudden contractions in a duct or pipe may also be dealt with in this way, provided that there is little or no loss between the upstream large section conduit and the vena contracta formed within the smaller conduit just downstream of the junction, as shown in Fig. 8.12.

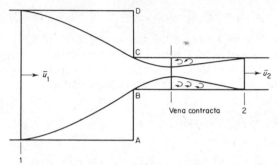

Figure 8.12   Sudden contraction. Loss approximated by consideration of sudden enlargement between the vena contracta and section 2

It is not possible to apply the momentum equation between sections 1 and 2 in Fig. 8.12, due to the uncertain pressure distribution across the face ABCD. However, it has been shown experimentally that the majority of the pressure loss occurs as a result of the eddies formed as the flow area expands from the vena contracta area up to the full cross-section of the downstream pipe. If the area of the vena contracta is A_c then accurate results may be achieved by applying the sudden enlargement expression between A_c and A_2 at section 2, thus

$$ h \simeq \frac{\bar{u}_2^2}{2g} \left\{ \frac{A_2}{A_c} - 1 \right\}^2 = \frac{\bar{u}_2^2}{2g} \left\{ \frac{1}{C_c} - 1 \right\}^2, \tag{8.66} $$

where C_c is the coefficient of contraction for the junction based on the smaller pipe entry diameter BC. The above equation indicates, since the expression in brackets is constant for any given area ratio, that it may be generalized into the form

$$ h = K \, \bar{u}_2^2/2g, \tag{8.67} $$

where K is known as the loss coefficient. Table 8.1 shows some experimental values of C_c and the corresponding values of K obtained with sharp pipe edges.

A_2/A_1	0·1	0·3	0·5	0·7	1·0
C_c	0·61	0·632	0·673	0·73	1·0
K	0·41	0·34	0·24	0·14	0

Table 8.1. Loss coefficients for sudden contraction

Example 8.7

In a water pipeline there is an abrupt change in diameter from 140 to 250 mm. If the head lost due to separation when the flow is from the smaller to the larger pipe is 0·6 m greater than the head lost when the same flow is reversed, determine the flow rate.

Solution

When the flow is from the smaller to the larger pipe the loss is due to sudden enlargement and is given by equation (8.65):

$$h = (\bar{u}_1^2/2g)(1 - A_1/A_2)^2 = (\bar{u}_1^2/2g)(1 - 0{\cdot}314)^2 = 0{\cdot}47\ \bar{u}_1^2/2g,$$

where $\bar{u}_1$ is the velocity in the smaller pipe. When the flow is reversed, the loss is due to sudden contraction. Area ratio,

$$A_2/A_1 = (140/250)^2 = 0{\cdot}314.$$

From Table 8.1,

$$K = 0{\cdot}33\ (\text{say}).$$

Therefore, the loss,

$$h' = 0{\cdot}33\ \bar{u}_2^2/2g,$$

where $\bar{u}_2$ is again the velocity in the smaller pipe. Since the flow rate is the same in both cases, then

$$\bar{u}_1 = \bar{u}_2 = \bar{u}$$

and $h - h' = 0{\cdot}6,$

so that

$$(0{\cdot}47 - 0{\cdot}33)(\bar{u}^2/2g) = 0{\cdot}6$$

and $\bar{u} = 9{\cdot}17\ \text{m s}^{-1}.$

8.8.2. Losses in pipe fittings, bends and at pipe entry

Losses in pipe fittings are usually expressed in the form already suggested for the loss at sudden contraction, namely,

$$h = K(\bar{u}^2/2g), \tag{8.67}$$

where K is the fitting loss coefficient. It is a non-dimensional constant and its value is obtained experimentally for any pipe fitting. Table 8.2 sets out some typical values. The major advantage of expressing losses due to separation in the above form is that it can easily be incorporated into Bernoulli's equation, as will be shown in Chapter 12.

Figure 8.13 illustrates the flow in a pipe bend, demonstrating the area of flow separation which results in the loss coefficients for bends listed in Table 8.2. As the bend becomes sharper, so the areas of separation become more extensive and the loss coefficient increases.

Loss at entry to a pipe from a reservoir is a special case of sudden contraction, in which the velocity in the reservoir is considered to be zero. Due to the fact that the fluid enters the pipe from all directions, a vena

Fitting	Loss coefficient, K
Gate valve (open to 75 per cent shut)	$0.25 \rightarrow 25$
Globe valve	10
Spherical plug valve (fully open)	0.1
Pump foot valve	1.5
Return bend	2.2
90° elbow	0.9
45° elbow	0.4
Large radius 90° bend	0.6
Tee junction	1.8
Sharp pipe entry	0.5
Radiused pipe entry	$\rightarrow 0.0$
Sharp pipe exit	0.5

Table 8.2. Head loss coefficients for a range of pipe fittings

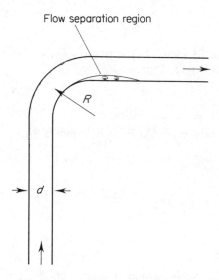

Flow separation region

Figure 8.13   Flow in a bend, illustrating separation

contracta is formed downstream of the pipe inlet and, consequently, the loss is associated with enlargement from the vena contracta to the full bore pipe. This is the same situation as in the case of sudden contraction.

Figure 8.14 illustrates various types of pipe entry conditions. The results tabulated in Table 8.2 and indicated in Fig. 8.14 may be explained by reference to the flow separation at entry to the pipe explained above, so that the sharper the entry corner, the smaller is the vena contracta and, hence, the greater the flow separation and the higher the value of K.

8.8.3. Equivalent length for pipe fitting loss calculations

Separation loss coefficients K may also be defined in terms of an equivalent length of straight pipe, of the same diameter as that including the fitting, that would result in the same frictional loss as that incurred by flow separation through the fitting. This is justified by consideration of

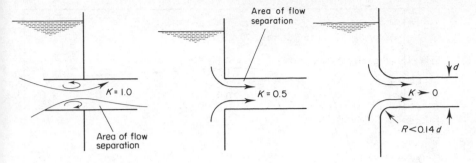

Figure 8.14 Pipe entry losses

Darcy's equation and equation (8.67),

$$h_f = 4 f l_e \bar{u}^2 / d\,2g = K\,\bar{u}^2 / 2g,$$

where l_e is the equivalent length of pipe, diameter d, that would yield a friction loss equivalent to the particular fitting. Thus,

$$l_e = Kd/4f \qquad\qquad (8.68)$$

and so l_e is normally calculated as a number of pipe diameters.

l_e may be the equivalent length for a single fitting or the summation of all the separation loss coefficients for a particular system. Hence, for the total pressure loss through a pipeline of length l and diameter d, the expression

$$h_f = 4f(l + l_e)\bar{u}^2 / 2dg \qquad\qquad (8.69)$$

may be employed.

8.8.4. Diffusers

In order to avoid the head losses incurred by the installation of sudden enlargements into pipe and duct flow, diffusers are commonly employed. Figure 8.15 illustrates a typical conical diffuser and the variation in pressure loss across it with diffuser included angle.

The loss experienced depends on the area ratio between which the diffuser operates and the included angle of the diffuser included angle θ. The total loss across the diffuser is made up of two components, the first due to fluid friction along the length of the diffuser, which, therefore, increases as θ decreases for a given area ratio, i.e. the diffuser length l increases as θ decreases and results in an increase in friction loss. The second contribution to the total loss is dependent on θ and is the separation loss, which increases with increasing included angle for a given area ratio, reaching a maximum when the diffuser approaches a sudden enlargement.

The minimum loss for any particular area ratio will, therefore, be a compromise, where the angle of the diffuser is sufficiently small to limit separation, or flow eddy, losses, but not so small as to increase the length of the diffuser to the point where the frictional losses become predominant. Normally 6 to 7° included angle is the minimum acceptable.

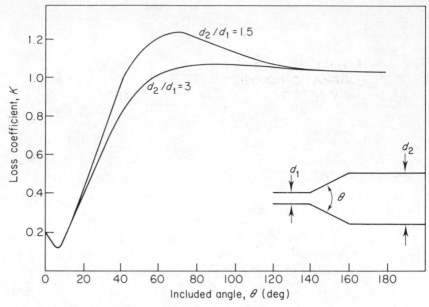

Figure 8.15   Loss of pressure in a conical diffuser

Diffusers are found in a wide range of applications where it is necessary to reduce flow velocity by means of an area change, without undue pressure loss. For example, wind tunnel return circuits are at one end of the size spectrum and venturi meter discharge diffusers at the other.

8.9 Computer program 'CBW'

(1) The program calculates the value of friction factor for full bore pipe or duct flow based on the Colebrook-White equation.

(2) The required input is:
    (i)   pipe diameter (m);
    (ii) pipe roughness (mm);
    (iii) steady flow rate in $m^3 s^{-1}$ or steady mean velocity ($m\ s^{-1}$);
    (iv) fluid density;
    (v) fluid kinematic or dynamic viscosity.

(3) The program utilizes the Colebrook-White expression (8.60),

$$1/\sqrt{f} = -4 \log_{10} (k/3.71\,D + 1.26/\text{Re}\sqrt{f}).$$

(4) Input example:

$$D = 1\ m; k = 0.4\ mm; Q = 2\ m^3 s^{-1};$$
$$\rho = 1000\ kg\ m^{-3}; \mu = 1 \times 10^{-3}\ kg\ ms^{-1}.$$

(5) Output:

    PIPE DATA

    PIPE DIAMETER D=   1 M.
    PIPE AREA A=   0.7854 M^2.

PIPE ROUGHNESS K=   0.4 MM.
ROUGHNESS RATIO K/D=   4E-4

FLOW DATA

FLUID DENSITY R=   1000 KG/M^3.
KINEMATIC VISCOSITY =   1E-6 M^2/S
FLOWRATE Q=   2 M^3/S.
MEAN FLOW VELOCITY V=   2.5465 M/S.
FLOW REYNOLDS NUMBER =   2.5465E6

FRICTION FACTOR F=   4.0527E-3

List:

```
10 REM CBW
20 MODE 0
30 @%=&50509
40 PRINT"CBW"
50 PRINT"SOLUTION OF THE COLEBROOK-WHITE EQUATION APPLIED TOCIRCULAR SECTION
DUCTS RUNNING FULL."
60 PRINT"PLEASE INPUT PIPE DATA AS FOLLOWS;-"
70 INPUT"INPUT PIPE DIAMETER (M.) D= ",D
80 A=PI*D^2/4.0
90 INPUT"INPUT PIPE ROUGHNESS (MM.) K= ",K
100 INPUT"INPUT FLOW RATE, M^3/S, OR MEAN FLOW VELOCITY, M/S. V= ",V
110 INPUT"INDICATE UNIT OF V, INPUT F FOR M^3/S OR U FOR M/S",A$
120 IF A$="F" THEN V=V/A
130 IF A$="U" THEN V=V
140 Q=V*A
150 PRINT"INPUT FLUID DATA:-"
160 INPUT"INPUT FLUID DENSITY (KG/M^3) ",R
170 INPUT"INPUT VISCOSITY AS MEW*10^N ","MEW= ",MEW," N= ",N
180 M=MEW*10.0^N
190 INPUT"TO INDICATER WHETHER KINEMATIC OR DYNAMIC VISCOSITY USED PLEASE INPU
T K OR D. ",A$
200 IF A$="D" THEN M=M/R
210 S=V*D/M
220 CLS
230 PRINT"PIPE DATA":PRINT:PRINT
240 PRINT"PIPE DIAMETER D- ",D," M."
250 PRINT"PIPE AREA A= ",A," M^2."
260 PRINT"PIPE ROUGHNESS K= ",K," MM."
270 PRINT"ROUGHNESS RATIO K/D= ",K/(D*1000.0):PRINT
280 PRINT:PRINT"FLOW DATA":PRINT
290 PRINT"FLUID DENSITY R= ",R," KG/M^3."
300 PRINT"KINEMATIC VISCOSITY = ",M," M^2/S"
310 PRINT"FLOWRATE Q= ",Q," M^3/S."
320 PRINT"MEAN FLOW VELOCITY V= ",V," M/S."
330 PRINT"FLOW REYNOLDS NUMBER = ",S
340 REM SET UP LIMITS FOR BISECTION SOLUTION.
350 T=0.05
360 B=0.0
370 F=(T+B)/2.0
380 Y=0.25/SQR(F)
390 X=((K/(1000.0*3.71*D))+1.26/(S*SQR(F)))
400 W=Y+LOG(X)
410 IF W<0.0 THEN T=F
420 IF W>0.0 THEN B=F
430 IF W=0.0 THEN GOTO 490
440 G=(T+B)/2.0
450 E=ABS((G-F)/F)
460 IF E<0.01 THEN GOTO 490
470 F=G
480 GOTO 380
490 PRINT:PRINT:PRINT"FRICTION FACTOR F= ",F
500 PRINT:PRINT:INPUT"DO YOU WISH TO REPEAT Y OR N ",A$
510 CLS
520 IF A$="Y" THEN GOTO 10
530 STOP
```

8.10 Further reading

Colebrook, C. F. (1939). Turbulent flow in pipes with particular reference to the transition region between smooth and rough pipe laws, J. I. C. E., **10**.

Moody, L. F. (1944). Friction factors for pipe flow, *Trans A.S.M.E.*, **66**.

Ward-Smith, A. J. (1980). *International Fluid Flow: The Fluid Dynamics of Flow in Pipes and Ducts*, Oxford University Press.

EXERCISES 8

8.1   Show that, for laminar flow between two, infinite, moving flat plates, distance z apart, the flow rate Q is given by expression of the form

$$Q = -\frac{1}{\mu}\frac{dp}{dl}\frac{z^3}{12} + \frac{z}{2}(U + V)$$

where μ is fluid viscosity, dp/dl is the pressure gradient in the flow direction and U and V are the absolute velocities of the two plates.

8.2   For the case set out in Exercise 8.1 above derive the shear stress expressions for each plate surface:

$$\tau_0 = \mu\left(U'/z - \frac{z}{2\mu}\,dp/dl\right)$$

$$\tau_z = \mu\left(U'/z + \frac{z}{2\mu}\,dp/dl\right) \text{ where } U' = (U + V)$$

8.3   A thin film of oil, thickness z and viscosity μ, flows down an inclined plate. Show that the velocity profile is given by

$$u = \frac{\rho g}{2\mu}(z^2 - y^2)\sin\theta$$

where u is the local velocity at a depth y below the free surface, θ is the plate inclination to the horizontal and ρ is the fluid density.

8.4   For the case in Exercise 8.3 above calculate the flow rate per unit plate width if fluid has a viscosity of 0.9 Ns/m^2, a density of 1260 kg/m^3, the plate is inclined at $30°$ and the depth of flow is 10 mm.
[0.137 litres/min/m]

8.5   A film of fluid, density 2400 kg/m^3, flows down a vertical plate with a free surface velocity of 0.75 m/s. If the film is 20 mm thick determine the fluid viscosity.
[6.28 Ns/m^2]

8.6   Fluid of density 1260 kg/m^3 and viscosity 0.9 Ns/m^2 passes between two infinite parallel plates, 2 cm separation. If the flow rate is 0.5 litres/sec/unit width calculate the pressure drop/unit length if both plates are stationary.
[0.68 $kN/m^2/m$]

8.7   The radial clearance between a hydraulic plunger and the cylinder wall is 0.15 mm, the length of the plunger 0.25 m and the diameter 150 mm. Calculate the leakage rate past the plunger at an instant when the pressure differential between the two ends of the plunger is 15 m of water. Viscosity of hydraulic fluids is 0.9 Ns/m^2.
[5.2×10^{-3} litres/min]

8.8   For laminar flow in a tube calculate the position of the average cross-sectional velocity.
[$0.293 \times$ radius from tube wall]

8.9   For laminar flow in a tube show that the mean velocity is half the centre line maximum velocity.

The centre line velocity is measured as 3 m/sec in a 0·1 m diameter tube. If the fluid flowing has a density of 1260 kg/m³ and a viscosity of 0·9 Ns/m² determine whether the flow is laminar and calculate the pressure gradient necessary.

[Re = 210, 4·3 kN/m²/m]

8.10   For the case set out in Exercise 8.9 calculate the centre line velocity corresponding to the accepted limit of laminar flow.

[32·8 m/s]

8.11   An oil having a viscosity of 0·048 kg/ms flows through a 50 mm diameter tube at an average velocity of 0·12 m/s. Calculate the pressure drop in 65 m of tube and the velocity 10 mm from the tube wall.

[4·8 kN/m², 0·153 m/s]

8.12   Oil of specific gravity 0·9 and kinematic viscosity 0·000 33 m²/s is pumped over a distance of 1·5 km through a 75 mm diameter tube at a rate of 25 x 10³ kg/hour. Determine whether the flow is laminar and calculate the pumping power required, assuming 70% mechanical efficiency.

[Re = 397, 48·8 kW]

8.13   For the flow conditions set out in Exercise 8.12 above calculate the shear stress at the tube walls.

[55·4 N/m²]

8.14   Show that for laminar flow in a tube the friction factor f is given by $f = 16/\text{Re}$ where Re is the flow Reynolds Number.

8.15   Define critical velocity and critical Reynolds Number as applied to flow in circular tubes.

8.16   Air at 20°C is drawn through a 0·5 m dia duct by a fan. If the volume flow rate is 4·5 m³/s and the duct is 12 m long, with a friction factor of 0·005, determine the fan shaft power necessary, assuming 80% mechanical efficiency. Take air density as 1·2 kg/m³ and viscosity as $1·8 \times 10^{-5}$ Ns/m².

[0·85 kW]

8.17   For the air flow in Exercise 8.16 calculate the Reynolds Number for the flow. If the flow is to be modelled by a water flow in a 0·1 m dia duct, calculate the water flow velocity. Take water density as 1000 kg/m³ and viscosity as $1·3 \times 10^{-3}$ Ns/m².

[76×10^4, 9·88 m/s]

8.18   Water at a density of 998 kg/m³ and kinematic viscosity 1×10^{-6} m²/s flows through smooth tubing at a mean velocity of 2 m/s. If the tube diameter is 30 mm calculate the pressure gradient per unit length necessary. Assume that friction factor for a smooth pipe is given by $16/\text{Re}$ for laminar flow and $0·079/\text{Re}^{1/4}$ for turbulent flow.

[1·34 kN/m²/m]

8.19   From the Colebrook–White formula for friction factor f

$$\frac{1}{\sqrt{f}} = -4 \log\left(\frac{k}{3·71D} + \frac{1·26}{\text{Re}\sqrt{f}}\right)$$

where Re is Reynolds No. based on tube diameter D and fluid mean velocity, and k is the tube roughness size, derive an expression linking pressure loss per unit length to flow rate, Q m³/s.

$$[Q = C_1 \Delta p^{1/2} D^{5/2} \log \left\{ \frac{k}{3 \cdot 71 D} + \frac{C_2}{\Delta p^{1/2} D^{3/2}} \right\}, \; C_1, C_2 \; \text{constants}]$$

8.20   In a laboratory the water supply is drawn from a roof storage tank 25 m above the water discharge point. If the friction factor is 0·008, the pipe diameter is 5 cm and the pipe is assumed vertical calculate the maximum volume flow achievable, if separation losses are ignored.
[0·001 m^3/s]

8.21   For the case set out in Exercise 8.20 above calculate the relative roughness of the pipe used, if the water is at 0°C.
[0·006]

8.22   Calculate the diameter of a spun cast iron waste water pipe that, when flowing full, is expected to carry 5100 litres/minute of water at 0°C, with a head loss of 4m per km of pipeline. Use the Colebrook–White equation and assume a wall roughness of 0·12 mm.
[300 mm diameter]

8.23   The friction factor applicable to turbulent flow in a smooth glass pipe is given by $f = 0 \cdot 079 / \text{Re}^{1/4}$. Calculate the pressure loss per unit length necessary to maintain a flow of 0·02 m^3/s of kerosene, s.g. = 0·82, viscosity $1 \cdot 9 \times 10^{-3}$ Ns/m^2, in a glass pipe of 8 cm diameter. If the tube is replaced by a galvanized steel pipeline, wall roughness 0·15 mm, calculate the increase in pipe diameter to handle this flow with the same pressure gradient.
[150 N/m^2/m, 17%]

8.24   Derive the expression for the loss of head at a sudden expansion and show how this may also be used to approximate to the loss through a sudden contraction.

8.25   Define static regain along a diffuser and show that it may be calculated as

$$\text{Static regain} = \tfrac{1}{2}\rho(V_1^2 - V_2^2) - \tfrac{1}{2}K\rho V_1^2$$

8.26   A 150 mm diameter pipe reduces in diameter abruptly to 100 mm diameter. If the pipe carries water at 30 litres/second calculate the pressure loss across the contraction and express this as a percentage of the loss to be expected if the flow was reversed. Take the coefficient of contraction as 0·6.
[3·2 kN/m^2, 141%]

8.27   Show that the K value for a fitting is linked to the equivalent length of the fitting L_e by an expression:

$$K = 4fL_e/D$$

where D is the diameter of the duct, and f is the applicable friction factor.

8.28   An air duct, carrying a volume Q of air per second, is abruptly changed in section. Deduce the diameter ratio for the two duct sections if the pressure loss is to be independent of flow direction. Assume a value of 0·6 for the contraction coefficient.
[1·66:1]

9 Boundary layer

The drag on a body passing through a fluid may be considered to be made up of two components, the *form drag*, which is dependent on the pressure forces acting on the body, and the *skin friction drag*, which depends on the shearing forces acting between the body and the fluid. Form drag will be dealt with in detail in Chapter 10, while the mechanics of skin friction will be covered in this chapter.

In Chapter 8, it was shown that in both laminar and turbulent flow in pipes the fluid velocity is not uniform but varies from zero at the wall to a maximum at the pipe centre. It was further shown that, in general, the velocity distribution is dependent upon the Reynolds number which defines the type of flow. This chapter will be concerned with the analysis of the effects the fluid viscosity has on the velocity gradient near a solid boundary and, hence, how it affects the skin friction. Such analysis is most conveniently carried out by consideration of flow over a flat plate of infinite width.

The shear stress on a smooth plate is a direct function of the velocity gradient at the surface of the plate. That a velocity gradient should exist in a direction perpendicular to the surface is evident, because the particles of fluid adjacent to the surface are stationary whilst those some distance above the surface move with some velocity. The condition of zero fluid velocity at the solid surface is referred to as 'no slip' and the layer of fluid between the surface and the free stream fluid is termed the boundary layer. Thus, it will be appreciated that any calculations of surface resistance or skin friction forces will obviously involve the integration of the shear stress at the surface over the whole fluid immersed area and will be directly concerned with the patterns of flow within the boundary layer.

Within this context, the importance of Reynolds number becomes self-evident as, with the dramatic change in particle motion consequent upon a transition from laminar to turbulent type of flow, considerable changes in boundary layer flow patterns and velocity gradients must be expected that will materially affect any calculations of surface resistance.

9.1. Qualitative description of the boundary layer

As mentioned above, the boundary layer is taken as that region of fluid close to the surface immersed in the flowing fluid. Figure 9.1 illustrates such a flat plate in a free fluid stream. Only the top surface boundary layer is shown but there will, in practice, be symmetry between the upper and lower surface boundary layers, provided both surfaces are identical in nature. The fluid in contact with the plate surface has zero velocity, 'no slip' and a velocity gradient exists between the fluid in the free stream and the plate surface. Now, shear stress may be defined (equation (8.3)) as

$$\tau = \mu \frac{\mathrm{d}u}{\mathrm{d}y}, \tag{9.1}$$

where τ is the shear stress, μ the fluid viscosity and $\mathrm{d}u/\mathrm{d}y$ the velocity gradient.

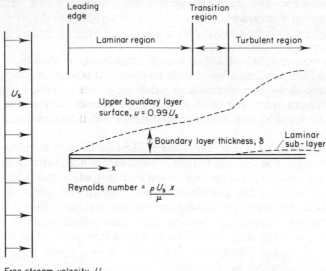

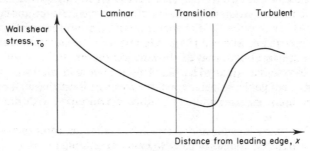

Figure 9.1   Development of the boundary layer along a flat plate, illustrating variations in layer thickness and wall shear stress

This shear stress acting at the plate surface sets up a shear force which opposes the fluid motion and fluid close to the wall is decelerated. Further along the plate, the shear force is effectively increased due to the increasing plate surface area affected, so that more and more of the fluid is retarded and the thickness of the fluid layer affected increases, as shown in Fig. 9.1. Returning to the Reynolds number concept, if the Reynolds number locally were based on distance from the leading edge of the plate, then it will be appreciated that, initially, the value is low so that the fluid flow close to the wall may be categorized as laminar. However, as the distance from the leading edge increases so does Reynolds number until a point must be reached where the flow régime becomes turbulent.

For smooth, polished plates the transition may be delayed until Re equals 500 000. However, for rough plates or for turbulent approach flows, transition may occur at much lower values. Again, the transition does not occur in practice at one well defined point but, rather, a transition zone is established between the two flow régimes, as shown in Fig. 9.1.

The random particle motion characterizing turbulent flow results in a far more rapid growth of the boundary layer in the turbulent region, so that the velocity gradient at the wall increases as does the corresponding shear force opposing motion.

Figure 9.1 also depicts the distribution of shear stress along the plate in the flow direction. At the leading edge, the velocity gradient is large, resulting in a high shear stress. However, as the laminar region progresses so the velocity gradient and shear stress decrease with thickening of the boundary layer. Following transition the velocity gradient again increases and the shear stress rises.

Theoretically, for an infinite plate, the boundary layer goes on thickening indefinitely. However, in practice, the growth is curtailed by other surfaces in the vicinity. This is particularly the case for boundary layers within ducts, as will be described in Section 9.2, where the growth is terminated when the boundary layers from opposite duct surfaces meet on the duct centre line.

It must be appreciated that the thickness of the boundary layer, δ in Fig. 9.1, is much smaller than x.

9.2. Dependence of pipe flow on boundary layer development at entry

The types of fluid flow considered in Chapter 8 have been confined to steady and uniform flow. The assumption of steady, uniform flow conditions led to the simplifying condition that the fluid elements were under no acceleration, either spatial or temporal, and allowed the development of the equations set out there. This assumption implies that the flow conditions are fully established and are not subject to any changes. This is, obviously, not the case in the initial length of a pipe where the boundary layer is still developing and growing in thickness up to its maximum, which for a closed pipe will be the pipe radius.

Initially, as the boundary layer develops, it will be laminar in form. However, as described earlier, the boundary layer will become turbulent, depending upon the ratio of inertial and viscous forces acting on the fluid, this condition being normally monitored by reference to the value of the flow Reynolds number.

For pipe flow, it is normal practice to base the Reynolds number, Re, on the mean flow velocity and the pipe diameter. Generally, for values of Re < 2000, the flow may be assumed to be laminar, although it has been shown possible to maintain laminar flow at higher values of Reynolds number under specialized laboratory conditions. Above Re = 2000 it is, however, reasonable to suppose that the flow will be turbulent and that the boundary layer development will include a transition and a turbulent region, as described for the flat plate. The only major difference is that, in the pipe flow case, there is a limit to boundary layer thickness growth, namely the pipe radius.

If, therefore, this limit is reached before transition occurs, i.e. if laminar boundary layers meet at the pipe centre, the flow in the remainder of the pipe will be laminar. On the other hand, if transition within the boundary layer occurs before they fill the pipe, the flow in the rest of the pipe will be turbulent. These two cases are illustrated in Fig. 9.2.

Once the boundary layer, whether laminar or turbulent in nature, has

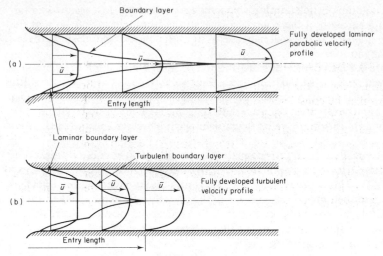

Figure 9.2   Development of fully-developed laminar and turbulent flow in a circular
pipe. (a) Laminar flow conditions, Re < 2000. (b) Turbulent flow
conditions, Re > 2000

grown to fill the whole pipe cross-section, the flow may be said to be fully
developed and no further changes in velocity profile are to be expected
downstream, provided that the pipeline characteristics (i.e. diameter, surface
roughness) remain constant.

Theoretically, the entry length for a particular pipe (i.e. the distance from
entry at which a laminar or turbulent boundary layer ceases to grow) is
infinite. However, it is normally assumed that the flow has become fully
developed when the maximum velocity, at the pipe centre line, becomes 0·99
of the theoretical maximum. Using this approximation, typical entry lengths
for establishment of fully developed laminar or turbulent flow may be taken
as 120 and 60 pipe diameters, respectively. The entry length characteristic of
turbulent flow is the shorter due to the higher growth rate of the turbulent
boundary layer.

Thus, the assumption of steady, uniform flow restricts the application of
the equations derived for pipe flow to that part of a conduit beyond the
entry length. Normally, this is not a serious restriction as the entry length is
usually small compared to the total length of the pipeline.

Similarly, all the equations derived for laminar flow have depended on the
shear stress–viscosity relation of equation (9.1), $\tau = \mu \, du/dy$. In turbulent
flow, due to the random nature of the motion of the fluid particles, the
apparent shear stress may be expressed as

$$\tau = (\mu + \epsilon) \frac{du}{dy},$$

(9.2)

where ϵ is the *eddy viscosity* and is often much larger than μ. Since eddy
viscosity is difficult to determine, equations dealing with the calculations of
pressure loss associated with turbulent flow in pipes, established in Chapter 8,
were developed, introducing the concept of an empirical friction factor.

However, it will be shown that the friction factor is related to the skin friction coefficient defined later in this chapter.

9.3. Factors affecting transition from laminar to turbulent flow régimes

As mentioned above, the transition from laminar to turbulent boundary layer conditions may be considered as Reynolds-number-dependent, $Re = \rho U_s x / \mu$, and a figure of 5×10^5 is often quoted. However, this figure may be considerably reduced if the surface is rough. For $Re < 10^5$, the laminar layer is stable, however, at Re near 2×10^5 it is difficult to prevent transition.

The presence of a pressure gradient dp/dx can also be a major factor. Generally, if dp/dx is positive, then transition Reynolds number is reduced, a negative dp/dx increasing transition Reynolds number. This effect forms the basis of suction high lift devices designed for aircraft wings. Figure 9.3

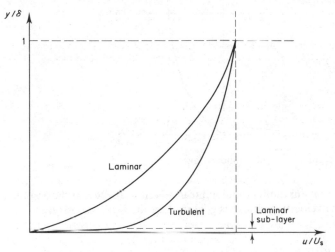

Figure 9.3   Typical velocity profiles in the laminar and turbulent boundary layer regions

illustrates typical velocity profiles through the boundary layer in both the laminar and turbulent regions, the increased velocity gradient du/dy being apparent. As mentioned, the growth of the boundary layer thickness is more rapid in the turbulent region, roughly varying as $x^{0.8}$ here compared to $x^{0.5}$ in the laminar region.

In calculations involving long plates, it is often reasonable to suppose that transition occurs close to the leading edge and, in such cases, the presence of the laminar section may be ignored.

The study of the turbulent boundary layer is the more important as in most engineering applications the flow Reynolds number is sufficiently high to ensure transition and the establishment of a turbulent boundary layer. However, it will be appreciated that the random motion of the fluid particles must die out very close to the surface to maintain the condition of no slip at the plane—fluid interface. To accommodate this, the presence of a laminar sub-layer in the turbulent region has been established, the thickness of this being small compared to the local boundary layer thickness, as shown in Fig. 9.1. The velocity profile across this sub-layer is assumed linear and tangential to the velocity profile up through the turbulent boundary layer.

9.4. Discussion of flow patterns and regions within the turbulent boundary layer

Figure 9.4 illustrates the velocity distribution through one particular section in a turbulent boundary layer. As mentioned above, very close to the plane

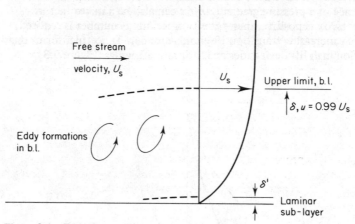

Figure 9.4   Eddy formation in the boundary layer

surface the flow remains laminar and a linear velocity profile may be assumed. In this region, the velocity gradient is governed by the fluid viscosity (equation (9.1))

$$\frac{du}{dy} = \tau_0/\mu.$$

Rearranging this expression yields, after integration,

$$u = (\tau_0/\mu)y \tag{9.3}$$

or $$\frac{u}{\sqrt{(\tau_0/\rho)}} = \frac{\sqrt{(\tau_0/\rho)}}{\nu}y \tag{9.4}$$

where ν is the fluid kinematic viscosity μ/ρ. The term $\sqrt{(\tau_0/\rho)}$ is common in boundary layer theory and termed the shear stress velocity due to the units of velocity applicable to the combination (*see* Chapter 8) and is denoted by u^*. Thus,

$$\frac{u}{u^*} = \frac{y}{\nu/u^*}. \tag{9.5}$$

Experimentally, it has been shown that the laminar sub-layer occurs in flows, where equation (9.5) has a value less than approximately 5, so that the thickness of the laminar sub-layer $y = \delta'$ becomes

$$\delta' = 5\nu/u^*, \tag{9.6}$$

indicating that the sub-layer thickness will be small for large shear stress flows, i.e. u^* large, and that it will increase in the downstream direction as shear stress decreases in this direction. Above the laminar sub-layer, the flow régime is turbulent and equation (9.1) no longer adequately represents the shear forces acting. It is appropriate here to describe the mechanism of flow within this upper region.

Due to the random motion of the fluid particles, eddy patterns are set up in the boundary layer which sweep small masses of fluid up and down through the boundary layer, moving in a direction perpendicular to the surface and the mean flow direction. Due to these eddies, fluid from the upper, higher velocity areas is forced into slower-moving stream above the laminar sub-layer, having the effect of increasing the local velocity here relative to its value in the laminar boundary layer. This increase in velocity of fluid close to the wall is shown in Fig. 9.3. Conversely, slow-moving fluid is lifted into the upper levels, slowing down the fluid stream and, by doing so, effectively thickening the boundary layer, explaining the more rapid growth of the turbulent boundary layer compared to the laminar.

The process described is, effectively, a momentum transfer phenomenon. However, the effect is analagous to a shear stress applied to the fluid as the overall deceleration is increased as boundary layer thickness increases. In order to explain this process, and to retain the useful form of equation (9.1), a new viscosity term may be introduced, the eddy viscosity, ϵ, and equation (9.1) rewritten as the relationship mentioned earlier (equation (9.2)),

$$\tau = (\epsilon + \mu) \frac{\mathrm{d}u}{\mathrm{d}y}.$$

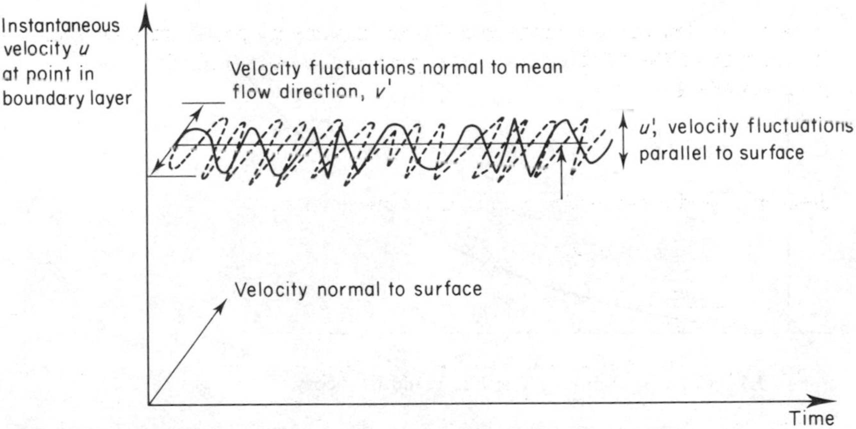

Figure 9.5   Velocity fluctuations in the mean flow direction and normal to the surface at a point in the turbulent boundary layer

Figure 9.5 illustrates the likely output from a velocity measuring device positioned within the turbulent boundary layer. Here it will be seen that, although the mean velocity is in the flow direction, there are fluctuations in velocity corresponding to the random particle motion.

If the fluid velocity is made up of a mean value $\bar{u}$ and fluctuating components u' and v' in the flow direction and perpendicular to it, respectively, then it may be assumed that the apparent shear stress required to duplicate the eddy effects discussed above would be

$$\tau = -\rho\,\overline{u'v'}, \tag{9.7}$$

i.e. the shear stress opposing motion is given by the product of fluid density and the average product of the normal velocity fluctuations over an incremental time period.

9.5. Prandtl mixing length theory

In the form of equation (9.7), little further may be done. However, Prandtl (1875–1953) – who, in 1904, was responsible for stating the basics of boundary layer theory and proposing that all viscous effects are concentrated within it – developed the necessary theory to relate the apparent shear stress to mean velocity distribution through the boundary layer. This theory, summarized below, is known as the Prandtl mixing length theory.

Prandtl defined the mixing length as that distance l in which a particle loses its excess momentum and assumes the mean velocity of its surroundings, an idea in some respects similar to the mean free path. In practice, the loss or transfer of momentum would be gradual over the length l. Assuming that the changes of velocity u' and v' following from this particle motion would be equal – a not unreasonable assumption – it may be seen that $v' = u' = l\,\mathrm{d}u/\mathrm{d}y$ (Fig. 9.6) and, from equation (9.7),

$$\tau = \rho l^2 \left(\frac{\mathrm{d}u}{\mathrm{d}y}\right)^2. \tag{9.8}$$

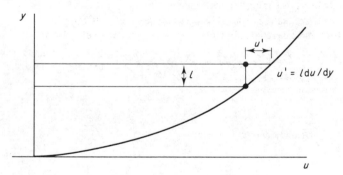

Figure 9.6   Concept of mixing length in Prandtl's theory

Close to the surface, Prandtl assumed that l became dependent on the distance from the surface, or $l = ky$. This allows for l to have zero value at the boundary where $y = 0$. Hence

$$\tau = \rho k^2 y^2 \left(\frac{\mathrm{d}u}{\mathrm{d}y}\right)^2,$$

where k was proposed as a universal constant having a value around 0·4. More recent work has shown distinct limitations in this approach and values varying from 0·4 have been recorded. However, the Prandtl mixing length theory was a major advance at the time and may still be of use in particular situations.

Close to the surface it may be assumed that the shear stress equals the surface value so

$$\tau_0 = \rho k^2 y^2 \left(\frac{du}{dy}\right)^2 \tag{9.9}$$

or $\quad du = \dfrac{\sqrt{(\tau_0/\rho)}}{k} \dfrac{dy}{y}$

which, on integration, yields

$$u/u^* = (1/k) \log_e y + C. \tag{9.10}$$

Values of the integration constant C have been experimentally determined in the form

$$C = 5·56 - (1/k) \log_e (v/u^*),$$

so that a velocity distribution

$$u/u^* = (1/k) \log_e \{y(u^*/v)\} + 5·56 \tag{9.11}$$

is obtained. Substituting 0·4 for k yields

$$u/u^* = 5·75 \log_{10} \{y(u^*/v)\} + 5·56 \tag{9.12}$$

in terms of log base 10.

Comparison of equation (9.12), which applies for $30 < yu^*/v < 500$, with equation (9.5) shows that there is a major change in velocity profile between the laminar sub-layer and the turbulent boundary layer region. However, both profiles are related through the yu^*/v term.

Figure 9.7 illustrates these profiles. However, above $yu^*/v = 500$, experimental results indicate that a better fit is obtained by a velocity defect

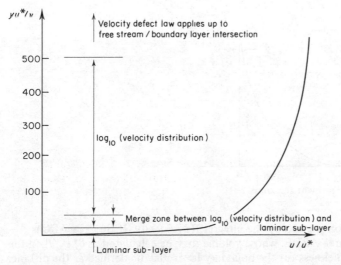

Figure 9.7   Overlap of velocity distributions

law of the form

$$(U_s - u)/u^* = f(y/\delta). \tag{9.13}$$

Thus, three zones of application of velocity distribution equations are apparent. These zones do not possess sharp boundaries, rather they merge into each other. In the intersection zones, experimental results straddle the predictions of each equation; however, the general boundaries of 30 and 500 for yu^*/v are adequate in this treatment.

In terms of experimental results, a simplified velocity profile, which applies to 90 per cent of the boundary layer thickness, but not to the 10 per cent close to the plane surface, was proposed by Prandtl,

$$u/U_s = (y/\delta)^n \tag{9.14}$$

The value of n for Reynolds numbers in the region $10^5 > Re > 10^7$ may be taken of $\frac{1}{7}$ and the expression (9.14) is known as the seventh power law. In addition, it is usually assumed that the velocity profile through the laminar sub-layer is linear and tangential to the seventh-power law.

9.6. Definitions of boundary layer thicknesses

So far the boundary layer thickness has been referred to only in physical terms; namely, boundary layer thickness is defined as that distance from the surface where the local velocity equals 99 per cent of the free stream velocity:

$$\delta = y_{(u = 0.99 \, U_s)}, \tag{9.15}$$

where U_s is the free stream velocity. It is possible, however, to define boundary layer thickness in terms of the effect on the flow.

9.6.1. Displacement thickness δ^*

Due to the presence of the boundary layer, the flow past a given point on the surface is reduced by a volume equivalent to the area ABC in Fig. 9.8. This

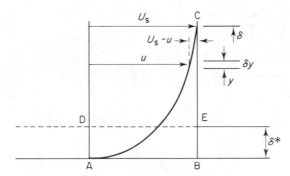

Figure 9.8   Displacement thickness

volume reduction is given by the integral $\int(U_s - u) \, dy$. If the area ABC is equated to an area ABDE, whose volume may be calculated as $\delta^* U_s$, then the displacement thickness for the boundary layer may be defined as the distance the surface would have to move in the y direction to reduce the flow passing by a volume equivalent to the real effect of the boundary layer:

$$\delta^* = \int_0^\infty (1 - u/U_s)\, dy. \tag{9.16}$$

9.6.2. Momentum thickness θ

The fluid passing through the element δy carries momentum at a rate $(\rho u \delta y)u$ per unit width, whereas, in the absence of the boundary layer, u would equal U_s, so that the total reduction in momentum flow is

$$\int_0^\infty \rho(U_s - u)u\, dy,$$

which may be equated to the momentum carried through a section θ deep per unit width at the free stream velocity (Fig. 9.9):

$$(\rho U_s \theta)\, U_s = \int_0^\infty \rho(U_s - u)u\, dy,$$

$$\theta = \int_0^\infty \frac{u}{U_s}\left(1 - \frac{u}{U_s}\right) dy. \tag{9.17}$$

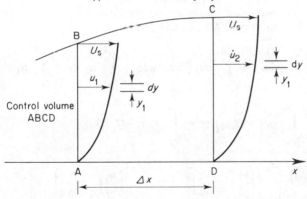

Figure 9.9   Control volume applied to a general section of boundary layer over a flat plate

9.7. Application of momentum equation to a general section of boundary layer

Von Kármán first applied the momentum equation to a general section of a boundary layer. Regardless of the position of the section in either the laminar or turbulent boundary layer regions, it is possible to equate the skin friction drag force to the product of rate of change of momentum and mass of fluid affected by the boundary layer. Figure 9.9 illustrates a control volume ABCD around a general section of boundary layer. It will be assumed that the flow continues to be incompressible and that as dp/dx, the pressure gradient in the flow direction, is also zero so will be dU_s/dx, i.e. any change in free stream velocity. Flow enters the control volume through AB and BC as shown and leaves via CD. Assuming a unit width of surface, then the momentum equation applied to this control volume in the flow direction becomes

$$-\tau_0 \Delta x = \int_0^{\delta_2} \rho u_2^2 \; \mathrm{d}y - \int_0^{\delta_1} \rho u_1^2 \; \mathrm{d}y - \rho U_s^2 (\delta_2 - \delta_1). \qquad (9.18)$$

Taking the terms in order: $\tau_0 \Delta x$ represents the shear force, opposing motion in the +ve flow direction, over the immersed area Δx;

$$\int_0^{\delta_2} \rho u_2^2 \; \mathrm{d}y = \int_0^{\delta_2} \rho u_2 \; \mathrm{d}y \, u_2$$

is the rate of momentum transfer through a section $\mathrm{d}y$ on CD at a height y above the surface, where the local velocity is u_2; the third term is identical in form to the preceding integral except that it applies to a section $\mathrm{d}y$ on AB; the last term is a measure of the momentum, in the flow direction, carried into the control volume across the boundary layer upper surface BC. But

$$U_s(\delta_2 - \delta_1) = \int_0^{\delta_2} u_2 \; \mathrm{d}y - \int_0^{\delta_1} u_1 \; \mathrm{d}y, \qquad (9.19)$$

i.e. the difference in flow rates past AB and CD. Hence,

$$-\tau_0 \Delta x = \int_0^{\delta_2} \rho u_2^2 \; \mathrm{d}y - \int_0^{\delta_1} \rho u_1{}^2 \; \mathrm{d}y - \rho U_s \left[\int_0^{\delta_2} u_2 \; \mathrm{d}y - \int_0^{\delta_1} u_1 \; \mathrm{d}y \right]$$

$$= \rho \left[\int_0^{\delta_2} (u_2^2 - U_s u_2) \; \mathrm{d}y - \int_0^{\delta_1} (u_1^2 - U_s u_1) \; \mathrm{d}y \right]$$

$$= \rho U_s^2 \left[\int_0^{\delta_2} \left\{ \left(\frac{u_2}{U_s} \right)^2 - \frac{u_2}{U_s} \right\} \mathrm{d}y - \int_0^{\delta_1} \left\{ \left(\frac{u_1}{U_s} \right)^2 - \frac{u_1}{U_s} \right\} \mathrm{d}y \right].$$

As Δx approaches zero in the limit, and multiplying both sides by -1, the equation above reduces to

$$\tau_0 = \rho U_s^2 \frac{\mathrm{d}}{\mathrm{d}x} \int_0^{\delta} \frac{u}{U_s} \left(1 - \frac{u}{U_s} \right) \mathrm{d}y \qquad (9.20)$$

or, by reference to the momentum thickness defined in equation (9.17),

$$\tau_0 = \rho U_s^2 \frac{\mathrm{d}\theta}{\mathrm{d}x}. \qquad (9.21)$$

Equation (9.20) is the momentum equation applied to a general boundary layer section and is of general use in deriving further relations in the boundary layer. It is also of use in the area of heat transfer through boundary layers, although these applications are outside the scope of this text.

In the following sections, more detailed relations applying to the laminar and turbulent boundary layers individually will be presented.

9.8. Properties of the laminar boundary layer formed over a flat plate in the absence of a pressure gradient in the flow direction

In practice, the laminar section of the boundary layer formed as a result of flow over a surface is short; however, it will always exist, even in flows that are nominally turbulent. For example, consider the inlet section of a circular cross-section duct. As discussed in Chapter 8, the flow in the duct will be considered turbulent if the Reynolds number based on mean fluid velocity and duct diameter $Re = \rho \bar{u} \, d/\mu > 2000$. However, as far as the boundary layer is concerned, the transition from laminar to turbulent occurs at Reynolds numbers above 10^5 based on mean fluid velocity and distance measured from the entry to the duct, $Re = \rho U_s x/\mu$, so that there will always be a finite length of laminar boundary layer.

Blasius developed a series of analytical solutions for the laminar boundary layer which will be quoted in this section and compared to approximate results that may be derived from the assumptions already discussed; namely, a linear relation between shear stress and vertical distance to the surface and the absence of a pressure gradient across the flat surface.

In all laminar flow, equation (9.1) applies, i.e.

$$\tau_0 = \mu \left(\frac{\mathrm{d}u}{\mathrm{d}y}\right)_{y=0}, \tag{9.22}$$

and, from equation (9.20), the momentum equation,

$$\mu\left(\frac{\mathrm{d}u}{\mathrm{d}y}\right)_{y=0} = \rho \frac{\mathrm{d}}{\mathrm{d}x} \int_0^\delta u\,(U_s - u)\,\mathrm{d}y. \tag{9.23}$$

If the assumption is made that velocity profiles through the boundary layer are geometrically similar along the whole length of the laminar section, then this may be expressed as

$$u = U_s f(\eta), \tag{9.24}$$

where $\eta = y/\delta$, $u = 0$ at $y = 0$, or $\eta = 0$, $u = U_s$ at $y = \delta$, or $\eta = 1$. Substituting into (9.23) yields

$$\frac{\mu}{\delta}\, U_s \left(\frac{\mathrm{d}f(\eta)}{\mathrm{d}\eta}\right)_{\eta=0} = \rho \frac{\mathrm{d}}{\mathrm{d}x} \left[U_s^2 \delta \int_0^1 (1 - f(\eta))f(\eta)\,\mathrm{d}\eta\right], \tag{9.25}$$

where the limits of integration have been changed as $\eta = 1$ at $y = \delta$.

Due to the geometric similarity of velocity profiles, $f(\eta)$ is independent of position along the laminar section x, so that

$$\int_0^1 (1 - f(\eta)) f(\eta) \, d\eta$$

is independent of x and may be regarded as a constant C_1 and $\partial f(\eta)/\partial \eta$ as a constant C_2. As a result, equation (9.25) becomes

$$\frac{\mu}{\delta} U_s C_2 = \rho U_s^2 C_1 \frac{d\delta}{dx}$$

or

$$\mu \frac{C_2}{C_1} = \rho U_s \delta \frac{d\delta}{dx} \, .$$

Integrating,

$$\delta^2/2 = (\mu/\rho U_s)(C_2/C_1) \, x + \text{constant},$$

so

$$\delta = \sqrt{\{(2C_2/C_1)(\mu/\rho U_s)\}} \, x = \sqrt{(2C_2/C_1)} x/\sqrt{(\text{Re})} \qquad (9.26)$$

if $\delta = 0$ at $x = 0$, i.e. zero boundary layer thickness at the leading edge, and

$$\text{Re}_x = \rho U_s x/\mu.$$

From (9.25),

$$\tau_0 = \rho U_s^2 C_1 \frac{d\delta}{dx} = \rho U_s^2 C_1 \sqrt{\left(\frac{2C_2}{C_1} \frac{\mu}{\rho U_s}\right)} \frac{d(x^{1/2})}{dx} \, ,$$

$$\tau_0 = \rho U_s^2 C_1 \sqrt{\left(\frac{2C_2 \mu}{C_1 \rho U_s}\right)} \frac{x^{-1/2}}{2} = \rho U_s^2 \sqrt{\left(\frac{C_1 C_2}{2 \, \text{Re}_x}\right)} \qquad (9.27)$$

and, from the integral of shear stress at the wall over the length l of the laminar boundary layer, the total skin friction force per unit width of surface may be written as

$$F = \int_0^l \tau_0 \, dx = \int_0^l \rho U_s^2 C_1 \frac{d\delta}{dx} \, dx = \left[\rho U_s^2 C_1 \delta \right]_0^l = \rho U_s^2 l \sqrt{\left(\frac{2 C_1 C_2}{\text{Re}_l}\right)},$$

when equations (9.27) and (9.26) are employed to substitute for τ_0 and δ. Simplifying yields

$$F = \rho U_s^2 \sqrt{(2C_1 C_2 \, \mu/\rho U_s l)} l$$

$$= \sqrt{(2C_1 C_2 \, \rho \mu U_s^3 \, l)} \qquad (9.28)$$

and the *skin friction coefficient* C_f may then be calculated as

$$C_f = F/\tfrac{1}{2}\rho U_s^2 l \text{ per unit width.} \qquad (9.29)$$

However, equation (9.28) is of little value in the form shown due to the presence of C_1 and C_2. If these constants could be evaluated, then the skin friction for a flat plane would be known. The values of C_1, C_2 depend on the assumptions made with respect to the variation of shear stress with distance above the plane, i.e. $\tau = f(y)$ or $f(\eta)$. Now, we have already mentioned that this function may be assumed to be linear and the boundary conditions at

$y = 0$ and $y = \delta$ are known to be $\tau = \tau_0$ and $\tau = 0$, respectively. This may be expressed by a relation of the form

$$\tau = C_3(\delta - y), \tag{9.30}$$

$$\mu \frac{du}{dy} = C_3(\delta - y)$$

$$\mu u = C_3(y\delta - y^2/2) + C_4.$$

Now $u = 0$ at $y = 0$. Therefore, $C_4 = 0$ and so

$$\mu u = C_3(y\delta - y^2/2).$$

As $u = U_s$ when $y = \delta$, $C_3 = 2\mu U_s/\delta^2$. Hence,

$$\mu u = 2\mu(U_s/\delta^2)(y\delta - y^2/2),$$

$$u/U_s = 2(y/\delta - y^2/2\delta^2) = 2(\eta - \eta^2/2),$$

and so $\quad u/U_s = 2\eta - \eta^2 \tag{9.31}$

is the resulting velocity profile from the linear shear stress vs. distance above surface assumption. This allows C_1 and C_2 to be evaluated as

$$C_1 = \int_0^1 (1 - f(\eta))f(\eta)\,d\eta = \int_0^1 [(1 - (2\eta - \eta^2)]\,(2\eta - \eta^2)\,d\eta$$

$$= 2/15$$

and $\quad C_2 = \left[\frac{\partial}{\partial \eta}f(\eta)\right]_{\eta=0} = \left[\frac{\partial}{\partial \eta}(2\eta - \eta^2)\right]_{\eta=0} = (2 - 2\eta)_{\eta=0}$

$$= 2.$$

Substituting back, yields

$$\delta = \sqrt{(2 \times 2 \times \tfrac{15}{2})}\{x/\sqrt{(Re_x)}\} = 5 \cdot 48x/\sqrt{(Re_x)} \tag{9.32}$$

and $\quad C_f = \dfrac{\sqrt{(2C_1 C_2 \rho\mu U_s^3 l)}}{\tfrac{1}{2}\rho U_s^2 l}$

$$= \sqrt{\left(\frac{2 \times \tfrac{2}{15} \times 2 \times \rho\mu U_s^3 l}{\tfrac{1}{4}\rho^2\,U_s^4 l^2}\right)}$$

$$= \sqrt{\frac{\left(\tfrac{32}{15}\right)}{Re_x}} \text{ per plate side} = 1\cdot4\,Re_x^{-1/2}. \tag{9.33}$$

Blasius was able, by reference to the general equations of motion for boundary layers, to plot the velocity distribution up through the laminar boundary layer in a form

$$u/U_s = f(y\,Re_x^{-1/2}/x) \tag{9.34}$$

and, from this plot, again making the assumptions that $y = \delta$ when $u = 0 \cdot 99 U_s$ and that the velocity profiles are geometrically similar along the surface, Blasius was able to show that, approximately,

$$\delta = 5x.\,Re_x^{-1/2}, \tag{9.35}$$

which is comparable with equation (9.32) above. Similarly, by taking the slope of the curve (9.34) at $y = 0$,

$$\left(\frac{\mathrm{d}u}{\mathrm{d}y}\right)_{y=0} = 0 \cdot 332 \frac{U_s}{x} \mathrm{Re}_x^{1/2}, \tag{9.36}$$

which, when substituted into

$$\tau_0 = \mu \left(\frac{\mathrm{d}u}{\mathrm{d}y}\right)_{y=0}$$

and integrated from $x = 0$ to l, gives a skin friction coefficient,

$$C_f = 1 \cdot 33 \ \mathrm{Re}_l^{-1/2} \tag{9.37}$$

which is comparable to equation (9.33) above.

A number of alternative velocity profiles have been suggested to replace (9.31). However, the results do not differ substantially from those shown above.

Values for the displacement and momentum thicknesses of the boundary layer may also be calculated in terms of δ, initially from equation (9.31) substituted into equations (9.16) and (9.17), respectively. Hence,

$$\delta^* = \delta \int_0^1 [1 - f(\eta)] \ \mathrm{d}\eta = \delta \int_0^1 (1 - 2\eta + \eta^2) \ \mathrm{d}\eta,$$

$$= \delta/3 = 1 \cdot 86x \ \mathrm{Re}_x^{-1/2} \tag{9.38}$$

and

$$\theta = \delta \int_0^1 f(\eta) [1 - f(\eta)] \ \mathrm{d}\eta$$

$$= \delta \int_0^1 (2\eta - \eta^2)(1 - 2\eta + \eta^2) \ \mathrm{d}\eta$$

$$= \tfrac{2}{15}\delta = 0 \cdot 73x \ \mathrm{Re}_x^{-1/2}. \tag{9.39}$$

Experimental results verify the Blasius solution except close to the leading edge of the surface, where the assumption of a zero velocity component normal to the surface is not strictly valid. However, the results above are usable for the calculation of skin friction forces and boundary layer thicknesses.

Typical laminar boundary layer thicknesses are of the order 0·75 mm in air at 100 m s^{-1}, Re $= 10^6$ and typical lengths, for a smooth flat plate, would be around 160 to 200 mm. Measurement of boundary layer velocity profiles is difficult and requires specialized information. The advent of the hot wire anemometer has made life a lot easier here, but great care is still necessary to ensure that the results obtained are not a direct function of the experimental set up. Again, it may be appreciated that the laminar section of the boundary layer is, generally, of secondary importance to the turbulent section, which will be dealt with in the next section.

Example 9.1

Oil with a free stream velocity of $3 \cdot 0$ m s^{-1} flows over a thin plate $1 \cdot 25$ m wide and 2 m long. Determine the boundary layer thickness and the shear stress at mid-length and calculate the total, double-sided resistance of the plate. ($\rho = 860$ kg m^{-3}, $\nu = 10^{-5}$ m^2 s^{-1}, $\nu = \mu/\rho$.)

Solution

Calculate the Reynolds number at $x = 1$ m.

$$\mathrm{Re}_x = U_s x/\nu = 3x/10^{-5}.$$

Therefore

$$\mathrm{Re}_x^{1/2} = 5 \cdot 48 \times 10^2.$$

Note, Re is low enough to allow laminar boundary layer to survive over whole plate.

$$\tau_0 = 0 \cdot 332 \, \mu(U_s/x) \, \mathrm{Re}_x^{1/2}$$

$$= 0 \cdot 332 \times \frac{10^{-5}}{860} \times \frac{3}{1} \times 5 \cdot 48 \times 10^2$$

$$= 6 \cdot 347 \times 10^{-5}.$$

The skin friction force is given by, double-sided,

$$F = 2 \times \tfrac{1}{2}\rho U_s^2 l \times b \times C_f,$$

where l is plate length and b is plate width,

$$F = 2 \times \tfrac{1}{2} \times 860 \times 3^2 \times 2 \times 1 \cdot 25 \times C_f,$$

where (from equation (9.37))

$$C_f = 1 \cdot 33 \, \mathrm{Re}_l^{-1/2}$$

$$= 1 \cdot 33/(6 \times 10^5)^{1/2}.$$

Therefore,

$$F = 860 \times 18 \times 1 \cdot 25 \times 1 \cdot 33/\{\sqrt{(60)} \times 10^2\}$$

$$= 33 \cdot 224 \text{ N}.$$

9.9. Properties of the turbulent boundary layer over a flat plate in the absence of a pressure gradient in the flow direction

The majority of boundary layers met in engineering practice are turbulent over most of their length, and so the study of this section of the development of the boundary layer is usually regarded as of greater fundamental importance than the laminar section. In many cases, the laminar section of the boundary layer is short enough, compared to the total length of the surface, to be ignored in calculations of skin friction forces.

The momentum equation (9.20) may be applied to the turbulent boundary layer as no limiting assumptions were made in its derivation. However, as mentioned in Chapter 5, a new relation for the velocity profile up through the boundary layer will have to be found and the shear stress will no longer be obtained simply from the product of fluid viscosity and the gradient of the velocity profile.

Due to the basic similarity between the development of boundary layers within circular cross-section pipes and over flat pipes, Prandtl suggested that results from the pipe case be applied to the analysis of flat plate turbulent boundary layers. As was mentioned in Section 9.2, the boundary layer growth in pipes is limited to the pipe radius R, so that $u = U_s$ at $y = R$, and the mean flow velocity in turbulent pipe flow is known to be about $0.8\ U_s$. The velocity distribution in such flow is adequately represented by the Prandtl power law,

$$u/U_s = (y/\delta)^n \tag{9.40}$$

where $n = \frac{1}{7}$ for $\mathrm{Re}_x < 10^7$. Obviously, this profile breaks down at the wall, where $y = 0$. However, the presence of a laminar sub-layer has already been discussed (Section 9.2) where the velocity decreases linearly to zero at the wall, this profile being tangential to the power law (Fig. 9.7).

To develop the analog between flat plates and pipe flow, it is necessary to appreciate that $\delta = R$ in the fully developed region and to develop some relation for τ_0 to replace equation (9.1), which no longer applies.

Blasius proposed that, for smooth pipes, the shear stress at the wall could be expressed by

$$\tau_0 = f\tfrac{1}{2}\rho\bar{u}^2, \tag{9.41}$$

where $\bar{u}$ is the mean fluid velocity $= 0.8\ U_s$ and f is an empirical constant known as the friction factor which is a function of flow Reynolds number ($\mathrm{Re} = \rho\bar{u}\,d/\mu$, d = pipe diameter) and the ratio of wall roughness to pipe diameter. Friction factors are covered in more detail in Chapter 8. Thus, $\tau_0 = \tfrac{1}{2}\rho(0.8\ U_s^2)f$ and, as Blasius developed the expression

$$f = 0.079/\mathrm{Re}^{1/4} = 0.079/(\rho\bar{u}d/\mu)^{1/4}$$

to apply to smooth pipes, substitution yields an expression

$$\tau_0 = \tfrac{1}{2}\rho(0.8\ U_s^2)\,0.079(\mu/\rho0.8U_s2R)^{1/4}$$

and, if $\delta = R$, then,

$$\tau_0 = 0.0225\,\rho U_s^2(\mu/\rho U_s\delta)^{1/4}. \tag{9.42}$$

As the assumption of zero pressure gradient has been made, equation (9.20) can be applied. Thus,

$$\tau_0 = \rho U_s^2\,\frac{\mathrm{d}}{\mathrm{d}x}\int \frac{u}{U_s}\left(1 - \frac{u}{U_s}\right)\mathrm{d}y$$

$$= \rho U_s^2\,\frac{\mathrm{d}\delta}{\mathrm{d}x}\int_0^1 (1 - \eta^{1/7})\eta^{1/7}\,\mathrm{d}\eta, \tag{9.43}$$

where

$$u/U_s = (y/\delta)^{1/7} = \eta^{1/7}.$$

Therefore,

$$\tau_0 = \frac{7}{72} \rho U_s^2 \frac{d\delta}{dx}. \tag{9.44}$$

Equating these two expressions (9.42) and (9.44) for τ_0 yields

$$\delta^{1/4} \, d\delta = 0.234 \, (\mu/\rho U_s)^{1/4} \, dx.$$

Integrating yields

$$\tfrac{4}{5} \, \delta^{5/4} = 0.234 \, (\mu/\rho U_s)^{1/4} x + C_5.$$

Now, if the turbulent boundary layer is assumed to extend to the plate leading edge, which is reasonable if the plate is long compared to the length of the laminar layer length, then $\delta = 0$ at $x = 0$ and $C_5 = 0$. Hence,

$$\delta^{5/4} = 0.292 \, (\mu/\rho U_s)^{1/4} x,$$

$$\delta = 0.37x/(\rho U_s x/\mu)^{1/5}$$

$$= 0.37x . \, \mathrm{Re}_x^{-1/5}. \tag{9.45}$$

Comparing equation (9.45) to (9.32), it may be seen that the turbulent boundary layer grows more rapidly than the laminar layer, the proportionality to distance along the plate being to the power $x^{4/5}$ and $x^{1/2}$, respectively. The skin friction force on the flat surface may be determined by eliminating δ between equations (9.42) and (9.45). Hence,

$$\tau_0 = 0.029 \, \rho U_s^2 (\mu/\rho U_s x)^{1/5}$$

and

$$F = \int_0^l \tau_0 \, dx \text{ per unit width,}$$

where l is plate length.

$$F = 0.036 \, \rho U_s^2 l (\mu/\rho U_s l)^{1/5}$$

$$= 0.036 \, \rho U_s^2 \, l \, \mathrm{Re}_l^{-1/5} \tag{9.46}$$

and the skin friction coefficient,

$$C_f = F/\tfrac{1}{2}\rho U_s^2 l \text{ per unit width}$$

$$= 0.072 \, \mathrm{Re}_l^{-1/5}. \tag{9.47}$$

The expression above is valid for Reynolds numbers up to 10^7, but experimental results indicate that a better approximation is given by

$$C_f = 0.074 \, \mathrm{Re}_l^{-1/5}. \tag{9.48}$$

Prandtl has suggested subtracting the length of the laminar layer, resulting in an expression

$$C_f = 0.074 \, \mathrm{Re}_l^{-1/5} - 1700 \, \mathrm{Re}_l^{-1}$$

to apply from

$$\mathrm{Re}_l = 5 \times 10^5 \text{ to } 10^7.$$

To extend the Reynolds number range further, Schlichting employed the logarithmic velocity distribution for pipes under turbulent flow conditions, which have already been mentioned in Chapter 8, resulting in a semi-empirical relation,

$$C_f = 0.455 \, (\log_{10} Re_l)^{-2.58} \qquad\qquad (9.49)$$

applying from $10^6 < Re_l < 10^9$.

Comparison of equation (9.47) with equation (9.37) shows that the skin friction is proportional to $\frac{9}{5}$-power of velocity of the main stream and the $\frac{4}{5}$-power of plate length for the turbulent layer, compared to the $\frac{3}{2}$- and $\frac{1}{2}$-powers, respectively, for the laminar layer. Generally, then, it may be seen that retention of a laminar boundary as long as possible is desirable from a

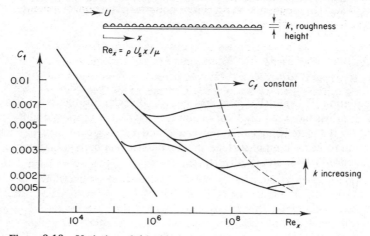

Figure 9.10 Variation of skin friction coefficient with Reynolds number

drag viewpoint. Figure 9.10, a plot of C_f vs. Re_l, illustrates the variations in skin friction coefficient.

Example 9.2

A smooth flat plate 3 m wide and 30 m long is towed through still water at 20 °C at a speed of 6 m s^{-1}. Determine the total drag on the plate and the drag on the first 3 m of the plate.

Solution
For the whole plate:

$$Re_l = 1000 \times 6 \times 30/10^{-3} = 180 \times 10^6,$$

$$\rho = 1000 \text{ kg m}^{-3}, \mu = 10^{-3} \text{ N s m}^{-3},$$

$$C_f = 0.455/[\log_{10}(1.8 \times 10^8)]^{2.58}, = 0.001\,96.$$

Drag on both sides of the plate,

$$F = 2\{\tfrac{1}{2}\rho U_s^2 \, C_f\}$$

$$= 2 \times \tfrac{1}{2} \times 1000 \times 36 \times 0.001\,96$$

$$= \mathbf{70.65 \ N.}$$

Considering the point at which the boundary layer becomes turbulent, assume transition at $\text{Re}_l = 10^5$:

$$10^5 = \rho U_s l_t / \mu,$$

where l_t is the transition length,

$$l_t = 10^5 \times 10^{-3}/10^3 = 0.033 \text{ m}.$$

Thus, it is reasonable to ignore the laminar layer compared to the 30 m plate length.

Drag on the first 3 m is then calculated in the same way as shown above for the full plate length.

9.10. Effect of surface roughness on turbulent boundary layer development and skin friction coefficients

Initially, the effect of surface roughness is to cause transition from laminar to turbulent conditions closer to the leading edge of the surface. Indeed, the method commonly used to trigger a turbulent boundary layer over model surfaces in wind tunnels is to fix a trip wire or a band of sandpaper or rough material along the surface leading edge to ensure correct drag readings from the models tested. Following transition, where the boundary layer is still thin, the value of k/δ may be significant and all the surface roughness protrudes above the boundary layer. In this case, all the drag is due to the eddies caused by the flow passing over the surface roughness and the drag is proportional to the square of the free stream velocity. As the boundary layer continues to develop so the layer depth increases, and laminar sub-layer eventually becomes thick enough to cover all the surface roughness, so that the eddy related losses mentioned above do not occur. In this case, which occurs for high Reynolds numbers, the degree of roughness of the surface becomes unimportant, i.e. a change in roughness height k would not affect the drag force. Under these special conditions the surface is said to have become hydraulically smooth.

9.11. Effect of pressure gradient on boundary layer development

So far, the assumption made of zero pressure gradients in the flow direction across the flat surfaces considered has been unquestioned. The presence of a pressure gradient $\partial p/\partial x$ effectively means a $\partial u/\partial x$ term, i.e. the flow stream velocity changes across the surface. If, for example, a curved surface is considered, then the velocity is seen to vary as shown in Fig. 9.11.

If the pressure *decreases* in the downstream direction, then the boundary layer tends to be reduced in thickness, and this case is termed a favourable pressure gradient.

If the pressure *increases* in the downstream direction, then the boundary layer thickens rapidly this case is referred to as an adverse pressure gradient. This adverse pressure gradient, together with the action of the shear forces described in the boundary layer, if they act for a sufficient length, will bring the boundary layer to rest and the flow separates from the surface. This flow separation has serious consequences in the design of aerofoils, as, once the flow breaks away from the surface, all lift is lost. Due to the continuing action of the adverse pressure gradient downstream of the separation point,

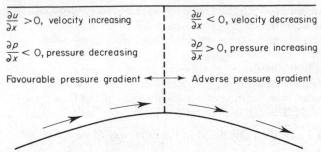

$\dfrac{\partial u}{\partial x} > 0$, velocity increasing $\dfrac{\partial u}{\partial x} < 0$, velocity decreasing

$\dfrac{\partial p}{\partial x} < 0$, pressure decreasing $\dfrac{\partial p}{\partial x} > 0$, pressure increasing

Favourable pressure gradient ← → Adverse pressure gradient

Figure 9.11 Variation of pressure and velocity over a curved surface

reversed flow eddies are formed which act to increase drastically the drag force acting on the surface. Figure 9.11 illustrates the changes in boundary layer velocity profile under the conditions described above.

Generally, then, for the design of aerofoils or other lift producing surfaces, such as pump and fan blades, the onset of separation should be avoided by design. In the particular case of aerofoil design, this has led to a number of ingenious lift-sustaining devices which act either to revitalize the slow moving air layer by introduction of a faster moving jet or the removal of the surface layer prior to separation by sucking away this low velocity layer. One of the earliest devices of the first type was the Handley Page leading edge slot which passed high velocity air from below the wing into the upper wing surface boundary layer prior to separation, thus preventing the change of shape of the velocity profile shown in Fig. 9.12. More recently, a French STOL transport relied on exhaust air from the turbines ducted and discharged along the wing leading edge to prevent separation and loss of lift at slow speed and high wing angles of attack.

The second method, sucking away the boundary layer, has been employed

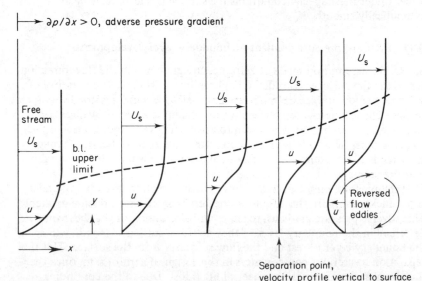

$\rightarrow \partial p / \partial x > 0$, adverse pressure gradient

Figure 9.12 Effect of an adverse pressure gradient on boundary layer development

in the study of laminar flow wings for long range transport aircraft where the marked reduction in skin friction drag that would follow from an entirely laminar boundary layer covering the wing would have obvious range and/or lifting capacity advantages.

The effects of wake formation are not solely concerned with aerofoils, but have resonant failure results in bridge design. Given certain wind speeds over and under a bridge span the alternate breaking away of the flow from the upper and lower surfaces can impose cyclic loads which, under special conditions, can correspond to the structure's natural frequency.

The examples above have all dealt with the formation of the boundary layer external to a flat or curved surface. However, as mentioned for the pipe case, boundary layers form within any duct and grow to fill the duct — imposing a velocity profile of some sort across the duct cross-section. Generally, this is of little concern. However, in the special case of aircraft engine intakes leading to the engine first stage compressors, the development of the boundary layer can be adverse to the performance of the engine. For the best output, the velocity profile at entry to the compressor, normally now of an axial design, should be as uniform as possible, which cannot occur with a fully developed boundary layer. To prevent this it is now quite common to bleed or suck the boundary layer away down the length of the intake.

A particular complication arises in the case of an aircraft engine designed to operate at supersonic speed. (With two exceptions these are all military aircraft.) It is necessary to decelerate the air prior to entry to the compressors. However, this requires the generation of a shock wave pattern in the intake. If this shock wave pattern is represented as a step increase in pressure, then it will be seen as a concentrated adverse pressure gradient which could cause boundary layer separation in the intake. This, in turn, would cause the formation of an eddy wake which would be likely to stall the compressor — with obvious consequences of loss of engine power. To avoid this boundary layer, bleeding is again employed.

With the few examples given above, the study of the boundary layer can be seen to be of fundamental importance in the understanding of fluid flow phenomena.

EXERCISES 9

9.1   Air at 20 °C and with a free stream velocity of 40 m/s flow past a smooth thin plate which is 3 m wide and 10 m long in the flow direction. Assuming a turbulent boundary layer from the leading edge determine the shear stress, laminar sub-layer thickness and the boundary layer thickness 6 m from the leading edge.

Take density = 1·2 kg/m^3 and kinematic viscosity as 1·49 × 10^{-5} m^2/s.
[1·89 N/m^2, 0·08 mm, 76 mm]

9.2   Determine the ratio of momentum and displacement thickness to the boundary layer thickness δ when the layer velocity profile is given by

$$u/U_s = (y/\delta)^{1/2}$$

where u is the velocity at a height y above the surface and the flow free stream velocity is U_s.
[0·166, 0·333]

9.3   Repeat Exercise 9.2 above if the velocity profile is given by

$$u/U_s = \sin\left(\frac{\pi}{2}\frac{y}{\delta}\right)$$

[0·136, 0·36]

9.4   Oil with a free stream velocity of 2 m/s flows over a thin plate 2 m wide and 3 m long. Calculate the boundary layer thickness and the shear stress at the mid length point and determine the total surface resistance of the plate.

Take density as 860 kg/m^3, kinematic viscosity as 10^{-5} m^2/s.

[13·6 mm, 2·8 x 10^{-5} N/m^2, 35·224 N]

9.5   A flat plate is drawn submerged through still water at a velocity of 9 m/s. If the plate is 3 m wide and 20 m long determine the laminar to turbulent transition position and the total drag force acting on the plate. Take water temp. as 20 °C.

[0·01 m, 9·5 kN]

9.6   An open rectangular box section, sides 3 m x 20 m and 1·5 m x 20 m, is drawn submerged through still water, at 20°, at a velocity of 9 m/s. Determine the overall drag force, neglecting any edge effects.

[28·5 kN]

9.7   Estimate the skin friction drag on an airship 92 m long, average diameter 18 m, being propelled at 130 km/h through air at 90 kN/m^2 absolute pressure and 27 °C.

[6·5 kN]

9.8   Assuming a velocity distribution defined by

$$u/U_s = \sin \pi y/2\delta$$

determine the general expressions for growth of the laminar boundary layer and for the surface shear stress for a smooth flat plate.

[$\delta = 4\cdot8/\mathrm{Re}_x^{1/2}, \tau_0 = 0\cdot33\sqrt{(\mu U_s^3 \rho/x)}$]

9.9   Air at 20 °C and 760 mm Hg absolute pressure flows past a smooth wind tunnel wall, with a free stream velocity of 160 km/h. Determine the position along the wall, in the flow direction, at which the boundary layer becomes turbulent and the distance to a boundary layer thickness of 25 mm. All wall measurements may be assumed to be taken from the working section entrance and edge/corner effects may be ignored.

[9·6 mm, 1·4 m]

9.10   Show that, if a flat plate, sides a, b in length, is towed through a fluid so that the boundary layer is entirely laminar, the ratio of towing speeds so that the drag force remains constant regardless of whether a or b is in the flow direction is given by

$$U_a/U_b = \sqrt[3]{(a/b)}$$

where U_a is the free stream velocity if side a is in the flow direction and U_b is the corresponding fluid velocity if b is in the flow direction.

9.11   Repeat Exercise 9.10 above if the boundary layer is considered fully turbulent.

[$U_a/U_b = \sqrt[5]{a/b}$]

10 Incompressible flow round a body

10.1. Régimes of external flow

In Chapter 7, flow around a cylinder was discussed and expressions enabling
the calculation of velocity and pressure in the flow field around the cylinder
were derived. Clearly, such a flow may be described as external as it is
concerned with the pattern of streamlines surrounding a solid body immersed
in a moving fluid. However, the treatment of Chapter 7 excluded any effects
which viscosity may have on the flow pattern because that chapter was
concerned with ideal flow only.

Chapter 9 introduced the concept of a boundary layer and dealt with the
effects viscosity has on a fluid adjacent to a solid surface and with the
calculation of forces acting on the surface due to fluid friction. We are,
therefore, now in a position to consider the external flows of real fluids,
namely taking into account viscous effects. The knowledge of potential flow
and of boundary layer theory makes it possible to treat an external flow
problem as consisting broadly of two distinct régimes: that immediately
adjacent to the body's surface, where viscosity is predominant and where
frictional forces are generated, and that outside the boundary layer, where
viscosity is neglected but velocities and pressures are affected by the physical
presence of the body together with its associated boundary layer. In this
outside zone, the theories of ideal flow may be used. In addition, there is the
stagnation point at the front of the body (which may stretch into a stagnation
region if the body is very blunt) and there is the flow region behind the body
(which is known as the wake). These flow régimes are shown in Fig. 10.1.

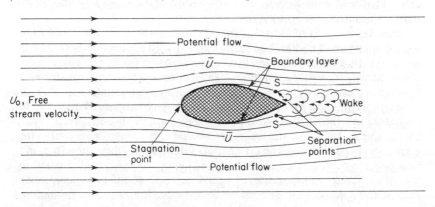

Figure 10.1   Flow régimes around an immersed body

The wake, which starts from points S at which the boundary layer
separation occurs, deserves a fuller description. It will be remembered from
Chapter 9 that separation occurs due to adverse pressure gradient ($\partial p/\partial x > 0$),
which, combined with the viscous forces on the surface, produces flow reversal,
thus causing the stream to detach itself from the surface. The same situation
exists at the rear edge of a body as it represents a physical discontinuity of

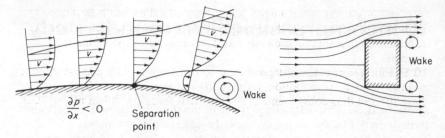

Figure 10.2   Formation of a vortex in a wake

the solid surface. In both cases the flow reversal produces a vortex, as shown in Fig. 10.2.

The flow in the wake is thus highly turbulent and consists of large-scale eddies. High-rate energy dissipation takes place there, with the result that the pressure in the wake is reduced. A situation is thus created whereby the pressure acting on the front of the body (the stagnation pressure) is in excess of that acting on the rear of the body, so that a resultant force acting on the body in the direction of the relative fluid motion exists. This force, arising from the pressure difference, or more generally from the non-uniform pressure distribution on the body, is called *pressure drag*.

It is worthwhile to recollect the findings of Chapter 7 dealing with ideal flow. There, in the absence of viscosity, the flow pattern over the rear part of a body, such as a cylinder, was symmetrical with respect to that over the front half of the body. There were two stagnation points – at the front and at the rear – and the pressure at these points was the same. Thus, there was no resultant force due to pressure acting on the body in the direction of the relative motion (*see* Section 7.10).

A situation very similar to this exists in the case of flow of the real fluids around streamlined bodies, but only at very low Reynolds numbers. There is no wake then and the pressure at the rear stagnation point is nearly equal to that at the front stagnation point. The degree to which the rear stagnation pressure approaches the front stagnation pressure is sometimes called *pressure recovery*. In the ideal flow, the pressure recovery is complete. When the flow separates from the surface and the wake is formed, the pressure recovery is not complete. The larger the wake, the smaller is the pressure recovery and greater the pressure drag. The art of streamlining a body lies, therefore, in shaping its contour so that separation, and hence the wake, is eliminated or at least confining the separation to a small rear part of the body and, thus, keeping the wake as small as possible. Such bodies are known as *streamlined* bodies. Otherwise a body is referred to as *bluff* and a significant pressure drag is associated with it.

10.2. Drag

Pressure drag was described in the preceding section, but asymmetry of pressure distribution, which is responsible for it, is not necessarily the only cause for the existence of a force acting on an immersed body in the direction of relative motion.

Thus, in general, when a body is immersed in a fluid and is in relative motion with respect to it, the *drag* is defined as that component of the resultant force acting on the body which is in the direction of the relative motion.

The force component perpendicular to the drag, i.e. acting in the direction normal to the relative motion, is called *lift* and was defined in Section 7.10. Both lift and drag components of the resultant force are shown in Fig. 10.3.

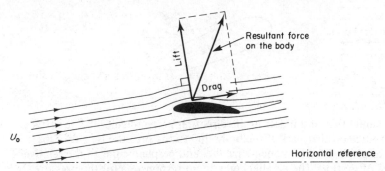

Figure 10.3   Lift and drag on a body

In Chapter 9, frictional drag was discussed in connection with the boundary layer theory. It is the force on the body acting in the direction of relative motion due to fluid shear stress at the surface. Thus, in external flow, the immersed body is subjected to frictional drag over its entire surface. *Total drag* on the body, often called *profile* drag, is, therefore, made up of two contributions, namely the pressure (or form) drag and the *skin friction* drag. Thus,

$$\text{Profile drag} = \text{Pressure drag} + \text{Skin friction drag.} \qquad (10.1)$$

The relative contribution of pressure drag and friction drag to the profile drag depends upon the shape of the body and its orientation with respect to the flow. Take, for example, a small rectangular flat plate. If it is held in a fluid stream 'edge on', as shown in Fig. 10.4(a), the pressure drag will be

(a)   (b)

Figure 10.4

negligible because even though the pressure recovery is incomplete, the resulting pressure difference will act on a very small frontal area (that perpendicular to the flow). The skin friction drag, however, will be substantial, due to the formation of the boundary layer on both sides of the plate.

If, however, the plate is held perpendicularly to the flow (as in Fig. 10.4(b)) the drag will be almost entirely due to pressure difference whereas the skin friction drag will be negligible.

The foregoing description of the two kinds of drag can now be formalized mathematically as follows. If (Fig. 10.5) p_s is the fluid pressure acting on the

Figure 10.5

surface element ds, and it acts in the direction perpendicular to the surface, then the force on that part of the body due to the pressure is p_s ds. This may be resolved into components parallel and perpendicular to the relative direction of motion. The parallel component responsible for the pressure drag is $p_s \cos \theta$ ds.

If this component is now integrated around the whole contour of the body, the pressure drag is obtained. Thus, pressure drag,

$$D_p = \oint p_s \cos \theta \; ds. \tag{10.2}$$

Similarly, the friction force on the body is manifested by the existence of the shear stress at the surface S. This also acts on the element ds and gives rise to a tangential force τ_0 ds, whose component in the direction of the motion is $\tau_0 \sin \theta$ ds. Performing, again, the integration around the body's contour, the skin friction drag is obtained. Thus, skin friction drag,

$$D_f = \oint \tau_0 \sin \theta \; ds. \tag{10.3}$$

Both contributions to the profile drag can, therefore, be theoretically calculated, but the first requires the knowledge of pressure distribution around the body and the other the knowledge of shear stress distribution on the surface. The determination of these could be very laborious and it is, therefore, usually simpler to measure the profile drag experimentally, as a force component in a wind tunnel. It is customary to relate the measured drag to the projected area of the body A, the fluid density ρ, and the free stream velocity U_0 by the expression

$$D = \tfrac{1}{2} C_D \rho U_0^2 A, \tag{10.4}$$

where C_D is known as the *drag coefficient* and A is the area of the body's projection on a plane perpendicular to the relative direction of motion.

A similar exercise of summation may be carried out for the force components normal to the direction of motion to give lift. This is also related to ρ, U_0 and A by an analogous expression,

$$L = \tfrac{1}{2} C_L \rho U_0^2 A. \tag{10.5}$$

The resultant force on the body is, of course, obtained by compounding lift and drag:

$$F = \sqrt{(L^2 + D^2)} = \tfrac{1}{2}\rho U_0^2 A \sqrt{(C_L^2 + C_D^2)}. \tag{10.6}$$

Example 10.1

A kite, which may be assumed to be a flat plate of face area $1 \cdot 2 \text{ m}^2$ and mass $1 \cdot 0$ kg, soars at an angle to the horizontal. The tension in the string holding the kite is 50 N when the wind velocity is 40 km h^{-1} horizontally and the angle of the string to the horizontal direction is $35°$. The density of air is $1 \cdot 2$ kg m^{-3}. Calculate the lift and the drag coefficients for the kite in the given position indicating the definitions adopted for these coefficients.

Solution

Since the wind is horizontal, the drag, by definiton, will also be horizontal and the lift vertical. The kite is in equilibrium and, therefore, lift and drag must be balanced by the string tension and the weight of the kite. Resolving forces into horizontal and vertical components,

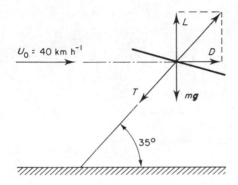

Figure 10.6

$$L = T \sin 35° + mg = 50 \sin 35° + 1 \cdot 0 \times 9 \cdot 81$$
$$= 38 \cdot 49 \text{ N},$$
$$D = T \cos 35° = 50 \cos 35° = 40 \cdot 95 \text{ N}.$$

But, lift,

$$L = \tfrac{1}{2} C_L \rho U_0^2 A$$

and, therefore,

$$C_L = 2L/\rho U_0^2 A = 2 \times 38 \cdot 49 \Big/ 1 \cdot 2 \left(\frac{40 \times 1000}{3600}\right)^2 1 \cdot 2$$

$$= 0 \cdot 432.$$

Similarly, the drag coefficient,

$$C_D = \frac{2D}{\rho U_0^2 A} = \frac{2 \times 40 \cdot 95}{1 \cdot 2 (40 \times 1000/3600)^2 \, 1 \cdot 2} = 0 \cdot 460.$$

Both coefficients have been based on the full area of the kite, because the projected area varies with incidence. This is also the accepted practice in the case of aerofoils.

10.3. Drag coefficient and similarity considerations

In order to obtain some idea of the nature of the drag coefficient, it is informative to carry out a dimensional analysis exercise (*see* Chapter 25) in which the drag on an immersed body is considered to be the dependent variable while the following are included as independent variables: the fluid density ρ, its viscosity μ, free stream velocity U_0, a linear dimension of the body l, the weight of the fluid per unit mass g (acceleration due to gravity), surface tension σ, and bulk modulus K. Thus,

$$D = f(\rho, \mu, U_0, l, \sigma, g, K)$$

or: $$D = \rho^a \mu^b U_0^c l^d g^e \sigma^f K^h$$

and, substituting the dimensions,

$$\frac{ML}{T^2} = \left(\frac{M}{L^3}\right)^a \left(\frac{M}{LT}\right)^b \left(\frac{L}{T}\right)^c (L)^d \left(\frac{L}{T^2}\right)^e \left(\frac{M}{T^2}\right)^f \left(\frac{M}{T^2 L}\right)^h.$$

Equating indices:

[M] $1 = a + b + f + h,$

therefore,

$$a = 1 - b - f - h;$$

[T] $-2 = -b - c - 2e - 2f - 2h$

therefore,

$$c = 2 - b - 2e - 2f - 2h;$$

[L] $1 = -3a - b + c + d + e - h,$

from which,

$$d = 1 + 3a + b - c - e + h = 1 + 3(1 - b - f - h) + b - 2 + b + 2e + 2f$$
$$+ 2h - e + h = 2 - b - f + e.$$

Therefore,

$$D = \rho^{(1-b-f-h)} \mu^b U_0^{(2-b-2e-2f-2h)} l^{(2-b-f+e)} g^e \sigma^f K^h$$

$$= \rho U_0^2 l^2 \left(\frac{\mu}{\rho U_0 l}\right)^b \left(\frac{\sigma}{\rho U_0^2 l}\right)^f \left(\frac{gl}{U_0^2}\right)^e \left(\frac{K}{\rho U_0^2}\right)^h.$$

But $\rho U_0 l/\mu$ = Re (Reynolds number), U_0^2/gl = Fr^2 (Froude number),

   $\rho l U_0^2/\sigma$ = We^2 (Weber number), $U_0^2/(K/\rho)$ = Ma^2 (Mach number),

so that

$$D = \rho U_0^2 l^2 \phi(\text{Re, Fr, We, Ma}). \tag{10.7}$$

Comparing this expression for drag with that of equation (10.4),

$$\tfrac{1}{2} C_D \rho U_0^2 A = \rho U_0^2 l^2 \phi \ (\text{Re, Fr, We, Ma}).$$

Since, for a body of a fixed shape,

$$A = \lambda l^2$$

where λ is a numerical constant and incorporating the constant $\tfrac{1}{2}$ as well as λ into the function ϕ', such that

$$\phi'[\ \ ] = \phi [\ \ ]/\tfrac{1}{2}\lambda,$$

we obtain

$$\tfrac{1}{2} \lambda C_D \rho U_0^2 l^2 = \rho U_0^2 l^2 \phi \ (\text{Re, Fr, We, Ma})$$

and, finally,

$$C_D = \phi' \ (\text{Re, Fr, We, Ma}). \tag{10.8}$$

Equation (10.8) demonstrates that the drag coefficient is not a numerical constant, but a proportionality coefficient whose numerical value is a function of a series of dimensionless groups. These groups, and also others which, for simplicity, were not incorporated into the analysis (such as relative roughness, free-stream turbulence level, cavitation number) come into play if the kind of forces represented by them are of significance. For example, Re will predominate in cases where viscous forces are dominant, Fr will only be significant in the presence of gravity waves (wave-making drag), Ma will dominate at high compressibility rates associated with high-speed gas flow, cavitation number will not be important unless cavitation occurs, etc.

It may, therefore, be said that, in general, the drag coefficients (and lift coefficients as well since an analogous expression may be derived for them) for two geometrically similar situations will be the same if the other para- meters are the same. For example, the drag on a smooth sphere in an incompressible fluid without cavitation will be such that

$$C_D = f(\text{Re}),$$

which means that so long as Re is the same the drag coefficient for any sphere of any size in any fluid will be the same provided other parameters are insignificant or irrelevant. For instance, if the free-stream level of turbulence is not the same C_D will vary. This is why the value of C_D for a sphere falling in a stationary liquid (zero turbulence) may be different from that obtained when the sphere is stationary and the fluid is moving past it. Similarly, the boundary layer transition and separation affect both lift and drag and, hence, their values for two situations may not be the same unless all the parameters involved are the same.

It must also be remembered that if one effect is absent from (say) the model situation of a dynamically similar system, it must also be absent from the prototype situation. For example, aerofoils tested in water must not cavitate if their performance in air is required or submarine hulls when tested in a wind tunnel should not be subjected to high velocities to avoid Mach number effects.

Although the values of drag coefficient vary with Re and other parameters described, they depend primarily upon the shape of the body and its orientation with respect to the fluid flow. Appendix 2 gives values of C_D at specified Re for a variety of commonly encountered shapes.

10.4. Resistance of ships

So far in this chapter, it was assumed that the body which is in relative motion with respect to the fluid is totally immersed in it. An important case, however, exists when the body is partly immersed in a liquid, the example being a ship. When it travels on the surface of water, two main sets of waves are produced, one originating at the bow and the other at the stern of the ship, both diverging from each side of the hull. Energy is required to generate these waves, and this energy originates from the propulsion system of the ship, which must therefore overcome not only the skin friction drag and the form drag but also the additional resistance in generating the waves. This additional resistance is known as *wave-making drag* or wave drag. (Note, however, that the term 'wave drag' is also used to describe the compressibility effects at supersonic velocities. *See* Section 11.1.)

It is not possible to measure the wave resistance directly. It is, therefore, normally obtained by measurement of the total drag and subtracting from it the calculated value of the skin friction drag:

$$\text{Wave-making resistance} = \text{Total drag} - \text{Skin friction drag.} \qquad (10.9)$$

In this equation the form drag (due to the wake at the stern) is included in the wave-making resistance.

The application of dimensional analysis to the problem carried out in Section 10.3 indicated that the friction drag is dependent upon Reynolds number and the wave-making resistance upon Froude number. The latter is the ratio of inertia forces to the gravity forces and, in the present context, is defined as

$$\text{Fr} = v/\sqrt{(gl)}, \qquad (10.10)$$

where v is the velocity of the ship and l is its length.

It may also be shown by dimensional analysis that the velocity of propagation of surface waves, sometimes called *celerity*, is given by

$$c = \sqrt{(gL)}\phi(d/L, h/L), \qquad (10.11)$$

where L is the wavelength, h is the height of the waves, d is depth of water. If the ratio h/L is small, the celerity is not affected by it and is given by

$$c = \sqrt{\{(gL/2\pi) \tan h(2\pi d/L)\}},$$

which, for deep-water waves, where $d \gg L$, reduces to

$$c = \sqrt{(gL/2\pi)}. \qquad (10.12)$$

Experiments which involve towing model ships indicate that the bow and stern waves produced by them travel at the same speed as the ship. This may be demonstrated by suddenly stopping the ship and measuring the wave velocity. Thus,

$$v = C = \sqrt{(gL/2\pi)}.$$

But, from (10.10),

$$v = \mathrm{Fr}\sqrt{(gl)},$$

so that, equating the two expressions, the Froude number may be written as

$$\mathrm{Fr} = 0{\cdot}4\sqrt{(L/l)} \qquad\qquad (10.13)$$

This expression is important because it indicates two things. First, it shows that the Froude number describes completely the interrelation between the ship's length and the wavelength produced by it and, hence, determines the wave-making flow pattern at a given speed; it, therefore, also demonstrates that, for dynamic similarity of the wave-making resistance, the Froude number must be the same for the model and for the prototype. Second, it indicates how many wavelengths there are in a ship's length and, hence, describes the interaction between the bow and the stern systems of waves which may be beneficial or detrimental to the ship's resistance. For example, at certain speeds the waves superpose in such a way that a travelling mound of water is built at the stern. The hydrostatic pressure of this mound acts on the ship pushing it forward and, hence, diminishing its wave-making resistance. However, at other speeds the superposition may produce a travelling trough at the stern, thus increasing the resistance. This is very pronounced at Fr of about 0·6, when the ship rides on the back of its first bow wave crest with the stern in the trough. This 'up-hill' ride means a very much increased wave-making resistance. The most dramatic situation occurs however at Fr = 1, attained by planing speed boats, at which the boat rides on top of the wave crest and the wave-making resistance is then reduced very considerably. Figure 10.7 shows the 'ups and downs' of the wave-making resistance due to

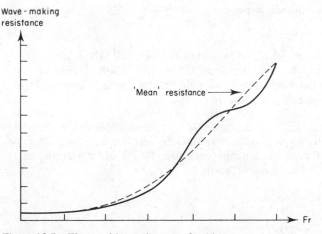

Figure 10.7   Wave-making resistance of a ship

the interaction of the wave systems as the Froude number is increased. A 'mean' resistance curve is also shown. A more extensive 'mean' wave-making resistance curve is drawn, together with the frictional resistance curve, on Fig. 10.8. The two curves add up to the total resistance of a ship. Note the drop of the wave-making resistance at Fr = 1 and the consequent effect upon the total resistance.

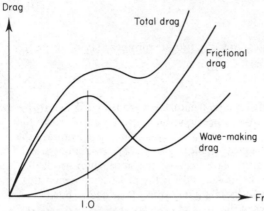

Figure 10.8 Ship's resistance

The procedure for predicting total resistance of a ship during its design stage is based on towing model tests which are aimed at determination of the wave-making resistance (including form drag) and on calculations of frictional drag related to the mean wetted area of the ship. Model towing tests are carried out at a corresponding speed based on Froude number as the criterion. Thus, if suffix 'm' refers to the model and suffix 'p' to the prototype,

$$(Fr)_m = (Fr)_p$$

or $$v_m/\sqrt{(gl_m)} = v_p/\sqrt{(gl_p)},$$

from which the corresponding speed

$$v_m = v_p \sqrt{(l_m/l_p)}. \tag{10.14}$$

The total drag of the model D_m is measured at this speed. The frictional drag of the model D_{fm} is calculated using the boundary layer theory or empirical formulae determined by towing thin plates. Hence, the model wave-making resistance R_m is obtained:

$$R_m = D_m - D_{fm}.$$

This is then scaled-up to predict the wave-making resistance of the prototype R_p using the general drag relationship (equation (10.7)), but neglecting all parameters except Froude number, so that

$$R_p = \rho_p v_p^2 l_p^2 \phi(Fr)_p$$

and

$$R_m = \rho_m v_m^2 l_m^2 \phi(Fr)_m.$$

Dividing one equation by the other, and since $(Fr)_m = (Fr)_p$ by design of the towing tests (corresponding speed), it follows that

$$R_p = R_m (\rho_p/\rho_m)(v_p/v_m)^2 (l_p/l_m)^2. \tag{10.15}$$

By adding to this the calculated skin friction drag for the prototype D_{fp}, the total drag is obtained:

$$D_p = R_p - D_{fp}. \tag{10.16}$$

Example 10.2

A ship is to be built having a wetted hull area of 2500 m^2 to cruise at 12 m s^{-1}. A $\frac{1}{40}$th full-size scale model is tested at the corresponding speed and the measured total resistance is found to be 32 N. From separate tests, the skin friction resistance for the model was found to be $3 \cdot 7 v^{1 \cdot 95}$ (newtons per square metre of wetted area) whereas for the prototype this is estimated to be $2 \cdot 9 v^{1 \cdot 8}$, where v is the velocity in metres per second.

Find the expected total resistance of the full-size ship if it operates in sea water of density 1025 kg m^{-3} whereas the model is tested in fresh water.

Solution

The corresponding speed at which the model must be tested is given by equation (10.14), obtained by equating the Froude number for the model and for the prototype:

$$v_m = v_p \sqrt{(l_m/l_p)} = 12\sqrt{(1/40)} = 1 \cdot 90 \text{ m s}^{-1}.$$

Now, the skin friction drag of the model at this test speed will be

$$D_{fm} = 3 \cdot 7 v_m^{1 \cdot 95} A_m = 3 \cdot 7 (1 \cdot 90)^{1 \cdot 95} (2500/40^2) = 20 \cdot 2 \text{ N}$$

and, hence, the model's wave-making resistance will be

$$R_m = D_m - D_{fm} = 32 - 20 \cdot 2 = 11 \cdot 8 \text{ N}.$$

Now, using equation (10.15), the wave-making resistance of the ship is obtained:

$$R_p = R_m \frac{\rho_p}{\rho_m} \left(\frac{l_p}{l_m}\right)^2 \left(\frac{v_p}{v_m}\right)^2 = 11 \cdot 8 \frac{1025}{1000} (40)^2 \left(\frac{12}{1 \cdot 90}\right)^2 = 771 \cdot 9 \text{ kN}.$$

Now the skin friction drag for the ship is calculated from

$$D_{fp} = 2 \cdot 9 v_p^{1 \cdot 8} \times A_p = 2 \cdot 9 (12)^{1 \cdot 8} 2500 \times 10^{-3} = 635 \cdot 13 \text{ kN}$$

and, hence, the total drag of the ship,

$$D_p = R_p + D_{fp} = 771 \cdot 9 + 635 \cdot 13 = \mathbf{1407 \cdot 06 \text{ kN}}.$$

10.5. Flow past a cylinder

In this section a thin, circular cylinder of infinite length, placed transversely in a fluid stream, will be used to discuss in greater detail the changes in flow pattern and in the drag coefficient which accompany the variation of Reynolds number. It must be remembered that for a given cylinder of a given diameter immersed in a given fluid the Reynolds number is directly proportional to the velocity and, therefore, the variation with Reynolds number could be imagined as the variation with velocity for a given cylinder. We assume also that there are no end effects and therefore that the flow is two-dimensional.

At very small values of Re, say below 0·5, the inertia effects are negligible and the flow pattern is very similar to that for ideal flow, the pressure recovery being nearly complete. Thus, pressure drag is negligible and the profile drag is nearly all due to skin friction. Figure 10.9(a) shows the flow pattern and associated pressure distribution for such a case. Figure 10.10 indicates a straight line relationship between C_D and Re in this range from which we conclude that the drag D is directly proportional to velocity U_0.

At increased Re, say between 2 and 30, separation of the boundary layer occurs at points S as indicated in Fig. 10.9(b). Two symmetrical eddies, rotating in opposition to one another, are formed. They remain fixed in position and the main flow closes behind them. The separation of the boundary layer is reflected in the variation of C_D graph by the curvature of the line indicating that the drag is now proportional to U_0^n, where $n \to 2$.

Further increase of Re tends to elongate the fixed eddies, which then begin to oscillate until at about Re = 90, depending upon the free-stream turbulence level, they break away from the cylinder as shown in Fig. 10.9(c). The breaking away occurs alternately from one and then the other side of the cylinder, the eddies being washed away by the main stream. This process is intensified by a further increase of Re, whereby the shedding of eddies from alternate sides of the cylinder is continuous, thus forming in the wake two discreet rows of vortices, as shown in Fig. 10.9(c). This is known as *vortex street* or von Kármán vortex street. At this stage the contribution of pressure drag to the profile drag is about three-quarters. von Kármán showed analytically, and confirmed experimentally, that the pattern of vortices in a vortex street follows a mathematical relationship, namely,

$$(h/l) = (1/\pi) \sinh^{-1}(l) = 0·281,$$ \hfill (10.17)

where h and l are indicated in Fig. 10.11.

It will be seen that shedding of each vortex produces circulation and, hence, gives rise to a lateral force on the cylinder. Since these forces are periodic following the frequency of vortex shedding, the cylinder may be subjected to a forced vibration. The familiar 'singing' of telephone wires is due to this phenomenon, caused by a lateral wind, whereas collapse of suspension bridges and 'flutter' of aerofoils are the result of a resonance between the natural frequency of the body and the frequency of forced vibration due to vortex shedding.

The frequency of such forced vibration, sometimes called self-induced vibration, may be calculated from an empirical formula due to Vincent Strouhal:

$$fd/U_0 = 0·198 \, (1 - 19·7/Re),$$ \hfill (10.18)

in which

$$fd/U_0 = \text{Str}$$ \hfill (10.19)

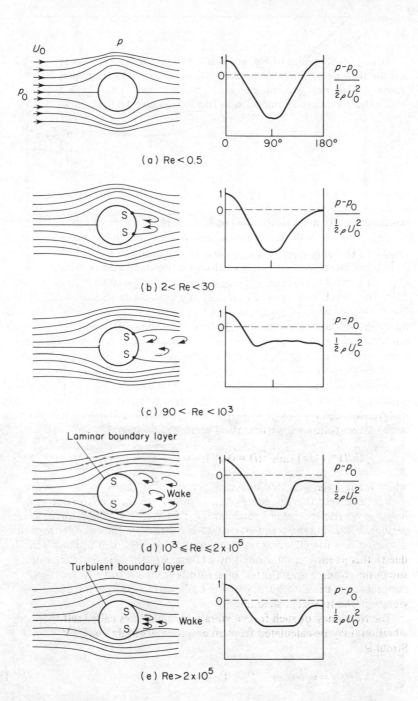

Figure 10.9   Flow past a cylinder

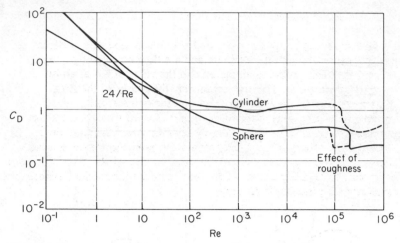

Figure 10.10   Drag coefficient for a sphere and a cylinder

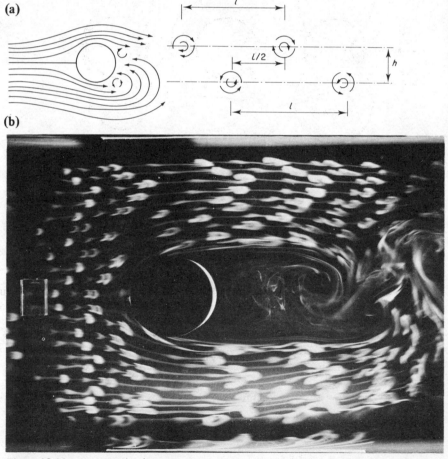

Figure 10.11   (a) von Kármán vortex street, (b) Smoke visualization showing von Kármán vortex street

and is known as *Strouhal number*. The formula applies to $250 < \text{Re} < 2 \times 10^5$.

It is fortunate that at higher values of Reynolds number the vortices disappear because of high rates of shear and are then replaced by a highly turbulent wake. This produces an increase in the value of C_D at about $\text{Re} = 3 \times 10^4$. Pressure drag is now responsible for nearly all the drag.

Up to $\text{Re} \simeq 2 \times 10^5$ the boundary layer on the cylinder is laminar, but at approximately that value, depending upon the intensity of free-stream turbulence, it changes to turbulent before separation, as indicated on Fig. 10.9(d). The effect of this is that separation points move further back and, hence, there is a marked drop in the value of C_D. At $\text{Re} > 10^7$ the value of C_D appears to be independent of Re, but there are insufficient experimental data available for this end of the range.

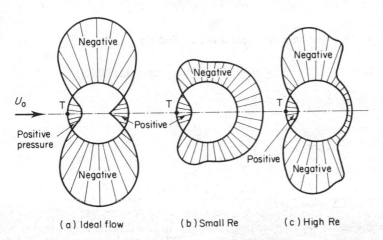

Figure 10.12   Pressure distribution around a cylinder

Figure 10.12 compares pressure distributions around a cylinder for ideal flow with the real flow at low and high values of Reynolds number.

Example 10.3

Electrical transmission towers are stationed at 500 m intervals and a conducting cable 2 cm in diameter is strung between them. If an 80 km h^{-1} wind is blowing transversely across the wires, calculate the total force each tower carrying 20 such cables is subjected to. Assume there is no interference between the wires and take air density as $1 \cdot 2$ kg m^{-3} and viscosity $1 \cdot 7 \times 10^{-5}$ N s m^{-2}.

Also establish whether the wires are likely to be subjected to self-induced vibrations and if so what would be the frequency.

Solution
Drag on one wire,

$$D = \tfrac{1}{2} \rho C_D U_0^2 A.$$

In order to establish the value of C_D, it is necessary to calculate the value of Re first.

$$Re = \frac{\rho\, U_0 d}{\mu} = \frac{1\cdot2 \times 80 \times 1000 \times 0\cdot02}{3600 \times 1\cdot7 \times 10^{-5}} = 3\cdot14 \times 10^4.$$

Now, from Fig. 10.10, for the above value of Re the drag coefficient is

$$C_D = 1\cdot2.$$

The projected area of a single wire between towers,

$$A = 0\cdot02 \times 500 = 10\ \text{m}^2.$$

Hence, drag on wire,

$$D = \tfrac{1}{2} \times 1\cdot2 \times 1\cdot2 \left(\frac{80 \times 1000}{3600}\right)^2 \times 10 = 3556\ \text{N}.$$

Therefore, the force on each tower due to 20 cables is

$$F = 20 \times 3556 = \mathbf{71\cdot11\ kN}.$$

Since $250 < Re < 10^5$, 'singing' may occur. Using formula (10.18),

$$f = 0\cdot198\, \frac{U_0}{d}\left(1 - \frac{19\cdot7}{Re}\right)$$

$$= 0\cdot198\, \frac{80 \times 1000}{3600 \times 0\cdot02}\left(1 - \frac{19\cdot7}{90\cdot67 \times 10^3}\right)$$

$$= \mathbf{219\cdot9\ Hz}.$$

10.6. Flow past a sphere

So far our discussion of drag has been confined to two-dimensional flow. We will now examine the flow past the simplest of all three-dimensional bodies, the sphere. There is a great similarity in the development of drag at increasing Re between the sphere and the cylinder, except that the vortex street associated with the latter and two-dimensional bodies such as aerofoils is not formed in the case of three-dimensional bodies. Instead, a vortex ring occurs, which for a sphere is formed at about Re = 10 and becomes unstable at $200 < Re < 2000$ when it tends to move downstream of the body, to be immediately replaced by a new ring. This process is not periodic, however, and does not give rise to vibrations of the sphere.

The study of flow past a sphere is of great practical importance because it is the foundation of a branch of fluid mechanics, namely *particle mechanics*. This subject concerns itself with all problems associated with flow of solid particles in a fluid or liquid particles in a gas and encompasses practical problems such as pneumatic conveying, particle separation, sedimentation, filtration, etc. In practice, most particles are not spheres and there are ways of classifying them in accordance to shape, but in the end they are always related to the sphere as the simplest theoretical shape and, therefore, most amenable to both the analytical as well as the experimental investigation.

At very low values of Re, during so called 'creeping' flow around a sphere, it may be assumed that the inertial effects are negligible and, hence, the steady flow Navier–Stokes equations may be greatly simplified by omission of the inertia term, thus enabling the calculation of viscous drag.

Stokes obtained the solution for drag by expressing the simplified Navier–Stokes equation together with the continuity equation in polar coordinates and using the boundary conditions that all velocity components are zero at the surface of the sphere. His solution is the well-known equation,

$$D = 6\pi\mu R U_0 \tag{10.20}$$

in which R is the radius of the sphere, U_0 is the free stream velocity of the fluid and μ is its absolute viscosity. This relationship holds true for $Re < 0.1$ but may be used with negligible error up to $Re = 0.2$. In this range, often referred to as *Stokes flow*, the drag coefficient may be calculated by equating the general drag equation (10.4) to the Stokes solution,

$$\tfrac{1}{2} C_D \rho U_0^2 A = 6\pi\mu R U_0,$$

but $A = \pi R^2$,

so that

$$C_D = 12\mu/\rho U_0 R = 24\mu/\rho U_0 d,$$

where $d = 2R$ is the diameter of the sphere. But $\rho U_0 d/\mu = Re$, based on the sphere diameter, and, therefore,

$$C_D = 24/Re \text{ for } Re < 0.2. \tag{10.21}$$

At larger values of Re, separation of the boundary layer occurs and the Navier–Stokes equations cannot be used. It is, therefore, necessary to rely on empirical expressions. One such formula extends Stokes' law to $Re < 100$ and is as follows:

$$C_D = (24/Re)(1 + \tfrac{3}{16} Re)^{1/2}. \tag{10.22}$$

Beyond $Re = 100$ it is necessary to use values of C_D as a function of Re from the graph, as in Fig. 10.13.

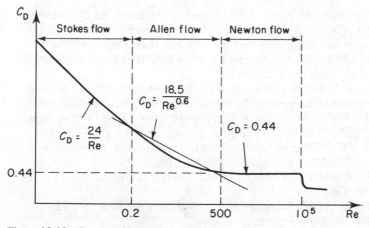

Figure 10.13   Drag coefficient for a sphere

Stokes' formula forms the basis for determination of viscosity of oils which consists in allowing a sphere of known diameter to fall freely in the oil. After initial acceleration, the sphere attains a constant velocity known as *terminal velocity* which is reached when the external drag on the surface and buoyancy, both acting upwards and in opposition to the motion, become equal to the downward force due to gravity. At this equilibrium condition,

$$6\pi\mu U_t R + \tfrac{4}{3}\pi R^3 \rho g = \tfrac{4}{3}\pi R^3 \rho_p g,$$

$$(\text{Drag}) + (\text{Buoyancy}) = (\text{Gravity})$$

where ρ = density of the fluid, ρ_p = density of the sphere material, U_t = terminal velocity. Thus,

$$6\mu U_t = \tfrac{4}{3}R^2(\rho_p - \rho)g$$

and $\qquad U_t = (2R^2/9\mu)\,(\rho_p - \rho)g$

or, in terms of the sphere's diameter d,

$$U_t = (d^2/18\mu)(\rho_p - \rho)g. \tag{10.23}$$

By timing the rate of fall of the sphere, the terminal velocity is measured and, hence, the viscosity of the fluid may be determined using the above equation. Alternatively, the method may also be used to determine the mean diameter of spherical particles by allowing them to settle freely in a liquid of known viscosity.

At large values of Reynolds number, the flow over the front half of a sphere may be divided into a thin boundary layer region, where viscosity effects are dominant, and an outer region, in which the flow corresponds to that of an inviscid fluid. The pressure is decreasing over the front half of the sphere from the stagnation point-onwards, thus having a stabilizing effect on the boundary layer, which remains laminar up to about Re = 5×10^5. Beyond the minimum pressure point on the sphere (at about $80°$) the boundary layer is subjected to an adverse pressure gradient and separation occurs. At low Re it begins at the rear stagnation point and with increasing Re it moves forward, reaching the $80°$ point from the front stagnation at a Reynolds number of about 1000. Pressure drag begins to dominate and C_D becomes independent of Re until, at about Re = 5×10^5, transition in the boundary layer occurs, it becoming turbulent before separation. This moves the separation point to the rear, making the wake smaller and abruptly reducing the value of C_D from about 0·5 to 0·2.

The experimental determination of C_D for a sphere is difficult because, first, the method of supporting the sphere in the wind tunnel affects the results and, second, because the results depend upon the free stream turbulence level, which is difficult to control, and upon the roughness of the sphere, which is difficult to reproduce. It is not surprising, therefore, that the early experimenters produced conflicting results. Any data which do not specify the method of support and free stream turbulence level should be viewed with caution.

However, for the purposes of particle mechanics the flow past a sphere is subdivided into three régimes as follows:

   (i) Stokes flow, $Re < 0.2$, $C_D = 24/Re$;

   (ii) Allen flow, $0.2 < Re < 500$, $C_D = f(Re)$;

  (iii) Newton flow, $500 < Re < 10^5$, $C_D = \text{constant} = 0.44$.

These régimes are shown in Fig. 10.13.

The calculation of terminal velocity is of great importance in particle mechanics because it forms the basis of such operations as settling or sorting. In sorting, for example, solid particles are introduced into a vertical stream of fluid, as shown in Fig. 10.14. If the fluid were stationary, that is

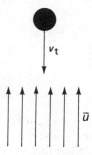

Figure 10.14   Spherical particle falling into a vertical fluid stream

$\bar{u} = 0$, the particle would attain a constant terminal descending velocity v_t. If, however, the fluid is moving vertically up with a velocity $\bar{u}$ there are three distinct possibilities:

    (i) when $\bar{u} < v_t$ the particle will be falling down with an absolute velocity $v = v_t - \bar{u}$, where v_t is now the relative velocity between the fluid and the particle and governs the drag on the particle;

    (ii) when $\bar{u} = v_t$ the particle will be suspended, having an absolute velocity of $v = 0$;

    (iii) when $\bar{u} > v_t$ the particle will move upwards with the fluid with an absolute velocity $v = \bar{u} - v_t$.

Thus, if particles of different size or weight are introduced into a vertical fluid stream, some will be carried over and some will descend, thus enabling *sorting* to be carried out.

The difficulty in deciding the correct sorting velocity $\bar{u}$ lies in the fact that it usually corresponds to the particle terminal velocity in the Allen flow régime, where C_D is a function of Reynolds number. This cannot be calculated until the velocity is known, which is precisely the variable we are trying to establish. To demonstrate the difficulty let us examine a general case for terminal velocity. It occurs when

       Drag + Buoyancy = Gravitational force

but drag,

$$D = \tfrac{1}{2}\rho C_D A v^2.$$

For a spherical particle,

$$A = \pi d^2/4$$

and $D = \tfrac{1}{8}\rho C_D \pi d^2 v^2,$

so that

$$\tfrac{1}{8}\rho C_D \pi d^2 v^2 + (\pi d^3/6)\rho g = (\pi d^3/6)\rho_p g.$$

Simplifying and rearranging,

$$\tfrac{1}{8}\rho C_D v^2 = (d/6)(\rho_p - \rho)g$$

and $v = v_t = \sqrt{\{\tfrac{4}{3} dg(\rho_p - \rho)/C_D \rho\}}.$ (10.24)

Thus,

$$v = f(C_D) = f_1(\text{Re}) = f_2(v).$$

For Stokes flow, $C_D = 24/\text{Re}$, and the substitution gives

$$v_t = d^2(\rho_p - \rho)g/18\mu$$

which is the equation (10.23) already derived.

For Allen flow two alternatives are possible.

(i) The $C_D = f(\text{Re})$ curve may be approximated to a straight line, giving

$$C_D = 18 \cdot 5/\text{Re}^{0 \cdot 6}$$ (10.25)

This yields a cumbersome and inaccurate equation for v_t.

(ii) A more satisfactory procedure is to eliminate v_t from equation (10.24) and to replace it by Re in the following manner. From (10.24),

$$C_D = 4d(\rho_p - \rho)g/3v_t^2\rho;$$

multiplying both sides by

$$\text{Re}^2 = (v_t \, d\rho/\mu)^2,$$

$$C_D \, Re^2 = 4d^3(\rho_p - \rho)\rho g/3\mu^2.$$ (10.26)

The right-hand side of this equation can be calculated for any given fluid and particle combination, since all relevant parameters are known. Thus, the value of $C_D \text{Re}^2$ becomes known. This is then referred to a graph relating $C_D \text{Re}^2$ to Re, shown in Fig. 10.15. This graph is simply a replot of the Allen part of the $C_D = f(\text{Re})$ graph.

Example 10.4

A particle of 1 mm diameter and density $1 \cdot 1 \times 10^3 \, \text{kg m}^{-3}$ is falling freely from rest in an oil of $0 \cdot 9 \, \text{kg m}^{-3}$ density and $0 \cdot 03 \, \text{N s m}^{-2}$ viscosity. Assuming that Stokes' law applies, how long will the particle take to reach 99 per cent of its terminal velocity? What is the Reynolds number corresponding to this velocity?

Solution

The equation of motion for the particle is

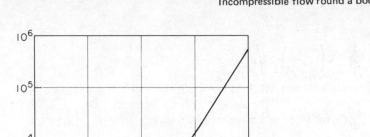

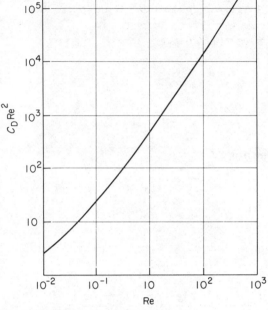

Figure 10.15   The C_D Re^2 vs. Re graph

Mass x Acceleration = Resultant force on the body in the direction of
motion

$$= \text{Gravity} - \text{Buoyancy} - \text{Drag},$$
$$m\frac{\mathrm{d}v}{\mathrm{d}t} = mg - m_0g - D,$$

where m = mass of particle, m_0 = mass of oil displaced by the particle,
D = drag on the particle. Thus

$$\frac{\mathrm{d}v}{\mathrm{d}t} = g - \frac{m_0}{m}\,g - \frac{D}{m},$$

but

$$D = 3\pi\mu dv \quad \text{and} \quad m = \tfrac{1}{6}\pi d^3\rho_\mathrm{p}.$$

Therefore,

$$D/m = 18\mu v/d^2\rho_\mathrm{p};$$

also $m_0/m = \rho_0/\rho_\mathrm{p},$

so that

$$\frac{dv}{dt} = g\left(1 - \frac{\rho_0}{\rho_p}\right) - \frac{18\mu v}{d^2\rho_p}.$$

To facilitate integration, let $A = g(1 - \rho_0/\rho_p)$ and $B = 18\mu/d^2\rho_p$, so that

$$\frac{dv}{dt} = A - Bv.$$

Hence,

$$t = \int_0^{0.99v_t} \frac{dv}{A - Bv} = \left[-\frac{1}{B}\log_e(A - Bv)\right]_0^{0.99v_t}$$

$$= -\frac{1}{B}\log_e(A - 0.99Bv_t) + \frac{1}{B}\log_e A$$

$$= \frac{1}{B}\log_e \frac{A}{A - 0.99Bv_t}.$$

But

$$v_t = d^2(\rho_p - \rho)g/18\mu,$$

therefore,

$$0.99Bv_t = 0.99\frac{18\mu}{d^2\rho_p} \times \frac{d^2(\rho_p - \rho)g}{18\mu} = 0.99\frac{(\rho_p - \rho)g}{\rho_p}$$

$$= 0.99(1 - \rho/\rho_p)g = 0.99A.$$

Hence,

$$t = \frac{1}{B}\log_e \frac{A}{A - 0.99A} = \frac{1}{B}\log_e \frac{1}{1 - 0.99} = \frac{1}{B}\log_e 100 = \frac{4.60}{B}.$$

But

$$B = 18\mu/d^2\rho_p = 18 \times 0.3/(0.1 \times 10^{-2})^2 \times 1.1 \times 10^3 = 490$$

and $t = 4.60/490 = \mathbf{0.0094}$ s.

Terminal velocity,

$$v_t = d^2(\rho_p - \rho_0)g/18\mu = 10^{-6}(1.1 - 0.9)10^3 \times 9.81/18 \times 0.03$$

$$= 3.63 \times 10^{-3} \text{ m s}^{-1}.$$

Reynolds number at this velocity,

$$\text{Re} = \rho v_t d/\mu = 0.9 \times 10^3 \times 3.63 \times 10^{-3} \times 10^{-3}/0.03$$

$$= \mathbf{0.1089}.$$

Example 10.5

A solid particle of specific gravity 2·4, when settling in oil of specific gravity 0·9 and viscosity 0·027 P, attains a terminal velocity of 3×10^{-3} m s^{-1}. What should be the velocity of an air stream (density 1.3 kg m^{-3}) blowing vertically up, in order to carry the particle at a velocity of 0.5 m s^{-1}? Viscosity of air may be taken as 1.7×10^{-5} N s m^{-2}.

Solution
It is first necessary to determine the diameter of the particle. This may be done from the settling data in oil, assuming Stokes' flow. This assumption will have to be checked. From equation (10.23),

$$v_t = d^2(\rho_p - \rho)g/18\mu.$$

Therefore,

$$d = \sqrt{\left\{\frac{18\mu v_t}{(\rho_p - \rho)g}\right\}} = \sqrt{\left\{\frac{18 \times 0.027 \times 10^{-1} \times 3 \times 10^{-3}}{(2.4 - 0.9)10^{-3} \times 9.81}\right\}} = 0.0995 \times 10^{-3}\,\text{m}.$$

To check the flow régime,

$$Re = \frac{vd\rho}{\mu} = \frac{3 \times 10^{-3} \times 0.0995 \times 10^{-3} \times 0.9 \times 10^3}{0.0027} = 0.0995.$$

Therefore Re < 0.1 and so the assumption of Stokes' flow was correct.

Now, for the particle to move vertically upwards with absolute velocity v in an air stream of absolute velocity u, the relative velocity v_t is

$$v_t = u - v.$$

It is this relative velocity which is responsible for the drag on the particle. Therefore,

$$v_t = \sqrt{\{4dg(\rho_p - \rho_{air})/3C_D \rho_{air}\}}.$$

Since we do not know C_D, we calculate

$$C_D\,Re^2 = 4d^3(\rho_p - \rho_{air})g\rho_{air}/3\mu^2$$

$$= 4(0.0995 \times 10^{-3})^3(2.4 \times 10^3 - 1.3)9.81 \times 1.3/3 \times (1.7 \times 10^{-5})^2$$

$$= 139$$

and from Fig. 10.15 we read that Re $= 5$ and hence the relative velocity v_t may be obtained from the expression for Re:

$$Re = \rho vd/\mu,$$

therefore,

$$v_t = \mu\,Re/\rho d = 1.7 \times 10^{-5} \times 5 / 1.3 \times 0.0995 \times 10^{-3} = 0.657\ \text{m s}^{-1}.$$

The upward air velocity required:

$$u = v_t + v = 0.657 + 0.5 = 1.157\ \text{m s}^{-1}.$$

10.7. Flow past an infinitely long aerofoil

An aerofoil may be defined as a streamlined body designed to produce lift.

There are other lift-producing surfaces such as hydrofoils or circular arcs. In general the following elementary aerofoil theory applies also to these surfaces. There is an accepted terminology concerning aerofoils and familiarization with it is necessary in order to understand the discussion of flow past aerofoils.

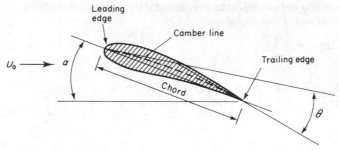

Figure 10.16   An aerofoil

Figure 10.16 shows an aerofoil section and the following are some of the most important terms relating to it:

Leading edge	the front, or upstream edge, facing the direction of flow;
Trailing edge	the rear, or downstream edge;
Chord line	a straight line joining the centres of curvature of the leading and trailing edges;
Chord, c	the length of chord line between the leading and trailing edges;
Camber line	the centre line of the aerofoil section;
Camber, δ	the maximum distance between the camber line and the chord line;
Percentage camber	$= 100\delta/c$ per cent is a measure of aerofoil curvature;
Span, b	the length of the aerofoil in the direction perpendicular to the cross-section;
Plan area, A	the area of the projection of the aerofoil on the plane containing the chord line. If the aerofoil is of constant cross-section, $A = c \times b$;
Mean chord, $\bar{c}$	$= A/b$;
Aspect ratio, AR	$= (\text{Span})/(\text{Mean chord}) = b/c = b^2/A$;
Deviation, θ	angle between the tangent to camber line at trailing edge and the tangent to camber line at leading edge;
Angle of attack (incidence)	the angle between the direction of the relative motion and the chord line;
Pressure coefficient, C_p	$= \dfrac{p - p_0}{\frac{1}{2}\rho U_0}$ where p is the local pressure and p_0 is the pressure far upstream of the aerofoil where velocity is V_0.

The primary purpose of an aerofoil is to produce lift when placed in a fluid stream. It will, of course, experience drag at the same time. In order to minimize drag, an aerofoil is a streamlined body. A measure of its usefulness

as a wing section of an aircraft or as a blade section of a pump or turbine is
the ratio of lift to drag. The higher this ratio is, the better the aerofoil, in the
sense that it is capable of producing high lift at a small drag penalty. In an
aircraft it is the lift on the wing surfaces which maintains the plane in the
air. At the same time it is the drag which absorbs all the engine power
necessary for the craft's forward motion. Similarly, in pumps, the head
generated is due to the lift produced by the impeller blades, whereas the
torque necessary to rotate the blades overcomes the drag on them. Thus,
the lift/drag ratio,

$$\frac{\text{Lift}}{\text{Drag}} = \frac{\frac{1}{2}\rho C_L U_0^2 A}{\frac{1}{2}\rho C_D U_0^2 A} = \frac{C_L}{C_D}.$$

(10.27)

The creation of lift is, therefore, of primary importance. How does an
aerofoil produce lift? How does it start when the motion of the aerofoil
begins and how is it maintained during the motion? We will attempt to answer
these questions by reference to potential flow theory, expanded in Chapter 7,
and by reference to the boundary layer theory.

The Kutta–Joukowski law, derived for a cylinder with circulation, relates
lift to circulation. It is not limited to cylinders, but may be shown to apply
to any two-dimensional section. The important point, however, is that lift
exists only if there is circulation around the section.

A rotating cylinder placed in a real fluid produces circulation by viscous
action of its rotating surface on the fluid. Aerofoils, however, do not rotate
and, hence, there must be a different mechanism of producing and maintaining
circulation. Let us first consider how the circulation starts. It was shown
earlier how a vortex is formed, either due to separation of a stream or at the
rear of a blunt body. Similarly, if two parallel streams of unequal velocity
meet, there is a discontinuity of velocity at their interface and that produces
a vortex. In the same manner, when a slightly inclined aerofoil starts motion
it splits the flow into two streams: one over the upper surface and one over
the lower surface. The velocities in these streams are not equal due to the
inclination of the aerofoil and, therefore, when they meet at the trailing edge
a starting vortex is produced as shown in Fig. 10.17. This vortex is cast-off
soon after the beginning of the motion. It does, however, give rise to circulation around the aerofoil which is equal in strength (but opposite in sign) to
the circulation of the starting vortex.

Let us now consider the situation during the motion of the aerofoil. For
the circulation to exist there must be vorticity in the stream which cuts

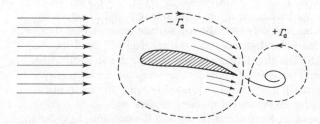

Figure 10.17   Starting vortex

across the circulation contour. The flow around the aerofoil may be considered as potential outside the boundary layer and, therefore, it is irrotational there. Hence, there is no vorticity and there cannot be any circulation associated with it. Within the boundary layer, however, the flow is viscous and due to the velocity gradient vorticity exists there.

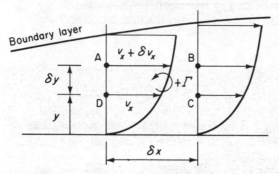

Figure 10.18   Circulation in the boundary layer

Consider an element of fluid ABCD in the boundary layer as shown in Fig. 10.18. Taking δx as small, the change of velocity in x direction $((\partial v_x/\partial x)\,\mathrm{d}x)$ may be neglected. The circulation for the element becomes

$$\Gamma_{\text{ABCD}} = -(v_x + \delta v_x)\delta x + v_x\delta x = -\delta v_x\delta x$$

and vorticity,

$$\zeta_{\text{ABCD}} = \delta v_x\delta x/\delta x\delta y = \delta v_x/\delta y.$$

Thus, there is vorticity in the boundary layer and its value depends upon the velocity gradient.

Figure 10.19 shows the flow around an aerofoil with an exaggerated boundary layer and vorticity within it. It will be noticed that, because the velocity gradients are of opposite sign on the top and bottom surfaces, the vorticity in the upper boundary layer is clockwise whereas the vorticity in the lower boundary layer is anticlockwise. If these two vorticities are equal in strength, they cancel each other and the resultant circulation around the contour is zero. This occurs, for example, in the case of a symmetrical aerofoil without camber placed in a fluid stream at zero angle of incidence. Because of complete flow symmetry, the growth of the boundary layer at top and bottom is identical and hence vorticities are the same in strength and of opposite rotation. However, if the vorticity over the top surface exceeds that over the bottom, the resultant circulation around the aerofoil will be clockwise, as shown in Fig. 10.18. The circulation contour may be drawn arbitrarily around the aerofoil and, provided it contains the whole of the boundary layer, the value of circulation will not be affected because the irrotational flow outside the boundary layer makes no contribution to it.

We therefore deduce that the circulation around the aerofoil, $\Gamma_{\text{a}} = \oint v\,\mathrm{d}s$, will be clockwise if the velocities over the upper surface are greater than the

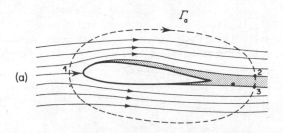

(a)

(b)

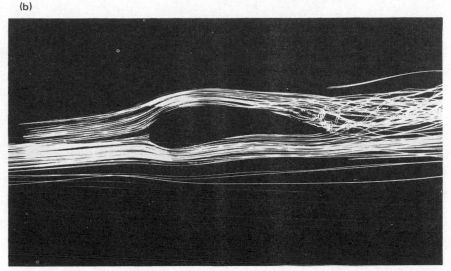

Figure 10.19   (a) Flow around an aerofoil, with circulation. (b) Helium bubble in air
flow visualization around an aerofoil

velocities over the lower surface, or, more exactly, if

$$\oint_1^2 v \, ds > \oint_3^1 v \, ds.$$

Such velocity distributions must, in accordance with Bernoulli's equation,
be accompanied by higher pressures on the bottom surface and lower pressures
on the top surface. Such a pressure distribution, as shown in Fig. 10.20,
gives rise to a resultant upward force, namely the lift.

Summarizing: at a small angle of incidence the fluid flowing over the
bottom surface of an aerofoil is slowed down, thus increasing the pressure,
which means that the pressure gradient there is favourable, the boundary
layer thickness is small and the anticlockwise vorticity in it is also small.
Over the upper surface the vorticity is greater, the pressure gradient adverse,
the boundary layer thicker and the clockwise vorticity in it greater. Thus,
the resulting pressure difference gives rise to lift, which may be related to
the circulation around the aerofoil. By the same argument, a negative lift
(downward force) may exist for negative values of angles of incidence.

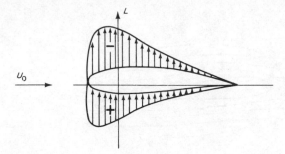

Figure 10.20   Pressure distribution around an aerofoil

The foregoing discussion indicates a strong dependence of lift upon the incidence angle. Let us consider this in greater detail by referring to the potential flow around a cylinder as our model. It will be remembered from our discussion of this topic in Chapter 7 that the increase of circulation around the cylinder alters the position of stagnation points, as shown in Fig. 7.29. Since lift (by Kutta-Joukowski, $L = \rho U_0 \Gamma$) depends upon circulation, it may therefore be related to the position of stagnation points. The analogy between the cylinder and the aerofoil is illustrated in Fig. 10.21.

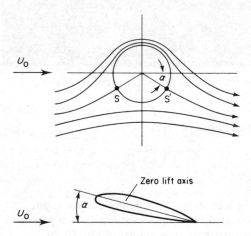

Figure 10.21   Relationship between the zero lift line on the aerofoil and position of stagnation points on a cylinder with circulation

The stream function for the flow around a cylinder with circulation is given by equation (7.64):

$$\Psi = U_0 \left(r - a^2/r\right) \sin \theta + (\Gamma/2\pi) \log_e r,$$

where U_0 = upstream velocity, a = radius of the cylinder, Γ = circulation around the cylinder, r, θ = cylindrical coordinates.

The velocity on the cylinder surface is the tangential velocity v_θ (since there is no velocity into or out of the cylinder) at $r = a$. Thus, since (from equation (7.13)

$$v_\theta = -\frac{\partial \Psi}{\partial r},$$

it follows that

$$v_\theta = -U_0(1 + a^2/r^2) \sin \theta - \Gamma/2\pi r$$

and, making $r = a$,

$$v_\theta = -2U_0 \sin \theta - \Gamma/2\pi a. \qquad (10.28)$$

At stagnation points, $v_\theta = 0$, and, therefore, $\Gamma/2\pi a = -2U_0 \sin \theta$, so that the location of the stagnation points is given by

$$\theta = \text{arc sin } (-\Gamma/4\pi U_0 a).$$

This equation results in two solutions, namely,

$$\theta_1 = -\alpha \quad \text{and} \quad \theta_2 = -(180 - \alpha).$$

The corresponding angle of incidence α for the aerofoil is, therefore, measured from the position of the aerofoil in the stream such that there is zero lift. The axis of the aerofoil parallel to the direction of flow under this condition and drawn through the trailing edge is known as the zero-lift axis and is shown in Fig. 10.22. Thus, α_0 is the negative angle of incidence corresponding to no lift.

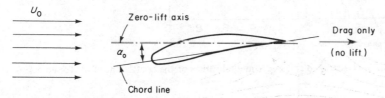

Figure 10.22   The zero lift axis of an aerofoil

Returning now to our analogy with the cylinder, it is possible to express the circulation in terms of α (equation (7.65)):

$$\Gamma = -4\pi U_0 a \sin \alpha$$

and, using Kutta–Joukowski's expression, which is for clockwise circulation defined as negative, the lift becomes

$$L = \rho U_0 \Gamma = 4\pi a \rho \sin \alpha U_0^2. \qquad (10.29)$$

Comparing this with equation (10.5) for lift, namely

$$L = \tfrac{1}{2} C_L \rho U_0^2 A,$$

the following relationship is obtained:

$$4\pi a \rho U_0^2 \sin \alpha = \tfrac{1}{2} C_L \rho U_0^2 A,$$

so that

$$C_L = (8\pi a/A) \sin \alpha, \qquad (10.30)$$

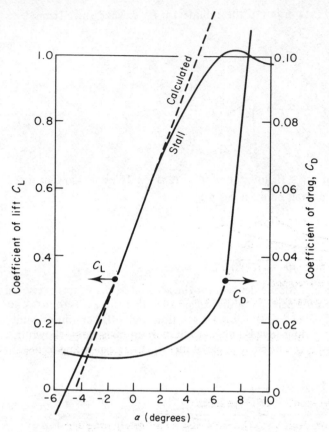

Figure 10.23 Calculated and experimental values of the coefficient of lift for an aerofoil

indicating that the coefficient of lift is directly proportional to sin α, which, for small angles of incidence, means that it is proportional to the angle of incidence. This theory is in good agreement with experimental results. Figure 10.23 shows the calculated values of C_L for a given aerofoil together with the measured values, as functions of the angle of attack. The drag coefficient is also shown.

The good agreement at small angles of attack is related to the fact that there is no separation of the boundary layer at these small angles. As the angle of incidence is increased, however, separation occurs at the top surface near the trailing edge, thus reducing slightly the rate of increase of the lift with the angle of attack. As the incidence is increased, the point of separation moves forward, as shown in Fig. 10.25. It will be seen that the wake widens and, hence, the drag increases until at some stage the separation point moves to a position such that any further increase of incidence no longer produces an increase of lift. This position is called *stall* and constitutes a critical angle of attack above which the lift drops rapidly, as indicated in Figs. 10.23 and 10.24. The stall is accompanied by a rapidly increasing drag, which is mainly due to the increasing wake.

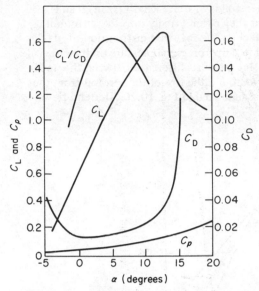

Figure 10.24   Typical characteristics of an aerofoil

A typical aerofoil characteristic is shown in Fig. 10.24. In addition to curves of lift and drag coefficients it also shows the lift/drag ratio and pressure coefficient C_p.

10.8.   Flow past an aerofoil of finite length

The previous section dealt with the flow past an infinitely long aerofoil or one which is bounded by parallel plates at the ends. Such conditions or assumptions mean that the flow is truly two-dimensional and there is no spanwise variation of flow patterns and forces for a constant chord aerofoil.

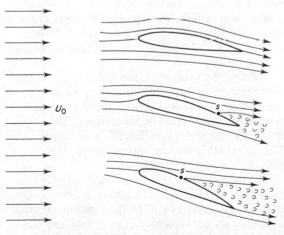

Figure 10.25   Separation due to increased angle of incidence of an aerofoil

In three-dimensional flow, the aerofoil is of finite length (span) b and without walls at the ends, so that they extend freely into the surrounding fluid. This has a considerable effect on the spanwise distribution of lift.

When an aerofoil is subjected to a lift force, the pressure on its underside is greater than that on the top. This pressure difference between the upper and the lower surface causes flow around the tips of the aerofoil from the underside to the upper surface, as indicated in Fig. 10.26. This end flow

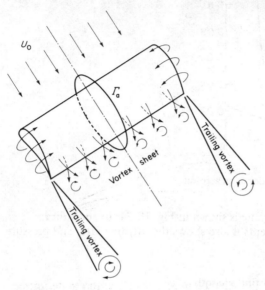

Figure 10.26   End effects on an aerofoil of finite span

affects the rest of the flow pattern in the following manner. The flow on the underside is deflected towards the tips of the aerofoil in order to supply the necessary end flow, whereas the flow at the top of the aerofoil is deflected from the tips towards the centre. This produces unstable flow at the trailing edge, causing a vortex sheet which rolls up into two vortices emanating from somewhere near the tips. It is the condensation of water vapour due to low pressure in these tip vortices that is sometimes seen at the tips of aircraft wings.

Since there is end flow at the tips, the pressure difference between the top and bottom surfaces of an aerofoil must decrease from a maximum at the mid-span towards the tips where it is zero. Consequently, the circulation around the aerofoil of finite span must also decrease from its maximum value Γ_a at the centre-line down to zero at the tips. The lift is, of course, affected in the same way. The distribution may be approximated to an ellipse, as shown in Fig. 10.27.

A further consequence of the tip vortices is that they induce a downward velocity component which is known as *downwash velocity* $\bar{v}_i$. Its presence means that the relative velocity of motion between the fluid and the aerofoil is no longer the free-stream velocity U_0 but velocity U, deflected from U_0 by an angle ϵ known as the induced angle of incidence. The resulting geometry

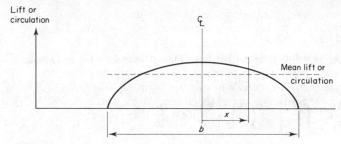

Figure 10.27   Distribution of lift along a wing's span

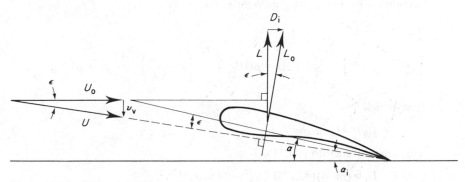

Figure 10.28   Induced drag

is shown in Fig. 10.28. What follows is that, in accordance with the definition of lift, which stipulates that it is perpendicular to the relative direction of motion, the true lift is normal to U. However, since it is more convenient and customary to relate lift and drag to the direction of the free stream relative to the aerofoil, the true lift L_0 is resolved into L, the component perpendicular to U_0, and D_i, the component parallel to U_0. This latter component, which is in the same direction as drag, is known as *induced drag* and is added to pressure drag and the skin friction drag to give the total drag on an aerofoil. The expression for induced drag is derived as follows. The true lift per unit length of span is given by

$$L_0 = \rho U \Gamma,$$

hence, the induced drag per unit span,

$$D_i' = L_0 \sin \epsilon = \rho U \Gamma \sin \epsilon.$$

But, $\sin \epsilon = v_v/U$ and, using Prandtl's approximation for elliptical spanwise lift distribution that $v_v = \Gamma_0/2b$, where Γ_0 is the maximum circulation at the centre-line,

$$\sin \epsilon = \Gamma_0/2bU$$

and

$$D_i' = \rho U \Gamma (\Gamma_0/2bU) = \rho \Gamma (\Gamma_0/2b).$$

Now, for elliptic spanwise distribution of Γ,

$$\Gamma = \Gamma_0 [1 - (2x/b)^2]^{1/2}$$

where x is the distance from the centre-line. Thus, the induced drag for the total span,

$$D_i = \frac{\rho}{2b} \Gamma_0^2 \int_{-b/2}^{+b/2} \left[1 - \left(\frac{2x}{b} \right)^2 \right]^{1/2} dx = \frac{\rho}{2b} \Gamma_0^2 \frac{b\pi}{4}$$

$$= \rho\pi\Gamma_0^2/8, \tag{10.31}$$

is obtained by substituting $2x/b = \sin\theta$.

But,

$$L_0 = \int_{-b/2}^{+b/2} \rho U \Gamma \, dx = \rho U \Gamma_0 \int_{-b/2}^{+b/2} \left[1 - \left(\frac{2x}{b} \right)^2 \right]^{1/2} dx = \rho U \Gamma_0 b \frac{\pi}{4},$$

from which

$$\Gamma_0 = 4L_0/\rho U b \pi$$

and, substituting into (10.31),

$$D_i = (\rho\pi/8)(4L_0/\rho U b\pi)^2 = 2L_0^2/\rho\pi U^2 b^2.$$

However, from similar triangles,

$$L_0/U = L/U_0$$

and, hence,

$$D_i = (2/\rho\pi b^2)(L/U_0)^2. \tag{10.32}$$

If the coefficient of induced drag is defined as

$$C_{D_i} = D_i/\tfrac{1}{2}\rho U^2 A$$

and, since $C_L = L/\tfrac{1}{2}\rho U^2 A$, by substitution

$$C_{D_i} = \frac{D_i}{L/C_L} = \frac{C_L}{L} \frac{2L^2}{\rho\pi b^2 U^2} = 2C_L \frac{L}{\rho\pi b^2 U^2}$$

$$= 2C_L \frac{\tfrac{1}{2}C_L\rho U^2 A}{\rho\pi b^2 U^2} = C_L^2 \frac{A}{\pi b^2} = C_L^2 \frac{cb}{\pi b^2}$$

$$= \frac{C_L^2}{\pi} \times \frac{c}{b}.$$

But

$$\frac{c}{b} = \frac{1}{\text{(Aspect ratio)}},$$

so that

$$C_{D_i} = \frac{C_L^2}{\pi \text{(Aspect ratio)}}. \tag{10.33}$$

This equation shows that a large aspect ratio minimizes the induced drag, as would be expected.

Example 10.6

A wing of an aeroplane of 10 m span and 2 m mean chord is designed to develop a lift of 45 kN at a speed of 400 km h^{-1}. A $\frac{1}{20}$th scale model of the wing section is tested in a wind tunnel at 500 m s^{-1} and ρ = 5·33 kg m^{-3}. The total drag measured is 400 N. Assuming that the wind tunnel data refer to a section of infinite span, calculate the total drag for the full-size wing. Assume elliptical lift distribution and take air density as 1·2 kg m^{-3}.

Solution
Wing area,

$$A = 2 \times 10 = 20 \text{ m}^2.$$

Coefficient of drag from the model data,

$$C_D = \frac{D}{\frac{1}{2}\rho U^2 A} = \frac{400}{\frac{1}{2} \times 5\cdot3(500)^2\, 20/(20)^2} = 0\cdot012.$$

For the prototype,

$$U = 400 \text{ km h}^{-1} = 111\cdot1 \text{ m s}^{-1}$$

and the lift coefficient,

$$C_L = \frac{L}{\frac{1}{2}\rho U^2 A} = \frac{45\,000}{\frac{1}{2} \times 1\cdot2(111\cdot1)^2\, 20} = 0\cdot304.$$

Now, assuming elliptical distribution, the coefficient of induced drag,

$$C_{D_i} = C_L^2/\pi(\text{AR}) = (0\cdot304)^2/\pi(\tfrac{10}{2}) = 0\cdot0059.$$

Hence, the total drag coefficient,

$$C_{D_t} = C_D + C_{D_i} = 0\cdot012 + 0\cdot0059 = 0\cdot0179$$

and the total drag on the wing,

$$D = \tfrac{1}{2}C_{D_t}\rho U^2 A = \tfrac{1}{2} \times 0\cdot0179 \times 1\cdot2(111\cdot1)^2 \times 20 = 2648\cdot9 \text{ N}.$$

Therefore,

$$D = 2\cdot65 \text{ N}.$$

10.9 Wakes and Drag

It was explained in Section 10.1 that pressure drag is closely related to boundary layer separation and the formation of a wake at the rear of the body. The size of the wake and the pressure within it are the two factors which determine the magnitude of the pressure drag. The wider the wake the greater is the area over which the pressure difference between the front and the rear of the body acts and hence the greater is the drag. Equally, the lower the pressure within the wake, the greater is the pressure difference acting on the body and hence the drag. The two effects are in fact related: as the width of the wake is reduced the pressure within it increases and approaches the free-stream pressure. This interdependence was first shown theoretically by Helmholtz, who assumed a stagnant wake region behind the body. For such theoretical flow (known as Helmholtz flow) the separation points move towards the rear of the body and the wake is reduced as the pressure within the wake is increased. Thus, when pressure recovery is assumed to be complete, that is when the pressure within wake is the same as the free-stream pressure, the wake disappears completely and there is no pressure drag. Empirical evidence of flow past cylinders, spheres and other bodies supports the above principles. In particular the work of Eisenberg and Reichardt provided strong evidence of linear correlation between the drag of various bodies and the pressure coefficient of the cavity behind them.

It follows therefore that in order to minimize pressure drag it is important to reduce the width of the wake as much as possible. This is achieved by preventing or delaying boundary layer separation from the surface of the body. In Chapter 9 it was shown that the turbulent boundary layer separates less easily than the laminar boundary layer and therefore in the former case the separation points are always further to the rear of the body, the wake is narrower and the drag coefficient is considerably smaller than for the laminar boundary layer. For example, for a cylinder and for a specific Re, laminar separation before transition occurs at $\theta = \pm 98°$ (measured from the rear stagnation point) and the drag coefficient $C_D = 1 \cdot 2$ whereas turbulent separation after transition occurs at $\theta = \pm 60°$ and $C_D = 0 \cdot 3$. Similarly for a sphere: for laminar separation $\theta = \pm 100°$ and $C_D = 0 \cdot 44$ but for turbulent separation $\theta = \pm 60°$ and $C_D = 0 \cdot 22$. (Figure 10.10 shows that C_D varies by up to 10% with Re over small Re ranges.) The level of free-stream turbulence has little effect on the value of the drag coefficient as such but it does affect the Reynolds number at which transition from laminar to turbulent boundary layer takes place (*see* Section 10.5). Generally higher levels of turbulence cause earlier transition.

Fluid velocity in the wake is greatly reduced compared with that upstream of the body. It is in general not uniform, unsteady and sometimes oscillatory. Much work has been done on the nature of this very complicated unsteady flow which can give rise to significant body forces, particularly on bluff bodies at subcritical Reynolds number. For most flows, however, it is sufficient to use time-averaged velocities, as measured by a Pitot static tube for example.

The drag force and hence the drag coefficient of a body, due to the relative motion of a fluid over it, may be determined either by direct force measurement or by calculation from detailed velocity and pressure distributions in the wake. The latter method is based on the application of Newton's second law of linear motion to an immersed body which causes fluid deceleration in the wake and hence a change of fluid linear momentum, but also a difference of pressure between the front and rear of the body.

Consider flow past a two-dimensional body mounted in a wind tunnel as shown in Fig. 10.29. Let the free-stream velocity and static pressure in front of the body be U_0 and p_0 respectively and the drag force on the body per unit span be D'. Let also the velocity and pressure profiles downstream of the body be as shown, so that at any distance y from the centre-line the velocity is $u_1 = f(y)$ and the pressure is $p_1 = f'(y)$. Now, if the half-width of the wake w is defined as that distance from the centre-line at which the

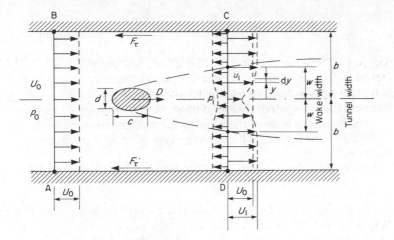

Figure 10.29

velocity becomes constant, then u_1 varies within the band $-w < y < +w$ and becomes equal to U_1 outside the wake where it is constant.

Taking the control volume ABCD such that AB is far upstream of the body where both the velocity and the pressure are constant, CD is downstream of the body and cuts through the wake, and BC and AD are coincident with the tunnel walls, the forces acting on the fluid in x-direction are: pressure forces due to pressure difference, the negative drag force D with which the body acts on the fluid and forces due to shear stresses F_τ along the tunnel walls resulting from the boundary layer there. The sum of all these forces must be equal to the rate of change of linear momentum in the x-direction.

Now,

Pressure force on fluid element dy, per unit span $= (p_0 - p_1) dy$,

Resultant force on the fluid within the control volume, per unit

$$\text{span} = \int_{-b}^{+b} (p_0 - p_1) \, dy - D' - F_\tau,$$

Mass flowrate through fluid element dy, per unit span $= \rho u_1 \, dy$,

Change of velocity $= (u_1 - U_0)$.

Therefore,

Rate of change of fluid momentum $= \int_{-b}^{+b} \rho u_1 (u_1 - U_0) \, dy,$

and equating the forces to the rate of change of momentum:

$$\int_{-b}^{+b} \rho u_1 (u_1 - U_0) \, dy = \int_{-b}^{+b} (p_0 - p_1) \, dy - D' - F_\tau,$$

and the drag force on the body per unit span:

$$D' = \int_{-b}^{+b} (p_0 - p_1)\, dy - \int_{-b}^{+b} \rho u_1 (u_1 - U_0)\, dy - F_\tau. \qquad (10.34)$$

The drag force includes both the skin friction drag and pressure drag because both produce the overall change of momentum. The force F_τ may be calculated using boundary layer analysis, but if the width of the tunnel b is large compared with the frontal width of the body d, then F_τ is small and may be neglected.

Thus, changing the order of velocities in the second integral:

$$D' = \int_{-b}^{+b} (p_0 - p_1)\, dy + \int_{-b}^{+b} \rho u_1 (U_0 - u_1)\, dy. \qquad (10.35)$$

However, the typical velocity and pressure distributions indicate that at distances greater than w from the centre-line the fluid is unaffected by the body and the velocity and pressure there are constant. If the wake is enclosed, that is within the walls of a tunnel, the constant velocity U_1 outside the wake is not equal to the free-stream velocity, because continuity demands that the velocity defect within the wake is made up outside it.

From the continuity equations, therefore,

$$U_0 b = U_1 (b - w) + \int_{-w}^{+w} u_1\, dy,$$

from which: $\quad U_1 = \dfrac{U_0 b}{(b - w)} - \dfrac{1}{(b - w)} \displaystyle\int_{-w}^{+w} u_1\, dy. \qquad (10.36)$

Similarly, the constant pressure p_1' outside of the wake may not be equal to p_0, so that in such a case the integration must include the whole tunnel width. So, to evaluate drag both the velocity and the pressure distributions are necessary if the traversing is carried out close behind the body. However, some simplifications and approximations are possible in appropriate cases. Firstly, for a free wake, such as that forming behind an aircraft wing during flight, there is no restriction of x-direction mass-flow continuity imposed by the tunnel walls and therefore the velocity outside the wake U_1 is equal to U_0. Similarly the pressure outside the wake p_1' is equal to the free-stream pressure p_0. Thus, both the integrals of equation (10.35) may have their limits changed to $\pm w$ because they become equal to zero outside these limits. Therefore, equation (10.35) for a free wake becomes:

$$D' = \int_{-w}^{+w} (p_0 - p_1)\, dy + \int_{-w}^{+w} \rho u_1 (U_0 - u_1)\, dy \qquad (10.37)$$

Now, since the drag is obviously independent of any traverse which may be carried out at any distance downstream of the body, it follows that the sum of the two integrals of equation (10.37) must remain constant and independent of x. So, as the wake diffuses with increasing distance from the body, its width will increase and the velocity defect $(U_0 - u)$ will decrease, as shown in Fig. 10.30. At some distance, sufficiently far away downstream of the body, the static pressure across the wake and flow will be constant

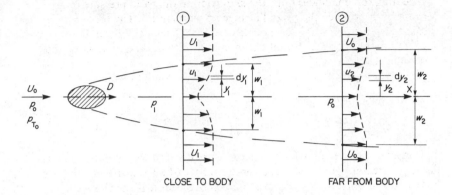

CLOSE TO BODY   FAR FROM BODY

Figure 10.30

and equal to the upstream free-stream static pressure p_0. Thus the pressure integral in equation (10.37) becomes equal to zero, and the drag becomes:

$$D' = \int_{-w_2}^{+w_2} \rho u_2 (U_0 - u_2) \, dy_2,$$

(10.38)

where suffix 2 refers to the far-away downstream section.

Using the definition of the coefficient of drag of equation (10.4),

$$D = \tfrac{1}{2} \rho C_D U_0^2 A,$$

and remembering that $D' = D/L$, we obtain:

$$C_D = \frac{D}{\frac{1}{2}\rho U_0^2 Lc} = \frac{D'}{\frac{1}{2}\rho U_0^2 c} = \frac{2}{c} \int_{-w_2}^{+w_2} \frac{u_2}{U_0} \left(1 - \frac{u_2}{U_0}\right) dy_2. \quad (10.39)$$

Note that in the above equation the value of C_D would be based on the span area, since it was taken that $A = cL$ which is usual for lifting surfaces such as plates or aerofoils. However, for bluff bodies the frontal area $A = dL$ is commonly used in which case equation (10.39) would take the form:

$$C_D = \frac{2}{d} \int_{-w_2}^{+w_2} \frac{u_2}{U_0} \left(1 - \frac{u_2}{U_0}\right) dy_2.$$

(10.39a)

To be consistent the numerical values of Reynolds number at which values of C_D are determined and quoted are customarily also based on c for lifting surfaces and on d for bluff bodies.

Equations (10.38) and (10.39) are convenient to use provided it is practicable to carry out the traversing at such a distance that $p_2 = p_0$. If this is not possible equation (10.37) must be used unless an approximation is acceptable. One such approximation was developed by B. M. Jones and has been used extensively together with a Pitot rake behind a body. The method assumes that within the wake along any given streamtube between sections 1 and 2 the total pressure remains constant. It is then possible to modify equation (10.38) so that it refers to section 1 using also continuity to account for larger wake width at section 2.

Thus: $p_T = p_1 + \frac{1}{2} \rho u_1^2$,

and also

$$p_T = p_2 + \frac{1}{2} \rho u_2^2.$$

Thus the velocities are given by:

$$u_1 = \sqrt{\left(\frac{2}{\rho} (p_T - p_1) \right)} \quad \text{and} \quad u_2 = \sqrt{\left(\frac{2}{\rho} (p_T - p_2) \right)},$$

but since $p_2 = p_0$,

$$u_2 = \sqrt{\left(\frac{2}{\rho} (p_T - p_0) \right)}.$$

Similarly the free-stream velocity U_0 may be expressed in terms of the total pressure there:

$$U_0 = \sqrt{\left(\frac{2}{\rho} (p_{T_0} - p_0) \right)}.$$

Continuity:

$$u_1 \, dy_1 = u_2 \, dy_2,$$

from which:

$$dy_2 = \frac{u_1}{u_2} \, dy_1.$$

Substituting into equation (10.38):

$$D' = \int_{-w_1}^{+w_1} \rho u_2 (U_0 - u_2) \frac{u_1}{u_2} \, dy_1 = \int_{-w_1}^{+w_2} \rho u_1 (U_0 - u_2) \, dy_1,$$

and replacing the velocities:

$$D' = 2 \int_{-w_1}^{+w_1} \sqrt{(p_T - p_1)} \, (\sqrt{(p_{T_0} - p_0)} - \sqrt{(p_T - p_0)}) \, dy_1. \quad (10.40)$$

Also, equation (10.39) becomes:

$$C_D = \frac{2}{c} \int_{-w_1}^{+w_1} \left(\frac{p_T - p_1}{p_{T_0} - p_0} \right)^{1/2} \left[1 - \left(\frac{p_T - p_0}{p_{T_0} - p_0} \right)^{1/2} \right] dy_1. \quad (10.41)$$

In order to evaluate the drag of a body and its drag coefficient using the above equations a traverse of total pressures and static pressures at some section downstream of the body is required together with the reading of total and static pressures upstream of the body. The integration may be performed graphically, the method was devised prior to the widespread availability of computers. It must be remembered, however, that the method is approximate and therefore it is preferable to use equation (10.37) in conjunction with a suitable numerical integration method and a computer.

10.10 Computer program 'WAKE'

(1) The program calculates the drag per unit span of a body in an airstream and its drag coefficient from pressure and stagnation pressure traverse across the wake downstream of the body, assuming two-dimensional, incompressible flow.

(2) The required input is as follows:

(i)   body characteristic dimension, i.e. frontal width or chord, C (mm);

(ii)   barometric pressure, B (mm of Hg);

(iii)   static pressure (gauge) upstream of the body, p_s (mm of H_2O);

(iv)   temperature upstream of the body, T ($^{\circ}C$);

(v)   either (a) free-stream air velocity, U_0 (m s^{-1}), or (b) stagnation pressure (gauge), upstream of the body, p_{T_0} (mm of H_2O);

(vi)   number of readings in the traverse, N;

(vii)   spacing between traverse points, H (mm);

(viii) either (a) up to 50 traverse readings of static pressure, P and stagnation pressure, p_T (mm of H_2O), or (b) the value of static pressure, p, at traverse section, if constant, and up to 50 traverse readings of stagnation pressure p_T (mm of H_2O).

(3) Use is made of equations (10.37) and (10.39) together with the equation of state (1.13) and Bernoulli's equation. The integrals of equation (10.37) are computed using the 'trapezoidal rule'.

(4) Input example:

$C = 20$ mm; $B = 770$ mm of Hg; $p_s = 78.5$ mm of water; $T = 20^{\circ}C$; $U_0 = 45$ m/s; $N = 15$; $H = 5$mm; at traverse section static pressure constant at: $p = 48.8$ mm of water; values of stagnation pressure, p_T (mm of water), at traverse points: 176, 176, 170, 150, 136, 110, 85, 70, 82, 111, 135, 155, 173, 175, 176.

(5) Output:

U	RO	D	CD
M/S	KG/M^3	KN/M	
45.00	1.23	41.08	1.65

List:

```
10 MODE7: @%=&2020A
20 REM: WAKE TRAVERSE
30 CLS: GOSUB 1040
40 PRINT "PROGRAM: WAKE"
50 DIM P(50),PT(50),Y(50),V(50),Z(50)
60 GOSUB 1040: GOSUB 1050
70 PRINT CHR$(132);"THIS PROGRAM CALCULATES THE DRAG ON A"
80 PRINT CHR$(132);"BODY IN AIR FROM PRESSURE & STAGNATION"
90 PRINT CHR$(132);"PRESSURE TRAVERSE ACROSS THE WAKE OF"
100 PRINT CHR$(132);"THE BODY, ASSUMING";CHR$(134);"TWO-DIMENSIONAL"
110 PRINT CHR$(132);"INCOMPRESSIBLE FLOW"
120 GOSUB 1040: GOSUB 1050
130  GOTO 150
140 CLS: GOSUB 1040
150 PRINT "WHAT IS THE ATMOSPHERIC (BAROMETRIC)"
160 INPUT "PRESSURE IN MM OF MERCURY";B
170 PA=13.6*9.81*B: REM: IN N/M^2
180 GOSUB1040: PRINT "WHAT IS THE STATIC (GAUGE) PRESSURE"
190 PRINT "IN MM OF WATER AND THE TEMPERATURE IN"
200 PRINT "DEGREES C IN THE FREE AIRSTREAM UPSTREAM"
210 INPUT "OF THE BODY, PS= ";PS
```

```
220 PO=PS*9.81+PA : REM:NOW IN N/M^2 ABS
230 INPUT "T=";TO
240 RO=PO/(287*(273+TO))
250 GOSUB 1040: INPUT"IS THE UPSTREAM AIR VELOCITY KNOWN (Y/N)";A$
260 IF A$="Y" THEN 290
270 IF A$="N" THEN 310
280 GOTO 250
290 GOSUB 1040: INPUT"WHAT IS IT (M/S)";UO
300 GOTO 340
310 GOSUB1030: PRINT "WHAT IS THE UPSTREAM STAGNATION PRESSURE"
320 INPUT "(PTO) IN MM OF WATER";PTO
330 UO=SQR(2*9.81*(PTO-PS)/RO)
340 GOSUB 1040: PRINT"WHAT IS THE FRONTAL OR CHORD DIMENSION"
350 PRINT"OF THE BODY (IN MM) ON WHICH THE DRAG"
360 INPUT "COEFFICIENT IS TO BE BASED, C= ";C
370 GOSUB 1040: PRINT"HOW MANY TRAVERSE POINTS ARE TO BE INPUT"
380 INPUT "N= ";N
390 GOSUB 1030: PRINT "WHAT IS THE DISTANCE (IN MM) BETWEEN ANY"
400 INPUT "ADJACENT TRAVERSE POINTS, H= ";H
410 GOSUB 1040: PRINT"IS THE STATIC PRESSURE ACROSS THE"
420 INPUT "TRAVERSE SECTION CONSTANT (Y/N)";B$
430 IF B$="Y" THEN 840
440 IF B$="N" THEN 460
450 GOTO 410
460 GOSUB 1040: PRINT"INPUT TRAVERSE VALUES OF STATIC PRESSURE"
470 PRINT "(P) AND STAGNATION PRESSURE (PT),"
480 PRINT "BOTH IN MM OF WATER COLUMN"
490 FOR X=1 TO N
500 INPUT "P, PT";P(X),PT(X)
510 Y(X)=9.81*(PS-P(X))
520 V(X)=SQR(9.81*2*(PT(X)-P(X))/RO)
530 Z(X)=V(X)*(UO-V(X))
540 NEXT
550 S1=Y(1)+Y(N)
560 FOR X=2 TO (N-1)
570 S1=S1+2*Y(X)
580 NEXT
590 I1=H*S1/2000
600 GOSUB 1040
610 S2=Z(1)+Z(N)
620 FOR X=2 TO (N-1)
630 S2=S2+2*Z(X)
640 NEXT
650 I2=RO*H*S2/2000
660 D=I1+I2
670 CD=2000*D/(RO*C*UO^2)
680 CLS: GOSUB 1040
690 PRINT TAB(6)"U",TAB(16)"RO",TAB(27)"D",TAB(35)"CD"
700 PRINT TAB(5)"M/S",TAB(14)"KG/M^3",TAB(26)"KN/M"
710 GOSUB 1040
720 PRINT UO,RO,D,CD
730 GOSUB1040: GOSUB1050
740 PRINT "DO YOU WANT TO INPUT"
750 INPUT "NEW TRAVERSE DATA (Y/N)";A$
760 IF A$="Y" THEN 410
770 IF A$="N" THEN 790
780 GOTO 740
790 GOSUB1040: PRINT "DO YOU WANT TO RUN THE"
800 INPUT "PROGRAM AGAIN (Y/N)";B$
810 IF B$="Y" THEN 140
820 CLS: PRINT TAB(15,15) CHR$(130)"THANK YOU"
830 END
840 GOSUB 1040: PRINT "IS THIS PRESSURE EQUAL TO "
850 INPUT "THE UPSTREAM PRESSURE (Y/N)";A$
860 IF A$="Y" THEN 890
870 IF A$="N" THEN 990
880 GOTO 840
890 I1=0
900 GOSUB1030
910 PRINT "INPUT TRAVERSE VALUES OF"
920 PRINT "STAGNATION PRESSURE IN MM OF WATER"
930 FOR X=1 TO N
940 INPUT "PT= ";PT(X)
950 V(X)=SQR(2*9.81*(PT(X)-P)/RO)
960 Z(X)=V(X)*(UO-V(X))
970 NEXT
980 GOTO 610
990 GOSUB1030: PRINT"WHAT IS IT IN "
1000 INPUT "MM OF WATER";P
1010 I1=9.81*(N-1)*H*(PS-P)/1000
1020 GOTO 900
1030 PRINT: RETURN
1040 PRINT: PRINT: RETURN
1050 FOR X=1 TO 4000: NEXT: RETURN
```

EXERCISES 10

10.1 A wing of a small aeroplane is rectangular in plan having a span of 10 m and a chord of 1·2 m. In straight and level flight at 240 km/h the total aerodynamic force acting on the wing is 20 kN. If the lift/drag ratio is 10 calculate the coefficient of lift and the total weight the aeroplane can carry. Assume air density to be 1·2 kg/m^3.
[0·622]

10.2 A screen across a pipe of rectangular cross-section 2 m by 1·2 m consists of well streamlined bars of 25 mm maximum width and at 100 mm centres, their coefficient of total drag being 0·30. A water stream of 5·5 m^3/s passes through the pipe. What is the total drag on the screen? If a rectangular block of wood 1 m by 0·3 m and about 25 mm thick is held by the screen, making suitable assumptions, estimate the increase of the drag.
[449 N, 726 N]

10.3 A parachute of 10 m diameter when carrying a load W descends at a constant velocity of 5·5 m/s in atmospheric air at a temperature of 18 °C and pressure of 1·0 10^5 N/m^2. Determine the load W if the drag coefficient for the parachute is 1·4.
[1·995 kN]

10.4 Prove that the viscous resistance F of a sphere of diameter d moving at constant speed v through a fluid of density ρ and viscosity μ may be expressed as

$$F = k \frac{\mu^2}{\rho} f \left(\frac{\rho v d}{\mu} \right) \quad \text{where } k \text{ is a constant}$$

Two balls made of steel and aluminium are allowed to sink freely in an oil of specific gravity 0·9. Determine the ratio of their diameters if dynamic similarity must be obtained when the balls attain their terminal sinking velocities. The specific gravities of steel and aluminium are 7·8 and 2·7 respectively.
[0·639]

10.5 The drag and bending moment on a structure in a 40 km/h wind is to be studied using a 1/20 scale model in a pressurized wind tunnel. If the tunnel and ambient temperatures are the same but the air density in the tunnel is 8 times that of the ambient air calculate the air speed in the tunnel and the bending moment for the structure if that measured on the model is 30 Nm.
[100 km/h, 4800 N m]

10.6 A submarine periscope is 0·15 m in diameter and is travelling at 15 km/h. What is the frequency of the alternating vortex shedding and the force per unit length of the periscope. Take density of water as 1·03 x 10^3 kg/m^3 and kinematic viscosity at 1·25 mm^2/s.
[5·5 Hz, 805 N/m]

10.7 A 1/20 model of a 120 m long cargo ship is towed in fresh water at a velocity of 2·5 m/s. The measured total drag is 105 N. The skin friction drag coefficient is 0·002 72 and the wetted area is 6·5 m^2. The estimated skin-friction drag coefficient for the prototype is 0·0018. Determine: (a) the wave drag for the model, (b) the wave drag coefficient for the model, (c) the wave drag and the total drag for the ship, and (d) the power required to tow the model and to propel the ship at its design cruising speed.
[49·75 N, 0·00245, 408 kN, 435 kN, 262·5 W, 4863 kW]

10.8   A spherical weather balloon of 2 m diameter is filled with hydrogen. The total mass of the balloon skin and the instruments it carries is 3·708 kg. If at a certain altitude the density of air is 1·0 kg/m^3 and is 10 times the density of hydrogen in the balloon determine the steady upward velocity of the balloon. Take viscosity of air to be 1·8 x 10^{-5} Ns/m^2.
[0·931 m/s]

10.9   A small spherical water droplet falls freely at a constant speed in air. An air bubble of the same diameter rises freely at a constant rate in water. Derive an expression for the ratio of the distances travelled by the droplet and the air bubble during the same time and calculate this ratio if $\rho_{air} = 1\cdot2$ kg/m^3, $\mu_{air} = 1\cdot7$ x 10^{-5}, $\mu_{water} = 10^{-3}$ Ns/m^2.
[58·8 for $d < 0\cdot0446$ mm; depends on d for $0\cdot0446 < d < 2\cdot04$ mm; 28·9 for $d > 2\cdot04$ mm]

10.10   (a) A submarine is deeply submerged and moving along a straight course. Describe the physical phenomena that give rise to resistance to its motion. The submarine now comes to the surface and continues on course. What changes occur in the resistance phenomena?

(b) The following data refer to a 1/20 scale model of a cargo vessel under test in a model basin:

Model speed	1·75 m/s
Total resistance	34·25 N
Model length	6·20 m
Wetted surface area	5·91 m^2
Basin water density	998 kg/m^3
Kinematic viscosity	0·1010 10^{-5} m^2/s

The I.T.T.C. coefficients may be calculated from:
$$C_F = 0\cdot075/(\log_{10}R - 2)^2$$
where R is the Reynolds number.

What will be the total resistance for the smooth ship at the corresponding speed in sea water of kinematic viscosity 0·1188 10^{-5} m^2/s?
[244·5 N]

10.11   When a slender body held transversely is tested in a wind tunnel it is found that the decrease in velocity in the wake is approximately linear. It decreases from the undisturbed velocity u_0 at double the solid width to $0\cdot2 \, U_0$ at the axis, the pressure in the wake being constant throughout and the same as that in the undisturbed stream. If such a body of 1·5 m width moves, under dynamically similar conditions, through still air at 150 m/s, calculate the drag on the solid per unit length and the drag coefficient. The air is at 5°C and a pressure of 510 mm of mercury. Take density of mercury as 13·6 x 10^3 kg/m^3 and gas constant for air as $R = 287$ J/kg K.
[10·75 kN, 0·747]

11 Compressible flow round a body

11.1. Effects of compressibility

In the previous chapter, the discussion of drag in external flow has been limited to the influence of Reynolds number and Froude number. The former involves the relative influence of two fluid properties, namely the density and the viscosity, whereas the latter is concerned with the effects of gravity. However, in Section 10.3 it was shown by dimensional analysis that Mach number, which is a measure of the importance of the elastic forces in the fluid, may be of significance. This occurs when changes of density are appreciable, and the flow is then called compressible. Mach number is also the ratio of the free stream velocity and the velocity of propagation of pressure waves, called the velocity of sound (*see* Section 5.13):

$$\text{Ma} = U_0/c \tag{11.1}$$

Since very significant changes occur at Ma = 1, the flows are classified into subsonic for Ma < 1 and supersonic for Ma > 1.

In subsonic flow, at relatively low velocities the viscous forces and hence Re are of predominant importance. The density changes are small, Ma is also small and its influence is negligible. As the velocity is increased we know from previous paragraphs that at some value of Re the drag coefficient becomes independent of it. This is, however, accompanied by a simultaneous increase of Mach number, whose influence becomes more and more pronounced, and cannot be neglected.

In supersonic flow, shock waves are formed. They not only affect the boundary layer and, hence, the skin friction drag and the position of separation, which controls the form drag, but also produce an abrupt change of pressure. This gives rise to additional drag known as *wave drag*. Since the wave drag is not related to viscosity, but to pressure change across the shock wave, it would be present in an ideal fluid at supersonic flow.

At supersonic flow, the wave drag constitutes the largest contribution to total drag and, therefore, streamlining the rear part of the body, which is so important in the subsonic flow, has little effect. As will be shown later, in order to reduce the wave drag in supersonic flow the nose of the body must be sharp and pointed. This confines the shock wave to only a small region. Thus, the streamlining requirements for the supersonic flow are completely the reverse of those for the subsonic flow. Whereas the latter requires a rounded nose and long, gradually pointed tail, the former requires a sharp, pointed nose and rounded, blunt tail.

The effect of Mach number on the coefficient of drag for projectiles is shown in Fig. 11.1, which also indicates the considerable reduction in C_D achieved by a pointed nose.

The effects of compressibility in external flow are not confined to drag. Since, fundamentally, they take into account the variations of fluid density, all parameters are affected. As an illustration, let us consider the very important conditions at the front stagnation point on a body. Let the pressure, temperature and density at the stagnation point be denoted by

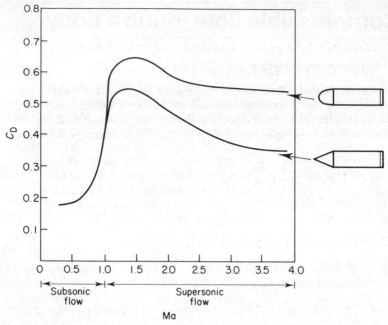

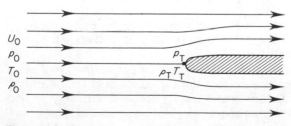

Figure 11.1   Effect of Mach number on the coefficient of drag for projectiles

suffix T and those in the free stream some distance upstream of the body by a suffix 0, as indicated in Fig. 11.2. The conditions at the stagnation point may be expressed in terms of those upstream by the application of Bernoulli's equation and by remembering that the velocity at the stagnation point is zero. First, assuming incompressible flow, we obtain

$$p_T = p_0 + \tfrac{1}{2}\rho U_0^2 \tag{11.2}$$

and, since $\rho_0 = \rho_T = \rho$, it follows from Boyle's law $(p/\rho = f(T))$ that

$$T_T/T_0 = p_T/p_0.$$

Hence, the stagnation temperature is obtained:

$$T_T = T_0(p_T/p_0) = T_0\{(p_0 + \tfrac{1}{2}\rho U_0^2)/p_0\}$$
$$= T_0\{1 + \tfrac{1}{2}(\rho/p_0)U_0^2\}. \tag{11.3}$$

Figure 11.2

But the equation of state (equation (1.13)) gives

$$p_0 = \rho R T_0, \tag{11.4}$$

from which $\rho/p_0 = 1/RT_0$ which, on substitution into (11.3), gives

$$T_T = T_0(1 + \tfrac{1}{2}U_0^2/RT_0) = T_0 + \tfrac{1}{2} U_0^2/R. \tag{11.5}$$

Now, assuming that the flow is compressible and the process by which it is brought to rest at the stagnation point is frictionless and adiabatic (no heat exchange) and, therefore, isentropic, the appropriate form of the Bernoulli equation, derived from equations (5.21) and (1.17) (*see* Example following) gives

$$\frac{\gamma}{\gamma - 1} \frac{p_T}{\rho_T} = \frac{\gamma}{\gamma - 1} \frac{p_0}{\rho_0} + \frac{U_0^2}{2}. \tag{11.6}$$

But, from the equation of state,

$$p/\rho = RT,$$

so that $\{\gamma/(\gamma - 1)\} R T_T = \{\gamma/(\gamma - 1)\} R T_0 + U_0^2/2$

and the stagnation temperature,

$$T_T = T_0 + \{(\gamma - 1)/\gamma R\}(U_0^2/2). \tag{11.7}$$

However, it was also shown in Section 5.13 that the velocity of sound is given by equation (5.31),

$$c = \sqrt{(\gamma R T)} = \sqrt{\{(\gamma - 1) c_p T\}} = \sqrt{(\gamma p/\rho)}, \tag{11.8}$$

from which $\gamma R = c^2/T_0$ may be substituted into equation (11.7), giving

$$T_T = T_0 + \{(\gamma - 1)/2\}(U_0^2/c^2)T_0.$$

But $U_0/c = \mathrm{Ma}_0,$

so that, finally,

$$T_T = T_0\{1 + \tfrac{1}{2}(\gamma - 1)\,\mathrm{Ma}_0^2\}. \tag{11.9}$$

Now, since, for isentropic processes,

$$T_T/T_0 = (p_T/p_0)^{(\gamma-1)/\gamma},$$

the stagnation pressure may be obtained by

$$p_T = p_0(T_T/T_0)^{\gamma/(\gamma-1)} = p_0\{1 + \tfrac{1}{2}(\gamma - 1)\,\mathrm{Ma}_0^2\}^{\gamma/(\gamma-1)} \tag{11.10}$$

In order to compare this expression with equation (11.2) for incompressible flow, it may be rearranged as a pressure ratio and then the expression in brackets expanded using the binomial theorem (justified because

$$\frac{\gamma - 1}{2}\,\mathrm{Ma}_0^2 < 1$$

for subsonic flow). Thus,

$$\frac{p_T}{p_0} = \left(1 + \frac{\gamma - 1}{2}\,\mathrm{Ma}_0^2\right)^{\gamma/(\gamma-1)}$$

$$= 1 + \frac{\gamma}{2}\,\mathrm{Ma}_0^2 + \frac{\gamma}{8}\,\mathrm{Ma}_0^4 + \frac{\gamma(2 - \gamma)}{48}\,\mathrm{Ma}_0^6 + \dots.$$

Rearranging and taking $(\gamma/2)\mathrm{Ma}_0^2$ outside the bracket,

$$\frac{p_T}{p_0} - 1 = \frac{\gamma}{2}\mathrm{Ma}_0^2 \left[1 + \frac{\mathrm{Ma}_0^2}{4} + \frac{(2-\gamma)}{24}\mathrm{Ma}_0^4 + \dots \right]$$

and

$$p_T - p_0 = \frac{\gamma}{2}p_0\mathrm{Ma}_0^2 \left[1 + \frac{\mathrm{Ma}_0^2}{4} + \frac{(2-\gamma)}{24}\mathrm{Ma}_0^4 + \dots \right].$$

But

$$\frac{\gamma}{2}p_0\mathrm{Ma}_0^2 = \frac{\gamma}{2}p_0\frac{U_0^2}{c^2} = \frac{\gamma}{2}p_0\frac{U_0^2}{\gamma p_0/\rho_0} = \frac{1}{2}\rho_0 U_0^2,$$

so that, finally,

$$p_T - p_0 = \tfrac{1}{2}\rho_0 U_0^2 \left[1 + \frac{\mathrm{Ma}_0^2}{4} + \frac{(2-\gamma)}{24}\mathrm{Ma}_0^4 + \dots \right] \qquad (11.11)$$

Now, if the flow is considered incompressible, the corresponding pressure difference $(p_T - p_0)_{\rho=\text{constant}}$ given by equation (11.2) is less than the correct pressure difference given above. The ratio between the two is called the *compressibility factor*. Thus,

$$\text{Compressibility factor} = \frac{p_T - p_0}{(p_T - p_0)_{\rho\,=\,\text{constant}}} = \frac{p_T - p_0}{\tfrac{1}{2}\rho_0 U_0^2}$$

$$= \left[1 + \frac{\mathrm{Ma}_0^2}{4} + \frac{(2-\gamma)}{24}\mathrm{Ma}_0^4 + \dots \right]. \qquad (11.12)$$

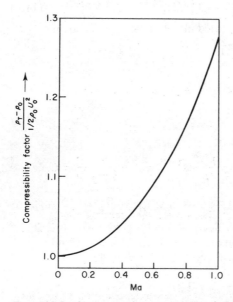

Figure 11.3   Variation of compressibility factor with Mach number

Figure 11.3 shows the variation of compressibility factor with Ma, indicating that for Ma < 0·2 the error in assuming the flow to be incompressible amounts to less than 1 per cent, but for Ma > 0·5 exceeds 5 per cent and becomes 27·6 per cent at Ma = 1.

Example 11.1

Show that for horizontal isentropic flow the Bernoulli's equation takes the form

$$\frac{\gamma}{\gamma-1}\frac{p}{\rho} + \frac{\bar{v}^2}{2} = \text{constant.}$$

Calculate, working from the above equation, the stagnation pressure, temperature and density for an air stream at Ma = 0·7 and density $\rho = 1·8$ kg m^{-3} and temperature of 75 °C. Take $R = 287$ J kg^{-1} K^{-1} and $\gamma = 1·4$.

Solution

Euler's equation (5.21) states

$$\frac{1}{\rho}\frac{dp}{ds} + \bar{v}\frac{d\bar{v}}{ds} + g\frac{dz}{ds} = 0,$$

which, upon integration, becomes

$$\int \frac{dp}{\rho} + \frac{\bar{v}^2}{2} + gz = \text{constant.}$$

Now, for horizontal flow, $z = 0$, and for isentropic flow, $p/\rho^{\gamma} = \text{constant}$ (equation (1.17)). Therefore,

$$\rho = \left(\frac{p}{\text{constant}}\right)^{1/\gamma} = \frac{p^{1/\gamma}}{G}$$

and

$$\int \frac{dp}{\rho} = G\int p^{-1/\gamma}\,dp = \frac{G}{1 - 1/\gamma}p^{(\gamma-1)/\gamma} + C.$$

But $G = p^{1/\gamma}/\rho$ and so, substituting,

$$\int \frac{dp}{\rho} = \frac{G\gamma}{\gamma-1}p^{(\gamma-1)/\gamma} + C = \frac{\gamma}{\gamma-1}\frac{p^{1/\gamma}}{\rho}p^{(\gamma-1)/\gamma} + C = \frac{\gamma}{\gamma-1}\frac{p}{\rho} + C.$$

Therefore, Euler's equation, after integration for isentropic conditions, becomes

$$\frac{\gamma}{\gamma-1}\frac{p}{\rho} + \frac{\bar{v}^2}{2} = \text{constant.}$$

At stagnation point, $v = 0$; therefore, applying Bernoulli's equation to a point in the free stream and the stagnation point (suffix T),

$$\frac{\gamma}{\gamma-1}\frac{p_T}{\rho_T} = \frac{\gamma}{\gamma-1}\frac{p_0}{\rho_0} + \frac{\bar{v}_0^2}{2},$$

$$\frac{p_T}{\rho_T} = \frac{p_0}{\rho_0} + \frac{\gamma-1}{\gamma}\frac{\bar{v}_0^2}{2}.$$

But, from equation (1.17),

$$\rho_T = \rho_0 (p_T/p_0)^{1/\gamma}. \tag{I}$$

Therefore

$$\frac{p_T}{\rho_0} \left(\frac{p_0}{p_T}\right)^{1/\gamma} = \frac{p_0}{\rho_0} + \frac{\gamma - 1}{\gamma} \frac{\bar{v}_0^2}{2},$$

$$p_T^{(\gamma-1)/\gamma} = p_0^{(\gamma-1)/\gamma} + \rho_0 \left(\frac{\gamma-1}{\gamma}\right) \frac{\bar{v}_0^2}{2p_0^{1/\gamma}} \tag{II}$$

Now, from the equation of state,

$$p_0 = \rho_0 RT = 1.8 \times 287 \,(273 + 75) = 179.8 \text{ kN m}^{-2};$$

velocity of sound,

$$c = (\gamma RT)^{\frac{1}{2}} = \sqrt{\{1.4 \times 287\,(273 + 75)\}} = 373.9 \text{ m s}^{-1};$$

stream velocity,

$$\bar{v}_0 = \text{Ma}\, c = 0.7 \times 373.9 = 261.7 \text{ m s}^{-1}.$$

Substituting into (II),

$$p_T^{0.4/1.4} = (179.8 \times 10^3)^{0.4/1.4} + 1.8 \left(\frac{0.4}{1.4}\right) \times \frac{(261.7)^2}{2(179.8 \times 10^3)^{0.714}}$$

$$= 31.72 + 3.12 = 34.84.$$

Hence,

$$p_T = (34.84)^{3.5} = 249.6 \text{ kN m}^{-2}.$$

Now, from (I),

$$\rho_T = 1.8 \left(\frac{249.6}{179.8}\right)^{0.714} = 2.275 \text{ kg m}^{-3}$$

and, therefore,

$$T_T = \frac{p_T}{\rho_T R} = \frac{249.6 \times 10^3}{2.275 \times 287} = 382.3 \text{ K}$$

$$= 109.3 \text{ }^{\circ}\text{C}.$$

11.2. Shock waves

Weak pressure change in the fluid is propagated through the fluid continuum with the velocity of sound, which is a function of the elastic properties of the fluid. Thus, if a periodic pressure disturbance occurs at a point S in Fig. 11.4 in a stationary fluid, the resulting pressure waves will travel radially outwards from point S as concentric spheres. If the period of the disturbance is Δt, then the distance travelled by a wave between the first and second disturbance will be $c\Delta t$. By the time the second wave covered the distance $c\Delta t$ the first wave, being $c\Delta t$ ahead of the second, would have travelled a total distance of $2c\Delta t$. Thus, all the successive waves are equidistant from each other in all directions, the distance being $c\Delta t$.

Now, consider a situation (as shown in Fig. 11.5) in which the source of

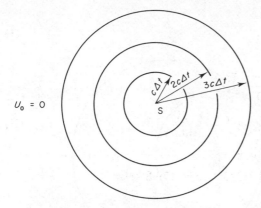

Figure 11.4 Wave propagation in a stationary fluid

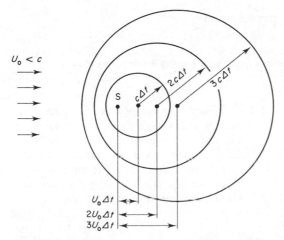

Figure 11.5 Wave propagation in a fluid moving with a velocity smaller than that of sound

periodic disturbance S is placed in a moving fluid whose velocity U_0 is less than the velocity of sound. The waves are still concentric spheres, but are being swept away by the moving fluid. The lateral distance over which each sphere moves during the periodic time Δt is $U_0 \Delta t$. Thus, the absolute velocity with which the disturbance is now propagated depends upon the direction, being $(U_0 + c)$ in the direction of fluid motion but only $(U_0 - c)$ in the opposite direction. The source remains within the spheres, but the distance between the consecutive waves will be large downstream of S and small upstream of it. This concentration of spheres' surfaces upstream will increase as the velocity U_0 approaches the velocity of sound c, until, when $U_0 = c$ and Ma = 1, all the spherical waves become tangential to each other at S. If the fluid velocity is increased further so that $U_0 > c$ and Ma > 1 the spheres are swept away faster than they are generated, the distance $U_0 \Delta t$ being greater than $c \Delta t$. Such a situation is shown in Fig. 11.6.

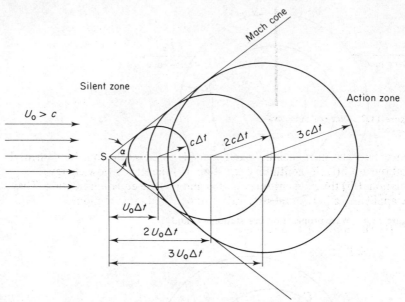

Figure 11.6   Wave propagation in a fluid moving with a velocity greater than that of sound

The surface tangential to all the spherical waves is, of course, a cone, known as the *Mach cone*, which contains within itself the subsonic region called the *action zone*. Outside of the cone, in the *silent zone*, the flow is supersonic and hence the disturbance generated at S is not 'communicated' to any part of the zone. This is the reason for it to be called silent. It follows from the geometry of the situation that the greater is U_0 the greater will be the distances travelled by the spheres ($U_0 \Delta t$) and, since $c\Delta t$ remains constant, the Mach angle α, defined as

$$\sin \alpha = c\Delta t / U_0 \Delta t = c/U_0 = 1/\text{Ma}$$

or $\alpha = \sin^{-1} (1/\text{Ma})$,   (11.13)

will decrease.

If the source of disturbance S is replaced by a thin wedge, as shown in Fig. 11.11, every point on the body becomes a source of disturbance and generates weak Mach waves. The pattern resulting from the superimposition of these waves yields a *shock wave* across which finite changes of flow parameters occur. If the plane of the shock wave is perpendicular to the direction of flow the shock wave is known as a *normal shock* wave. Consider such a shock wave: let the parameters upstream of the shock wave, in the silent zone where Ma > 1, be denoted by a suffix 1 and those downstream of the shock, in the action zone where Ma < 1 be denoted by suffix 2, as indicated in Fig. 11.7. The flow is considered to be adiabatic and frictionless, but not isentropic. This is because there is dissipation of mechanical energy across the shock which results in an increase of entropy. The process is, thus, an irreversible one. A perfect gas is also stipulated. The derivation of the relationship between the upstream and downstream Mach numbers, and,

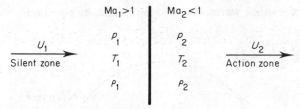

Figure 11.7   Normal shock wave

hence, between the remaining parameters, is based on the four fundamental relationships: (i) the continuity equation, (ii) the steady flow energy equation, (iii) the momentum equation and (iv) the equation of state. They are applied to a horizontal streamtube of constant cross-section.

(i) The continuity equation

$$\rho_1 U_1 A = \rho_2 U_2 A,$$

therefore,

$$\rho_1 U_1 = \rho_2 U_2.$$

But, for adiabatic flow,

$$Ma = U/c = U/\sqrt{(\gamma RT)},$$

so that

$$U = Ma \sqrt{(\gamma RT)},$$

which gives

$$\rho_1 Ma_1 \sqrt{(\gamma RT_1)} = \rho_2 Ma_2 \sqrt{(\gamma RT_2)}.$$

Therefore,

$$\rho_1 Ma_1 \sqrt{T_1} = \rho_2 Ma_2 \sqrt{T_2}. \tag{11.14}$$

However, from the equation of state,

$$p_1/\rho_1 T_1 = p_2/\rho_2 T_2,$$

so that

$$\rho_1/\rho_2 = (p_1/p_2)(T_2/T_1).$$

Substituting into (11.14) in order to eliminate the density ratio:

$$(p_1/p_2)(T_2/T_1) Ma_1 \sqrt{T_1} = Ma_2 \sqrt{T_2},$$

which gives

$$p_1 Ma_1/\sqrt{T_1} = p_2 Ma_2/\sqrt{T_2}. \tag{11.15}$$

(ii) Steady flow energy equation (6.10):

$$U_1^2/2 + H_1 = U_2^2/2 + H_2 = H_T,$$

where H_1 and H_2 are the enthalpies and H_T is the stagnation enthalpy, which remains constant across the shock waves. But $H_T = c_p T_T$ and, by equation (11.9),

$$T_T = T_0 \left(1 + \frac{\gamma - 1}{2} \ Ma_0^2 \right),$$

so that

$$T_1 \left(1 + \frac{\gamma - 1}{2} \ Ma_1^2 \right) = T_2 \left(1 + \frac{\gamma - 1}{2} \ Ma_2^2 \right). \tag{11.16}$$

(iii) Momentum equations:

$$p_1 A - p_2 A = \rho_1 A U_1 (U_2 - U_1),$$

$$p_1 - p_2 = \rho_1 U_1 U_2 - \rho_1 U_1^2.$$

But $\rho_1 U_1 = \rho_2 U_2$, so that

$$p_1 - p_2 = \rho_2 U_2^2 - \rho_1 U_1^2$$

or

$$p_1 + \rho_1 U_1^2 = p_2 + \rho_2 U_2^2.$$

However,

$$U^2 = \gamma R T Ma^2,$$

so that

$$p_1 + \gamma R T_1 Ma_1^2 \rho_1 = p_2 + \gamma R T_2 Ma_2^2 \rho_2$$

and

$$p_1 (1 + \gamma Ma_1^2 \ R T_1 \rho_1 / p_1) = p_2 (1 + \gamma Ma_2^2 \ R T_2 \rho_2 / p_2).$$

In addition,

$$R T \rho / p = 1$$

which gives

$$p_1 (1 + \gamma Ma^2) = p_2 (1 + \gamma Ma_2^2) \tag{11.17}$$

In order to obtain the relationship between Ma_1 and Ma_2, it is necessary to eliminate pressures and temperatures from equations (11.15), (11.16) and (11.17). The former objective is realized by dividing equation (11.17) by equation (11.15), which gives

$$\left(\frac{1 + \gamma Ma_1^2}{Ma_1} \right) \sqrt{T_1} = \left(\frac{1 + \gamma Ma_2^2}{Ma_2} \right) \sqrt{T_2}.$$

This equation is now divided by the square root of equation (11.16). The result is

$$\frac{1 + \gamma Ma_1^2}{Ma_1 \{ 1 + \frac{1}{2} (\gamma - 1) Ma_1^2 \}^{1/2}} = \frac{1 + \gamma Ma_2^2}{Ma_2 \{ 1 + \frac{1}{2} (\gamma - 1) Ma_2^2 \}^{1/2}} = f(Ma, \gamma). \tag{11.18}$$

It shows that the above particular function of Mach number and γ is constant across the shock and determines the relationship between the upstream and downstream values of the Mach number. It is plotted in Fig. 11.8 for air ($\gamma = 1 \cdot 4$).

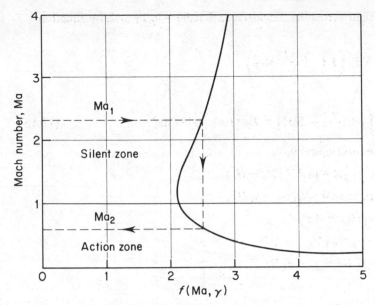

Figure 11.8   Change of Mach number across a shock wave

By solving equation (11.18), the expression for Ma_2 is obtained:

$$\mathrm{Ma}_2 = \left[\frac{\mathrm{Ma}_1^2 + 2/(\gamma - 1)}{\{2\gamma/(\gamma - 1)\}\,\mathrm{Ma}_1^2 - 1} \right]^{1/2}. \tag{11.19}$$

This equation can now be substituted into equations (11.15), (11.16) and (11.17) to give the following ratios across the shock wave:

$$\frac{T_2}{T_1} = \frac{\gamma(\gamma - 1)}{(\gamma + 1)^2 \mathrm{Ma}_1^2} \left(1 + \frac{\gamma - 1}{2}\,\mathrm{Ma}_1^2 \right) \left(\frac{2\gamma}{\gamma - 1}\,\mathrm{Ma}_1^2 - 1 \right), \tag{11.20}$$

$$\frac{p_2}{p_1} = \frac{2\gamma}{\gamma + 1}\,\mathrm{Ma}_1^2 - \frac{\gamma - 1}{\gamma + 1}, \tag{11.21}$$

$$\frac{\rho_2}{\rho_1} = \frac{U_1}{U_2} = \frac{\gamma + 1}{2}\,\frac{\mathrm{Ma}_1^2}{1 + \{(\gamma - 1)/2\}\,\mathrm{Ma}_1^2}. \tag{11.22}$$

The above three ratios are functions of Ma and γ only and are plotted in Fig. 11.9 for $\gamma = 1 \cdot 4$.

The *strength* of a shock wave is defined as the ratio of the pressure rise across the shock to the upstream pressure. Thus,

$$\text{Shock strength} = (p_2 - p_1)/p_1 = p_2/p_1 - 1. \tag{11.23}$$

By substitution from equation (11.21),

$$\text{Shock strength} = 2\gamma(\mathrm{Ma}_1^2 - 1)/(\gamma + 1). \tag{11.24}$$

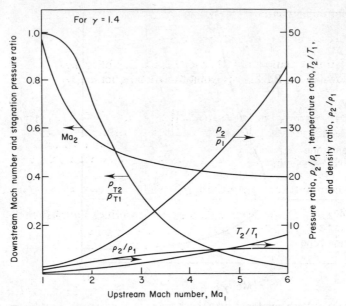

Figure 11.9 Changes of parameters of state across a shock wave

It is also useful to have an expression for the ratio of the stagnation pressures which may be obtained using equation (11.10),

$$p_T = p_0 \left[1 + \{(\gamma - 1)/2\} \, \text{Ma}_0^2 \right]^{\gamma/(\gamma-1)},$$

together with (11.21). This procedure, although using an isentropic equation, is justified because the stagnation pressure is defined as resulting from a reversible adiabatic and, hence, isentropic process and, in any case, would take place either upstream or downstream of the shock. Thus, from (11.10),

$$\frac{p_{T_1}}{p_{T_2}} = \frac{p_1}{p_2} \left[\frac{1 + \{(\gamma - 1)/2\} \, \text{Ma}_1^2}{1 + \{(\gamma - 1)/2\} \, \text{Ma}_2^2} \right]^{\gamma/(\gamma-1)}.$$

Substituting for p_1/p_2 from (11.21),

$$\frac{p_{T_1}}{p_{T_2}} = \left(\frac{2\gamma}{\gamma + 1} \, \text{Ma}_1^2 - \frac{\gamma - 1}{\gamma + 1} \right)^{-1} \left[\frac{1 + \{(\gamma - 1)/2\} \, \text{Ma}_1^2}{1 + \{(\gamma - 1)/2\} \, \text{Ma}_2^2} \right]^{\gamma/(\gamma-1)}$$

Now, eliminating Ma_2^2 using equation (11.19) and simplifying gives

$$\frac{p_{T_2}}{p_{T_1}} = \left[\frac{(\gamma + 1) \, \text{Ma}_1^2}{(\gamma - 1) \, \text{Ma}_1^2 + 2} \right]^{\gamma/(\gamma-1)} \left[\frac{\gamma + 1}{2\gamma \text{Ma}_1^2 - (\gamma - 1)} \right]^{1/(\gamma-1)} \qquad (11.25)$$

It is now possible to show that the flow across the shock is irreversible, and (hence) accompanied by an increase of entropy, by obtaining the relationship between pressure and density and comparing it with the isentropic relationship (equation (1.17)), namely

$$p/\rho^\gamma = \text{constant}.$$

Eliminating Ma_1^2 from equations (11.21) and (11.22), the following relationship, known as the Rankine–Hugoniot relation, is obtained:

$$\frac{\rho_2}{\rho_1} = \left\{ \left(\frac{\gamma + 1}{\gamma - 1} \right) \frac{p_2}{p_1} + 1 \right\} \bigg/ \left\{ \frac{p_2}{p_1} + \left(\frac{\gamma + 1}{\gamma - 1} \right) \right\}. \tag{11.26}$$

It is evidently different from

$$p/\rho^\gamma = \text{constant} \quad \text{or} \quad \rho_2/\rho_1 = (p_2/p_1)^{1/\gamma}.$$

Figure 11.10 demonstrates the deviation of the Rankine–Hugoniot relation from the isentropic equation for $\gamma = 1\cdot4$.

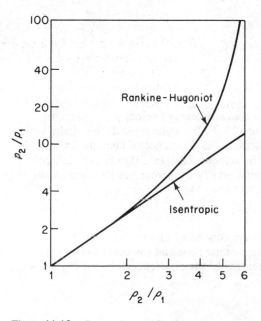

Figure 11.10   Comparison of Rankine–Hugoniot and isentropic curves for $\gamma = 1\cdot4$

The increase of entropy across a shock is obtained from:

$$S_2 - S_1 = \int_1^2 \frac{dQ}{T} = c_v \int_1^2 \frac{dT}{T} + \int_1^2 \frac{p}{T} d\left(\frac{1}{\rho} \right) = c_v \log_e \frac{T_2}{T_1} + \int_1^2 R\rho \, d\left(\frac{1}{\rho} \right)$$

$$= c_v \log_e \left(\frac{T_2}{T_1} \right) - R \log_e \left(\frac{\rho_2}{\rho_1} \right) \tag{11.27}$$

An alternative expression for the specific entropy in terms of pressure ratio may be obtained as follows:

$$\frac{S_2 - S_1}{c_v} = \log_e \frac{T_2}{T_1} - \frac{R}{c_v} \log_e \frac{p_2}{p_1},$$

but $\dfrac{R}{c_v} = \dfrac{c_p - c_v}{c_v} = (\gamma - 1)$ and $\dfrac{p_2}{p_1} = \dfrac{T_1}{T_2} \times \dfrac{p_2}{p_1},$

so that

$$\frac{S_2 - S_1}{c_v} = \log_e \frac{T_2}{T_1} - (\gamma - 1)\left(\log_e \frac{T_1}{T_2} + \log_e \frac{p_2}{p_1}\right)$$

$$= \log_e \frac{T_2}{T_1} - \gamma \log_e \frac{T_1}{T_2} + \log_e \frac{T_1}{T_2} - (\gamma - 1)\log_e \frac{p_2}{p_1}$$

$$= \log_e \frac{T_2}{T_1} + \gamma \log_e \gamma \frac{T_2}{T_1} - \log_e \frac{T_2}{T_1} - (\gamma - 1)\log_e \frac{p_2}{p_1}.$$

Finally,

$$\frac{S_2 - S_1}{c_v} = \gamma \log_e \frac{T_2}{T_1} - (\gamma - 1)\log_e \frac{p_2}{p_1}. \qquad (11.27a)$$

Example 11.2

A Pitot–static tube is inserted into an airstream of velocity U_0, pressure $1 \cdot 02 \times 10^5$ N m^{-2} and temperature 28 °C. It is connected differentially to a mercury U-tube manometer. Calculate the difference of mercury levels in the two limbs of the manometer if the velocity U_0 is (a) 50 m s^{-1}, (b) 250 m s^{-1} and (c) 420 m s^{-1}. Take the specific gravity of mercury as 13·6 and for air $\gamma = 1 \cdot 4$ and $R = 287$ J kg^{-1} K^{-1}.

Solution

The two limbs of the manometer are connected one to the total (or stagnation) connection of the Pitot–static tube and the other to the static connection. Thus, the manometer 'reads' the difference between the two so that

$$p_T - p = \rho_{Hg}gh,$$

where ρ_{Hg} is the density of mercury and h is the difference between the mercury levels. Thus,

$$h = (p_T - p)/\rho_{Hg}g$$

It is, therefore, necessary to obtain $(p_T - p)$ for the three cases. First, the value of the Mach number must be calculated in order to establish the type of flow taking place, which will govern the choice of appropriate equations.

(a)

$$Ma = \frac{U_0}{c} = \frac{U_0}{\sqrt{(\gamma RT)}} = \frac{50}{\sqrt{(1 \cdot 4 \times 287 \times 301)}} = \frac{50}{347 \cdot 77} = 0 \cdot 14.$$

Therefore the flow may be considered as incompressible and equation (11.2) may be used:

$$p_T - p_0 = \tfrac{1}{2}\rho U_0^2.$$

But, $p/\rho = RT$,

from which

$$\rho = p/RT = 1{\cdot}02 \times 10^5/287 \times 301 = 1{\cdot}18 \text{ kg m}^{-3}$$

and $p_T - p = \tfrac{1}{2}\rho U_0^2 = \tfrac{1}{2} \times 1{\cdot}18(50)^2 = 1475 \text{ N m}^{-2}$,

$$h = 1475/13{\cdot}6 \times 10^3 \times 9{\cdot}81 = 11{\cdot}06 \times 10^{-3} \text{ m of mercury}$$

= 11·06 mm of mercury

(b)

$$\text{Ma} = 250/347{\cdot}77 = 0{\cdot}719$$

Compressibility effects must be taken into account and, therefore, either (i) equation (11.10) is used or (ii) the value of the compressibility factor is obtained from Fig. 11.3.

(i)

$$p_T = p_0 [1 + \{(\gamma - 1)/2\}\text{Ma}_0^2]^{\gamma/(\gamma-1)}$$
$$= 1{\cdot}02 \times 10^5 [1 + (0{\cdot}4/2)(0{\cdot}719)^2]^{1{\cdot}4/0{\cdot}4} = 1{\cdot}44 \times 10^5 \text{ N m}^{-2}.$$

Therefore,

$$p_T - p_0 = (1{\cdot}44 - 1{\cdot}02)\,10^5 = 0{\cdot}42 \times 10^5 \text{ N m}^{-2}.$$

(ii) From Fig. 11.3, for Ma = 0·719, $(p_T - p_0)/\tfrac{1}{2}\rho_0 U_0^2 = 1{\cdot}135$. Therefore,

$$p_T - p_0 = 1{\cdot}135 \times \frac{1{\cdot}18}{2}(250)^2 = 0{\cdot}418 \times 10^5 \text{ N m}^{-2}.$$

Taking $0{\cdot}42 \times 10^5$ as more accurate,

$$h = \frac{0{\cdot}42 \times 10^5}{13{\cdot}6 \times 10^3 \times 9{\cdot}81} = 31{\cdot}5 \times 10^{-3} \text{ m of mercury}$$

= 315 mm of mercury.

(c)

$$\text{Ma} = 420/347{\cdot}77 = 1{\cdot}208.$$

The flow is supersonic and, therefore, a shock wave will be formed due to the disturbance created by the Pitot-static tube. As the nose of the tube is rounded, it is reasonable to assume that the shock will be detached and a section of it just upstream of the Pitot-static tube will be normal to it. Thus, the pressure downstream of the shock and upstream of the tube will be given by equation (11.21):

$$p_2 = p_1\left(\frac{2\gamma}{\gamma + 1}\,\text{Ma}_1^2 - \frac{\gamma - 1}{\gamma + 1}\right) = 1{\cdot}02 \times 10^5\left[\frac{2{\cdot}8}{2{\cdot}4}(1{\cdot}208)^2 - \frac{0{\cdot}4}{2{\cdot}4}\right]$$
$$= 1{\cdot}567 \times 10^5 \text{ N m}^{-2}.$$

Now, in order to calculate the stagnation pressure there, using equation (11.10), it is necessary first to determine the Mach number in the action zone between the shock wave and the Pitot–static tube. This may be obtained from equation (11.19):

$$\mathrm{Ma}_2 = \left[\frac{\mathrm{Ma}_1^2 + 2/(\gamma - 1)}{\{2\gamma/(\gamma - 1)\}\,\mathrm{Ma}_1^2 - 1}\right]^{1/2}$$

or $\quad \mathrm{Ma}_2^2 = \dfrac{(1\cdot208)^2 + 2/0\cdot4}{(2\cdot8/0\cdot4)(1\cdot208)^2 - 1} = 0\cdot70$

Hence,

$$p_{T_2} = p_2\left(1 + \frac{\gamma - 1}{2}\,\mathrm{Ma}_2^2\right)^{\gamma/(\gamma-1)} = 1\cdot567 \times 10^5 \left(1 + \frac{0\cdot4}{2} \times 0\cdot70\right)^{1\cdot4/0\cdot4}$$

$$= 2\cdot479 \times 10^5 \ \mathrm{N\,m}^{-2}.$$

Therefore,

$$h = \frac{(2\cdot479 - 1\cdot567)\,10^5}{13\cdot6 \times 10^3 \ \times 9\cdot81} = 683\cdot4 \times 10^{-3} \ \mathrm{m} = \textbf{683 mm of mercury.}$$

11.3. Oblique shock waves

When a shock wave is not perpendicular to the direction of flow, it is called an oblique shock wave (Fig. 11.11). It occurs during flow past a wedge or sharp

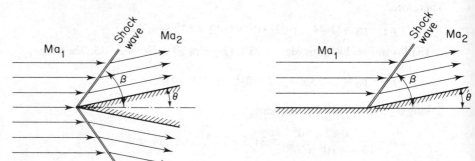

Figure 11.11   Oblique shock wave

object or when the supersonic flow is forced to change direction by a solid boundary, as shown in Fig. 11.12.

One way of treating an oblique shock wave is to consider its normal and tangential components. The normal component undergoes changes associated with the normal shock wave, whereas the tangential component remains unchanged. Thus, only the normal velocity component is reduced causing the deflection of the flow.

It is important to note that although $u_{2\mathrm{n}}$ must be subsonic, being downstream of the normal shock, the resultant velocity downstream of the oblique shock, namely

$$U_2 = \sqrt{(u_{2\mathrm{n}}^2 + u_{2\mathrm{t}}^2)},$$

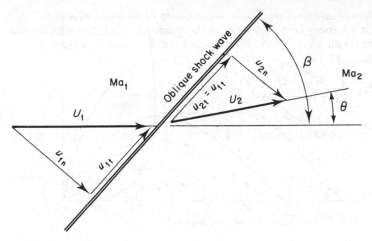

Figure 11.12   Flow deflection due to an oblique shock wave

may be supersonic, provided u_{2t} is large enough. Thus, Ma_2 is always smaller than Ma_1, but it may be greater than one.

The equations derived for the normal shock wave are valid provided they are applied to the normal velocity components. Since

$$u_{1n} = U_1 \sin \beta$$

and $$u_{2n} = U_2 \sin (\beta - \theta)$$

where β = shock angle (with respect to upstream flow direction), θ = deflection angle, it is sufficient to substitute these expressions — as well as $Ma_1 \sin \beta$ for Ma_1 and $Ma_2 \sin (\beta - \theta)$ for Ma_2 — in the normal shock equations. The angles β and θ are related by

$$\frac{u_{2n}}{u_{1n}} = \frac{\tan (\beta - \theta)}{\tan \beta} = \frac{\rho_1}{\rho_2} = \frac{p_1 T_1}{p_2 T_2}. \tag{11.28}$$

Using equations previously derived, it may be shown that

$$\frac{\tan (\beta - \theta)}{\tan \beta} = \frac{2 + (\gamma - 1) Ma_1^2 \sin^2 \beta}{(\gamma + 1) Ma_1^2 \sin^2 \beta} \tag{11.29}$$

and $$\tan \theta = \frac{2 \cot \beta (Ma_1^2 \sin \beta - 1)}{Ma_1^2 (\gamma + \cos 2\beta) + 2}, \tag{11.30}$$

from which the deflection angle may be determined. This equation has two real roots, giving two values of β for each value of θ and Ma_1 as shown by the plot in Fig. 11.13. The two values correspond to a strong and a weak wave, respectively. For the strong shock wave, the downstream flow is always subsonic and the shock angle is large; for the weak wave, the downstream flow is usually supersonic and the shock angle is smaller. The chain-dotted curve in Fig. 11.13 separates the region of $Ma_2 < 1$ from that in which $Ma_2 > 1$. The heavy line, however, joins the maximum values of θ (= θ_{max}) and thus separates the weak shock from the strong.

The plot also indicates that, for the shock to occur, θ must be smaller than

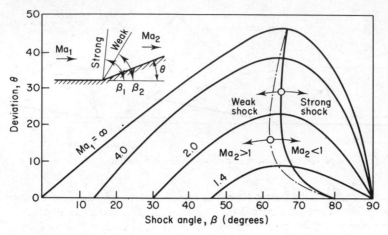

Figure 11.13   Oblique shock angles for $\gamma = 1 \cdot 4$

θ_{max} for a given value of the upstream Mach number. If the physical situation is such that this condition is not satisfied — for example, if the wedge angle is greater than θ_{max} for the flow Mach number — the shock will detach itself from the wedge, thus creating a subsonic space just in front of the wedge. Such a shock wave is always curved, as shown in Fig. 11.14. It, thus, extends further and significantly increases the wave drag on the body. Hence a sharp, pointed nose is a better shape for supersonic flow.

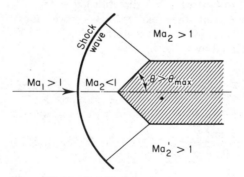

Figure 11.14   Detached shock wave

11.4. Supersonic expansion and compression

Consider supersonic flow round an infinitesimal corner, which may be convex or concave as shown, with the angle $\delta\theta$ greatly exaggerated (Fig. 11.15). The corner constitutes a disturbance and, since $\delta\theta$ is very small, the disturbance is small, thus generating a very weak shock wave. Such a wave of infinitesimal strength is called a *Mach wave*. It may be represented by a Mach line whose angle μ is given by

$$\sin \mu = 1/\text{Ma}. \qquad (11.31)$$

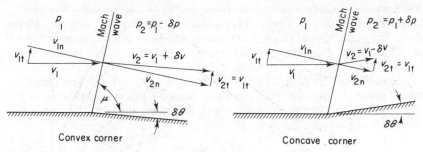

Figure 11.15   Supersonic expansion and compression at a corner

When the supersonic flow is forced to negotiate a corner, the flow remains parallel to both the upstream and the downstream solid boundary surfaces. Since the tangential velocity component remains unaltered, it follows from the velocity triangles of Fig. 11.15 that there must be a change in the resultant velocity. This change is positive, i.e. there is an increase of velocity for the convex corner and there is a negative change or a decrease of velocity

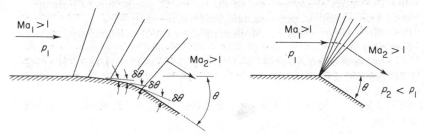

Figure 11.16   Prandtl–Mayer expansion

for a concave corner. These changes must be accompanied by the corresponding changes in pressure. Thus, there is a pressure drop or expansion at a convex corner and a pressure rise or compression at a concave corner.

Any convex corner of finite deflection θ may be regarded as a series of consecutive infinitesimal corners of deflections $\delta\theta$. This gives rise to a fan of

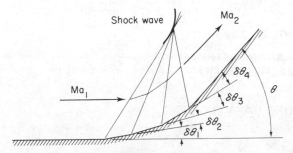

Figure 11.17   Shock wave at a concave corner

Mach waves (or characteristics), as shown in Fig. 11.16, through which smooth and isentropic expansion takes place. This is known as Prandtl–Mayer flow. The evaluation of changes of pressure and Mach number is carried out in steps through each successive Mach line. Such a step-by-step method is called the method of characteristics and is beyond the scope of this book.

A concave corner of finite deflection gives rise to a series of Mach lines which converge into an envelope and thus form a shock wave as shown in Fig. 11.17. Such a compression process is, therefore, not isentropic, since the changes occur across a shock wave.

11.5 Computer program 'NORSH'

(1) The program calculates and lists the Mach number, celerity, gas velocity and its parameters of state downstream of a normal shock together with entropy change across the shock. It also calculates those quantities upstream of the shock which are not input and all upstream and downstream quantities in a tabular form.

(2) The required input is:

(i) the values of the gas constant and of the ratio of specific heats;

(ii) the values of any two out of the following parameters of state upstream of the shock: pressure, temperature, density;

(iii) the value of either the Mach number or the gas velocity upstream of the shock.

(3) Use is made of equations (1.13), (5.31), (11.1), (11.10), (11.19), (11.21), (11.22) and (11.27).

(4) Input example:

$R = 287$ J kg^{-1}K^{-1}; $\gamma = 1\cdot4$; $p_1 = 102$ kNm^{-2}; $T_1 = 301$ K; $v_1 = 420$ m s^{-1}.

(5) Output:

	UPSTREAM CONDITIONS	DOWNSTREAM CONDITIONS
MACH NUMBER =	1.21	0.84
PRESSURE KPA =	102	156.57
STAGN. PRES. KPA =	249.85	247.86
TEMPERATURE K =	301	340.98
DENSITY KG/CU.M =	1.181	1.6
CELERITY M/S =	347.77	370.14
VELOCITY M/S =	420	309.96

THE ENTROPY CHANGE IS: 2.29 J/KG K

DO YOU WANT TO CHANGE UPSTREAM
GAS VELOCITY OR MACH NUMBER?
(Y/N)?Y

DO YOU KNOW THE NEW GAS VELOCITY?
(Y/N)?N

NEW MACH NUMBER = ?4.0

List:
```
 10 MODE 7
 20 CLS: PRINT: PRINT
 30 PRINT "PROGRAM: NORSH"
 40 GOSUB 1160: GOSUB 1140
 50 PRINT "WHAT IS THE VALUE OF"
 60 PRINT "THE GAS CONSTANT R IN J/KG K"
 70 INPUT "R =";R
 80 GOSUB 1140: PRINT "WHAT IS THE VALUE OF THE RATIO"
 90 PRINT "OF SPECIFIC HEATS GAMMA?"
100 INPUT "G =";G
110 GOSUB 1140: PRINT "IS THE PRESSURE UPSTREAM OF"
120 PRINT "THE NORMAL SHOCK KNOWN?"
130 INPUT "(Y/N)";A$
140 IF A$ = "N" THEN 240
150 GOSUB 1140: PRINT "WHAT IS IT IN KPA?"
160 INPUT "P=";P1
170 GOSUB 1120: GOSUB 1140: PRINT "IS THE TEMPERATURE UPSTREAM"
180 PRINT "OF THE NORMAL SHOCK KNOWN?"
190 INPUT "(Y/N)";B$
200 IF B$ = "N" THEN 320
210 GOSUB 1140: PRINT "WHAT IS IT IN DEGREES K?"
220 INPUT "T=";T1
230 GOSUB 1140: RO1=1000*P1/(R*T1): GOTO 360
240 GOSUB 1120: GOSUB 1140: PRINT "WHAT IS THE ABSOLUTE GAS"
250 PRINT "TEMPERATURE IN DEGREES K"
260 PRINT "UPSTREAM OF THE NORMAL SHOCK?"
270 INPUT "T=";T1
280 GOSUB 1140: PRINT "WHAT IS THE GAS DENSITY IN"
290 PRINT "KG/CU.M UPSTREAM OF THE SHOCK?"
300 INPUT "RO=";RO1
310 P1=RO1*R*T1/1000: GOTO 360
320 GOSUB 1140: PRINT "WHAT IS THE GAS DENSITY IN"
330 PRINT "KG/CU.M UPSTREAM OF THE SHOCK?"
340 INPUT "RO=";RO1
350 T1=P1*1000/(RO1*R)
360 GOSUB 1140: C1=SQR(G*R*T1)
370 GOSUB 1140: PRINT "IS THE GAS VELOCITY UPSTREAM"
380 PRINT "OF THE SHOCK KNOWN?"
390 INPUT "(Y/N)";C$
400 IF C$ = "N" THEN 450
410 GOSUB1140: PRINT "WHAT IS IT IN M/S?"
420 INPUT "V=";V1
430 IF V1<C1 GOTO 490
440 M1=V1/C1: GOTO 540
450 GOSUB 1140: PRINT "WHAT IS THE VALUE OF MACH"
460 PRINT "NUMBER UPSTREAM OF THE SHOCK?"
470 INPUT "M=";M1
480 IF M1>1 GOTO 530
490 GOSUB 1120: GOSUB 1140: PRINT "THE UPSTREAM CONDITIONS"
500 PRINT "SUBSONIC. THERE IS NO SHOCK."
510 PRINT "CHECK YOUR DATA!": GOSUB 1170
520 GOSUB 1140: GOTO 1190
530 V1=C1*M1
540 GOSUB 1140: K=G/(G-1)
550 PT1=P1*(1+(G-1)*(M1)^2/2)^K
560 MI1=INT(M1*100+.5)/100
570 PI1=INT(P1*100+.5)/100
580 PTI1=INT(PT1*100+.5)/100
590 TI1=INT(T1*100+.5)/100
600 ROI1=INT(RO1*1000+.5)/1000
610 CI1=INT(C1*100+.5)/100
620 VI1=INT(V1*100+.5)/100
630 P2=P1*(2*G*(M1)^2/(G+1)-(G-1)/(G+1))
640 A=(G+1)*(M1)^2/2
650 B=(G-1)*(M1)^2/2
660 RO2=RO1*(A/(1+B))
670 V2=V1*RO1/RO2
680 T2=P2*1000/(R*RO2)
690 E=M1^2+2/(G-1):F=2*G*(M1)^2/(G-1): M2=SQR(E/(F-1))
700 C2=V2/M2
710 PT2=P2*(1+(G-1)*(M2)^2/2)^K
720 S=R*LN(T2/T1)/(G-1)-R*LN(RO2/RO1)
730 MI2=INT(M2*100+.5)/100
740 PI2=INT(P2*100+.5)/100
750 TI2=INT(T2*100+.5)/100
760 ROI2=INT(RO2*100+.5)/100
770 D=M1^2*(G-1)/2
780 CI2=INT(C2*100+.5)/100
790 VI2=INT(V2*100+.5)/100
800 PTI2=INT(PT2*100+.5)/100
810 SI=INT(S*100+.5)/100
820 C2=V2/M2
830 GOSUB 1120: PRINT SPC(15)"UPSTREAM","DOWNSTREAM"
```

```
840 PRINT SPC(14)"CONDITIONS","CONDITIONS": GOSUB 1150
850 PRINT "MACH NUMBER =";TAB(18)MI1;TAB(32)MI2
860 PRINT "PRESSURE KPA =";TAB(18)PI1;TAB(32)PI2
870 PRINT "STAGN.PRES. KPA =";TAB(18)PTI1;TAB(32)PTI2
880 PRINT "TEMPERATURE K =";TAB(18)TII;TAB(32)TI2
890 PRINT "DENSITY KG/CU.M =";TAB(18)ROI1;TAB(32)ROI2
900 PRINT "CELERITY M/S =";TAB(18)CI1;TAB(32)CI2
910 PRINT "VELOCITY M/S =";TAB(18)VI1;TAB(32)VI2
920 GOSUB 1130: PRINT "THE ENTROPY CHANGE IS: ";SI;" J/KG K"
930 GOSUB 1170: GOSUB1140: PRINT "DO YOU WANT TO CHANGE UPSTREAM"
940 PRINT "GAS VELOCITY OR MACH NUMBER?"
950 INPUT "(Y/N)";A$
960 IF A$="Y" GOTO 1040
970 GOSUB 1140: PRINT "DO YOU WANT TO CHANGE UPSTREAM"
980 PRINT "PARAMETERS OF STATE?"
990 INPUT "(Y/N)";B$
1000 IF B$="Y" THEN CLS: GOTO110
1010 GOSUB 1120: GOSUB 1140: GOSUB 1140
1020 PRINT "THANK YOU"
1030 GOTO 1110
1040 GOSUB 1140: PRINT "DO YOU KNOW THE NEW GAS VELOCITY?"
1050 INPUT "(Y/N)";C$
1060 IF C$="Y" GOTO 1090
1070 GOSUB 1140: INPUT "NEW MACH NUMBER =";M1
1080 GOTO 480
1090 GOSUB 1140: INPUT "WHAT IS IT?";V1
1100 GOTO 430
1110 END
1120 CLS:RETURN
1130 PRINT:RETURN
1140 PRINT:PRINT:RETURN
1150 STAR$=STRING$(40,"*"):PRINT STAR$:RETURN
1160 FOR X=1 TO 3000:NEXT:RETURN
1170 FOR X=1 TO 10000:NEXT:RETURN
```

EXERCISES 11

11.1   A Pitot static tube is inserted into an air stream and the mercury manometer connected differentially to it shows a difference in levels of 300 mm. The free stream temperature and pressure are 40 °C and 150 kN/m² abs. Calculate the air velocity and the percentage error which would have been committed if the flow was considered as incompressible. (Specific gravity of mercury = 13·6.)
[219 m/s, 4·6%]

11.2   If the difference between static and stagnation pressure in standard air (p = 101·3 kN/m²; T = 288 K) is 600 mm of mercury compute the air velocity assuming (*a*) the air is incompressible, (*b*) the air is compressible and hence calculate the compressibility factor.
[361 m/s, 324 m/s, 1·02]

11.3   An air stream issues from a nozzle into the atmosphere where the barometric pressure is 750 mm of mercury and the temperature is 20 °C. Assuming that for air the difference between the stagnation temperature and the free stream temperature is given by

$$T_T - T_0 = \left(\frac{V_0}{45}\right)^2 \quad \text{degrees Centigrade}$$

where V_0 = 250 m/s is the free stream velocity, calculate the stagnation temperature, pressure density and the Mach number of the flow. For air R = 287 J/kg K and γ = 1·4.
[324 K, 142·3 kN/m², 1·53 kg/m³, 0·729]

11.4   A Pitot static tube is inserted into the test section of a subsonic wind tunnel. It indicates a static pressure of 80 kN/m² while the difference

between stagnation and static pressure is shown as 120 mm of mercury. The barometric pressure is 760 mm of mercury and the stagnation temperature is 40 °C. Calculate the Mach number and the air velocity.
[0·4, 124 m/s]

11.5   Starting from the differential form of Euler's equation:

$$\frac{1}{\rho}\frac{\mathrm{d}p}{\mathrm{d}x} + v\frac{\mathrm{d}v}{\mathrm{d}x} + g\frac{\mathrm{d}z}{\mathrm{d}x} = 0$$

show that for air ($R = 287$ J/kg K; $\gamma = 1·4$) assuming horizontal, isentropic flow the difference between stagnation temperature and free stream temperature is approximately given by

$$T_T - T_0 = \left(\frac{V_0}{45}\right)^2 \quad \text{degrees Centigrade}$$

where V_0 is the free stream velocity in m/s. Calculate also the percentage error involved in the above approximation.
[0·8%]

11.6   An air stream with velocity 500 m/s, static pressure 60 kN/m² and temperature −18 °C undergoes a normal shock. Determine the air velocity and the static and stagnation conditions after the wave.
[252 m/s, 160 kN/m², 223 kN/m²]

11.7   Given that the Mach number downstream of a normal shock is expressed in terms of the Mach number upstream of the shock as follows:

$$\mathrm{Ma}_2^2 = \frac{\mathrm{Ma}_1^2 + \dfrac{2}{\gamma - 1}}{\dfrac{2\gamma}{\gamma - 1}\mathrm{Ma}_1^2 - 1}$$

derive an expression for the pressure ratio across the shock wave and hence an expression for the density ratio in terms of pressure ratio and Ma_1.

11.8   A normal shock moves into still air with a velocity of 1500 m/s. The still air is at 10 °C and 80 kN/m². Calculate the velocity of air behind the wave and the static and stagnation pressures and temperatures behind the wave.
[700 °C, 2·37 kN/m²]

11.9   The Mach angle as measured from a Schlieren photograph of a bullet has a magnitude of 30°. Estimate the speed of the bullet if the temperature and pressure of the atmosphere were 5 °C and 90 kN/m² respectively. What Mach angle would indicate the same velocity in air at 15 °C and 101 kN/m²?
[668·4 m/s, 30·6°]

11.10   Show that for a normal shock wave:

$$p_1(1 + \gamma\mathrm{Ma}_1^2) = p_2(1 + \gamma\mathrm{Ma}_2^2)$$

where p_1 and p_2 are pressures upstream and downstream of a shock wave respectively, Ma_1 and Ma_2 are the Mach numbers upstream and downstream of the shock respectively and γ is the ratio of specific heats.

A projectile with a rounded nose moves through still air at Ma = 5. The air pressure is 60 kN/m^2 and the temperature is $-10\,°C$. Assuming that the shock wave formed at the nose of the projectile is detached and normal to it determine the stagnation pressure and temperature at the nose. Take $\gamma = 1\cdot4$ and use the relationship given in Exercise 11.8.

[1959 kN/m^2, 1524 K]

11.11 A two-dimensional wedge is used to measure the Mach number of the flow in a supersonic wind tunnel using air. If the total wedge angle is 20° and the shock wave angle is 60° calculate the Mach number in the tunnel and downstream of the shock. Also determine the minimum Mach number for which this wedge could be used.

[1·46, 1·14, 1·42]

11.12 (a) Starting from the momentum considerations and given that Mach number downstream of a normal shock Ma_2 is related to the Mach number upstream of the shock Ma_1 by the equation:

$$Ma_2^2 = \frac{Ma_1^2 + \dfrac{2}{\gamma - 1}}{\dfrac{2\gamma}{\gamma - 1}\,Ma_1^2 - 1}$$

show that for air the shock strength is given by:

$$\frac{p_2 - p_1}{p_1} = 1\cdot167\,(Ma_1^2 - 1)$$

(b) A supersonic aircraft flies horizontally overhead at 3000 m through still air. The time interval between the instant the aircraft is directly overhead an observer on the ground and the instant the shock wave is detected by him is 7·0 s.

If the velocity of sound in air is 335 m/s calculate the velocity of the aircraft and the stagnation pressure on its nose. Take the atmospheric pressure at 3000 m to be 70 kN/m^2. Note that for normal shock in air:

$$Ma_2^2 = \frac{Ma_1^2 + 5}{7Ma_1^2 - 1}$$

Take $\gamma = 1\cdot4$ for air.

[537·1 m/s; 268·6 kN/m^2]

11.13 Air at pressure 85 kN/m^2 and temperature 10°C flows at a supersonic velocity over a symmetrical wedge of 16° included angle. When the pointed edge of the wedge is pointed forward a weak shock wave is formed making an angle of 25° with the axis of symmetry of the wedge. Calculate the pressure, temperature and air velocity downstream of the shock wave.

The wedge is now turned so that the blunt face is forward and exposed to the same air stream velocity. What is the stagnation pressure on the blunt face of the wedge and the air velocity downstream of the shock wave?

For an oblique shock the relationship between the deflection angle θ and the shock angle β at an upstream Mach number Ma_1 in air is given by:

$$\frac{\tan(\beta - \theta)}{\tan\beta} = \frac{2 + 0\cdot4Ma_1^2\,\sin^2\beta}{2\cdot4Ma_1^2\,\sin^2\beta}$$

[1085 kN/m^2; 264·5 m/s]

PART IV
Steady flow in pipelines and open channels

In the previous section, the behaviour of real fluids has been examined and, in particular, the energy losses which occur due to friction and other causes. In the following chapters, consideration is given to the practical design of pipelines and channels. It is usual to treat liquids under steady flow conditions as if they were incompressible, since the changes of pressure are not large enough to produce significant changes of density. This permits the use of the simple constant density form of the continuity and energy equations, as shown in Chapter 12, which also covers the analysis of pipe networks under such conditions.

Pipelines can have two different functions. The first is to convey fluids from point to point, in which case almost the whole of the head available to produce flow is used in overcoming resistance in the pipeline. The second is to transmit power from a pump, pressure vessel or high level reservoir, the fluid travelling through the pipeline so that it arrives at the point of use under pressure or at high velocity, a comparatively small proportion of the head being lost in overcoming pipeline resistance, as explained in Chapter 13.

When gases flow through pipelines it is, usually, necessary to take changes of density and temperature along the length of the pipe into account. In Chapter 14, the basic equations of compressible flow are considered and first applied to frictionless flow through orifices, venturi contractions and nozzles. The formation of a normal shock wave in a diffuser is discussed. For pipelines of constant cross-section with frictional resistance, an analysis is made for both adiabatic and isothermal conditions.

Flow of liquids through open channels is dealt with in two parts. In Chapter 15 we consider uniform flow and the design of channel cross-sections for optimum performance, while Chapter 16 is concerned with non-uniform flow phenomena and the water surface profiles which can occur under these conditions.

12 Steady, incompressible flow in pipelines and pipe networks

12.1. General approach

This section is concerned with the analysis of steady flow of a fluid in closed or open conduits. A *closed conduit* is a pipe or duct through which the fluid flows while completely filling the cross-section. Since the fluid has no free surface, it can be either a liquid or a gas, its pressure may be above or below atmospheric pressure and this pressure may vary from cross-section to cross-section along its length. An *open conduit* is a duct or open channel along which a liquid flows with a free surface. At all points along its length the pressure at the free surface will be the same, usually atmospheric. An open conduit may be covered providing that it is not running full and the liquid retains a free surface; a partly filled pipe would, for example, be treated as an open channel.

In either case, as the fluid flows over the solid boundary a shear stress will be developed at the surface of contact (as discussed in Chapter 8) which will oppose motion. This so-called frictional resistance results in a loss of energy from the system, and, since we are dealing with a continuous stream, this can be expressed most conveniently as a loss of energy per unit weight (measured, for example, in newton-metres per newton), which is the same as a head loss expressed in terms of the fluid flowing. In addition to the energy loss due to friction, losses of energy can also occur in the form of separation losses at changes of section, bends and fittings, as described in Section 8.8. The first approach to the analysis of bounded systems is, therefore, to consider the

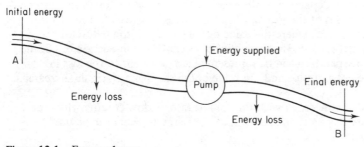

Figure 12.1 Energy change

energy balance between two cross-sections of the flow. In Fig. 12.1, for flow from section A to section B, working in terms of energy per unit weight,

| Total energy per unit weight at A | = | Total energy per unit weight at B | + | Loss of energy per unit weight due to friction and separation | − | Energy supplied per unit weight between A and B. |

Also, for steady flow to be maintained it is necessary that

$$\frac{\text{Mass per unit time}}{\text{passing A}} = \frac{\text{Mass per unit time}}{\text{passing B}}.$$

For incompressible flow the density of the fluid remains constant and the continuity of flow equation can be reduced to

$$\frac{\text{Volume per unit time}}{\text{passing A}} = \frac{\text{Volume per unit time}}{\text{passing B}}.$$

Analysis of all steady flow problems in pipes and channels is based on the application of the energy equation and the continuity of flow equation between suitable points in the system.

12.2 Incompressible flow through pipes

For incompressible flow, since there is no change of density with pressure, the energy equation reduces to Bernoulli's equation with the addition of terms for the energy losses due to friction and separation, for work done by the fluid in driving a turbine and for energy supplied by a pump where appropriate (*see* equation (6.6)). All these terms represent energy per unit weight or heads of the fluid concerned.

The head lost h_f, or energy loss per unit weight, due to friction can be conveniently expressed in terms of the velocity head of the fluid by using the Darcy formula

$$h_f = \frac{4fl}{d} \cdot \frac{\bar{v}^2}{2g}$$

(equation (8.30)), which can be applied to both turbulent and laminar flow, although in the latter case f is a function of Reynolds number ($f = 16/\mathrm{Re}$) and is, therefore, dependent on the velocity $\bar{v}$. Separation losses can also be expressed in terms of the velocity head, since (as shown in Section 8) the head lost, $h = k\bar{v}^2/2g$, where the value of k depends on the nature of the bend or fitting (Table 8.3). Alternatively, as explained in Section 8.8.3, the losses due to separation can be represented as equivalent to the loss in additional lengths of pipe and the value of l in the Darcy formula increased accordingly.

In problems concerned with the volume rate of flow Q rather than the mean velocity $\bar{v}$, an alternative form of the Darcy formula can be used, obtained by writing

$$\bar{v} = \frac{Q}{\text{Pipe area}} = \frac{Q}{(\pi/4)d^2}.$$

Substituting in the Darcy formula,

$$h_f = \frac{4fl}{d} \cdot \frac{16Q^2}{2g\pi^2 d^4}.$$

In SI units, putting $g = 9.81 \text{ m s}^{-2}$, this reduces to

$$h_f = flQ^2/3.03d^5$$

or, within an error of 1 per cent,

$$h_f = flQ^2/3d^5. \tag{12.1}$$

In general, for all pipes and fittings, the loss of head, which is the loss of energy per unit weight, can be expressed in the form

$$h = kQ^2, \tag{12.2}$$

where k is a resistance coefficient which, for a pipe, would be $fl/3d^5$ in SI units.

Example 12.1

Water discharges from a reservoir A (Fig. 12.2) through a 100 mm pipe 15 m long which rises to its highest point at B, 1·5 m above the free surface of the

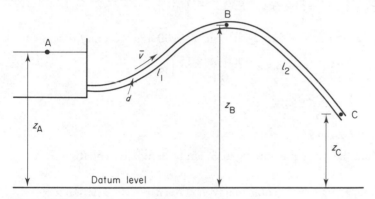

Figure 12.2 Flow through a siphon

reservoir, and discharges direct to the atmosphere at C, 4 m below the free surface at A. The length of pipe l_1 from A to B is 5 m and the length of pipe l_2 from B to C is 10 m. Both the entrance and exit of the pipe are sharp and the value of f is 0·08. Calculate (a) the mean velocity of the water leaving the pipe at C and (b) the pressure in the pipe at B.

Solution

(a) To determine the velocity $\bar{v}$, first apply Bernoulli's equation to the point A on the free surface and to the point C at the exit from the pipe, since the pressure and elevation of these points are known:

$$\text{Total energy per unit weight at A} = \text{Total energy per unit weight at C} + \text{Losses.} \tag{I}$$

Since the entrance to the pipe is sharp, there will be a loss of $0·5\ \bar{v}^2/2g$ (see Section 8.8.2). The loss due to friction in the length of pipe AC is given by the Darcy formula as

$$\frac{4f(l_1 + l_2)}{d} \cdot \frac{\bar{v}^2}{2g}.$$

There will be no loss of energy at the exit because, although the pipe exit is sharp, the water emerges into the atmosphere without any change of the cross-section of the stream.

At both A and C the pressure is atmospheric, so that $p_A = p_C$ = zero gauge pressure. Also, if the area of the free surface of the reservoir is large, the velocity at A is negligible. Thus,

Total energy per unit weight at A = z_A,

Total energy per unit weight at C = $z_C + \bar{v}^2/2g$.

Substituting in (I),

$$z_A = \left(z_C + \frac{\bar{v}^2}{2g}\right) + 0.5\frac{\bar{v}^2}{2g} + \frac{4f(l_1 + l_2)}{d}\frac{\bar{v}^2}{2g},$$

$$z_A - z_C = \frac{\bar{v}^2}{2g}\left\{1 + 0.5 + \frac{4f(l_1 + l_2)}{d}\right\}.$$

Putting $z_A - z_C = 4$ m, $l_1 = 5$ m, $l_2 = 15$ m, $d = 100$ mm $= 0.1$ m, $f = 0.08$

$$4 = \frac{\bar{v}^2}{2 \times 9.81}\left(1 + 0.5 + \frac{4 \times 0.08 \times 15}{0.1}\right) \text{ m,}$$

$$\bar{v}^2 = \frac{4 \times 2 \times 9.81}{49.5} = 1.585,$$

$$\bar{v} = 1.26 \text{ m s}^{-1}.$$

(b) To find the gauge pressure p_B at B, apply Bernoulli's equation to A and B,

Total energy per unit weight at A	=	Total energy per unit weight at B	+	Loss per unit weight at entry	+	Friction loss per unit weight in AB,

$$\left(\frac{p_A}{\rho g} + \frac{\bar{v}_A^2}{2g} + z_A\right) = \left(\frac{p_B}{\rho g} + \frac{\bar{v}^2}{2g} + z_B\right) + 0.5\frac{\bar{v}^2}{2g} + \frac{4fl_1}{d}\frac{\bar{v}^2}{2g}.$$

Since A is at atmospheric pressure, $p_A = 0$ (gauge) and, if the reservoir is large, $\bar{v}_A = 0$, so that

$$z_A = \frac{p_B}{\rho g} + z_B + \frac{\bar{v}^2}{2g}\left(1 + 0.5 + \frac{4fl_1}{d}\right),$$

$$p_B = \rho g(z_A - z_B) - \rho\frac{\bar{v}^2}{2}\left(1.5 + \frac{4fl_1}{d}\right).$$

Substituting $(z_A - z_B) = -1.5$ m, $\bar{v} = 1.26$ m s^{-1}, $f = 0.08$, $l_1 = 5$ m, $d = 100$ m $= 0.1$ m, $\rho = 10^3$ kg m^{-3},

$$p_B = 10^3 \times 9.81 \times (-1.5) - \frac{10^3 \times 1.26^2}{2}\left(1.5 + \frac{4 \times 0.08 \times 5}{0.1}\right)$$

$$= -14.71 \times 10^3 - 13.87 \times 10^3 \text{ N m}^{-2}$$

$$= -28.58 \times 10^3 \text{ N m}^{-2}$$

$$= \textbf{28.58 kN m}^{-2} \textbf{ below atmospheric pressure.}$$

12.3. Computer program 'SIPHON'

(1) The program considers the case of flow between two reservoirs or discharging from a constant level reservoir through a multisection series pipeline that may have a high point in its profile. The reservoirs need not be open to atmosphere but may be subjected to some constant pressure.

The program calculates the theoretical maximum flowrate between supply tank and discharge and then calculates the local pressure at a series of points along the pipe system. Pressure levels at each section are compared to the fluid vapour pressure and an impractical flow situation is thus recognized and indicated in the program output.

(2) The required input is:

(i)   pressure above fluid in supply tank, p_1 (kN m^{-2} abs.);
(ii)   pressure above fluid in receiving tank, or at final pipe discharge, p_2 (kN m^{-2} abs.);
(iii)   fluid level in supply tank, z_1 (m);
(iv)   discharge level or fluid level in receiving tank z_2 (m);
(v)   number of pipes in series, n;
(vi)   fluid density ρ (kg m^{-3});
(vii)   fluid vapour pressure (N m^{-2} abs.);
(viii) For each pipe in turn:
          level of entry above datum (m),
          pipe length (m),
          pipe diameter (m),
          pipe friction factor,
          presence of any separation losses (yes or no).
        For each separation loss in turn:
          value of k,
          position of loss along this pipe length as a percentage (0 – 100%);
(ix)   discharge level for final pipe (m).

(3) Use is made of Bernoulli's equation, Darcy's friction loss equation and the separation loss at each pipe fitting.

(4) Input example:
$p_1 = 100 \cdot 0$ kN m^{-2} abs.; $p_2 = 100 \cdot 0$ kN m^{-2} abs.; $z_1 = 10$ m; $z_2 = 6$ m; $n = 2$; $\rho = 1000$ kg m^{-3}; vapour pressure = 100 N m^{-2} abs.

For pipe 1:
$z_1 = 9$ m, $l_1 = 5$ m, $D_1 = 0 \cdot 1$ m, $f_1 = 0 \cdot 08$, 1 separation loss $k = 0 \cdot 5$ at $0.0\% l_1$.

For pipe 2:
$z_2 = 11 \cdot 5$ m, $l_2 = 10$ m, $D_2 = 0 \cdot 1$ m, $f_2 = 0 \cdot 08$, 1 separation loss $k = 1 \cdot 0$ at $100.0\% l_2$.
Discharge level for pipe 2 = 5m.

(5) Output:

THEORETICAL MAX. FLOWRATE Q = 0.01 M^3/S.

PRESSURE = 100.00 KN/M^2. AT ELEVATION = 10.00 M.

PIPE   1.00 RESULTS AT   0.50 M.   INTERVALS.
FLOW VELOCITY =   1.26 M/S.

```
PRESSURE = 109.81 KN/M^2. AT ELEVATION =   9.00 M.
PRESSURE = 105.69 KN/M^2. AT ELEVATION =   9.25 M.
PRESSURE = 101.97 KN/M^2. AT ELEVATION =   9.50 M.
PRESSURE =  98.25 KN/M^2. AT ELEVATION =   9.75 M.
PRESSURE =  94.53 KN/M^2. AT ELEVATION = 10.00 M.
PRESSURE =  90.81 KN/M^2. AT ELEVATION = 10.25 M.
PRESSURE =  87.09 KN/M^2. AT ELEVATION = 10.50 M.
PRESSURE =  83.37 KN/M^2. AT ELEVATION = 10.75 M.
PRESSURE =  79.65 KN/M^2. AT ELEVATION = 11.00 M.
PRESSURE =  75.93 KN/M^2. AT ELEVATION = 11.25 M.
PRESSURE =  72.20 KN/M^2. AT ELEVATION = 11.50 M.
```

```
PIPE   2.00 RESULTS AT   1.00 M.  INTERVALS.
FLOW VELOCITY =    1.26 M/S.
```

```
PRESSURE =  72.20 KN/M^2. AT ELEVATION = 11.50 M.
PRESSURE =  76.04 KN/M^2. AT ELEVATION = 10.85 M.
PRESSURE =  79.88 KN/M^2. AT ELEVATION = 10.20 M.
PRESSURE =  83.72 KN/M^2. AT ELEVATION =  9.55 M.
PRESSURE =  87.56 KN/M^2. AT ELEVATION =  8.90 M.
PRESSURE =  91.40 KN/M^2. AT ELEVATION =  8.25 M.
PRESSURE =  95.24 KN/M^2. AT ELEVATION =  7.60 M.
PRESSURE =  99.08 KN/M^2. AT ELEVATION =  6.95 M.
PRESSURE = 102.92 KN/M^2. AT ELEVATION =  6.30 M.
PRESSURE = 106.76 KN/M^2. AT ELEVATION =  5.65 M.
PRESSURE = 110.60 KN/M^2. AT ELEVATION =  5.00 M.
```

List:

```
10 REM SIPHON
20 MODE 3
30 @%=&20206
40 PRINT"PROGRAM SIPHON"
50 PRINT"PROGRAM DESIGNED TO CALCULATE FLOW THROUGH A SIPHON AND THE ASSOCIAT
ED PRESSURE - DISTANCE PROFILE."
60 PRINT:PRINT"THE SYSTEM CONSISTS OF TWO TANKS LINKED BY A SERIES PIPE NETWO
RK."
70 PRINT:PRINT"THE PROGRAM ASSUMES THAT PIPE LENGTHS,FRICTION FACTORS,ELEVATI
ONS,SEPARATION LOSS COEFFS. ETC ARE KNOWN."
80 PRINT:PRINT"THE DOWNSTREAM TANK MAY BE REPLACED BY AN OPEN DISCHARGE WHOSE
ELEVATION IS KNOWN."
90 INPUT"PRESSURE OVER SUPPLY RESERVOIR (KN/M^2 ABS.) PIN = ",PIN
100 PIN=PIN*1000.0
110 IP=0
120 INPUT"PRESSURE OVER EXIT RESERVOIR OR AT OPEN DISCHARGE POINT (KN/M^2 ABS.)
POUT= ",POUT
130 POUT=POUT*1000.0
140 DIM L(5):DIM Z(5):DIM D(5):DIM F(5)
150 DIM J(5):DIM T(5,5)
160 DIM S(5):DIM K(5,5,2)
170 DIM H(60):DIM V(60)
180 INPUT"SUPPLY WATER LEVEL ABOVE DATUM (M.) ZIN= ",ZIN
190 INPUT"DISCHARGE WATER LEVEL OR PIPE EXIT LEVEL FOR OPEN DISCHARGE (M.) ZOU
T= ",ZOUT
200 INPUT"NUMBER OF PIPES N= ",N
210 INPUT"FLUID DENSITY (KG/M^3) R= ",R
220 INPUT"FLUID VAPOUR PRESSURE (N/M^2 ABS.) VAP= ",VAP
230 G=9.81
240 CLS:PRINT"INPUT SYSTEM DATA IN RESPONSE TO THE FOLLOWING PROMPTS..."
250 ZTOP=0.0
260 FOR I=1 TO N
270 PRINT:PRINT"PIPE NUMBER = ",I
280 INPUT"PIPE ENTRY LEVEL ABOVE DATUM (M.) Z(I)= ",Z(I)
290 IF I=1 THEN GOTO 310
300 IF Z(I)>Z(I-1) AND Z(I)>ZTOP THEN ZTOP=Z(I)
310 INPUT"PIPE LENGTH (M.) L(I)= ",L(I)
320 INPUT"PIPE DIAMETER (M.) D(I)= ",D(I)
330 INPUT"PIPE FRICTION FACTOR F(I)= ",F(I)
```

```
340 INPUT"ARE THERE ANY SEPARATION LOSS COEFFS. APPLICABLE TO THIS PIPE? Y OR
N",Z$
350 IF Z$="N" THEN GOTO 430
360 INPUT"NUMBER OF SEPARATION LOSS COEFFS FOR THIS PIPE J(I)= ",J(I)
370 FOR M=1 TO J(I)
380 PRINT"PLEASE INPUT THE FOLLOWING DATA FOR EACH LOSS IN TURN..."
390 INPUT"INPUT VALUE OF SEPARATION COEFF K(I,M,1)= ", K(I,M,1)
400 INPUT"INPUT POSITION OF SEPARATION LOSS ALONG PIPE AS A % OF PIPE LENGTH K
(I,M,2)= ",K(I,M,2)
410 K(I,M,2)=K(I,M,2)/100.0
420 NEXT M
430 NEXT I
440 IF (PIN+R*G*(ZIN-ZTOP))<0.0 THEN GOTO 1070
450 INPUT"INPUT DISCHARGE LEVEL OF LAST PIPE Z(N+1)= ",Z(N+1)
460 REM CALCULATE SUM OF FRICTION AND LOSS COEFFS...
470 SUMF=0.0
480 FOR M=1 TO N
490 SUMF=SUMF+4.0*R*F(M)*L(M)/(2.0*D(M)*((PI*D(M)^2)/4.0)^2)
500 NEXT M
510 SUMK=0.0
520 FOR M=1 TO N
530 IF J(M)=0 THEN GOTO 570
540 FOR P=1 TO J(M)
550 SUMK=SUMK+R*K(M,P,1)/(2.0*((PI*D(M)^2/4.0)^2))
560 NEXT P
570 NEXT M
580 Q=SQR(((PIN-POUT)+R*G*(ZIN-ZOUT))/(SUMF+SUMK))
590 CLS:PRINT"THEORETICAL MAX. FLOWRATE Q = ",Q," M^3/S."
600 REM CALCULATION OF LOCAL PRESSURE ALONG THE PIPE SYSTEM...
610 FOR M=1 TO N
620 S(M)=4.0*R*F(M)*L(M)*Q^2/((2.0*D(M)*(PI*D(M)^2/4.0)^2))
630 IF J(M)=0.0 THEN GOTO 670
640 FOR P=1 TO J(M)
650 T(M,P)=K(M,P,1)*0.5*R*Q^2/((PI*D(M)^2/4.0)^2)
660 NEXT P
670 NEXT M
680 H(1)=PIN
690 V(1)=ZIN
700 M1=1
710 FOR I=1 TO N
720 LZ=L(I)/10.0
730 DZ=(Z(I+1)-Z(I))/10.0
740 M1=M1+1
750 IF I=1 THEN H(M1)=PIN+R*G*(ZIN-Z(1))-Q^2/(2.0*(PI*D(1)^2/4.0)^2)
760 IF I>1 THEN H(M1)=H(M1-1)
770 V(M1)=Z(I)
780 FOR M2=2 TO 11
790 SL=0.0
800 IF J(I)=0 THEN GOTO 850
810 FOR M3=1 TO J(I)
820 IF (LZ*(M2-1))>(K(I,M3,2)*L(I)) AND (LZ*(M2-2))<(K(I,M3,2)*L(I)) THEN SL=S
L+T(I,M3)
830 IF (LZ*(M2-1))>(K(I,M3,2)*L(I)) AND (LZ*(M2-2))=(K(I,M3,2)*L(I)) THEN SL=S
L+T(I,M3)
840 NEXT M3
850 M1=M1+1
860 V(M1)=Z(I)+(M2-1)*DZ
870 H(M1)=H(M1-1)+R*G*(V(M1-1)-V(M1))-S(I)/10.0-SL
880 NEXTM2
890 NEXT I
900 M6=0
910 I1=1:PRINT
920 FOR M3=1 TO M1
930 IF M3=2 OR M3=2+I1*11 THEN GOTO 950
940 GOTO 990
950 IF M3=2+I1*11.0 THEN I1=I1+1
960 PRINT:PRINT"PIPE ",I1," RESULTS AT ",L(I1)/10.0," M. INTERVALS."
970 PRINT"FLOW VELOCITY = ",Q/(PI*D(I1)^2/4.0)," M/S."
980 PRINT
990 PRINT"PRESSURE = ",H(M3)/1000.0," KN/M^2. AT ELEVATION = ",V(M3)," M."
1000 IF M3-1 THEN GOTO 1030
1010 IF H(M3)<VAP AND H(M3-1)>VAP THEN M6=M3
1020 IF H(M3)>VAP AND H(M3-1)<VAP THEN M7=M3
1030 NEXT M3
1040 PRINT:IF M6>0 THEN PRINT"PRESSURE BELOW VAPOUR PRESSURE FROM ELEVATION ",V
(M6)," TO ",V(M7)," M. ABOVE DATUM."
1050 PRINT:IF M6>0 THEN PRINT"TRY INTRODUCING AN EXTRA SEPARATION LOSS AT DISCH
ARGE FROM THE LAST PIPE IN THE SYSTEM. THIS WILL REDUCE FLOWRATE AND RAISE PRESS
URE IN THE PIPES UPSTREAM..."
1060 GOTO 1080
1070 PRINT:PRINT"NO FLOW POSSIBLE DUE TO MAXIMUM HEIGHT OF PIPELINE ...."
1080 PRINT:INPUT"DO YOU WISH TO TRY AGAIN? Y OR N",Z$
1090 IF Z$="Y" THEN GOTO 10
1100 PRINT:PRINT"CALCULATION COMPLETE"
```

12.4 Incompressible flow through pipes in series

When pipes of different diameters are connected end to end to form a pipe-line, so that the fluid flows through each in turn, the pipes are said to be in series. Since losses are calculated per unit weight and each unit weight of the fluid passes through each pipe in succession, the total loss of energy per unit weight over the whole pipeline will be the sum of the losses for each pipe together with any separation losses such as might occur at the junctions, entrance or exit.

Example 12.2

Two reservoirs A and B (Fig. 12.3) have a difference of level of 9 m and are

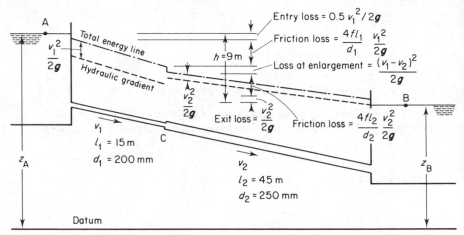

Figure 12.3   Pipes in series, showing head losses and the total energy line and hydraulic gradient

connected by a pipeline 200 mm in diameter over the first part AC, which is 15 m long, and then 250 mm diameter for CB, the remaining 45 m length. The entrance to and exit from the pipe are sharp and the change of section at C is sudden. The friction coefficient f is 0·01 for both pipes.

(a) List the losses of head (energy per unit weight) which occur, giving an expression for each. (b) Use program 'SIPHON' 12.3 to calculate system flowrate and hydraulic gradient.

Solution

(a) The losses of head which will occur are as follows.

(i) Loss at entrance to pipe AC. This is a separation loss and, since the entrance is described as sharp and is below the free surface of the reservoir (from Section 8.8.2), the value of k will be 0·5:

Loss of head at entry, $h_1 = 0.5\, \bar{v}_1^2/2g$.

(ii) Friction loss in AC. Using the Darcy formula, we have

$$\text{Loss of head in friction in AC,} = h_{f_1} = \frac{4fl_1}{d_1}\frac{\bar{v}_1^2}{2g}.$$

(iii) Loss at change of section at C. There will be a separation loss at the sudden change of section. From section 8.8.1, the loss at a sudden enlargement will be

Loss of head at sudden enlargement, $h_2 = (\bar{v}_1 - \bar{v}_2)^2/2g$.

(iv) Friction loss in CB. Using the Darcy formula,

Loss of head in friction in CB, $= h_{f_2} = \dfrac{4fl_2}{d_2} \dfrac{\bar{v}_2^2}{2g}$.

(v) Loss at exit. Since the exit is described as sharp and is beneath the surface of the reservoir B, there will be a separation loss as explained in Section 8.8:

Loss of head at exit, $h_3 = \bar{v}_2^2/2g$.

(b) Volume flowrate, $Q = 0 \cdot 158 \text{ m}^3 \text{s}^{-1}$.

12.5. Incompressible flow through pipes in parallel

When two reservoirs are connected by two or more pipes in parallel, as shown in Fig. 12.4, the fluid can flow from one to the other by a number of

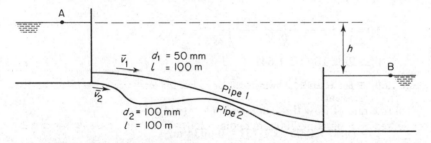

Figure 12.4   Pipes in parallel

alternative routes. The difference of head h available to produce flow will be the same for each pipe, and, since this head represents the energy per unit weight of fluid that can be used to overcome resistance to flow, it will be available for each unit weight irrespective of the route which that element of fluid follows. Thus, each pipe can be considered separately, entirely independently of any other pipes running in parallel. For incompressible flow, Bernoulli's equation can be applied for flow by each route and the total volume rate of flow will be the sum of the volume rates of flow in each pipe.

Example 12.3

Two sharp ended pipes of diameter $d_1 = 50$ mm and $d_2 = 100$ mm, each of length $l = 100$ m, are connected in parallel between two reservoirs which have a difference of level $h = 10$ m, as in Fig. 12.4. If the Darcy coefficient $f = 0 \cdot 008$ for each pipe calculate: (a) the rate of flow for each pipe, (b) the diameter D

of a single pipe 100 m long which would give the same flow if it was substituted for the original two pipes.

Solution
(a) Since the two pipes are in parallel, we can follow unit weight of the fluid through each pipe independently and apply Bernoulli's equation to points A and B on the free surfaces of the upper and lower reservoirs, respectively.

For flow by way of pipe 1,

$$
\begin{array}{llll}
\text{Total energy} & \text{Total energy} & \text{Losses per unit weight} \\
\text{per unit} & = \text{per unit} & + \text{ in pipe 1 for} \\
\text{weight at A} & \text{weight at B} & \text{entry, friction and exit,}
\end{array}
$$

$$
\left(\frac{p_A}{\rho g} + \frac{\bar{v}_A^2}{2g} + z_A\right) = \left(\frac{p_B}{\rho g} + \frac{\bar{v}_B^2}{2g} + z_B\right) + \left(0.5\frac{\bar{v}_1^2}{2g} + \frac{4fl}{d_1}\frac{\bar{v}_1^2}{2g} + \frac{\bar{v}_1^2}{2g}\right).
$$

Since $p_A = p_B$ = atmospheric pressure and, if the reservoirs are large, $\bar{v}_A$ and $\bar{v}_B$ will be negligible,

$$
z_A - z_B = \left(1.5 + \frac{4fl}{d_1}\right)\frac{\bar{v}_1^2}{2g}.
$$

Putting $z_A - z_B = h = 10$ m, $f = 0.008$, $l = 100$ m, $d_1 = 50$ mm $= 0.05$ m,

$$
10 = \left(1.5 + \frac{4 \times 0.008 \times 100}{0.05}\right)\frac{\bar{v}_1^2}{2g},
$$

$$
\bar{v}_1^2 = 2g \times 10/(1.5 + 64),
$$

$$
\bar{v}_1 = 1.731 \text{ m s}^{-1}.
$$

Volume rate of flow through pipe 1, $Q_1 = (\pi/4)d_1^2\,\bar{v}_1$

$$
= (\pi/4) \times 0.05^2 \times 1.731 = \mathbf{0.0034 \text{ m}^3 \text{ s}^{-1}}.
$$

For flow by way of pipe 2,

$$
\begin{array}{llll}
\text{Total energy} & \text{Total energy} & \text{Losses per unit weight} \\
\text{per unit} & = \text{per unit} & + \text{ in pipe 2 for entry,} \\
\text{weight at A} & \text{weight at B} & \text{friction and exit,}
\end{array}
$$

$$
\left(\frac{p_A}{\rho g} + \frac{\bar{v}_A^2}{2g} + z_A\right) = \left(\frac{p_B}{\rho g} + \frac{\bar{v}_B^2}{2g} + z_B\right) + \left(0.5\frac{\bar{v}_2^2}{2g} + \frac{4fl}{d_2}\frac{\bar{v}_2^2}{2g} + \frac{\bar{v}_2^2}{2g}\right).
$$

Since $p_A = p_B$ and both $\bar{v}_A$ and $\bar{v}_B$ can be assumed negligible,

$$
z_A - z_B = \left(1.5 + \frac{4fl}{d_2}\right)\frac{\bar{v}_2^2}{2g}.
$$

Putting $z_A - z_B = h = 10$ m, $f = 0.008$, $l = 100$ m, $d_2 = 100$ mm $= 0.10$ m,

$$
10 = \left(1.5 + \frac{4 \times 0.008 \times 100}{0.10}\right)\frac{\bar{v}_2^2}{2g}
$$

$$\bar{v}_2^2 = 2g \times 10/(1\cdot5 + 32),$$

$$\bar{v}_2 = 2\cdot42 \text{ m s}^{-1}.$$

Volume rate of flow through pipe 2,

$$Q_2 = (\pi/4)d_2^2\bar{v}_2$$

$$= (\pi/4) \times 0\cdot10^2 \times 2\cdot42 = 0\cdot0190 \text{ m}^3 \text{ s}^{-1}.$$

(b) Replacing the two pipes by the equivalent single pipe which will convey the same total flow,

Volume rate of flow through single pipe,

$$Q = Q_1 + Q_2$$

$$= 0\cdot0034 + 0\cdot0190 = 0\cdot0224 \text{ m}^3 \text{ s}^{-1}.$$

If $\bar{v}$ is the velocity in the single pipe, $Q = (\pi/4)D^2\bar{v}$. Therefore,

$$\bar{v} = \frac{4Q}{\pi D^2} = \frac{4 \times 0\cdot0224}{\pi D^2} = \frac{0\cdot028\ 52}{D^2}.$$

Applying Bernoulli's equation to A and B

| Total energy per unit weight at A | = | Total energy per unit weight at B | + | Losses per unit weight in equivalent pipe for entry friction and exit, |

$$\left(\frac{p_A}{\rho g} + \frac{\bar{v}_A^2}{2g} + z_A\right) = \left(\frac{p_B}{\rho g} + \frac{\bar{v}_B^2}{2g} + z_B\right) + \left(0\cdot5\ \frac{\bar{v}^2}{2g} + \frac{4fl}{D}\frac{\bar{v}^2}{2g} + \frac{\bar{v}^2}{2g}\right).$$

Making the same assumptions as before,

$$z_A - z_B = \left(1\cdot5 + \frac{4fl}{D}\right)\frac{\bar{v}^2}{2g}.$$

Putting $z_A - z_B = h = 10$ m, $f = 0\cdot008$, $l = 100$ m, $\bar{v} = 0\cdot028\ 52/D^2$,

$$10 = \left(1\cdot5 + \frac{4 \times 0\cdot008 \times 100}{D}\right) \times \frac{(0\cdot028\ 52)^2}{2gD^4}$$

$$= (1\cdot5\ D + 3\cdot2)(0\cdot028\ 52)^2/2gD^5.$$

Therefore,

$$241\ 212\ D^5 - 1\cdot5D - 3\cdot2 = 0. \tag{I}$$

This equation can be solved graphically or by successive approximations. An approximate answer can be obtained by omitting the second term; then,

$$241\ 212\ D^5 = 3\cdot2 \quad \text{and} \quad D = 0\cdot1058 \text{ m}.$$

To obtain a more precise answer, let the left-hand side of (I) be called $f(D)$, then, if $D = 0\cdot1058$ m,

$$f(D) = 3\cdot198 - 0\cdot159 - 3\cdot2 = -0\cdot161.$$

The negative value of $f(D)$ suggests that the value chosen for D was too small. If $D = 0.1070$ m,

$$f(D) = 3.383 - 0.161 - 3.2 = + 0.022.$$

Comparing these two results, the correct value of D will be a little less than 0.107 m. This result is sufficiently accurate for practical purposes.

Diameter of equivalent single pipe = 0.107 m = **107 mm**.

12.6. Incompressible flow through branching pipes: the three-reservoir problem

If the flow from the upper reservoir passes through a single pipe which then divides and the two branch pipes lead to two separate reservoirs with different surface levels, as shown in Fig. 12.5, the problem is more complex, particularly

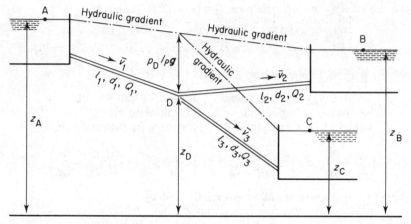

Figure 12.5   The three-reservoir problem

as it is sometimes difficult to decide the direction of flow in one of the pipes. Thus, in Fig. 12.5, if we draw the hydraulic gradient lines as shown, flow will be from D to B if the level of the hydraulic gradient at D is above the level of the free surface at B, but if it is below the level of B then flow will be in the reverse direction from B to D. Unfortunately, the hydraulic gradient cannot be drawn until the problem has been solved and so its value, $(z_D + p_D/\rho g)$, at D cannot be determined initially. In many cases, the direction of flow is reasonably obvious, but if it is doubtful, e.g. in DB, imagine that this branch is closed and calculate the value of $(z_D + p_D/\rho g)$ when there is flow from A to C only. If $(z_D + p_D/\rho g)$ is greater than z_B for this condition, flow will initially be from B to D when branch DB is opened. In some cases, conditions at D might then change sufficiently for the flow to reverse, but, if the correct assumption has been made, the continuity requirement that the sum of the flows into the junction is equal to the sum of the flows leaving the junction will be satisfied. If this is not the case, the assumed direction of flow must be reversed and a new solution calculated.

Example 12.4

Water flows from a reservoir A (Fig. 12.5) through a pipe of diameter $d_1 = 120$ mm and length $l_1 = 120$ m to a junction at D, from which a pipe of diameter $d_2 = 75$ mm and length $l_2 = 60$ m leads to reservoir B in which the water level is 16 m below that in reservoir A. A third pipe, of diameter $d_3 = 60$ mm and length $l_3 = 40$ m, leads from D to reservoir C, in which the water level is 24 m below that in reservoir A. Taking $f = 0.01$ for all the pipes and neglecting all losses other than those due to friction, determine the volume rates of flow in each pipe.

Solution

In this case, the levels of reservoir B and C are such that flow is obviously from D to B and D to C, as indicated in Fig. 12.5. There are three unknowns, $\bar{v}_1$, $\bar{v}_2$ and $\bar{v}_3$, and the necessary three equations are obtained by applying Bernoulli's equation first for flow from A to B, then for flow from A to C and finally writing the continuity of flow equation for the junction D.

For flow from A to B,

| Total energy per unit weight at A | = | Total energy per unit weight at B | + | Friction loss per unit weight in AD | + | Friction loss per unit weight in BD, |

$$\left(\frac{p_A}{\rho g} + \frac{\bar{v}_A^2}{2g} + z_A\right) = \left(\frac{p_B}{\rho g} + \frac{\bar{v}_B^2}{2g} + z_B\right) + \frac{4fl_1}{d_1}\cdot\frac{\bar{v}_1^2}{2g} + \frac{4fl_2}{d_2}\cdot\frac{\bar{v}_2^2}{2g}.$$

Putting $p_A = p_B$ and treating $\bar{v}_A$ and $\bar{v}_B$ as negligibly small,

$$z_A - z_B = \frac{4fl_1}{d_1}\cdot\frac{\bar{v}_1^2}{2g} + \frac{4fl_2}{d_2}\cdot\frac{\bar{v}_2^2}{2g}.$$

Substituting $z_A - z_B = 16$ m, $f = 0.01$, $l_1 = 120$ m, $d_1 = 0.120$ m, $l_2 = 60$ m, $d_2 = 0.075$ m,

$$16 = \frac{4 \times 0.01 \times 120\,\bar{v}_1^2}{0.120 \times 2g} + \frac{4 \times 0.01 \times 60\,\bar{v}_2^2}{0.075 \times 2g},$$

$$16 = 2.0387\,\bar{v}_1^2 + 1.6310\,\bar{v}_2^2. \tag{I}$$

For flow from A to C,

| Total energy per unit weight at A | = | Total energy per unit weight at C | + | Friction loss per unit weight in AD | + | Friction loss per unit weight in DC, |

$$\left(\frac{p_A}{\rho g} + \frac{\bar{v}_A^2}{2g} + z_A\right) = \left(\frac{p_C}{\rho g} + \frac{\bar{v}_C^2}{2g} + z_C\right) + \frac{4fl_1}{d_1}\cdot\frac{\bar{v}_1^2}{2g} + \frac{4fl_3}{d_3}\cdot\frac{\bar{v}_3^2}{2g},$$

giving

$$z_A - z_C = \frac{4fl_1}{d_1}\cdot\frac{\bar{v}_1^2}{2g} + \frac{4fl_3}{d_3}\cdot\frac{\bar{v}_3^2}{2g}.$$

Putting $z_A - z_C = 24$ m, $f = 0.01$, $l_1 = 120$ m, $d_1 = 0.120$ m, $l_3 = 40$ m, $d_3 = 0.060$ m,

$$24 = \frac{4 \times 0.01 \times 120\,\bar{v}_1^2}{0.120 \times 2g} + \frac{4 \times 0.01 \times 40\,\bar{v}_3^2}{0.060 \times 2g},$$

$$24 = 2.0387\,\bar{v}_1^2 + 1.3592\,\bar{v}_3^2. \tag{II}$$

For continuity of flow at D,

Flow through AD = Flow through DB + Flow through DC,

$$Q_1 = Q_2 + Q_3,$$

$$(\pi/4)d_1^2\bar{v}_1 = (\pi/4)d_2^2\bar{v}_2 + (\pi/4)d_3^2\bar{v}_3,$$

$$\bar{v}_1 = (d_2/d_1)^2\bar{v}_2 + (d_3/d_1)^2\bar{v}_3.$$

Substituting numerical values,

$$\bar{v}_1 = (0.075/0.120)^2\bar{v}_2 + (0.060/0.120)^2\bar{v}_3$$

$$\bar{v}_1 - 0.3906\,\bar{v}_2 - 0.2500\,\bar{v}_3 = 0. \tag{III}$$

Values of $\bar{v}_1$, $\bar{v}_2$ and $\bar{v}_3$ are found by solution of the simultaneous equations (I), (II) and (III). From (I),

$$\bar{v}_2 = \sqrt{(9.81 - 1.25\,\bar{v}_1^2)}. \tag{IV}$$

From (II)

$$\bar{v}_3 = \sqrt{(17.657 - 1.5\,\bar{v}_1^2)}. \tag{V}$$

Substituting in equation (III),

$$\bar{v}_1 - 0.3906\,\sqrt{(9.81 - 1.25\,\bar{v}_1^2)} - 0.25\,\sqrt{(17.657 - 1.5\,\bar{v}_1^2)} = 0. \tag{VI}$$

Equation (VI) can be solved graphically or by successive approximation. In the latter case, if the square roots are to be real, the value of $\bar{v}_1$ cannot exceed the lowest value that will make one of the terms under the square root signs equal to zero, this will be given by $\bar{v}_1^2 = 9.81/1.25 = 7.848$, so that $\bar{v}_1$ must be less than $\sqrt{(7.848)} = 2.80$ m s^{-1}. Calling the left-hand side of equation (VI) $f(\bar{v}_1)$ and choosing, as a first approximation, a value of $\bar{v}_1$ less than 2.80 m s^{-1} which, by inspection, will make $f(\bar{v}_1)$ approximately zero, calculate $f(\bar{v}_1)$. If this is not zero, choose further values of $\bar{v}_1$ until a value is found that makes $f(\bar{v}_1)$ sufficiently close to zero to be acceptable. Thus, if

$$\bar{v}_1 = 1.9 \text{ m s}^{-1}, f(\bar{v}_1) = 1.9 - 0.8990 - 0.8747 = +0.1263;$$

$$\bar{v}_1 = 1.8 \text{ m s}, f(\bar{v}_1) = 1.8 - 0.9374 - 0.8943 = -0.0317;$$

$$\bar{v}_1 = 1.82 \text{ m s}, f(\bar{v}_1) = 1.82 - 0.9300 - 0.8905 = -0.0005.$$

Taking $\bar{v}_1 = 1.82$ m s^{-1} as a sufficiently accurate result,

Volume rate of flow in AD, $Q_1 = (\pi/4)d_1^2\bar{v}_1$

$$= (\pi/4)(0.120)^2 \times 1.82 = 0.0206 \text{ m}^3 \text{ s}^{-1}.$$

From equation (IV),

$$\bar{v}_2 = \sqrt{(9.81 - 1.25 \times 1.82^2)} = 2.381 \text{ m s}^{-1},$$

Volume rate of flow in DB, $Q_2 = (\pi/4)d_2^2\bar{v}_2$

$$= (\pi/4)(0.075)^2 \times 2.381$$

$$= 0.0105 \text{ m}^3 \text{ s}^{-1}.$$

From equation (V),

$$\bar{v}_3 = \sqrt{(17 \cdot 657 - 1 \cdot 5 \times 1 \cdot 82^2)} = 3 \cdot 562 \text{ m s}^{-1}.$$

Volume rate of flow in DC, $Q_3 = (\pi/4) d_3^2 \bar{v}_3$

$$= (0 \cdot 060)^2 \times 3 \cdot 562 = \mathbf{0 \cdot 0101 \text{ m}^3 \text{ s}^{-1}}.$$

Checking for continuity at D,

$$Q_2 + Q_3 = 0 \cdot 0105 + 0 \cdot 0101 = 0 \cdot 0206 = Q_1.$$

12.7. Resistance coefficients for pipelines in series and in parallel

Both the friction and separation losses in a pipeline are functions of the mean velocity of flow $\bar{v}$ and, since $\bar{v} = Q/A$ where Q is the volume rate of flow and A the cross-sectional area of the pipe,

Total head loss, $h = KQ^n$,   (12.3)

where n is some power which depends on the type of flow. For turbulent flow n will be equal to two and, if separation losses are negligible, equation (12.3) becomes $h = KQ^2$, which is identical with equation (12.2). The form $h = KQ^2$ gives a misleading impression that the loss of head between two points in a pipeline is the same irrespective of the sign of Q, i.e. that h is independent of the direction of flow, which is of course absurd. It would be preferable to write $h = KQ|Q|$, where $|Q|$ means the numerical value of Q without regard to sign. Similarly, equation (12.3) could be written $h = KQ(|Q|)^{n-1}$.

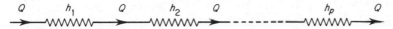

Figure 12.6   Resistances in series

For pipes in series (Fig. 12.6), Q is the same for each pipe and the losses of head $h_1, h_2, \ldots$ in each pipe are additive:

$$\text{Total loss of head} = h_1 + h_2 + \ldots + h_p$$

$$= K_1 Q^n + K_2 Q^n + \ldots + K_p Q^n$$

$$= (K_1 + K_2 + \ldots + K_p) Q^n. \qquad (12.4)$$

If several pipes are connected in parallel, as shown in Fig. 12.7, the loss of head between A and B must be the same for each pipe. Thus,

$$h = K_1 Q_1^n = K_2 Q_2^n = \ldots = K_p Q_p^n. \qquad (12.5)$$

Also, for continuous flow,

$$\text{Total flow, } Q = \text{Sum of the flows through each pipe}$$

$$= Q_1 + Q_2 + \ldots + Q_p.$$

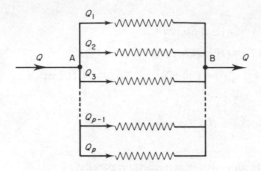

Figure 12.7   Resistances in parallel

Substituting for $Q_1, Q_2, \ldots Q_p$ from equation (12.5),

$$Q = \sqrt[n]{(h/K_1)} + \sqrt[n]{(h/K_2)} + \ldots + \sqrt[n]{(h/K_p)}. \tag{12.6}$$

The set of parallel pipes can be considered as equivalent to a single pipe with a resistance coefficient K carrying the total flow Q for which

$$h = KQ^n \quad \text{or} \quad Q = \sqrt[n]{(h/K)}.$$

Substituting in equation (12.6),

$$\sqrt[n]{(h/K)} = \sqrt[n]{(h/K_1)} + \sqrt[n]{(h/K_2)} + \ldots + \sqrt[n]{(h/K_p)}$$

or, assuming that $n = 2$,

$$1/\sqrt{K} = 1/\sqrt{K_1} + 1/\sqrt{K_2} + \ldots + 1/\sqrt{K_p}. \tag{12.7}$$

Example 12.5

A system of pipes conveying water is connected in parallel and in series, as shown in Fig. 12.8. The section DE represents the resistance of a valve for

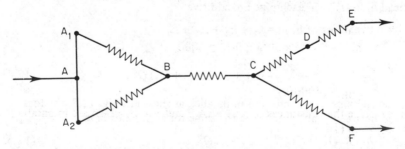

Figure 12.8   Resistance network

controlling the flow which has a resistance coefficient $K_{DE} = (4000/n)^2$, where n is the percentage valve opening.

The friction factor f in the Darcy formula is 0·006 for all pipes, and their lengths and diameters are given by

Pipe	Length l (m)	Diameter d (m)
$AA_1 B$	30	0·1
$AA_2 B$	30	0·125
BC	60	0·15
CD	15	0·1
CF	30	0·1

The head at A is 100 m, at E is 40 m and at F is 60 m. If the valve is adjusted to give equal discharge rates at E and F, calculate the head at C, the total volume rate of flow through the system and the percentage valve opening. Neglect all losses except those due to friction.

Solution
For any pipe, neglecting separation losses and working in SI units,

$$\text{Head lost in friction, } h = \frac{4fl}{d} \cdot \frac{\bar{v}^2}{2g} = \frac{flQ^2}{3d^5} = KQ^2,$$

Therefore,

$$K = fl/3d^5.$$

For pipe AA_1B,

$$K_{AA_1 B} = 0 \cdot 006 \times 30/3 \times 0 \cdot 1^5 = 6000.$$

For pipe AA_2B,

$$K_{AA_2 B} = 0 \cdot 006 \times 30/3 \times 0 \cdot 125^5 = 1966.$$

For pipe BC,

$$K_{BC} = 0 \cdot 006 \times 60/3 \times 0 \cdot 15^5 = 1580.$$

For pipe CD,

$$K_{CD} = 0 \cdot 006 \times 15/3 \times 0 \cdot 1^5 = 3000.$$

For pipe CF,

$$K_{CF} = 0 \cdot 006 \times 30/3 \times 0 \cdot 1^5 = 6000.$$

For the valve,

$$K_{DE} = (4000/n)^2.$$

First, combine the resistances of pipes AA_1B and AA_2B, which are in parallel, using equation (12.5):

$$1/\sqrt{K_{AB}} = 1/\sqrt{K_{AA_1 B}} + 1/\sqrt{K_{AA_2 B}} = (K_{AA_1 B}^{1/2} + K_{AA_2 B}^{1/2})/ (K_{AA_1 B} \times K_{AA_2 B})^{1/2},$$

$$K_{AB} = (K_{AA_1 B} \times K_{AA_2 B})/(K_{AA_1 B}^{1/2} + K_{AA_2 B}^{1/2})^2$$

$$= (6000 \times 1966)/(6000^{1/2} + 1966^{1/2})^2 = 795.$$

Now combine K_{AB} with K_{BC} to find the equivalent resistance when BC is added in series with the pipes between A and B. From equation (12.12),

$$K_{AC} = K_{AB} + K_{BC} = 795 + 1580 = 2375.$$

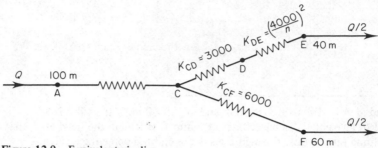

Figure 12.9   Equivalent pipeline

If Q is the total volume rate of flow entering at A,

Loss of head between A and C = $H_A - H_C = K_{AC}Q^2 = 2375 \, Q^2$.   (I)

Since it is required that the discharges at E and F should be equal, the flows through CE and CF will each be $\frac{1}{2}Q$:

Loss of head between C and F = $H_C - H_F = K_{CF}(\frac{1}{2}Q)^2$

$$= 6000 \, Q^2/4 = 1500 \, Q^2.$$ (II)

From equations (I) and (II),

$$(H_A - H_C)/2375 = Q^2 = (H_C - H_F)/1500,$$
$$1500 \, H_A - 1500 \, H_C = 2375 \, H_C - 2375 \, H_F,$$
$$3875 \, H_C = 1500 \, H_A + 2375 \, H_F.$$

Putting $H_A = 100$ m and $H_F = 60$ m,

$$3875 \, H_C = 1500 \times 100 + 2375 \times 60$$
$$H_C = 75 \cdot 48 \text{ m}.$$

From equation (I),

$$2375 \, Q^2 = H_A - H_C = 100 - 75 \cdot 48 = 24 \cdot 52 \text{ m},$$
$$Q = 0 \cdot 1016 \text{ m}^3 \text{ s}^{-1}.$$

Since pipe CD and valve DE are in series,

Loss of head between C and E = $H_C - H_E = (K_{CD} + K_{DE})(\frac{1}{2}Q)^2$.

Substituting numerical values,

$$75 \cdot 48 - 40 = \{3000 + (4000/n)^2\}(0 \cdot 0508)^2,$$
$$35 \cdot 52 = 7 \cdot 74 + 41 \, 290/n^2,$$
$$27 \cdot 78 \, n^2 = 41 \, 290,$$
$$n = 38 \cdot 55 \text{ per cent}.$$

12.8. Incompressible flow in a pipeline with uniform draw-off

If a pipeline has a large number of tappings along its length from which the fluid is discharged, as in the case of a perforated pipe used as a sprinkler, the problem can be treated as if fluid was being drawn off at a uniform rate per unit length. Under these circumstances, the volume rate of flow across successive cross-sections will decrease as the distance from the point of input increases. If the pipe is of constant diameter, the velocity and, therefore, the frictional loss of head per unit length will also decrease. Such problems are solved by considering a short length of pipe and then integrating to obtain the result for the whole pipe.

12.9. Incompressible flow through a pipe network

A pipe network is a set of pipes which are interconnected so that the flow

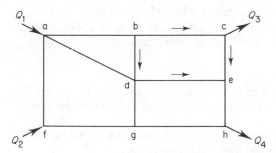

Figure 12.10   Pipe network

from a given input or to a given outlet may come through several different routes. Thus, in Fig. 12.10, the input at a may be divided between pipes ab, ad, and af; part of this may leave at c and part at h, each combining with part of the input Q_2. An attempt to apply Bernoulli's equation and the continuity of flow equation to the various elements in the network would lead to a very large number of simultaneous equations which would be cumbersome to solve. The alternative approach is to use a method of successive approximations, assuming values for the flow in each pipe, or the heads at each junction, and checking whether the values chosen satisfy the requirements that

(i) the loss of head between any two junctions must be the same for all routes between these junctions (e.g. in loop bced (Fig. 12.10),

$$\frac{\text{Loss of head}}{\text{in pipe bc}} + \frac{\text{Loss of head}}{\text{in pipe de}} = \frac{\text{Loss of head}}{\text{in pipe bd}} + \frac{\text{Loss of head}}{\text{in pipe ce}}$$

(ii) the inflow to each junction must equal the outflow from that junction.

If the values chosen do not satisfy these conditions throughout the network, they must be corrected by successive approximations until they do so within the required degree of accuracy. The Hardy–Cross method provides a system for calculating the value of the correction to be made, each loop or junction being considered in turn and corrected assuming that conditions in the remainder of the network remain unaltered. Obviously, corrections to one element will affect conditions elsewhere and the required balance of heads and flows will not be reached as a result of the first correction. However, each successive repetition of the process will bring the system nearer to the final balanced condition.

12.10. Head balance method for pipe networks

The head balance method is used when the total volume rate of flow through the network is known, but the heads or pressures at junctions within the network are unknown. For each pipe, an assumption must first be made of the direction and volume rate of flow so as to satisfy condition (ii) (that the inflow to each junction must equal the outflow from that junction). In the loop bced in Fig. 12.10, the directions of flow might be as indicated by the arrows, thus the flow in bc and ce is clockwise round this loop and the flow in bd and de is anticlockwise. To satisfy condition (i), the loss of head between b and e must be the same by either the clockwise or anticlockwise route. Neglecting losses other than friction for any pipe,

$$\text{Head lost, } h = KQ^n,$$

where Q = volume rate of flow in the pipe, K = resistance coefficient which, in SI units, would be $fl/3\,D^5$ and n is a constant which, for turbulent flow, would be 2.

If Σ_c and Σ_{cc} represent summations of quantities in the clockwise and anticlockwise (counterclockwise) directions, respectively,

$$\begin{matrix}\text{Loss of head in pipes in} \\ \text{which flow is clockwise}\end{matrix} \quad = \Sigma_c h \; = \Sigma_c KQ^n,$$

$$\begin{matrix}\text{Loss of head in pipes in} \\ \text{which flow is anticlockwise}\end{matrix} = \Sigma_{cc} h = \Sigma_{cc} KQ^n.$$

The values initially chosen for the volume rate of flow in each pipe Q are unlikely to meet the requirement that

$$\Sigma_c h = \Sigma_{cc} h.$$

If it is assumed that $\Sigma_c h > \Sigma_{cc} h,$

$$\text{Out of balance head} = \Sigma_c h - \Sigma_{cc} h = \Sigma_c KQ^n - \Sigma_{cc} KQ^n.$$

To remove this out of balance head, while keeping the total flow through the loop constant, the clockwise flow must be *reduced* by an amount δQ and the anticlockwise flow increased by δQ, so that

$$\Sigma_c h - \Sigma_{cc} h = \Sigma_c K(Q - \delta Q)^n - \Sigma_{cc} K(Q + \delta Q)^n = 0.$$

Expanding the terms in brackets and neglecting all terms involving the second or higher orders of δQ, which is a small quantity compared to Q,

$$\Sigma_c K(Q^n - nQ^{n-1} \delta Q) = \Sigma_{cc} K(Q^n + nQ^{n-1} \delta Q),$$

from which

$$\delta Q = \frac{\Sigma_c KQ^n - \Sigma_{cc} KQ^n}{n(\Sigma_c KQ^{n-1} + \Sigma_{cc} KQ^{n-1})}.$$

Now $KQ^n = h$ and $KQ^{n-1} = h/Q$, therefore,

$$\delta Q = \frac{\Sigma_c h - \Sigma_{cc} h}{n(\Sigma_c(h/Q) + \Sigma_{cc}(h/Q))}.$$

Adopting a sign convention that values of h and Q are to be regarded as positive in pipes in which the flow is clockwise *with regard to the loop under consideration* and negative if anticlockwise,

$$\delta Q = -\frac{\Sigma h}{n \Sigma(h/Q)}.$$

The negative sign indicates that the positive (clockwise) values of Q are to be reduced and the negative (anticlockwise) values of Q are to be increased. When a system has a number of loops, corrections to one loop will unbalance adjoining loops which will require further correction. Also pipes common to two loops will receive corrections for each loop. The process is therefore iterative and must be continued until the desired degree of accuracy is achieved.

12.11. Computer program 'HARDYC'

(1) The computer program calculates the distribution of known inflows and outflows within a network consisting of a number of loops. The program is based on the Hardy-Cross method and allows for frictional and separation losses within each pipe length.

The program allows a network design to be input in response to a number of prompts. The sign convention used is that flow into a node is always positive, outflow is negative. Initial flow values in each pipe are assigned by equally distributing unaccounted for outflow between the available pipes.

(2) The required input is:

   (i)   fluid density;
   (ii)   number of pipes in system;
   (iii) length of pipe joining each node, 1 to n (note input is 0 if no such pipe exists);
   (iv) diameter of pipe if it exists, D(m);
   (v)   assumed friction factor f for each pipe;
   (vi) separation loss for this pipe (repeat until all combinations covered);
   (vii) inflow at each node 1 to n (note sign convention, and 0 if no flow);
   (viii) number of loops in system;
   (ix) for each loop, clockwise, enter node number.

A clear sketch will help in entering these data.

(3) The program is based on the Hardy-Cross method that assumes net flow at a node to be zero and net pressure loss around a loop to be zero.

(4) Input example:

Density = 1000 kg m^{-3}; Number of pipes = 8; Number of loops = 3; Loops: 1278, 2367, 3456.

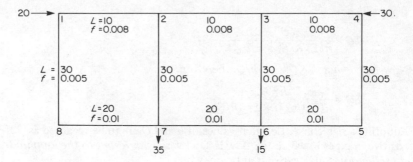

Figure 12.11 Network example for 'HARDYC'

All diameters = 1·0 m.

(5) Output:

FINAL FLOW DISTRIBUTION. . . . .

INFLOW/OUTFLOW FROM NETWORK NODE   1 =   20 M^3/S.
INFLOW/OUTFLOW FROM NETWORK NODE   2 =   0 M^3/S.
INFLOW/OUTFLOW FROM NETWORK NODE   3 =   0 M^3/S.
INFLOW/OUTFLOW FROM NETWORK NODE   4 =   30 M^3/S.
INFLOW/OUTFLOW FROM NETWORK NODE   5 =   0 M^3/S.
INFLOW/OUTFLOW FROM NETWORK NODE   6 =   -15 M^3/S.
INFLOW/OUTFLOW FROM NETWORK NODE   7 =   -35 M^3/S.
INFLOW/OUTFLOW FROM NETWORK NODE   8 =   0 M^3/S.

FINAL FLOW DISTRIBUTION (−VE OUT OF NODE) . . . .
FLOW FROM NODE   1 TO NODE   2 Q=   −8.41 M^3/S.
FLOW FROM NODE   1 TO NODE   8 Q=   −11.6 M^3/S.
FLOW FROM NODE   2 TO NODE   3 Q=   6.33 M^3/S.
FLOW FROM NODE   2 TO NODE   7 Q=   −14.7 M^3/S.
FLOW FROM NODE   3 TO NODE   4 Q=   17.3 M^3/S.
FLOW FROM NODE   3 TO NODE   6 Q=   −10.9 M^3/S.
FLOW FROM NODE   4 TO NODE   5 Q=   −12.7 M^3/S.
FLOW FROM NODE   5 TO NODE   6 Q=   −12.7 M^3/S.
FLOW FROM NODE   6 TO NODE   7 Q=   −8.67 M^3/S.
FLOW FROM NODE   7 TO NODE   8 Q=   11.6 M^3/S.

List:

```
10 REM HARDYC
20 MODE 7:@%=&30305
30 PRINT"PROGRAM HARDY"
40 PRINT:PRINT"PROGRAM DESIGNED TO USE THE HARDY CROSS TECHNIQUE TO DETERMINE
THE DISTRIBUTION OF FLOW WITHIN A PIPE NETWORK."
50 PRINT:PRINT"THE PROGRAM ASSUMES THAT ALL PIPE AND FLUID DATA IS KNOWN, EG.
 PIPE LENGTHS, DIAMETERS, ASSUMED FRICTION FACTORS, SEPARATION LOSS COEFFS, FLUI
D DENSITY AND VISCOSITY."
60 PRINT:PRINT"IN ADDITION ALL INFLOWS AND OUTFLOWS FROM THE NETWORK ARE ASSU
MED KNOWN."
80 PRINT:PRINT:PRINT"PLEASE INPUT DATA IN RESPONSE TO THE FOLLOWING PROMPTS.
A CLEARLY LABELLED SKETCH MAY AID YOU IN ENTERING THIS DATA..."
90 DIM Z(10,10,4):DIM Q(10): DIM X(10,10): DIM NN(10):DIM QP(10,10):DIM PR(10
,10)
95 PRINT:PRINT:
100 INPUT"FLUID DENSITY (KG/M^3) = ",RO
```

```
105 Z$="Y"
120 INPUT"TOTAL NUMBER OF NODES IN SYSTEM N= ",N
130 FOR K=1 TO 4
140 FOR I=1 TO N
150 FOR J=1 TO N
160 IF K=1 THEN Z(I,J,K)=-1.0 ELSE Z(I,J,K)=0.0
170 PR(I,J)=0.0
180 NEXT J
190 NEXT I
200 NEXT K
210 FOR I=1 TO N
220 FOR J=1 TO N
230 IF J=I THEN GOTO 280
240 IF Z(J,I,1)>0.0 OR Z(J,I,1)=0.0 THEN GOTO 340
250 PRINT"INPUT LENGTH OF PIPE (M.) DESIGNATED AS LINKING NODE ",I," TO NODE "
,J," (ENTER 0.0 TO INDICATE NO LINKING PIPE.)"
260 INPUT Z(I,J,1)
270 IF Z(I,J,1)=0.0 THEN PRINT "NO SUCH PIPE..."
280 IF J=I THEN Z(I,J,1)=0.0
290 IF Z(I,J,1)=0.0 THEN GOTO 380
300 INPUT"DIAMETER OF PIPE (M.) = ",Z(I,J,2)
310 INPUT"ASSUMED FRICTION FACTOR FOR THIS PIPE = ",Z(I,J,3)
320 INPUT"TOTAL SEPARATION LOSS COEFF. FOR THIS PIPE LENGTH 'K' = ",Z(I,J,4)
330 GOTO 380
340 Z(I,J,1)=Z(J,I,1)
350 Z(I,J,2)=Z(J,I,2)
360 Z(I,J,3)=Z(J,I,3)
370 Z(I,J,4)=Z(J,I,4)
380 NEXT J
390 NEXT I
400 REM INFLOW ALLOCATION
410 PRINT"INPUT INFLOW AT EACH NODE IN TURN. 0.0 = NO INFLOW, -VE VALUES INDIC
ATE OUTFLOW AT THAT NODE. (NOTE TOTAL = ZERO FOR NETWORK.)"
420 SUMQ=0.0
430 FOR I=1 TO N
440 PRINT"INFLOW AT NODE ",I," IN M^3/S.":INPUT Q(I)
450 SUMQ=SUMQ+Q(I)
460 NEXT I
470 IF ABS(SUMQ)>0.0 THEN PRINT "ERROR IN INFLOW ENTRIES PLEASE TRY AGAIN..."
480 IF ABS(SUMQ)>0.0 THEN GOTO 410
490 REM ALLOCATE NODES TO LOOPS.
500 INPUT"INPUT NUMBER OF LOOPS TO BE CONSIDERED NL= ",NL
510 FOR L=1 TO NL
520 PRINT"LOOP NUMBER =",L
530 M=0
540 M=M+1
550 INPUT"NODE NUMBER (NOTE MOVE CLOCKWISE ROUND EACH LOOP INCLUDING RETURN TO
1ST. NODE) = ",X(L,M)
560 IF M>1 AND X(L,M)=X(L,1) THEN GOTO 580
570 GOTO 540
580 NN(L)=M
590 NEXT L
600 REM CALC. FLOW IN EACH PIPE LEADING FROM EACH NODE AS 1ST APPROXIMATION.
610 REM AT NODE 1 ASSUME INFLOW SPLITS EQUALLY INTO ALL PIPES LEADING FROM 1.
620 I=1
630 NF=0
640 FOR J=1 TO N
650 IF Z(1,J,1)=0.0 THEN GOTO 670
660 NF=NF+1
670 NEXT J
680 FOR J=1 TO N
690 IF Z(1,J,1)=0.0 THEN GOTO 720
700 QP(1,J)=-Q(1)/NF
710 QP(J,1)=-QP(1,J)
720 NEXT J
730 FOR I=2 TO N
740 NF=0.0
750 INFLO=Q(I)
760 FOR J=1 TO N
770 IF Z(I,J,1)=0.0 THEN GOTO 820
780 IF ABS(QP(I,J))>0.0 THEN GOTO 810
790 NF=NF+1
800 GOTO 820
810 INFLO=INFLO+QP(I,J)
820 NEXT J
830 FOR J=1 TO N
840 IF Z(I,J,1)=0.0 OR ABS(QP(I,J))>0 THEN GOTO 870
850 QP(I,J)=-INFLO/NF
860 QP(J,I)=-QP(I,J)
870 NEXT J
880 NEXT I
890 PRINT"INITIAL FLOW DISTRIBUTION...."
900 FOR I=1 TO N
910 FOR J=1 TO N
```

```
920 IF J=I THEN GOTO 940
930 PRINT "NODE ",I," TO NODE ",J," FLOW = ",QP(I,J)," M^3/S."
940 NEXT J
950 NEXT I
960INPUT"PRESS RETURN TO CONTINUE",Z$
970 REM FOR EACH LOOP IN TURN CALCULATE LOSS IN EACH PIPE, SUM DP=0 ETC.
980 DQ=0.0
990 FOR I1=1 TO NL
1000 SUMDP=0.0
1010 SUMDPQ=0.0
1020 FOR I2=1 TO NN(I1)-1
1040 L=Z(X(I1,I2),X(I1,I2+1),1)
1050 D=Z(X(I1,I2),X(I1,I2+1),2)
1060 F=Z(X(I1,I2),X(I1,I2+1),3)
1070 S=Z(X(I1,I2),X(I1,I2+1),4)
1080 QZ=QP(X(I1,I2),X(I1,I2+1))
1090 DP=4.0*F*L*RO*QZ^2/(2.0*D*(PI*D^2/4)^2)+S*0.5*RO*QZ^2/(PI*D^2/4.0)^2
1100 PRINT"I=",X(I1,I2),"J=",X(I1,I2+1)
1110 PRINT "Q=",QZ
1120 SIGN=QZ/ABS(QZ)
1130 DP=DP*SIGN
1140 PRINT"DP=",DP
1150 SUMDP=SUMDP+DP
1160 DPQ=ABS(DP/QZ)
1170 SUMDPQ=SUMDPQ+ABS(DP/QZ)
1180 PRINT"DPQ=",DPQ
1190 NEXT I2
1200 DQ1=-SUMDP/(2.0*SUMDPQ)
1210 FOR I2=1 TO NN(I1)-1
1220 QP(X(I1,I2),X(I1,I2+1))=QP(X(I1,I2),X(I1,I2+1))+DQ1
1230 QP(X(I1,I2+1),X(I1,I2))=-QP(X(I1,I2),X(I1,I2+1))
1240 NEXT I2
1250 PRINT DQ1
1255 IF Z$="N" THEN GOTO 1270
1260 INPUT"DO YOU WISH TO VIEW THESE STEPS AT EACH ITERATION ? Y OR N.",Z$
1270 IF ABS(DQ1)<0.0001*ABS(QP(1,2)) THEN GOTO 1310
1280 DQ=DQ1
1290 NEXT I1
1300 GOTO 990
1310 CLS:PRINT"FINAL FLOW DISTRIBUTION....."
1320 PRINT
1330 FOR I=1 TO N
1340 PRINT"INFLOW/OUTFLOW FROM NETWORK NODE ",I," = ",Q(I)," M^3/S."
1350 NEXT I
1360 PRINT:PRINT"FINAL FLOW DISTRIBUTION (-VE OUT OF NODE).... "
1370 FOR I=1 TO N
1380 FOR J=1 TO N
1390 IF Z(I,J,1)=0 THEN GOTO 1430
1400 IF PR(J,I)=1.0 THEN GOTO 1430
1410 PRINT"FLOW FROM NODE ",I," TO NODE ",J," Q= ",QP(I,J)," M^3/S."
1420 PR(I,J)=1.0
1430 NEXT J
1440 NEXT I
1450 INPUT" DO YOU WISH TO REPEAT? Y OR N.",Z$
1460 IF Z$="Y" THEN GOTO 10
1470 PRINT "CALCULATIONS COMPLETE."
```

12.12. The quantity balance method for pipe networks

When the heads at various points in a pipe network are known and it is necessary to calculate the quantities flowing in each pipe, the quantity balance method can be used. An estimate is made of the head at each junction (or node) in the network and the volume rate of flow Q is calculated for each pipe from the difference of head h between the junctions at each end of the pipe and its resistance coefficient K, using the formula $h = KQ^n$. If the resulting inflow does not equal the outflow at each junction, the original estimates of head must be corrected.

If, in Fig. 12.12, the head at b has been overestimated by an amount δh relative to the head at a, c and d, the values of Q_{ab} and Q_{cb} will be too small and the value of Q_{bd} will be too great. Differentiating the equation $h = KQ^n$, we have

$$\delta h = KnQ^{n-1}\delta Q$$

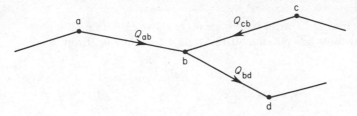

Figure 12.12   Pipe network — quantity balance network

or, since $K = h/Q^n$,

$$\delta Q = (Q/nh)\,\delta h.$$

If, therefore, the original estimate of the head at b is reduced by an amount δh, the flows in ab and cb will be increased to $Q_{ab} + \delta Q_{ab}$ and $Q_{cb} + \delta Q_{cb}$. If the flows are now correct, inflow and outflow at b will balance:

$$\text{Flow in ab} + \text{Flow in cb} \quad = \text{Flow in bd},$$

$$(Q_{ab} + \delta Q_{ab}) + (Q_{cb} + \delta Q_{cb}) = (Q_{bd} - \delta Q_{bd}). \tag{12.8}$$

If h_{ab}, h_{cb} and h_{bd} are the assumed losses of head in pipes ab, cb and bd used to calculate Q_{ab}, Q_{cb} and Q_{bd},

$$\delta Q_{ab} = \frac{Q_{ab}}{nh_{ab}}\,\delta h, \quad \delta Q_{cb} = \frac{Q_{cb}}{nh_{cb}}\,\delta h \quad \text{and} \quad \delta Q_{bd} = \frac{Q_{bd}}{nh_{bd}}\,\delta h.$$

Substituting in equation (12.8),

$$\left(Q_{ab} + \frac{Q_{ab}}{nh_{ab}}\,\delta h \right) + \left(Q_{cb} + \frac{Q_{cb}}{nh_{cb}}\,\delta h \right) = \left(Q_{bd} - \frac{Q_{bd}}{nh_{bd}}\,\delta h \right),$$

$$\delta h = -\frac{(Q_{ab} + Q_{cb} - Q_{bd})}{Q_{ab}/nh_{ab} + Q_{cb}/nh_{cb} + Q_{bd}/nh_{bd}}.$$

Using the sign convention that for flow towards the junction b both Q and h are positive,

$$\delta h = -\frac{\Sigma Q}{\Sigma(Q/nh)},$$

where $\Sigma Q = Q_{ab} + Q_{cb} - Q_{bd}$ = Algebraic sum of the flows towards the junction.

When δh has been calculated, the new value of the head at b can be used to determine revised values of Q_{ab}, Q_{cb} and Q_{bd}. The process is repeated until ΣQ has been reduced to a negligible quantity. The following simple example shows the application of this method to the three-reservoir problem.

Example 12.6
A reservoir A (Fig. 12.13) with its surface 60 m above datum supplies water to a junction D through a 300 mm diameter pipe 1500 m long. From the

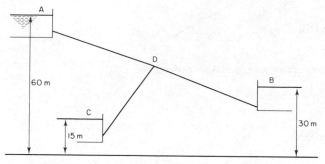

Figure 12.13   The three-reservoir problem

junction, a 250 mm diameter pipe 800 m long feeds reservoir B, in which the surface level is 30 m above datum, while a 200 mm diameter pipe 400 m long feeds reservoir C, in which the surface level is 15 m above datum. Calculate the volume rate of flow to each reservoir. Assume that the loss of head due to friction is given by $h = flQ^2/3d^5$ for each pipe and that $f = 0.01$.

Solution
First choose, by inspection, a value for the head at D. Since the elevation of A is much greater than that of B or C, flow is likely to be from D to B and C (as indicated in the question) and the head at D would, therefore, be greater than that at B. Assume a trial value for the head at D of 35 m. Then, initially,

Head loss in AD, $h_{AD} = 60 - 35 = 25$ m,

Head loss in DB, $h_{DB} = 35 - 30 = 5$ m,

Head loss in DC, $h_{DC} = 35 - 15 = 20$ m.

					First correction			
Pipe	K	Assumed value of h (m)	$Q = \sqrt{(h/K)}$ (m³ s⁻¹)	$\dfrac{Q}{2h}$	$\delta h = \dfrac{-\Sigma Q}{\Sigma(Q/2h)}$	$h_1 = h + \delta h$	$Q_1 = \sqrt{(h_1/K)}$ (m³ s⁻¹)
AD	2058	25	+0·1102	0·00220	+0·23	+25·23	+0·1107
DB	2730	− 5	−0·0428	0·00428	+0·23	− 4·77	−0·0418
DC	4167	−20	−0·0693	0·00173	+0·23	−19·77	−0·0689
			Σ −0·0019	0·00821			0

Table 12.1. The calculations for junction D

For each pipe, $h = KQ^2$ where $K = fl/3d^5$. Therefore, the flow Q for any value of h is $Q = \sqrt{(h/K)}$.

For pipe AD, $K_{AD} = \dfrac{0.01 \times 1500}{3 \times 0.300^5} = 2058$.

For pipe DB, $K_{DB} = \dfrac{0.01 \times 800}{3 \times 0.250^5} = 2730$.

For pipe DC, $K_{DC} = \dfrac{0.01 \times 400}{3 \times 0.200^5} = 4167$.

The calculations for junction D can now be set out as shown in Table 12.1. After the first correction it can be seen from Table 12.1 that ΣQ is zero four places of decimals and no further correction is necessary. The flows in the pipes are

$Q_{AD} = 0.1107$ m^3 s^{-1} from reservoir A,

$Q_{DB} = 0.0418$ m^3 s^{-1} to reservoir B,

$Q_{DC} = 0.0689$ m^3 s^{-1} to reservoir C.

12.13. Further reading

Jeppson R. W. (1979) *Analysis of flow in pipe networks*, Ann Arbor Science.

EXERCISES 12

12.1   Two vessels in which the difference of surface levels is maintained constant at 2.4 m are connected by a 75 mm diameter pipeline 15 m long. If the frictional coefficient f may be taken as 0.008, determine the volume rate of flow through the pipe.
[11.9 litre/s]

12.2   The difference in surface levels in two reservoirs connected by a syphon is 7.5 m. The diameter of the syphon is 300 mm and its length 750 m. The friction coefficient f is 0.0064. If air is liberated from solution when the absolute pressure is less than 1.2 m of water, what will be the maximum length of the inlet leg of the syphon to run full, if the highest point is 5.4 m above the surface level in the upper reservoir. What will be the discharge.
[360 m, 107 dm^3/s]

12.3   Two reservoirs whose difference of level is 15 m are connected by a pipe ABC whose highest point B is 2 m below the level in the upper reservoir A. The portion AB has a diameter of 200 mm and the portion BC a diameter of 150 mm, the friction coefficient being the same for both portions. The total length of the pipe is 3 km.

Find the maximum allowable length of the portion AB if the pressure head at B is not to be more than 2 m below atmospheric pressure. Neglect the secondary losses.
[2030 m]

12.4   A pipeline 30 m long connects two tanks which have a difference of

water level of 12 m. The first 10 m of pipeline from the upper tank is 40 mm diameter and the next 20 m is 60 mm diameter. At the change in section a valve is fitted. Calculate the rate of flow when the valve is fully opened assuming that its resistance is negligible and that f for both pipes is 0·0054. In order to restrict the flow the valve is then partially closed. If k for the valve is now 5·6, find the percentage reduction in flow.
[0·00738 m³/s, 25·8 per cent]

12.5   Oil of specific gravity 0·9, kinematic viscosity 4×10^{-5} m²/s flows from one reservoir to another by gravity along a 200 m long pipe. If the flow rate is 0·028 m³/s in a 150 mm smooth pipe which contains 2 x 90° bends, having k values of 0·19 and a smooth entry and abrupt exit at the lower reservoir. Calculate the vertical separation of the reservoir surface levels.
[252 m]

12.6   A smooth walled tube is used in a 3000 m long pipeline carrying water at 15 °C between two reservoirs whose surface elevations are 6 m apart. Entry is sharp edged and the outlet is also abrupt to the downstream reservoir. The pipeline contains 6 x 45° bends and two globe valves. Determine the necessary pipe diameter so that the discharge should be 28 litres/s to the lower reservoir.

Take the equivalent length of each bend as 26·5 diameters, the valves as 75 diameters and the entry as 30 diameters.
[70 mm]

12.7   When the pressure in a pipe falls below the vapour pressure local boiling occurs with consequent disruption and possible stoppage of flow. Two large reservoirs have surface levels which differ by an amount H. A hill which is higher than the level of the upper reservoir is situated between them. The two reservoirs are to be joined by a pipe as shown in Fig. 12.14. The lengths

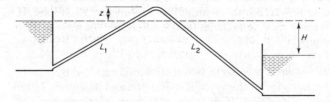

Figure 12.14

L_1 and L_2 depend upon the maximum height z which the pipe reaches above the level of the upper reservoir and are given by

$$L_1 = 400 - 54z \text{ metres and } L_2 = 3100 - 46z \text{ metres}$$

Neglecting the kinetic energy of the fluid in the pipe derive an expression for the gauge pressure at the highest point reached by the pipe in terms of H, z, ρ, and g where ρ is the density of the fluid and g is the acceleration due to gravity.

Taking $H = 70$ m and the vapour gauge pressure = 0·8 bar, determine L_1, L_2 and z for that pipe route which gives maximum flow of water for a given diameter.

$$\left[\rho g \left\{ \frac{H(4 - 0.54z)}{35 - z} + z \right\}, 184 \text{ m}, 2916 \text{ m}, 4 \text{ m} \right]$$

12.8   A horizontal duct system draws atmospheric air into a circular duct 0·3 m diameter, 20 m long, then through a centrifugal fan and discharges it to atmosphere through a rectangular duct 0·25 m by 0·20 m, 50 m long. Assuming that the friction factor for each duct is 0·01 and accounting for an inlet loss of one half of the velocity head and also for the kinetic energy at outlet, find the total pressure rise across the fan to produce a flow of 0·5 m³/s.

Sketch also the total energy and hydraulic gradient lines putting in the most important values. Assume the density of air to be 1·2 kg/m³.
[694 N/m²]

12.9   Liquid F12 flows through a tube 10 m long, 500 mm diameter, at a rate of 18·4 kg/h. The average temperature of the liquid is 23 °C. Calculate the pressure drop due to friction. For turbulent flow assume that $f = 0·08/$ $Re^{0·25}$ and that for laminar flow the usual assumption applies. At 23 °C, $\rho = 1304$ kg/m³ and $\mu = 254 \times 10^{-6}$ kg/m s.
[480 kN/m²]

12.10   For flow through pipes at high Reynolds number, the coefficient of friction is given by the following relation,

$$\frac{1}{\sqrt{f}} - 4 \log_{10}\left(\frac{r}{\epsilon}\right) = 3·48$$

where r = pipe radius, ϵ = mean height of roughness projections. A pipe of internal diameter 0·15 m is formed of a material for which ϵ is 0·000 38 m. The pipe is 1524 m long and it connects two water reservoirs whose surface levels are maintained at the same height. Water may be pumped along the pipe and the maximum pumping power available is 82 kW. Calculate the maximum rate of flow in the pipe.
[0·06 m³/s]

12.11   Reservoir A feeds water to a lower reservoir B, the water surface in B being 30 m below that in A. A pipe 50 m long, 0·5 m diameter, leaves A and is series connected to a 100 m long, 0·25 m diameter pipe which discharges below the surface level in B. This second pipe is fitted with an adjustable valve at its exit with the following loss characteristic:

Valve loss = k × velocity head in pipe

Valve opening per cent	100	75	50	25
k	3	10	25	60

Neglecting all minor losses and assuming a constant friction factor of 0·01 for both pipes, calculate the flow rate from A to B if the valve is fully open and also the valve setting to restrict this flow to one half of this value.
[0·273 m³/s, 25 per cent]

12.12   A pipeline conveying water between reservoirs A and B is 30·5 cm diameter and 366 m long. The difference of head between the two surfaces is 4·12 m. Determine the flow rate if $f = 0·005$.

It is required to increase the flow by 50 per cent by duplicating a portion of the pipe. If the head and friction factor are unchanged and minor losses are ignored, find the length of the second pipe which is of the same diameter as the first.
[268 m]

12.13   There is a pressure loss of 300 kN/m^2 when water is pumped through pipeline A at a rate of 2 m^3/s and there is a pressure loss of 250 kN/m^2 when water is pumped at a rate of 1·4 m^3/s through pipeline B. Calculate the pressure loss which will occur when 1·5 m^3/s of water are pumped through pipes A and B jointly if they are connected (*a*) in series, (*b*) in parallel, assuming that junction losses may be neglected. In the latter case calculate the volume rate of flow through each pipe.
[455 kN/m^2, 54·4 kN/m^2, 0·853 m^3/s, 0·652 m^3/s]

12.14   A complex ventilation system for a coalmine may be reduced to the system shown in Fig. 12.15 where R_1, R_2 and R_3 represent the equivalent

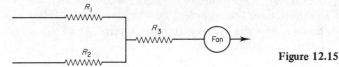

Figure 12.15

resistances of the three main sections of the mine. Assuming an air density of 1·17 kg/m^3 these resistances are

R_1 = 49 mm of water total pressure at 100 m^3/s

R_2 = 73 mm of water total pressure at 100 m^3/s

R_3 = 10 mm of water total pressure at 200 m^3/s

The fan characteristic at a density of 1·2 kg/m^3 is:

Discharge Q m^3/s	0	100	150	200	250	300	350
Fan total pressure mm of water		175	180	175	160	135	100	60

(*a*) Determine the volume rate of flow handled by the fan and the fan total pressure. (*b*) If due to the increased length of workings the resistance of the whole system changes and is found to be 150 mm of water total at 200 m^3/s and density 1·17 kg/m^3, determine the percentage increase of fan speed required to maintain the same flow through the fan.
[265 m^3/s, 125 mm of water; 29 per cent]

12.15   Water flows in the parallel pipe system shown in Fig. 12.16 for which the following data are available:

Pipe	Diameter	Length	f
AaB	0·10 m	300 m	0·0060
AbB	0·15 m	250 m	0·0055
AcB	0·20 m	500 m	0·0050

The supply pipe to point A is 0·30 m diameter and the mean velocity of water in it is 3 m/s. If the elevation of point A is 100 m and the elevation of point B is 30 m above datum, calculate the pressure at point B if that at A is 200

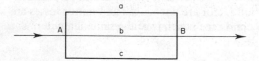

Figure 12.16

kN/m^2. What is the discharge in each pipe. Neglect all minor losses.
[538 kN/m^2, 0·024 m^3/s, 0·075 m^3/s, 0·144 m^3/s]

12.16   Water is handled by a system of pipes as shown in Fig. 12.17 the details being as follows:

Pipe	Length (m)	Diameter (m)	f
$A_1 B = A_2 B$	100	0·50	0·0055
BC	300	0·75	0·0050
CD	500	0·30	0·0060
CE	400	0·25	0·0060
CF	500	0·30	0·0060

The elevation of outlets D, E and F is 100 m above the elevation of inlets A_1 and A_2. All outlets and inlets are open to atmosphere. If the mean velocity in

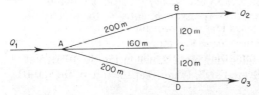

Figure 12.17

the pipes A_1B and A_2B is 2·5 m/s, calculate the flow rate through the pump P, the pressure difference across the pump and the power consumed. Take the pump efficiency as 76%.
[0·98 m^3/s, 1682 kN/m^2, 1650 kW]

12.17   For the system of pipes shown in Fig. 12.18 determine (a) Q_1, Q_2, Q_3 and the pressure at C, (b) Q_2 and the pressure at C when $Q_3 = 0$. Assume for

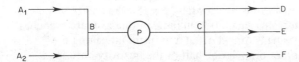

Figure 12.18

both cases that the pressure at A is 280 kN/m^2 and that the discharge at B and D is to atmosphere. All pipes are 0·30 m diameter except AD which is 0·40 m diameter and f for all pipes is 0·005.
[1·86 m^3/s, 0·69 m^3/s, 1·17 m^3/s, 43·6 kN/m^2; 0·96 m^3/s, 208·9 kN/m^2]

12.18   Three reservoirs A, B and C are interconnected by three pipes which all meet at junction J. The water surface of reservoir B is 20 m above the surface of C whilst the surface of A is 40 m above the surface of B. A flow control valve is fitted just before junction J in pipe AJ.

The head loss h_L through pipes and components can be written as $h_L = rQ^2$ where r is the resistance coefficient and Q is the volume rate of flow. The values of r for the valve and the pipes are

$$r_{AJ} = 150, r_{BJ} = 200, r_{CJ} = 300, r_{valve} = (400/n)^2$$

where n is the percentage valve opening. Find the value of n which will make the discharge into reservoir C twice that into reservoir B.
[56·6 per cent]

12.19   Two evaporative condensers are supplied with water through a horizontal pipe system. The main pipe is 50 mm in diameter and 70 m long, and branches into a 25 mm diameter pipe 30 m long and a 37·5 mm pipe 45 m long. The total head producing flow in the system is 0·1 m of water. Taking the coefficient of friction f for all pipes as 0·005 and neglecting all other losses, calculate the velocity and discharge in each pipe.
[5·75 m/s, 0·114 m³/s; 7·08 m/s, 0·0344 m³/s; 0·708 m/s, 0·0696 m³/s]

12.20   A horizontal water main comprises 1500 m of 150 mm diameter pipe followed by 900 m of 100 mm diameter pipe, the friction factor f for each pipe being 0·007. All the water is drawn off at a uniform rate per unit length along the pipe. If the total input to the system is 25 dm³/s, find the total pressure drop along the main, neglecting all losses other than pipe friction. Also draw the hydraulic gradient taking the pressure head at inlet as 54 m.
[20·68 m]

12.21   A 675 mm water main runs horizontally for 1500 m and then branches into two 450 mm mains each 3000 m long. In one of these branches the whole of the water entering is drawn off at a uniform rate along the length of the pipe. In the other branch one half of the quantity entering is drawn off at a uniform rate along the length of the pipe. If f = 0·006 throughout, calculate the total difference of head between inlet and outlet when the inflow to the system is 0·28 m³/s. Consider only frictional losses and assume atmospheric pressure at the end of each branch.
[4·78 m]

12.22   Water entering a 150 mm diameter pipe 1300 m long is all drawn off at a uniform rate per metre of length along the pipe. Neglecting all losses other than pipe friction, find the volume rate of flow entering the pipe when the pressure drop along the pipe is 2·55 bar. Take f = 0·008. Draw the hydraulic gradient for the system if the pressure at entry to the pipe is 2·8 bar.
[0·0416 m³/s]

12.23   Water enters the four-sided ring main shown in Fig. 12.19 at A at the rate of 0·40 m³/s and is delivered at B, C and D at the rate of 0·15, 0·10

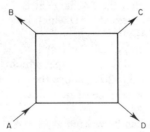

A   D   **Figure 12.19**

and 0·15 m³/s. All pipes are 0·6 m in diameter with a friction coefficient f of 0·0078 and their lengths are AB and CD 150 m, BC 300 m and DA 240 m. Determine the flow through each pipe and the pressures at B, C and D if that at A is 105 kN/m².
[Q_{AB} = 0·216 m³/s, Q_{BC} = 0·066 m³/s, Q_{DC} = 0·034 m³/s, Q_{AD} = 0·184 m³/s, p_A = 105 kN/m², p_B = 102·7 kN/m², p_C = 102·3 kN/m², p_D = 102·4 kN/m²]

12.24   Using the Hardy-Cross method, estimate the flow rate in each of the

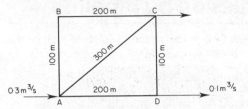

Figure 12.20

pipes in the network shown in Fig. 12.20. The pipe diameter is 0·5 m throughout and the friction factor f may be taken as 0·005.

[AB = BC = 0·0831 m³/s, AC = 0·0825 m³/s, AD = 0·1344 m³/s,

DC = 0·0344 m³/s]

12.25   The head loss for flow in a duct can be written as $h_f = rQ^n$ where r
is the pipe resistance and Q is the volume rate of flow. The fuel gallery for a
small gas turbine is shown in Fig. 12.21. Each injection nozzle passes 5 litres

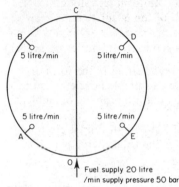

Figure 12.21

of kerosine per minute. The relationships between the pipe resistances are as
follows

$$r_{BC} = r_{CD} = r_{AO} = r_{OE},\quad r_{AB} = r_{DE},$$

$$r_{AB} = 2r_{BC},\quad r_{OC} = 3r_{BC}$$

If the pipe OC is 2 m long, 0·01 m diameter and the friction factor f is 0·010
in the formula

$$h_f = \frac{4fL}{d} \cdot \frac{v^2}{2g}$$

find the pressure drop between O and C.
[3380 N/m²]

13 Power transmission by pipeline

13.1. Transmission of power by pipeline

In addition to conveying fluids, pipelines can be used to transmit power from point to point. In principle, any fluid can be used for this purpose, but normally the working fluids will be oil (or similar liquid) in 'oil hydraulic' machinery and water as, for example, in hydro-electric power plants. In the examples of pipelines discussed so far, when the flow has been between one reservoir and another the difference of head H between entry and exit of the pipeline has been used wholly to overcome the loss of head in friction h_f. Even when the fluid has been discharged to atmosphere, the fraction of the original total energy H which is available in the form of kinetic energy of the outflowing fluid is usually small. For a pipeline to be able to deliver power at its exit, only a fraction of the total head H at the inlet can be used to overcome the loss of head due to friction h_f and the remainder will then be available as the power head h_p, which is the energy per unit weight available at the outlet. Thus,

$$H = h_f + h_p. \tag{13.1}$$

The energy delivered at the outlet may be conveyed either in the form of fluid under pressure but at low velocity or, using a suitable nozzle, it can take the form of the kinetic energy of a high speed jet at atmospheric or ambient pressure.

As was seen in Section 6.13, the power P of a stream of fluid is given by ρgQH, where Q is the volume rate of flow and H the total energy per unit weight. Thus, for a pipeline transmitting power,

Power supplied at inlet = ρgQH,
Power used in friction = ρgQh_f,
Power available at outlet = ρgQh_p,

therefore,

$$\text{Efficiency of transmission} = \frac{\text{Outlet power}}{\text{Inlet power}} = \frac{\rho gQh_p}{\rho gQH} = \frac{h_p}{H}.$$

Alternatively, since $H = h_p + h_f$,

$$h_p = H - h_f,$$

Efficiency of transmission, $\eta = 1 - h_f/H$ \hfill (13.2)

and

$$\text{Head lost in friction} = (1 - \eta)H.$$

For turbulent flow, the loss of head in friction h_f for a given volume rate of flow Q would be given by $h_f = KQ^2$, which corresponds to the parabolic curve in Fig. 13.1(a). For a given total head H at inlet, Fig. 13.1(a) also shows the variation with Q of the head available for power h_p, since $h_p = H - h_f$.

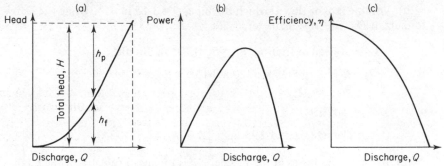

Figure 13.1   Power and efficiency in a pipeline

The relation between the actual power transmitted, $\rho g Q h_p$, and the volume rate of flow Q is shown in Fig. 13.1(b). Clearly, when Q is zero h_p is a maximum, but the power transmitted involves the product of these two quantities and is, therefore, zero. Similarly, when Q is a maximum h_p is zero and, again, the power transmitted is zero. For intermediate values of Q, both h_p and Q have positive values and power will be transmitted, rising from zero to a maximum and then falling to zero again as shown.

From equation (13.2), the efficiency of power transmission η has a maximum value of unity when h_f is zero, which corresponds with a value for Q of zero. Since $h_f = KQ^2$, the efficiency of transmission is given by

$$\eta = (H - KQ^2)/H$$

and will decrease with increasing values of Q (as shown in Fig. 13.1(c)) reaching zero when Q is a maximum and $h_f = H$.

Example 13.1

A pipeline 3220 m long conveys water from a reservoir which is discharged through a nozzle at its lower end to drive a Pelton wheel. The surface level of the reservoir is maintained constant at 220 m above the nozzle and the rate of flow is $3 \cdot 15$ m³ s⁻¹.

If 85 per cent of the potential energy per unit weight of the water in the reservoir is to be available as kinetic energy at the jet, calculate (a) the efficiency of transmission, (b) the pipe diameter, (c) the nozzle diameter and (d) the power of the jet. The value of f in the Darcy formula for the pipe is $0 \cdot 0075$ and all losses except friction may be ignored.

Solution

Gross head available = 220 m. If 85 per cent of this energy is to be available as kinetic energy of the jet,

Power head, $h_p = 0 \cdot 85 \times 220 = 187$ m

and   Head lost in friction, $h_f = H - h_p = 33$ m.

(a)

Efficiency of transmission $= h_p/H = 0 \cdot 85 H/H$

$= $ **85 per cent.**

(b) Volume rate of flow through the pipe, $Q = 3 \cdot 15 \text{ m}^3 \text{s}^{-1}$. Using the Darcy formula, if D is the pipe diameter,

Loss of head in friction, $h_f = flQ^2/3D^5$ in SI units,

$$D^5 = \frac{flQ^2}{3h_f} = \frac{0 \cdot 0075 \times 3220 \times 3 \cdot 15^2}{3 \times 33}$$

$$= 2 \cdot 4205,$$

Pipe diameter, $D = 1 \cdot 193$ m.

(c) Assuming no losses in the nozzle, if $\bar{v}$ is the mean velocity of the jet,

Power head, h_p = Kinetic energy per unit weight of jet = $\bar{v}^2/2g$,

Jet velocity, $\bar{v} = \sqrt{(2gh_p)}$.

But the flow through the jet is equal to the flow Q through the pipe. If d is the jet diameter,

$$Q = (\pi/4)d^2\,\bar{v} = (\pi/4)d^2\,\sqrt{(2gh_p)},$$

$$d^2 = 4Q/\sqrt{(2gh_p)} = 4 \times 3 \cdot 15/\sqrt{(2g \times 187)} = 0 \cdot 2080,$$

Jet diameter, $d = 0 \cdot 456$ m.

(d)

Power of jet = $\rho g Q h_p$ = 1000 x 9·81 x 3·15 x 187 W

$$= 5779 \text{ kW}.$$

13.2 Conditions for transmission of maximum power through a given pipeline

Looking at Fig. 13.1, it is clear that the power available from a given pipeline for a total input head H rises to a maximum and then falls back to zero as Q increases to its maximum value. Comparing Fig. 13.1(b) with Fig. 13.1(c), maximum discharge will not coincide with maximum efficiency but will occur at a substantially lower efficiency. It is of interest to establish the conditions under which a pipeline will transmit the maximum power for a given size, since this also enables us to determine the smallest pipe diameter which will transmit the required power under the given conditions. For a given pipeline of diameter d and length l, if the total head available is H, the head lost in friction is h_f and the power head available at delivery is h_p:

$$h_p = H - h_f.$$

Using the Darcy formula,

$$h_f = \frac{4fl}{d} \frac{\bar{v}^2}{2g} = k\bar{v}^2$$

where $k = 4fl/2gd$ is a constant for a given pipeline. Thus

$$h_p = H - k\bar{v}^2.$$

Power transmitted, P = Volume rate of flow × Power head,

$$P = \rho g A \bar{v} h_{\mathrm{p}} = \rho g A \bar{v}(H - k\bar{v}^2), \tag{13.3}$$

where A = area of cross-section of the pipe and ρ is the mass density of the fluid. For any given pipeline ρ, A, H and k are constant. Thus the power transmitted will be a maximum when $\mathrm{d}P/\mathrm{d}v = 0$. Differentiating equation (13.3),

$$\frac{\mathrm{d}P}{\mathrm{d}v} = \rho g A H - 3 \rho g A k \bar{v}^2 = 0.$$

Thus, for maximum power,

$$H = 3k\bar{v}^2 = 3h_{\mathrm{f}}$$

or $h_{\mathrm{f}} = \frac{1}{3}H$ and $h_{\mathrm{p}} = \frac{2}{3}H$.

Efficiency for maximum power = $(H - h_{\mathrm{f}})/H = (H - \frac{1}{3}H)/H$

$$= 0.6667 = 66\frac{2}{3} \text{ per cent.}$$

Note that this value has been obtained using the Darcy formula for the loss of head due to friction, which assumes that h_{f} is proportional to $\bar{v}^2$. For any other power relationship, the condition for maximum power transmission would be different.

Example 13.2

A pipeline 1500 m long conveys water to a turbine, the difference of level between the surface of the reservoir and the turbine outlet being 141 m. If the shaft power output of the turbine is 350 kW and the turbine efficiency is 70 per cent, calculate the smallest diameter of pipe which could be used, assuming that $f = 0.008$.

Solution

The smallest pipe will be one that is working under conditions for maximum power transmission and, if it is assumed that h_{f} is proportional to $\bar{v}^2$, we have

Head lost in friction, $h_{\mathrm{f}} = \frac{1}{3}H$,
Power head delivered, $h_{\mathrm{p}} = \frac{2}{3}H$,

where H is the total head available.
    If Q is the volume rate of flow,

Power supplied by the pipeline = $\rho g Q h_{\mathrm{p}}$,

Shaft power output from turbine = $\eta_{\mathrm{t}} \rho g Q h_{\mathrm{p}}$,

where η_{t} = turbine efficiency. Thus, putting $H = 141$ m, $h_{\mathrm{p}} = 94$ m and $h_{\mathrm{f}} = 47$ m.

$$Q = \frac{\text{shaft power}}{\eta_{\mathrm{t}} \rho g Q h_{\mathrm{p}}} = \frac{350 \times 10^3}{0.70 \times 10^3 \times 9.81 \times 94}$$

$$= 0.5422 \text{ m}^3 \text{ s.}$$

Using the Darcy formula and working in SI units,

$$h_f = flQ^2/3d^5,$$

where d = pipe diameter; and so

$$d^5 = flQ^2/3h_f$$

$$= \frac{0\cdot008 \times 1500 \times 0\cdot5422^2}{3 \times 47} = 0\cdot0250.$$

Therefore,

Pipe diameter required, $d = 0\cdot4783$ m.

13.3. Relationship of nozzle diameter to pipe diameter for maximum power transmission

When the power transmitted by a pipeline is delivered in the form of a high velocity jet from a nozzle, the proportion of the total input head H which is converted into kinetic energy will depend upon the relation between the diameter of the nozzle d_n and the diameter of the pipe d. Assuming that, for maximum power transmission, $h_f = \frac{1}{3}H$ and $h_p = \frac{2}{3}H$, we have

$$h_f = \tfrac{1}{2} h_p. \tag{13.4}$$

If $\bar{v}$ = mean velocity in the pipe and $\bar{v}_n$ = mean velocity in the nozzle, using the Darcy formula,

$$h_f = \frac{4fl}{d} \cdot \frac{\bar{v}^2}{2g}$$

and if there is no loss in the nozzle, the whole of the power head is converted into kinetic energy, so that

$$h_p = \bar{v}_n^2/2g.$$

Substituting in equation (13.4),

$$\frac{4fl}{d} \cdot \frac{\bar{v}^2}{2g} = \frac{1}{2} \cdot \frac{\bar{v}_n^2}{2g},$$

$$\bar{v}/\bar{v}_n = (d/8fl)^{1/2}.$$

For continuity of flow,

$$(\pi/4)d^2\bar{v} = (\pi/4)d_n^2\,\bar{v}_n,$$

giving

$$d_n/d = (\bar{v}/\bar{v}_n)^{1/2} = (d/8fl)^{1/4}.$$

This result applies to a single pipeline with friction losses calculated in accordance with the Darcy formula. Other conditions can be treated similarly.

EXERCISES 13

13.1 Water is conveyed to a turbine through a pipe 1200 m long from a reservoir. There is a fall of 126 m between the reservoir surface level and the discharge from the turbine. If the output power is to be 300 kW and the efficiency of the turbine is 70 per cent, calculate the smallest size pipe which could be employed. Take f as 0·008.
[0·46 m]

13.2 Power is to be transmitted hydraulically a distance of 8000 m by means of a number of 100 mm diameter pipes laid horizontally in parallel. The pressure at the inlet to the pipes is maintained constant at 6450 kN/m². Determine the minimum number of pipes required to ensure an efficiency of transmission of at least 92 per cent when the power delivered is 150 kW. Take $f = 0·0075$.
[5].

13.3 Calculate the power which can be delivered to a factory 6·4 km distant from a hydraulic power station through three horizontal pipes each 150 mm diameter laid in parallel, if the inlet pressure is maintained constant at 5000 kN/m² and the efficiency of transmission is 94 per cent.

If one of the pipes becomes unavailable, what increase in pressure at the power station would be required to transmit the same power at the same delivery pressure as before and what would be the efficiency of transmission under these conditions. Take $f = 0·0075$.
[170·5 kW, 326 kN/m², 88·3 per cent]

13.4 A number of hydraulically operated machines are supplied with water under pressure by a pipe 300 m long, the pressure at the machines being 4140 kN/m². The power available at the machines is 220 kW and the frictional loss of power in the pipes is 88 kW. Find the necessary diameter of the pipes if $f = 0·03$, and also the pressure at the pumping end. What is the maximum power at which the machines could be worked for the same pump pressure.
[137 mm, 5790 kN/m², 222 kW]

13.5 The water supply to a turbine in a power station is 1·27 m³/s under a total head of 285 m and is transmitted through a pipe 720 m long. If the efficiency of transmission is 95 per cent and $f = 0·0075$, find the necessary diameter of pipe. What should be the diameter of the jet driving the turbine.
[726 mm, 149 mm]

13.6 A long pipe of diameter d and length l supplies a nozzle of diameter d_0 at the end. The total head is H, the friction coefficient of the pipe is f and the jet loss $k u_0^2/2g$ where u_0 is the jet speed. Determine the ratio d_0/d for maximum jet momentum and the efficiency of transmission for this condition.

$$\left[\left(\frac{(k+1)d}{4fL} \right)^{1/4}, \frac{1}{2(k+1)} \right]$$

13.7 Water is supplied from a reservoir through a 300 mm diameter pipe 600 m long to a nozzle which is situated 108 m below the free surface of the water in the reservoir. The friction coefficient f of the pipe is 0·005. Find the greatest possible power of the issuing jet.
[210 kW]

13.8   A pump feeds water to a hose 50 m long which is fitted with a nozzle which has a coefficient of discharge of 0·98 and discharges a 40 mm diameter jet at 30 m/s when the nozzle is at the same level as the pump. The pump efficiency is 70 per cent and it draws water from a level 3 m below the nozzle. The friction coefficient for the hose is $f = 0.007$. If the efficiency of hydraulic power transmission through the hose is 75 per cent, calculate the hose diameter and the power required to drive the pump.
[100·8 mm, 35·2 kW]

13.9   Water is supplied under a head of 270 m to a pipeline of length 1050 m, the diameter of the pipe being 200 mm and the coefficient of friction f being 0·0075. The pipeline terminates in a nozzle of circular section. Find from first principles the best diameter for the nozzle if the reaction of the jet is to be a maximum. Calculate also the power of the jet. Neglect all losses other than pipe friction.
[58·2 mm, 114 kW]

13.10   A Pelton wheel is supplied by four equal pipes in parallel connected to a short common pipe leading to the nozzle. If the losses in the short pipe and the nozzle are small in comparison with the friction losses in the four pipes, show that the power delivered to the wheel by the jet is a maximum when $d = (2D^5/fL)^{1/4}$ in which d is the nozzle diameter, D the pipe diameter, L the length of each pipe and f the friction coefficient.

Hence find the maximum power which can be delivered to a Pelton wheel of suitable jet diameter if $D = 0.6$ m and $L = 3$ km when the level of the water in the reservoir is 300 m above the centre of the jet. What is the diameter of the nozzle and the velocity of the jet. Take $f = 0.005$.
[9860 kW, 319 mm, 62·7 m/s]

13.11   Show that if η_p is the efficiency of power transmission through a pipeline conveying water to a reaction turbine which has an efficiency η_t and H is the head from reservoir level to the turbine tailrace, then the power developed by the turbine will be proportional to

$$\eta_t \eta_p \sqrt{(1 - \eta_p)H\sqrt{H}}$$

A pipeline 3000 m long conveys water to a turbine which develops 3750 kW with an efficiency of 90 per cent. The efficiency of power transmission through the pipeline is 91 per cent and the friction coefficient for the pipeline is 0·004. The head from reservoir level to turbine tailrace is 24 m. Find the diameter of the pipeline.
[3·71 m]

13.12   A single jet Pelton wheel is to be supplied by a number of pipes in parallel connected to a short common pipe leading to a nozzle. Show that the power delivered to the wheel by the jet is a maximum when the diameter of the nozzle is $d = n^{1/2}(D^5/8fL)^{1/4}$ where n is the number of pipes and D, L and f are the diameter, length and friction coefficient of each pipe. Neglect losses in the short pipe and nozzle.

Hence find the flow of water and number of pipes required if D is 600 mm, L is 2800 m, the power of the jet is 10 000 kW and the level of the water in the reservoir is 300 m above the nozzle. Take $f = 0.005$.
[5·1 m³/s, 4]

14 Compressible flow in pipes

14.1. Compressible flow: the basic equations

When considering flow in ducts and pipes, in Chapters 8 and 12, it was assumed that the fluid could be treated as if it were incompressible and, therefore, of constant density. For a wide range of fluids employed in engineering this assumption is valid because the pressure changes which occur are normally too small to cause an appreciable change in density. For gases, however, this assumption cannot be made, since large variations of density can be produced as a result of the changes of pressure which occur in normal engineering applications: compressibility must be taken into account except where such pressure changes are very small. Thus, in considering the continuous flow of a compressible fluid, the relationship between density and the other factors affecting fluid flow must be considered. This will be the equation of state relating the absolute pressure p, absolute temperature T and the mass density ρ which was given in Chapter 1, equation (1.13) for a perfect gas as $p = \rho RT$, where R is the gas constant for the particular gas concerned: hence, we have

$$\rho = p/RT \tag{14.1}$$

The continuity of flow equation, which arises from the principle of conservation of mass as discussed in Section 4.12, must also be used, in the form

$$\dot{m} = \rho A \bar{v} = \text{constant}, \tag{14.2}$$

where $\dot{m}$ is the mass flow rate through a cross-section of area A at which the velocity and the density of the fluid are $\bar{v}$ and ρ, respectively.

The steady flow energy equation was discussed in Section 6.2. For compressible flow in a horizontal plane, equation (6.10) becomes

$$\tfrac{1}{2}v_1^2 + H_1 + q - w = \tfrac{1}{2}v_2^2 + H_2 \tag{14.3}$$

where H is the enthalpy, q the heat added per unit mass and w the work done per unit mass. If q and w are zero, equation (14.3) reduces to

$$\tfrac{1}{2}v^2 + H = \text{constant} = H_0, \tag{14.4}$$

where $H_0 = $ total or stagnation enthalpy $= c_p T_0$ where T_0 is the stagnation temperature.

For frictionless flow, the Euler equation, as derived in Section 5.12, is also applicable. Equation (5.21), which states that along a streamline

$$\frac{\mathrm{d}p}{\rho} + v\,\mathrm{d}v + g\mathrm{d}z = 0, \tag{14.5}$$

can be integrated when the relationship between p and ρ is known. If friction and other forces act, the momentum equation can be applied in its basic form, as given in equation (5.5).

14.2. Steady, isentropic flow in non-parallel-sided ducts neglecting friction

Although isentropic flow, which is frictionless flow under adiabatic conditions, is an ideal which cannot be fully realized in practice, the assumption of isentropic conditions gives a satisfactory approximation for the analysis of flow through short transitions, orifices, venturi meters and nozzles in which friction and heat transfer are minor effects which can be neglected.

Since, for an incompressible fluid, ρ was constant, it was possible to write equation (14.2) as $A\bar{v}$ = constant, indicating that for steady flow the velocity must increase if the area of the stream decreases. For compressible flow, this need not be the case since ρ is also variable. Considering a horizontal stream, from equation (14.5),

$$\frac{\mathrm{d}p}{\rho} + \bar{v}\,\mathrm{d}\bar{v} = 0, \tag{14.6}$$

but, from equation (5.31), $\mathrm{d}p/\mathrm{d}\rho = c^2$, where c is the velocity of sound, so that

$$\frac{\mathrm{d}p}{\rho} = c^2\,\frac{\mathrm{d}\rho}{\rho}.$$

Substituting in equation (14.6),

$$c^2\,\frac{\mathrm{d}\rho}{\rho} + \bar{v}\,\mathrm{d}\bar{v} = 0. \tag{14.7}$$

Differentiating equation (14.2) and dividing by $\rho A \bar{v}$,

$$\frac{\mathrm{d}\rho}{\rho} + \frac{\mathrm{d}\bar{v}}{\bar{v}} + \frac{\mathrm{d}A}{A} = 0. \tag{14.8}$$

Eliminating $\mathrm{d}\rho/\rho$ between equations (14.7) and (14.8),

$$\frac{\bar{v}\mathrm{d}\bar{v}}{c^2} - \frac{\mathrm{d}\bar{v}}{v} - \frac{\mathrm{d}A}{A} = 0.$$

Dividing through by $\mathrm{d}\bar{v}/A$,

$$\frac{\mathrm{d}A}{\mathrm{d}\bar{v}} = \frac{A}{\bar{v}}\left(\frac{v^2}{c^2} - 1\right)$$

or, since v/c is the Mach number Ma,

$$\frac{\mathrm{d}A}{\mathrm{d}\bar{v}} = \frac{A}{\bar{v}}\,(\mathrm{Ma}^2 - 1). \tag{14.9}$$

From equation (14.9), it can be seen that, for steady frictionless flow with no restriction on heat transfer:

(i) if Ma $<$ 1 (subsonic flow), $\mathrm{d}A/\mathrm{d}\bar{v}$ is always negative, indicating that the velocity must increase as the cross-sectional area of the duct decreases;

(ii) if Ma $>$ 1 (supersonic flow), $\mathrm{d}A/\mathrm{d}\bar{v}$ is always positive, indicating that for the velocity to increase the area of the duct must also increase;

(iii) If Ma = 1 (sonic velocity), $dA/d\bar{v}$ is zero. Since

$$\frac{dA}{d\bar{v}} = \frac{dA}{dx} \bigg/ \frac{d\bar{v}}{dx},$$

and $d\bar{v}/dx$ cannot be infinite, the value of dA/dx must be zero, indicating that the cross-sectional area must be a minimum, since the second derivative is positive, when the velocity reaches the velocity of sound, as in Fig. 14.1.

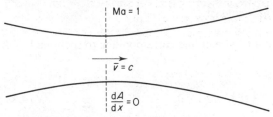

Figure 14.1 Convergent-divergent nozzle

The effect of a convergent–divergent nozzle on the flow of a compressible fluid will therefore depend upon the Mach number. On the upstream side and initially, supersonic flow will be decelerated towards Ma = 1 by the convergent section, while a subsonic flow will be accelerated towards Ma = 1. Once the throat (or minimum area of cross-section) is passed, the flow will accelerate for the supersonic case. For the subsonic case, if Ma = 1 is not attained in the throat, the flow will decelerate in the divergent section. It is only at a throat or minimum area of cross-section that the velocity can be sonic and the Mach number unity. To obtain supersonic steady flow of a compressible fluid flowing initially at subsonic velocity or contained at rest in a reservoir, the fluid must pass through a convergent–divergent nozzle.

14.3. Mass flow through a venturi meter

When a gas flows through a venturi meter (Fig. 14.2), the mass flow rate can be determined using the method explained in Section 6.7, but the form of Bernoulli's equation will be that obtained by integrating equation (14.5), which is

$$\int \frac{dp}{\rho} + \frac{\bar{v}^2}{2} + gz = \text{constant}. \tag{14.10}$$

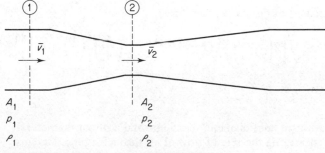

Figure 14.2 Flow through a venturi meter

For the short distance between the full bore and throat sections, conditions can be considered as adiabatic and the relationship between pressure and density will be $p/\rho^\gamma = \text{constant} = k$. Putting $\rho = (p/k)^{1/\gamma}$ in equation (14.10),

$$k^{1/\gamma} \int p^{-1/\gamma} \, dp + \tfrac{1}{2} \bar{v}^2 + gz = \text{constant.}$$

Integrating and putting $k = p/\rho^\gamma$,

$$\left(\frac{\gamma}{\gamma - 1}\right) \frac{p}{\rho} + \tfrac{1}{2} \bar{v}^2 + gz = \text{constant.}$$

or, for two points on a horizontal streamline corresponding to sections 1 and 2 in Fig. 14.2,

$$\left(\frac{\gamma}{\gamma - 1}\right) \left(\frac{p_1}{\rho_1} - \frac{p_2}{\rho_2}\right) + \tfrac{1}{2} (\bar{v}_1^2 - \bar{v}_2^2) = 0. \tag{14.11}$$

Also, since for adiabatic flow

$$p_1/\rho_1^\gamma = p_2/\rho_2^\gamma,$$

$$\rho_2 = \left(\frac{p_2}{p_1}\right)^{1/\gamma} \rho_1 \quad \text{and} \quad \frac{p_2}{\rho_2} = \frac{p_1}{\rho_1} \left(\frac{p_2}{p_1}\right)^{(\gamma - 1)/\gamma}$$

or, putting $p_2/p_1 = r$,

$$\frac{p_2}{\rho_2} = \frac{p_1 r^{(\gamma - 1)/\gamma}}{\rho_1}. \tag{14.12}$$

For continuity of flow by mass, $\rho_1 A_1 \bar{v}_1 = \rho_2 A_2 \bar{v}_2$,

$$\bar{v}_2 = \frac{A_1}{A_2} \left(\frac{\rho_1}{\rho_2}\right) \bar{v}_1 = \frac{A_1}{A_2} \left(\frac{1}{r}\right)^{1/\gamma} \bar{v}_1. \tag{14.13}$$

Substituting from equations (14.12) and (14.13) in equation (14.11),

$$\left(\frac{\gamma}{\gamma - 1}\right) \frac{p_1}{\rho_1} (1 - r^{(\gamma-1)/\gamma}) = \frac{\bar{v}_1^2}{2} \left\{ \left(\frac{A_1}{A_2}\right)^2 \left(\frac{1}{r}\right)^{2/\gamma} - 1 \right\},$$

$$\bar{v}_1 = \sqrt{\left\{ 2 \left(\frac{\gamma}{\gamma - 1}\right) \frac{p_1}{\rho_1} (1 - r^{(\gamma-1)/\gamma}) \Big/ \left[\left(\frac{A_1}{A_2}\right)^2 \left(\frac{1}{r}\right)^{2/\gamma} - 1 \right] \right\}}.$$

Mass flow per unit time, $\dot{m} = C_d A_1 \bar{v}_1 \rho_1$,

where C_d is a coefficient of discharge, therefore,

$$\dot{m} = C_d A_1 \rho_1 \sqrt{\left\{ 2 \left(\frac{\gamma}{\gamma - 1}\right) \frac{p_1}{\rho_1} (1 - r^{(\gamma-1)/\gamma}) \Big/ \left[\left(\frac{A_1}{A_2}\right)^2 \left(\frac{1}{r}\right)^{2/\gamma} - 1 \right] \right\}}.$$

Example 14.1

A venturi meter having an inlet diameter of 75 mm and a throat diameter of 25 mm is used for measuring the rate of flow of air through a pipe. Mercury U-tube gauges register pressures at the inlet and throat equivalent to 250 mm and 150 mm of mercury, respectively. Determine the volume of air flowing through the pipe per unit time in $m^3 s^{-1}$. Assume adiabatic conditions ($\gamma = 1.4$). The density of the air at the inlet is 1.6 kg m^{-3} and the barometric pressure is 760 mm of mercury.

Solution

$$p_1 = \frac{760 + 250}{1000} \times 13.6 \times 10^3 \times 9.81 = 134\ 750\ \text{N m}^{-2},$$

$$p_2 = \frac{760 + 150}{1000} \times 13.6 \times 10^3 \times 9.81 = 121\ 408\ \text{N m}^{-2},$$

$$\rho_1 = 1.6\ \text{kg m}^{-3}, \rho_2 = \rho_1 \left(\frac{p_2}{p_1}\right)^{1/\gamma} = 1.6 \left(\frac{121\ 408}{134\ 750}\right)^{1/1.4} = 1.485\ \text{kg m}^{-3}.$$

For continuous flow, $A_1 \bar{v}_1 \rho_1 = A_2 \bar{v}_2 \rho_2$. Therefore,

$$\bar{v}_2 = \frac{A_1}{A_2} \cdot \frac{\rho_1}{\rho_2} \cdot \bar{v}_1 = \frac{d_1^2}{d_2^2} \cdot \frac{\rho_1}{\rho_2} \cdot \bar{v}_1 = \left(\frac{75}{25}\right)^2 \left(\frac{1.6}{1.485}\right) \bar{v}_1 = 9.697 \bar{v}_1.$$

Applying Bernoulli's equation for adiabatic conditions,

$$\left(\frac{\gamma}{\gamma - 1}\right) \left(\frac{p_1}{\rho_1} - \frac{p_2}{\rho_2}\right) = \frac{\bar{v}_2^2 - \bar{v}_1^2}{2},$$

$$\frac{1.4}{0.4} \left(\frac{134\ 750}{1.6} - \frac{121\ 408}{1.485}\right) = \frac{\bar{v}_1^2}{2}(9.697^2 - 1),$$

$$\bar{v}_1 = 13.6\ \text{m s}^{-1}.$$

Volume of flow $= A_1 \bar{v}_1 = (\pi/4)(0.075)^2 \times 13.6$

$$= 0.060\ \text{m}^3 \text{s}^{-1}.$$

14.4. Mass flow from a reservoir through an orifice or convergent–divergent nozzle

Conditions for flow through an orifice or a nozzle, as shown in Fig. 14.3(a) and (b), can be taken as adiabatic and Bernoulli's equation in the form of equation (14.11), will apply. But, if the reservoir is large, $\bar{v}_1 = 0$ and $\bar{v}_2 = \bar{v}$ so that equation (14.11) reduces to

$$\left(\frac{\gamma}{\gamma - 1}\right) \left(\frac{p_0}{\rho_0} - \frac{p}{\rho}\right) = \frac{\bar{v}^2}{2}. \tag{14.14}$$

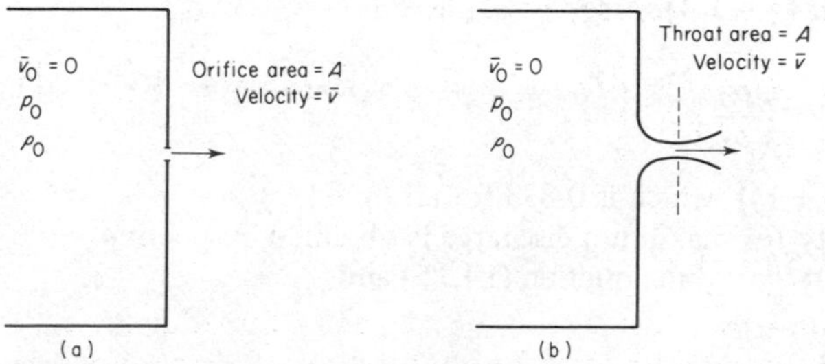

$\bar{v}_0 = 0$
p_0
ρ_0

Orifice area $= A$
Velocity $= \bar{v}$

$\bar{v}_0 = 0$
p_0
ρ_0

Throat area $= A$
Velocity $= \bar{v}$

(a) (b)

Figure 14.3 Mass flow from a large reservoir

Since, for adiabatic conditions, $p_0/\rho_0^\gamma = p/\rho^\gamma$,

$$\rho = \rho_0 r^{1/\gamma}$$

where $r = (p/p_0)$ and $p/\rho = p_0 r^{(\gamma-1)/\gamma}/\rho_0$. Substituting in equation (14.14),

$$\left(\frac{\gamma}{\gamma-1}\right) \frac{p_0}{\rho_0} (1 - r^{(\gamma-1)/\gamma}) = \frac{\bar{v}^2}{2}$$

$$\bar{v} = \sqrt{\left\{2\left(\frac{\gamma}{\gamma-1}\right) \frac{p_0}{\rho_0} (1 - r^{(\gamma-1)/\gamma})\right\}}. \tag{14.15}$$

Mass flow per unit time, $\dot{m} = A\bar{v}\rho = A\rho_0 r^{1/\gamma}\bar{v}$

$$= A\rho_0 \sqrt{\left\{2\left(\frac{\gamma}{\gamma-1}\right) \frac{p_0}{\rho_0} r^{2/\gamma} (1 - r^{(\gamma-1)/\gamma})\right\}}. \tag{14.16}$$

In practice, the actual discharge will be $C_d\dot{m}$, where C_d is a coefficient of discharge.

14.5. Conditions for maximum discharge from a reservoir through a convergent–divergent duct or orifice

For the throat section, where $p = p_t$ and $\rho = \rho_t$, it can be seen from equation (14.16) that for maximum discharge under given initial conditions p_0 and ρ_0, the quantity $r^{2/\gamma}(1 - r^{(\gamma-1)/\gamma})$ must be a maximum. This will occur for the value of r which makes

$$\frac{d}{dr}\{r^{2/\gamma}(1 - r^{(\gamma-1)/\gamma})\} = 0$$

$$(2/\gamma)r^{(2-\gamma)/\gamma} - \{(\gamma+1)/\gamma\}r^{1/\gamma} = 0$$

$$r^{(\gamma-1)/\gamma} = 2/(\gamma+1)$$

where $r = p_t/p_0$. Thus, for the throat section, the ratio of the throat pressure p_t to the upstream pressure p_0 is

$$p_t/p_0 = \{2/(\gamma+1)\}^{\gamma/(\gamma-1)}$$

which is 0·528 for air ($\gamma = 1·4$). For adiabatic conditions, $p/\rho^\gamma = $ constant and $\rho_t/\rho_0 = (p_t/p_0)^{1/\gamma}$. Thus,

$$\rho_t/\rho_0 = \{2/(\gamma+1)\}^{1/(\gamma-1)},$$

which is 0·634 for air ($\gamma = 1·4$). Also,

$$\frac{T_t}{T_0} = \frac{p_2}{p_1} \cdot \frac{\rho_1}{\rho_2} = \left(\frac{p_2}{p_1}\right)^{(\gamma-1)/\gamma},$$

so that $T_t/T_0 = \{2/(\gamma+1)\}$ which is 0·833 for air ($\gamma = 1·4$).

The throat velocity for maximum discharge is obtained by putting $r = p_t/p_0 = \{2/(\gamma+1)\}^{\gamma/(\gamma-1)}$ in equation (14.15) and

$$\frac{p_0}{\rho_0} = \frac{p_t}{\rho_t}\left(\frac{1}{r}\right)^{(\gamma-1)/\gamma}$$

Then, for maximum discharge,

$$\text{Throat velocity, } \bar{v} = \sqrt{\left\{ 2 \left(\frac{\gamma}{\gamma-1} \right) \frac{p_t}{\rho_t} \left[\left(\frac{1}{r} \right)^{(\gamma-1)/\gamma} - 1 \right] \right\}}$$

$$= \sqrt{\left\{ 2 \left(\frac{\gamma}{\gamma-1} \right) \frac{p_t}{\rho_t} \left(\frac{\gamma+1}{2} - 1 \right) \right\}}$$

$$= \sqrt{\left(\frac{\gamma p_t}{\rho_t} \right)} = c_t,$$

where c_t is the local velocity of sound in the throat or orifice.

14.6. The Laval nozzle

Named after its Swedish inventor, de Laval (1845-1913), this nozzle is designed to produce supersonic flow. It takes the form of a convergent-divergent nozzle with subsonic flow in the converging section, critical or trans-sonic conditions in the throat and supersonic flow in the diverging section.

If ρ, $\bar{v}$ and A are the density, velocity and cross-sectional area at any section of the nozzle and ρ_t, $\bar{v}_t$, A_t are the critical values at the throat, then, since the mass flow rate is the same at each cross-section,

$$\rho \bar{v} A = \rho_t \bar{v}_t A_t,$$

$$A/A_t = \rho_t \bar{v}_t / \rho \bar{v}. \tag{14.17}$$

The velocity at any point can be expressed in terms of the Mach number at that point and the local speed of sound,

$$\bar{v} = \text{Ma } c = \text{Ma} \sqrt{(\gamma R T)}$$

for adiabatic conditions.

At the throat, Ma = 1 and $T = T_t$, so that $\bar{v}_t = \sqrt{(\gamma R T_t)}$. Substituting in equation (14.17),

$$\frac{A}{A_t} = \frac{\rho_t}{\rho} \left(\frac{T_t}{T} \right) \frac{1}{\text{Ma}}. \tag{14.18}$$

Now, for isentropic flow from a large reservoir in which the conditions are given by p_0, ρ_0 and T_0 and $\bar{v}_0$ is zero, from Bernoulli's equation at any section of the nozzle,

$$\frac{\bar{v}^2}{2} = \left(\frac{\gamma}{\gamma-1} \right) R (T_0 - T).$$

Dividing by c^2, where $c = \sqrt{(\gamma R T)}$, the local velocity of sound, and rearranging,

$$\frac{\bar{v}^2}{c^2} = \text{Ma}^2 = \frac{2}{\gamma-1} \left(\frac{T_0}{T} - 1 \right),$$

$$\frac{T_0}{T} = 1 + \left(\frac{\gamma-1}{2} \right) \text{Ma}^2 \tag{14.19}$$

and since, for isentropic flow,

$$\frac{T_0}{T} = \left(\frac{p_0}{p}\right)^{(\gamma-1)/\gamma} = \left(\frac{\rho_0}{\rho}\right)^{(\gamma-1)},$$

$$\frac{p_0}{p} = \left(1 + \frac{\gamma-1}{2}\,\mathrm{Ma}^2\right)^{\gamma/(\gamma-1)} \qquad (14.20)$$

and

$$\frac{\rho_0}{\rho} = \left(1 + \frac{\gamma-1}{2}\,\mathrm{Ma}^2\right)^{1/(\gamma-1)}. \qquad (14.21)$$

we have,

$$\frac{\rho_t}{\rho} = \frac{\rho_t}{\rho_0} \times \frac{\rho_0}{\rho} = \left\{\frac{1 + [(\gamma-1)/2]\,\mathrm{Ma}^2}{(\gamma+1)/2}\right\}^{1/(\gamma-1)}$$

and

$$\frac{T_t}{T} = \frac{T_t}{T_0} \times \frac{T_0}{T} = \left\{\frac{1 + [(\gamma-1)/2]\,\mathrm{Ma}^2}{(\gamma+1)/2}\right\}.$$

Substituting these values in equation (14.18),

$$\frac{A}{A_t} = \frac{1}{\mathrm{Ma}} \left\{\frac{1 + [(\gamma-1)/2]\,\mathrm{Ma}^2}{(\gamma+1)/2}\right\}^{(\gamma+1)/2(\gamma-1)}, \qquad (14.22)$$

in which A is the area at the section at which the Mach number is Ma and A_t is the area at the throat. The value of A/A_t will never be less than unity and, for any given value of A/A_t, there will be two values of the Mach number, one less than unity and the other greater than unity.

The maximum mass flow rate $\dot{m}_{\max}$ will be given by

$$\dot{m}_{\max} = \rho_t A_t \bar{v}_t,$$

which can be expressed in terms of the reservoir conditions ρ_0, p_0 and T_0 since $\bar{v}_t = \sqrt{(\gamma R T_t)}$ and, from Section 14.5, $\rho_t/\rho_0 = [2/(\gamma+1)]^{1/(\gamma-1)}$ and $T_t/T_0 = 2/(\gamma+1)$, giving

$$\dot{m}_{\max} = \rho_0 \left(\frac{2}{\gamma+1}\right)^{1/(\gamma-1)} A_t \sqrt{\left(\frac{2\gamma R T_0}{\gamma-1}\right)}.$$

Putting $\rho_0 = p_0/R T_0$,

$$\dot{m}_{\max} = \frac{A_t p_0}{\sqrt{T_0}} \sqrt{\left\{\frac{\gamma}{R}\left(\frac{2}{\gamma+1}\right)^{(\gamma+1)/(\gamma-1)}\right\}}. \qquad (14.23)$$

If $\gamma = 1{\cdot}4$, this becomes

$$\dot{m}_{\max} = 0{\cdot}686\, A_t p_0 / \sqrt{(R T_0)}.$$

indicating that the mass flow varies linearly with A_t and p_0, but inversely as the square root of the absolute temperature.

Example 14.2

A supersonic wind-tunnel consists of a large reservoir containing gas under high pressure which is discharged through a convergent–divergent nozzle to a test section of constant cross-sectional area. The cross-sectional area of the throat of the nozzle is 500 mm^2 and the Mach number in the test section is 4. Calculate the cross-sectional area of the test section assuming $\gamma = 1 \cdot 4$.

Solution
From equation (14.22), putting $\gamma = 1 \cdot 4$ and $Ma = 4$,

$$\frac{A}{A_t} = \frac{1}{4} \left\{ \frac{1 + 0 \cdot 2 \times 4^2}{1 \cdot 2} \right\}^{2 \cdot 4/0 \cdot 8} = 10 \cdot 72,$$

Area of test section $= 10 \cdot 72 \times 500 = \textbf{5360 mm}^2$.

Equation (14.23) shows that the maximum mass flow is a function only of the reservoir conditions and the throat area, and cannot be affected by reducing the outlet pressure. Such a change could only be propagated upstream at the velocity of sound and, therefore, could not pass through the throat where the fluid velocity is sonic. Under these conditions, the nozzle is said to be choked.

When the mass flow rate is a maximum, the flow downstream of the throat can be either supersonic or subsonic depending on the downstream pressure. From equations (14.16) and (14.23) at any section of area A,

$$\dot{m} = A\rho_0 \sqrt{\left\{ 2 \frac{\gamma}{(\gamma - 1)} \frac{p_0}{\rho_0} \left(\frac{p}{p_0}\right)^{2/\gamma} \left[1 - \left(\frac{p}{p_0}\right)^{(\gamma-1)/\gamma}\right] \right\}}$$

$$= \frac{A_t p_0}{\sqrt{T_0}} \sqrt{\left\{ \frac{\gamma}{R} \left(\frac{2}{\gamma + 1}\right)^{(\gamma+1)/(\gamma-1)} \right\}},$$

Eliminating $\dot{m}$

$$\left(\frac{p}{p_0}\right)^{2/\gamma} \left\{1 - \left(\frac{p}{p_0}\right)^{(\gamma-1)/\gamma}\right\} = \frac{\gamma - 1}{2} \left(\frac{2}{\gamma + 1}\right)^{(\gamma+1)/(\gamma-1)} \left(\frac{A_t}{A}\right)^2.$$

Thus, for a given value of A_t/A, which must be less than unity, in the diverging duct there will be two possible values of p/p_0 between zero and unity, the upper value corresponding to subsonic flow and the lower to supersonic flow. For all other values of p/p_0 less than the upper value, isentropic flow is impossible and shock waves occur.

Flow through a nozzle can be classified by reference to the exit conditions. Figure 14.4 shows the variations of pressure and Mach number through a Laval nozzle. In the converging section, the pressure falls from the stagnation pressure p_0 at the entry, where the Mach number is small, to the value corresponding to the critical pressure ratio $p_t/p_0 = [2/(\gamma + 1)]^{\gamma/(\gamma-1)}$ at the throat, where $Ma = 1$. The pressure then continues to decrease until it reaches the exit and the Mach number increases correspondingly as shown. The exit pressure of the fluid issuing from the nozzle will not necessarily be the same as the back pressure of the fluid outside into which the nozzle is discharging. If the exit pressure is higher than the back pressure, the nozzle is *under-expanded*, since the flow could have expanded further, and, therefore, expansion waves form at the nozzle exit (Fig. 14.5(a)).

If the exit pressure is less than the back pressure, shock waves occur and the nozzle is said to be *over-expanded*. If the difference is small, oblique

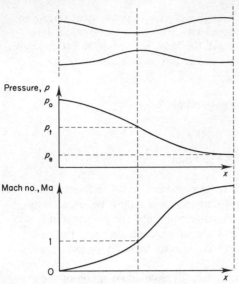

Figure 14.4   Variation of static pressure and Mach number in a Laval nozzle

shock waves form at the exit (Fig. 14.5(b)), but for larger differences of pressure, a normal shock wave will form in the nozzle (Fig. 14.5(c)). Figure 14.6 summarizes the variations of pressure ratio and Mach number.

14.7. Normal shock wave in a diffuser

When a normal shock wave forms in a diffuser, a supersonic flow is decelerated to a subsonic flow with consequent increase in stagnation temperature, pressure and density. An analogy can be drawn to the hydraulic jump discussed in Chapter 16.

Taking a control volume enclosing the wave (Fig. 14.6) of cross-sectional area A, for steady flow

$$\rho_1 \bar{v}_1 A = \rho_2 \bar{v}_2 A. \tag{14.24}$$

Putting $\rho_1 = p_1/RT_1$,

$$\bar{v}_1 = \text{Ma}_1 \sqrt{(\gamma R T_1)},$$

and if $\rho_2 = p_2/RT_2$,

$$\bar{v}_2 = \text{Ma}_2 \sqrt{(\gamma R T_2)},$$

where Ma_1 and Ma_2 are the Mach numbers upstream and downstream of the shock wave. Dividing through by A, equation (14.24) becomes

$$(p_1/RT_1)\text{Ma}_1 \sqrt{(\gamma R T_1)} = (p_2/RT_2)\,\text{Ma}_2 \sqrt{(\gamma R T_2)}. \tag{14.25}$$

From the momentum equation,

$$\left. \begin{aligned} (p_1 - p_2)A &= \dot{m}(\bar{v}_2 - \bar{v}_1), \\ p_1 - p_2 &= \rho_2 \bar{v}_2^2 - \rho_2 \bar{v}_1^2 \end{aligned} \right\} \tag{14.26}$$

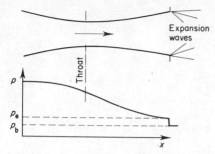

(a) Under-expanded, expansion waves

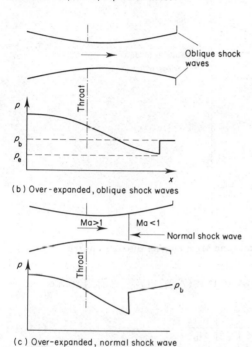

(b) Over-expanded, oblique shock waves

(c) Over-expanded, normal shock wave

Figure 14.5   Flow through a nozzle

Putting $\rho = p/RT$ and $\bar{v} = \mathrm{Ma}\,\sqrt{(\gamma RT)}$,

$$p_1 + (p_1/RT_1)\bar{v}_1^2 = p_2 + (p_2/RT_2)\bar{v}_2^2,$$

$$p_1(1 + \gamma\,\mathrm{Ma}_1^2) = p_2(1 + \gamma\,\mathrm{Ma}_2^2), \qquad (14.27)$$

which is the same as equation (11.17). Thus, the static pressure ratio across a shock wave is given by

$$p_2/p_1 = (1 + \gamma\,\mathrm{Ma}_1^2)/(1 + \gamma\,\mathrm{Ma}_2^2), \qquad (14.28)$$

and, since $\mathrm{Ma}_1 > 1$ and $\mathrm{Ma}_2 < 1$, static pressure increases across the shock wave $(p_2 > p_1)$.

Assuming adiabatic conditions, there will be no change in the stagnation

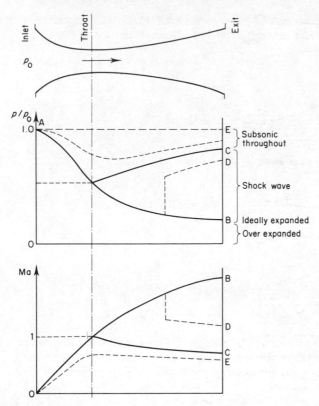

Figure 14.6   Variation of pressure ratio and Mach number through a nozzle

temperature across the shock wave, so that $(T_0)_1 = (T_0)_2 = T_0$. From equation (14.19),

$$\frac{T_2}{T_1} = \frac{T_2}{T_0} \times \frac{T_0}{T_1} = \frac{1 + [(\gamma - 1)/2] \; Ma_1^2}{1 + [(\gamma - 1)/2] \; Ma_2^2} \tag{14.29}$$

which corresponds with equation (11.16). Substituting in equation (14.25) from equations (14.28) and (14.29),

$$\frac{Ma_1}{1 + \gamma Ma_1^2} \left(1 + \frac{\gamma - 1}{2} \; Ma_1^2\right)^{1/2} = \frac{Ma_2}{1 + \gamma Ma_2^2} \left(1 + \frac{\acute{\gamma} - 1}{2} \; Ma_2^2\right)^{1/2}.$$

If this is solved for Ma_2 in terms of Ma_1, there are two solutions. The first is $Ma_1 = Ma_2$, which is the case for no shock wave. The second is

$$Ma_2^2 = \frac{(\gamma - 1) \; Ma_1^2 + 2}{2\gamma \; Ma_1^2 - (\gamma - 1)} \tag{14.30}$$

which corresponds with equation (11.19).

Example 14.3

Air is flowing through a duct and a normal shock wave is formed at a cross-section at which the Mach number is 2·0. If the upstream pressure and temperature are 105 bar and 15 °C, respectively, find the Mach number, pressure and temperature immediately downstream of the shock waves. Take $\gamma = 1\cdot4$.

Solution
From equation (14.30),

$$\mathrm{Ma}_2^2 = \frac{(1\cdot4 - 1)\, 2^2 + 2}{2 \times 1\cdot4 \times 2^2 - (1\cdot4 - 1)} = 0\cdot333,$$

$$\mathrm{Ma}_2 = \mathbf{0\cdot577}.$$

From equation (14.28),

$$p_2 = p_1 \frac{1 + 1\cdot4 \times 2^2}{1 + 1\cdot4 \times 0\cdot577^2} = 4\cdot5\, p_1 = 4\cdot5 \times 105$$

$$= \mathbf{473\ bar}.$$

From equation (14.29),

$$T_2 = T_1 \frac{1 + 0\cdot2 \times 2^2}{1 + 0\cdot2 \times 0\cdot577^2} = 1\cdot687\, T_1 = 1\cdot687 \times 288$$

$$= \mathbf{486\ K\ or\ 213\ ^{\circ}C}.$$

An insight into the nature of the changes in flow conditions which occur across a shock wave, where the area can be considered to be constant, can be obtained by examining the relationship graphically. If the upstream conditions

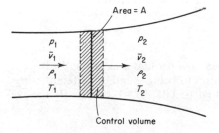

Figure 14.7   Normal shock wave

(Fig. 14.7) are taken as fixed, curves can be drawn showing all the corresponding possible conditions downstream of the shock wave. It is possible to draw one set of curves, known as *Fanno lines*, in which each curve represents conditions which, for a particular mass flow, satisfy the continuity and energy equations, which are

$$\text{Mass flow per unit area, } G = \dot{m}/A = \rho\bar{v} = \text{constant} \tag{14.31}$$

and

$$\text{Stagnation enthalpy, } H_0 = H + \bar{v}^2/2 = \text{constant.} \tag{14.32}$$

It is instructive to plot the Fanno lines as a graph of enthalpy H against entropy S. The entropy equation for a perfect gas is

$$S - S_1 = c_v \log_e \left\{ \frac{p}{p_1} \left(\frac{\rho_1}{\rho} \right)^\gamma \right\}$$ (14.33)

and

$$H = c_p T = c_p p / R\rho.$$ (14.34)

Combining equations (14.31), (14.32), (14.33) and (14.34), we have

$$S = S_1 + c_v \log_e \{ H(H_0 - H)^{(\gamma-1)/2} \} + \text{constant},$$ (14.35)

where the constant is determined by the mass flow per unit area G and the upstream conditions.

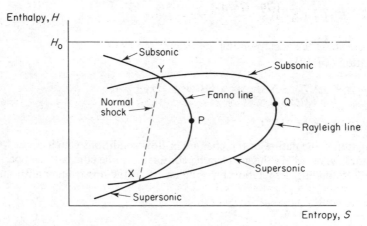

Figure 14.8   Fanno and Rayleigh lines

This is shown plotted for a given mass flow per unit area in Fig. 14.8. Maximum entropy occurs at P, the conditions being found by differentiating equation (14.35) with respect to H and putting $dS/dH = 0$ for $H = H_P$ the value at the point P.

$$\frac{dS}{dH} = \frac{1}{H_P} - \frac{(\gamma-1)}{2} \frac{1}{H_0 - H_P} = 0,$$

$$H_P = \{ 2/(\gamma+1) \} H_0,$$

$$H_0 = \{ (\gamma+1)/2 \} H_P = H_P + v_P^2/2,$$

$$v_P^2 = (\gamma-1) H_P = (\gamma-1) c_p T_P = (\gamma-1) \{ \gamma R/(\gamma-1) \} T_P$$

$$v_P = \sqrt{(\gamma R T_P)} = \text{Velocity of sound}.$$

Thus, the Fanno line shows that maximum entropy occurs at the point P for which the Mach number is 1 and conditions are sonic. If $H > H_P$, the flow is subsonic and, if $H < H_P$, the flow is supersonic. For a shock wave, the conditions before and after the shock must both lie on the Fanno line for the mass flow and area of the section at which the shock occurs.

To determine the position of these points, we now consider the requirement that both the continuity and momentum equations must be satisfied for a given mass flow. A curve known as the *Rayleigh line* can be drawn showing all the points on the *H-S* diagram which satisfy these requirements. The momentum equation is

$$p_1 - p = (m/A)(\bar{v} - \bar{v}_1)$$

and, combining this with equation (14.31),

$$p + \rho\bar{v}^2 = p_1 + \rho_1\bar{v}_1^2 = \text{constant}$$

or $p + G^2/\rho = \text{constant} = B.$

Substituting for p in equation (14.33),

$$S = S_1 + c_v \log_e \left\{ \frac{(B - G^2/\rho)}{\rho^\gamma} \right\} + \text{constant}, \tag{14.36}$$

where the constant depends upon the upstream conditions.

Now,

$$H = c_p T = \frac{c_p}{R} \cdot \frac{1}{\rho} \left(B - \frac{G^2}{\rho} \right). \tag{14.37}$$

From equations (14.36) and (14.37), the Rayleigh line can be plotted (Fig. 14.8) for the given mass flow.

The conditions corresponding to the point Q for maximum entropy are found by differentiating equations (14.36) and (14.37) to give $dS/d\rho$ and $dH/d\rho$, then, dividing and equating to zero,

$$\frac{dS}{dH} = \frac{dS}{d\rho} \cdot \frac{d\rho}{dH} = \frac{c_v}{c_p} R\rho_Q \frac{[G^2/\{\rho_Q(B - G^2/\rho_Q)\}] - \gamma}{2G^2/\rho_Q - B} = 0.$$

If the denominator is not zero, this gives

$$\frac{G^2}{\rho_Q(B - G^2/\rho_Q)} - \gamma = 0$$

or, substituting for G in terms of $\bar{v}$ from equation (14.31),

$$\bar{v}_Q = \sqrt{(\gamma p_Q/\rho_Q)} = \text{Velocity of sound}.$$

Thus, sonic conditions occur at the point of maximum entropy. The upper limb of the curve corresponds to subsonic flow and the lower limb to supersonic flow.

For a shock wave the continuity, energy and momentum equations must be satisfied and it is, therefore, clear that conditions before and after the shock wave must lie on both the appropriate Fanno and Rayleigh lines, namely points X and Y on Fig. 14.8. The value of the entropy S will be greater at Y than at X since the shock occurs from supersonic to subsonic conditions.

Note an alternative plot for Fanno and Rayleigh lines is a *T-S* diagram, but, since $H = c_p T$, the diagrams are similar.

14.8. Compressible flow in a duct with friction under adiabatic conditions: Fanno flow

The flow of a liquid through a duct against resistance due to friction was discussed in Chapter 8. The analysis of the flow of a compressible fluid under similar circumstances is, fundamentally, the same, but is complicated by the interdependence of density, pressure and temperature, all of which will change from point to point along the length of the duct. It is necessary to make some assumptions that will simplify the problem.

One such assumption is that the duct or pipe is perfectly insulated and that conditions in the fluid are adiabatic. This is known as Fanno flow. For steady flow in a duct of constant cross-sectional area, the continuity equation can be written $\rho\bar{v}$ = constant, where ρ is the density and $\bar{v}$ the velocity at any cross-section. Differentiating,

$$\frac{d\bar{v}}{v} + \frac{d\rho}{\rho} = 0. \tag{14.38}$$

The energy equation will be $H + \bar{v}^2/2$ = constant where H is the enthalpy and conditions are adiabatic. Now, $H = c_p T$ and, for a perfect gas, it can be shown that $c_p = \gamma R/(\gamma - 1)$, so that the energy equation reduces to

$$\frac{\gamma R}{\gamma - 1} T + \frac{\bar{v}^2}{2} = \text{constant}$$

or differentiating,

$$\frac{\gamma R}{\gamma - 1} dT + \bar{v}\, d\bar{v} = 0. \tag{14.39}$$

The force momentum equation is derived from consideration of the control volume shown in Fig. 14.9. Neglecting gravitational forces, which are small,

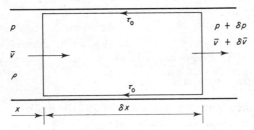

Figure 14.9   Friction in a duct

and assuming a shear stress τ_0 at the wall of the pipe, equate the forces in the direction of motion to the rate of change of momentum in that direction across the system boundaries:

$$A\{p - (p + \delta p)\} - \tau_0 P\, dx = \rho\bar{v}A\{(\bar{v} + \delta\bar{v}) - \bar{v}\}, \tag{14.40}$$

where P is the perimeter of the duct cross-section.

From equation (8.25), $\tau_0 = f\rho\bar{v}^2/2$, where f is the resistance coefficient in the Darcy equation

$$h_{\mathrm{f}} = \frac{fl}{m} \cdot \frac{\bar{v}^2}{2g}.$$

Simplifying, equation (14.40) becomes

$$- \mathrm{d}p - f\rho\frac{\bar{v}^2}{2} \cdot \frac{P}{A} = \rho\bar{v}\,\mathrm{d}\bar{v} \tag{14.41}$$

or, putting $(A/P) = m$,

$$\rho\bar{v}\,\mathrm{d}\bar{v} + \mathrm{d}p + \frac{f\rho}{m} \cdot \frac{\bar{v}^2}{2}\,\mathrm{d}x = 0, \tag{14.42}$$

which is the general equation for flow with friction in ducts.

Since both velocity and temperature will be changing as the gas flows along the duct, the value of the Mach number will also vary from point to point. The amount of change can be found by combining the equation of state with the momentum equation. From equation (14.1), $p/\rho = RT = c^2/\gamma$, where c is the sonic velocity, which will be $\sqrt{(\gamma RT)}$ for adiabatic conditions. Putting $\rho = \gamma p/c^2$ and dividing by p, equation (14.42) becomes

$$\gamma\frac{\bar{v}^2}{c^2} \cdot \frac{\mathrm{d}\bar{v}}{\bar{v}} + \frac{\mathrm{d}p}{p} + \frac{f}{m} \cdot \frac{\gamma}{c^2} \cdot \frac{\bar{v}^2}{2}\,\mathrm{d}x = 0$$

or

$$\gamma\,\mathrm{Ma}^2\frac{\mathrm{d}\bar{v}}{\bar{v}} + \frac{\mathrm{d}p}{p} + \gamma\frac{f}{m} \cdot \frac{\mathrm{Ma}^2}{2}\,\mathrm{d}x = 0, \tag{14.43}$$

where Ma is the Mach number $(\bar{v}/c)$,

Differentiating equation (14.1), which is the equation of state,

$$\frac{\mathrm{d}p}{p} = \frac{\mathrm{d}\rho}{\rho} + \frac{\mathrm{d}T}{T}. \tag{14.44}$$

Substituting in equation (14.44) for ρ in terms of $\bar{v}$ from equation (14.38) and for T in terms of $\bar{v}$ from equation (14.39), we have

$$\frac{\mathrm{d}p}{p} = -\frac{\mathrm{d}\bar{v}}{\bar{v}} - (\gamma - 1)\,\mathrm{Ma}^2\frac{\mathrm{d}\bar{v}}{\bar{v}}. \tag{14.45}$$

Putting this value of $(\mathrm{d}p/p)$ in equation (14.43),

$$(\mathrm{Ma}^2 - 1)\frac{\mathrm{d}\bar{v}}{\bar{v}} + \frac{\gamma f}{m}\frac{\mathrm{Ma}^2}{2}\,\mathrm{d}x = 0. \tag{14.46}$$

Now, the Mach number is defined as $\mathrm{Ma} = \bar{v}/c = \bar{v}/\sqrt{(\gamma RT)}$ which, on differentiation, gives

$$\frac{\mathrm{dMa}}{\mathrm{Ma}} = \frac{\mathrm{d}\bar{v}}{\bar{v}} - \frac{1}{2}\frac{\mathrm{d}T}{T}. \tag{14.47}$$

From equation (14.39),

$$\frac{\mathrm{d}T}{T} = -\frac{\bar{v}\,\mathrm{d}\bar{v}\,(\gamma - 1)}{\gamma RT} = (\gamma - 1)\,\mathrm{Ma}^2\frac{\mathrm{d}\bar{v}}{\bar{v}},$$

so that equation (14.47) becomes

$$\frac{d\text{Ma}}{\text{Ma}} = \frac{d\bar{v}}{\bar{v}} \left\{ 1 - \frac{(\gamma - 1)}{2} \text{Ma}^2 \right\},$$ (14.48)

from which $d\bar{v}/\bar{v}$ can be eliminated from equation (14.46) to obtain

$$\frac{(1 - \text{Ma}^2)\, d\text{Ma}}{\text{Ma}^3 \{1 + [(\gamma - 1)/2]\, \text{Ma}^2\}} = \frac{\gamma f}{2m}\, dx.$$ (14.49)

From equation (14.49), it can be seen that if flow is subsonic and $\text{Ma} < 1$, then $(d\text{Ma}/dx) > 0$ and the Mach number increases with distance along the duct.

If the flow is supersonic, $\text{Ma} > 1$ and $(d\text{Ma}/dx) < 0$, the Mach number will decrease along the duct. Thus, the effect of pipe friction is always to cause the Mach number to approach unity. It is impossible for the Mach number of a compressible flow to change from subsonic to supersonic in a duct of constant cross-section and, consequently, the maximum Mach number that can be attained by an initially subsonic flow is unity, reached at the exit from the duct. Conversely, supersonic flow may only become subsonic due to the occurrence of shock waves in the duct.

In order to integrate equation (14.49) to obtain values of Ma against length, some reasonable assumption must be made about the variation of friction factor f along the duct. The continuity equation states that $\rho \bar{v}$ is constant along the length of the duct therefore, any variation of Reynolds number, which governs the friction factor f, can only occur as a result of a change in the viscosity of the fluid. Viscosity is dependent on temperature. For example, a change in temperature of 20 per cent, which can frequently occur in compressible flow, would produce a 10 per cent change in viscosity in the case of air. However, the resulting 10 per cent change in Reynolds number would normally give rise to a much smaller change in the friction factor f and, so, it is reasonable to assume a constant value for f when integrating equation (14.49). This value is equal to the average value of f along the duct. Reducing the left-hand side of equation (14.49) to partial fractions yields

$$\left(\frac{1}{\text{Ma}^3} + \frac{\gamma + 1}{2\text{Ma}} + \frac{(\gamma + 1)(\gamma - 1)\,\text{Ma}}{4(1 + [(\gamma - 1)/2]\text{Ma}^2)} \right) d\text{Ma} = \frac{\gamma f}{2m}\, dx.$$ (14.50)

Integrating both sides,

$$-\frac{1}{2\text{Ma}^2} - \frac{\gamma + 1}{2} \log_e \text{Ma} + \frac{\gamma + 1}{4} \log_e \left\{ 1 + \frac{(\gamma - 1)}{2} \text{Ma}^2 \right\} = \frac{\gamma f x}{2m} + C.$$ (14.51)

To determine the constant of integration C, let x_1 be the distance along the pipe at which the Mach number becomes unity. Then,

$$C = -\frac{\gamma f x_1}{2m} - \frac{1}{2} + \frac{(\gamma + 1)}{4} \log_e \frac{(\gamma + 1)}{2}.$$ (14.52)

Substitution of C into equation (14.51) yields an expression linking Mach number to distance along the duct

$$\frac{1 - \text{Ma}^2}{\gamma \text{Ma}^2} + \frac{\gamma + 1}{2\gamma} \log_e \left\{ \frac{(\gamma + 1) \text{Ma}^2}{2 + (\gamma - 1) \text{Ma}^2} \right\} = \frac{f(x_2 - x)}{m}. \qquad (14.53)$$

Example 14.4

Air (for which $\gamma = 1 \cdot 4$) flows along a circular pipe with a diameter d of 50 mm. Assuming that conditions are adiabatic and that the Mach number at the entrance to the pipe is $0 \cdot 2$, calculate the distance from the entrance of the pipe to the section at which the Mach number will be (a) $1 \cdot 0$, (b) $0 \cdot 6$. Take $f = 0 \cdot 003$ 75.

Solution

(a) The distance x_1 at which the Mach number is unity can be found from equation (14.53), since $\gamma = 1 \cdot 4$, $m = d/4$, $f = 0 \cdot 003$ 75, and when $x = 0$, $M = 0 \cdot 2$. Substituting these values,

$$\frac{1 - 0 \cdot 2^2}{1 \cdot 4 \times 0 \cdot 2^2} + \frac{2 \cdot 4}{2 \cdot 8} \log_e \frac{2 \cdot 4 \times 0 \cdot 04}{2 + 0 \cdot 4 \times 0 \cdot 04} = \frac{4 \times 0 \cdot 003 \ 75 \ (x_1 - 0)}{d},$$

$$x_1 = 968 \cdot 9 d. \qquad (14.54)$$

Putting $d = 50$ mm $= 0 \cdot 05$ m,

$$x = 48 \cdot 44 \text{ m.}$$

(b) The distance from the entrance, where Ma $= 0 \cdot 2$ and $x = x_{0 \cdot 2} = 0$, to the section at which Ma $= 0 \cdot 6$ and $x = x_{0 \cdot 6}$ cannot be found directly. First, find the distance from $x_{0 \cdot 6}$ to x_1 from equation (14.53)

$$\frac{1 - 0 \cdot 6^2}{1 \cdot 4 \times 0 \cdot 6^2} + \frac{2 \cdot 4}{2 \cdot 8} \log_e \frac{2 \cdot 4 \times 0 \cdot 36}{2 + 0 \cdot 4 \times 0 \cdot 36} = \frac{4 \times 0 \cdot 003 \ 75 \ (x_1 - x_{0 \cdot 6})}{d},$$

$$x_1 - x_{0 \cdot 6} = 32 \cdot 7 d.$$

Substituting for x_1 from equation (14.54),

$$x_{0 \cdot 6} = (968 \cdot 9 - 32 \cdot 7)d = 936 \cdot 2 d.$$

Putting $d = 0 \cdot 05$ m,

$$x_{0 \cdot 6} = 46 \cdot 81 \text{ m.}$$

The variation of pressure along the length of the duct can also be obtained from equation (14.45)

$$\frac{dp}{p} = -\frac{d\bar{v}}{\bar{v}} \{ 1 + (\gamma - 1) \text{ Ma}^2 \}.$$

Substituting this value in equation (14.48) and rearranging,

$$\frac{dp}{p} = -\frac{d\text{Ma}}{\text{Ma}} \left\{ \frac{1 + (\gamma - 1) \text{ Ma}^2}{1 + [(\gamma - 1)/2] \text{ Ma}^2} \right\}, \qquad (14.55)$$

from which it can be seen that $(dp/d\text{Ma})$ is negative, indicating that the pressure decreases with increasing Mach number. Thus, the observed pressure decrease for subsonic flow along a duct corresponds to an increase of Mach number. From equation (14.55),

$$\frac{dp}{p} = \left\{ -\frac{1}{Ma} - \frac{[(\gamma - 1)/2]\ Ma}{1 + [(\gamma - 1)/2]\ Ma^2} \right\} d\,Ma.$$

Integrating,

$$\log_e p = \log_e Ma - \tfrac{1}{2} \log_e \left\{ 1 + \left(\frac{\gamma - 1}{2}\right) Ma^2 \right\} + C. \qquad (14.56)$$

The constant of integration C can be evaluated in terms of the pressure p_1 corresponding to $Ma = 1$, giving

$$\log_e p_1 = -\tfrac{1}{2} \log_e \left(\frac{\gamma + 1}{2}\right) + C$$

and so, from equation (14.56),

$$\frac{p}{p_1} = \frac{1}{Ma} \left\{ \frac{\gamma + 1}{2 + (\gamma - 1)Ma^2} \right\}, \qquad (14.57)$$

where p is the pressure corresponding to Mach number Ma.

14.9. Isothermal flow of a compressible fluid in a pipeline

When a gas flows at low velocities in a long duct through which heat transfer can occur readily, conditions may be approximately isothermal so that the temperature T can be considered constant. From the equation of state, $p/\rho = $ constant $= p_1/\rho_1$, where p_1 and ρ_1 are the values of pressure and density at a given point and p and ρ the corresponding values at any other point, or

$$\rho = (p/p_1)\rho_1. \qquad (14.58)$$

For a duct of constant cross-sectional area A, the continuity equation $\rho A \bar{v} = \rho_1 A_1 \bar{v}_1 = $ constant, so that $\rho \bar{v} = \rho_1 \bar{v}_1$ and, hence,

$$\bar{v} = \bar{v}_1\ (\rho/\rho_1).$$

Substituting for ρ from equation (14.58), the velocity at any section is

$$\bar{v} = \bar{v}_1\ (p_1/p). \qquad (14.59)$$

For flow with frictional resistance, from equation (14.42),

$$\bar{v}\,d\bar{v} + \frac{dp}{\rho} + \frac{f}{m} \cdot \frac{\bar{v}^2}{2}\,dx = 0. \qquad (14.60)$$

By integration, the pressure drop along the duct can be determined as follows.

14.9.1. Approximate solution neglecting velocity change

Under these conditions, $d\bar{v} = 0$ and equation (14.60) becomes

$$\frac{dp}{\rho} + \frac{f}{m}\,\frac{\bar{v}^2}{2}\,dx = 0.$$

Substituting for ρ and $\bar{v}$ from equations (14.58) and (14.59),

$$\frac{dp}{p}\frac{p_1}{\rho_1} + \frac{f}{2m}\bar{v}_1^2\left(\frac{p_1}{p}\right)^2 dx = 0,$$

$$p\,dp = -\frac{f}{2m}\rho_1 p_1 \bar{v}_1^2\,dx. \qquad (14.61)$$

Integrating and putting $\rho_1 = p_1/RT$,

$$p^2 - p_1^2 = -fp_1^2\bar{v}_1^2(x - x_1)/mRT,$$

$$p = p_1\sqrt{\{1 - f(x - x_1)\bar{v}_1^2/mRT\}}, \qquad (14.62)$$

where p is the pressure at a distance x downstream from the point at which the pressure is p_1 and m is the hydraulic mean depth.

14.9.2. Solution allowing for velocity changes

Since for continuity of flow $\rho\bar{v}$ = constant and for a perfect gas $p = \rho \times$ constant, for isothermal conditions,

$$\frac{d\bar{v}}{\bar{v}} = -\frac{d\rho}{\rho} = -\frac{dp}{p}. \qquad (14.63)$$

Substituting for dp in equation (14.60) and dividing by v^2,

$$\frac{f\,dx}{2m} = \frac{p}{\rho}\cdot\frac{d\bar{v}}{\bar{v}^3} - \frac{d\bar{v}}{\bar{v}}. \qquad (14.64)$$

Now, as T is constant and as viscosity may be assumed to be a function only of T at normal pressures, Reynolds number is constant and, for uniform roughness along the duct, the friction factor f may also be treated as constant.

As $p/\rho = RT$ = constant, equation (14.64) may be integrated directly:

$$\frac{f}{2m}(x_1 - x) = \frac{RT}{2}\left(\frac{1}{\bar{v}_1^2} - \frac{1}{\bar{v}^2}\right) - \log_e\frac{\bar{v}}{\bar{v}_1},$$

$$\frac{f}{2m}(x_1 - x) = \frac{1}{2\gamma}\left(\frac{1}{Ma_1^2} - \frac{1}{Ma^2}\right) - \log_e\frac{Ma}{Ma_1}, \qquad (14.65)$$

where suffix 1 refers to conditions at a known point in the duct.

There is a limitation arising from these results in respect of the maximum attainable Mach number for a subsonic isothermal flow. Substituting equation (14.63) into equation (14.42),

$$\frac{dp}{dx} = \left(\frac{f\rho v^2}{2m}\right)\Big/\left(\frac{\rho\bar{v}^2}{p} - 1\right) = \frac{f\rho\bar{v}^2}{2m(\gamma\,Ma^2 - 1)}. \qquad (14.66)$$

Equation (14.66) shows that, for $Ma < (1/\gamma)^{1/2}$, the value of dp/dx is negative but, when $Ma = (1/\gamma)^{1/2}$, the value of dp/dx becomes infinite and discontinuities arise in both pressure and velocity variations. Thus, there is a maximum flow length for isothermal flow, obtained by putting $Ma = (1/\gamma)^{1/2}$ in equation (14.65), comparable to the limiting flow length for adiabatic flow which was shown to be limited by $Ma = 1$. In practice, $Ma = (1/\gamma)^{1/2}$ is never achieved, as dp/dx would have to be infinite. If the actual length of the pipe exceeds the limiting flow length, choking would occur and the rate of flow would adjust until conditions were such that the value Ma = $(1/\gamma)^{1/2}$ would not be reached until the end of the actual pipe.

From equation (14.65), substituting for $\bar{v}$ from equation (14.59),

$$\frac{dp}{dx} = \left(\frac{f}{2m} \cdot \frac{p}{p_1} \cdot \frac{\rho_1 \bar{v}_1^2 p_1^2}{p^2}\right) \bigg/ \left(\frac{p}{p_1}\frac{\rho_1 \bar{v}_1^2 p_1^2}{p^3} - 1\right)$$

$$= \frac{f}{2m} \rho_1 \bar{v}_1^2 p_1^2 \frac{p}{(\rho_1 \bar{v}_1^2 p_1 - p^2)},$$

$$p \, dp - \rho_1 \bar{v}_1^2 p_1 \frac{dp}{p} = -\frac{f}{2m}\rho_1 \bar{v}_1^2 p_1^2 \, dx. \tag{14.67}$$

This equation can be compared to the approximate solution of equation (14.61), to which it will reduce if $\rho_1 \bar{v}_1^2 p_1 \, (dp/p)$ is small. Integrating equation (14.67) and putting $\rho_1 = p_1/RT$,

$$\frac{p^2 - p_1^2}{2} - \bar{v}_1^2 \frac{p_1^2}{RT}\log_e\left(\frac{p}{p_1}\right) = -\frac{fp_1^2 \bar{v}_1^2 (x - x_1)}{2mRT}. \tag{14.68}$$

Example 14.5

Air flows along a pipe 100 mm in diameter under isothermal conditions. At the entrance the pressure is 200 kN m^{-2}, the volume rate of flow is 28 m^3 min^{-1} and the temperature is constant at 15 °C. If the pipe is 60 m long and the value of f is 0·004, calculate the pressure at the outlet assuming $R = 287$ J kg^{-1} K^{-1} (a) neglecting changes of velocity, (b) allowing for velocity changes.

Solution

(a) At entrance,

$$\bar{v}_1 = Q/(\pi/4d^2) = \frac{28}{(\pi/4) \times 0\cdot01 \times 60} = 59\cdot40 \text{ m s}^{-1}.$$

Substituting in equation (14.62), $p_1 = 200 \times 10^3$ N m^{-2}, $f = 0\cdot004$, $(x - x_1) = 60$ m, $m = d/4 = 0\cdot025$ m, $R = 287$ J kg^{-1} K^{-1}, $T = 15$ °C $= 288$ K,

$$p = 200 \times 10^3 \sqrt{\left(1 - \frac{0\cdot004 \times 60 \times (59\cdot4)^2}{0\cdot025 \times 287 \times 288}\right)} \text{ N m}^{-2}$$

$$= 153\cdot6 \text{ kN m}^{-2}.$$

(b) In this case use equation (14.68), which can be solved by trial. Substituting the numerical values:

$$\frac{p^2}{2} - \frac{(200 \times 10^3)^2}{2} - 59\cdot4^2 \frac{(200 \times 10^3)^2}{287 \times 288}\log_e\left(\frac{p}{200 \times 10^3}\right)$$

$$= -\frac{0\cdot004(200 \times 10^3)^2 \times 59\cdot4^2 \times 60}{2 \times 0\cdot025 \times 287 \times 288}$$

$$\frac{1}{2}\left(\frac{p}{10^3}\right)^2 - 20\,000 - 1707\log_e\left(\frac{p}{200 \times 10^3}\right) = -8196.$$

Try $p = 152$ kN m^{-2}:

> Left-hand side = $11\ 552 - 20\ 000 + 468 = 7980$.

Try $p = 150$ kN m^{-2}:

> Left-hand side = $11\ 250 - 20\ 000 + 491 = -8259$.

Try $p = 150 \cdot 5$ kN m^{-2}.

> Left-hand side = $11\ 325 - 20\ 000 + 485 = -8190$,

which is approximately equal to the right-hand side. Therefore,

> Pressure at outlet = **$150 \cdot 5$ kN m^{-2}**.

EXERCISES 14

14.1 Air at 5 bar and 560 K is expanded in steady flow in a horizontal convergent–divergent duct to an exit velocity of 640 m/s. The walls of the duct are heated so as to keep the temperature drop to one half the value of an isentropic expansion to the same velocity and pressure from the same initial conditions. Calculate, assuming negligible initial velocity (a) the heat supplied, (b) the final temperature, (c) the mass flow rate per square metre of exit area.
[102·4 kJ/kg, 458 K, 588 kg/m^{2}s]

14.2 A venturi meter which has a throat diameter of 25 mm is installed in a horizontal pipeline 75 mm diameter conveying air. The pressure at the inlet to the metre is 133·3 kN/m^2 and that at the throat is 100 kN/m^2, both pressures being absolute. The temperature of the air at the inlet is 15 °C. Assuming isentropic flow, determine the mass flow rate in kg/s. For air $\gamma = 1·4$ and $R = 287$ J/kgK.
[0·14 kg/s]

14.3 A sharp-edged circular orifice of 45 mm diameter is used to measure the flow of air from the atmosphere into a large tank. Barometric pressure is 735 mm of mercury, temperature 17 °C and the difference of pressure between the atmosphere and the inside of the tank is equivalent to a head of 20 mm of water. Determine the mass of air in kg/min passing into the tank if the coefficient of discharge for the orifice is 0·6. Take $R = 287$ J/kgK.
[1·2 kg/min]

14.4 A convergent-divergent nozzle is fitted into the side of a reservoir. The temperature and pressure are kept constant at 80°C and 10 bar. It is found that the thrust exerted by the jet of air issuing from the nozzle is 11·12 kN. Assuming that the expansion of air in the nozzle is isentropic and takes place down to atmospheric pressure which may be assumed to be 1 bar, calculate (a) the nozzle throat and exit areas, (b) the Mach number of the jet issuing from the nozzle. For air $R = 287$ J/kg/K and $\gamma = 1·4$.
[88·2 cm^2, 171 cm^2, 2·16]

14.5 Calculate the maximum mass flow possible through a frictionless, heat insulated, convergent nozzle if the entry or stagnation conditions are 5 bar and 15 °C and the throat area is 6·5 cm^2. Also calculate the temperature of the air at the throat. Take $c_p = 1·00$ kJ/kg K, $\gamma = 1·4$.
[0·765 kg/s, 241 K]

14.6 Air flows through a converging conical nozzle. The length of the

nozzle is 0·350 m and its diameter changes linearly from 0·102 m at the entry
to 0·051 m at the throat. At a section half-way along the nozzle a Pitot–static
tube is mounted on the centreline. The Pitot–static tube shows a dynamic
pressure of 614 N/m^2. Calculate the velocity of flow at this section and also
the longitudinal gradient of static pressure. Assume that the density of the air
is constant at 1·42 kg/m^3.
[29·4 m/s, 5·35 kN/m^3]

14.7   A convergent–divergent nozzle is supplied with compressed air from a
reservoir at a pressure of 1 MN/m^2 abs. The throat area is 8 cm^2 and the
nozzle expands to a parallel section of area 13·5 cm^2 before discharging into
a region where the absolute pressure is 100 kN/m^2. Calculate the exit Mach
number.
[2·0]

14.8   Air initially at standard temperature and pressure flows into an
evacuated tank through a convergent nozzle contracting to a diameter of
4 cm. What pressure must be maintained in the tank to produce a sonic jet.
What is the mass flow.
[53·5 kN/m^2, 0·3 kg/s]

14.9.   Carbon dioxide discharges through an orifice 10 mm diameter from a
large container in which the gas has a temperature of 15 °C and is at a pressure
of 7 atmospheres (abs). Calculate the speed of the jet and the mass flow rate.
Take atmospheric pressure as 101·3 kN/m^2 and $\gamma = 1·3$ for carbon dioxide.
[248 m/s, 0·16 kg/s]

14.10   A converging–diverging nozzle fed from a reservoir has an exit area
3 times the throat area. What is the ratio of exit pressure to reservoir pressure
for isentropic flow of air if the Mach number at exit is greater than unity.
[$p/p_0 = 0·047$]

14.11   An air stream at Mach 2·0, pressure 60 kN/m^2 and temperature
217 K enters a diverging channel with a ratio of exit area to inlet area of 3·0.
Determine the back pressure necessary to produce a normal shock wave in
the channel at an area equal to twice the inlet area. Assume isentropic flow
except for the normal shock.

14.12   Show that if the Mach number upstream of a normal shock wave in
air is large the density ratio across the shock wave is 6 and the downstream
Mach number is 0·378.

14.13   A shock wave occurs in a duct carrying air where the upstream Mach
number is 2 and the upstream temperature and pressure are 15 °C and 20
kN/m^2 abs. Calculate the Mach number, pressure, temperature and velocity
after the shock wave.
[0·577, 90 kN/m^2, 213 °C, 255 m/s]

14.14   Air flows through a parallel passage in which a shock wave is formed
if the subscripts 1 and 2 refer to conditions just before and just after the wave,
show that

$$\frac{u_1^2}{2} + \frac{\gamma}{\gamma-1}\frac{p_1}{\rho_1} = \frac{u_2^2}{2} + \frac{\gamma}{\gamma-1}\left(\frac{u_2}{u_1}\right)\left(u_1^2 - u_1 u_2 + \frac{p_1}{\rho_1}\right)$$

If $p_1 = 690$ kN/m^2 abs, $\rho_1 = 5·45$ kg/m^3 and $u_1 = 450$ m/s, calculate the
values of p_2 and u_2 immediately after the shock wave given that $\gamma = 1·4$. Also
calculate the Mach number before and after the wave.
[400 m/s, 825 kN/m^2, 1·09, 0·927]

14.15   A normal shock wave forms in an air stream at a static temperature

of 22 K, the total temperature being 400 K. Estimate the Mach number and static temperature behind the shock wave.
[0·577, 101 °C]

14.16   Fanno flow (adiabatic flow with friction) prevails as air moves through a pipe of 50 mm diameter. At a certain point along the pipe the Mach number is 0·2. Find the maximum distance from this point to the exit from the pipe if choking is avoided. Assume that the friction factor f is 0·006. Start from the energy equation and the momentum equation in differential form.
[30·28 m]

14.17   Air flows adiabatically at the rate of 2·7 kg/s through a horizontal 100 mm diameter pipe for which a mean value of f is 0·006. If the initial pressure and temperature are 1·8 bar abs and 50 °C, what is the maximum length of the pipe for which choking will not occur. What are then the temperature and pressure at the exit end and half way along the pipe.
[4·75 m; 9·2 °C, 82·7 kN/m²; 44·2 °C, 150·8 kN/m²]

14.18   At a particular section of a duct air with a static temperature of 32 °C and static pressure 80 kN/m² flows with a velocity of 365 m/s. Assuming reversible adiabatic flow calculate the velocity and temperature at a section where the static pressure is 120 kN/m² and estimate the Mach number at both sections. Take R = 287 J/kgK.
[241 m/s, 71 °C, 1·033, 0·63]

14.19   A horizontal pipe of length L and diameter D conveys air. Assuming the air to expand according to the law p/ρ = constant and that acceleration effects are small, prove the equation

$$\frac{p_1}{\rho_1}(\rho_1^2 - \rho_2^2) = \frac{64fL\dot{m}^2}{\pi^2 D^5}$$

where $\dot{m}$ is the mass flow rate and 1 and 2 refer to the inlet and discharge ends of the pipe respectively.

Calculate the value of $\dot{m}$ if the length is 135 m, the diameter 15 cm, inlet conditions 10 bar abs and 65°C; discharge pressure 8·85 bar abs and f = 0·005, R = 287 J/kgK.
[6·25 kg/s]

14.20   Air flows through a pipe isothermally which is 50 mm in diameter and 1200 m long. Calculate the flow, in m³/min of free air at 15 °C and 101·3 kN/m² abs, if the initial pressure is 1 MN/m² abs and the final pressure 0·7 MN/m² abs. The temperature is constant at 5 °C and the friction coefficient f = 0·004.
[5·49 m³/min]

14.21   Air passes steadily through a horizontal duct of diameter 15 cm and length 300 m. The mass flow rate is 4·5 kg/s, the pressure at entry is 5 bar abs and at exit is 1·25 bar abs. Assuming the flow to be shock free and isothermal at 60 °C determine (*a*) the friction factor for the duct which may be assumed to be constant, (*b*) the heat transfer rate in watts to the air in the duct, (*c*) the Mach number at exit.

What percentage error would have occurred in your answer to (*a*) had the velocity term been neglected. Take R = 287 J/kgK and γ = 1·4.
[0·0043, 60·5 kW, 0·535, 6·1 per cent].

15 Uniform flow in open channels

15.1. Flow with a free surface in open channels and ducts

As explained in Section 12.1, flow in an open channel or a duct in which the liquid has a free surface differs from flow in pipes in so far as the pressure at the free surface is constant (normally atmospheric) and does not vary from point to point in the direction of flow, as the pressure can do in a pipeline. A further difference is that the area of cross-section is not controlled by the fixed boundaries, since the depth can vary from section to section without restraint.

The types of flow occurring in open channels can be classified as steady if conditions do not vary with time and uniform if they do not vary from cross-section to cross-section. Thus *steady uniform flow* will occur in long channels of constant cross-section and slope over that portion which is far enough from entry or exit for the flow to have reached its terminal velocity. Such a situation will occur when the energy loss due to friction is exactly supplied by the reduction in potential energy which occurs due to the fall in bed level. Under these conditions the depth is constant and known as the *normal depth*.

At entry and exit where the depth is varying and wherever the cross-section is changing, as would be the case in most rivers or natural channels, *steady non-uniform flow* will occur if conditions do not change with time. This is also referred to as *varied flow*. Since the flow in an open channel has a free surface, gravity waves can be formed which are an example of *unsteady non-uniform flow*, a result of the fact that conditions of depth and velocity change with time relative to a fixed point on the bed of the channel.

Both laminar and turbulent flow can occur depending on the value of the Reynolds number. In a pipe, laminar flow occurs when the value of the Reynolds number $\rho \bar{v} d/\mu < 2000$, ρ and μ being the mass density and dynamic viscosity of the fluid, $\bar{v}$ the mean velocity and d the pipe diameter. This relation can also be applied to a channel if the diameter d is replaced by the *hydraulic mean depth* m, which is defined as the ratio A/P of the cross-sectional area A of the liquid flowing to the *wetted perimeter* P (the length of the line of contact between the liquid and the channel boundary at that section). Thus, for a rectangular channel of width B in which the depth of the liquid is D,

Cross-sectional area $= BD$,

Wetted perimeter $= B + 2D$,

Hydraulic mean depth $= BD/(B + 2D)$.

For a pipe of diameter d running full, $A = (\pi/4)d^2$ and $P = \pi d$, so that $m = d/4$. Replacing m by $d/4$ in the Reynolds number, the criterion for the type of flow in channels will be

Laminar flow $\quad \rho \bar{v}(4m)/\mu < 2000 \quad$ or $\quad \rho \bar{v} m/\mu < 500$.

For values of $\rho \bar{v} m/\mu$ between 500 and 2000, flow will be transitional and, if $\rho \bar{v} m/\mu > 2000$, flow is generally turbulent.

In practice, laminar flow is rare in channels and will only occur if the kinematic viscosity μ/ρ is very high or m is very small, as, for example, in the flow of a thin film of liquid over an inclined surface. Normally, flow is turbulent and in this section this will be assumed to be the case.

The continuity, momentum and energy equations can be applied to channel flow in the same way as for pipe flow. Thus, in Fig. 15.1, since there

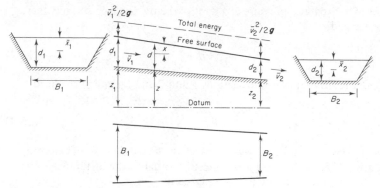

Figure 15.1   Channel flow

is no change of density between sections 1 and 2, for continuity of steady flow the volume rate of flow Q must be the same at both sections:

$$Q = B_1 D_1 \bar{v}_1 = B_2 D_2 \bar{v}_2, \tag{15.1}$$

where $\bar{v}_1$ and $\bar{v}_2$ are the mean velocities at the two sections. For wide channels of approximately rectangular section it is sometimes convenient to consider the flow per unit width q, so that

$$q_1 = Q/B_1 = \bar{v}_1 D_1 \quad \text{and} \quad q_2 = Q/B_2 = \bar{v}_2 D_2.$$

In travelling from section 1 to section 2 there will be a change of momentum per second of the liquid corresponding to the change of velocity:

Rate of change of momentum = Mass per second x Change of velocity

$$= \rho Q(\bar{v}_2 - \bar{v}_1).$$

This change is produced by the difference in the hydrostatic forces at sections 1 and 2. From equation (3.2),

Force in direction of motion at section 1 $= \rho g A_1 \bar{x}_1$,

Force opposing motion at section 2 $= \rho g A_2 \bar{x}_2$,

where $\bar{x}_1$ and $\bar{x}_2$ are the depths from the free surface to the centroids of the cross-sections.

Resultant force in the direction of motion $= \rho g(A_1 \bar{x}_1 - A_2 \bar{x}_2)$. By Newton's second law

Force = Rate of change of momentum

$$\rho g(A_1 \bar{x}_1 - A_2 \bar{x}_2) = \rho Q(\bar{v}_2 - \bar{v}_1),$$
$$(A_1 \bar{x}_1 - A_2 \bar{x}_2) = Q(\bar{v}_2 - \bar{v}_1)/g. \tag{15.2}$$

For the energy equation, Bernoulli's equation with a term for loss of energy can be used, since the fluid flowing in the channel can be assumed to be incompressible. Considering conditions at a point on any streamline at a depth x below the free surface (Fig. 15.1),

$$\text{Total energy per unit weight, } H = \frac{p}{\rho g} + \frac{\bar{v}^2}{2g} + (z + d - x).$$

Now p is the hydrostatic pressure at a depth x below the free surface, therefore $p/\rho g = x$ and

$$\begin{matrix} \text{Total energy at any point} \\ \text{per unit weight, } H \end{matrix} = z + d + \frac{v^2}{2g}. \qquad (15.3)$$

Applying Bernoulli's equation to sections 1 and 2 and including the head loss h

$$z_1 + d_1 + \frac{v_1^2}{2g} = z_2 + d_2 + \frac{v_2^2}{2g} + h. \qquad (15.4)$$

In the special case of uniform, steady flow $\bar{v}_1 = \bar{v}_2$ and $d_1 = d_2$. Therefore, from equation (15.4), the head loss h must be equal to the difference of bed level: $h = z_1 - z_2$ and equation (15.4) reduces to

$$d_1 + \bar{v}_1^2/2g = d_2 + \bar{v}_2^2/2g,$$

which corresponds to Bernoulli's equation for the frictionless flow of a liquid in a channel with a horizontal bed.

The total energy per unit weight measured above bed level, $(d + \bar{v}^2/2g)$, is termed the *specific energy*. For problems involving steady flow it is often easier to analyse the situation using specific energy instead of total energy and omitting losses of energy due to friction, which correspond to the difference between specific energy and total energy.

The hydraulic gradient line for steady, uniform or non-uniform flow will coincide with the free surface (as shown in Fig. 15.1), while the total energy line will lie $\bar{v}^2/2g$ above the free surface. The gradients of the total energy line, the water surface and the bed will normally differ although they are interrelated. They will only be the same for steady uniform flow in open channels.

15.2. Resistance formulae for steady uniform flow in open channels

The analysis of the resistance to flow of a liquid in a channel is the same as that for flow in a pipe, as shown in Section 8.5. For steady flow, the resistance due to the shear force on the channel boundaries is exactly equal and opposite to the component of the force due to gravity acting in the direction of flow. The resistance laws for pipes can be applied to channels if they are written in terms of the hydraulic mean depth m. Thus, the Darcy formula for the loss of head h_f in a length l can be written

$$h_f = \frac{fl}{m} \cdot \frac{\bar{v}^2}{2g}, \qquad (15.5)$$

since, for a pipe running full, $m = A/P = (\pi/4) d^2/\pi d = d/4$ and, in the form of equation (15.5), the Darcy formula can be applied directly to open channels of any form.

Many problems involve steady flow at uniform depth and constant cross-sectional area. Under these conditions, the bed slope s is equal to the slope of the total energy line i, which is the loss of energy per unit weight per unit length h_f/l. It is, therefore, more convenient for resistance formulae for channels to be written in terms of i. From equation (15.5),

$$\bar{v}^2 = \frac{2g}{f} m \frac{h_f}{l}$$

so that

$$\bar{v} = \sqrt{(2g/f)} \sqrt{(mi)}$$

or, putting $\sqrt{(2g/f)} = C$,

$$\bar{v} = C\sqrt{(mi)}. \tag{15.6}$$

This is the *Chezy formula*. Unlike the Darcy coefficient f, which is dimensionless, the Chezy coefficient C has dimensions $L^{1/2} T^{-1}$ and, therefore, its numerical value will vary with the system of units employed. In Section 8.3 it was seen that, for pipes, the Darcy friction coefficient f and, therefore, the Chezy coefficient C varied with the Reynolds number and relative roughness of the boundary. The same relationship will hold for open channels, but the Reynolds number will be calculated as $\bar{v}m/\nu$, so that C will depend on the mean velocity $\bar{v}$, the hydraulic mean depth m, the kinematic viscosity ν and the relative roughness. There is experimental evidence that the value of the resistance coefficient does vary with the shape of the channel and therefore with m and possibly also with the bed slope s, which for uniform flow will be equal to i, the relationship for velocity being of the form $\bar{v} = K m^x i^y$, where K, x and y are constants.

A number of empirical formulae have been proposed for the determination of the value of C in the Chezy formula. In 1869 Ganguillet and Kutter proposed the following formula based on an analysis of the behaviour of rivers and open channels which, stated in SI units, is

$$C = \frac{23 + 0\cdot00155/s + 1/n}{1 + (23 + 0\cdot0015/s)\, n/\sqrt{m}}, \tag{15.7}$$

where s is the bed slope and n is a roughness coefficient, which increases with increasing roughness of the channel boundary. Typical values of n are given in Table 15.1. Equation (15.7) is usually referred to as the *Kutter formula*. It is not very convenient to use unless presented in the form of tables giving values of C for values of m, s and n. There is some doubt whether the terms involving s are justified.

The simplest formula, and one very widely used, is that published in 1890 by Robert Manning, who found from the experimental data then available that C varied as $m^{1/6}$ and was dependent on the roughness coefficient n of the channel boundaries. The *Manning formula*, obtained by putting Manning's value of C in the Chezy formula, stated in SI units, is usually written in the form

$$\bar{v} = (1/n)\, m^{2/3} i^{1/2}, \tag{15.8}$$

where n has the same value as in the Kutter formula. Equation (15.8) is also

Surface of channel	Condition	
	Good	Poor
Neat cement	0·010	0·013
Cement mortar	0·011	0·015
Concrete, *in situ*	0·012	0·018
Concrete, precast	0·011	0·013
Cement rubble	0·017	0·030
Dry rubble	0·025	0·035
Brick with cement mortar	0·012	0·017
Plank flumes, planed	0·010	0·014
unplaned	0·011	0·015
Metal flumes, semicircular, smooth	0·011	0·015
corrugated	0·022	0·030
Cast iron	0·013	0·017
Steel, rivetted	0·017	0·020
Canals, earth straight and uniform	0·017	0·025
dredged earth	0·025	0·033
rock cuts, smooth	0·025	0·035
rock cuts, jagged	0·035	0·045
rough beds with weeds on sides	0·025	0·040
Natural streams, clean smooth and straight	0·025	0·035
rough	0·045	0·060
very weedy	0·075	0·150

Table 15.1. Values of n in Manning's formula for flow in open channels

known as the *Strickler formula* and $1/n$ as the Strickler coefficient. The dimensions of n in the Manning formula are $L^{-1/3} T$.

The *Bazin formula*, published in 1897, does not relate C to the bed slope s. Stated in SI units,

$$C = \frac{86·9}{1 + k/\sqrt{m}},$$
(15.9)

where k depends on the surface roughness. Typical values of k are given in Table 15.2.

A number of other formulae have been put forward, but the experimental investigation of open channel flow is complicated by the effects of the free surface and of cross-currents, as well as the variety of bed conditions which can occur. It should also be remembered that such formulae apply, strictly, only to uniform flow and, if applied to non-uniform flow, it must be assumed that the loss of energy per unit weight at a given section is the same as for

Surface of channel	k
Smooth cement or planed wood	0·060
Planks, ashlar and brick	0·160
Rubble masonry	0·460
Earth channels of very regular surface	0·850
Ordinary earth channels	1·303
Exceptionally rough channels	1·750

Table 15.2. Values of k in Bazin's formula (SI units)

uniform flow at the same depth. In practice, if the flow is diverging, this loss will be greater because of increased turbulence, while for converging flow the loss will be decreased. The results obtained provide a means of estimating flow in channels, but their relation to the actual flow will depend upon the experience of the user in selecting the most suitable formula and appropriate resistance coefficient (*see* Section 16.3).

Example 15.1

An open channel has a cross-section in the form of a trapezium (Fig. 15.2)

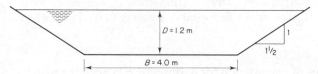

Figure 15.2

with a bottom width B of 4 m and side slopes of 1 vertical to $1\frac{1}{2}$ horizontal. Assuming that the roughness coefficient n is 0·025, the bed slope is 1 in 1800 and the depth of the water is 1.2 m find the volume rate of flow Q using (a) the Chezy formula with C determined from the Kutter formula, and (b) the Manning formula.

Solution

$$\text{Width of water surface} = B + 2 \times 1\cdot5D$$

$$= 4 + 3 \times 1\cdot2 = 7\cdot6 \text{ m},$$

$$\text{Area of cross-section, } A = \{(7\cdot6 + 4)/2\} \times 1\cdot2 = 6\cdot96 \text{ m}^2,$$

$$\text{Wetted perimeter, } P = B + 2\sqrt{(D^2 + 2\cdot25D^2)}$$

$$= 4 + 2 \times 1\cdot2\sqrt{(1 + 2\cdot25)} = 8\cdot33 \text{ m},$$

$$\text{Hydraulic mean depth, } m = A/P = 6\cdot96/8\cdot33 = 0\cdot836 \text{ m}.$$

For uniform steady flow,

$$\text{Total energy gradient } i = \text{Bed slope } s = 1/1800.$$

(a) From the Kutter formula,

$$C = \frac{23 + 0\cdot001\ 55/i + 1/n}{1 + (23 + 0\cdot001\ 55/i)n/\sqrt{m}} \quad \text{in SI units}$$

$$= \frac{23 + 0\cdot001\ 55 \times 1800 + 1/0\cdot025}{1 + (23 + 0\cdot001\ 55 \times 1800)\ 0\cdot025/\sqrt{(0\cdot836)}} = 38\cdot6$$

Volume rate of flow, $Q = CA\sqrt{(mi)}$

$$= 38\cdot8 \times 6\cdot96\sqrt{(0\cdot836/1800)} = \mathbf{5\cdot82 \ m^3 s^{-1}}.$$

(b) Using the Manning formula,

$$\text{Volume rate of flow, } Q = A(1/n)\, m^{2/3} i^{1/2}$$

$$= 6 \cdot 96 \times 0 \cdot 836^{2/3}/0 \cdot 025 \times (1/1800)^{1/2} = 5 \cdot 82 \text{ m}^3 \text{ s}^{-1}.$$

In addition to energy losses due to friction, there will be separation losses wherever the flow is disturbed, as, for example, at a reservoir entrance or exit, a change of section or a bend. These are similar to those occurring in pipe flow and can be expressed in the form $k(\bar{v}^2/2g)$, where $\bar{v}$ is the mean velocity and k is a coefficient of the same order as for pipes. Such separation losses are normally small compared with overall friction losses, but can be of local importance, since the loss of head will appear as a change in level of the free surface.

15.3. Optimum shape of cross-section for uniform flow in open channels

The shape of a channel or, more precisely, the cross-section of flow, will affect the ratio of the area of flow A to the wetted perimeter P and, therefore, the value of the hydraulic mean depth m. For uniform flow with a given bed slope, the mean velocity and discharge depend on m and so the shape of a channel will affect its hydraulic effectiveness. Given complete freedom of choice of cross-section, the optimum shape, hydraulically, would be that producing a maximum discharge for a given area, bed slope and surface roughness, which would be that with the smallest wetted perimeter. Such a channel would also have the smallest cross-section of flow for a required discharge and, since its wetted perimeter is a minimum, would require the least amount of lining material or surface finishing. The optimum cross-section would, therefore, also tend to be the cheapest.

In practice, the choice of cross-section may be dictated by other factors. Of all sections with an open surface, a semicircle has the smallest wetted perimeter for a given area, but semicircular channels are not easy to construct in many materials. Channels excavated in the ground are usually trapezoidal in cross-section, but although the optimum side slopes can be calculated, the nature of the ground will determine the slope that can be used. Unlined earth banks will normally not stand at slopes steeper than $1\frac{1}{2}$ horizontal to 1 vertical and, in sandy soil, the side slopes may be as flat as 3 to 1. Channels cut in rock, lined with concrete or constructed from timber or metal can be built with vertical sides if required. The following example shows how the optimum shape of trapezoidal channels can be calculated.

Example 15.2

Find the proportions of a trapezoidal channel (Fig. 15.3) which will make the discharge a maximum for a given cross-sectional area of flow and given side slopes. Show also that if the side slopes can be varied the most efficient of all trapezoidal sections is a half hexagon.

A trapezoidal channel has side slopes of 3 horizontal to 4 vertical and the slope of its bed is 1 in 2000. Determine the optimum dimensions of the channel if it is to carry water at $0 \cdot 5$ m^3 s^{-1}. Use the Chezy formula assuming that $C = 80$ m$^{1/2}$ s^{-1}.

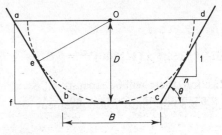

Figure 15.3

Solution

Using the Chezy formula,

$$Q = ACm^{1/2}i^{1/2} = AC(A/P)^{1/2}i^{1/2}.$$

Maximum discharge for given values of A, C and i will, therefore, occur when P is a minimum.

In Fig. 15.3, base width = B, depth = D and the side slope is 1 vertical to n horizontal.

$$\text{Area of flow, } A = (B + nD)D, \tag{I}$$

from which

$$B = A/D - nD. \tag{II}$$

Wetted perimeter, $P = bc + 2\,cd$

$$= B + 2D\sqrt{(n^2 + 1)}.$$

Substituting from equation (II),

$$P = A/D + (2\sqrt{(n^2 + 1)} - n)D. \tag{III}$$

If A and n are fixed, P will be a minimum when $dP/dD = 0$. Differentiating equation (III),

$$\frac{dP}{dD} = -A/D^2 + 2\sqrt{(n^2 + 1)} - n = 0,$$

$$A = D^2 (2\sqrt{(n^2 + 1)} - n).$$

Substituting for A from equation (I),

$$BD + nD^2 = D^2(2\sqrt{(n^2 + 1)} - n).$$

For maximum discharge,

$$B = 2D(\sqrt{(n^2 + 1)} - n). \tag{IV}$$

(Note special case of rectangular sections, i.e. $n = 0$.)

We now find the side slopes for the section which will have the greatest possible efficiency. The value of B will be that for optimum efficiency, given by equation (IV),

$$\text{Area, } A = BD + nD^2 = 2D^2 (\sqrt{(n^2 + 1)} - n) + nD^2$$

$$= D^2 (2\sqrt{(n^2 + 1)} - n).$$

So, for maximum efficiency,

$$D = A^{1/2}/(2\sqrt{(n^2 + 1)} - n)^{1/2}.$$

Substituting in equation (III),

$$P = 2A^{1/2}(2\sqrt{(n^2 + 1)} - n)^{1/2}. \tag{V}$$

Since P^2 will also be a minimum when P is a minimum, it is convenient to square equation (V):

$$P^2 = 4A(2\sqrt{(n^2 + 1)} - n).$$

Differentiating and equating to zero,

$$\frac{d(P^2)}{dn} = 4A\left(\frac{2n}{\sqrt{(n^2 + 1)}} - 1\right) = 0,$$

$$2n = \sqrt{(n^2 + 1)}.$$

Squaring,

$$4n^2 = n^2 + 1 \quad \text{and} \quad n = 1/\sqrt{3}:$$

If θ is the angle of the side of the horizontal, $\tan\theta = 1/n = \sqrt{3}$ and $\theta = 60°$. Thus, the cross-section of flow of greatest possible efficiency will be a half-hexagon.

For the given cross-section in this problem, $n = \frac{3}{4}$; therefore, substituting in equation (IV) for maximum discharge,

$$B = 2D\left(\sqrt{\left(\frac{3}{4}\right)^2 + 1} - \frac{3}{4}\right) = 2D\left(\frac{5}{4} - \frac{3}{4}\right) = D,$$

$$\text{Area of cross-section, } A = BD + \tfrac{3}{4}D^2$$

$$= D^2 + \tfrac{3}{4}D^2 = \tfrac{7}{4}D^2,$$

$$\text{Wetted perimeter, } P = B + 2 \times \tfrac{5}{4}D = \tfrac{7}{2}D,$$

$$\text{Hydraulic mean depth, } m = A/P = \tfrac{1}{2}D.$$

Substituting in the Chezy formula,

$$Q = ACm^{1/2}i^{1/2} = \tfrac{7}{4}D^2 \times C(\tfrac{1}{2}D)^{2}i^{1/2}.$$

Putting $Q = 0.5 \text{ m}^3 \text{ s}^{-1}$, $C = 80 \text{ m}^{1/2} \text{ s}^{-1}$ and $i = 1/2000$,

$$0.5 = \tfrac{7}{4} \times \frac{80D^{5/2}}{(2 \times 2000)^{1/2}}$$

$$\text{Depth, } D = \left(\frac{4 \times 0.5 \times 63.25}{7 \times 80}\right)^{2/5} = 0.552 \text{ m},$$

$$\text{Base width, } B = D = 0.552 \text{ m}.$$

It can also be shown that for a channel of optimum proportions the sides and base are tangential to a semicircle with its centre O in the free surface.

Drawing Oe perpendicular to ab from the midpoint of the free surface:

$$\sin \widehat{Oae} = \frac{Oe}{Oa} = \frac{Oe}{\frac{1}{2}(B + 2nD)},$$

$$\sin \widehat{abf} = \frac{af}{ab} = \frac{D}{D\sqrt{(n^2 + 1)}}.$$

But $\widehat{Oae} = \widehat{abf}$ and so

$$\frac{Oe}{\frac{1}{2}(B + 2nD)} = \frac{D}{D\sqrt{(n^2 + 1)}},$$

$$Oe = \frac{B + 2nD}{2\sqrt{(n^2 + 1)}}.$$

From equation (IV) in Example 15.2,

$$B = 2D(\sqrt{(n^2 + 1)} - n) = 2D\sqrt{(n^2 + 1)} - 2nD \qquad (15.10)$$

and so

$$Oe = \frac{2D\sqrt{(n^2 + 1)} - 2nD + 2nD}{2\sqrt{(n^2 + 1)}} = D.$$

As ab is perpendicular to Oe it will be a tangent to a semicircle of centre O to which bc and cd will also be tangential.

The hydraulic mean depth for an optimum trapezoidal section will be

$$m = \frac{BD + nD^2}{B + 2D\sqrt{(n^2 + 1)}} = \frac{B + nD}{B/D + 2\sqrt{(n^2 + 1)}}.$$

Substituting for B from equation (15.10),

$$m = \frac{2D\sqrt{(n^2 + 1)} - nD}{2\sqrt{(n^2 + 1)} - 2n + 2\sqrt{(n^2 + 1)}} = \frac{D}{2}.$$

15.4. Optimum depth for flow with a free surface in covered channels

For covered channels which are not flowing full, there will be optimum depths of flow for maximum velocity and for maximum discharge. This arises because, as the level of the free surface rises, a point is reached beyond which the wetted perimeter increases very rapidly in comparison with the area of flow, causing the hydraulic mean depth and, therefore, the velocity, to decrease. Since the discharge is the product of the area and velocity, there will also come a point at which the discharge will start to diminish as the depth continues to increase, because the increase in area is more than offset by the reduction in mean velocity. The method of determining these optimum conditions is similar to that used in Section 15.3 but, since the shape of the conduit is fixed, the area A can no longer be treated as constant.

For the steady, uniform flow in the circular conduit shown in Fig. 15.4 the Chezy formula gives

$$\text{Mean velocity, } \bar{v} = Cm^{1/2}i^{1/2} = C(A/P)^{1/2}i^{1/2}$$

and so, for constant values of C and i, the maximum value of $\bar{v}$ will occur when A/P is a maximum.

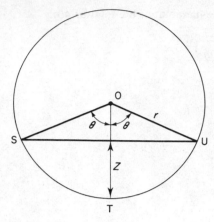

Figure 15.4

If the free surface subtends an angle 2θ at the centre O for any depth Z,

Area of flow, A = Sector OSTU − Triangle OSU

$$= \tfrac{1}{2}r^2 \times 2\theta - r^2 \sin \theta \cos \theta$$

$$= r^2(\theta - \tfrac{1}{2} \sin 2\theta),$$

Wetted perimeter, $P = 2r\theta$.

For maximum velocity,

$$\frac{\mathrm{d}(A/P)}{\mathrm{d}\theta} = \frac{1}{P^2}\left(P\frac{\mathrm{d}A}{\mathrm{d}\theta} - A\frac{\mathrm{d}P}{\mathrm{d}\theta}\right) = 0$$

or $$P\frac{\mathrm{d}A}{\mathrm{d}\theta} = A\frac{\mathrm{d}P}{\mathrm{d}\theta}.$$

Substituting for P, A, $\mathrm{d}A/\mathrm{d}\theta$ and $\mathrm{d}P/\mathrm{d}\theta$,

$$2r\theta \times r^2(1 - \cos 2\theta) = r^2(\theta - \tfrac{1}{2} \sin 2\theta) \times 2r,$$

$$\theta\,(1 - \cos 2\theta) = \theta - \tfrac{1}{2} \sin 2\theta,$$

$$2\theta = \tan 2\theta,$$

giving $2\theta = 257 \cdot 5°$.

Depth of flow, $Z = r - r \cos \theta$

$$= r(1 + 0 \cdot 62) = 1 \cdot 62r$$

$$= 0 \cdot 81 \times \textbf{Pipe diameter.}$$

For maximum discharge, the result will depend on the choice of resistance formula. Using the Chezy formula,

Discharge, $Q = ACm^{1/2}i^{1/2} = AC(A/P)^{1/2}i^{1/2} = C(A^3/P)^{1/2}i^{1/2}$.

For given values of C and i, the discharge Q will be a maximum when (A^3/P)

is a maximum. Differentiating with respect to θ and equating to zero,

$$\frac{d(A^3/P)}{d\theta} = \frac{1}{P^2}\left(3PA^2\frac{dA}{d\theta} - A^3\frac{dP}{d\theta}\right) = 0,$$

$$3P\frac{dA}{d\theta} - A\frac{dP}{d\theta} = 0.$$

Substituting for P, A, $dA/d\theta$ and $dP/d\theta$,

$$3 \times 2r\theta \times r^2(1 - \cos 2\theta) - r^2(\theta - \tfrac{1}{2}\sin 2\theta) \times 2r = 0.$$

Dividing by r^3 and simplifying,

$$4\theta - 6\theta \cos 2\theta + \sin 2\theta = 0,$$

from which

$$2\theta = 308°,$$

$$\theta = 154° = 2·68 \text{ rad.}$$

Depth for maximum discharge, $Z = r(1 - \cos\theta)$

$$= r(1 + 0·90)$$

$$= \textbf{0·95 x Pipe diameter.}$$

Since any slight increase in depth will cause a reduction in the volume rate of flow that the channel can carry, it is usual to design on the assumption that the section will run full.

Hydraulic mean depth for maximum discharge, m_1
$$= \frac{r^2(2·68 \times \tfrac{1}{2}\sin 308°)}{2r \times 2·68}$$

$$= 0·574r.$$

Hydraulic mean depth running full, m_2
$$= 0·5r.$$

$$\frac{\text{Discharge running full}}{\text{Maximum discharge}} = \left(\frac{m_2}{m_1}\right)^{1/2} = \left(\frac{0·5}{0·574}\right)^{1/2}$$

Discharge running full = **0·933 x Maximum discharge.**

15.5. Further reading

Anon. (March 1963) A.S.C.E. Report of task force on Friction Factors in Open Channels, *J. Hydraulics Division A.S.C.E.*, **89** (HY2).

Chow, V. T. (1959). *Open Channel Hydraulics*, McGraw-Hill, New York.

Henderson, F. M. (1966). *Open Channel Flow*, Macmillan.

Rouse, H. (June 1965) Critical analysis of open channel roughness, *J. Hydraulics Division A.S.C.E.*, 1–15.

EXERCISES 15

15.1   A rectangular channel is 2·5 m wide and has a uniform bed slope of 1 in 500. If the depth of flow is constant at 1·7 m calculate (a) the hydraulic mean depth, (b) the velocity of flow, (c) the volume rate of flow. Assume that the value of the coefficient C in the Chezy formula is 50 in SI units.
[0·72 m, 1·9 m/s, 8·1 m³/s]

15.2   An open channel has a Vee-shaped cross-section with sides inclined at an angle of 60° to the vertical. If the rate of flow is 80 dm³/s when the depth at the centre is 0·25 m, what must be the slope of the channel assuming C = 45 SI units.
[1 in 401]

15.3   A channel 5 m wide at the top and 2 m deep has sides sloping 2 vertically in 1 horizontally. The slope of the channel is 1 in 1000. Find the volume rate of flow when the depth of water is constant at 1 m. Take C as 53 in SI units.

What would be the depth of water if the flow were to be doubled.
[4·79 m³/s, 1·6 m]

15.4   Water is conveyed in a channel of semi-circular cross section with a slope of 1 in 2500. The Chezy coefficient C has a value of 56 SI units. If the radius of the channel is 0·55 m what will be the volume in dm³/s flowing when the depth is equal to the radius.

If the channel had been rectangular in form with the same width of 1·1 m and depth of flow of 0·55 m, what would be the discharge for the same slope and value of C.
[280 dm³/s, 355 dm³/s]

15.5   A 900 mm diameter conduit 3600 m long is laid at a uniform slope of 1 in 1500 and connects two reservoirs. When the levels in the reservoirs are low the conduit runs partly full and it is found that a normal depth of 600 mm gives a rate of flow of 0·322 m³/s.

The Chezy coefficient C is given by Km^n where K is a constant, m is the hydraulic mean depth and $n = \frac{1}{6}$. Neglecting losses of head at entry and exit obtain (a) the value of K, (b) the discharge when the conduit is flowing full and the difference in level between the two reservoirs is 4·5 m.
[67·6, 0·562 m³/s]

15.6   An earth channel is trapezoidal in cross section with a bottom width of 1·8 m and side slopes of 1 vertical to 2 horizontal. Taking the friction coefficient k in the Bazin formula as 1·3 and the slope of the bed as 0·57 m per km, find the discharge in m³/s when the depth of water is 1·5 m.
[5·64 m³/s]

15.7   The water supply for a turbine passes through a conduit which for convenience has its cross-section in the form of a square with one diagonal vertical. If the conduit is required to convey 8·5 dm³/s under conditions of maximum discharge at atmospheric pressure when the slope of the bed is 1 in 4900, determine its size assuming that the velocity of flow is given by

$$v = 80i^{1/2}m^{2/3}$$

[2·99 m side]

15.8   A trapezoidal channel is to be designed to carry 280 m³ per minute of water. Determine the cross-sectional dimensions of the channel if the slope

is 1 in 1600, side slopes $45°$ and the cross section is to be a minimum. Take
$C = 50$ SI units.
[$D = 1·53$ m, $B = 1·27$ m]

15.9   A circular section open conduit conveys liquid under maximum
velocity conditions. Show that the depth of liquid is 81 per cent of the
diameter. Show, without complete calculation, that this will not be the
maximum discharge condition.

Such a conduit having a diameter of 0·8 m is to discharge 0·6 m³/s at
maximum velocity. Find the required channel slope if the Chezy constant
is 90 SI units.
[1 in 1050]

15.10   It is required to excavate a canal out of rock. It is to be of rectangular
cross-section and to bring 14·2 m³ of water from a distance of 6·5 km with a
velocity of 2·25 m/s. Determine the gradient and the most suitable section.
[1 in 1200, $D = 1·78$ m, $B = 3·56$ m]

15.11   An egg-shaped sewer has a section formed by circular arcs the top
being a semicircle of radius R. The area and wetted perimeter of the section
below the horizontal diameter of the semicircle are $3R^2$ and $4·82R$ respectively.
Prove that, if C in the Chezy formula is constant, the maximum flow will
occur when the water surface subtends an angle of approximately $55°$ at the
centre of curvature of the semicircle.

15.12   The upper portion of the cross-section of an open channel is a semicircle of radius a, the lower portion is a semiellipse of width $2a$, depth $2a$ and
perimeter $4·847a$, whose minor axis coincides with the horizontal diameter
of the semicircle. The channel is required to convey 14 m³/s when running
three-quarters full (i.e. with three-quarters of the vertical axis of symmetry
immersed), the slope of the bed i being 0·001. Assuming that the mean
velocity of flow is given by Manning's formula $v = 80i^{1/2}m^{2/3}$, determine the
dimensions of the section and the depth under maximum flow conditions.
[$a = 1·29$ m, 3·68 m]

15.13   The cross-section of a closed channel is a square with one diagonal
vertical, s is the side of the square and y is the depth of the water line below
the apex. Show that for maximum discharge $y = 0·127$ s and that for
maximum velocity $y = 0·414$ s.

15.14   Find an expression for the theoretical depth for maximum velocity
in a closed circular channel in terms of the diameter d.

Compare the discharge at maximum velocity with that when the channel is
running full, assuming that the Chezy coefficient is unaltered, and that the
pressure remains atmospheric.
[$0·81d$, $0·964$]

15.15   An open channel of economic trapezoidal cross-section with sides
inclined at $60°$ to the horizontal is required to give a discharge of 10 m³/s
when the slope of the bed is 1 in 1600. Calculate the dimensions of the cross-
section assuming $v = 74i^{1/2}m^{2/3}$.
[$D = 1·82$ m, $B = 2·11$ m]

16 Non-uniform flow in open channels

16.1. Specific energy and alternative depths of flow

As explained in Section 15.1, specific energy E is defined as the energy per unit weight of the liquid at a cross-section measured above bed level at that point. If D is the depth and $\bar{v}$ is the mean velocity

$$E = D + \frac{\bar{v}^2}{2g} \tag{16.1}$$

An examination of this equation shows that, for a given specific energy, the possible depths of flow are limited. Considering a wide rectangular channel, width B, cross-sectional area A, through which there is a volume rate of flow Q,

$$\bar{v} = Q/A = Q/BD.$$

Substituting in equation (16.1),

$$E = D + \frac{1}{2g}\left(\frac{Q}{BD}\right)^2$$

or, putting $Q/B = q$ = volume rate of flow per unit width,

$$E = D + q^2/2gD^2 \tag{16.2}$$

$$D^3 - ED^2 + q^2/2g = 0 \tag{16.3}$$

This equation has three roots, of which two are positive and real and the other is negative and unreal. For a constant value of specific energy E, there are normally two, and only two, alternative depths for a given discharge q, as can be seen from Fig. 16.1(a). Similarly, as shown in Fig. 16.1(b), for a constant value of the discharge per unit width q, there will normally be two, and only two, alternative depths for a given value of specific energy.

The larger of these two values corresponds to the condition of deep slow flow which is known as *tranquil* or *streaming* flow. The smaller value is that for shallow fast flow which is known as *shooting* flow.

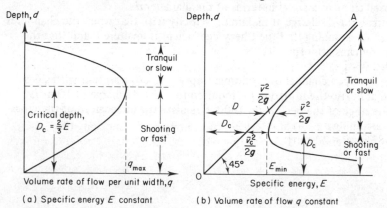

(a) Specific energy E constant   (b) Volume rate of flow q constant

Figure 16.1   Alternative depths of flow

From Fig. 16.1(a) and (b) it can be seen that there is a *critical depth* D_c at which the two roots coincide, when the discharge for a given specific energy is a maximum and the energy required for a given discharge is a minimum. To find this value, differentiate equation (16.2) assuming that q is constant:

$$\frac{\mathrm{d}E}{\mathrm{d}D} = 1 - 2\,q^2/2gD^3.$$

When $\mathrm{d}E/\mathrm{d}D$ is zero, flow will be at the critical depth D_c. Thus,

$$\text{Critical depth, } D_c = (q^2/g)^{1/3} = (Q^2/gB^2)^{1/3}. \tag{16.4}$$

The corresponding value of the specific energy will be E_{min} and is obtained by substituting the value of $q^2 = gD_c^3$ from equation (16.4) in equation (16.2) giving

$$E = D_c + gD_c^3/2gD_c^2 = \tfrac{3}{2}D_c.$$

Thus, the critical depth of flow D_c in a rectangular channel will be $\tfrac{2}{3}E$.

The same result could have been obtained by differentiating equation (16.3), assuming that E is constant, since

$$q = D[2g(E-D)]^{1/2}, \tag{16.5}$$

$$\frac{\mathrm{d}q}{\mathrm{d}D} = \sqrt{(2g)}\{(E-D)^{1/2} - \tfrac{1}{2}D/(E-D)^{1/2}\}.$$

For maximum discharge, when $D = D_c$ at the critical depth, $\mathrm{d}q/\mathrm{d}D = 0$ and so $(E - D_c) - \tfrac{1}{2}D_c = 0$, from which $D_c = \tfrac{2}{3}E$.

The maximum discharge per unit width for a given value of E is found by substituting this value in equation (16.5):

$$q_{max} = \tfrac{2}{3}E[2g(E - \tfrac{2}{3}E)]^{1/2} = g^{1/2}(\tfrac{2}{3}E)^{3/2}$$

or $\qquad q_{max} = \sqrt{(gD_c^3)}. \tag{16.6}$

The velocity of flow corresponding to critical depth is known as the *critical velocity* v_c. From equation (16.1), putting $E = \tfrac{3}{2}D_c$ and $D = D_c$ for critical flow conditions,

$$\tfrac{3}{2}D_c = D_c + v_c^2/2g,$$

$$v_c = \sqrt{(gD_c)}.$$

Referring to Section 5.14, it will be seen that this is the velocity of propagation of a surface wave. Thus, for critical flow conditions, the Froude number $\bar{v}/\sqrt{(gD)} = v_c/\sqrt{(gD_c)} = 1$.

For tranquil flow, the velocity will be less than the critical velocity v_c and this is sometimes termed *sub-critical* flow. Similarly, for shooting flow, the velocity will be greater than v_c and may be termed *super-critical* flow. An important difference between them is that for tranquil flow the mean velocity $\bar{v}$ is less than the velocity of propagation of a disturbance relative to the stream and, therefore, disturbances can be propagated both up and down-stream, so enabling downstream conditions to determine the behaviour of the flow. For shooting flow, the velocity of the stream exceeds the velocity of

propagation and, therefore, disturbances cannot travel upstream and downstream conditions cannot control the behaviour of the flow.

Examination of Fig. 16.1 shows that, when flow is in the region of the
critical depth, small changes of energy or flow rate are associated with
relatively large changes of depth. Small surface waves are, therefore, easily
formed but, since the velocity of propagation is equal to the critical velocity
these waves will be stationary or *standing waves*, and their presence is an
indication of critical flow conditions.

In Fig. 16.1(b) a line OA can be drawn at 45° through the origin.
Assuming that the scales for E and D are the same, horizontal distances from
the vertical axis to this line will be equal to the depth D and, since
$E = D + \bar{v}^2/2g$, the distance from OA to the specific energy curve will
represent $\bar{v}^2/2g$. For tranquil flow, $\bar{v}$ will decrease as D increases if q is
constant so that the specific energy curve is asymptotic to OA. Similarly, as
D decreases $\bar{v}$ increases and the specific energy curve will be asymptotic to
the E axis.

Although at a cross-section there can be two alternative depths for a given
specific energy, the maintenance of uniform flow at one or other of these
depths is dependent on the slope of the channel. Energy losses are a function
of velocity. For shooting flow, therefore, the slope must be greater than for
tranquil flow, since energy losses will be greater. The slope of the channel
which will just maintain flow at the critical depth is known as the *critical
slope*. For uniform tranquil flow the slope is said to be *mild* and for uniform
shooting flow it is termed *steep*.

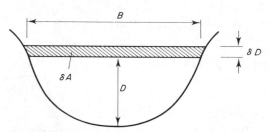

Figure 16.2

16.2. Critical depth in non-rectangular channels

For a channel of any shape and cross-sectional area A (Fig. 16.2) the specific
energy for any depth D is

$$E \doteq D + \bar{v}^2/2g$$

or, since $\bar{v} = Q/A$,

$$E = D + Q^2/2A^2g. \tag{16.7}$$

For flow at critical depth and velocity the specific energy is a minimum for a
given value of Q and $\mathrm{d}E/\mathrm{d}D$ is zero. Differentiating equation (16.7),

$$1 - \frac{2Q^2 A^{-3}}{2g} \frac{\mathrm{d}A}{\mathrm{d}D} = 0. \tag{16.8}$$

Referring to Fig. 16.2, a change of depth δD will produce a change in cross-sectional area of $\delta A = B\delta D$; therefore, $dA/dD = B$. Substituting in equation (16.8) under critical flow conditions,

$$Q^2 B/A^3 g = 1, \qquad (16.9)$$

where B and A are the breadth and area of the flow under these conditions.

Critical velocity, $v_c = Q/A$

and from equation (16.9)

$$v_c = (Ag/B)^{1/2} \quad \text{or} \quad v_c = (g\bar{D})^{1/2} \qquad (16.10)$$

where $\bar{D}$ is the average depth, defined as A/B for critical flow conditions.

Example 16.1

Find, in terms of the specific energy E, the critical velocity and the critical depth in a trapezoidal channel with bottom width B and the side slopes, $\dot{n}:1$.

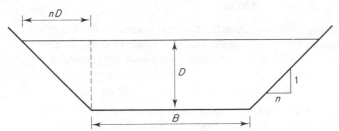

Figure 16.3

Solution The cross-section most frequently used for large open channels is trapezoidal. Thus, in Fig. 16.3

Area of section, $A = (B + nD)D$

Specific energy, $E = D + \bar{v}^2/2g$

giving $\bar{v} = \sqrt{\{2g(E-D)\}}$.

Volume rate of flow, $Q = A\bar{v} = D(B + nD)\sqrt{\{2g(E-D)\}}$ \qquad (I)

Under critical flow conditions for a constant value of E, the discharge Q is a maximum and $dQ/dD = 0$. From (I),

$$\log_e Q = \log_e D + \log_e (B + nD) + \tfrac{1}{2} \log_e 2g + \tfrac{1}{2} \log_e (E-D).$$

Differentiating with respect to D,

$$\frac{1}{Q}\frac{dQ}{dD} = \frac{1}{D} + \frac{n}{B + nD} + \frac{-1}{2(E-D)} \cdot$$

Since $dQ/dD = 0$ when $D = $ critical depth D_c,

$$\frac{1}{D_c} + \frac{n}{BnD_c} - \frac{1}{2(E-D_c)} = 0,$$

$$5nD_c^2 + (3B - 4nE)\,D_c - 2RE = 0. \tag{II}$$

If n is not zero

$$D_c = \frac{-(3B - 4nE) \pm \sqrt{(9B^2 - 24BnE + 16n^2E^2 + 40BnE)}}{10n}$$

$$= \frac{4nE - 3B + \sqrt{(16n^2E^2 + 16nEB + 9B^2)}}{10n}. \tag{III}$$

This is a general solution. If $B = 0$, the channel is triangular and (III) gives $D_c = \frac{4}{5}E$. If $n = 0$, (III) does not apply, but (II) gives $D_c = \frac{2}{3}E$ for a rectangular section.

In practice tables and curves are available for the determination of the critical depth in trapezoidal channels having any of a number of bottom widths and side slopes.

16.3. Computer program 'CRITNOR'

(1) The program deals with steady uniform open channel flow depths in circular or rectangular section channels. The critical depth separating sub-critical from super-critical flow is found and a choice of expressions for normal depth is presented.

(2) The required input is:

(i)   choice of circular or rectangular section channels;
(ii) channel diameter or width and maximum depth allowable (m);
(iii) choice of expression for the loss coefficient C in Chezy's equation: four options are offered based on Manning's n, the Bazin and Kutter expressions and the Colebrook-White equation modified for partially filled pipe flow by expressing hydraulic mean depth as $D/4$;
(iv) Manning's n or surface roughness k as appropriate following choice in (iii);
(v)   channel slope S;
(vi) flowrate Q in $m^3\,s^{-1}$.

(3) The program utilizes equations (15.6) (Chezy), (15.7), (15.8), (15.9) and (8.60) (Colebrook-White) with $D = 4m$, where $m = A/P$.

(4) Input example:
Circular section $I = 1$; $D = 0.1$ m; Manning equation $J = 1$; $n = 0.009$; $S = 0.01$; $Q = 0.0002\ m^3 s^{-1}$.

(5) Output:

CHANNEL DATA

PARTIALLY FILLED PIPE OF DIAMETER =   0.1 M.

FLOWRATE =   2E-4 M^3/S.

CRITICAL DEPTH =   1.3867E-2 M.

CHANNEL SLOPE =   1E-2

MANNING EQUATION USED WITH N =   9E-3

NORMAL DEPTH=   1.1133E-2 M.

DO YOU WISH TO REPEAT, Y OR N. ?N

List:

```
10 PRINT"PROGRAM CRITNOR"
20 MODE 0
30 PRINT:PRINT"PROGRAM DESIGNED TO CALCULATE FLOW CRITICAL AND NORMAL DEPTHS
IN EITHER PARTIALLY FULL PIPES OR RECTANGULAR CHANNELS."
40 PRINT:PRINT"CRITICAL DEPTH BASED ON SPECIFIC ENERGY RELATIONSHIPS."
50 PRINT:PRINT"NORMAL DEPTH BASED ON THE CHEZY EQUATION FOR STEADY FLOW WITH
A CHOICE OF GOVERNING LOSS COEFFS."
60 PRINT:INPUT"INPUT GEOMETRY CODE:- ";"I=1 CIRCULAR SECTION CHANNELS, I=2 RE
CTANGULAR SECTION CHANNELS. ",I
70 INPUT"INPUT PIPE DIAMETER OR CHANNEL MAX FLOW DEPTH (M.) D= ",D
80 IF I=2 THEN INPUT "CHANNEL WIDTH (M.) B= ",B
90 PRINT:PRINT"CHOICE OF C VALUES IN CHEZY EQUATION AS FOLLOWS;-"
100 PRINT"MANNING J=1"
110 PRINT"BAZIN J=2"
120 PRINT"KUTTER J=3"
130 PRINT"COLEBROOK WHITE J=4"
140 INPUT"J= ",J
150 IF J=1 THEN INPUT"MANNING'S N= ",N
160 IF J=2 THEN INPUT"BAZIN. SURFACE ROUGHNESS K= ",K
170 IF J=3 THEN INPUT"KUTTER. VALUE OF MANNING'S N =",N
180 IF J=4 THEN INPUT"COLEBROOK-WHITE. SURFACE ROUGHNESS K= ",K
190 INPUT "CHANNEL SLOPE S= ",S
200 REM CALC. OF CRITICAL DEPTH FOR ALL CASES.
210 INPUT"FLOWRATE Q= ",Q
220 U=D
230 L=0.0
240 C=(U+L)/2.0
250 GOSUB 820
260 F=1.0-Q^2*T/(9.81*A^3)
270 IF F<0.0 THEN L=C
280 IF F>0.0 THEN U=C
290 IF F=0.0 THEN GOTO 340
300 E=(U+L)/2.0
310 IF ABS((E-C)/C) < 0.01 THEN GOTO 340
320 C=E
330 GOTO 250
340 CLS:PRINT"CHANNEL DATA"
350 PRINT:IF I=1 THEN PRINT"PARTIALLY FILLED PIPE OF DIAMETER = ",D," M."
360 IF I=2 THEN PRINT"OPEN RECTANGULAR CHANNEL OF WIDTH = ",B," M."
370 PRINT:PRINT"FLOWRATE = ",Q," M^3/S."
380 PRINT:PRINT"CRITICAL DEPTH = ",C," M."
390 REM CALC NORMAL DEPTH USING BISECTION METHOD.
400 U=D
410 L=0.0
420 C=(U+L)/2.0
430 GOSUB 820
440 IF J=1 THEN Z=(A/P)^0.1667/N
450 IF J=2 THEN Z=86.9/(1.0+K/SQR(A/P))
460 IF J=3 THEN Z=(23.0+0.00155/S+1.0/N)/(1.0+(23+0.00155/S)*N/SQR(A/P))
470 IF J=4 THEN GOSUB 660
480 F=Z*SQR((A/P)*S)-Q/A
490 IF F<0.0 THEN L=C
500 IF F>0.0 THEN U=C
510 IF F=0.0 THEN GOTO 560
520 E=(U+L)/2.0
530 IF ABS((E-C)/C) < 0.01 THEN GOTO 560
540 C=E
550 GOTO 430
560 PRINT:PRINT"CHANNEL SLOPE = ",S
570 PRINT:IF J=1 THEN PRINT"MANNING EQUATION USED WITH N = ",N
580 IF J=2 THEN PRINT"BAZIN EQUATION USED WITH K = ",K
590 IF J=3 THEN PRINT"KUTTER EQUATION USED WITH N = ",N
600 IF J=4 THEN PRINT"COLEBROOK WHITE EQUATION USED WITH K = ",K
610 PRINT:PRINT"NORMAL DEPTH = ",C," M."
620 PRINT:INPUT"DO YOU WISH TO REPEAT, Y OR N. ",Z$
630 IF Z$="Y" THEN GOTO 10
640 GOTO 960
650 REM CALC OF FRICTION FACTOR FROM COLEBROOK WHITE WITH M REPLACED BY D/4
660 V=0.05
670 G=0.0
680 X=(V+G)/2.0
690 GOTO 700
700 F=1.0/SQR(X)+4.0*LOG((K/(3.71*4.0*(A/P)))+1.26/((Q/A)*4.0*(A/P)*SQR(X)/(1.
41*10.0^(-6))))
710 IF F>0.0 THEN G=X
720 IF F<0.0 THEN V=X
730 IF F=0.0 THEN GOTO 800
```

```
740 PRINT "F= ",F
750 PRINT"X=",X
760 W=(V+G)/2.0
770 IF ABS((W-X)/X)<0.01 THEN GOTO 800
780 X=W
790 GOTO 690
800 Z=SQR(2.0*9.81/X)
810 RETURN
820 IF I=1 THEN GOTO 890
830 REM RECTANGULAR CHANNELS
840 T=B
850 P=B+2.0*C
860 A=T*C
870 GOTO 950
880 REM PARTIALLY FILLED PIPES.
890 IF C<D/2.0 THEN O=2.0*ATN(SQR(C*(D-C))/(D/2.0-C))
900 IF C=D/2.0 THEN O=PI
910 IF C>D/2.0 THEN O=PI+2.0*ATN((C-D/2.0)/(SQR(C*(D-C))))
920 A=((D^2.0)/8.0)*(O-SIN(O))
930 T=2.0*((C*(D-C))^0.5)
940 P=D*O/2.0
950 RETURN
960 PRINT:PRINT"CALCULATION COMPLETE."
```

16.4. Non-dimensional specific energy curves

The curve in Fig. 16.1(a) is drawn for a single value of E and is, therefore, one of a family of curves for other values of E which would all be similar in shape. These can be reduced to a single curve in non-dimensional form applying to all values of E by dividing equation (16.2) by q^2_{max}, the square of the maximum discharge occurring at critical depth:

$$\frac{E}{q^2_{max}} = \frac{D}{q^2_{max}} + \frac{1}{2gd^2}\left(\frac{q}{q_{max}}\right)^2.$$

Now, from equation (16.6), $q_{max} = (gd_c^3)^{1/2}$. Hence,

$$\frac{E}{gD_c^3} = \frac{D}{gd_c^3} + \frac{1}{2gD^2}\left(\frac{q}{q_{max}}\right)^2,$$

$$\left(\frac{q}{q_{max}}\right)^2 = \frac{2E}{D_c}\left(\frac{D}{D_c}\right)^2 - 2\left(\frac{D}{D_c}\right)^3.$$

For a rectangular channel, $E = \frac{3}{2}d_c$; therefore,

$$\left(\frac{q}{q_{max}}\right)^2 = 3\left(\frac{D}{D_c}\right)^2 - 2\left(\frac{D}{D_c}\right)^3.$$

Similarly, Fig. 16.1(b) can be presented in a non-dimensional form applicable to all values of q by dividing equation (16.2) by D_c.

$$\frac{E}{D_c} = \frac{D}{D_c} + \frac{1}{2gD_c}\left(\frac{q}{D}\right)^2$$

or, since for minimum energy $q = gD_c^3$,

$$\frac{E}{D_c} = \frac{D}{D_c} + \frac{1}{2}\left(\frac{D_c}{D}\right)^2.$$

16.5. Occurrence of critical flow conditions

Since, at the critical depth, the volume rate of flow is a maximum for a given specific energy, cross-sections at which the flow passes through the critical depth are known as *control sections*. Such sections are a limiting factor in the design of a channel and can be expected to occur under the following circumstances:

(i) *Transition from tranquil to shooting flow*. This may occur as shown in Fig. 16.4 where there is a change of bed slope *s*. Upstream the slope is mild

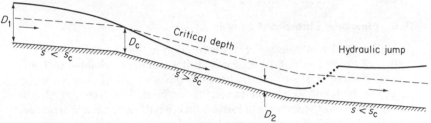

Figure 16.4

and *s* is less than the critical slope s_c. Over a considerable distance the depth will change smoothly from D_1 to D_2 and at the break in the slope the depth will pass through the critical depth forming a control section which regulates the depth upstream. The reverse transition from shooting to tranquil flow occurs abruptly by means of a hydraulic jump (*see* page 462).

(ii) *Entrance to a channel of steep slope from a reservoir*. If the depth of flow in the channel is less than the critical depth channel the water surface must pass through the critical depth in the vicinity of the entrance (Fig. 16.5) since conditions in the reservoir correspond to tranquil flow.

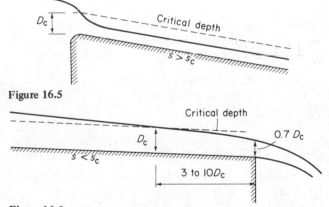

Figure 16.5

Figure 16.6

(iii) *Free outfall from a channel with a mild slope*. In Fig. 16.6 if the slope *s* of the channel is less than s_c the upstream flow will be tranquil. At the outfall there is no resistance to flow so that, theoretically, it will be a maximum and the depth should be critical. In practice, the gravitational acceleration creates a curvature of the streamlines and an increase of velocity at the brink so that the depth is less than critical. Experiments indicate that critical depth occurs at a distance of $3\,D_c$ to $10\,D_c$ from the brink and that the depth at the brink is approximately $0.7\,D_c$. If the slope of the channel is steep, s being greater than s_c, the upstream flow will be shooting and the depth will everywhere be less than the critical depth.

(iv) *Change of bed level or channel width*. Under certain circumstances flow will occur at critical depth if a hump is formed in the bed of the channel or the width of the channel is reduced. These cases are discussed in Sections 16.6 and 16.7.

16.6. Flow over a broad-crested weir

A broad-crested weir consists of an obstruction in the form of a raised portion of the bed extending across the full width of the channel with a flat upper surface or crest sufficiently broad in the direction of flow for the surface of the liquid to become parallel to the crest. The upstream edge is rounded to avoid the separation losses which would occur at a sharp edge.

In Fig. 16.7 the flow upstream is tranquil and the conditions downstream

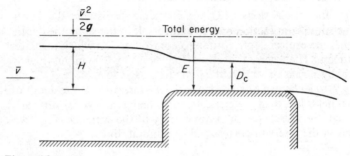

Figure 16.7

allow a free fall over the weir. Since there is no restraining force on the liquid, the discharge over the weir will be the maximum possible and flow over the weir will take place at the critical depth. For a rectangular channel, from equation (16.4),

$$D_c = (Q^2/gB^2)^{1/3}$$

so that $Q = B(gD_c^3)^{1/2}$.

Since $D_c = \tfrac{2}{3}E$,

$$Q = B(g \times \tfrac{8}{27}E^3)^{1/2} = 1.705\,BE^{3/2} \text{ in SI units.} \tag{16.11}$$

The specific energy E measured above the crest of the weir will, assuming no losses, be equal to $H + \bar{v}^2/2g$, where H is the height of the upstream water level above the crest and $\bar{v}$ is the mean velocity at a point upstream where the flow is uniform. If the depth upstream is large compared with the depth over the weir, $\bar{v}^2/2g$ is negligible and equation (16.11) can be written

$$Q = 1.705\,BH^{3/2} \text{ in SI units.} \tag{16.12}$$

A single measurement of the head H above the crest of the weir would then be sufficient to determine the discharge Q.

Since the critical depth $D_c = (Q^2/gB^2)^{1/3}$, the depth over the crest of the weir is fixed, irrespective of its height. Any increase of height of the weir will not alter D_c but will cause an increase in the depth of flow upstream.

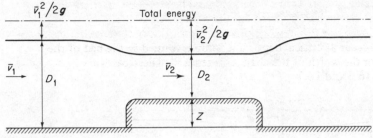

Figure 16.8

If, as in Fig. 16.8, the level of the flow downstream is raised, the surface level will be drawn down over the hump, but the depth may not fall to the critical depth. The rate of flow can be calculated by applying Bernoulli's equation and the continuity of flow equations and will depend upon the difference in surface level upstream and over the weir.

16.7. Effect of lateral contraction of a channel

When the width of a channel is reduced while the bed remains flat (Fig. 16.9), the discharge per unit width increases. If losses are neglected,

Figure 16.9

the specific energy remains constant and so, from Fig. 16.1(a), for tranquil flow the depth will decrease while for shooting flow the depth will increase as the channel narrows.

A lateral contraction followed by an expansion can be used for flow measurement as an alternative to the broad-crested weir. If the conditions are such that the free surface does not pass through the critical depth, the arrangement forms a *venturi flume* analogous to the venturi meter used for flow in pipes. Referring to Figure 16.10(a), for continuity of flow,

$$B_1 D_1 \bar{v}_1 = B_2 D_2 \bar{v}_2 \qquad\qquad (16.13)$$

Applying Bernoulli's equation to upstream and throat sections and ignoring losses,

$$D_1 + \bar{v}_1^2/2g = D_2 + \bar{v}_2^2/2g.$$

Substituting for $\bar{v}_1$ from equation (16.13),

$$\frac{\bar{v}_2^2}{2g}\left(1 + \frac{B_2^2 D_2^2}{B_1^2 D_1^2}\right)^2 = D_1 - D_2 = h,$$

$$\bar{v}_2 = \sqrt{\left\{\frac{2gh}{1 - (B_2 D_2/B_1 D_1)^2}\right\}}$$

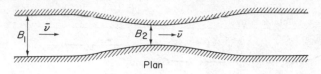

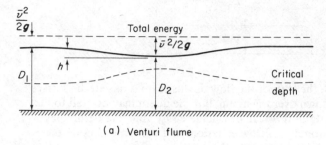

(a) Venturi flume

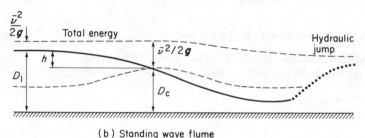

(b) Standing wave flume

Figure 16.10

Volume rate of flow, $Q = B_2 D_2 \bar{v}_2$

$$= B_2 D_2 \sqrt{\left\{\frac{2gh}{1 - (B_2 D_2 / B_1 D_1)^2}\right\}} . \quad (16.14)$$

Due to energy losses, the actual discharge in practice will be slightly less than this value and is given by

$$Q = C_d B_2 D_2 \sqrt{\left\{\frac{2gh}{1 - (B_2 D_2 / B_1 D_1)^2}\right\}},$$

where C_d is a coefficient of discharge of the order of 0·95 to 0·99.

If the degree of contraction and the flow conditions are such that the upstream flow is tranquil, and the free surface passes through the critical depth in the throat as shown in Fig. 16.10(b),

$$Q = B_2 D_c v_c = B_2 D_c \sqrt{\{2g(E - D_c)\}},$$

where E is the specific energy measured above the bed level at the throat and, since the critical depth $D_c = \frac{2}{3}E$,

$$Q = B_2 \times \tfrac{2}{3}E \sqrt{(2g \times \tfrac{1}{3}E)} = 1·705\, BE^{3/2} \text{ in SI units.}$$

Assuming, as for the broad-crested weir, that the upstream velocity head is negligible,

$$Q = 1·705\, BH^{3/2} \text{ in SI units,} \quad (16.15)$$

where H is the height of the upstream free surface above bed level at the throat. In some cases, in addition to the lateral contraction, a hump is formed in the bed (as shown in Fig. 16.11), in which case $H = D_1 - Z$.

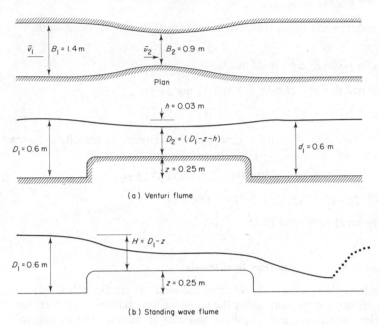

Plan

$h = 0.03$ m

$D_1 = 0.6$ m

$D_2 = (D_1 - z - h)$

$d_1 = 0.6$ m

$z = 0.25$ m

(a) Venturi flume

$H = D_1 - z$

$D_1 = 0.6$ m

$z = 0.25$ m

(b) Standing wave flume

Figure 16.11

If the upstream conditions are tranquil and the bed slope is the same downstream as upstream, it will not be possible for shooting flow to be maintained for any great distance from the throat. It will revert to tranquil flow downstream by means of an hydraulic jump or standing wave. A venturi flume operating in this mode is known as a *standing wave flume*.

Example 16.2

A venturi flume is formed in a horizontal channel of rectangular cross-section 1·4 m wide by constricting the width to 0·9 m and raising the floor level in the constricted section by 0·25 m above that of the channel. If the difference in levels of the free surface between the throat and upstream is 30 mm and both upstream and downstream depths are 0·6 m, calculate the volume rate of flow.

If the downstream conditions are changed so that a standing wave forms clear of the constriction, what will be the volume rate of flow if the upstream depth is maintained at 0·6 m?

Solution
The venturi flume is shown in Fig. 16.11 and, from equation (16.14),

$$Q = B_2 D_2 \sqrt{\left\{ \frac{2gh}{1 - (B_2 D_2 / B_1 D_1)^2} \right\}}$$

$$= 0 \cdot 9 \times 0 \cdot 32 \sqrt{\left\{ \frac{2 \times 9 \cdot 81 \times 0 \cdot 3}{1 - (0 \cdot 9 \times 0 \cdot 32/1 \cdot 4 \times 0 \cdot 6)^2} \right\}}$$

$$= 0 \cdot 2352 \ \mathrm{m^3 \ s^{-1}}.$$

The standing wave flume is shown in Fig. 16.10(b) and, from equation (16.15),

$$Q = 1 \cdot 705 \ B_2 E^{3/2} \ \text{in SI units.}$$

If it is assumed that $E = H = D_1 - Z = 0 \cdot 35$ m, then

$$Q = 1 \cdot 705 \times 0 \cdot 9 \times 0 \cdot 35^{3/2} = 0 \cdot 3177 \ \mathrm{m^3 \ s^{-1}}.$$

Checking to determine the effect of neglecting the upstream velocity $\bar{v}_1$, since $Q = B_1 \bar{v}_1 D_1$,

$$\bar{v}_1 = Q/B_1 D_1 = 0 \cdot 3177/1 \cdot 4 \times 0 \cdot 6 = 0 \cdot 3782 \ \mathrm{m \ s^{-1}},$$

$$\bar{v}_1^2/2g = (0 \cdot 3782)^2/2 \times 9 \cdot 81 = 0 \cdot 0073 \ \mathrm{m}.$$

This is very small compared to H.

16.8. Non-uniform, steady flow in channels

In non-uniform flow, the depth and cross-sectional area of the actual flow may vary from point to point along the length of the channel in contrast to uniform flow, in which the slope, shape and area of cross-section are constant and flow takes place at the *normal* depth. True uniform flow is found only in artificial channels and, even then, flow will be non-uniform near the entrance and exit. In natural streams, the slope of the bed and the shape and size of the cross-section vary considerably and true uniform flow does not exist. Nevertheless, this type of flow can be analysed using the equations for uniform flow, such as the Chezy or Manning formulae, by dividing its length into sections, or *reaches*, within which conditions are approximately constant.

Non-uniform flow may be either *gradually varied flow*, in which the change in conditions extends over a long distance, or it can be local non-uniform flow, known as *rapidly varied flow*, in which changes take place suddenly, as at an hydraulic jump, or over a short distance, as at the entrance to a channel or at an obstruction such as a bridge pier or weir. For the purpose of analysis, the distinction between these types is that, for gradually varied flow, it is assumed that changes take place sufficiently slowly for the effects of acceleration to be negligible. In the analysis of rapidly varied flow, the acceleration of the fluid and the resulting rate of change of momentum cannot be overlooked.

It was seen in Section 15.1 that, for uniform flow, the total energy line, the water surface and the bed are all parallel, so that the bed slope s is equal to the slope of the total energy line i and the depth is constant. In non-uniform flow, the depth is variable, s may be greater or less than i for any particular reach and the flow may be accelerating or decelerating. In practice, the variation in depth may be of considerable importance, e.g. in determining the possibility of the occurrence of flooding.

16.9. Equations for gradually varied flow

For channels of regular cross-section it is possible to derive analytically an expression for the variation of depth from point to point and so to determine, theoretically, the profile of the free surface if it is assumed that:

   (i) the channel is rectangular, straight and of constant roughness;
   (ii) the bed slope is small, so that the depth measured normal to the bed can be assumed equal to the vertical depth;
   (iii) flow is steady and streamlines are approximately parallel, so that pressure distribution is hydrostatic.

Consider a length of channel δl (Fig. 16.12) of rectangular cross-section and bed slope s. At section A the depth is D and the velocity v, while at

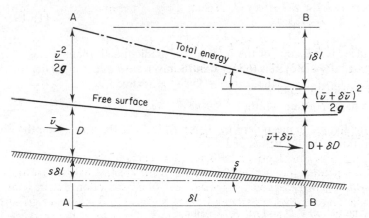

Figure 16.12

section B the depth is $D + \delta D$ and the velocity $v + \delta v$. The loss of energy between A and B will be $i\delta l$, where i is the slope of the total energy line, and the fall in the bed level will be $s\delta l$. Applying Bernoulli's equation to A and B,

$$s\delta l + D + v^2/2g = (D + \delta D) + (v + \delta v)^2/2g + i\delta l,$$

$$s\delta l = \delta D + (v\delta v/g) + i\delta l$$

(neglecting second order of small quantities),

$$\delta D/\delta l = s - (v\delta v/g\delta l) - i. \tag{16.16}$$

Assuming a constant width of channel, for continuity of flow the discharge per unit width is constant from section to section and so

$$vD = (v + \delta v)(D + \delta D),$$

$$vD = vD + v\delta D + D\delta v + \text{second order terms},$$

$$\delta v = -v\delta D/D.$$

Substituting in equation (16.16)

$$\frac{\delta D}{\delta l} = s - i + \frac{v^2}{gD}\frac{\delta D}{\delta l},$$

$$\frac{\delta D}{\delta l} = \frac{s - i}{1 - v^2/gD}. \tag{16.17}$$

This is the basic equation for non-uniform flow and has a number of alternative forms. Since $v/\sqrt{(gD)}$ is the Froude number Fr, equation (16.17) becomes

$$\delta D/\delta l = (s - i)/(1 - \mathrm{Fr}^2). \tag{16.18}$$

Also, since $v = q/D$,

$$v^2/gD = q^2/gD^3 = (D_c/D)^3,$$

where D_c is the critical depth. Substituting in equation (16.17),

$$\frac{\delta D}{\delta l} = \frac{s - i}{1 - (D_c/D)^3} \tag{16.19}$$

Moreover, since i is the slope of the total energy gradient and, therefore, is equal to the bed slope required to maintain the given flow at a normal depth D, if a resistance formula of the form $v = Km^a i^b$ is assumed,

Discharge per unit width, $q = DKm^a i^b$,

where m is the hydraulic mean depth at depth D. If D_n is the normal depth for the given flow q on the actual bed slope s,

$$q = D_n K m_n^a s^b$$

and so $Dm^a i^b = D_n m_n^a s^b$.

For a wide channel, $m = D$ and $m_n = D_n$. Hence,

$$(i/s)^b = (D_n/D)^{1+a}$$

or $i/s = (D_n/D)^c$,

where $c = (1 + a)/b$. Substituting in equation (16.19),

$$\frac{\delta D}{\delta l} = s \frac{1 - (D_n/D)^c}{1 - (D_c/D)^3}. \tag{16.20}$$

If the width of the channel is not constant, but changes from B to $B + \delta B$, the continuity of flow equation will be

$$Q = B\bar{v}D = (B + \delta B)(\bar{v} + \delta\bar{v})(D + \delta D)$$

and, neglecting second-order terms,

$$\delta\bar{v} = \frac{\bar{v}}{D}\delta D - \frac{\bar{v}}{B}\delta B.$$

Substituting in equation (16.16),

$$\frac{\delta D}{\delta l} = s - i + \frac{\bar{v}^2}{gD}\frac{\delta D}{\delta l} + \frac{\bar{v}^2}{gB}\frac{\delta B}{\delta l}$$

$$= \frac{s - i + (\bar{v}^2/gB)(\delta B/\delta l)}{1 - \bar{v}^2/gD}. \tag{16.21}$$

From this equation, it can be seen that, if $s = i$, dD/dl is zero and so flow is uniform. Also, if $\bar{v}^2/gD = 1$ or $Fr = 1$, dD/dl is infinite, so that, theoretically, the slope of the water surface is vertical. For any other conditions, these equations give the slope of the water surface at any point and so, using a step by step method of integration, the profile of the free surface can be constructed.

16.10. Classification of water surface profiles

It can be seen from equation (16.20) that the gradient of the free surface dD/dl may be positive or negative, depending on the signs of the numerator and denominator, which in turn will depend on the relative values of the actual depth D, the critical depth D_c, and the normal depth D_n. Whether the normal depth is above or below the critical depth will depend on the classification of the bed slope. Surface profiles are classified by a letter and a number. The letter refers to the bed slope, which may be one of the following categories:

M	Mild slope	$D_n > D_c$;
C	Critical slope	$D_n = D_c$;
S	Steep slope	$D_n < D_c$;
H	Horizontal	$s = 0$;
A	Adverse	bed slope s is negative.

The number refers to the relation between the actual depth of flow D, the normal depth D_n and the critical depth D_c as follows:

1   free surface of stream lies above both normal and critical depth lines;
2   free surface of stream lies between normal and critical depth lines;
3   free surface of stream lies below both normal and critical depth lines.

Although there are five categories of slope and three categories of depth only twelve combinations are possible. For horizontal and adverse slopes, uniform flow is impossible, thus eliminating the H1 and A1 curves; while for critical slope, critical depth and normal depth coincide, thus eliminating the C2 curve. The water profiles corresponding to the various categories are shown in Fig. 16.13 with an indication of how they may occur. Note that the vertical scale has been greatly exaggerated and that even a steep slope, in this context, would amount to only a few degrees inclination to the horizontal. Water surface curves in which the depth increases in the direction of flow are known as *backwater* curves. If the depth decreases in the direction of flow they are called *drop down* or *draw down* curves. Backwater curves have a positive value of dD/dl and draw down curves have a negative value of dD/dl. The type of curve corresponding to each of the classifications can, therefore, be determined by examining whether the numerator and denominator are positive or negative in equation (16.20) for the M, S and C curves, or in equation (16.19) for the H and A curves, as shown in Table 16.1.

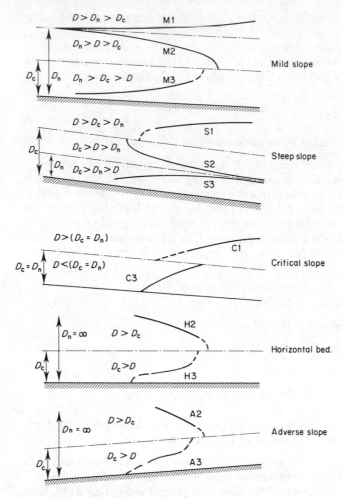

Figure 16.13   Water surface profiles

For all type 1 curves, the surface will approach the horizontal asymptotically since, as the depth increases, the velocity decreases and tends to zero. All curves approach the normal depth line asymptotically since the steady state is one of uniform flow at a distance remote from the disturbance. In theory, curves cross the critical depth line vertically, since the denominator in equation (16.20) becomes zero, making dD/dl infinite. However, in this region the theory does not apply, since it conflicts with the assumption (that changes will be gradual) on which it is based.

Example 16.3

A wide canal has a bed slope of 1 in 1000 and conveys water at a normal depth of 1·2 m. A weir is to be constructed at one point to increase the depth of flow to 2·4 m. How far upstream of the weir will the depth be 1·35 m. Take C in the Chezy formula as 55 in SI units.

Class	Relative depths	Equation (16.20) $1-(D_n/D)^c$	$1-(D_c/D)^3$	Free surface slope, dD/dl
M1	$D > D_n > D_c$	+ve	+ve	+ve
M2	$D_n > D > D_c$	−ve	+ve	−ve
M3	$D_n > D_c > D$	−ve	−ve	+ve
S1	$D > D_c > D_n$	+ve	+ve	+ve
S2	$D_c > D > D_n$	+ve	−ve	−ve
S3	$D_c > D_n > D$	−ve	−ve	+ve
C1	$D > (D_n = D_c)$	+ve	+ve	+ve
C2		Not feasible since D_n and D_c coincide		
C3	$D < (D_n = D_c)$	−ve	−ve	+ve
		Equation (16.19) $s - i$	$1-(D_c/D)^3$	
H1		Not feasible since D_n is indeterminate		
H2	$D > D_c$	−ve	+ve	−ve
H3	$D < D_c$	−ve	−ve	+ve
A1		Not feasible since D_n is indeterminate		
A2	$D > D_c$	−ve	+vc	−ve
A3	$D < D_c$	−ve	−ve	+ve

Table 16.1

Solution
From equation (16.20), using the Chezy formula and putting $a = \frac{1}{2}$, $b = \frac{1}{2}$ so that $c = (1 + \frac{1}{2})/\frac{1}{2} = 3$,

$$\delta l = \delta D \{1 - (D_c/D)^3\}/s\{1 - (D_n/D)^3\}, \tag{I}$$

normal depth, $D_n = 1 \cdot 2$ m, bed slope, $s = 1/1000$. The discharge per unit width Q for uniform flow is found from the Chezy formula, taking the hydraulic mean depth m as equal to the depth, since the channel is wide.

$$Q/B = q = D_n C (D_n s)^{1/2} = 1 \cdot 2 \times 55 (1 \cdot 2/1000)^{1/2} = 2 \cdot 286 \text{ m}^2 \text{ s}^{-1},$$

$$\text{Critical depth, } D_c = (q^2/g)^{1/3} = 0 \cdot 811 \text{ m}.$$

Using a step by step method, the distance δl corresponding to a change in depth δD can be calculated working from the known conditions at the weir back upstream. The general rule is to work from the cause of the change towards the point at which flow at normal depth occurs, i.e. upstream for tranquil flow and downstream for shooting flow.

Dividing the range of depths from 2·4 to 1·35 m into seven steps of 0·15 m, the length for each step can be calculated from (I) using the mean depth $\bar{D}$ for each step. Putting $\delta D = 0 \cdot 15$ m and $s = 1/1000$,

$$\delta l = 150 \frac{\{1 - (0 \cdot 811/\bar{D})^3\}}{\{1 - (1 \cdot 2/\bar{D})^3\}} \text{ m}.$$

The work is best carried out in tabular form:

Depth (m)	Mean depth, D_m	$1 - (0.811/\bar{D})^3$	$1 - (1.2/\bar{D})^3$	L (m)
2.4				
2.25	2.325	0.9576	0.8625	166.54
2.10	2.175	0.9482	0.8321	170.93
1.95	2.025	0.9358	0.7919	177.26
1.80	1.875	0.9191	0.7379	186.83
1.65	1.725	0.8961	0.6634	202.62
1.50	1.575	0.8627	0.5577	232.03
	1.425	0.8157	0.4028	303.76

Total 1439.97

Distance upstream at which depth is 1.35 m = **1440 m approx.**

16.11. The hydraulic jump

The hydraulic jump is an important example of local non-uniform flow. As can be seen from the water surface profiles in Fig. 16.14, there is no possibility of a smooth transition from shooting to tranquil flow since, theoretically, the slope of the water surface should be vertical as it passes through the critical depth. In practice this cannot occur, and the transition takes the form of the hydraulic jump with a steep upward sloping water surface and violently turbulent conditions accompanied by a substantial loss of energy. As for steady flow, the mass per unit time flowing upstream and downstream of the jump (Fig. 16.14) will be equal and, since the velocity upstream is greater than the velocity downstream, there will be a change in the momentum of the stream per unit time as it passes through the jump. For a channel with a moderate bed slope, the force slowing down the stream and

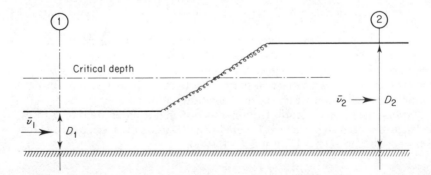

Figure 16.14

producing this rate of change of momentum is due to the difference in the resultant forces caused by the hydrostatic pressure at the downstream and the upstream cross-sections.

If Q is the volume rate of flow, B the width of the channel, assumed to be rectangular, D_1 and D_2, $\bar{v}_1$ and $\bar{v}_2$ the depths and mean velocities at the upstream and downstream sections, respectively, for continuity of flow,

$$Q = B\bar{v}_1 D_1 = B\bar{v}_2 D_2 \quad \text{and so} \quad \bar{v}_2 = \bar{v}_1 (D_1/D_2),$$

$$\begin{array}{r}\text{Rate of change of momentum} \\ \text{between sections 1 and 2}\end{array} = \rho Q(\bar{v}_1 - \bar{v}_2)$$

$$= \rho B\bar{v}_1^2 D_1 (1 - D_1/D_2),$$

$$\begin{array}{r}\text{Hydrostatic force acting downstream} \\ \text{at section 1}\end{array} = \tfrac{1}{2}\rho g D_1^2 B,$$

$$\begin{array}{r}\text{Hydrostatic force acting upstream} \\ \text{at section 2}\end{array} = \tfrac{1}{2}\rho g D_2^2 B,$$

$$\text{Resultant force acting upstream} = \tfrac{1}{2}\rho g B(D_2^2 - D_1^2).$$

By Newton's second law,

$$\rho B\bar{v}_1^2 D_1 (1 - D_1/D_2) = \tfrac{1}{2}\rho g B(D_2^2 - D_1^2),$$

$$D_2^2 - D_1^2 = (2\bar{v}_1^2 D_1/gD_2)(D_2 - D_1),$$

$$D_2 + D_1 = 2\bar{v}_1^2 D_1/gD_2,$$

$$D_2^2 + D_1 D_2 - 2\bar{v}_1^2 D_1/gD_2 = 0,$$

$$D_2 = \tfrac{1}{2}D_1 \left(-1 + \sqrt{\{1 + 8\bar{v}_1^2/gD_1\}}\right). \tag{16.22}$$

From equation (16.22), the conjugate depths D_1 and D_2 before and after the jump can be determined. Since $\bar{v}_1/\sqrt{(gD_1)} = \mathrm{Fr}_1$ the Froude number at the upstream section,

$$D_2 = \tfrac{1}{2}D_1 \left(-1 + \sqrt{\{1 + 8\mathrm{Fr}_1^2\}}\right).$$

The loss of energy in the jump will be equal to the difference of specific energies at the upstream and downstream sections.

$$\text{Loss of head} = (D_1 + \bar{v}_1^2/2g) - (D_2 + \bar{v}_2^2/2g).$$

16.12. Location of an hydraulic jump

It is often desirable to be able to determine the position at which an hydraulic jump will occur. For example, in the design of a spillway over a dam (Fig. 16.15), the energy of the fast flowing stream must be partially dissipated to prevent erosion of the bed downstream. This can be done by arranging for the formation of an hydraulic jump, but to prevent damage this must occur on the apron. Shooting flow down the face of the dam is retarded by the flatter slope of the apron, which is insufficient to maintain its original high velocity, and a jump will occur. An obstruction can be introduced on the apron to force the jump to form at the desired position but, if this is not done, the position at which the jump will occur naturally

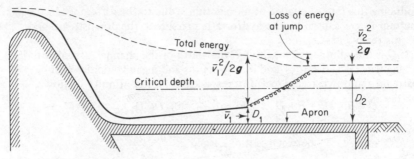

Figure 16.15

can be estimated using equation (16.22) to determine the possible conjugate upstream and downstream depths for the known upstream and downstream conditions.

If the discharge and velocity at the foot of the spillway are known and the downstream depth D_2 is fixed, the conjugate depth D_1 can be determined. From equation (16.22), if q is the discharge per unit width,

$$D_1 = \tfrac{1}{2}D_2 (-1 + \sqrt{\{1 + 8q^2 D_2/g\}}). \tag{16.23}$$

Starting from the known conditions at the foot of the dam, the distance along the apron at which the depth of flow will have increased to D_1 can be calculated as explained in Section 16.10.

If the downstream conditions are not fixed, the position of the jump can still be determined by applying equation (16.22) to find the conjugate depths which are compatible with the surface profiles upstream and downstream. For example, if the slope of the channel changes from steep to mild (as in Fig. 16.16), the jump may occur either upstream or downstream of the break in the slope. To decide which of the two alternatives is possible, first determine the normal depth for the upstream and downstream slopes. Considering the possibility of the jump occurring upstream of the break, calculate the conjugate depth corresponding to the normal depth on the upstream slope. If this conjugate depth is less than the normal depth on the downstream slope, the jump will form on the upstream slope and be followed by an S1 curve leading to the normal depth downstream. On the other hand, if it is greater than the normal depth on the downstream slope, the jump cannot occur upstream of the break. The jump must, therefore, occur downstream of the break (as in Fig. 16.16(a)), the depth after the jump being normal depth on the downstream slope and the corresponding conjugate depth D'_1 occurring immediately before the jump. An M3 curve is formed upstream of the jump.

The position of the jump can also be determined graphically by plotting the water surface profiles for the upstream and downstream flow AA and BB, respectively (Fig. 16.17); in this case an accelerating flow upstream and a backwater curve starting from an obstacle downstream which is backing up the flow. If, for a number of points such as C, D and E on the upstream profile, the conjugate depths C', D' and E' are plotted, the position of the hydraulic jump will be the intersection of the line joining C', D' E' with the downstream backwater curve.

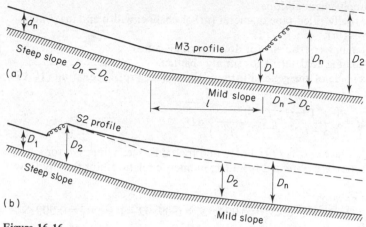

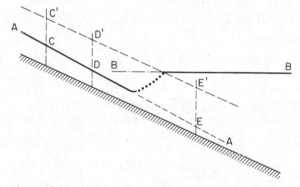

Figure 16.16

Figure 16.17

16.13. Computer program 'CHANNEL'

(1) The program calculates the gradually varied flow depth profiles for a
range of cases featuring a change in open channel slope.

The following cases are dealt with:
(a)   steep slope to steep slope;
(b)   mild slope to mild slope;
(c)   mild slope to steep slope;
(d)   steep slope to mild slope.

In cases (a), (b) and (c), the water surface profile in the downstream channel
is calculated. In case (d) the presence of an hydraulic jump in either of the
channels is covered, its position calculated and the water surface profiles
determined.

(2) The required input is:
(i)   choice of channel section;
(ii)   steady flow rate ($m^3 s^{-1}$);
(iii) downstream channel slope;

(iv) supply channel slope;
(v)   partially filled pipe diameter (m) or channel width and maximum
depth valves (m);
(vi) Manning coefficient for downstream channel;
(vii) Manning coefficient for supply channel.
(3) Use is made of the general form of the surface profile equation (16.17),
namely,

$$dl = \left\{ \frac{1 - Q^2\, T/g\, A^3}{S - (Q^2 m^{4/3})/n^2 A^2} \right\} dD$$

Values of surface width T, and of A and hydraulic mean depth m for a
circular section are obtained from trigonometric relationships for any flow
depth.
(4) Input example:
Code = 1, $Q = 1\ \mathrm{m}^3\,\mathrm{s}^{-1}$, $S_1 = 0{\cdot}001$; $S_s = 0{\cdot}09$, $D = 1$ m; $n_1 = 0{\cdot}009$,
$n_s = 0{\cdot}009$.
(5) Output:

```
CRITICAL DEPTH AT                    1 M^3/S IS              570 MM.

NORMAL DEPTH IN SUPPLY CHANNEL =          207 MM.
SUPERCRITICAL SUPPLY FLOW
NORMAL DEPTH IN DOWNSTREAM CHANNEL =              758 MM.
SUBCRITICAL DOWNSTREAM FLOW
SUPPLY CHANNEL SLOPE =           9E-2
DOWNSTREAM CHANNEL SLOPE =                1E-3
SUPPLY CHANNEL MANNING COEFF. =           9E-3
DOWNSTREAM CHANNEL MANNING COEFF. =                9E-3
HYDRAULIC JUMP POSITIONED IN DOWNSTREAM CHANNEL.
CONJUGATE DEPTH =                 392 MM.
HYDRAULIC JUMP PRESENT
PRESS RETURN TO CONTINUE...?

DEPTH PROFILE FROM SLOPE CHANGE TO JUMP.

DEPTH =      207 MM. AT          0 M. FROM ENTRY.
DEPTH =      219 MM. AT       6.59 M. FROM ENTRY.
DEPTH =      232 MM. AT       13.2 M. FROM ENTRY.
DEPTH =      244 MM. AT       19.8 M. FROM ENTRY.
DEPTH =      256 MM. AT       26.5 M. FROM ENTRY.
DEPTH =      269 MM. AT       33.1 M. FROM ENTRY.
DEPTH =      281 MM. AT       39.7 M. FROM ENTRY.
DEPTH =      293 MM. AT       46.2 M. FROM ENTRY.
DEPTH =      305 MM. AT       52.7 M. FROM ENTRY.
DEPTH =      318 MM. AT       59.2 M. FROM ENTRY.
DEPTH =      330 MM. AT       65.6 M. FROM ENTRY.
DEPTH =      342 MM. AT       71.9 M. FROM ENTRY.
DEPTH =      355 MM. AT       78.2 M. FROM ENTRY.
DEPTH =      367 MM. AT       84.3 M. FROM ENTRY.
DEPTH =      379 MM. AT       90.3 M. FROM ENTRY.
DEPTH =      392 MM. AT       96.2 M. FROM ENTRY.
DO YOU WISH TO CALCULATE THE DRAWDOWN PROFILE AT FREE DISCHARGE FROM THE DOWNSTR
EAM CHANNEL, Y OR N. ?Y
DRAWDOWN PROFILE FOR DOWNSTREAM CHANNEL ASSUMING FREE DISCHARGE.
DISTANCES MEASURED -VE UPSTREAM FROM CHANNEL EXIT.

DEPTH =      570 MM. AT         ·0 M. FROM EXIT.
DEPTH =      583 MM. AT     -0.524 M. FROM EXIT.
DEPTH =      595 MM. AT      -2.14 M. FROM EXIT.
DEPTH =      608 MM. AT      -5.06 M. FROM EXIT.
DEPTH =      620 MM. AT      -9.54 M. FROM EXIT.
DEPTH =      633 MM. AT      -15.9 M. FROM EXIT.
DEPTH =      645 MM. AT      -24.7 M. FROM EXIT.
DEPTH =      658 MM. AT      -36.5 M. FROM EXIT.
DEPTH =      670 MM. AT      -52.2 M. FROM EXIT.
DEPTH =      683 MM. AT      -73.5 M. FROM EXIT.
DEPTH =      695 MM. AT       -103 M. FROM EXIT.
DEPTH =      708 MM. AT       -144 M. FROM EXIT.
DEPTH =      720 MM. AT       -207 M. FROM EXIT.
DEPTH =      733 MM. AT       -321 M. FROM EXIT.
DEPTH =      745 MM. AT       -690 M. FROM EXIT.
```

List:

```
  10 REM CHANNEL
  20 MODE 3: @%=&30308
  30 DIM DEP(50):DIM X(50):DIM H(3):DIM DL(3)
  40 CLS:PRINT"PROGRAM CHANNEL."
  50 PRINT"PROGRAM DESIGNED TO CALCULATE THE WATER SURFACE PROFILES FOR DEVELOP
ING GRADUALLY VARIED FLOW"
  60 PRINT:PRINT"BOTH RECTANGULAR AND PARTIALLY FILLED PIPE FLOW ARE CATERED FO
R BY REFERENCE TO THEIR RESPECTIVE GEOMETRIES."
  70 INPUT"INPUT CHANNEL GEOMETRY CODE, I=1 FOR PARTIALLY FILLED PIPE FLOW AND
I=2 FOR A RECTANGULAR SECTION CHANNEL. I= ",I
  80 PRINT:PRINT"THE FOLLOWING CASES ARE COVERED..."
  90 PRINT:PRINT"FLOW FROM AN UPSTREAM CHANNEL INTO A SECOND CHANNEL HAVING THE
SAME DIMENSIONS BUT DIFFERENT SLOPE AND MANNING COEFF. IS TREATED."
 100 PRINT:PRINT"THE DOWNSTREAM SLOPE MAY BE STEEPER OR MILDER THAN THAT OF THE
SUPPLY CHANNEL."
 110 PRINT:PRINT"SUB OR SUPERCRITICAL FLOW MAY OCCUR IN THE SUPPLY CHANNEL AND
THE ENSUING FLOW DEPTH PROFILES IN THE DOWNSTREAM CHANNEL ARE CALCULATED."
 120 PRINT:INPUT"STEADY FLOW RATE  (M^3/S) Q= ",Q
 130 INPUT"DOWNSTREAM CHANNEL SLOPE S= ",S1
 140 INPUT"SUPPLY CHANNEL SLOPE SS= ",SS
 150 IF I=1 THEN INPUT"PARTIALLY FILLED PIPE DIAMETER (M) D= ",D1
 160 IF I=2 THEN INPUT"RECTANGULAR CHANNEL WIDTH (M) B= ",B
 170 IF I=2 THEN INPUT"MAXIMUM DEPTH IN RECTANGULAR CHANNEL (M) D=",D1
 180 IF I=2 THEN BS=B
 190 DS=D1
 200 INPUT"DOWNSTREAM CHANNEL MANNING'S N= ",N1
 210 INPUT"SUPPLY CHANNEL MANNING'S N= ",NS
 220 REM CALC. OF CRITICAL DEPTH FOR ALL CASES.
 230 D=D1
 240 S=S1
 250 N=N1
 260 U=D
 270 L=0.0
 280 C=(U+L)/2.0
 290 GOSUB 1460
 300 F=1.0-Q^2*T/(9.81*A^3)
 310 IF F<0.0 THEN L=C
 320 IF F>0.0 THEN U=C
 330 IF F=0.0 THEN GOTO 380
 340 E=(U+L)/2.0
 350 IF ABS((E-C)/C) < 0.01 THEN GOTO 380
 360 C=E
 370 GOTO 290
 380 HC1=C
 390 HCS=C
 400 REM CALC NORMAL DEPTH USING BISECTION METHOD.
 410 IR=0
 420 D=D1
 430 S=S1
 440 N=N1
 450 IR=IR+1
 460 IF IR=2 THEN D=DS
 470 IF IR=2 THEN S=SS
 480 IF IR=2 THEN N=NS
 490 U=D
 500 L=0.0
 510 C=(U+L)/2.0
 520 GOSUB 1460
 530 Z=(A/P)^0.1667/N
 540 F=Z*SQR((A/P)*S)-Q/A
 550 IF F<0.0 THEN L=C
 560 IF F>0.0 THEN U=C
 570 IF F=0.0 THEN GOTO 620
 580 E=(U+L)/2.0
 590 IF ABS((E-C)/C) < 0.01 THEN GOTO 620
 600 C=E
 610 GOTO 520
 620 IF IR=1 THEN HN1=C
 630 IF IR=1 THEN GOTO 450
 640 IF IR=2 THEN HNS=C
 650 IF I=1 THEN PRINT"PIPE DIAMETER = ",D," M."
 660 IF I=2 THEN PRINT"CHANNEL WIDTH = ",B," M. AND MAX DEPTH = ",D," M."
 670 PRINT:PRINT"STEADY FLOWRATE = ",Q," M^3/S."
 680 PRINT:PRINT"CRITICAL DEPTH AT ",Q," M^3/S IS ",HC1*1000.0," MM."
 690 PRINT:PRINT"NORMAL DEPTH IN SUPPLY CHANNEL =",HNS*1000.0," MM."
 700 IF HNS>HCS THEN PRINT"SUBCRITICAL SUPPLY FLOW" ELSE PRINT"SUPERCRITICAL SU
PPLY FLOW"
 710 PRINT"NORMAL DEPTH IN DOWNSTREAM CHANNEL =",HN1*1000.0," MM."
 720 IF HN1>HC1 THEN PRINT"SUBCRITICAL DOWNSTREAM FLOW" ELSE PRINT"SUPERCRITICA
L DOWNSTREAM FLOW"
 730 PRINT"SUPPLY CHANNEL SLOPE = ",SS
 740 PRINT"DOWNSTREAM CHANNEL SLOPE = ",S1
 750 PRINT"SUPPLY CHANNEL MANNING COEFF. = ",NS
```

```
 760 PRINT"DOWNSTREAM CHANNEL MANNING COEFF. = ",N1
 770 IF HNS>HCS AND HN1<HC1 THEN J=1
 780 IF HNS<HCS AND HN1<HC1 THEN J=2
 790 IF HNS>HCS AND HN1>HC1 THEN J=3
 800 IF HNS<HCS AND HN1>HC1 THEN J=4
 810 IF J=1 THEN H(1)=HC1
 820 IF J=1 THEN DH=(HN1-HC1)/30.0
 830 IF J=2 OR J=3 THEN H(1)=HNS
 840 IF J=2 AND HNS>HN1 THEN DH=-(HNS-HN1)/30.0
 850 IF J=2 AND HNS<HN1 THEN DH=(HN1-HNS)/30.0
 860 IF J=3 AND HNS>HN1 THEN DH=-(HNS-HN1)/30.0
 870 IF J=3 AND HNS<HN1 THEN DH=(HN1-HNS)/30.0
 880 IF J=4 THEN GOTO 1670
 890 GOTO 950
 900 IF J=4 AND HCDS>HN1 THEN H(1)=HNS
 910 IF J=4 AND HCDS>HN1 THEN DH=(HCD1-HNS)/30.0
 920 IF J=4 AND HCDS<HN1 THEN H(1)=HN1
 930 IF J=4 AND HCDS<HN1 THEN DH=(HCDS-HN1)/30.0
 940 IF J=4 AND HCDS<HN1 THEN N=NS
 950 DEP(1)=H(1)
 960 K=0
 970 X(1)=0.0
 980 IF J=1 THEN PRINT"FLOW ENTERS DOWNSTREAM CHANNEL AT CRITICAL DEPTH."
 990 IF J=3 THEN PRINT"FLOW ENTERS DOWNSTREAM CHANNEL AT SUPPLY CHANNEL NORMAL
DEPTH."
1000 IF J=4 THEN PRINT"HYDRAULIC JUMP PRESENT"
1010 INPUT"PRESS RETURN TO CONTINUE...",X$
1020 PRINT:IF J<4 THEN PRINT"DEPTH PROFILE IN DOWNSTREAM PIPE..."
1030 IF J=4 THEN PRINT"DEPTH PROFILE FROM SLOPE CHANGE TO JUMP."
1040 IF J=4 AND HCDS<HN1  THEN PRINT"NOTE -VE. DISTANCES INDICATE JUMP LOCATED
UPSTREAM OF SLOPE CHANGE IN SUPPLY CHANNEL."
1050 PRINT:IF K>0 AND J>2 THEN PRINT"DEPTH =",DEP(1)*1000.0," MM. AT ",X(1)," M
 FROM EXIT."
1060 IF K=0 THEN PRINT"DEPTH =",DEP(1)*1000.0," MM. AT ",X(1)," M. FROM ENTRY."
1070 IS=1
1080 FOR IZ=1 TO 29 STEP 2
1090 IS=IS+1
1100 H(2)=DEP(1)+DH*(IZ+1)
1110 H(3)=DEP(1)+DH*IZ
1120 FOR IJ=1 TO 3
1130 C=H(IJ)
1140 GOSUB 1460
1150 DL(IJ)=DLX
1160 NEXT IJ
1170 DXP=DH*(DL(1)+4.0*DL(2)+DL(3))/3.0
1180 H(1)=H(2)
1190 IF J=4 AND HCDS<HN1 THEN SIGN=-1.0 ELSE SIGN=1.0
1200 IF K>0 THEN SIGN=-1.0
1210 X(IS)=X(IS-1)+ABS(DXP)*SIGN
1220 DEP(IS)=H(1)
1230 IF K>0 AND J>2 THEN PRINT"DEPTH = ",DEP(IS)*1000.0," MM. AT ",X(IS)," M. F
ROM EXIT. "
1240 IF K=0 THEN PRINT"DEPTH = ",DEP(IS)*1000.0," MM. AT ",X(IS)," M. FROM ENTR
Y. "
1250 NEXT IZ
1260 IF K=1 THEN GOTO 1360
1270 IF J>1 THEN GOTO 1360
1280 IF J=1 THEN INPUT"DO YOU WISH TO CALCULATE THE DRAWDOWN PROFILE IN THE SUP
PLY CHANNEL, Y OR N. ",Z$
1290 IF Z$="N" THEN GOTO 1360
1300 H(1)=HCS
1310 DEP(1)=HCS
1320 DH=(HNS-HCS)/30.0
1330 PRINT"DRAWDOWN PROFILE FOR SUPPLY CHANNEL, DISTANCES MEASURED -VE UPSTREAM
. "
1340 K=1
1350 GOTO 1050
1360 IF J<3 OR  K=2 THEN GOTO 2140
1370 IF J=3 OR J=4 THEN INPUT"DO YOU WISH TO CALCULATE THE DRAWDOWN PROFILE AT
FREE DISCHARGE FROM THE DOWNSTREAM CHANNEL, Y OR N. ",Z$
1380 IF Z$="N" THEN GOTO 2150
1390 H(1)=HC1
1400 DEP(1)=HC1
1410 K=2
1420 DH=-(HC1-HN1)/30.0
1430 PRINT"DRAWDOWN PROFILE FOR DOWNSTREAM CHANNEL ASSUMING FREE DISCHARGE."
1440 PRINT"DISTANCES MEASURED -VE UPSTREAM FROM CHANNEL EXIT."
1450 GOTO 1050
1460 IF I=1 THEN GOTO 1540
1470 REM RECTANGULAR CHANNELS
1480 T=B
1490 P=B+2.0*C
1500 A=T*C
1510 HBAR=C/2.0
```

```
1520 GOTO 1620
1530 REM PARTIALLY FILLED PIPES.
1540 IF C<D/2.0 THEN O=2.0*ATN(SQR(C*(D-C))/(D/2.0-C))
1550 IF C=D/2.0 THEN O=PI
1560 IF C>D/2.0 THEN O=PI+2.0*ATN((C-D/2.0)/(SQR(C*(D-C))))
1570 A=((D^2.0)/8.0)*(O-SIN(O))
1580 T=2.0*((C*(D-C))^0.5)
1590 P=D*O/2.0
1600 XO=0.666*D/2.0*((3.0*SIN(O/2)-3.0*SIN(O/2))/(4.0*(O/2-0.5*SIN(O))))
1610 HBAR=C-D/2+XO
1620 HCRIT=1.0-Q^2*T/(9.81*A^3)
1630 HNORM=S*(1.0-Q^2*(N^2*S)/(A^3.333/P^1.333))
1640 DLX=HCRIT/HNORM
1650 RETURN
1660 REM CALC. OF JUMP LOCATION.
1670 C=HNS
1680 D=DS
1690 N=NS
1700 S=SS
1710 GOSUB 1460
1720 HB=HBAR
1730 AB=A
1740 Z1=1000.0*(Q^2/AB+9.81*HB*AB)
1750 U=DS
1760 L=HCS
1770 C=(U+L)/2.0
1780 IR=0
1790 GOSUB 1460
1800 Z2=1000.0*(Q^2/A+9.81*HBAR*A)
1810 IF IR=0 THEN F=Z1-Z2 ELSE  F=Z2-Z1
1820 IF F>0.0 THEN L=C
1830 IF F<0.0 THEN U=C
1840 IF F=0.0 THEN GOTO 1890
1850 E=(L+U)/2.0
1860 IF ABS((E-C)/C)<0.005 THEN GOTO 1890
1870 C=E
1880 GOTO 1790
1890 IF IR=1 THEN GOTO 2090
1900 IF IR=0 THEN HCDS=C
1910 IF HCDS<HN1 THEN PRINT"HYDRAULIC JUMP POSITIONED IN SUPPLY CHANNEL"
1920 IF HCDS<HN1 THEN PRINT"CONJUGATE DEPTH = ",HCDS*1000.0," MM."
1930 IF HCDS>HN1 THEN PRINT"HYDRAULIC JUMP POSITIONED IN DOWNSTREAM CHANNEL."
1940 IF HCDS<HN1 THEN GOTO 2110
1950 REM CALCULATE CONJUGATE DEPTH UPSTREAM OF JUMP IF IT FORMS IN THE CHANNEL
DOWNSTREAM OF THE SLOPE CHANGE...
1960 IR=1
1970 C=HN1
1980 D=D1
1990 S=S1
2000 N=N1
2010 GOSUB 1460
2020 HB=HBAR
2030 AB=A
2040 Z1=1000.0*(Q^2/AB+9.81*HB*AB)
2050 U=HCS
2060 L=HNS
2070 C=0.5*(U+L)
2080 GOTO 1790
2090 HCD1=C
2100 PRINT"CONJUGATE DEPTH = ",HCD1*1000.0," MM."
2110 GOTO 900
2120 PRINT:INPUT"CALCULATION COMPLETE. DO YOU WISH TO REPEAT? Y OR N.",Z$
2130 IF Z$="Y" THEN GOTO 10
2140 PRINT:INPUT"DO YOU WISH TO REPEAT? Y OR N.",Z$
2150 IF Z$="Y"THEN GOTO 40 ELSE PRINT"END"
```

16.14. Further reading

Henderson, F. M. (1966). *Open channel flow*, Macmillan.
Stephenson, D. (1981). *Stormwater hydrology and drainage*. Elsevier Scientific Publishing Co., Amsterdam.

EXERCISES 16

16.1   A channel has a trapezoidal cross section with a base width of 0·6 m
and sides sloping at 45°. When the flow along the channel is 20 m³/min determine the critical depth.
[0·27 m]

16.2   Determine the critical depth for a channel of trapezoidal section
conveying 1·33 m³/s. The base width of the channel is 2·4 m and the sides
slope at 60° to the horizontal.
[0·31 m]

16.3   Contrast and relate the Chezy and Manning formulae for the mean
velocity of flow in open channels. A rectangular channel is 6 m wide and will
carry a discharge of 22·5 m³/s of water. Determine the necessary slopes to
achieve uniform flow at (i) a depth of 3 m, (ii) a depth of 0·6 m, (iii) the
critical depth. Assume that for this channel $n = 0·012$ in the Manning
equation.
[1/7631, 1/70, 1/482]

16.4   Water flows across a broad crested weir in a rectangular channel
400 mm wide. The depth of the water just upstream of the weir is 70 mm
and the crest of the weir is 40 mm above the channel bed. Calculate the fall
of the surface level and the corresponding discharge assuming that the velocity
of approach is negligible.
[10 mm, 3·54 x 10⁻³ m³/s]

16.5   A venturi flume with a level bed is 12 m wide and 1·5 m deep
upstream with a throat width of 6 m. Assuming that a standing wave forms
downstream calculate the rate of flow of water if the discharge coefficient is
0·94. Correct for the velocity of approach.
[18·56 m³/s]

16.6   Show that the equation

$$Q = a_1 \left[\left(\frac{2g(h_1 - h_2)}{r^2 - 1} \right) \right]$$

where $r = a_1/a_2$ commonly derived for the frictionless flow of water with a
rate of discharge Q through a venturi meter, is also applicable to the frictionless flow of water over zero bed slope (a) through a venturi flume, (b) over a
broad-crested weir, (c) under a sluice gate, (d) over a rounded crest of a spillway to the horizontal bed at the toe of a dam. Define a_1 and a_2 in each of
the above cases and state for which of these cases and for what special
conditions the equation $Q = 1·704 B E^{3/2}$ is valid. Derive this equation for the
relevant conditions carefully specifying E.

16.7   A venturi flume is placed in an open channel 2 m wide in which the
throat width is 1·2 m. The upstream depth is 1 m and the floor is effectively
horizontal. Calculate the flow when (a) the depth at the throat is 0·9 m and
(b) a standing wave is produced beyond the throat.
[1·8 m³/s, 2·04 m³/s]

16.8   A channel of rectangular cross-section is 1·2 m wide and the normal
flow is 0·7 m³/s with a depth of 0·6 m. If a streamlined hump x metres high is
installed on the bed, draw a curve on squared paper showing how the theoretical ratio of the depth over the hump to the total energy immediately

upstream varies with x. Take the datum for the energy head as passing through the top of the hump.

Hence state whether a standing wave can form downstream when (a) $x = 0.1$ m, (b) $x = 0.3$ m, giving reasons.

[No, Yes]

16.9 A venturi flume in a rectangular channel of width B has a throat width b. The depth of liquid at entry is H and at the throat is h. Derive an expression for the theoretical volume flow rate of the liquid in terms of H, b and the ratio h/H. Develop also a relationship between the ratios h/H and b/B. State what assumptions you make regarding the downstream flow.

$$\left[Q = 5.69 b H^{3/2} \left(\frac{h}{H} \right)^{3/2} ; \quad \frac{b}{B} = (\sqrt{3}) \left(\frac{H}{h} \right) - (\sqrt{2}) \left(\frac{H}{h} \right)^{3/2} \right]$$

16.10 A standing wave or venturi flume having a throat width of 0.375 m is installed in a channel 0.6 m wide, the bed of the channel being horizontal. By successive approximation, or otherwise, determine (i) the still water head upstream, (ii) the rate of discharge. The depth of water in the approach channel is 0.36 m. Neglect friction losses.

[0.385 m, 0.152 m³/s]

16.11 A rectangular prismatic channel 1.2 m wide has a uniform slope of 1 in 1600 and a normal depth of 0.6 m when the flow rate is 0.72 m³/s. When a sluice is lowered the upstream depth is increased to 1 m. Determine the distance upstream from the sluice where the depth of water is 0.8 m. Using a step by step method to solve the problem, divide the range of depth into two equal parts.

[472 m]

16.12 A wide rectangular channel has a slope of 0.0003, the normal depth of flow in it being 0.6 m. A weir is placed across the channel increasing the local depth to 0.8 m. Find approximately how far upstream of the weir the depths will be 0.75 m and 0.725 m given that the Chezy constant is 55.5 in SI units.

[151 m, 227 m]

16.13 A rectangular channel of slope 0.001 carries 100 m³/s of water and is 5 m wide. If an overflow weir is installed across the channel which raises the water level at the weir to a depth of 6 m (a) compute the normal depth of flow, (b) compute in two steps the distance upstream to the point where the depth of water is 5.8 m and (c) classify the surface profile. Take the value of n in Manning's formula as 0.02.

[4.07 m, 143 m, M1]

16.14 A sluice across a channel 6 m wide discharges a stream 1.2 m deep. What will be the flow if the upstream depth is 6 m.

The conditions downstream cause an hydraulic jump to occur at a place where concrete blocks have been placed in the bed. What will be the force on the blocks if the downstream depth is 3.06 m.

[78 m³/s, 124.2 N]

16.15 The stream issuing from beneath a vertical sluice gate is 0.3 m deep at the vena contracta. Its mean velocity is 6 m/s. A standing wave is created on the level bed below the sluice gate. Find the height of the jump, the loss of head and the power dissipated per unit width of sluice.

[1.04 m, 0.7 m, 12.36 kW]

16.16   At the foot of a 30 m wide spillway from a dam where the discharge velocity is 28·2 m/s and the depth is 0·96 m a hydraulic jump is formed on a horizontal apron. Calculate the height of the jump and the total power dissipated in it.
[12 m, 237 MW]

16.17   Water issuing from a sluice enters a horizontal rectangular channel with uniform velocity v and depth y. Show that if y is less than the critical depth an hydraulic jump will be formed.

If the velocity when the water enters the channel is 4 m/s and the Froude number is 1·4 obtain (a) the depth of·flow after the jump, (b) the loss of specific energy due to the formation of the jump.
[1·28 m, 0·02 m]

16.18   In a rectangular channel 0·6 m wide a jump occurs where the Froude number is 3. The depth after the jump is 0·6 m. Estimate the total loss of head and the power dissipated at the jump.
[0·225 m, 0·786 kW]

16.19   Derive the relationship

$$h_2^2 + h_1 h_2 = \frac{2Q^2}{gh_1 b^2}$$

connecting the depths h_1 and h_2 before and after an hydraulic jump in a rectangular section channel of width b where Q is the volume rate of flow.

A rectangular section channel of width 0·52 m has a flow depth at a certain section of 0·61 m when the flow rate is 0·44 m³/s. The channel bed slopes at 1 in 110. If a hydraulic jump has just occurred upstream, find the depth before the jump and the Chezy constant C for the channel. Find also the change in specific energy across the jump.
[0·27 m, 34·1 SI units, 0·06 m loss]

16.20   A wide channel with uniform rectangular section has a change of slope from 1 in 95 to 1 in 1420 and the flow is 3·75 m³/s per metre width. Determine the normal depth of flow corresponding to each slope and show that a hydraulic jump will occur in the region of the junction. Calculate the height of the jump and sketch the surface profiles between the upstream and downstream regions of uniform flow. Manning's coefficient $n = 0·013$ and it may be assumed that the channel is wide in comparison to the depth of flow so that the hydraulic mean depth is approximately equal to the depth of flow.
[0·576 m]

PART V
Unsteady flow in bounded systems

Unsteady flow may be defined as a state in which the flow parameters are time-dependent, governed by partial differential equations requiring, in their complete form, numerical methods of solution using digital computers. By considering the rates of change of the various flow parameters it is possible to place most unsteady flow phenomena into one of three categories.

(i) *Quasi-steady flows* in which the rate of change of mass flow is continuous with time, but the fluid acceleration and the forces responsible for acceleration are negligible. In such cases the steady flow equations may be applied with reasonable accuracy (e.g. continuous filling and emptying of reservoirs and tanks).

(ii) *Mass oscillation* in which the rate of change of fluid velocity is sufficient for the forces causing fluid acceleration to be important, but still so slow as to permit the compressibility of the fluid to be ignored. The pressures generated within the affected system are often termed *surge pressures*. Examples include reciprocating machinery and oscillatory fluid motion such as that found in pipe systems with more than one free surface.

(iii) Flows in which the time taken to change fluid velocity is comparable to the period of the system based on the wave propagation velocity through the fluid, modified by the pipe properties, and the pipe length. If these times are comparable then the compressibility of the fluid becomes significant and the solution requires graphical or computer-based numerical techniques. These unsteady flow conditions, historically referred to as *waterhammer*, may result from rapid valve operation, pump shutdown or turbine load rejection and are commonly termed *pressure* or *fluid transients*.

Unsteady flow conditions may also occur in open channels, generally characterized by the attenuation of surface waves during their passage along the channel. These conditions in channels or partially filled pipes may be analysed utilizing similar techniques to transient theory, although the transients are now in depth rather than in pressure and the wave propagation velocities are independent of any fluid or pipe elasticity.

Examples of each category of transient will be dealt with in this section, including those in free surface flow.

17 Quasi-steady flow

Let us consider the case of a tank emptying through an orifice into the atmosphere. It may be assumed that the rate of change of fluid discharge as the fluid level in the tank drops is sufficiently slow to permit application of the steady-state relationship at any instant during the discharge.

Hence, application of the quasi-steady approximation results in the following relationship between the instantaneous rate of change of the head H above the orifice and the outflow through the orifice:

$$A_s\left(-\frac{dH}{dt}\right) = C_d A_o \sqrt{(2gH)}, \tag{17.1}$$

where the surface area $A_s \gg$ the cross-sectional area A_o of the orifice. Note that the surface area A_s will be a function $f(H)$ of the head and that a negative sign is included in the rate of change of head since the fluid level in the tank is falling.

Rearrangement and incorporation of the fact that $A_s = f(H)$ yields an expression which may be integrated to give the time taken by the fluid

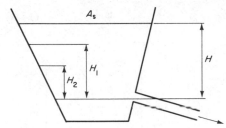

Figure 17.1 Discharge from a tank through an orifice of area A_o and discharge coefficient C_d

level to fall from $H = H_1$ to $H = H_2$ (Fig. 17.1):

$$t = -\int_{H_1}^{H_2} \{f(H)\, H^{-1/2}/C_d A_o \sqrt{(2g)}\}\, dH \tag{17.2}$$

The method of integration of equation (17.2) will be dependent upon the form of $f(H)$. Normally, for this type of example, it may be assumed that the coefficient of discharge at the orifice C_d is a constant.

If the tank or reservoir has a constant inflow Q_{in} during the discharge considered above, then equation (17.1) becomes

$$A_s\left(-\frac{dH}{dt}\right) = C_d A_o \sqrt{(2gH)} - Q_{in} \tag{17.3}$$

and equation (17.2) may be re-expressed as

$$t = -\int_{H_1}^{H_2} \{f(H)/[C_d A_o \sqrt{(2gH)} - Q_{in}]\}\, dH. \tag{17.4}$$

Let us now consider a system in which the orifice has been replaced by a pipe of length l and diameter d. Then, for quasi-steady conditions, it follows that the instantaneous head H balances the total losses in the discharge pipeline. Hence,

$$H = (4fl/d + k)\,(\bar{u}^2/2g) \qquad (17.5)$$

where k is the sum of all the separation loss coefficients for the discharge pipeline and could include the loss through a partially shut valve and $\bar{u}$ is the mean flow velocity in the pipe.

Consideration of the continuity of flow between the fluid surface and the pipe discharge yields

$$A_s\left(-\frac{dH}{dt}\right) = A_o\bar{u} = C\sqrt{H}, \qquad (17.6)$$

where $\quad C = A_o \sqrt{\{2g/(4fl/d + k)\}}.$ $\qquad (17.7)$

Rearrangement again results in an expression for the time taken for the surface to fall from H_1 to H_2:

$$t = -\int_{H_1}^{H_2} \{f(H)/C\sqrt{H}\}\,dH \qquad (17.8)$$

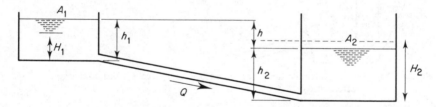

Figure 17.2   Unsteady flow transfer between two steady reservoirs

Integration of equation (17.8) will depend upon $f(H)$, but, if the tank is assumed to have a constant cross-section over the depth range under consideration, then the solution for t is given by

$$t = 2A_s(H_2^{1/2} - H_1^{1/2})/C. \qquad (17.9)$$

Strictly, the friction factor f which appears in equation (17.7) is a function of both the discharge flow Reynolds number and the discharge pipe roughness. However, it is usually sufficient to assume a constant value for f based on the initial flow rate. If it is required to introduce friction factor variation into the solution, this may be done by relating the friction factor to the instantaneous head through the Reynolds number relations and equation (17.5) and integrating the resulting expression by an incremental method.

Quasi-steady flow may also occur when a tank or reservoir is connected to a higher level tank and fluid transfer occurs under gravity (Fig. 17.2). The major difference between this case and those previously described is the presence of two free surfaces. Consider the case of two reservoirs connected by a single pipeline of length l and diameter d.

The instantaneous head difference h for quasi-steady conditions balances the total losses in the connecting pipeline. Hence,

$$h = (4fl/d + k)\,(\bar{u}^2/2g) \tag{17.10}$$

where, again, k represents the sum of all the separation losses in the connecting pipeline.

Now, from the application of continuity of flow between the two reservoirs at any instant, it follows that the rate of change of the fluid surface levels are linked, by the volume flow rate Q:

$$A_1\left(-\frac{dh_1}{dt}\right) = Q = A_2\left(\frac{dh_2}{dt}\right). \tag{17.11}$$

Note the difference in sign as one surface level rises while the other falls.

The rate of change of h, the difference in the two levels, is given by the sum of the rate of change of the fluid levels in the individual reservoirs. Hence,

$$-\frac{dh}{dt} = -\frac{dh_1}{dt} + \frac{dh_2}{dt}\,. \tag{17.12}$$

Substituting from equation (17.11),

$$\frac{dh_1}{dt} = \frac{dh}{dt}\bigg/\left(1 + \frac{A_1}{A_2}\right). \tag{17.13}$$

From equations (17.10) and (17.11),

$$\frac{dh_1}{dt} = A_0\bar{u} = C\sqrt{h}, \tag{17.14}$$

where C is given by equation (17.7). However, equation (17.14) is still not solvable since it is necessary to express t in terms of one head only. Therefore, from equation (17.13), we obtain the final expression

$$-A_1\frac{dh}{dt}\bigg/\left(1 + \frac{A_1}{A_2}\right) = C\sqrt{h}$$

and, thus,

$$t = -\int_{H_1}^{H_2} \{A_1/C\sqrt{h}\,(1 + A_1/A_2)\}\,dh \tag{17.15}$$

It has, of course, been assumed that the surface areas of the two reservoirs remain constant over the depth range H_1 to H_2. If this is not the case, then the solution of equation (17.15) may become difficult and it may be necessary to employ numerical or graphical methods of integration.

EXERCISES 17

17.1   A vertical cylindrical tank, 0·4 m diameter and 3 m high is used as part of a flow calibration diverter unit. If the water collected in the tank, up to a depth of 2·5 m, is discharged through an orifice and valve in the tank base, that may be represented by a 50 mm diameter orifice of discharge coefficient 0·6, calculate the time to empty half the collected volume and express as a % of the time to fully empty.
[22·15 s, 29·1%]

17.2   A vertical axis tank is conical in shape, the diameter increasing uniformly from 1 m at the base to 1·75 m diameter at a height of 3 m. The tank is to be emptied by means of a 50 mm orifice in the base having a discharge coefficient of 0·6. Calculate the time to reduce the water level from 2 m to 1 m above the base.
[232·6 s]

17.3   For the case set out in Exercise 17.2 above calculate the inflow necessary to hold the liquid level at 1·5 m above the base.
[383 litres/min]

17.4   For the tank in Exercise 17.2 calculate the time to discharge the full contents of the tank if the discharge is carried away by a 25 mm diameter pipe, length 4 m, friction factor 0·005. Assume that the effect of the orifice to pipe connection can be represented by a separation loss having a K value of 2, and that final discharge is at tank base level.
[5044 s]

17.5   A rectangular cross-section tank, 2 m x 3 m, is filled with water up to a depth of 2 m. Calculate the time to reduce the volume in the tank by 50% if the discharge is via a 40 mm diameter pipe, 6 m long, for which a friction factor of 0·005 may be assumed and the separation losses may be represented by a K value of 0·9. Assume final discharge 2 m below tank base level.
[1141·6 s]

17.6 ·   A 1·2 m deep rectangular tank is 2 m x 1 m in area and has a V-notch in one side. The lowest point of the V-notch is 770 mm above the base of the tank. A water supply to the tank of 1036 litres/minute establishes a steady depth of 1 m above the base of the tank. If the water inflow ceases calculate the time needed for the level to fall to 150 mm.
[68·8 s]

17.7   A cylindrical tank is 1·8 m in diameter and 3 m long and is mounted horizontally. Oil of s.g. 0·87 is stored in the tank is drawn off through an orifice, 20 mm diameter, 0·6 discharge coefficient, at the tank's lowest point. Calculate the time taken to reduce the level in the tank from 0·9 m to 0·8 m above the orifice.
[699·6 s]

17.8   A 2 m deep tank is 2 m x 3 m in area and is divided into two equal halves by a vertical separation plate. Flow from one tank to the other takes place through a square orifice, 1 cm side, having a discharge coefficient of 0·6. If water is initially at 1·5 m depth on one side of the plate and 0·5 m depth on the other calculate the time taken for the depths in both tanks to be equal.
[6773 s]

18 Unsteady flow in closed pipeline systems

Whenever the steady-state operating conditions of a fluid system are changed, either intentionally by a planned valve or pump operation or inadvertently due to some system failure, then the change in the prevailing equilibrium flow conditions will be communicated to the system as a whole by pressure waves, travelling at the appropriate sonic velocity and propagating away from the point in the system where the change in the steady flow conditions was imposed. After a short time the system will attain a new equilibrium condition, assuming, of course, that the pressures generated during the change in steady conditions have not been sufficient to damage the system. For example, if the flow into a pipe network is increased during operation, then, after some time has elapsed, the flow will again become steady, with a new flow distribution throughout the pipe network. The transition from one set of steady conditions to the next will always be accompanied by the propagation of pressure and velocity waves throughout the system and a consequent variation of pressure at all points in the pipe network. The severity of these transient pressures, which may be of destructive proportions, will, for any particular system, depend on the rate of change of the flow velocity imposed locally at the item of equipment whose operation introduced the fluid acceleration. The potentially destructive nature of the pressure waves that are necessary to convey a change in operating conditions through a pipe system, although they may have a very short duration, explains the considerable practical importance of unsteady flow analysis.

The rate of change of conditions imposed on the system is of prime importance and governs, for any particular system, the method to be employed in calculating the effects of the pressure wave propagation. If the rate of change of the flow velocity is slow in comparison with the time taken for a pressure wave to pass through the system, then it is possible to consider the fluid as incompressible, i.e. a change in flow conditions is instantaneously transmitted through the system as the fluid is implied to have an infinite sonic velocity. The analysis of unsteady flow conditions from this assumption is known as rigid column theory and is considerably simpler and cheaper than a full solution involving fluid compressibility.

In recent years, the trend in engineering fluid systems has been towards higher flow rates with systems incorporating ever larger pumps and turbines, this being particularly the case in the power generation industry. The need to control such systems within a reasonable time has meant that the propagation of severe pressure transients as a result of system control has become more common, with the majority of cases falling within the third unsteady flow category defined (page 473), where fluid compressibility and system elasticity can no longer be neglected. Historically, this category has been referred to as waterhammer, a title which in no way reflects the occurrence of phenomena in all fluid systems, whether the moving fluid is liquid ammonia, water, crude oil or natural gas. The established method of solution has been the graphical method proposed in the 1930s by Schnyder and Bergeron. However, since the early 1960s, the advent of the digital computer has meant that solutions by numerical methods can be attempted, particularly those involving the method of characteristics, where the major contribution since 1960 has been due to Professor Victor Streeter and his associates.

It is worth mentioning here that, although the rigid column theory may only be applied to transient conditions displaying slow rates of flow acceleration, the analysis based on the wave theory, incorporating the fluid compressibility, may be universally applied. Effectively, the rigid column theory is a limiting case of the wave theory. However, considerations of time and cost dictate that rigid column theory be applied whenever possible.

Rigid column theory will be outlined first, followed by an introduction to pressure transient wave theory and the numerical solution based on the method of characteristics.

18.1. Rigid column theory

Consider a simple pipeline from an open reservoir discharging into the atmosphere through a valve (Fig. 18.1). Friction is neglected, so that all

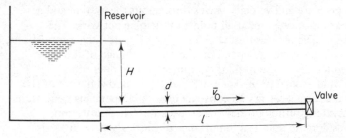

Figure 18.1   Simple pipeline of constant cross-section A terminated by a valve

points along the pipe are at a head H. The valve is now shut, with the result that after time t the velocity in the pipe has been reduced by $\Delta \bar{v}$ and the head at the valve has risen by ΔH. The total force accelerating the flow is $-\rho g A \Delta H$ and the acceleration of the flow is $-\mathrm{d}\bar{v}/\mathrm{d}t$. Note the negative signs here as the flow is retarded. Hence, from Newton's second law,

$$-\rho g A \Delta H = -\frac{\mathrm{d}\bar{v}}{\mathrm{d}t}\rho A l,$$

$$\Delta H = \frac{l}{g}\frac{\mathrm{d}\bar{v}}{\mathrm{d}t}. \tag{18.1}$$

ΔH in equation (18.1) is referred to as the surge or inertia pressure resulting from the closure of the valve.

Equation (18.1) may also be employed to describe the establishment of flow on valve opening in the pipeline illustrated in Fig. 18.1, provided that the flow acceleration involved is low. The flow is accelerated from rest, following valve opening, by the difference in head between the reservoir surface and the valve discharge level. Initially, the full reservoir head H is available to accelerate the flow. However, as the flow builds up, so this accelerating head is decreased because of the increasing frictional and separation losses along the pipeline. The final velocity $\bar{v}_0$ attained by the flow is given by the energy balance expression

$$H = 4fl_e\bar{v}_0^2/2dg, \tag{18.2}$$

where f is the friction coefficient, l_e is the equivalent length of pipe (and so represents all the separation losses in the system) and d is the pipe diameter. Thus, at any instant during acceleration, the accelerating head ΔH is

$$\Delta H = H - 4fl_e\bar{v}^2/2dg,$$

where $\bar{v}$ is the mean instantaneous fluid velocity at that time. From equation (18.1) it follows that

$$H - \frac{4fl_e\bar{v}^2}{2dg} = \frac{l}{g}\frac{d\bar{v}}{dt} \tag{18.3}$$

and, by substitution from equation (18.2),

$$H(1 - \bar{v}^2/\bar{v}_0^2) = \frac{l}{g}\frac{d\bar{v}}{dt},$$

therefore,

$$t = \frac{l\bar{v}_0^2}{gH}\int_0^{\bar{v}_0} \frac{1}{\bar{v}_0^2 - \bar{v}^2}\,d\bar{v} \tag{18.4}$$

or $$t = \tfrac{1}{2}(l\bar{v}_0^2/gH)\log_e\{(\bar{v}_0 + \bar{v})/(\bar{v}_0 - \bar{v})\} \tag{18.5}$$

which implies an infinite time for $\bar{v}_0$ to be achieved. However, the flow velocity does attain 99 per cent of $\bar{v}_0$ in a finite time:

$$t_{0.99\bar{v}_0} = (l\bar{v}_0^2/gH)\log_e(1{\cdot}99/0{\cdot}01)$$

$$= 2{\cdot}646\ l\bar{v}_0^2/gH.$$

The apparent anomaly suggested by the infinite time required to establish flow, as indicated by equation (18.5), arises from the assumption of fluid incompressibility. In practice, the propagation of pressure waves through a fluid results in equilibrium conditions being established in a shorter time than indicated by equation (18.5).

18.2. Surge tanks and shafts

It has been mentioned that rigid column theory may only be applied to unsteady flow conditions where the rate of change of the flow boundary conditions is slow compared to the transient pipe periods within the system. There is one very important area of unsteady flow where rigid column theory may be applied successfully and this is to the analysis of waterhammer in hydroelectric installations and, particularly, the study of surge tanks for pressure surge control.

Figure 18.2 illustrates a typical hydroelectric scheme, where both the tunnels supplying water to the turbines and the surge shafts designed to limit surge pressures are cut into the rock.

If the electrical output taken from the generators varies, there is a

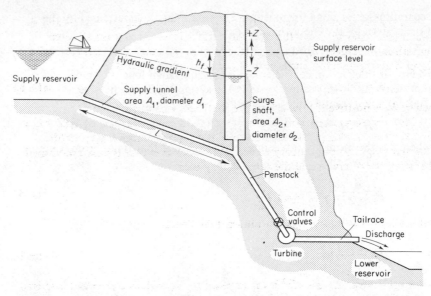

Figure 18.2   Schematic of a surge shaft layout in a hydroelectric scheme. (Note that in a pump storage scheme, water is returned to the upper lake during low electricity demand periods.)

tendency for the turbines to change speed; the turbine governors then operate control valves in the penstock approach to the turbines to stabilize the turbine speed. In the limiting case of load rejection, the penstock valves close, resulting in a pressure rise in the penstock that propagates towards the supply reservoir. If no surge control devices are provided, the system may then suffer severe damage, so the rate of deceleration of the water supply is reduced by, effectively, using the pressure surge in the penstock to divert flow into the surge tank, as shown.

Conversely, as the generators come back on load, the surge tank water re-enters the penstock, effectively increasing the flow acceleration and minimizing any negative surges that would otherwise occur. Apart from preventing damage to the installation, the provision of a surge tank also aids in the design of the turbine governors, as the presence of periodic pressure waves in the system following every load change would cause the governors to hunt. It is worth noting that the surge tank should be as close as possible to the turbines, as the penstock will suffer the unreduced surge pressure and an adjacent surge tank will reduce the penstock pipe period and the consequent surges following load change. The penstock has, therefore, to be designed to take the full surge pressure, a point sometimes not fully stressed.

Returning to Fig. 18.2, it will be seen that the surge tank or shaft extends up above the reservoir surface to prevent overflow during operation, although this has been known to happen inadvertently in some installations due to bad design or system flow rate increase since the shaft was designed. Under normal conditions, the level in the tank will be below the reservoir level by an amount h_f equal to the friction and separation losses in the tunnels from the reservoir to the surge shaft.

As the penstock valves shut, so flow is diverted up the surge shaft and the water level rises to above the reservoir surface level. At any time t after initiation of surge by load change, the level in the surge shaft has reached a level Z above reservoir surface, then the head opposing h is

$$h = Z \pm h_f, \tag{18.6}$$

where h_f is positive if flow is into the shaft and vice versa.

The force opposing motion of water in the supply tunnel is $\rho g A_1 (Z \pm h_f)$ and the rate of change of momentum in the supply tunnel is $\rho A_1 l \, d\bar{v}/dt$. Applying rigid column theory to the mass oscillation that is now established between the reservoir and the surge shaft yields

$$\rho g A_1 (Z \pm h_f) = -\rho A_1 l \frac{d\bar{v}}{dt},$$

where A_1 is the supply tunnel cross-sectional area

$$\frac{l}{g} \frac{d\bar{v}}{dt} + Z \pm h_f = 0, \tag{18.7}$$

where $h_f = 4 f l \bar{v}^2 / 2 d_1 g = K \bar{v}^2$ if l is assumed to include, as equivalent lengths, any separation losses in the tunnel.

At the entrance to the surge shaft, continuity of flow must apply, i.e. the flow along the supply tunnel at time t must equal the flow Q to the turbines plus the flow up into the surge shaft. Thus, as the surface velocity in the surge shaft is dZ/dt, it follows that

$$A_1 \bar{v} = A_2 \frac{dZ}{dt} + Q. \tag{18.8}$$

In order to solve equations (18.7) and (18.8), it is necessary to make assumptions about the variation of friction factor and the rate of flow reduction to the turbine. Even under total load rejection, when $Q = 0$, the variable friction factor still prevents an analytical solution and so it is now usual to use a step by step numerical integration, preferably utilizing a digital computer. If friction is ignored, then an approximation to maximum surge level and its period can be made. Neglecting f, the equations become

$$\frac{l}{g} \frac{d\bar{v}}{dt} + Z = 0 \tag{18.9}$$

and, if $Q = 0$ for full load rejection, then

$$A_2 \frac{dZ}{dt} = A_1 \bar{v}.$$

When $t = 0$, $Z = 0$ in the frictionless case as $h_f = 0$ and $dZ/dt = Q_0/A_2$, where Q_0 is the steady flow before load rejection.

A solution to equations (18.7) and (18.8) under these conditions is

$$Z = (Q_0/A_2) \sqrt{(A_2 l/A_1 g)} \sin \{\sqrt{(A_1 g/A_2 l)}\} t,$$

so that the maximum level is

$$Z_{max} = (Q_0/A_2) \sqrt{(A_2 l/A_1 g)}$$

and the period of the mass oscillation is

$$T = 2\pi \sqrt{(A_2 l / A_1 g)}.$$

If friction is included, then the maximum level reached is lower and the water level oscillation damps at a calculatable frequency about the steady-state value of $-h_f$. Care should be taken to ensure that the turbine governors cannot respond to this frequency or that it does not correspond to a system resonant frequency. One important point here is that the minimum extreme of the oscillation, represented by $-Z_{max}$ in the frictionless case, should not be greater than the distance to the penstock entry, as this would result in air entrainment into the penstock.

In order to solve for the friction case, equations (18.7) and (18.8) can be treated numerically as shown by the following example. The various designs of surge shaft will be discussed later, in Chapter 20.

Example 18.1

In a hydroelectric scheme the supply tunnel is 1·25 m in diameter and has a friction factor of 0·01. At 200 m along the tunnel there is an open surge shaft of 4 m diameter. The steady flow to the turbines is 2 m^3 s^{-1}. Illustrate how to obtain peak water level in the surge shaft by both a frictionless flow assumption and by a step by step numerical integration following a sudden full load rejection.

Solution
If $f = 0$, then

$$Z = (Q_0 / A_2) \sqrt{(A_2 l / A_1 g)},$$

where $A_1 = (\pi/4) \times 1·25^2 = 1·227 \text{ m}^2$,

$A_2 = (\pi/4) \times 4^2 = 12·566 \text{ m}^2$,

$l = 200 \text{ m}.$

Thus, $Z_{max} = 2·29 \text{ m}.$

For a numerical integration, replace dZ, dt and $d\bar{v}$ by ΔZ, Δt and $\Delta \bar{v}$, so that

$$\frac{l}{g} \frac{d\bar{v}}{dt} + Z \pm h_f = 0$$

becomes

$$20·39 \, \Delta \bar{v} / \Delta t + Z \pm 0·326 \, \bar{v}^2 = 0,$$

$$\Delta \bar{v} = -\Delta t \, (0·049 Z + 0·0159 \bar{v}^2) \qquad \text{(I)}$$

and $A_1 \bar{v} = A_2 \dfrac{dZ}{dt} + Q$

becomes

$$\bar{v} = 10·24 \, \Delta Z / \Delta t$$

as $Q = 0$, or

$$\Delta Z = \Delta t \cdot \bar{v}/10.24. \tag{II}$$

The equation for $\bar{v}$ and Z may now be solved. At $t = 0$,

$$\bar{v} = \bar{v}_0 = Q_0/A_1 = 1.63 \text{ m s}^{-1} \tag{III}$$

$$Z_0 = -h_f = -0.326\bar{v}_0^2 = -0.866 \text{ m}. \tag{IV}$$

The choice of Δt is based on experience; 5 s will be taken here. On a digital computer there is obviously no restriction. It is perhaps worth noting here that these equations may also be solved on an analog computer with obvious advantages as far as operator interaction is concerned. Thus, for the period $t = 0$ to $t = 5$ s:

First approximation. Put $\bar{v}$ = velocity at $t = 0$, Z as water level at $t = 5$ s, thus $\bar{v} = 1.63 \text{ m s}^{-1}$. Then

$$\Delta Z = 5 \times 1.63/10.24 = 0.796 \text{ m from (II) and (III)},$$

$$Z = -0.866 + 0.796 = -0.07 \text{ m from (IV)},$$

$$\Delta\bar{v}_1 = -5(0.049 \times -0.07 + 0.159 \times 1.63^2) = -0.194 \text{ m s}^{-1},$$

$$\bar{v}_{t=5s} = 1.63 - 0.194 = 1.436 \text{ m s}^{-1}.$$

Second approximation. Put $\bar{v} = \frac{1}{2}(1.436 + 1.63) = 1.533 \text{ m s}^{-1}$. Then,

$$\Delta Z = 5 \times 1.533/10.24 = 0.748 \text{ m},$$

$$Z = -0.866 + 0.748 = -0.118 \text{ m},$$

$$\Delta\bar{v} = -5(0.049 \times -0.118 + 0.0159 \times 1.533^2) = -0.157 \text{ m s}^{-1},$$

$$\bar{v}_{t=5s} = 1.63 - 0.157 = 1.473 \text{ m s}^{-1}.$$

Third approximation. Put $\bar{v} = \frac{1}{2}(1.473 + 1.63) = 1.55 \text{ m s}^{-1}$. Then,

$$\Delta Z = 5 \times 1.55/10.24 = 0.757 \text{ m},$$

$$Z = -0.866 + 0.757 = -0.109 \text{ m},$$

$$\Delta\bar{v} = -5(0.049 \times -0.109 + 0.0159 \times 1.55^2) = -0.164 \text{ m s}^{-1},$$

$$\bar{v}_{t=5s} = 1.63 - 0.164 = 1.466 \text{ m s}^{-1}.$$

Fourth approximation. Put $\bar{v} = \frac{1}{2}(1.63 + 1.466) = 1.548 \text{ m s}^{-1}$. Then,

$$\Delta Z = 5 \times 1.548/10.24 = 0.756 \text{ m},$$

$$Z = -0.866 + 0.756 = -0.11 \text{ m},$$

$$\Delta\bar{v} = -5(0.049 \times -0.11 + 0.0159 \times 1.548^2) = -0.1635 \text{ m s}^{-1},$$

$$\bar{v}_{t=5s} = 1.63 - 0.1635 = 1.47 \text{ m s}^{-1}.$$

The agreement between the third and fourth approximation is close enough to allow the values of $Z = -0.11$ and $\bar{v} = 1.47 \text{ m s}^{-1}$ to be taken as the values at $t = 5$ s. The period from 5 to 10 s is then handled in the same way. This case may be dealt with by use of program SHAFT, Section 18.3.

The calculations show that the maximum level in the surge shaft was reached around 25 s after load rejection and that the level reached was 1·5 to 1·6 m above the reservoir surface level. In order that these values be more accurately defined, the time step should be reduced to 1 or 2 s from the 20 s point in the calculations onwards, until the time at which the surge shaft surface velocity becomes zero.

18.3. Computer program 'SHAFT'

(1) The program calculates the water level position within a vertical, constant cross-section surge shaft following total load rejection. In addition to determining the maximum change in water level, the program will also display water levels at each time step and the frictionless surge, amplitude and period.

(2) The required input is as follows:

  (i)   Tunnel length, l (m);
  (ii)   Tunnel diameter, d_1 (m);
  (iii) Tunnel friction factor, f;
  (iv)   Initial steady flow in the turbines, $(m^3 s^{-1})$;
  (v)   Surge shaft diameter, d_2 (m);
  (vi)   Chosen time step, Δt (s);
  (vii) Total simulation time, T (s).

(3) Use is made of rigid column surge theory and in particular of the momentum and continuity equations for a simple system, (18.7), (18.8) and (18.9).

(4) Input example:
$D = 200$ m; $d_1 = 1·25$ m; $f = 0·01$; $Q = 2 m^3 s^{-1}$; $d_2 = 4$m; $\Delta t = 5$s; $T = 65$s.

(5) Output:

```
TUNNEL DATA.
TUNNEL LENGTH =    200.000       M.
TUNNEL DIAMETER =    1.250 M.
TUNNEL AREA =        1.227 M^2.
FRICTION FACTOR =    0.010
INITIAL FLOW =       2.000 M^3/S

SHAFT DATA.
SHAFT DIAMETER =     4.000 M.
SHAFT AREA =        12.566 M^2

FRICTIONLESS MAXIMUM SURGE =    2.300 M.
FRICTIONLESS OSCILLATION PERIOD =   90.784 S.

TIME STEP =  2.000 SEC.
PRESS P TO OUTPUT RESULTS AT EACH TIMESTEP OR O TO DISPLAY MAXIMUM VALUES ONLY.?
O
MAXIMUM  SURGE =    1.653 M. AT TIME =      26.000 S.
DO YOU WISH TO REPEAT? Y OR N.  ?N
```

List:

```
10 REM SHAFT
20 @%=&20306
30 PRINT"PROGRAM SHAFT":PRINT
40 PRINT"PROGRAM DESIGNED TO CALCULATE SURGE SHAFT WATER LEVEL OSCILLATIONS F
OLLOWING LOAD REJECTION."
50 INPUT"INPUT TUNNEL LENGTH (M.) L= ",L
60 INPUT"TUNNEL DIAMETER (M.) D= ",D
70 INPUT"TUNNEL FLOW FRICTION FACTOR F= ",F
80 G=9.81:A=PI*D^2/4.0
```

```
 90 INPUT"INPUT INITIAL FLOW TO TURBINES (M^3/S.) Q= ",Q
100 V=Q/A
110 H=4.0*F*L*V^2/(2.0*D*G)
120 K=4.0*F*L/(2.0*G*D)
130 INPUT"INPUT SURGE SHAFT DIAMETER (M.) E= ",E
140 B=PI*E^2/4.0
150 INPUT"INPUT TIME STEP SIZE (SEC.) C= ",C
160 INPUT"INPUT TOTAL SIMULATION TIME (SEC.) ",J
170 CLS
180 PRINT"TUNNEL DATA."
190 PRINT"TUNNEL LENGTH = ",L," M."
200 PRINT"TUNNEL DIAMETER = ",D," M."
210 PRINT"TUNNEL AREA = ",A," M^2."
220 PRINT"FRICTION FACTOR = ",F
230 PRINT"INITIAL FLOW = ",Q," M^3/S"
240 PRINT:PRINT"SHAFT DATA."
250 PRINT"SHAFT DIAMETER = ",E," M."
260 PRINT"SHAFT AREA = ",B," M^2"
270 P=(Q/B)*SQR((B*L)/(A*G))
280 R=2.0*PI*SQR((B*L)/(A*G))
290 PRINT:PRINT"FRICTIONLESS MAXIMUM SURGE = ",P," M."
300 PRINT"FRICTIONLESS OSCILLATION PERIOD = ",R," S."
310 PRINT:PRINT"TIME STEP = ",C," SEC."
320 INPUT"PRESS P TO OUTPUT RESULTS AT EACH TIMESTEP OR O TO DISPLAY MAXIMUM V
ALUES ONLY.",Z$
330 IF Z$="P" THEN CLS
340 IF Z$="P" THEN PRINT"SURGE SHAFT WATER LEVEL VS. TIME PROFILE."
350 T=0.0
360 V=Q/A
370 Z=-H
380 I=0
390 IF Z$="P" THEN PRINT"TIME= ",T," SEC. Z= ",Z," M. V= ",V," M/S. "
400 T=T+C
410 M=V
420 W=Z
430 S=V
440 X=V*A*C/B
450 Z=W+X
460 Y=-C*(G*Z/L+G*K*V*ABS(V)/L)
470 U=S+Y
480 R=ABS((U-M)/U)
490 IF R<0.005 OR R=0.005 THEN V=U
500 IF R<0.005 OR R=0.005 THEN GOTO 390
510 M=U
520 V=(S+U)/2.0
530 IF Z$="P" THEN GOTO 560
540 IF I=0 AND W>Z THEN PRINT"MAXIMUM SURGE = ",W," M. AT TIME = ",T-C," S."
550 IF I=0 AND W>Z THEN I=1
560 IF T>J THEN GOTO 580
570 GOTO 440
580 INPUT"DO YOU WISH TO REPEAT? Y OR N. ",Z$
590 IF Z$="Y" THEN GOTO 10
600 STOP
```

EXERCISES 18

18.1 The outlet valve on a 25 m pipeline is closed in 2 seconds. If the initial velocity of flow was 8 m/s calculate the pressure rise on valve closure if the pipe is assumed to be rigid.
[100 kN/m^2]

18.2 Show that the time taken to establish flow in a pipeline from a large reservoir by opening a valve on the pipe discharge is infinite if rigid column theory is applied.

Further show that the velocity–time curve following valve operation is given by an equation of the form

$$\text{Time to } nV_0 = 0.102l\,V_0^2 \ln[(1+n)/(1-n)]$$

where V_0 is final flow velocity after an infinite time, l is the ratio of pipe length to head available in the reservoir, and n is the % of final velocity attained at any time.

18.3 A valve positioned at the discharge end of a 30 m pipe is opened at a time when the level in the tank supplying the pipe is 6 m above the pipe inlet. Calculate the time taken to accelerate the flow to 50% of its final value and plot the curve of discharge velocity against time following valve operation.

Assume the friction factor for the pipe is 0·01, its diameter 0·05 m and any separation losses may be represented by 50 diameters of equivalent length.
[1·26 s]

18.4 The pressure rise on closure of a valve on the discharge from a 100 m long pipe supplied by a constant head reservoir whose fluid surface level is 6 m above the valve is to be limited to 190 kN/m², this figure calculated by rigid column theory. If the pipe diameter is 25 mm, the pipe friction factor applicable is 0·01, and if the separation losses may be represented by the valve fully open K value of 10, calculate minimum valve closure time to comply with the design pressure limitation.
[0·43 s]

18.5 For the case set out in Exercise 18.4 above determine the time taken to re-establish flow if the valve is re-opened at some later time. Assume 99% flow time to be sufficiently accurate.
[3·1 s]

18.6 In a small hydroelectric plant the supply tunnel is 1·2 m in diameter and the friction factor is estimated as 0·01. 150 m downstream of the supply reservoir a simple surge shaft of 3·5 m diameter is positioned. The initial flow to the turbines is 2·25 m³/s. Calculate the maximum upsurge in the shaft relative to the initial steady state level.
[2·9 m, value dependent on time step chosen]

18.7 In a hydroelectric scheme the water flow in the low pressure supply tunnel under full load is 12 m³/sec. The low pressure tunnel is 980 m long, of 2·133 m diameter, and is protected by a simple surge shaft of 6 m diameter. The friction factor applicable to the supply tunnel is 0·005 and the entrance to the surge shaft from the supply tunnel is at a level 21·5 m below the water surface in the supply reservoir.

On start up the penstock valves are suddenly opened to full load condition and the surge shaft level falls to supply water to the penstock. Calculate by step by step integration the minimum surge shaft water level.
[9 m]

18.8 Show that, if friction loss is proportional to the square of flow velocity, the mass oscillation in a simple surge shaft following a sudden load rejection on flow shut down is given by an expression of the form

$$\frac{d^2z}{dt^2} \pm 2f \frac{D^2}{d^3} \left(\frac{dz}{dt}\right)^2 + \frac{ag}{AL} z = 0$$

where D, A, d, a, are the diameter and cross-sectional areas of supply tunnel and shaft respectively, f is the friction factor for the supply tunnel, L is the length of the supply tunnel and z is the shaft water surface level relative to the supply reservoir.

19 Pressure transient theory

Before embarking on the analysis of pressure transient phenomena and the derivation of the appropriate wave equations, it will be useful to describe the general mechanism of pressure propagation by reference to the events following the instantaneous closure of a valve positioned at the mid-length point of a frictionless pipeline carrying fluid between two reservoirs. The two pipeline sections upstream and downstream of the valve are identical in all respects. Transient pressure waves will be propagated in both pipes by valve operation and it will be assumed that the rate of valve closure precludes the use of rigid column theory.

As the valve is closed, so the fluid approaching its upstream face is retarded with a consequent compression of the fluid and an expansion of the pipe cross-section. The increase in pressure at the valve results in a pressure wave being propagated upstream which conveys the retardation of flow to the column of fluid approaching the valve along the upstream pipeline. This pressure wave travels through the fluid at the appropriate sonic velocity, which will be shown to depend on the properties of the fluid and the pipe material.

Similarly, on the downstream side of the valve the retardation of flow results in a reduction in pressure at the valve, with the result that a negative pressure wave is propagated along the downstream pipe which, in turn, retards the fluid flow. It will be assumed that this pressure drop in the downstream pipe is insufficient to reduce the fluid pressure to either its vapour pressure or its dissolved gas release pressure, which may be considerably different.

Thus, closure of the valve results in the propagation of pressure waves along both pipes and, although these waves are of different sign relative to the steady pressure in the pipe prior to valve operation, the effect is to retard the flow in both pipe sections. The pipe itself is affected by the wave propagation as the upstream pipe swells as the pressure rise wave passes along it, while the downstream pipe contracts due to the passage of the pressure reducing wave. The magnitude of the deformation of the pipe cross-section depends on the pipe material and can be well demonstrated if, for example, thin-walled rubber tubing is employed. The passage of the pressure wave through the fluid is preceded, in practice, by a strain wave propagating along the pipe wall at a velocity close to the sonic velocity in the pipe material. However, this is a secondary effect and, while knowledge of its existence can explain some parts of a pressure–time trace following valve closure, it has little effect on the pressure levels generated in practical transient situations.

Following valve closure, the subsequent pressure–time history will depend on the conditions prevailing at the boundaries of the system. In order to describe the events following valve closure in the simple pipe system outlined above, it will be easier to refer to a series of diagrams illustrating conditions in the pipe at a number of time steps (Fig. 19.1).

Assuming that valve closure was instantaneous, the fluid adjacent to the valve in each pipe would have been brought to rest and pressure waves conveying this information would have been propagated at each pipe at the

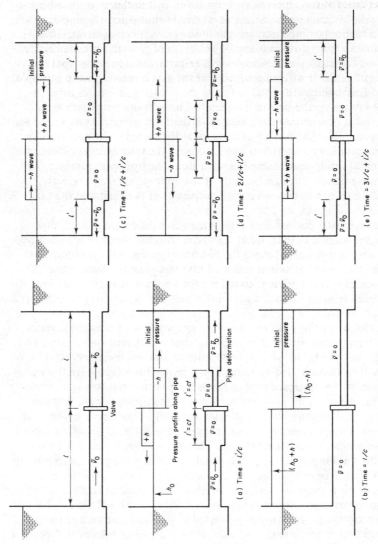

Figure 19.1 Pressure and pipe diameter profiles at a number of instants following an instantaneous valve closure. Frictional effects have been neglected

appropriate sonic velocity c. At a later time t, the situation is as shown in
Fig. 19.1(a), the wavefronts having moved a distance $l' = ct$ in each pipe. The
deformation of the pipe cross-section will also have travelled a distance l'
as shown.

The pressure waves reach the reservoirs terminating the pipes at a time
$t = l/c$ following valve closure (Fig. 19.1(b)). At this instant, an unbalanced
situation arises at the pipe–reservoir junction, as it is clearly impossible for
the layer of fluid adjacent to the reservoir inlet to maintain a pressure different from that prevailing at that depth in the reservoir. Hence, a restoring
pressure wave having a magnitude sufficient to bring the pipeline pressure
back to its value prior to valve closure is transmitted from each reservoir at
a time l/c. For the upstream pipe, this means that a pressure wave is propagated towards the closed valve, reducing the pipe pressure to its original
value and restoring the pipe cross-section. The propagation of this wave
also produces a fluid flow from the pipe into the reservoir as the pipe ahead
of the moving wave is at a higher pressure than the reservoir. Now, as the
system is assumed to be frictionless, the magnitude of this reversed flow
will be the exact opposite of the original flow velocity, as shown in Fig.
19.1(c).

At the downstream reservoir, the converse occurs, resulting in the
propagation of a pressure rise wave towards the valve and the establishment
of a flow from the downstream reservoir towards the valve (Fig. 19.1(c)).

For the simple pipe considered here, the restoring pressure waves in both
pipes reach the valve at a time $2l/c$. The whole of the upstream pipe has,
thus, been returned to its original pressure and a flow has been established
out of the pipe. At time $2l/c$, as the wave has reached the valve, there remains
no fluid ahead of the wave to support the reversed flow. A low pressure region,
therefore, forms at the valve, destroying the flow and giving rise to a pressure
reducing wave which is transmitted upstream from the valve, once again
bringing the flow to rest along the pipe and reducing the pressure within the
pipe (as shown in Fig. 19.1(d)). It is assumed that the pressure drop at the
valve is insufficient to reduce the pressure to the fluid vapour pressure. As the
system has been assumed to be frictionless, all the waves will have the same
absolute magnitude and will be equal to the pressure increment, above steady
running pressure, generated by the closure of the valve. If this pressure increment is h, then all the waves propagating will be $\pm h$, as shown in Fig. 19.1.
Thus, the wave propagating upstream from the valve at time $2l/c$ has a value
$-h$, and reduces all points along the pipe to $-h$ below the initial pressure
by the time it reaches the upstream reservoir at time $3l/c$.

Similarly, the restoring wave from the downstream reservoir that reached
the valve at time $2l/c$ had established a reversed flow along the downstream
pipe towards the closed valve. This is brought to rest at the valve, with a
consequent rise in pressure which is transmitted downstream as a $+h$ wave
arriving at the downstream reservoir at $3l/c$, at which time the whole of the
downstream pipe is at pressure $+h$ above the initial pressure with the fluid at rest.

Thus, at time $3l/c$ an unbalanced situation similar to the situation at
$t = l/c$ again arises at the reservoir–pipe junctions with the difference that it
is the upstream pipe which is at a pressure below the reservoir pressure and
the downstream pipe that is above reservoir pressure. However, the
mechanism of restoring wave propagation is identical with that at $t = l/c$,

resulting in a $-h$ wave being transmitted from the upstream reservoir, which effectively restores conditions along the pipe to their initial state (as shown in Fig. 19.1(e)), and a $+h$ wave being propagated upstream from the downstream reservoir, which establishes a flow out of the downstream pipe. Thus, at time $t = 4l/c$ when these waves reach the closed valve, the conditions along both pipes are identical to the conditions at $t = 0$, i.e. the instant of valve closure. However, as the valve is still shut, the established flow cannot be maintained and the cycle described above repeats.

The pipe system chosen to illustrate the cycle of transient propagation was a special case as, for convenience, the pipes upstream and downstream of the valve were identical. In practice, this would be unusual. However, the cycle described would still apply, except that the pressure variations in the two pipes would no longer show the same phase relationship. The period of each individual pressure cycle would be $4l/c$, where l and c took the appropriate values for each pipe. It is important to note that once the valve is closed the two pipes will respond separately to any further transient propagation.

The period of the pressure cycle described is $4l/c$. However, a term often met in transient analysis is 'pipe period'; this is defined as the time taken for a restoring reflection to arrive at the source of the initial transient propagation and, thus, has a value $2l/c$. In the case described, the pipe period for both pipes was the same and was the time taken for the reflection of the transient wave propagated by valve closure to arrive at the closed valve from the reservoirs.

From the description of the transient cycle above (Fig. 19.1), it is possible to draw the pressure–time records at points along the pipeline (as shown in Fig. 19.2). These variations are arrived at simply by calculating the time at which any one of the $\pm h$ waves reaches a point in the system assuming a constant propagation velocity c. The major interest in pressure transients lies in methods of limiting excessive pressure rises and one obvious method is to reduce valve speeds. However, reference to Fig. 19.2 illustrates an important point: no reduction in generated pressure will occur until the valve closing time exceeds one pipe period. The reduction in peak pressure achieved by slowing the valve closure arises as a result of the arrival of negative waves from the upstream reservoir at the valve prior to valve closure and, as no reflection can return to the valve before a time $2l/c$ from the start of valve motion, no beneficial pressure relief can be achieved if the valve is not open beyond this time. Generally, valve closures in less than a pipe period are referred to as *rapid* and those taking longer than $2l/c$ are *slow*.

In the absence of friction, the cycle would continue indefinitely. However, in practice, friction damps the pressure oscillations within a short period of time. In systems where the frictional losses are high, the neglect of frictional effects can result in a serious underestimate of the pressure rise following valve closure. In these cases, the head at the valve is considerably lower than the reservoir head. However, as the flow is retarded, so the frictional head loss is reduced along the pipe and the head at the valve increases towards the reservoir value. As each layer of fluid between the valve and the reservoir is brought to rest by the passage of the initial $+h$ wave, so a series of secondary positive waves each of a magnitude corresponding to the friction head recovered is transmitted towards the valve, resulting in

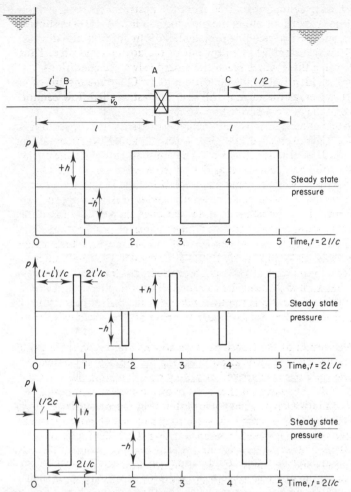

Figure 19.2   Pressure variation following an instantaneous valve closure at points along the two identical pipes linking the reservoirs. Note that the closer to the reservoir the recording point, the shorter is the duration of the pressure change

the full effect being felt at time $2l/c$ (as shown in Fig. 19.3). As the flow reverses in the pipe during time $2l/c$ to $4l/c$, the opposite effect is recorded at the valve because of the re-establishment of a high friction loss, these variations being shown by lines AB and CD (Fig. 19.3). In certain cases, such as long distance oil pipelines, this effect may contribute the larger part of the pressure rise following valve closure.

In addition to the assumptions made with regard to friction in the cycle description, mention was also made of the condition that the pressure drop waves at no time reduced the pressure in the system to the fluid vapour pressure. If this had occurred, then the fluid column would have separated and the simple cycle described would have been disrupted by the formation of a vapour cavity at the position where the pressure was reduced to vapour

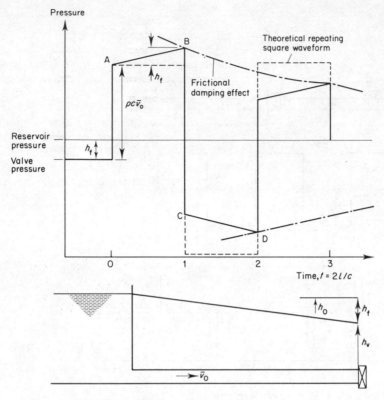

Figure 19.3   Effect of friction on the pressure variation recorded at the valve following instantaneous valve closure

level. In the system described, this could happen on the valve's downstream face at time 0 or on the upstream face at time $2l/c$. The formation of such a cavity is followed by a period of time when the fluid column moves under the influence of the pressure gradients between the cavity and the system boundaries. This period is normally terminated by the generation of excessive pressures on the final collapse of the cavity. This phenomena is generally referred to as *column separation* and is frequently made more complex by the release of dissolved gas in the vicinity of the cavity (*see* Section 19.7).

19.1. Differential equations defining transient propagation

Pressure transient propagation may be defined in any closed pipe application by two basic equations, namely the equations of motion and continuity applied to a short segment of the fluid column. The dependent variables are the fluid's average pressure and velocity at any pipe cross-section and the independent variables are time and distance, normally considered positive in the steady flow direction. Friction will be assumed proportional to velocity squared and steady flow friction relationships will be assumed to apply to the unsteady flow cases considered.

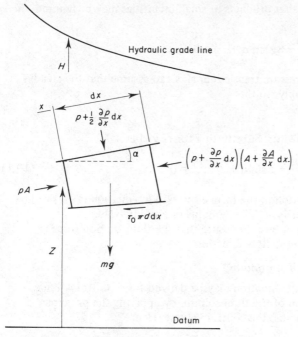

Figure 19.4   Element of fluid in an inclined pipeline, considered in the derivation of the equations of motion and continuity

19.1.1. The equation of motion

Figure 19.4 illustrates the forces acting on an element of fluid in an inclined pipeline. To develop the equation of motion it is necessary to equate the total resolved force in the flow direction to the product of the elements mass and acceleration.

Referring to Fig. 19.4 and resolving parallel to the axis of flow yields an expression for the total force:

$$pA - \left(p + \frac{\partial p}{\partial x}\,dx\right)\left(A + \frac{\partial A}{\partial x}\,dx\right) + \left(p + \frac{1}{2}\frac{\partial p}{\partial x}\,dx\right)\frac{\partial A}{\partial x}\,dx$$

$$-\tau_0 \pi d\,dx - mg\sin\alpha = m\left(\frac{v\partial v}{\partial x} + \frac{\partial v}{\partial t}\right)$$

made up of two opposing pressure forces, a component of the pressure force due to the change in pipe cross-section, friction force and the component of weight, which can be equated to the mass times acceleration term as shown above.

It is reasonable to assume that changes in fluid density are small compared to the density ρ, so that

$$m = \rho\left(A + \frac{1}{2}\frac{\partial A}{\partial x}\,dx\right)dx$$

Similarly, by assuming that products of small quantities may be ignored, the rearranged terms are

$$\frac{1}{\rho}\frac{\partial p}{\partial x} + \frac{\partial \bar{v}}{\partial t} + \frac{\bar{v}\partial \bar{v}}{\partial x} + g \sin \alpha + \frac{4\tau_0}{\rho d} = 0.$$

It is customary in pressure transient analysis to assume that the steady-state friction factors apply, so that

$$\tau_0 = \tfrac{1}{2}\rho f \bar{v}|\bar{v}|,$$

where f is the friction factor. Substituting for τ_0 yields

$$\frac{1}{\rho}\frac{\partial p}{\partial x} + \frac{\partial \bar{v}}{\partial t} + \frac{\bar{v}\partial \bar{v}}{\partial x} + g \sin \alpha + 2f\frac{\bar{v}|\bar{v}|}{d} = 0. \tag{19.1}$$

The equation of motion in the form expressed in equation (19.1) will be used throughout this analysis. The modulus or absolute-value sign is introduced in the friction term to ensure that the fluid friction force is always in opposition to the flow direction.

19.1.2. The equation of continuity

The unsteady continuity equation may be derived for a control volume enclosing a short section of the fluid column by applying the principle of continuity of flow across the control volume. Hence,

$$\text{Net mass flow rate into the control volume across its ends} = \text{Rate of increase of mass of fluid within the control volume}$$

or, in terms of the fluid element shown in Fig. 19.4,

$$\rho A \bar{v} - \left\{\dot{\rho} A \bar{v} + \frac{\partial}{\partial x}(\rho A \bar{v})\,\mathrm{d}x\right\} = \frac{\partial}{\partial t}(\rho A)\,\mathrm{d}x. \tag{19.2}$$

Expanding equation (19.2) yields

$$-\left\{\rho \bar{v}\frac{\partial A}{\partial x} + A\bar{v}\frac{\partial \rho}{\partial x} + \rho A\frac{\partial \bar{v}}{\partial x}\right\} = A\frac{\partial \rho}{\partial t} + \rho\frac{\partial A}{\partial t}$$

or
$$\left(\frac{\bar{v}}{A}\frac{\partial A}{\partial x} + \frac{1}{A}\frac{\partial A}{\partial t}\right) + \left(\frac{\bar{v}}{\rho}\frac{\partial \rho}{\partial x} + \frac{1}{\rho}\frac{\partial \rho}{\partial t}\right) + \frac{\partial \bar{v}}{\partial x} = 0,$$

which may be reduced to

$$\frac{1}{A}\cdot\frac{\mathrm{d}A}{\mathrm{d}t} + \frac{1}{\rho}\cdot\frac{\mathrm{d}\rho}{\mathrm{d}t} + \frac{\partial \bar{v}}{\partial x} = 0. \tag{19.3}$$

The first term in equation (19.3) expresses the deformation of the pipe under internal pressure. The second term expresses the fluid's compressibility.

Returning to the pipe deformation, it will be assumed that the pipe is subject to elastic deformations and that the pipe is sufficiently 'thin' (based on its diameter to wall thickness ratio) to allow thin cylinder theory to be applied. In this case, the circumferential strain for a pressure increment $\mathrm{d}p$ is

$$\frac{\mathrm{d}d}{d} = \frac{d}{2Ee}\,\mathrm{d}p,$$

where d is the pipe diameter, e is the wall thickness and E is the Young's modulus of the pipe material. The associated change in cross-sectional area is

$$dA = (\pi d/2)\, dd.$$

Hence,

$$\frac{1}{A}\frac{dA}{dt} = \frac{d}{Ee}\frac{d\rho}{dt}.$$

Now the bulk modulus of the fluid may be expressed as $K = \rho\, dp/d\rho$. Thus, the second term in equation (19.3) may be replaced by

$$\frac{1}{\rho} \cdot \frac{d\rho}{dt} = \frac{1}{K}\frac{dp}{dt}$$

and so the continuity equation may be written in the form

$$\left(\frac{1}{K} + \frac{d}{Ee}\right)\frac{dp}{dt} + \frac{\partial \bar{v}}{\partial x} = 0.$$

It is convenient to write $1/c^2 = \rho(1/K + d/Ee)$, where c has the dimensions of velocity and is the velocity of wage propagation appropriate to the pipe–fluid combination being considered. It is interesting to compare this value of c to the expression for the sonic velocity in an infinite expanse of the fluid given by $\sqrt{(K/\rho)}$. From this comparison, it may be seen that if a pipe–fluid equivalent bulk modulus term K' is introduced, such that

$$1/K' = 1/K + d/Ee,$$

then the wave speed in the fluid–pipe system is given by

$$c^2 = \frac{K'}{\rho} = \frac{1}{\rho}\left(\frac{K}{1 + dK/Ee}\right), \tag{19.4}$$

an expression directly analogous to equation (5.31). The equivalent bulk modulus effectively expresses the combined elasticity of the fluid and the pipe material.

The final form of the continuity equation can, therefore, be written as

$$\frac{\partial p}{\partial t} + \bar{v}\frac{\partial p}{\partial x} + \rho c^2\,\frac{\partial \bar{v}}{\partial x} = 0, \tag{19.5}$$

where dp/dt has been expanded into its partial differential form. It will be noted that longitudinal strain has been neglected in this derivation; however, this is reasonable for the majority of transient conditions. Inclusion of longitudinal strain would have resulted in the inclusion of a Poisson's ratio term in the equivalent bulk modulus expression, but the effect remains small unless $E \ll K$, i.e. the pipe Young's modulus becomes much smaller than the fluid bulk modulus — as may occur with plastic or rubber tubing.

The two equations defining transient propagation may now be solved either by graphical or numerical techniques. Examples of both types of solution will be presented.

19.2. The effect of pipe elasticity and free gas on wave propagation velocity

If even a small quantity of free gas is present in a liquid flow, the change in mixture compressibility can be large enough to cause considerable reductions in the wave speed.

Consider a flow of a liquid of density ρ_f and bulk modulus K_f along a pipe of diameter d and wall thickness e, where the Young's modulus of the

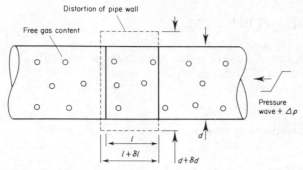

Figure 19.5   Effect of a pressure wave in a gas-fluid pipeline

pipe wall is E and the Poisson's ratio is ν (Fig. 19.5). The mean density $\bar{\rho}$ of the fluid–gas mixture is given by

$$\bar{\rho} = y\rho_g + (1 - y)\rho_f, \tag{19.6}$$

where y is the proportion of free gas by volume and ρ_g is the gas density.

Now, detailing the volumetric changes that occur if a pressure wave compresses a volume V_f of fluid by dV_f yields

$$dV_f = -(V_f/K_f) \, dp \text{ for the fluid,}$$

$$dV_g = -(V_g/K_g) \, dp \text{ for the gas,}$$

where V_g is the gas volume present, K_g is the gas bulk modulus and $V_f + V_g = $ total element volume, V_t. The pipe section containing the initial volume V_f can be distorted both radially and longitudinally.

If the original volume $V_t = \pi d^2 l/4$, where l is the length of pipe element chosen in the direction of flow, then the distorted volume increase due to the passage of a pressure wave dp is given by

$$dV_p = \pi d(\delta d/2)l \tag{19.7}$$

> Volume
> increase due
> to radial
> distortion
> δd

where products of small quantities (e.g. $\delta d\delta l$) are neglected.

From thin walled pipe theory, the longitudinal stress F_L and the circumferential stress F_C are given by

$$F_L = (d/4e) \, dp \quad \text{and} \quad F_C = (d/2e) \, dp,$$

while the longitudinal strain is given by

$$\delta l/l = (F_L - \nu F_C)/E$$

and the circumferential strain by

$$\delta d/d = (F_C - \nu F_L)/E,$$

where ν is Poisson's ratio of the pipe material.
Thus,

$$dV_p = \pi d^2 l(F_C - \nu F_L)/2E.$$

The final form of the volumetric strain expression above will depend upon the assumptions made about pipe restraint and, hence, longitudinal stress and strain.

(i) If the pipe is fully restrained at one end only, then both longitudinal and circumferential stresses occur. Hence,

$$dV_p/V = \{(d/Ee)(1 - \nu/2)\}\, dp. \tag{19.8}$$

(ii) If the pipe is fully restrained against axial movement along its whole length, then longitudinal strain $\delta l/l = 0$ and $F_L = \nu F_C$. Hence,

$$dV_P/V = \{(d/Ee)(1 - \nu^2)\}\, dp \tag{19.9}$$

(iii) The pipe is supplied with expansion joints at regular intervals along its length, which may be necessary to take up thermal expansion. These are standard in many systems, including aircraft fuel systems. In this case, the longitudinal stress $F_L = 0$ and

$$dV_p/V = (d/Ee)\, dp \tag{19.10}$$

A general form of the expression for pipe distortion is, therefore,

$$dV_p/V = (d/Ee)\, C'\, dp. \tag{19.11}$$

The total change in volume for the fluid, gas and pipe section is then

$$dV_t = dV_p - dV_f - dV_g$$
$$= \{(d/Ee)C'V_t + V_f/K_f + V_g/K_g\}\, dp,$$

where

$$V_f = (1 - y)\, V_t,$$
$$V_g = y V_t$$
$$dV_t = \{dC'/Ee + (1 - y)/K_f + y/K_g\}V_t\, dp.$$

As a result, an overall effective bulk modulus for the pipe, fluid and gas combination can be written

$$K_{eff} = \{(1 - y)/K_f + y/K_g + dC'/Ee\}^{-1}$$

and an expression for the wave speed c may be deduced as

$$c = \sqrt{(K_{eff}/\bar{\rho})}$$
$$= \sqrt{\{(y\rho_g + (1 - y)\rho_f)^{-1}((1 - y)/K_f + y/K_g + dC'/Ee)^{-1}\}} \tag{19.12}$$

At low temperatures and pressures, the gas bulk modulus K_g can be approximated by the initial absolute pressure of the gas p_a. This is an approximation, as the passage of the pressure wave compresses the gas and changes its pressure; however, this is acceptable as a first approximation.

In the absence of free gas the wave speed expression reduces to

$$c = \sqrt{\left/\left\{K/\rho\left(1 + \frac{dK_f}{Ee}C'\right)\right\}\right.} \qquad (19.13)$$

and Fig. 19.6 illustrates the effect of the pipe bore to wall thickness ratio d/e on wave speed through water in steel, cast iron and glass pipelines. It will be seen that the effect of increasing d/e, effectively increasing the elasticity of the pipe wall, reduces the wave speed through the fluid.

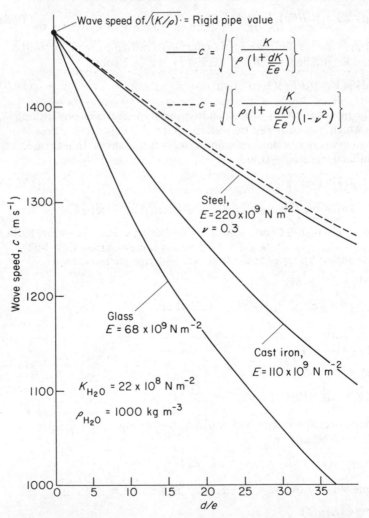

Figure 19.6   Influence of the pipe diameter/pipe wall thickness ratio on the wave speed in water

It will also be seen that incorporating the effects of longitudinal strain has little effect over the normal working range; indeed, for glass and cast iron, as the value of Poisson's ratio is around 0·25, the value of C' tends to unity anyway. If the pipe wall is assumed rigid, i.e. $E \to \infty$, then the wave speed expression reduces to

$$c = \sqrt{(K_f/\rho)}$$

and has the value corresponding to the acoustic velocity in an expanse of fluid. Therefore, the value is the same irrespective of pipe material (as shown in Fig. 19.6). Similarly, for the rigid column theory the fluid bulk modulus tends to infinity as the fluid is assumed incompressible and pressure wave propagation may therefore be assumed to be instantaneous.

The presence of even a small quantity of free gas becomes important in reducing the wave speed through the mixture. Figure 19.7 illustrates the predicted reduction in wave speed through a kerosene and nitrogen mixture flowing in a polythene pipeline. The results of a series of wave speed measurements are also included as a comparison to the predicted figures. It will be seen that the effect of neglecting pipe elasticity is only important at low gas contents. As the percentage of free gas increases, so the wave speed reducing effect of the pipe elasticity is numerically swamped by the effect of the gas terms, see Section 19.3.

It is sometimes stated that reduction in wave speed can be beneficial in reducing transient pressures. However, this is not automatically true, as will be seen later. Reducing wave speed increases the pipe period for any system, so that any valve operation taking a fixed time becomes effectively faster in terms of pipe periods, which is the only time measurement having any relevance in transient pressure prediction.

Table 19.1 presents Young's Modulus and Poisson's ratio values for some common pipe materials while Table 19.2 details some common fluid bulk modulus and density values.

Material	$E \times 10^{-9}$ (N m^{-2})	ν	Fluid	$K \times 10^{-8}$ (N m^{-2})	ρ (kg m^{-3})
Steel	200–214	0·3	Water at 0 °C	20·5	1000
Cast iron	80–110	0·25	at 20 °C	20·5	998
Aluminium	70	0·3	at 80 °C	20·5	972
Concrete	20–30	0·1–0·3	Oil s.a.e. 10	16·7	880–940
Copper	107–130	0·34	s.a.e. 30	18·6	880–940
Rubber	0·7–7	0·46–0·49	Kerosene	13·6	814
PVC plastic	2·4–2·8	—	Sea water		
PTFE plastic	0·35	—	at 10°C	22	1026
Glass	68	0·24	Ethyl alcohol		
Polythene	3·1	—	at 15°C	12	799
			Carbon tetra-		
			chloride at 15°C	11·2	1590

Table 19.1. Values of Young's modulus $E(\times 10^{-9})$ and Poisson's ratio ν

Table 19.2. Values of bulk modulus $K(\times 10^{-8})$ and density ρ for some common fluids

19.3. Computer program 'WAVESPD'

(1) The computer program calculates the value of transient pressure wave propagation velocity in both rigid and elastic walled pipes of circular cross-section. In addition the wave velocity is also calculated if free gas is present in the fluid, provided the gas pressure, density and % volume is known. Values of pipe period corresponding to this wave speed are also calculated.

(2) The required input is:

(i)   Pipe Young's modulus, diameter, wall thickness and Poisson's ratio;

(ii) fluid bulk modulus and density;

(iii) gas density and absolute pressure, and free gas content as a ratio of total volume of gas and fluid;

(iv) pipe length if the pipe period is required.

(3) The program is based on wave speed equations (19.12) and (19.13) and Figs 19.6 and 19.7.

(4) Input example:

Pipe   Young's modulus $70 \cdot 0 \times 10^9$ N m^{-2};
diameter $0 \cdot 05$ m.; thickness $0 \cdot 006$ m;
Poisson's ratio $0 \cdot 3$. Expansion joints, code 3.

Fluid   Bulk modulus $13 \cdot 6 \times 10^8$ N m^{-2};
Density 800 kg m^{-3};

Gas   Density $1 \cdot 2$ kg m^{-3};
pressure 175 kN m^{-2};
content $0 \cdot 004$ by volume.

Pipe   Length 23 m.

(5) Output:

WAVE SPEED =   230.66 M/SEC

DO YOU WISH TO CALCULATE PIPE PERIOD FOR THIS WAVE SPEED? (Y/N). ?Y

INPUT PIPE LENGTH IN M.?23

PIPE PERIOD =   0.20 S. FOR   23.00 M. AT C=   230.66 M/S.

List:

```
   10 REM WAVESPD
   20 MODE 7
   30 @%=&20208
   40 CLS:PRINT"PROGRAM DESIGNED TO CALCULATE PRESSURE TRANSIENT PROPAGATION VEL
OCITY."
   50 PRINT:PRINT
   60 INPUT"DO YOU WISH TO INCLUDE PIPE ELASTICITY IN WAVE SPEED CALCULATION? (Y
/N) ",U$
   70 IF U$="N" THEN C1=0.0 ELSE C1=1.0
   80 IF U$="N" THEN GOTO 190
   90 PRINT:INPUT"PIPE DIAMETER (M.)",D
  100 PRINT:INPUT"PIPE WALL THICKNESS (M.)",W
  110 PRINT:PRINT"PIPE MATERIAL YOUNGS MODULUS  AS E*10^n (UNITS-N/M^2)":INPUT"E
=",E:INPUT"n=",N
  120 PRINT:INPUT"PIPE POISSON'S RATIO = ",PR
  130 PRINT:PRINT"VALUE OF CONSTANT C'=(1.0-0.5*PR) OR (1.0-PR^2) OR 1.0 DEPENDA
NT ON PIPE RESTRAINT"
  140 PRINT:INPUT"INPUT CODE 1,2,OR 3 FOR EXPRESSION CHOOSEN. ",CO
  150 IF CO=1 THEN CC=(1.0-0.5*PR)
  160 IF CO=2 THEN CC=(1.0-PR^2)
  170 IF CO=3 THEN CC=1.0
  180 E=E*10.0^N
  190 PRINT:INPUT"FLUID DENSITY KG/M^3",R
  200 PRINT:PRINT"FLUID BULK MODULUS AS B*10^n (UNIT N/M^2)":INPUT"B=",B:INPUT"n
```

```
=",N
 210 B=B*10.0^N
 220 PRINT:INPUT"ANY FREE GAS PRESENT? (Y/N)."Z$
 230 IF Z$="N" THEN GOTO 290
 240 PRINT:INPUT"FREE GAS DENSITY KG/M^3 = ",RG
 250 PRINT:INPUT"BULK MODULUS FREE GAS TAKEN AS ITS ABS.PRESSURE IN KN/M^2 BG=
",BG
 260 BG=BG*1000.0
 270 PRINT:INPUT"RATIO FREE GAS BY VOLUME Y= ",Y
 280 REM CALCULATION OF WAVE SPEED
 290 IF C1=0.0 THEN C=SQR(B/R) ELSE C=SQR((B/R)/(1.0+D*CC*B/(E*W)))
 300 IF Z$="N" THEN GOTO 350
 310 BF=C^2*R
 320 IF C1=0.0 THEN BM=1.0/((1.0-Y)/B+Y/BG) ELSE  BM=1.0/((1.0-Y)/BG+D*CC/E
*W)
 330 RM=Y*RG+(1.0-Y)*R
 340 C=SQR(BM/RM)
 350 CLS
 360 IF U$="N" THEN PRINT"PIPE ASSUMED RIGID..."
 370 IF U$="N" THEN GOTO 420
 380 PRINT:PRINT"PIPE DIAMETER =",D," M. WALL THICKNESS =",W," M."
 390 PRINT:PRINT"YOUNGS MODULUS =",E/10.0^6," MN/M^2."
 400 PRINT"PIPE POISSON'S RATIO = ",PR
 410 PRINT"PIPE RESTRAINT COEFF. C'= ",CC
 420 PRINT:PRINT"FLUID DATA...."
 430 PRINT:PRINT"DENSITY =",R," KG/M^3. BULK MODULUS =",B/10.0^6," MN/M^2."
 440 IF Z$="N" THEN GOTO 500
 450 PRINT
 460 PRINT"GAS DATA..."
 470 PRINT"GAS DENSITY KG/M^3 = ",RG
 480 PRINT"GAS BULK MODULUS N/M^2 = ",BG
 490 PRINT"% FREE GAS BY VOLUME, % = ",Y*100.0
 500 PRINT:PRINT"WAVE SPEED =",C," M/SEC"
 510 PRINT:INPUT"DO YOU WISH TO CALCULATE PIPE PERIOD FOR THIS WAVE SPEED? (Y/N
).",N$
 520 IF N$="N" THEN GOTO 560
 530 PRINT:INPUT"INPUT PIPE LENGTH IN M.",L
 540 PP=2.0*L/C
 550 PRINT:PRINT"PIPE PERIOD = ",PP," S. FOR ",L," M. AT C= ",C," M/S."
 560 PRINT:INPUT"DO YOU WISH TO REPEAT FOR ANOTHER AIR VOLUME. (Y/N).",X$
 570 IF X$="Y" THEN GOTO  270
 580 PRINT:INPUT"DO YOU WISH TO REPEAT FOR ANOTHER CASE? (Y/N). ",M$
 590 IF M$="Y" THEN GOTO 10
 600 PRINT:PRINT"CALCULATION COMPLETE"
```

19.4. Simplification of the basic pressure transient equations

In order to describe some basic aspects of transient propagation and to introduce the graphical method of solution, it is necessary to rewrite the equations of motion and continuity in the simplified form below:

$$\text{Motion,} \quad \frac{\partial p}{\partial x} + \rho \frac{\partial \bar{v}}{\partial t} = 0; \tag{19.14}$$

$$\text{Continuity,} \quad \frac{\partial p}{\partial t} + \rho c^2 \frac{\partial \bar{v}}{\partial x} = 0. \tag{19.15}$$

In this form, the equations apply to a frictionless, horizonal pipeline, where the convective terms $\bar{v}\partial\bar{v}/\partial x$ and $\bar{v}\partial p/\partial x$ may be ignored in comparison to $\partial v/\partial t$ and $\partial p/\partial t$. The general solution of these two equations, due to D'Alembert, is:

$$p - p_0 = F(t + x/c) + f(t - x/c), \tag{19.16}$$

$$\bar{v} - \bar{v}_0 = -(1/\rho c)\{F(t + x/c) - f(t - x/c)\}. \tag{19.17}$$

The f and F functions are entirely arbitrary and may be selected to satisfy the conditions imposed at the system boundaries. Referring to the simple pipe of Fig. 19.1, the F function may be interpreted as a pressure wave

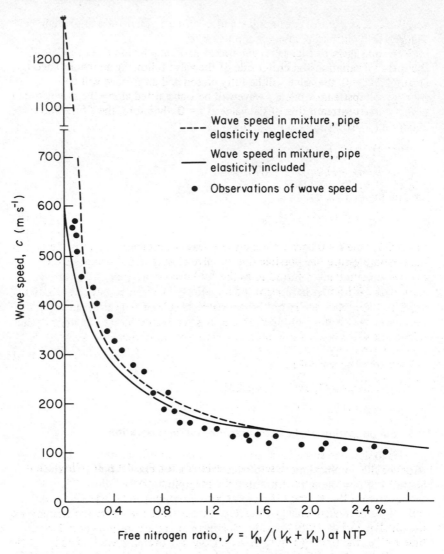

Figure 19.7   Influence of the free gas content on wave speed through a nitrogen-kerosene mixture in a polythene pipeline, of 50 mm diameter, 6 mm wall thickness and 15 m length. The free gas pressure has been taken as 175 kN m^{-2}

moving in the $-x$ direction, i.e. upstream, as x must decrease with time. From (19.16), the dimensions of both functions are those of pressure. Similarly, the f function may be interpreted as a pressure wave moving in the $+x$ direction, i.e. downstream. Both waves propagate at the sonic velocity c. The significance of equation (19.16) now becomes clear; it implies that at any time t following the initial disturbance, the pressure at any point in the pipe may be found from a summation of the F and f waves that have passed that point in the time t. It may be assumed that the pressure waves travel at a uniform velocity c and do not attenuate or change their shape as they propagate along the pipeline.

Referring again to Fig. 19.1, equation 19.16 may be used to calculate the pressure variations on either side of the valve following an instantaneous closure. At $t = 0$, the valve will be fully closed and an F wave will be propagated upstream while a f wave will be transmitted along the downstream pipe. On the upstream side of the valve at $t = 0$, therefore, the f function in equation (19.16) will be zero, so that

$$p - p_0 = F(t + x/c),$$

$$\bar{v} - \bar{v}_0 = (1/\rho_c)F(t + x/c).$$

Eliminating $F(\ )$ yields

$$\Delta p = \rho c(\bar{v}_0 - \bar{v}) = \rho c \bar{v}_0 \tag{19.18}$$

as $\bar{v} = 0$ at time $t = 0$ because the closure was instantaneous. This is the maximum pressure rise possible due to valve closure and the expression (19.18) is commonly referred to as the Joukowsky pressure rise after Joukowsky, who first demonstrated its validity in 1897. Equations (19.16) and (19.17) apply equally on the downstream side of the valve. However, at $t = 0$ here, it is the $F(\ )$ function that has zero value and so the magnitude of the initial $f(\ )$ wave propagated downstream will be given by

$$p - p_0 = f(t - x/c),$$

$$\bar{v} - \bar{v}_0 = -(1/\rho c)\{-f(t - x/c)\}.$$

Hence,

$$\Delta p = -\rho c \bar{v}_0.$$

Thus, for the closure case described previously for Fig. 19.1, it follows that the pressure waves referred to were of a magnitude $\pm \rho c \bar{v}_0$.

Equations (19.16) and (19.17) can also be employed to calculate the reflections produced when an incident wave reaches the reservoirs terminating the pipeline in Fig. 19.8. Consider the upstream reservoir-pipe junction: at time $t = l/c$, the first $F(\ )$ wave arrives. Now, the overall pressure change at the reservoir-pipe junction must be zero. Hence, from equation (19.16),

$$\Delta p = 0 = F(t + x/c) + f(t - x/c)$$

and so

$$f(t - x/c) = -F(t + x/c).$$

Hence, the reflected wave that is transmitted downstream from the reservoir is equal in magnitude but of opposite sign to the incident wave. If the above analysis is applied to the downstream reservoir at time $t = l/c$, then the reflected $F(\ )$ wave produced will be equal to $-f(\ )$, the initial wave propagating downstream from the valve. If the concept of a reflection coefficient is introduced at this stage, defined as the ratio of the reflected wave to the incident wave, i.e. $C_R = f(\ )/F(\ )$ for the upstream reservoir, then it will be seen that the reservoir has a reflection coefficient of $-1\cdot0$. This is true of all constant pressure boundary conditions and, for example, applies equally to vapour cavities formed during column separation as long as the cavity pressure remains a constant at vapour level.

Similarly, it is possible to calculate the appropriate reflection coefficient for the closed valve. At time $t = 2l/c$, the $f()$ and $F()$ waves in the two pipes of Fig. 19.8 arrive at the closed valve. However, so long as the pressure remains above vapour level, there can be no change in the velocity of the fluid adjacent to the valve, i.e. it must remain at rest. Hence, from equation (19.17),

$$\Delta\bar{v} = 0 = -(1/\rho c)\{F(t + x/c) - f(t - x/c)\}$$

and so

$$F(t + x/c) = f(t - x/c)$$

and the reflection coefficient may be shown to be $+1\cdot0$.

The same analysis applies to the case of a transient arriving at the end of a closed pipe, and it is to be noted that this implies that the pressure recorded at the dead end will be twice the value of the incident pressure wave — a very important consideration if the dead ended pipe happens to be a transducer connection.

The use of equations (19.16) and (19.17), together with known boundary conditions, allows reflection and transmission coefficients to be calculated for a range of cases likely to be met in any reasonably complex pipe network. However, this approach is limited to the propagation of fairly sharp transients, which may be approximated by step functions. In order to consider more gradual pressure changes it is necessary to employ the graphical method of solution of equations (19.16) and (19.17).

Example 19.1

An outlet control valve on a long water distribution main is partially shut in 20 ms to restrict delivery. Calculate the length of the wavefront so propagated if the wave propagation speed in the pipe is 1250 m s^{-1}.

If the change in velocity produced by the valve action is $0\cdot5$ m s^{-1}, show that the convective terms $\bar{v}\partial\bar{v}/\partial x$ and $\bar{v}\partial p/\partial x$ may be ignored in equations (19.1) and (19.5). Under what circumstances would this simplification be unacceptable.

Solution

Small pressure waves may be imagined to propagate from the closing valve throughout its closing motion. Thus, pressure waves leave the valve for a time of 20 ms and so the length of the wavefront will be given by

$$c\,\Delta t = 1250 \times 0\cdot02 = 25\text{ m}$$

If the change in flow velocity is $0\cdot5$ m s^{-1}, then the associated pressure rise is given by equation (19.18):

$$\Delta p = \rho c(\bar{v}_0 - \bar{v}) = 1000 \times 1250 \times \Delta\bar{v},$$

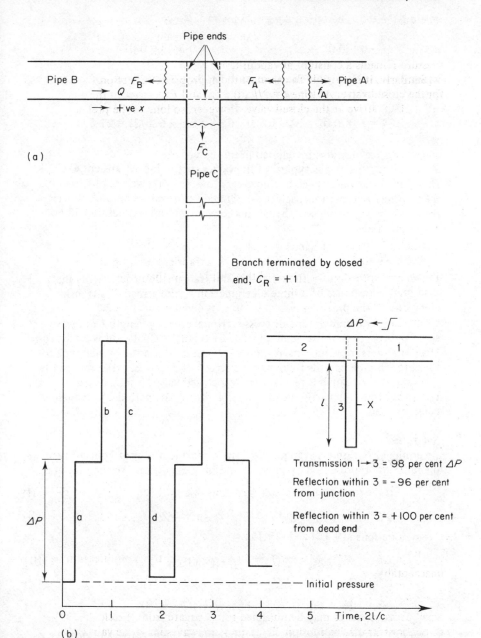

Figure 19.8   (a) Pipe layout. (b) Pressure variation at mid-length point X of closed-ended branch pipe.
In (b), ΔP in pipe 1 reaches junction at $t = 0$; a = partial transmission of ΔP into branch; b = total positive reflection of pipe end of pipe 1 at dead end; c = partial negative reflection of pipe end of pipe 2 at junction; d = total positive reflection of pipe end of pipe 3 at dead end

where $\Delta \bar{v} = 0.5$. Therefore, $\Delta p = 625$ kN m^{-2}. Now,

$$\frac{\partial \bar{v}}{\partial t} = 0.5/0.02 = 25, \qquad \bar{v}\frac{\partial \bar{v}}{\partial x} = 0.5 \times 0.5/25 = 0.01$$

where ∂x is the length of the wavefront, i.e. $\partial x = 25$ m. Thus, $\bar{v}\partial\bar{v}/\partial x$ may be ignored with respect to $\partial\bar{v}/\partial t$. Similarly,

$$\frac{\partial p}{\partial t} = 625/0.02 = 31\ 250, \qquad \bar{v}\frac{\partial p}{\partial x} = 0.5 \times 625/25 = 12.5$$

and so $\bar{v}\partial p/\partial x$ may also be ignored in this case.

As stated, this case is typical of the values met in most transient examples. However, if the wave speed becomes very low, then the values obtained for the convective terms approach the $\partial\bar{v}/\partial t$ and $\partial p/\partial t$ values because of the reduction in wavefront length and this convenient approximation will no longer be valid.

Example 19.2

Derive an expression for the reflection and transmission coefficients that describe the response of a three pipe junction to the arrival of a transient along one of the pipelines.

A small bore branch pipe of cross-sectional area A_2, length l and wave speed c, is connected to a main pipeline of area $A_1 = 25A_2$ and wave speed c. Draw the pressure variations at the mid-length point on the branch pipe for four branch pipe periods following the arrival of a step ΔP, transient at the junction, if the branch is terminated by a closed valve. It may be assumed that no further transients arrive at the junction during the period of time under consideration.

Solution

Assuming a frictionless system, then, from considerations of continuity at the junction, represented by pressure and flow conditions on the pipe ends,

$$p_A = p_B = p_C \quad \text{and} \quad \Delta p_A = \Delta p_B = \Delta p_C \tag{I}$$

$$Q_A = Q_B + Q_C \quad \text{and} \quad \Delta Q_A = \Delta Q_B + \Delta Q_C \tag{II}$$

From equations (19.16) and (19.17),

$$\Delta p_A = F_A + f_A, \quad \Delta Q_A = -(A_A/\rho c_A)(F_A - f_A) \tag{III}$$

$$\Delta p_B = F_B + f_B, \quad \Delta Q_B = -(A_B/\rho c_B)(F_B - f_B) \tag{IV}$$

$$\Delta p_C = F_C + f_C, \quad \Delta Q_C = -(A_C/\rho c_C)(F_C - f_C) \tag{V}$$

where F, f refer to the pressure waves moving in the negative and positive x directions, respectively.

If the incident wave moves in the negative x direction in pipe A, then this will generate waves in pipes B and C moving in the negative x direction, i.e. F-type transmission waves, and it will also produce a reflection of itself in pipe A moving in the positive x direction, i.e. an f type wave. Thus, the f_B, f_C pressure waves included in the above equations are zero as there are no waves present in pipes B and C of this type.

Substituting into (I) and (II)

$$f_A + F_A = F_B + F_C \tag{VI}$$

$$F_A - f_A = (c_A/A_A)\{(A_B/c_B)F_B + (A_C/c_C)F_C\}. \tag{VII}$$

Let the junction reflection coefficient be defined as $C_R = f_A/F_A$ and the junction transmission coefficient by $C_T = F_B/F_A = F_C/F_A$, then, from (VI) and (VII),

$$C_R = (A_A/c_A - A_B/c_B - A_C/c_C)/(A_A/c_A + A_B/c_B + A_C/c_C),$$

$$C_T = (2A_A/c_A)/(A_A/c_A + A_B/c_B + A_C/c_C).$$

In order to draw the pressure variations at the mid-length point of the branch, it is necessary to calculate the reflection and transmission coefficients for the junction for the arrival of a transient along either the main pipe or along the branch, as the latter will occur when reflections return to the junction from the end of the branch.

For a transient arriving along pipe A, $A_1 = A_A = A_B = 25A_C = 25A_2$, $c_1 = c_2 = c_A = c_B = c_C$. Hence, from the two expressions above,

$$C_R = (-A_2/c_2)/(51A_2/c_2) = -1.96 \text{ per cent,}$$

$$C_T = (50A_2/c_2)/(51A_2/c_2) = 98.04 \text{ per cent.}$$

For a transient arriving along pipe C, it is necessary for the suffix A to refer to the pipe bearing the transient. Hence $A_2 = A_B = A_C = 25A_A = 25A_1$ and $c_1 = c_2 = c_A = c_B = c_C$, with the result that the values for the reflection and transmission coefficients are

$$C_R = -49/51 = -96.08 \text{ per cent,}$$

$$C_T = 2/51 = 3.92 \text{ per cent.}$$

The pressure variations over four branch pipe periods may now be drawn by considering the individual waves travelling within the branch.

As the branch is terminated by a closed valve, the reflection coefficient at the end of the branch is +1. This means that any wave arriving at the closed end is reflected as an equal magnitude wave with no change of sign.

19.5. The Schnyder–Bergeron graphical method

This method is based on the D'Alembert solution of the transient equations.

$$p_{(x,t)} - p_0 = F(t + x/c) + f(t - x/c),$$

$$\bar{v}_{(x,t)} - \bar{v}_0 = -(1/\rho c)\{F(t + x/c) - f(t - x/c)\},$$

where $\bar{v}$ and p are the flow velocity and pressure at a section x at a time t, $F(\ ), f(\ )$ are pressure waves moving in the negative and positive x directions, respectively, at the wave speed c. Eliminating $f(\ )$ from the equations above yields

$$p_{(x,t)} - p_0 = \rho c(\bar{v}_{(x,t)} - \bar{v}_0) + 2F(t + x/c). \tag{19.19}$$

Referring to Fig. 19.9, if particular values X, T for the variables x, t are

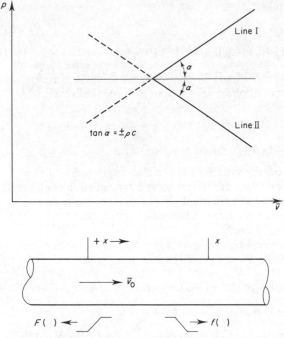

Figure 19.9   Characteristic transient lines in the p–v plane used in the Schnyder–Bergeron graphical method

introduced then (19.19) becomes

$$p_{(X,T)} - p_0 = \rho c (\bar{v}_{(X,T)} - \bar{v}_0) + 2F(T + X/c) \qquad (19.20)$$

If an imaginary observer is assumed to travel in the negative x direction with velocity c, then the $F(\ )$ function will appear constant as he travels with the wave; thus,

$$F(T + X/c) = F(t + x/c)$$

and the $F(\ )$ function may be eliminated from (19.19) and (19.20):

$$p_{(X,T)} - p_{(x,t)} = +\rho c (\bar{v}_{(X,T)} - \bar{v}_{(x,t)}). \qquad (19.21)$$

Referring to Fig. 19.9, this is the equation defining line I.

By an identical process, the second transient line on Fig. 19.9, line II, can be shown to represent an observer travelling in the positive x direction at velocity c, i.e. an observer travelling with the $f(\ )$ wave. Line II is represented by the equation

$$p_{(X,T)} - p_{(x,t)} = -\rho c (\bar{v}_{(X,T)} - \bar{v}_{(x,t)}). \qquad (19.22)$$

In order to calculate the $p, \bar{v}$ unknowns, it is necessary to solve a pair of equations. Two possibilities exist:

(i) Either equation (19.19) or (19.20) may be solved graphically with a boundary condition known in terms of a $p - \bar{v}$ relation. On Fig. 19.1, for example, this solution corresponds to the arrival of a wave at the reservoir terminating the pipe. Common boundary conditions that may be solved in this way include valve discharge relationships, closed-ended pipes where $\bar{v} = 0$, reservoirs where $p = $ a constant and pumps where the p–$\bar{v}$ relation is supplied by the pump characteristic. As previously mentioned, column separation can be dealt with by the graphical method; the appropriate p–$\bar{v}$ relation during the existence of the vapour cavity being $p = $ vapour pressure. However, the duration of the cavity has to be calculated by a summation of the flow across the cavity-fluid interface and errors in this summation resulting from the use of too coarse a time step can result in the prediction of transient pressures considerably different from those experienced. Because of this restriction, column separation is more accurately dealt with by the computer-based methods. The appropriate p–$\bar{v}$ relation in each case may be drawn onto the p–$\bar{v}$ diagram and the required solution is indicated by the point of intersection of this line and the appropriate $F(\ )$ or $f(\ )$ line.

(ii) Internal points along the pipeline may be solved by the intersection of lines I and II on Fig. 19.9. As described, the solution is restricted to frictionless, horizontal pipes. It is possible to incorporate friction losses by the introduction of a number of friction joints along the pipe. However, the number of such 'joints', each representing a discrete friction loss, has to be kept to a minimum to reduce the complexity of the final graphical construction.

The graphical method is best described by a number of worked examples.

Example 19.3

A simple pipe 500 m long is used to pass oil of relative density 0·8 from a storage reservoir to a loading jetty. The pipe is terminated by a valve which is closed linearly in 4 pipe periods. If the reservoir pressure is maintained constant at 250 kN m^{-2} during pipe discharge at a steady rate corresponding to a flow velocity of 1·0 m s^{-1} and if the wave speed in the pipe-fluid

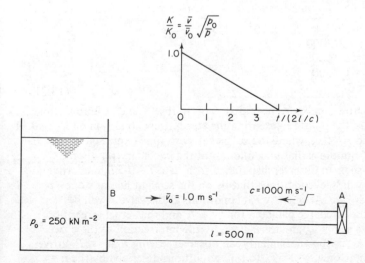

Figure 19.10   Layout of a simple pipeline

combination is 1000 m s^{-1}, calculate the pressure rise at the valve on closure and the difference between this rise and that suffered if the valve was inadvertently closed in one pipe period. The pipe may be assumed frictionless and horizontal; the valve loss characteristic may be assumed to have a form $v = K\sqrt{p_0}$ at any degree of opening, K/K_0 varying linearly from 1 to 0 with valve closure position.

Solution

$$\text{Pipe period} = 2l/c = 2 \times 500/1000 = 1 \text{ s},$$

$$\text{Pressure rise following closure in one pipe period} = \rho c v_0$$

$$= 800 \times 1000 \times 1 \cdot 0$$

$$= 800 \text{ kN m}^{-2}.$$

Boundary conditions for the slow closure may be extrapolated as follows.

At B, the pressure remains constant at 250 kN m^{-2} at all times. At the valve, the pressure has a value equal to the reservoir pressure up to the start of valve closure at time $t = 0$. The valve discharges to atmosphere.

At the valve A the velocity $\bar{v} = \bar{v}_0$ up to time $t = 0$ and $\bar{v} = 0$ at all times later than $t = 4T_p$, where $T_p = 2l/c$, the pipe period. During valve closure, conditions at the valve will be governed by the valve discharge characteristic $\bar{v} = K\sqrt{p}$. It is accepted practice to assume that the steady-state values of the valve discharge loss coefficient K apply during transient propagation. Therefore, parabolae may be drawn in the p–$\bar{v}$ plane to represent the valve discharge characteristic at any time during closure and the intersections of the appropriate transient line with these parabolae yield the p–$\bar{v}$ conditions at the valve at that instant in time. Initially, only the parabolae for $t = 0$, T_p, $2T_p$, $3T_p$ need be drawn, as the condition for $t = 4T_p$ and all later times is supplied by the $\bar{v} = 0$ axis.

Values of K for the four parabolae are as follows:

t	K
0	$K_0 = \bar{v}/\sqrt{p} = 1/\sqrt{(25 \times 10^4)} = 1/500$
$2l/c$	$3K_0/4 = 3/2000$
$4l/c$	$K_0/2 = 1/1000$
$6l/c$	$K_0/4 = 1/2000$
$8l/c$	$0\quad = 0$

The steady flow conditions prevail at A up to time 0 and at B up to time $t = 0 \cdot 5 T_p$. Thus, the point representing the steady flow condition on Fig. 19.11 may be labelled A_0, $B_{0.5}$, where the suffix refers to time expressed as t/T_p.

It follows from the earlier introduction of the imaginary waveborne observer, that such an observer dispatched from B at $t = 0 \cdot 5 T_p$ and arriving at A at $t = T_p$ will 'see' conditions that lie on the straight line of slope $-\rho c$ passing through $B_{0.5}$. At time $t = T_p$, conditions at the valve A will be described by the discharge parabola for $t = 2l/c$ and the actual values of p–$\bar{v}$ at the valve at $t = T_p$ are, thus, given by the intersection of these two lines at point A_1, as shown. Similarly, an observer leaving A at $t = T_p$ arrives at the reservoir at time $t = 1 \cdot 5 T_p$ to find conditions described by the $p =$ constant line, the resulting intersection of this boundary with the transient

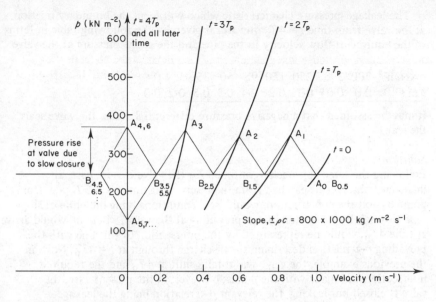

Figure 19.11 Schnyder–Bergeron construction for slow valve closure

line of slope $+\rho c$ representing this observer's travels yields point $B_{1.5}$ on the diagram.

By following the same procedure for successive observers, it is possible to establish the points of intersection A_{2-4} and $B_{2.5-4.5}$, as shown. A final observer dispatched from B at $t = 4·5\,T_p$ arrives at the valve at time $t = 5\,T_p$ to find the valve shut and the conditions there governed by the zero flow line. Thus, point A_5 may be determined.

The above analysis yields pressure values at the valve at 1 pipe period intervals. However, more frequent values may be calculated by plotting more parabolae to represent the valve discharge at more frequent intervals.

Similarly, pressure variations at points along the pipe may be calculated by the dispatch of observers in opposite directions from known points, so that they meet at the required intermediate point. This is the equivalent, in graphical terms, of solving the two equations represented by the transient lines.

From the diagram, the pressure rise recorded at the valve following a 4 pipe period closure was 120 kN m^{-2}, and the decrease in the transient due to slowing the valve motion was 680 kN m^{-2}.

Example 19.4

For the simple pipe employed in Example 19.3, assume that, instead of the valve becoming fully shut at a time of four pipe periods following initiation of closure, a leak develops at the valve. Due to the design of the valve seat, this leakage decreases with increasing pressure, above a certain pressure level, due to distortion of the seals. Starting from the point $B_{3.5}$ on the Schnyder–Bergeron construction presented with Example 19.3, derive the pressure-time curve at the valve following its supposed closure.

The leakage-pressure characteristic which will form the boundary relation at the valve from time $t = 4T_p$ onwards is given by the following table in terms of the equivalent flow velocity in the pipe and the gauge pressure at the valve.

p kN m^{-2}	700	375	250	150	95	50	35	5	0
$\bar{v}$ m s^{-1}	0·0	0·05	0·1	0·2	0·4	0·5	0·5	0·2	0·0

It may be assumed that a negative pressure differential across the valve seals the leak.

Solution
Following the same approach as outlined for the valve closure case, an imaginary observer dispatched from the reservoir at time $t = 3·5T_p$, i.e. from point $B_{3.5}$ on the construction, would 'see' conditions lying on a line of slope $-\rho c$ through $B_{3.5}$. The conditions prevailing at the valve, where he would arrive at time $4T_p$, would be represented by the intersection of this line with the prevailing p–v relation describing the discharge through at $t = 4T_p$. Now, in the previous example, the valve was totally shut at $4T_p$ and the required intersection was made with the p-axis; however, in the present case, the valve is closed but leaking, the relevant p–v relation being the leakage-pressure line shown on the construction. Thus, the conditions at the valve at $t = 4T_p$ are given by point A_4, the intersection of the $-\rho c$ line and the

Similarly, an observer leaving the valve at $t = 4T_p$ 'sees' conditions along the $+\rho c$ line A_4–$B_{4.5}$, thus yielding the point $B_{4.5}$ on the diagram by intersection with the reservoir pressure line. By following the same approach, points A_5, $B_{5.5}$, A_6 and $B_{6.5}$ may be determined.

Finally, consider the observer dispatched from the reservoir at $t = 6·5T_p$, point $B_{6.5}$ on Fig. 19.12. This observer will experience conditions along a line of slope $-\rho c$ as he or she travels towards the valve. It is clear from Fig. 19.12 that the point of intersection will lie below the zero gauge pressure line and so, at first, it appears that the correct leakage point would be A_7, because the leakage is cured by a negative pressure differential across the valve. However, at A_7, the $-\rho c$ line crosses the $v = 0$ axis and so A'_7 cannot be the correct leakage point because it would imply a pressure at the valve face below the fluid vapour pressure. In this case, column separation would occur at the valve and a vapour cavity would form. Throughout the duration of this cavity, the boundary condition at the valve would be the p = vapour pressure line. Subsequent collapse of the cavity would result in appreciable pressure rise at the valve and pressure variations would continue until damped out by the action of friction – which has been neglected in the constructions.

19.6. The method of characteristics

The graphical method of solution of the transient equations has two major limitations. First, the inability to represent friction losses accurately without greatly increasing the complexity of the construction; second, the difficulty in representing graphically transients in any complex pipe system. The advent of the digital computer encouraged the use of numerical methods of solution and, at present, there are two established methods available: a computerized version of the Schnyder–Bergeron method and a numerical method based on

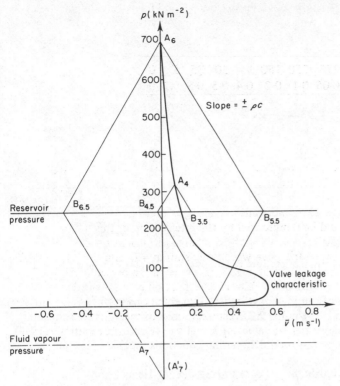

Figure 19.12   Schnyder–Bergeron construction following valve closure and subsequent leakage from the valve

the method of characteristics. Both methods represent frictional losses by a number of friction joints along the pipeline; however, the number of such discrete pressure drops can be very large so that frictional losses can be accurately represented. In practice, the number of 'joints' depends only on computer run time costs and not on construction complexity, as was the case for the graphical method.

The choice of numerical method is one of personal preference, there being little difference in the cost for the analysis of any particular system. The method of characteristics solution will be introduced as there is probably more published work employing this method to aid the reader who wishes to take the subject further.

The starting points for the solution are the equations of motion and continuity, (19.1) and (19.2). These equations may be combined linearly by writing

$$\psi = \psi_1 + \lambda\psi_2 = 0,$$

where $\psi_1 =$ equation (19.1) and $\psi_2 =$ equation (19.2):

$$\rho\left\{(\bar{v} + \lambda c^2)\frac{\partial \bar{v}}{\partial x} + \frac{\partial \bar{v}}{\partial t}\right\} + \lambda\left\{\left(\frac{1}{\lambda} + \bar{v}\right)\frac{\partial p}{\partial x} + \frac{\partial p}{\partial t}\right\} + \rho g \sin\alpha + 2\rho f \bar{v}|\bar{v}|/d = 0.$$

$$(19.23)$$

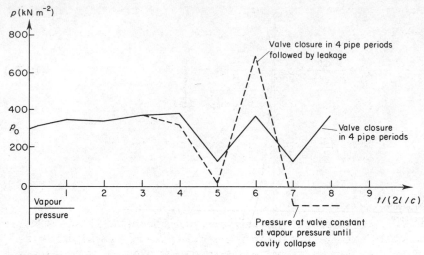

Figure 19.13   Comparison of pressure variations at the valve for Examples 19.3 and 19.4

Any distinct pair of values of λ will yield two equations in p, $\bar{v}$ equivalent in all respects to (19.1) and (19.2). The method of characteristics solution consists of choosing a pair of values of λ that will transform equations (19.1) and (19.2) into a pair of total differential equations capable of numerical solution.

As stated previously, $p = p(x, t)$ and $\bar{v} = \bar{v}(x, t)$. Hence,

$$\frac{\mathrm{d}p}{\mathrm{d}t} = \frac{\partial p}{\partial t} + \frac{\partial p}{\partial x} \cdot \frac{\mathrm{d}x}{\mathrm{d}t}, \tag{19.24}$$

$$\frac{\mathrm{d}\bar{v}}{\mathrm{d}t} = \frac{\partial \bar{v}}{\partial t} + \frac{\partial \bar{v}}{\partial x} \cdot \frac{\mathrm{d}x}{\mathrm{d}t}. \tag{19.25}$$

By comparing these relations with equation (19.23) it may be seen that writing

$$\frac{\mathrm{d}x}{\mathrm{d}t} = \bar{v} + \lambda c^2 = \bar{v} + 1/\lambda$$

and substituting into (19.23) yields

$$\frac{\mathrm{d}p}{\mathrm{d}t} = \left\{ \frac{1}{\lambda} + \bar{v} \right\} \frac{\partial \rho}{\partial x} + \frac{\partial p}{\partial t} \tag{19.26}$$

$$\frac{\mathrm{d}\bar{v}}{\mathrm{d}t} = \left\{ \bar{v} + \lambda c^2 \right\} \frac{\partial \bar{v}}{\partial x} + \frac{\partial \bar{v}}{\partial t}, \tag{19.27}$$

which implies that equation (19.23) may be replaced by the ordinary differential equation

$$\frac{\mathrm{d}\bar{v}}{\mathrm{d}t} + \frac{\lambda}{\rho} \frac{\mathrm{d}p}{\mathrm{d}t} + g \sin \alpha + 2f \frac{\bar{v}}{d} |\bar{v}| = 0, \tag{19.28}$$

where $\lambda = \pm 1/c$ \hfill (19.29)

and $dx/dt = \bar{v} \pm c.$ (19.30)

Substituting for λ in equation (19.28) yields the following pair of equations:

$$\left.\begin{array}{l} \dfrac{d\bar{v}}{dt} + \dfrac{1}{\rho c}\dfrac{dp}{dt} + g\sin\alpha + 2f\dfrac{\bar{v}}{d}|\bar{v}| = 0 \\[4mm] \dfrac{dx}{dt} = \bar{v} + c \end{array}\right\}$$

(19.31)

(19.32)

$$\left.\begin{array}{l} \dfrac{d\bar{v}}{dt} - \dfrac{1}{\rho c}\dfrac{dp}{dt} + g\sin\alpha + 2f\dfrac{\bar{v}}{d}|\bar{v}| = 0 \\[4mm] \dfrac{dx}{dt} = \bar{v} - c \end{array}\right\}$$

(19.33)

(19.34)

Equations (19.31) and (19.33) only apply if equations (19.32) and (19.34) are satisfied; this implies that they only apply along certain lines drawn in an $x\text{-}t$ plane. These lines are referred to as *characteristic curves* and, in the majority of cases, they have a slope of $\pm c$. This is because typical flow velocity values, $\bar{v}$, are small compared to the wave speeds in a particular pipe. Figure 19.14 illustrates these characteristic lines usually referred to as C^+ and C^- lines. It will be seen that, if conditions at points R, S at time t are known, then conditions at point P, lying at the intersection of the C^+, C^- lines, may be calculated by solving equations (19.31) to (19.33).

Application of a first-order finite difference approximation to equations (19.31) to (19.33) of the form

$$\int_{x_0}^{x_1} f(x)\,dx \triangleq f(x_0)(x_1 - x_0)$$

yields the following relations, where it has been assumed that $c \gg v$:

$$\left.\begin{array}{l} \bar{v}_P - \bar{v}_R + \dfrac{1}{\rho c}(p_P - p_R) + \left(g\sin\alpha_R + \dfrac{2}{d}f_R\bar{v}_R|\bar{v}_R|\right)(t_P - t_R) = 0, \\[4mm] x_P - x_R = c_R(t_P - t_R), \end{array}\right\}$$

(19.35)

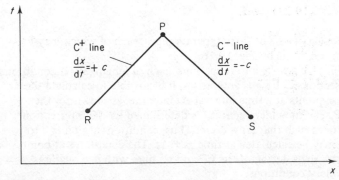

Figure 19.14   Characteristic lines in the $x\text{-}t$ space

$$\bar{v}_P - \bar{v}_S - \frac{1}{\rho c}(p_P - p_S) + \left(g \sin \alpha_S + \frac{2}{d} f_S \bar{v}_S |\bar{v}_S|\right)(t_P - t_S) = 0, \Big\}$$

$$x_P - x_S = c_S(t_P - t_S).$$

$$(19.36)$$

where suffixes P, R, S refer to points in the x, t plane (Fig. 19.14).
The equations derived above indicate that any transient condition in a pipe
may be solved by dividing the pipe into a number of sections and solving
equations like (19.35) and (19.36). The boundary conditions are dealt with
in the same way as outlined for the graphical method, i.e. the applicable C^+
or C^- equation is solved with a p–$\bar{v}$ relation defining the boundary. The best
way of describing the method is by means of a worked example that will
take the solution to the point where a computer program could be written.

Example 19.5

A simple pipe system consists of a pump, fitted with a non-return valve,
which supplies a chemical to two tanks as shown in Fig. 19.15. Derive the
necessary equations to allow a program to be written to predict pressure
variations in the system following the closure of either of the tank inlet

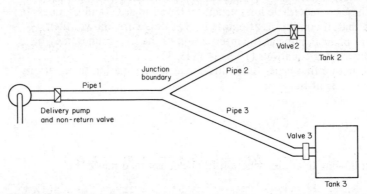

Figure 19.15 Layout of the system. The pump draws directly from a supply tank

valves. It may be assumed that the pump continues to run at constant speed
throughout.

Solution

The system as shown consists of three separate pipes, each of which may be
represented in the x–t plane by Fig. 19.16.

In general, each pipe is split into N sections, each of length Δx. If conditions
at each of these sections at time t_0 are known, it is possible to calculate the
conditions at these points at a time interval Δt later, i.e. as shown for the
points R, S and P. The time increment Δt is determined by the intersection
of the C^+ and C^- lines such that $\Delta t = \Delta x/c$. Thus, conditions at points 2 to
N along the pipe may be calculated at time $t_0 + \Delta t$. The conditions at points
1 and $N + 1$ require the solution of the C^- and C^+ lines with p–$\bar{v}$ relations
defining the boundary conditions.

The system in question consists of three pipes which may have different

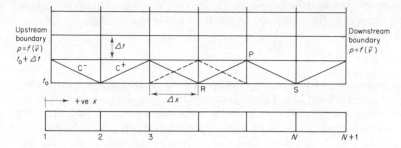

Figure 19.16   Characteristic solution in the $x-t$ plane for a general pipe

properties. However, for an orderly solution, it is necessary for the time step employed to be the same for all the pipes, i.e.

$$\Delta t = \Delta x_1/c_1 = \Delta x_2/c_2 = \Delta x_3/c_3 \text{ etc.}$$

This may be achieved in any system by correct choice of N for each pipe.

Assuming that the pipes have been split up into the correct number of sections, it is now necessary to review the equations to be solved at each time step.

The p-$\bar{v}$ values at each internal pipe section, i.e. section numbers 2 to N in each pipe, may be calculated at time $(t + \Delta t)$ by the solution of equations (19.35) and (19.36), representing the C^+, C^- lines intersecting at that grid point. Thus, for a horizontal pipe system, the general equations to be solved for each point in the 2 to N range are:

$$[\bar{v}_{j,i}] = \bar{v}_{j,i-1}\left(1 - \frac{2}{d_j}\, f_{j,i-1}|\bar{v}_{j,i-1}|\Delta t\right) - \frac{1}{\rho c_j}\left([p_{j,i}] - p_{j,i-1}\right),$$

$$[\bar{v}_{j,i}] = \bar{v}_{j,i+1}\left(1 - \frac{2}{d_j}\, f_{j,i+1}|\bar{v}_{j,i+1}|\Delta t\right) + \frac{1}{\rho c_j}\left([p_{j,i}] - p_{j,i+1}\right).$$

(Note that the p-$\bar{v}$ terms in [ ] refer to conditions at time $t + \Delta t$ and that it has been assumed that steady-state friction factor relations apply with sufficient accuracy during transient propagation.) Suffix j refers to the pipe and suffix i refers to the pipe section number. It will be seen that the solution requires the knowledge of p-$\bar{v}$ conditions at all points along the pipe one time step earlier; hence, in order for the solution to advance further, conditions at points 1 and $N + 1$ must be calculated.

The relevant p-$\bar{v}$ relation applying at the boundary between the pump and the non-return valve at the inlet to pipe 1 is the pump discharge characteristic normally presented as a polynomial in terms of pump discharge Q and the pressure rise across the pump:

$$\Delta p = C_1 + C_2 Q + C_3 Q^2.$$

This expression may be solved with the C^- equation between conditions at point $(1, 2)$ at time t and point $(1, 1)$ at time $t + \Delta t$:

$$[\bar{v}_{1,1}] = \bar{v}_{1,2}\left(1 - \frac{2}{d_1}\, f_{1,2}|\bar{v}_{1,2}|\Delta t\right) + \frac{1}{\rho c_1}\left([p_{1,1}] - p_{1,2}\right).$$

If, during the propagation of transients, the flow reverses in the pipe adjacent to the pump, then the pump characteristic is replaced by the non-return valve condition

$$[\bar{v}_{1,1}] = 0 \cdot 0$$

until such time as the flow again establishes itself in the positive x direction. The choice of solution in a real program would simply depend on the result of monitoring the value of flow velocity at the pump outlet at each time step.

The valve boundary condition terminating pipes 2 and 3 must be presented in two parts, either for a closing valve or for a valve that is fully shut. In the first case, the boundary equation is supplied by the valve steady-state discharge characteristic, expressed as

$$[\bar{v}_{j,N+1}] = \tau\bar{v}_0([p_{j,N+1}]/p_0)^{1/2},$$

where the suffix 0 refers to the flow velocity and associated pressure drop, p_0 through the valve when fully open and τ is the valve discharge coefficient whose value varies from unity to zero during valve closure. If the valve discharges to a pressurized tank then $[p_{j,N+1}]$ is replaced by $([p_{j,N+1}] - p_{\text{tank}})$. In practice, τ for a valve is treated as a function of the valve open area or open angle. During valve closure the valve position–time curve may be monitored or estimated by reference to other similar valve performance data and this allows τ to be calculated as a function of time. Thus, the valve equation above may be solved at any time step during closure with the appropriate C^+ equation. In this case, the appropriate C^+ expression is that between point j, N at time t and the valve coordinate $j, N + 1$ at time $t + \Delta t$:

$$[\bar{v}_{j,N+1}] = \bar{v}_{j,N}\left(1 - \frac{2}{d_j}\, f_{j,N}\, |\bar{v}_{j,N}|\Delta t\right) - \frac{1}{\rho c_j}([p_{j,N+1}] - p_{j,N}).$$

Following valve closure, the boundary equation is supplied by

$$\bar{v}_{j,N+1} = 0 \cdot 0.$$

Values representing the curves of τ vs. valve position and valve position vs. time would be included in the input data for the program and so any combination of valve actions could be studied, or the effect of changing valve design could be predicted illustrating the advantage of the computer-based solution as a design tool.

Let us now consider the boundary condition at the junction, which is common to all three pipes. At the junction there are, effectively, six unknowns to be calculated: $[p, \bar{v}_{1,N+1}]$, $[p, \bar{v}_{2,1}]$ and $[p, \bar{v}_{3,1}]$. The available equations are

(i) continuity of flow at the junction,

$$Q_1 = Q_2 + Q_3$$

or, in the [ ] notation for conditions at $t + \Delta t$,

$$A_1 [\bar{v}_{1,N+1}] = A_2 [\bar{v}_{2,1}] + A_3 [\bar{v}_{3,1}];$$

(ii) assuming that the junction displays zero separation loss,

$$[p_{1,N+1}] = [p_{2,1}] = [p_{3,1}];$$

(iii) the characteristic equations linking conditions at

point $1, N$ at time t to conditions at point $1, N + 1$
  at time $t + \Delta t(C^+)$

point $2, 2$ at time t to conditions at point $2, 1$
  at time $t + \Delta t(C^-)$

point $3, 2$ at time t to conditions at point $3, 1$
  at time $t + \Delta t(C^-)$.

These six equations may be solved easily and the algebraic solution included in the program, providing values of pressure and velocity at each time step at the three-pipe junction. Thus, the general solution is allowed to progress.

The solution requires knowledge of the initial conditions along the system and this may be input as data or a section may be included in the program to calculate steady-state flow distributions by (see Chapter 12). Values of the friction factor at each pipe section may be calculated at each time step from the usual Reynolds number relations.

19.7. Column separation

Referring back to Fig. 19.1 it will be seen that, following valve closure, the negative reflections from the upstream tank could reduce the pressure at the valve face to the fluid vapour pressure. If this occurs, and a simple check on calculated pressure values at the $(N + 1)$th section will confirm this, then the fluid column breaks away from the valve and moves towards the reservoir, continuing to travel in this direction until brought to rest by the adverse pressure gradient between the tank pressure p_{air_1} and the vapour pressure at the valve face. This same pressure gradient accelerates the flow back towards the closed valve, closing the cavity and generating a large pressure rise when the column is brought to rest by the closed valve.

The boundary condition $[\bar{v}_{N+1}] = 0$ applied at the valve following closure obviously breaks down during the presence of the cavity and is replaced by the pressure boundary equation

$$[p_{N+1}] = \text{Vapour pressure} = p_{\text{vap}}. \tag{19.37}$$

Solving this with the C^+ characteristic yields

$$[\bar{v}_{N+1}] = K_3 - K_2 p_{\text{vap}}. \tag{19.38}$$

from equation (19.35). In order to determine the duration of application of this pressure boundary condition, it is necessary to calculate and monitor the volume of the cavity V. This can be done on a time step basis as

$$V_{t+\Delta t} = V_t - (A/2)\,\Delta t([\bar{v}_{N+1}] + \bar{v}_{N+1}), \tag{19.39}$$

where the term in square brackets represents velocity at time $t + \Delta t$, so that the mean velocity over the time step is used in cavity volume summation, and A is pipe cross-sectional area. If the value of V is monitored, the collapse of the cavity is indicated by $V = 0$. The negative sign in the above equation is included so that the cavity volume increases in a positive manner for

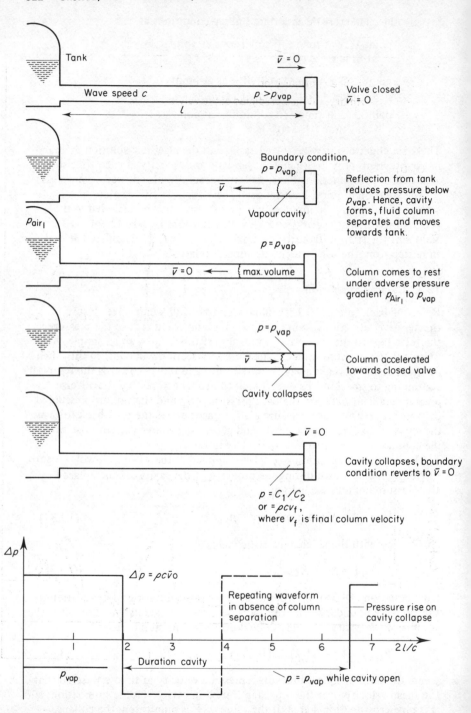

Figure 19.17 Δp vs time at a closed valve

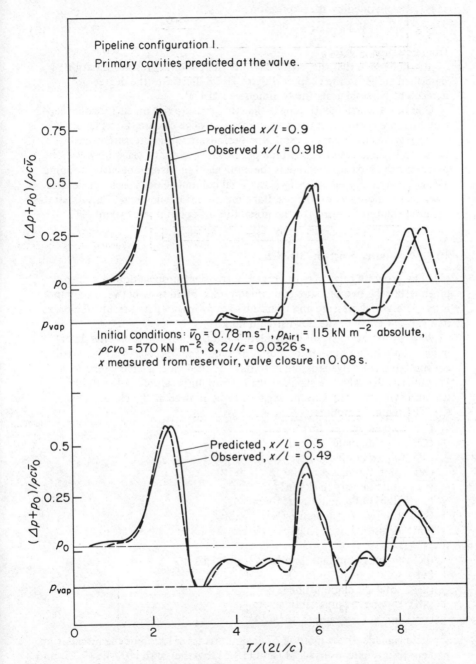

Figure 19.18 Predicted and observed pressure variations at two points upstream of the valve following closure

reverse flow in the pipe. The sequence of events is illustrated by Fig. 19.17. The boundary at the upstream tank, where the tank pressure is constant at p_{air_1} is provided by the expression:

$$[V_{1,1}] = K_1 + K_2\, p_{\text{air}_1} \tag{19.40}$$

from equation (19.36).

It will be seen that the cavity disrupts the periodic pressure variation described earlier in the chapter. Figure 19.18 illustrates the degree of agreement possible using the technique described.

One point worth mentioning is that the pressure rise on cavity collapse may be more severe than the pressure rise on valve closure. Consider a slow valve closure, taking several pipe periods, then the pressure rise is considerably less than the one pipe period closure rise of $\rho c \bar{v}_0$; however, the pressure rise on cavity collapse is the equivalent of instantaneously stopping a flow at a velocity $\bar{v}_f$, where $\bar{v}_f$ is the final column velocity as it accelerates towards the closed valve, the resultant pressure rise being $\rho c \bar{v}_f$. This illustrates the importance of testing for the possibility of column separation.

19.8. Computer program 'SURGE'

(1) The program calculates the pressure transient propagation in a simple pipeline linking two pressure controlled tanks. Fluid flow between the tanks is by differential pressure and transients are propagated as a result of closure of a valve at entry to the downstream tank.

Wave propagation speed is calculated and the pressure transients determined at a number of sections along the pipe. Tank pressures are assumed constant and the valve is assumed to have a linear closing characteristic. Pressures at the valve are checked against vapour pressure for the fluid and vapour cavity growth; 'column separation' is allowed at the closed valve.

(2) The required input is:
    (i)   pipe length, L (m);
    (ii)   pipe diameter, D (m);
    (iii) pipe wall thickness, w (m);
    (iv) pipe Young's modulus, E (N m^{-2});
    (v)   fluid density, ρ (kg m^{-3});
    (vi) fluid bulk modulus, K (N m^{-2});
    (vii) fluid vapour pressure (N m^{-2} abs.);
    (viii) pressure loss at fully open valve, Δ_p (N m^{-2}) for a steady flow ($\text{m}^3\,\text{s}^{-1}$);
    (ix) downstream tank pressure (N m^{-2} abs.);
    (x)   valve closure time (s);
    (xi) total simulation time (s);
    (xii) number of pipe sections, n;
    (xiii) initial flow rate, Q ($\text{m}^3\,\text{s}^{-1}$).

(3) Use is made of the method of characteristics solution of the momentum and continuity equations (19.1) and (19.5) together with (19.4), (19.35) and (19.36).

(4) Input example:
$L = 100$ m; $D = 1\!\cdot\!0$ m; $w = 0\!\cdot\!1$ m; $E = 1000$ MN m^{-2}; $f = 0\!\cdot\!01$;

$\rho = 1000$ kg m^{-3}; $K = 2000$ MN m^{-2}; Tank $p = 2000$ N m^{-2} abs.;
Valve $\Delta p = 1234$ N m^{-2} at $0 \cdot 2$ m^3 s^{-1}; t (closure) = $0 \cdot 2$s;
t (total) = 3s; $n = 4$ sections; $Q = 12$ m^3 s^{-1}.

(5) Output:

```
PIPE DATA.....

PIPE LENGTH =     100.00 M. PIPE DIAMETER =          1.00 M. WALL THICKNESS =
0.10 M.

YOUNGS MODULUS = 1000.00 MN/M^2.
PIPE FRICTION FACTOR =    0.01

FLUID DATA....

DENSITY =      1000.00 KG/M^3. BULK MODULUS = 2000.00 MN/M^2.
VAPOUR PRESSURE =     100.00 N/M^2 ABS.

VALVE AND TANK DATA...

DOWNSTREAM TANK PRESSURE =     2000.00 N/M^2 ABS.
VALVE PRESSURE LOSS =   1234.00 N/M^2 AT THROUGHFLOW =    0.20CU.M/S.
VALVE CLOSURE TIME =    0.20 SEC.

CALCULATION PARAMETERS...

WAVE SPEED =     308.61 M/SEC. PIPE PERIOD =     0.65 SEC
VALVE CLOSURE TIME =    0.31 PIPE PERIODS.
JOUKOWSKY MAX PRESSURE SURGE =    4715.16 KN/M^2.

PRESS P TO CONTINUE?P
MAXIMUM PRESSURE AT VALVE DURING CLOSURE =    9466.10 KN/M^2.
AT TIME =    0.65 SEC.
```

List:

```
     10 REM SURGE
     20 MODE 0
     30 @%=&20208
     40 CLS:PRINT"PROGRAM SURGE DESIGNED TO CALCULATE PRESSURE TRANSIENTS UPSTREAM
OF A CLOSING VALVE."
     50 PRINT:PRINT"PIPELINE MODELLED CONSISTS OF A SINGLE PIPE LINKING TWO PRESSU
RISED TANKS WITH A VALVE AT ENTRY TO THE DOWNSTREAM TANK."
     60 PRINT:PRINT"THE PROGRAM ALLOWS FOR COLUMN SEPARATION UPSTREAM OF THE CLOSE
D VALVE AND THEREFORE REQUIRES ABSOLUTE PRESSURE VALUES TO BE USED."
     70 DIM P(21):DIM PP(21):DIM V(21):DIM VV(21):DIM VOL(2)
     80 PRINT:PRINT"PLEASE INPUT SYSTEM DESCRIPTION IN RESPONSE TO THE FOLLOWING P
ROMPTS.....":PRINT
     90 INPUT"PIPE LENGTH (M.)",L
    100 INPUT"PIPE DIAMETER (M.)",D
    110 INPUT"PIPE WALL THICKNESS (M.)",W
    120 INPUT"PIPE MATERIAL YOUNGS MODULUS  AS E*10^n (UNITS-N/M^2)":INPUT"E=",E:I
NPUT"n=",N
    130 E=E*10.0^N
    140 INPUT"FLUID DENSITY KG/M^3",R
    150 PRINT"FLUID BULK MODULUS AS B*10^n (UNIT N/M^2)":INPUT"B=",B:INPUT"n=",N

    160 B=B*10.0^N
    170 INPUT "FLUID VAPOUR PRESSURE N/M^2 ABSOLUTE.",VAP
    180 INPUT"PIPE FRICTION FACTOR",F
    190 INPUT"VALVE STEADY STATE PRESSURE LOSS N/M^2",P0
    200 INPUT"VALVE STEADY THROUGHFLOW (CU.M/S.) TO GENERATE LOSS P0",Q0
    210 INPUT"DOWNSTREAM TANK PRESSURE N/M^2 ABSOLUTE.",PR2
    220 INPUT"VALVE CLOSURE TIME SEC.",TC
    230 INPUT"TOTAL SIMULATION TIME SEC.";TMAX
    240 INPUT"NUMBER OF PIPE SECTIONS CONSIDERED",N
    250 INPUT"INITIAL PIPE FLOWRATE CU.M/S",Q
    260 INPUT"PLEASE CHOOSE DATA OUTPUT FORMAT BY INPUTING VALUE JO = 1 FOR ALL RE
SULTS AT EACH TIME STEP, = 2 FOR VALVE PRESSURE ONLY AT EACH TIME STEP, =3 FOR M
AXIMUM PRESSURE AT VALVE ONLY.",JO
    280 REM CALCULATION OF MODEL PARAMETERS SUCH AS WAVE SPEED AND PIPE PERIOD
    290 VOL(1)=0.0:VOL(2)=0.0
    300 A=PI*D^2/4: VO=Q0/A
    310 C=SQR((B/R)/(1.0+D*B/(E*W)))
    320 PJ=R*C*(Q/A)/1000.0
    330 DX=L/N: PER=2.0*L/C:DT=DX/C
    340 PV=PR2+P0*((Q/A)/VO)^2
    350 DF=R*2*F*DX*(Q/A)^2/D
    360 CLS:PRINT"PIPE DATA....."
    370 PRINT:PRINT"PIPE LENGTH =",L," M. PIPE DIAMETER =",D," M. WALL THICKNESS =
",W," M."
    380 PRINT:PRINT"YOUNGS MODULUS =",E/10.0^6," MN/M^2."
```

```
 390 PRINT"PIPE FRICTION FACTOR =",F
 400 PRINT:PRINT"FLUID DATA...."
 410 PRINT:PRINT"DENSITY =",R," KG/M^3. BULK MODULUS =",B/10.0^6," MN/M^2."
 420 PRINT"VAPOUR PRESSURE =",VAP," N/M^2 ABS."
 430 PRINT:PRINT"VALVE AND TANK DATA..."
 440 PRINT:PRINT"DOWNSTREAM TANK PRESSURE =",PR2," N/M^2 ABS."
 450 PRINT"VALVE PRESSURE LOSS =",P0," N/M^2. AT THROUGHFLOW =",QO, "CU.M/S."
 460 PRINT"VALVE CLOSURE TIME =",TC," SEC."
 470 PRINT:PRINT"CALCULATION PARAMETERS..."
 480 PRINT:PRINT"WAVE SPEED =",C," M/SEC. PIPE PERIOD =",PER," SEC"
 490 PRINT"VALVE CLOSURE TIME =",TC/PER," PIPE PERIODS."
 500 PRINT"JOUKOWSKY MAX PRESSURE SURGE =",PJ," KN/M^2."
 510 PRINT:INPUT"PRESS P TO CONTINUE",Z$
 520 REM SET UP CONDITIONS AT TIME ZERO.
 530 T=0.0:PMAX=0.0
 540 FOR I=1 TO N+1
 550 J=N+2-I
 570 V(J)=Q/A
 580 P(J)=PV+(I-1)*DF
 600 NEXT I
 610 PR1=P(1)
 620 IF J0>1 THEN GOTO 680
 630 CLS:PRINT "TIME=",T
 640 FOR I=1 TO N+1
 650 PRINT "NODE=",I,"  PRESSURE =",P(I)/1000.0," KN/M^2.  VELOCITY =",V(I)," M
/S."
 655 IF VOL(2)>0.0 THEN PRINT"VAPOUR POCKET AT CLOSED VALVE. VOLUME =",VOL(2),"
M^3."
 660  NEXT I
 670 INPUT "PRESS P TO CONTINUE",Z$
 680 IF J0=2 THEN PRINT "TIME= ",T," S. VALVE PRESSURE= ",P(N+1)/1000.0," KN/M^
2. P.VALVE/J'SKY= ",(P(N+1)-PV)/(1000.0*PJ)
 682 IF J0=2 AND VOL(2)>0.0 THEN PRINT"VAPOUR POCKET AT CLOSED VALVE. VOLUME ="
,VOL(2)," M^3"
 685 T=T+DT
 690 KC=1.0/(R*C)
 700 REM UPSTREAM TANK BOUNDARY
 710 PP(1)=PR1
 720 K3=V(2)-KC*P(2)-2.0*F*V(2)*ABS(V(2))*DT/D
 730 VV(1)=K3+KC*PP(1)
 740 REM INTERNAL NODES
 750 FOR I=2 TO N
 760 K1=V(I-1)+KC*P(I-1)-2.0*F*V(I-1)*ABS(V(I-1))*DT/D
 770 K3=V(I+1)-KC*P(I+1)-2.0*F*V(I+1)*ABS(V(I+1))*DT/D
 780 VV(I)=0.5*(K1+K3)
 790 PP(I)=(K1-K3)/(2.0*KC)
 795 IF PP(I)<VAP THEN PP(I)=VAP
 800 NEXT I
 810 REM VALVE BOUNDARY CONDITION
 820 K1=V(N)+KC*P(N)-2.0*F*V(N)*ABS(V(N))*DT/D
 830 IF T>TC THEN GOTO 890
 840 TAU=1.0-T/TC
 850 K5=(TAU*V0)^2/P0
 860 VV(N+1)=-K5/(2.0*KC)+SQR((K5/(2.0*KC))^2-K5*(PR2-K1/KC))
 870 PP(N+1)=(K1-VV(N+1))/KC
 880 GOTO 980
 890 IF VOL(1)>0.0 THEN GOTO 930
 900 VV(N+1)=0.0
 910 PP(N+1)=K1/KC
 920 IF PP(N+1)>VAP THEN GOTO 980
 930 PP(N+1)=VAP
 940 VV(N+1)=K1-KC*VAP
 950 VOL(2)=VOL(1)-A*DT*0.5*(VV(N+1)+V(N+1))
 960 VOL(1)=VOL(2)
 970 IF VOL(2)<0.0 THEN VOL(1)=0.0
 980 IF PMAX=0.0 AND P(N+1)>PP(N+1) THEN TPMAX=T-DT
 982 IF PMAX=0.0 AND P(N+1)>PP(N+1) THEN PMAX=P(N+1)
 985 FOR I=1 TO N+1
1000 P(I)=PP(I):V(I)=VV(I)
1010 NEXT I
1030 IF T<TMAX THEN GOTO 620
1060 IF J0=3 THEN PRINT"MAXIMUM PRESSURE AT VALVE DURING CLOSURE = ",PMAX/1000.
0," KN/M^2."
1065 IF J0=3 THEN PRINT"AT TIME = ",TPMAX," SEC. "
1090 STOP
```

19.9. Open channels and partially filled pipes

Unsteady flow in open channels may be characterized by the changes in an inflow hydrograph shape as it propagates along the channel. In general an attenuation in the maximum depth and flow rate recorded at any down-stream station is observed, with trailing edges of the waves flattening while leading edges may steepen.

These effects are illustrated in Fig. 19.19, where the hydrograph is thought of as a series of incremental waves each of individual depth and thus

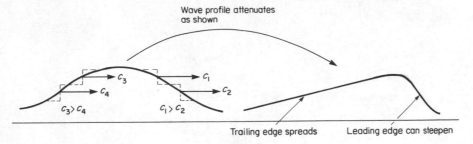

Figure 19.19 Unsteady flow wave attenuation in open channels or partially filled pipeflows

possessing an individual wave propagation velocity, c, which increases with depth. For trailing slope the deeper waves travel faster than the shallow trailing edge, thus 'stretching' the profile. Conversely the leading edge steepens. However, frictional forces, dependent on velocity squared, act to limit this so that all waves do not become steep-fronted surges automatically. The overall effect, then, is one of attenuation in depth as the wave progresses downstream. The phenomenon is complex and depends upon channel para-meters such as roughness, gradient and cross-sectional size and shape as well as the base flow upon which the wave travels, i.e. the greater the base flow depth the less attenuation suffered by a surface wave.

Accurate prediction of this attenuation would be advantageous in pipe sizing for varying inflows, such as urban stormwater systems, flood-routing or building drainage schemes. Analytical solutions are limited but a numerical approach based on finite difference methods is practical. One of these is the method of characteristics applied earlier to pressure transients.

19.9.1. Differential equations defining unsteady open channel flow

The basic equations of motion and continuity required are similar to those utilized in the study of pressure transients. From Fig. 19.20 the equation of motion becomes:

$$- \rho g\, \Delta x\, A\, \frac{\partial h}{\partial x} - \tau_0\, P\, \Delta x + \rho g\, A\, \Delta x \sin \alpha = \frac{\partial}{\partial x}\left(\rho\, V^2 A\right) \Delta x$$

$$+ \frac{\partial}{\partial t}\left(\rho\, A V\right) \Delta x. \tag{19.41}$$

Introducing $m = A/P$ and dividing by $\rho A\, \Delta x$:

$$g\, \frac{\partial h}{\partial x} + \frac{\tau_0}{\rho m} - g \sin \alpha + 2\, V\, \frac{\partial V}{\partial x} + \frac{V^2}{A}\, \frac{\partial A}{\partial x} + \frac{V}{A}\, \frac{\partial A}{\partial t} + \frac{\partial V}{\partial t} = 0. \tag{19.42}$$

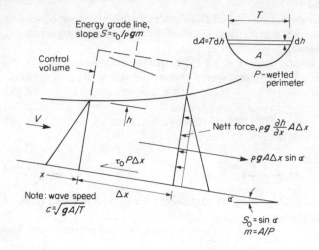

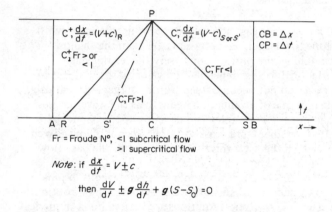

Figure 19.20 Method of characteristics solution of the unsteady partially
filled pipe flow equations

The continuity equation applied to the element of flow, Fig. 19.20, yields

$$-\frac{\partial}{\partial x}\,(\rho\,A\,V) = \frac{\partial}{\partial t}\,(\rho\,A\,\Delta x);$$

expanding:

$$V\,\frac{\partial A}{\partial x} + \frac{\partial A}{\partial t} + A\,\frac{\partial V}{\partial x} = 0. \tag{19.43}$$

Multiplying equation (19.43) by V/A and subtracting from (19.42) yields:

$$g\,\frac{\partial h}{\partial x} + \frac{\tau_0}{\rho m} - g\,\sin + \alpha\,\frac{\partial V}{\partial t} + \frac{\partial V}{\partial t} = 0. \tag{19.44}$$

Let $S_0 = \sin\alpha$ and $\tau_0 = \frac{1}{2}\,\rho f V^2$ as previously.

For open channel flow we may assume Chezy's equation to apply for unsteady flow resistance provided the rate of change of the flow parameters is gradual: hence $\tau = \frac{1}{2}\,\rho f.\,C^2\,mS$ where C is the Chezy coefficient and S is the slope of the energy grade line.

Hence $\tau_0/\rho m = gS$ and equation (19.44) becomes:

$$g\,\frac{\partial h}{\partial x} + g\,(S - S_0) + V\,\frac{\partial V}{\partial x} + \frac{\partial V}{\partial t} = 0 \tag{19.45}$$

Returning to Fig. 19.20 and the continuity equation, it may be seen that $\partial A = \partial hT$ and thus equation (19.43) becomes

$$V\,T\,\frac{\partial h}{\partial x} + T\,\frac{\partial h}{\partial t} + A\,\frac{\partial V}{\partial x} = 0. \tag{19.46}$$

Equations (19.45) and (19.46) may be combined linearly and reduced to a total differential expression,

$$\frac{dV}{dt} \pm \frac{g}{\sqrt{(g\,A/T)}}\,\frac{dh}{dt} + g\,(S - S_0) = 0, \tag{19.47}$$

if

$$\frac{dx}{dt} = V \pm \sqrt{(g\,A/T)} \tag{19.48}$$

For the general channel shape of Fig. 19.20, the wave propagation velocity,

$$c = \sqrt{(g\,A/T)}, \tag{19.49}$$

reducing for a rectangular channel to $\sqrt{(gh)}$, an expression derived earlier).

Thus equations (19.47) and (19.48) provide a basis for a finite difference solution similar in form to that developed previously for transient propagation. Referring to Fig. 19.20, if the variables V and h are known at R and S then four equations may be written in terms of the unknowns at P,

$$V_P - V_R + g \int_{h_R}^{h_P} \frac{1}{c}\,dh + \int_{t_R}^{t_P} g\,(S - S_0)\,dt = 0, \tag{19.50}$$

$$x_P - x_R = \int_{t_R}^{t_P} (V + c)\,dt, \tag{19.51}$$

$$V_P - V_S + g \int_{h_S}^{h_P} \frac{1}{c}\,dh + \int_{t_S}^{t_P} g\,(S - S_0)\,dt = 0 \tag{19.52}$$

$$x_P - x_S = \int_{t_S}^{t_P} (V - c)\,dt \tag{19.53}$$

These equations, forming two pairs of relationships, are the C^+ and C^- characteristics for unsteady flow prediction in open channels. As such the expressions may be related to a fixed grid in Δx and Δt as shown in Fig.

19.2. Two major differences may be seen between the open channel application and the transient analysis presented earlier.

In subcritical flow the local wave speed, $c = \sqrt{(g\,A/T)}$, exceeds the local mean flow velocity; thus characteristics intersecting at P originate in both adjacent grid sections. However in supercritical flow $V > c$; thus downstream information cannot propagate upstream and the slopes of both characteristics are positive, i.e. both C^{+} and C^{-} lines originate in the upstream grid section. The second main difference is that the relative magnitudes of V and c are such that V cannot be ignored relative to c and the characteristic base points R and S shown do not lie at the nodes upstream and downstream of P (Fig. 19.20). In general in transient analysis this is the case as unless the pipe walls are very elastic or the free gas content of the fluid is high, the wave speed c is several orders of magnitude greater than the local flow velocity.

For a rectangular, regular grid, the length Δx may be fixed. However, the value of Δt must be such that R and S lie as shown in Fig. 19.20. Thus Δt is based on

$$\Delta t = \frac{\Delta x}{(V+c)_{\max}}$$

where the maximum values of V and c at the time t, are used to assure stability.

As R and S cannot fall at the nodes on either side of P, interpolation for the base values is necessary.

Referring to Fig. 19.20 for the subcritical case,

$$\frac{c_C - c_R}{c_C - c_A} = \frac{V_C - V_R}{V_C - V_A} = \frac{x_C - x_R}{x_C - x_A} = (V_R + c_R)\frac{\Delta t}{\Delta x} = \frac{h_C - h_R}{h_C - h_A}\,.$$

As

$$x_P = x_C \text{ and } x_P - x_R = (V_R + c_R)\Delta t,\, \theta = \Delta t/\Delta x,$$

$$V_R = \frac{V_C + \theta\,(-V_C c_A + c_C V_A)}{1 + \theta\,(V_C - V_A + c_C - c_A)}\,,$$

$$c_R = \frac{c_C(1 - V_R\,\theta) + c_A V_R\,\theta}{1 + c_C\,\theta - c_A\,\theta}\,,$$

$$h_R = h_C - (h_C - h_A)\,(\theta(V_R + c_R)).$$

Similarly,

$$V_S = \frac{V_C - \theta(V_C c_B - c_C V_B)}{1 - \theta(V_C - V_B - c_C + c_B)}\,,$$

$$c_S = \frac{c_C + V_s\,\theta(c_C - c_B)}{1 + \theta(c_C - c_B)}\,,$$

$$h_S = h_C + \theta(V_S - c_S)\,(h_C - h_B).$$

For the supercritical flow regime the equations relating to R remain unchanged, those for S' may be shown to be:

$$V_{S'} = \frac{V_C(1 - c_A\theta) - V_A c_C\theta}{1 + \theta(V_C - V_A + c_A - c_C)},$$

$$c_{S'} = \frac{c_C + V_{S'}\theta(c_A - c_C)}{1 + c_A\theta - c_C\theta}.$$

In representing equations (19.50) to (19.53) by a first-order finite difference approximation a number of points need to be expanded.

Frictional representation is via the energy line slope S, determined from the Chezy equation. For large open channels it is acceptable to use the Manning equation where C is given by $m^{1/6}/n$. However, for partially filled pipe flows, certainly for pipe diameters $< 1 \cdot 0$ m, the Manning equation does not adequately represent the frictional losses and it is preferable to use a modified version of the Colebrook–White equation where D is replaced by $m = D/4$ (*see* Section 8.6 and equation (8.60)). The value of friction factor f may then be utilized in $C = \sqrt{(2g/f)}$.

In equations (19.50) and (19.52), $\int(1/c)dh$ may be approximated by $(1/c)\int dh$ if the variation in c, the local wave speed, is small over the depth change likely in the calculation. If this is not the case, and particularly for non-rectangular section channels, such as circular or elliptic cross-section sewers, Henderson has suggested replacing the depth by a 'stage' variable linked to depth by the equivalence:

$$d\omega = \sqrt{\left(\frac{gT}{A}\right)} \; dh = \frac{1}{c} \; dh.$$

For any cross-section, values of ω corresponding to depth can be determined by evaluating for each depth the integral

$$\omega = \int_0^h \sqrt{\left(\frac{gT}{A}\right)} \; dh$$

where T and A are geometric functions of h.

While it is relatively simple to incorporate this, in the example given below it will be assumed that c remains sensibly constant over dh at each time step.

Base values of h, V and c at the start of the calculation may be provided by assuming an initial steady flow through the pipe or channel. This may be assumed to be at normal depth for the channel slope, roughness and flow rate with the appropriate development profiles at entry and exit from the channel (*see* Section 16.5), calculated using gradually varied flow theory.

Once the initial values are set up, the calculation time can be progressed and depth and velocity values at the new time calculated. As with transient analysis this introduces the most important component of the method of characteristics solution, namely the choice of suitable boundary conditions.

Within the scope of this presentation, channel exit conditions will be limited to free discharge for either sub- or super-critical flow. In the super-critical case, the presence of the exit cannot be communicated upstream, hence the flow leaves at a depth and velocity simply calculated from the $C^+ C^-$ characteristics for that node.

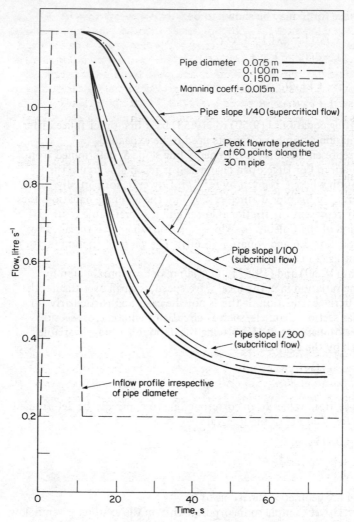

Figure 19.21 Peak flow rate predicted along a 30 m pipe for pipe gradients 1/40, 1/100, 1/300 and pipe diameters of 0·15, 0·1 and 0·075 m

In the subcritical case the flow exhibits a drawdown profile with the depth at exit normally taken as being the appropriate critical depth. Hence the exit conditions are solved by combining the C^+ characteristic with the critical depth relationship,

$$1 - \frac{Q^2 T}{gA^3} = 0,$$

equation (16.9) developed earlier. As both T and A are functions of h and $Q = VA$, solution with equation (19.50) is possible by an iterative technique.

Entry conditions to the pipe are more diverse, depending on supply channel or pipe geometry. One simplistic boundary would be to assume normal flow depth at entry. More realistically inflow energy may be utilized, for example where flow enters a near-horizontal pipe from a vertical stack that featured an annular flow, such as in the case of building drainage.

Therefore, given the initial steady state and the variation in inflow with time it is possible to apply the techniques described to trace the progress of the flow variation downstream. Figure 19.21 presents results generated by a simulation similar to that presented in Section 19.10 which clearly demonstrates the attenuation of flow maximum depth and calculated local flow rate along a partially filled pipe.

The technique described may be utilized in the study of flood routing or of flows in urban or building drainage systems. It can provide valuable guidance to designers in determining pipe sizing and gradients for such systems. Quite obviously the study requires further consideration of boundary conditions which is beyond the scope of this presentation, particularly relating to the effect of junctions and non-free discharge exits where the presence of the exit is transmitted upstream by the generation of a subcritical backwater profile, terminated by an upstream hydraulic jump in the supercritical flow case. At junctions such jumps will be free to move in the supply channels. However this can be dealt with by the techniques described allowing a network analysis program to be developed.

19.10 Computer program 'WAVES'

(1)   The program calculates wave attenuation in an open channel of either circular or rectangular cross-section. Flow is assumed to enter the channel at normal depth and discharges at critical depth if the channel is at a subcritical flow slope. A wave entering the channel is represented by a time-dependent inflow profile and the subsequent attenuation of this wave is traced at a number of sections along the channel.

(2)   The required input is:

(i)   channel shape code;
(ii)   channel width and maximum depth or partially filled pipe diameter (m);
(iii)   channel length (m).
(iv)   Manning coefficient;
(v)   channel slope;
(vi)   number of sections;
(vii)   number of points on inflow hydrograph;
(viii)   for each hydrograph point, time and flow rate.

(3)   Use is made of the method of characteristics solution of the open channel unsteady flow momentum and continuity equations.

(4)   Input example:

```
>RUN
PROGRAM DESIGNED TO PREDICT WAVE ATTENUATION IN AN OPEN CHANNEL OR PARTIALLY FIL
LED PIPE.

THIS PROGRAM DOES REQUIRE A FAIRLY LONG RUN TIME SO YOU SHOULD EXPERIMENT WITH S
HORT SIMULATION TIMES PREFERABLY IN THE MAX. VALUE OUTPUT MODE....

INPUT GEOMETRY CODE:-I=1 FOR CIRCULAR PIPES, I=2 FOR RECTANGULER CHANNELS.I= ?1
INPUT PIPE DIAMETER OR MAXIMUM OPEN CHANNEL DEPTH D (M.) =?0.1
CHANNEL LENGTH L (M.) =?10
MANNING'S N =?0.009
```

```
CHANNEL SLOPE S =?0.01
INPUT NUMBER OF CHANNEL SECTIONS TO BE CONSIDERED (MAX =5) N =?5
TOTAL SIMULATION TIME (SEC.) TMAX= ?15
INPUT INFLOW PROFILE:-
INPUT NUMBER OF INFLOW-TIME COORDINATES (NOTE THAT FIRST MUST BE INITIAL FLOW AT
 TIME=0.0) K =?5
TIME (SEC.)=?0
FLOW (M^3/S.) =?0.00011
TIME (SEC.)=?1
FLOW (M^3/S.) =?0.0029
TIME (SEC.)=?2
FLOW (M^3/S.) =?0.0029
TIME (SEC.)=?3
FLOW (M^3/S.) =?0.00011
TIME (SEC.)=?25
FLOW (M^3/S.) =?0.00011
PLEASE CHOOSE OUTPUT FORMAT BY INPUTING VALUE J9=1 FOR RESULTS AS CALC. PROGRESS
ES OR J9=0 FOR MAX VALUES ONLY. ?0


CHANNEL DATA...

PARTIALLY FILLED PIPE FLOW, PIPE DIAMETER, D=        0.1 M.
CHANNEL LENGTH, L=          10 M.
NUMBER OF SECTIONS CONSIDERED N=          5

CHANNEL SLOPE, S=    1E-2 M.

CHANNEL MANNING'S N=        9E-3
INITIAL FLOW Q0=  1.1E-4 M^3/S.
INITIAL FLOW CRITICAL DEPTH HC=        1.04E-2      M.
INITIAL FLOW NORMAL DEPTH HN= 8.5E-3 M.
SUPERCRITICAL FLOW.
PRESS RETURN TO CONTINUE.?
MAX DEPTH AND FLOW ARE AS FOLLOWS:
NODE =      1 DEPTH =     4.3E-2 M. QMAX.=  2.9E-3 M^3/S. TIME =  1.15      S.
NODE =      2 DEPTH =     2.41E-2     M. QMAX.=  1.05E-3     M^3/S. TIME =  3.08
       S.
NODE =      3 DEPTH =     1.61E-2     M. QMAX.=  4.59E-4     M^3/S. TIME =  4.36
       S.
NODE =      4 DEPTH =     1.39E-2     M. QMAX.=  3.21E-4     M^3/S. TIME =  7.49
       S.
NODE =      5 DEPTH =     1.29E-2     M. QMAX.=  2.69E-4     M^3/S. TIME =  10.9
       S.
NODE =      6 DEPTH =     1.23E-2     M. QMAX.=  2.41E-4     M^3/S. TIME =  14.5
       S.
SIMULATION COMPLETE....
```

List:

```
   10 REM WAVES
   20 MODE 3:@%=&30306
   30 DIM HH(6), H(6),W(6),V(6),VV(6),TF(5),Q(5),HMAX(6),THMAX(6),VHMAX(6)
   40 PRINT"PROGRAM DESIGNED TO PREDICT WAVE ATTENUATION IN AN OPEN CHANNEL OR P
ARTIALLY FILLED PIPE."
   50 PRINT:PRINT"THIS PROGRAM DOES REQUIRE A FAIRLY LONG RUN TIME SO YOU SHOULD
 EXPERIMENT WITH SHORT SIMULATION TIMES PREFERABLY IN THE MAX. VALUE OUTPUT MODE
....."
   60 PRINT:INPUT"INPUT GEOMETRY CODE:-","I=1 FOR CIRCULAR PIPES, I=2 FOR RECTAN
GULER CHANNELS.","I= ",I
   70 INPUT"INPUT PIPE DIAMETER OR MAXIMUM OPEN CHANNEL DEPTH D (M.) =",D
   80 IF I=2 THEN INPUT "INPUT CHANNEL WIDTH B (M.) =",B
   90 INPUT"CHANNEL LENGTH L (M.) =",L
  100 INPUT"MANNING'S N =",MN
  110 INPUT"CHANNEL SLOPE S =",S
  120 INPUT"INPUT NUMBER OF CHANNEL SECTIONS TO BE CONSIDERED (MAX =5) N =",N

  130 INPUT"TOTAL SIMULATION TIME (SEC.) TMAX= ",TMAX
  140 PRINT"INPUT INFLOW PROFILE:-"
  150 INPUT"INPUT NUMBER OF INFLOW-TIME COORDINATES (NOTE THAT FIRST MUST BE INI
TIAL FLOW AT TIME=0.0) K =",K
  160 FOR KK=1 TO K
  170 INPUT"TIME (SEC.)=",TF(KK),"FLOW (M^3/S.) =",Q(KK)
  180 NEXT KK
  190 INPUT"PLEASE CHOOSE OUTPUT FORMAT BY INPUTING VALUE J9=1 FOR RESULTS AS CA
LC. PROGRESSES OR J9=0 FOR MAX VALUES ONLY. ",J9
  200 Q0=Q(1)
  210 REM CALCULATION OF INITIAL FLOW NORMAL AND CRITICAL DEPTHS.
  220 ICALL=1
  230 QQ=Q0
  240 GOSUB 1440
  250 HC=C:AC=A:TC=T
  260 ICALL=2
  270 GOSUB 1440
```

```
280 HN=C:AN=A:TN=T
290 CLS:PRINT"CHANNEL DATA..."
300 PRINT:IF I=1 THEN PRINT"PARTIALLY FILLED PIPE FLOW, PIPE DIAMETER, D= ",D,
" M."
310 IF I=2 THEN PRINT"RECTANGULAR OPEN CHANNEL OF WIDTH, B= ",B," M."
320 PRINT"CHANNEL LENGTH, L= ",L," M."
330 PRINT"NUMBER OF SECTIONS CONSIDERED N= ",N
340 PRINT:PRINT"CHANNEL SLOPE, S= ",S," M."
350 PRINT:PRINT"CHANNEL MANNING'S N= ",MN
360 PRINT"INITIAL FLOW QO= ",QO," M^3/S."
370 PRINT"INITIAL FLOW CRITICAL DEPTH HC= ",HC," M."
380 PRINT"INITIAL FLOW NORMAL DEPTH HN= ",HN," M."
390 IF HN<HC THEN PRINT"SUPERCRITICAL FLOW." ELSE PRINT"SUBCRITICAL FLOW."
400 INPUT"PRESS RETURN TO CONTINUE.",Z$
410 TIM=0.0
420 REM SET UP BASE CONDITIONS AT TIME 0.0
430 FOR J=1 TO N+1
440 HMAX(J)=0.0
450 H(J)=HN
460 IF HN>HC AND J=N+1 THEN H(J)=HC
470 V(J)=Q(1)/AN
480 IF HN>HC AND J=N+1 THEN V(J)=Q(1)/AC
490 W(J)=SQR(9.81*AN/TN)
500 IF HN>HC AND J=N+1 THEN W(J)=SQR(9.81*AC/TC)
510 NEXT J
520 REM CALCULATION TIMESTEP.
530 WMAX=W(1):VMAX=V(1)
540 FOR J= 2 TO N+1
550 IF W(J)>W(J-1) THEN WMAX=W(J)
560 IFV(J)>V(J-1) THEN VMAX=V(J)
570 NEXT J
580 DT=L/(N*3.0*(VMAX+WMAX))
590 DX=L/N
600 TIM=TIM+DT
610 REM DETERMINE INFLOW AT TIME T
620 FOR KK=2 TO K
630 IF TF(KK-1)<TIM AND TF(KK)>TIM THEN KZ=KK
640 NEXT KK
650 QIN=Q(KZ-1)+(TIM-TF(KZ-1))*(Q(KZ)-Q(KZ-1))/(TF(KZ)-TF(KZ-1))
660 REM SET INLET DEPTH AT NORMAL FOR QIN
670 ICALL=2:QO=QIN
680 GOSUB 1440
690 HH(1)=C
700 VV(1)=QIN/A
710 REM INTERNAL NODES
720 TH=DT/DX
730 FOR J=2 TO N+1
740 VR=(V(J)+TH*(W(J)*V(J-1)-V(J)*W(J-1)))/(1.0+TH*(V(J)-V(J-1)+W(J)-W(J-1)))
750 WR=(W(J)*(1.0-VR*TH)+W(J-1)*VR*TH)/(1.0+W(J)*TH-W(J-1)*TH)
760 HR=H(J)-(H(J)-H(J-1))*TH*(VR+WR)
770 IF HN<HC THEN GOTO 830
780 IF J=N+1 THEN GOTO 860
790 VS=(V(J)-TH*(V(J)*W(J+1)-W(J)*V(J+1)))/(1.0-TH*(V(J)-V(J+1)-W(J)+W(J+1)))
800 WS=(W(J)+VS*TH*(W(J)-W(J+1)))/(1.0+TH*(W(J)-W(J+1)))
810 HS=H(J)+TH*(VS-WS)*(H(J)-H(J+1))
820 GOTO 860
830 VS=(V(J)*(1.0+TH*W(J-1))-V(J-1)*TH*W(J))/(1.0+TH*(V(J)-V(J-1)+W(J-1)-W(J))
)
840 WS=(W(J)+VS*TH*(W(J-1)-W(J)))/(1.0+TH*(W(J-1)-W(J)))
850 HS=H(J)-(H(J)-H(J-1))*TH*(VS-WS)
860 C=HR
870 GOSUB 1570
880 SR=((VR*ABS(VR)*MN^2))/(A/P)^1.333
890 IF HN>HC AND J=N+1 THEN GOTO 1000
900 C=HS
910 GOSUB 1570
920 SS=((VS*ABS(VS)*MN^2))/(A/P)^1.333
930 REM CALCULATION OF CONDITIONS AT EACH NODE AT TIME T
940 K1=9.81/WR
950 K3=9.81/WS
960 K2=K1*HR+VR-9.81*(SR-S)*DT
970 K4=VS-K3*HS-9.81*(SS-S)*DT
980 HH(J)=(K2-K4)/(K1+K3)
990 VV(J)=K4+K3*HH(J)
1000 NEXT J
1010 IF HN<HC THEN GOTO 1190
1020 REM EXIT CONDITIONS BASED ON CRITICAL FLOW DEPTH AT CURRENT FLOW.
1030 K1=9.81/WR
1040 K2=K1*HR+VR-9.81*(SR-S)*DT
1050 UP=D:LO=0.0
1060 C=(UP+LO)/2.0
1070 GOSUB 1570
1080 F=1.0-(ABS(K2-K1*C)*(K2-K1*C))*T/(9.81*A)
1090 IF F<0.0 THEN LO=C
```

```
1100 IF F=0.0 THEN GOTO 1160
1110 IF F>0.0 THEN UP=C
1120 E=(UP+LO)/2.0
1130 IF ABS((E-C)/C)<0.01 THEN GOTO 1160
1140 C=E
1150 GOTO 1070
1160 HH(N+1)=C
1170 VV(N+1)=K2-K1*HH(N+1)
1180 REM SET UP BASE CONDITIONS AT NEXT TIME STEP.
1190 IFJ9=1 THEN  PRINT"TIME= ",TIM," SEC. INFLOW= ",QIN," M^3/S."
1200 FOR J=1 TO N+1
1220 IF HH(J)>H(J)  THEN HMAX(J)=HH(J)
1230 IF HH(J)>H(J)  THEN THMAX(J)=TIM
1240 IF HH(J)>H(J)  THEN VHMAX(J)=VV(J)
1250 H(J)=HH(J)
1260 V(J)=VV(J)
1270 C=H(J)
1280 GOSUB 1570
1290 W(J)=SQR(9.81*A/T)
1300 QL=V(J)*A
1310 IF J9=1 THEN PRINT"NODE= ",J," DEPTH= ",H(J)," M. Q.LOCAL= ",QL," M^3/S."
1320 NEXT J
1330 IF J9=1 THEN   INPUT"PRESS RETURN TO CONTINUE",Z$
1340 IF TIM<TMAX THEN GOTO 530
1350 IF J9=0 THEN PRINT"MAX DEPTH AND FLOW ARE AS FOLLOWS:"
1360 FOR J=1 TO N+1
1370 IF HMAX(J)=0.0 THEN THMAX(J)=TMAX
1375 IF HMAX(J)=0.0 THEN VHMAX(J)=Q(1)/AN
1380 IF HMAX(J)=0.0 THEN HMAX(J)=HN
1390 C=HMAX(J)
1400 GOSUB 1570
1405 IF HMAX(J)=HN THEN QM=Q(1) ELSE QM=VHMAX(J)*A
1410 PRINT "NODE =",J," DEPTH =",HMAX(J)," M. QMAX.= ",QM," M^3/S. TIME ="THMAX
(J)," S."
1420 NEXT J
1430 GOTO 1730
1440 UP=D
1450 LO=0.0
1460 C=(UP+LO)/2.0
1470 GOSUB 1570
1480 IF ICALL=1 THEN F=1.0-QQ^2*T/(9.81*A^3)
1490 IF ICALL=2 THEN F=((A/P)^(0.1667)/MN)*SQR((A/P)*S)-QQ/A
1500 IF F<0.0 THEN LO=C
1510 IF F>0.0 THEN UP=C
1520 IF F=0.0 THEN GOTO 1560
1530 E=(UP+LO)/2.0
1540 IF ABS((E-C)/C)<0.01 THEN GOTO 1560
.1550 C=E:GOTO 1470
1560 RETURN
1570 IF I=1 GOTO 1650
1580 REM RECTANGULAR CHANNELS
1590 T=B
1600 P=B+2.0*C
1610 A=B*C
1620 GOTO 1710
1630 REM PARTIALLY FILLED PIPES
1640 PRINT"C=",C
1650 IF C<D/2.0 THEN O=2.0*ATN(SQR(C*(D-C))/(D/2.0-C))
1660 IF C=D/2.0 THEN O=PI
1670 IF C>D/2.0 THEN O=PI+2.0*ATN((C-D/2.0)/(SQR(C*(D-C))))
1680 A=((D^2)/8.0)*(O-SIN(O))
1690 P=D*O/2.0
1700 T=2.0*SQR(C*(D-C))
1710 RETURN
1720 STOP
1730 PRINT"SIMULATION COMPLETE...."
```

19.11. Further reading

Bergeron, L. (1961). *Waterhammer in Hydraulics and Wave Surges in Electricity* John Wiley, New York.

BHRA International Conferences on Pressure Surge, Nos. 1 to 4 (1972, 1976, 1980, 1983), British Hydromechanics Research Association, Cranfield.

Chow, V. T. (1959). *Open Channel Hydraulics*, McGraw-Hill, New York.

Finite Elements in Water Resources, Proc. 4th International Conference, Hanover. (June 1982).

Fox, J. A. (1977). *Hydraulic Analysis of Unsteady Flow in Pipe Networks*
Macmillan, London.
Henderson, F. M. (1966). *Open Channel Flow*, Macmillan Publishing, New
York.
Lister, M. (1960). The numerical solution of hyperbolic partial differential
equations by the method of characteristics. In *Mathematical Methods for
Digital Computers*, Wiley, New York.
Parmakian, J. (1963). *Waterhammer Analysis*. Dover, New York.
Streeter, V. L. and Wylie, E. B. (1984). *Fluid Mechanics*. McGraw-Hill, New
York.
Symposium on Surges in Pipelines. Institute of Mechanical Engineers,
London (1965).
Urban Drainage Systems. Proc. 1st International Conference, Southampton,
1982. Pitman Advanced Publishing Program.
Watters, G. Z. (1979) *Modern Analysis and Control of Unsteady Flow in
Pipelines,* Ann Arbor Science Publishers, Michigan.
Wylie, E. B. and Streeter, V. L. (1983). *Fluid Transients*, FEB Press, Ann
Arbor.

EXERCISES 19

19.1   Water at a temperature of 20 °C flows through a pipe system made up,
over a considerable length, of a range of different pipe sizes and materials.
For each of the sections detailed below calculate the appropriate wave propagation velocity. Assume that the effects of longitudinal strain, as represented
by the inclusion of Poisson's ratio, may be neglected.

   (1) 50 mm diameter steel, 5 mm wall,
        Young's modulus $E = 204 \times 10^9 \, \text{N/m}^2$
   (2) 50 mm diameter aluminium, 3 mm wall,
        $E = 70 \times 10^9 \, \text{N/m}^2$
   (3) 25 mm diameter glass, 5 mm wall,
        $E = 1 \times 10^9 \, \text{N/m}^2$
   (4) 75 mm diameter copper, 3 mm wall,
        $E = 2 \cdot 5 \times 10^9 \, \text{N/m}^2$
   (5) 900 mm diameter cast iron 15 mm wall,
        $E = 2 \cdot 0 \times 10^9 \, \text{N/m}^2$
   (6) 1·2 m diameter concrete, 75 mm wall,
        $E = 0 \cdot 5 \times 10^9 \, \text{N/m}^2$

Assume density of water 998 kg/m^3 and bulk modulus as $2 \times 10^9 \, \text{N/m}^2$.
[1344 m/s, 1160 m/s, 426 m/s, 308 m/s, 181 m/s, 175 m/s]

19.2   For the aluminium pipe in Exercise 2 above, determine the pipe
diameter to wall thickness ratio above which the effects of longitudinal strain,
as expressed through the inclusion of Poisson's ratio $\mu = 0 \cdot 3$, should be
included in wave speed calculations. Take 10% error as the boundary.
[$D/e > 30$]

19.3   Show that for a fluid containing a proportion of free gas, the wave propagation speed c becomes

$$c = \sqrt{\left[\left(\frac{Kp}{(1-y)p} + yK\right)\middle/\left(y\rho_a + (1-y)\rho\right)\right]}$$

where K = bulk modulus of fluid, ρ = density of fluid, p = absolute pressure of gas of density ρ_a, and y is the proportion of free gas by volume.

For the flow of kerosene mean pressure 180 kN/m², density 800 kg/m³, bulk modulus $1{\cdot}26 \times 10^9$ N/m² in an aluminium pipe, diameter 50 mm, wall thickness 1 mm determine the wave propagation velocity for a range of free air contents from 0 to 1% at atmospheric pressure.
[300 m/s at 0·4%]

19.4   For each of the pipelines described in Exercise 19.1 determine the physical length of the wavefront propagated by a valve closure in 0·8 seconds. Assume each pipe is infinitely long.
[1142 m, 986 m, 362 m, 262 m, 154 m, 149 m]

19.5   If the flow velocity in each case referred to in Exercise 19.4 was 2·58 m/s calculate the pressure rise recorded at the closing valve.
[3470 kN/m², 2990 kN/m², 1099 kN/m², 795 kN/m², 466 kN/m², 450 kN/m²]

19.6   A simple test rig incorporating a constant head reservoir and each of the pipes detailed in Exercise 19.6, connected radially, is constructed. Each of the pipes is 60 m long. Determine the pipe period for each pipeline.
[89 ms, 103 ms, 282 ms, 390 ms, 663 ms, 686 ms]

19.7   A 20 m long, 75 mm diameter steel pipeline, wall thickness 6 mm, carries water from a large reservoir tank, held at a constant head of 6 m. Discharge is 0·022 m²/s through a variable speed valve positioned 10 m from the supply tank. Discharge is to a second constant head tank held at 2 m head.

If the valve closure is instantaneous determine the theoretical magnitudes of the pressure waves propagated away from the valve, under frictionless conditions.

Comment on the downstream pressure variation and draw pressure–time curves at points 5 m, 2·5 m and 0·5 m from the upstream tank.
[±670 m, flow separates downstream of valve]

19.8   A 10 m long aluminium pipe, 50 mm diameter, 3 mm wall thickness carries 0·01 m³/s kerosene, 800 kg/m³ density, bulk modulus $1{\cdot}26 \times 10^9$ N/m² from a constant head reservoir held at 300 kN/m² abs. If the valve positioned at the discharge end of the pipe is closed instantaneously calculate the pressure rise recorded. If the valve closing time is then changed to $\frac{1}{4}$ and then $\frac{3}{4}$ pipe periods in turn determine the maximum pressure recorded at the valve and at a point 1 m from the supply tank. Assume that the rate of rise of pressure at the valve during closure is linear with time.
[4480 kN/m², 4480 kN/m²; 4480 kN/m², 4480 kN/m², 1810 kN/m², 603 kN/m²]

19.9   Explain why the assumption of linear pressure rise with time during valve closure made in Exercise 19.8 would probably result in an underestimate, in practice, of the pressures recorded at the measuring point 1 m from the supply tank.

19.10   For the case set out in Exercise 19.8 determine the point closest to the supply reservoir where the maximum pressure rise recorded at the valve could also be measured, for both the 1/4 and 3/4 pipe period valve closures.

Also calculate the fastest valve closure possible with a resultant pressure rise lower than the maximum recorded in Exercise 19.8.

[2·5 m, 7·5 m, $t > 2L/c$]

19.11   Explain why, in cases where the valve closure time is fixed and longer than one pipe period, the belief that reducing the wave speed, either by free air or changing the pipe material, will reduce peak pressures is not necessarily correct.

19.12   Show that the transient reflection and transmission coefficient for a junction of n pipes is given by an expression:

$$C_R = \left(\frac{2A_j}{c_j} - \sum_{i=1}^{n}\frac{A_i}{c_i}\right)\bigg/ \sum_{i=1}^{n} A_i c_i$$

$$C_T = \left(\frac{2A_j}{c_j}\right)\bigg/ \left(\sum_{i=1}^{n}\frac{A_i}{c_i}\right)$$

where the transient approaches the junction along pipe j.

19.13   Show that the reflection coefficient at a closed end is +1 and at a constant pressure boundary is −1. Fluid flows between two reservoir tanks, held at constant pressures p_1 and p_2 absolute. A valve is placed midway between the tanks. Assuming an instantaneous valve closure plot the theoretical pressure variations at the downstream face of the valve and midway between the valve and the downstream reservoir. Assume frictionless flow with all the pressure loss $p_1 - p_2$ occurring at the valve. If the pressure drop on valve closure, $\rho c V_0$, is greater than (p_2 fluid vapour pressure) sketch the resultant pressure variation on the valve downstream face if the 1st and 2nd cavities last for 6 and 4 pipe periods respectively. Show that during the existence of the first vapour cavity at the valve, the pressure variations recorded at the mid downstream pipe position oscillate between fluid vapour pressure and downstream reservoir pressure at a frequency of c/L where L is pipe length and c is the wave speed.

19.14   A pressure transducer is attached to an infinite pipe, area A, wave speed C, by means of a connection tube, area a, wave speed c. Show that if $A \gg a$ and $c \to C$ and if the pressure front approaching the transducer junction is short in comparison with the length of the transducer connection, the error in recording pipeline pressure will approach 100%.

If the pressure front is linear and has a length equal to the 6 x length of the transducer connection tube, show that the error in recording pipeline pressure is of the order of 50%.

19.15   A water pipeline 425 m long, wave speed 1125 m/s, has a valve at its downstream end that closes in 3 seconds; the valve loss characteristic may be assumed to have the form $V = k\sqrt{/p}$ at any degree of opening, the ratio k/k_0, i.e. ratio of k to its value for the fully open valve, reducing from 1 to zero linearly during the valve closure. If the head in the supply reservoir is 300 kN/m² and the initial flow is at 1 m/s, determine graphically the maximum pressure recorded at the valve.

[46 m]

19.16   Kerosene, density 820 kg/m^3, flows along an aluminium pipe, 50 mm diameter, 15·24 m long, 3 mm wall thickness. The valve terminating the pipe is closed in 0·03 seconds. For an initial fluid velocity of 0·3 m/s determine graphically the pressure variations at the valve and the duration of the 1st vapour cavity formed at the valve following closure. Take supply reservoir pressure constant at 130 kN/m^2 abs.

[0·03 s, 370 kN/m^2 abs on collapse]

19.17   Write down the necessary characteristic equations to allow Exercise 19.16 to be solved on a digital computer.

19.18   Write down the necessary characteristic equations to allow Exercise 19.13 to be solved on a digital computer.

19.19   Write a short computer program that will allow the relation between transducer connection tubing length and diameter and incoming wave front length to be investigated for the type of installation described in Exercise 19.14.

20 Surge control

Pressure fluctuations arising under unsteady flow conditions were described in some detail in Chapter 19 and it might be concluded that, following suitable analysis, surge problems could be designed out of any pipe–machine system. However, in many cases, the potentially dangerous surge pressure only occurs as a result of some other system failure, such as mains power failure to a pumping station, and it would be impossible to so restrict the performance of the pumping system as to avoid the generation of surge pressures in the event of power failure. In cases such as this, the best approach is to include some form of surge suppression device in the system. However, this should only come into action on an abnormal pump shut down and should not operate as a normal procedure.

A wide range of surge suppression devices exist, all having the common factor that they tend to slow down the rate of change of flow in the system.

(i) *Valve closure control*. By introducing controls onto the valve closure mechanism, particularly to slow down the final stages of closure, worthwhile reductions in surge pressure can be achieved. Alternatively, closing a valve in a system concurrently with some negative surge pressure generating phenomena can propagate a positive transient that will maintain the system pressure above vapour or gas release levels.

(ii) *Increasing pump inertia*. As mentioned, one of the major problems with pumped systems is the prevention of column separation following power failure. If the rate of change of flow can be decreased by keeping the pump turning longer, then the danger of column separation is reduced. This can be achieved by incorporating a flywheel into the pump design – not a good solution as it has to be driven while the pump starts up.

(iii) *Surge shafts or air vessels*. Positive surge pressures can be allieviated by allowing fluid to leave the pipeline and enter vertical surge shafts, thus absorbing some of the excess energy. If the pressures are very high, air chambers or accumulators where gas or air is compressed may be used.

(iv) *Air or fluid admission valves*. During flow stoppage it is possible for low pressure regions to form and column separation to occur in the fluid in the pipeline. Examples of this are downstream of a valve or pump on flow stoppage. An inflow valve placed near to the likely separation site can allow air or fluid to enter the system, thereby filling the cavity, restoring system pressure and alleviating the surge pressure rise likely on cavity collapse. In many cases, introduction of air into the system by such valves is troublesome when the system is restarted, so that fluid inflow valves should be used if practical.

(v) *Relief valves*. An outflow valve can be arranged to blow off excess fluid as the pressure rises. This is a cheap but rather dangerous solution as jamming open of such valves could cause substantial loss of fluid from the system.

(vi) *Bypass systems*. These are, effectively, extensions to the inflow fluid relief valve devices and merely allow water from a sump to bypass the pump and enter the downstream pipe section, if the pressure here, following pump shut down, falls below sump pressure.

In a study of available surge suppression devices, it is probably best to consider the range of solutions available for a particular problem. Here it is best to take, as an example, the negative pressure surge problems caused by flow stoppage, which can lead to column separation and large resurge pressures on cavity collapse as this type of surge problem is common.

Consider the case of a mains pipeline pumping fluid from one reservoir to another over some intervening high ground. The potential pressure surge problems arise in the following manner:

(i) excessive positive pressure rise on terminal valve closure;
(ii) column separation at the pump discharge during emergency shut down;
(iii) column separation at a high point in the system following pump shut down.

Figure 20.1 illustrates the pipe network involved.

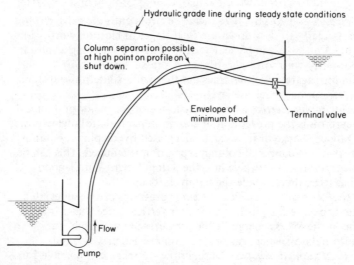

Figure 20.1   Layout of a pump supply system prone to surge on pump shutdown

20.1. Control of surge following valve closure with pump running

Valve closure will produce a positive pressure surge that will propagate through the system, as described in Chapter 19. The simplest solution is to decrease the valve closure rate. However, merely increasing the operation time of the valve is not the most efficient technique. The majority of the pressure surge generated will be produced by the final stages of valve closure (perhaps the last 15 per cent of the operation accounting for 80 per cent of the flow reduction), so that the most efficient method of reducing surge generation is to introduce a two-speed valve closure operation, the last 15 per cent of travel taking proportionately longer than the first 85 per cent (*see* Fig. 20.2).

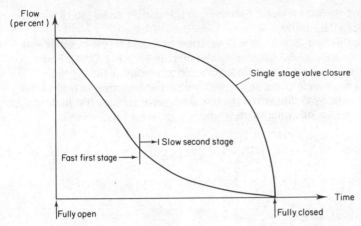

Figure 20.2   Flow–time relation for one- and two-stage valve closure

The boundary equation at the closing valve is supplied by the valve characteristic:

$$\tau = (\bar{v}/\bar{v}_0)\sqrt{(\Delta p_0/\Delta p)},$$

which varies from $1 \geqslant \tau \geqslant 0$ during valve closure. Now τ is known as a function of valve position; say $\tau = f(\theta)$, where θ is the angle of closure of the valve or a linear distance moved by a gate type valve. Thus, as position may be monitored with time and may be expressed for each stage of closure as

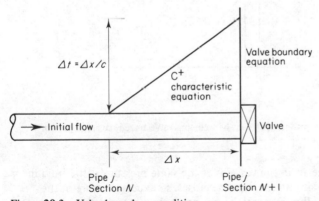

Figure 20.3   Valve boundary condition

$\theta = f(t)$, then τ is known as a function of time. Referring to Fig. 20.3, the resulting equation,

$$\bar{v}_{j,N+1} = \{\bar{v}_0 \sqrt{(p_{j,N+1}/p_0)}\} \times f(t),$$

can be solved with the C^+ characteristic line joining conditions at section (j, N) at time t to conditions at the valve section $(j, N + 1)$ at time $t + \Delta t$.

An alternative solution here is to provide a bypass valve adjacent to the closing valve which will open to pass a small quantity of flow as the main valve closes and then shuts itself. However, this is merely an adaptation of the method described above.

A common solution to this surge generation case, and to the negative surge situation, is to install an air chamber on the pipeline (*see* Fig. 20.4). The purpose of this device is to allow fluid to be diverted out of the pipeline under high pressure conditions. In many large civil engineering projects, open surge tanks may be used, but often the pressures generated are too high for these to be practical. Figure 20.4 illustrates a typical air vessel–pipeline

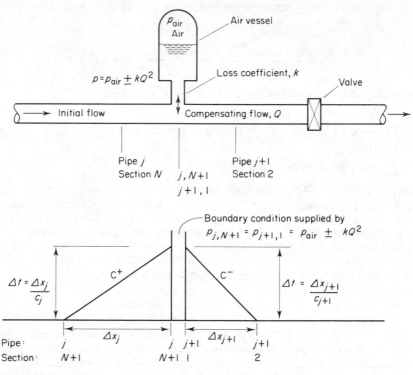

Figure 20.4   Air vessel installed upstream of a closing valve to reduce surge pressures

junction. Initially, the air in the vessel is at the same pressure as the fluid in the pipeline. As the valve on the pipeline closes, so excess pressure in the main forces fluid into the air vessel, compressing the gas and effectively delaying and reducing the surge pressure wave propagated through the pipeline fluid. The gas compression may be assumed to follow the relation

$$p \cdot V^n = \text{constant},$$

where p is gas pressure, V is the volume it occupies and n is an exponent (normally taken as 1·2 as a compromise between isothermal and adiabatic gas compression).

The air volume at any time t can be calculated from the air volume at the start of the time step and the inflow, or outflow, through the tank orifice over that time step Δt. Thus, the general volume equation is

$$V_t = V_{t-\Delta t} + Q \times \Delta t.$$

Considering the air vessel connection as a junction, then six equations are to be solved at each time step. Following the principles outlined in Chapter 19, the available equations are: C^+ characteristic joining section (j, N) at time $(t - \Delta t)$ to $(j, N + 1)$ at time t,

$$\bar{v}_{j,1} = K_3 - K_2 p_{j,1};$$

C^- characteristic joining section $(j + 1, 2)$ at time $(t - \Delta t)$ to $(j + 1, 1)$ at time t,

$$\bar{v}_{j+1,1} = K_1 + K_4 p_{j+1,1};$$

pressure change in air vessel from initial conditions,

$$p_{\text{air}} = p_{\text{initial}} \, (V_{\text{initial}}/V_t)^{1.2};$$

pressure drop across orifice,

$$p_{j,N+1} = p_{\text{air}} \pm kQ^2;$$

pressure continuity at the junction,

$$p_{j,N+1} = p_{j+1,1};$$

and continuity of flow

$$\bar{v}_{j,N+1} A_j = \bar{v}_{j+1,1} A_{j+1} \pm Q.$$

Solution of these equations, obviously achieved numerically, at each time step allows calculation of the volume–time curve for the vessel under a particular surge condition. The vessel should be resized to ensure that it does not empty either when pressure in the pipe falls or when the air volume becomes too small as the pressure rises in the pipeline. A compressor is necessary to make up any air losses due to absorption or leakage.

An alternative solution, which may be applied in certain circumstances, is a simple pressure relief valve which opens as soon as the pressure reaches a present level and allows fluid to blow out of the system. As soon as the pressure falls, the valve reseats. The obvious problem is the relief valve that does not close after use and allows fluid to drain out of the pipeline.

As mentioned, in large scale projects, such as hydroelectric power stations, surge tanks or shafts may be included in the station tunnelling.

Surge tanks are examples of fluid mass oscillation and, as described earlier (Fig. 18.2), may be analysed by means of rigid column theory. This provides considerable savings in computer time and cost and, in addition, it is possible to use a frictionless approximation to gain a first approximation to the period and amplitude of the mass oscillation, namely

$$T = 2\pi \sqrt{(lA_2/gA_1)}$$

and $\quad Z = (Q_0/A_2) \sqrt{(lA_2/gA_1)}.$

In general, surge tanks or shafts, as they are commonly cut out of rock in hydroelectric schemes, may be classified as simple, orifice or differential. Figure 20.5 illustrates some common designs. The simple surge shaft has an unrestricted entry from the penstock–supply tunnel junction and must be so

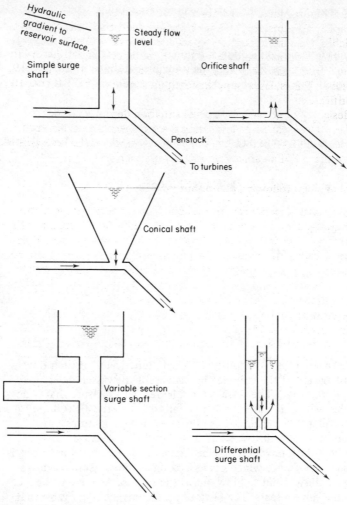

Figure 20.5   Schematics of some common surge shaft designs

designed that it will not overflow on an upsurge, unless overspill provision is made, nor empty during downsurge, caused by turbine acceleration or by fluid oscillation. If the shaft empties, then air will be drawn into the tunnel-penstock system, and this should be avoided wherever possible. The period of mass oscillation is relatively long for this type of shaft.

The orifice surge tank has a restricted opening from the tunnel system. The orifice losses aid in dissipating the upsurge pressures caused by turbine load rejection. Also, during the downsurge cycle, the orifice damps the oscillation and helps to prevent the shaft emptying and reduces the air influx problem.

Conical or variable section surge tanks may be of either the simple or orifice type. These types of shaft again aid in reducing the probability of air entry on downsurge.

The differential surge tank is a combination of the orifice and simple designs. For rapid valve closures, on load rejection, the benefits of the orifice shaft are available, the central tank being designed to overspill into the larger outer shaft. In the valve opening case, on load acceptance water is directly available in the central shaft with no orifice restriction to assist in accelerating the penstock flow. This is, later, supplemented by flow from the outer shaft via the base orifices.

The final design of surge tank for a given situation depends on the frequency of any particular surge-producing system operation and in each case an analysis of the likely advantages of each design should be investigated, preferably with the aid of an analog or digital computer.

20.2. Control of surge following pump shut down

Referring to Fig. 20.1, negative surge problems arise if the pump fails, thus generating a negative pressure wave relative to the steady-state pressure of the system. The effect of this is to reduce the pressure of the system and, if this falls to gas release or vapour pressure, then column separation may occur with potentially dangerous resurge pressures being generated on collapse of the vapour cavity. In order to limit these pressure rises, the principle followed is to increase the fluid pressure by passing fluid from some control device into the pipeline as soon as the pressure in the pipeline falls low enough to indicate possible column separation. There are a number of possible methods here. Obviously, the air vessel discussed in Section 20.1 can be employed if it is positioned close to the potential low pressure region; normally, such a vessel is mounted at the pump discharge. The equations outlined for the application of the air vessel to relief of a positive surge apply without modification. An alternative solution is to introduce an inwards relief valve, which could pass either air or fluid into the pipeline as soon as pipeline pressure falls. Figure 20.6 illustrates such a valve employed on an aircraft refuelling system to pass

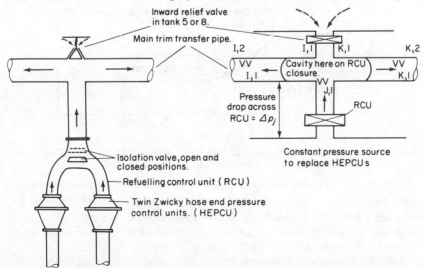

Figure 20.6   Comparison between the twin hose supply to the Concorde RCU and the layout assumed in the analysis

relief air or fuel into the pipeline following the emergency shut down of the main refuelling valve, which could occur under aircraft power failure conditions. The necessary equations are similar to those for the air vessel, the main divergence, or simplification, being that the pressure outside the relief valve can usually be taken as a constant, normally atmospheric pressure. Figure 20.7 illustrates a comparison between measured pressure traces for

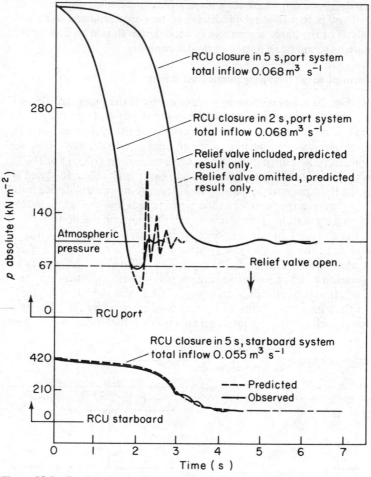

Figure 20.7   Predicted and observed pressure variations downstream of the Concorde RCU during and after its closure in an all-tanks-refuelling case

such a valve and an analysis based on the techniques described in Chapter 19, including the effects of air release. In this application such a valve is acceptable because there is no danger of contamination, as there would be if such a system were used to introduce air into a water supply main.

A similar device for use in conjunction with pump shut down would be a pump bypass pipe, linking the pump discharge to the pump supply reservoir and incorporating a non-return valve (*see* Fig. 20.8).

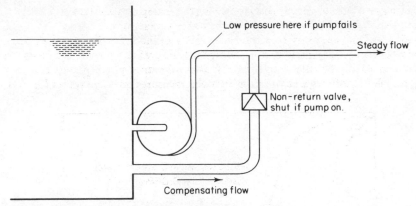

Figure 20.8   Pump bypass system to supply compensating flow if pump fails and low pressure region forms on pump discharge

Column separation, caused by main pipeline pressure falling to gas release or fluid vapour pressure, can occur at any point in the pipe system. It is more likely, however, to be sited close to the pump or closing valve. One prime site for column separation is a high point in the pipe profile (Fig. 20.1), e.g. where the pipe goes over the top of a hill, thus reducing the fluid static pressure. If analysis of the system indicates that separation may occur at such positions, then an air vessel or inwards relief valve may be positioned at these locations. The introduction of air through such a relief valve, or from a badly sized air vessel, can lead to system priming problems when the pumps are restarted so that, as a general rule, the introduction of air should be avoided.

Thus, it may be seen that there are a variety of surge suppression devices available. Obviously there are conditions that suit certain devices and care must be taken to choose the correct installations.

20.3. Further reading

Thorley, A. R. D. and Enever, K. J. E. (1979). *Control and Suppression of pressure surges in pipelines and tunnels*, CIRIA Report 84, London.

PART VI
Fluid machinery

Energy may exist in various forms. Hydraulic energy is that which may be possessed by a fluid. It may be in the form of kinetic, pressure, potential, strain or thermal energy. Mechanical energy is that which is associated with moving or rotating parts of machines, usually transmitting power. It is thus the purpose of hydraulic machines to transform energy either from mechanical to hydraulic or from hydraulic to mechanical. This distinction, based on the 'direction' of energy transfer, forms the basis of grouping hydraulic machines into two distinct categories. All machines in which hydraulic energy forms the input and is transformed into mechanical energy, so that the output is in the form of a rotating shaft or a moving part of a machine, are known as turbines or motors. In the other category, the input is mechanical, the transfer is from mechanical into hydraulic energy and the output is in the form of a moving fluid, sometimes compressed and at elevated temperature. Such machines are called pumps, fans and compressors. Thus, in the first category, work is done by the fluid and energy is subtracted from it, whereas, in the second category of machines, the work is done on the fluid and energy is added to it.

However, sometimes fluids because of their characteristic properties are used by some machines as media to form a link in the energy transfer chain. In an hydraulic coupling, for example, mechanical energy is transformed into hydraulic, only to be changed back into mechanical in the other half of the coupling. There is no gain in mechanical advantage, but, because of the fluid properties and the type of fluid flow in the coupling, a smooth and gradual transfer of power is made possible.

The action of an hydraulic coupling is *rotodynamic* as distinct from positive displacement, which is characteristic of, say, an hydraulic jack. Thus, the principle of machine operation affords further means of its classification and is quite independent of the direction of energy transfer.

In *positive displacement* machines, fluid is drawn or forced into a finite space bounded by mechanical parts and is then sealed in it by some mechanical means. The fluid is then forced out or allowed to flow out from the 'space' and the cycle is repeated. Thus, in positive displacement machines,

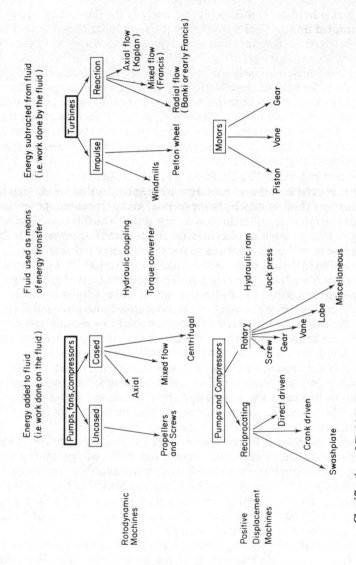

Classification of Fluid Machines

the fluid flow is intermittent or fluctuating to a greater or lesser extent and the flow rate of the fluid is governed by the dimensions of the 'space' in the machine and by the frequency with which it is filled and emptied.

In rotodynamic machines, there is a free passage of fluid between the inlet and outlet of the machine without any intermittent 'sealing' taking place. All rotodynamic machines have a rotating part called a runner, impeller or rotor, which is able to rotate continuously and freely in the fluid, allowing an uninterrupted flow of fluid through it at the same time. Thus, the transfer of energy between the rotor and fluid is continuous and is a result of the rate of change of angular momentum.

These two criteria, namely the direction of energy transfer and the type of action, form the basis of classification of hydraulic machines, as shown opposite. From it, it will be seen that pumps and compressors increase the energy of the fluid and may be either rotodynamic or positive displacement. Fans are always rotodynamic. In turbines the work is done by the fluid and the action is rotodynamic, whereas motors are positive displacement machines also receiving energy from the fluid.

Note that in this Part, all fluid velocities are, in fact, mean velocities. However, for convenience we have not included the bars over the symbols which had been conventional in the previous Parts of this text.

21 Theory of rotodynamic machines

21.1. Introduction

This chapter is concerned with flow through the rotodynamic machines and the relationships between the rate of fluid flow and the difference in total head across the impeller. Both are related to the type of machine under consideration and, hence, the geometric parameters of the impeller. However, certain fundamental relationships may be arrived at by considerations of angular momentum applied to a simplified, or idealized, impeller.

All rotodynamic machines, as previously stated, have a rotating part called the impeller, through which the fluid flow is continuous.

The direction of fluid flow in relation to the plane of impeller rotation distinguishes different classes of rotodynamic machines. One possibility is for the flow to be perpendicular to the impeller and, hence, along its axis of rotation, as shown in Fig. 21.1(a). Machines of this kind are called *axial flow*

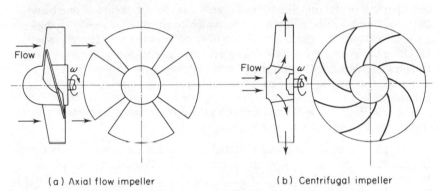

(a) Axial flow impeller   (b) Centrifugal impeller

Figure 21.1   Axial flow and centrifugal impellers

machines. In *centrifugal* machines (sometimes called 'radial flow'), although the fluid approaches the impeller axially, it turns at the machine's inlet so that the flow through the impeller is in the plane of the impeller rotation. This is shown in Fig. 21.1(b). *Mixed flow* machines constitute a third category. They derive their name from the fact that the flow through their impellers is partly axial and partly radial. Figure 21.2(a) shows a mixed flow fan impeller looking at it from the discharge side. It should be noticed that the hub is conical, thus the direction of flow leaving the impeller is somewhere between the axial and radial. Figure 21.2(b) and (c) show two other types of impeller.

Both pumps and turbines can be axial flow, mixed flow or radial flow. In the case of pumps, the latter are normally referred to as centrifugal. All impellers consist of a supporting disc or cylinder and blades attached to it. It is the motion of the blades which is related to the motion of the fluid, one doing work on the other or vice versa. In any case, there are forces exerted on the blades and, since they rotate with the impeller, torque is transmitted because of the rate of change of angular momentum.

It was shown in Section 5.7 that forces on moving vanes may be determined by considering the velocity triangles which represent, vectorially, the absolute velocity of the fluid, the velocity of the vane and the relative velocity between the two. However, it was justifiably assumed in Section 5.7 that the fluid velocity is the same across the jet. Also, the analysis was confined to the changes of kinetic energy of the fluid only, as the free jet is under atmospheric pressure throughout and, hence, there is no change in the pressure energy of the fluid. This is not the case when the fluid flows through blade passages of an impeller. Nevertheless, the understanding of the analysis of Section 5.7 is a necessary prerequisite to the considerations of this chapter.

21.2. One-dimensional theory

The real flow through an impeller is three-dimensional, that is to say the velocity of the fluid is a function of three positional coordinates, say, in the cylindrical system, r, θ and z, as shown in Fig. 21.3. Thus, there is a variation of velocity not only along the radius but also across the blade passage in any plane parallel to the impeller rotation, say from the upper side of one blade to the underside of the adjacent blade, which constitutes an abrupt change — a discontinuity. Also, there is variation of velocity in the meridional plane, i.e. along the axis of the impeller. The velocity distribution is, therefore, very complex and dependent upon the number of blades, their shapes and thicknesses, as well as on the width of the impeller and its variation with radius.

The one-dimensional theory simplifies the problem very considerably by making the following assumptions.

(i) The blades are infinitely thin and the pressure difference across them is replaced by imaginary body forces acting on the fluid and producing torque.

(ii) The number of blades is infinitely large, so that the variation of velocity across blade passages is reduced and tends to zero. This assumption is equivalent to stipulating *axisymmetrical* flow, in which there is perfect symmetry with regard to the axis of impeller rotation. Thus,

$$\frac{\partial v}{\partial \theta} = 0.$$

(iii) Over that part of the impeller where transfer of energy takes place (blade passages) there is no variation of velocity in the meridional plane, i.e. across the width of the impeller. Thus,

$$\frac{\partial v}{\partial z} = 0.$$

The result of these assumptions is that whereas, in reality,

$$v = f(r, \theta, z),$$

for the one-dimensional flow

$$v_\infty = f(r) \text{ only.}$$

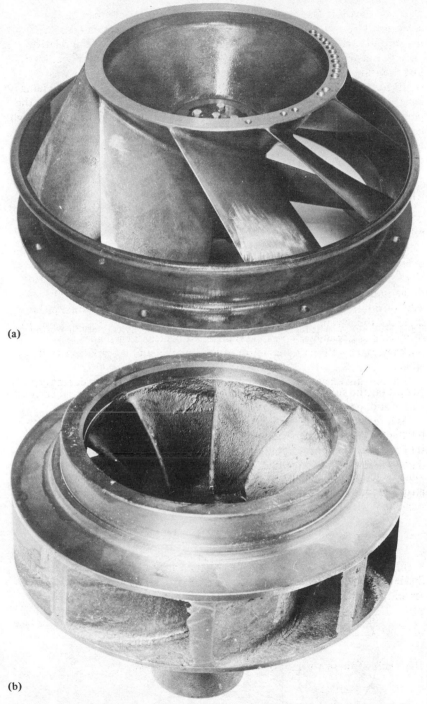

(a)

(b)

Figure 21.2   (a) A mixed flow fan impeller. (By courtesy of Airscrew-Howden Ltd)
(b) A centrifugal pump impeller (shrouded). (By courtesy of Worthington-Simpson Ltd)

Figure 21.2 (c) A centrifugal pump impeller (unshrouded). (By courtesy of Worthington-Simpson Ltd)

Note, that the suffix ∞ stipulates the assumption of an infinite number of blades and, hence, axisymmetry.

As a result, the flow through, say, a centrifugal impeller may be represented by a diagram such as Fig. 21.4. Although finite blades are shown, they are not taken into account in the theory. Furthermore assumption (ii) implies that the fluid streamlines are confined to infinitely narrow interblade passages, and, hence, their paths are congruent with the shape of the interblade centreline, shown by a chain-dotted line. Thus, the flow of fluid

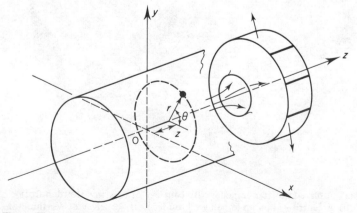

Figure 21.3   A centrifugal impeller in relation to cylindrical coordinates

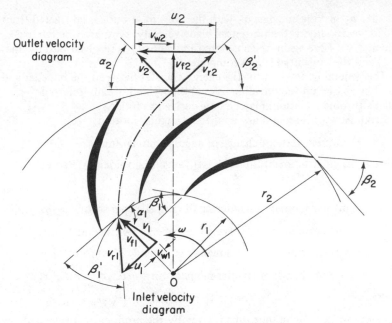

Figure 21.4   One-dimensional flow through a centrifugal impeller

through an impeller passage may be regarded as a flow of fluid particles along the centreline of the interblade passage.

The assumptions of the theory enable us to limit our analysis to changes of conditions which occur between impeller inlet and impeller outlet without reference to the space in between, where the real transfer of energy takes place. This space is treated as a 'black box' having an input in the form of an inlet velocity triangle and an output in the form of the outlet velocity triangle. Such velocity triangles for a centrifugal impeller rotating with a constant angular velocity ω are shown in Fig. 21.4.

At inlet, the fluid moving with an absolute velocity v_1 enters the impeller through a cylindrical surface of radius r_1 and may make an angle α_1 with the tangent at that radius. At outlet, the fluid leaves the impeller through a cylindrical surface of radius r_2, absolute velocity v_2 inclined to the tangent at outlet by the angle α_2.

The velocity triangles shown in Fig. 21.4 are obtained as follows. The inlet velocity triangle is constructed by firstly drawing the vector representing the absolute velocity v_1 at an angle α_1. The tangential velocity of the impeller, u_1 is then subtracted from it vectorially in order to obtain v_{r1}, the relative velocity of the fluid with respect to the impeller blades at the radius r_1. In this basic velocity triangle, the absolute velocity v_1 is resolved into two components: one in the radial direction, called velocity of flow v_{f_1}, and the other, perpendicular to it and, hence, in the tangential direction v_{w_1}, sometimes called velocity of whirl. These two components are useful in the analysis and, therefore, they are always shown as part of the velocity triangles.

Similarly, the outlet velocity triangle consists of the absolute fluid velocity v_2 making an angle α_2 with the tangent at outlet, subtracted from which, vectorially, is the tangential blade velocity u_2 to give the relative velocity v_{r_2}. Here again, the absolute fluid velocity is resolved into the radial (v_{f_2}) and the tangential (v_{w_2}) components.

The general expression for the energy transfer between the impeller and the fluid, based on the one-dimensional theory and usually referred to as Euler's turbine equation, may be now derived as follows.

From Newton's second law applied to angular motion,

Torque = Rate of change of angular momentum.

Now,   Angular momentum = (Mass) (Tangential velocity) (Radius).

Therefore,

Angular momentum entering the impeller per second = $\dot{m} v_{w1} r_1$,

Angular momentum leaving the impeller per second = $\dot{m} v_{w2} r_2$,

in which $\dot{m}$ is the mass of fluid flowing per second. Therefore,

Rate of change of angular momentum = $\dot{m} v_{w2} r_2 - \dot{m} v_{w1} r_1$,

so that   Torque transmitted = $\dot{m}(v_{w2} r_2 - v_{w1} r_1)$.

Since the work done in unit time is given by the product of torque and angular velocity,

Work done per second = (Torque) $\omega = \dot{m}(v_{w2} r_2 - v_{w1} r_1)\, \omega$,

but $\omega = u/r$, so that $\omega r_2 = u_2$ and $\omega r_1 = u_1$. Hence, on substitution,

Work done per second, $E_t = \dot{m}(u_2 v_{w2} - u_1 v_{w1})$.   (21.1)

The SI units of the above expression are joules per second or watts.

Since the work done per second by the impeller on the fluid, such as in this case, is the rate of energy transfer then:

Rate of energy transfer/Unit mass of fluid flowing, $Y = gE = E_t/\dot{m}$,

The product $gE = Y$ known as specific energy is of significance in the case of pumps and fans. The units of Y are joules per kilogramme.

From the specific energy, the Euler's head E is given by:

$$E = (1/g)(u_2 v_{w2} - u_1 v_{w1}).$$ (21.2)

The units of this equation are joules per kilogramme divided by m/s^2. This, of course, simplifies to metres and, therefore, is the same as all the terms of Bernoulli's equation and, consequently, E may be used in conjunction with it. Equation (21.2) is known as Euler's equation. From its mode of derivation it is apparent that Euler's equation applies to a pump (as derived) and to a turbine. In the latter case, however, since $u_1 v_{w1} > u_2 v_{w2}$, E would be negative, indicating the reversed direction of energy transfer. It is, therefore, common for a turbine to use the reversed order of terms in the brackets to yield positive E. Since the units of E reduce to metres of the fluid handled, it is often referred to as Euler's head and, in the case of pumps or fans, it represents the ideal theoretical head developed H_{th}.

It is useful to express Euler's head in terms of the absolute fluid velocities rather than their components. From velocity triangles of Fig. 21.4,

$$v_{w1} = v_1 \cos \alpha_1 \quad \text{and} \quad v_{w2} = v_2 \cos \alpha_2,$$

so that $E = (1/g)(u_2 v_2 \cos \alpha_2 - u_1 v_1 \cos \alpha_1),$ (21.3)

but, using the cosine rule,

$$v_{r1}^2 = u_1^2 + v_1^2 - 2u_1 v_1 \cos \alpha_1,$$

so that $u_1 v_1 \cos \alpha_1 = \tfrac{1}{2}(u_1^2 - v_{r1}^2 + v_1^2).$

Similarly,

$$u_2 v_2 \cos \alpha_2 = \tfrac{1}{2}(u_2^2 - v_{r2}^2 + v_2^2).$$

Substituting into (21.3),

$$E = (1/2g)(u_2^2 - u_1^2 + v_2^2 - v_1^2 + v_{r1}^2 - v_{r2}^2)$$

and $E = (v_2^2 - v_1^2)/2g + (u_2^2 - u_1^2)/2g + (v_{r1}^2 - v_{r2}^2)/2g.$ (21.4)

In this expression, the first term denotes the increase of the kinetic energy of the fluid in the impeller. The second term represents the energy used in setting the fluid into a circular motion about the impeller axis (forced vortex). The third term is the regain of static head due to reduction of relative velocity in the fluid passing through the impeller.

Let us now consider the application of Euler's equation to centrifugal and axial flow machines.

In the former case, the velocity triangles are as shown in Fig. 21.4 and, in addition, the following relationships hold. Since, in general, $u = \omega r$, it follows that the tangential blade velocities at inlet and outlet are given by:

$$\left.\begin{aligned} u_1 &= \omega r_1, \\ u_2 &= \omega r_2. \end{aligned}\right\}$$ (21.5)

Since the flow at inlet and outlet is through cylindrical surfaces and the velocity components normal to them are v_{f_1} and v_{f_2}, the continuity equation applied to inlet and outlet for the mass flow $\dot{m}$ and infinitely thin blades gives:

$$\dot{m} = \rho_1 2\pi r_1 b_1 v_{f_1} = \rho_2 2\pi r_2 b_2 v_{f_2},$$ (21.6)

where b_1 and b_2 are the impeller widths, as shown in Fig. 21.5, and ρ_1 and ρ_2 are the inlet and outlet densities, respectively. For incompressible flow, equation (21.6) simplifies to

$$r_1 b_1 v_{f_1} = r_2 b_2 v_{f_2}.$$ (21.7)

Now, assuming that m, ω, r_1 and r_2 are known, the following arguments are usually employed in order to draw the velocity triangles.

At inlet the usual assumptions are as follows.

(i) The absolute velocity is radial, therefore

$$v_1 = v_{f_1} \quad \text{and} \quad v_{w1} = 0,$$

hence, v_1 is calculated from equation (21.6) and $\alpha_1 = 90°$. If this condition

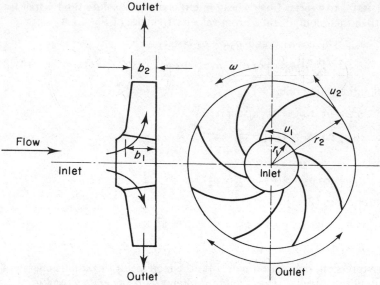

Figure 21.5   A centrifugal pump or fan impeller

does not apply, which only occurs if there is a prewhirl (v_{w1}) component present, perhaps due to inlet vanes or unfavourable inlet conditions, then v_{f_1} is calculated from equation (21.6) and α_1 can be determined only if v_{w1} is known.

(ii) The blade angle at inlet β_1 is such that the blade meets the relative velocity tangentially. Thus, $\beta_1 = \beta'_1$. This assumption is known as the 'no shock' condition and is applied in determining the blade inlet angle during design in order to minimize the entry loss.

Thus the inlet triangles may be drawn.

For the outlet triangles, it is assumed that the fluid leaves the impeller with a relative velocity tangential to the blade at outlet. Thus, $\beta'_2 = \beta_2$ and, in order to draw the outlet velocity triangles, β_2 must be known. The direction of v_{r_2} is then drawn, as well as the v_{f_2} vector, which is radial and whose magnitude is calculated from equation (21.6). It is, thus, possible to draw the u_2 ($= \omega r_2$) vector perpendicular to v_{f_2} and starting from the intersection with the direction of v_{r_2}. The absolute velocity v_2 is then obtained by completing the triangle, from which,

$$\cot \beta_2 = (u_2 - v_{w2})/v_{f_2},$$

so that $v_{w2} = u_2 - v_{f_2} \cot \beta_2.$

Substituting this into Euler's equation, and remembering that $v_{w1} = 0$, the following expression is obtained:

$$E = \left(\frac{u_2}{g}\right)(u_2 - v_{f_2} \cot \beta_2). \tag{21.8}$$

The total amount of energy transferred by the impeller is, thus,

$$E_t = \dot{m}gE = \dot{m}u_2(u_2 - v_{f_2} \cot \beta_2) \tag{21.9}$$

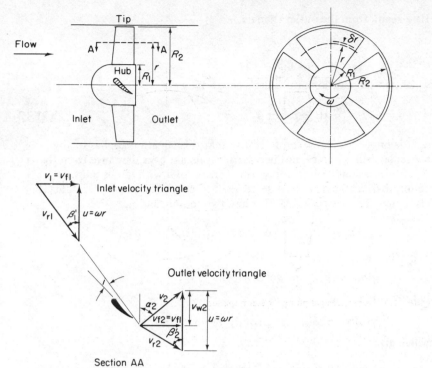

Figure 21.6   Axial flow impeller and velocity triangles

Consider now an axial flow machine, as shown in Fig. 21.6. The important difference between the axial flow machine and the centrifugal is that since, in the former, the flow is axial, the changes from inlet to outlet take place at the same radius and, hence,

$$u_1 = u_2 = u = \omega r. \tag{21.10}$$

Also, since the flow area is the same at inlet and outlet,

$$v_{f_1} = v_{f_2} = v_f$$

and is obtained from

$$\dot{m} = \rho v_f \pi (R_2^2 - R_1^2). \tag{21.11}$$

The following assumptions are made with regard to the velocity triangles.

(i) There is no prewhirl at inlet and, hence,

$$\alpha_1 = 90°, \quad v_{w_1} = 0, \quad v_1 = v_f.$$

(ii) For 'no shock' condition, the blade is set at an angle such that it meets the relative fluid velocity tangentially or at an appropriate angle of incidence for an aerofoil section.

(iii) At outlet, the relative velocity leaves the blade tangentially and a similar procedure to that for a centrifugal impeller is used to complete the velocity triangles.

Here again, from the outlet triangle,

$$\cot \beta_2' = (u - v_{w2})/v_f,$$

so that $v_{w2} = u - v_f \cot \beta_2',$

which, on substitution into Euler's equation, gives

$$E = \left(\frac{u}{g}\right)(u - v_f \cot \beta_2'). \tag{21.12}$$

It is important, however, to realize that this equation applies to any particular radius r and is not necessarily constant over the range from R_1 to R_2. For this condition to apply, the increase of u with radius must be counterbalanced by an equal decrease of $v_f \cot \beta_2'$. Since v_f = constant, the blade must be twisted so that, for any two radii r_a and r_b,

$$u_a^2 - u_a v_f \cot \beta_{2a}' = \text{constant} = u_b^2 - u_b v_f \cot \beta_{2b}'.$$

Rearranging,

$$v_f(u_b \cot \beta_{2b}' - u_a \cot \beta_{2a}') = u_b^2 - u_a^2,$$

but $u = \omega r$, so that

$$v_f(\omega r_b \cot \beta_{2b}' - \omega r_a \cot \beta_{2a}') = \omega^2 (r_b^2 - r_a^2),$$

which gives

$$r_b \cot \beta_{2b}' - r_a \cot \beta_{2a}' = (\omega/v_f)(r_b^2 - r_a^2). \tag{21.13}$$

However, this condition, known as the 'free vortex' design, is not always met. It is then necessary to apply Euler's equation to an element dr and to integrate from R_1 to R_2 as follows.

The energy transfer for the element

$$dE_t = (u/g)(u - v_f \cot \beta_2') \, dW.$$

However,

$$dW = 2\pi \rho g v_f r \, dr$$

and $u = \omega r$

so that $$E_t = 2\pi \rho \omega v_f \int_{R_1}^{R_2} r^2 (\omega r - v_f \cot \beta_2') \, dr, \tag{21.14}$$

in which $\beta_2' = f(r)$ must be known.

Example 21.1

An axial flow fan has a hub diameter of 1·50 m and a tip diameter 2·0 m. It rotates at 18 rad s^{-1} and, when handling 5·0 m^3 s^{-1} of air, develops a theoretical head equivalent to 17 mm of water. Determine the blade outlet and inlet angles at the hub and at the tip. Assume that the velocity of flow is independent of radius and that the energy transfer per unit length of blade (δr) is constant. Take the density of air as 1·2 kg m^{-3} and the density of water as 10^3 kg m^{-3}.

Solution

Velocity of flow, $v_f = \dfrac{Q}{A} = \dfrac{Q}{\pi(R_2^2 - R_1^2)} = \dfrac{5}{\pi(1 - 0.5625)} = 3.64 \text{ m s}^{-1}$.

Blade velocity at tip is given by

$$u_t = \omega R_2 = 18 \times 1 = 18 \text{ m s}^{-1}$$

and, at hub,

$$u_h = \omega R_1 = 18 \times 0.75 = 13.5 \text{ m s}^{-1}.$$

Since, for 'no shock' condition, $\cot \beta_1 = u/v_f$ the inlet blade angle at tip is given by

$$\beta_{1t} = \text{arc cot } \dfrac{18}{3.64} = 11.4°$$

and, at hub,

$$\beta_{1h} = \text{arc cot } \dfrac{13.5}{3.64} = 15.1°.$$

Since the head generated by the tip and hub sections is the same, the outlet angles may be obtained by applying Euler's equation to these sections. From equation (21.12),

$$E = (u/g)(u - v_f \cot \beta_2),$$

but $E = H_{th} = 17 \text{ mm of water} = 0.017 \times 10^3/1.2 = 14.16 \text{ m of air.}$

Therefore at tip,

$$14.16 = (18/9.81)(18 - 3.64 \cot \beta_{2t}),$$

$$\beta_{2t} = 19.5°$$

and at hub,

$$14.16 = (13.5/9.81)(13.5 - 3.64 \cot \beta_{2h}),$$

$$\beta_{2h} = 48.6°.$$

21.3 Isolated blade and cascade considerations

The major assumption underlying the considerations of the previous section was that of an infinite number of blades in the impeller. In practice, of course, the number of blades is finite and their spacing depends on a particular impeller design and, therefore, may vary considerably. The distance between the adjacent blades, s is known as *pitch*, whereas the ratio of blade chord to the pitch,

$$\sigma = c/s, \tag{21.15}$$

is called *blade solidity* and is a measure of the closeness of blades. If the blades are very close to each other, the passages between them may be treated as conduits and the flow through them treated accordingly as bounded flow. If, however, the blades are very far apart, they must be treated as bodies in an external flow, provided their mutual interference is negligible. Thus, the two extremes of high solidity and zero solidity provide well-defined situations requiring reasonably clear-cut treatment, but in real machines neither of them applies. Furthermore, what makes the treatment more difficult is the fact that in many impellers the blade solidity varies with radius. This is the case in axial flow impellers, where blades near the tip are much further apart than.they are near the hub.

In this section we will first consider relationships which result from assuming that blades are far apart, so that $s \to \infty$ and, hence, $\sigma = 0$. For such cases the theories expanded in Chapter 10 become directly applicable. We will then look at blades close together and arranged into cascades, for which a modified approach will be necessary.

It was shown in Chapter 10 that the lift is dependent upon circulation around the lifting surface or body and, also, that it is related to the pressure distribution around it:

$$L = \int_0^{2\pi} p \sin \theta \ d\theta = \tfrac{1}{2} C_L \rho U_0^2 A,$$

but also $L' = \rho U_0 \Gamma$ (from equation (7.67)),

where L' is the lift per unit length of the body, so that

$$L = \rho U_0 \Gamma l,$$

where l is the length of the body. Thus, there is a direct relationship between the pressure distribution and, hence, the resultant force on the body, which may be an impeller blade, and the circulation around it, namely

$$\rho U_0 \Gamma l = \int_0^{2\pi} p \sin \theta \ d\theta,$$

which leads to

$$\Gamma = \tfrac{1}{2} C_L U_0 A/l. \tag{21.16}$$

It is, therefore, possible to relate Euler's equation to the circulation as follows.

Consider the circulation around a single blade as shown in Fig. 21.7:

$$\Gamma_{ABCD} = \oint v \ ds = \int_A^B v \ ds + \int_B^C v \ dl + \int_C^D v \ ds + \int_D^A v \ dl.$$

However,

$$\int_B^C v \ dl = -\int_D^A v \ dl, \qquad \int_A^B v \ ds = -v_{w1} s_1 \quad \text{and} \quad \int_C^D v \ ds = v_{w2} s_2.$$

Therefore, if we denote the circulation around the blade by $\Gamma_b = \Gamma_{ABCD}$, then,

$$\Gamma_b = s_2 v_{w2} - s_1 v_{w1}. \tag{21.17}$$

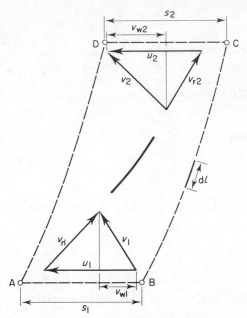

Figure 21.7   Circulation around the blade

If we now consider two adjacent blades, the circulation around them may be obtained by considering Fig. 21.8:

$$\Gamma = \Gamma_{ACDF} = \int_A^B v\ ds + \int_B^C v\ ds + \int_C^D v\ ds + \int_D^E v\ ds + \int_E^F v\ ds + \int_F^A v\ ds.$$

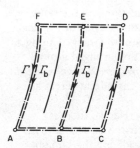

Figure 21.8   Circulation around two blades

However,

$$\int_C^D v\ ds = -\int_F^A v\ ds \quad \text{and} \quad \int_A^B v\ ds + \int_E^F v\ ds = \Gamma_b.$$

Also, $$\int_B^C v\ ds + \int_D^E v\ ds = \Gamma_b.$$

Therefore,

$$\Gamma = \Gamma_b + \Gamma_b = 2\Gamma_b.$$

This result may be generalized for a number of blades z into:

$$\Gamma = z\Gamma_b \qquad (21.18)$$

and, substituting from equation (21.18), we obtain

$$\Gamma = z(s_2 v_{w2} - s_1 v_{w1}),$$

but $zs_1 = 2\pi r_1$ and $zs_2 = 2\pi r_2$, so that

$$\Gamma = 2\pi(r_2 v_{w2} - r_1 v_{w1}). \qquad (21.19)$$

However, Euler's equation states that

$$E = (1/g)(u_2 v_{w2} - u_1 v_{w1}) = (\omega/g)(r_2 v_{w2} - r_1 v_{w1}),$$

so that, comparing the two equations, we obtain

$$\Gamma/2\pi = Eg/\omega$$

or $\qquad E = (\omega/g)(\Gamma/2\pi).$ $\qquad (21.20)$

However, in terms of individual blade circulation, this equation becomes

$$E = (\omega/g)(z\Gamma_b/2\pi). \qquad (21.21)$$

This equation may be used in conjunction with equation (21.16), which relates circulation to the coefficient of lift in an ideal situation, since the latter equation is based on Kutta–Joukowski's potential flow analysis.

The isolated blade approach, as described above, has some application to axial flow impellers in which solidity is small, such as propellers or axial flow fans used in cooling towers.

When solidity is significant, a different approach is required. An arrangement of geometrically identical blades such that they are the same distance from one another and are positioned in the same way with respect to the direction of flow is called a *cascade*. If the blades are arranged along a straight line, the cascade is called *straight* and, if they are arranged around the circumference of a circle, they are referred to as *circular*. A development of an axial flow impeller constitutes a moving straight cascade, whereas a centrifugal impeller is a rotating circular cascade.

The main purpose of cascades is to deflect the flow. Hence, there is always a change of momentum across a cascade and a force associated with it. If the velocities upstream and downstream of a cascade are the same in magnitude, there will be a change of momentum due to change in direction, but it follows from Bernoulli's equation that there will be no pressure difference between the upstream and the downstream sides of the cascade. It is then known as an *impulse* cascade. If the pressure difference exists due to absolute velocities not being the same, the cascade is called *reaction*. Those reaction cascades in which fluid is accelerated (fall of pressure) are usually used in turbines, whereas those in which the fluid is decelerated and, hence, there is an increase of pressure, are used in pumps and compressors.

Consider, now, a case of a stationary straight cascade of height Z. Let the upstream fluid velocity v_1 making an angle α_1 with the line of the

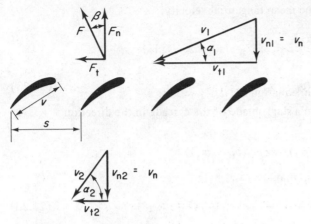

Figure 21.9   A straight cascade

cascade be deflected so that the downstream velocity v_2 makes an angle α_2, as shown in Fig. 21.9. Now the flow *deflection* is

$$\epsilon = \alpha_2 - \alpha_1$$

and it is an important characteristic of a cascade.

The fluid velocities v_1 and v_2 may be resolved into components parallel and normal to the cascade, v_t and v_n, respectively. Since there is a difference of velocity, the pressure change (assuming no loss and using Bernoulli's equation) is given by

$$p_1 - p_2 = \tfrac{1}{2}\rho(v_2^2 - v_1^2). \tag{21.22}$$

However, since, by the continuity equation, the mass flow through the cascade is

$$\dot{m} = sZ\rho_1 v_{n1} = sZ\rho_2 v_{n2},$$

where Z is the height of the cascade, it follows that for incompressible flow,

$$v_{n1} = v_{n2} = v_n.$$

Thus, the change in velocity is entirely due to the change of the tangential velocity component v_t, which also follows from substitution of the following relationships into equation (21.22):

$$v_2^2 = v_{t2}^2 + v_n^2,$$
$$v_1^2 = v_{t1}^2 + v_n^2,$$

from which,

$$v_2^2 - v_1^2 = v_{t2}^2 - v_{t1}^2$$

and equation (21.22) becomes

$$p_1 - p_2 = \tfrac{1}{2}\rho(v_{t2}^2 - v_{t1}^2).$$

If we now introduce the mean tangential velocity,

$$v_t = \tfrac{1}{2}(v_{t_1} + v_{t_2}),$$

we obtain

$$p_1 - p_2 = \rho v_t (v_{t_2} - v_{t_1}) \tag{21.23}$$

and the force acting on a single blade of the cascade in the direction perpendicular to it is

$$F_n = sZ(p_1 - p_2) = sZ\rho v_t(v_{t_2} - v_{t_1}),$$

but $s(v_{t_2} - v_{t_1}) = \Gamma_b$, so that

$$F_n = \rho Z v_t \Gamma_b. \tag{21.24}$$

Now the rate of change of momentum across the cascade is again due to a change in v_t and, therefore, gives rise to a force in the direction of the cascade, F_t. So,

$$F_t = \dot{m}(v_{t_2} - v_{t_1}) = sZ\rho v_n(v_{t_2} - v_{t_1}),$$

from which

$$F_t = \rho Z v_n \Gamma_b. \tag{21.25}$$

The resultant force on the blade, therefore, is

$$F = \sqrt{(F_n^2 + F_t^2)} = \rho Z \Gamma_b \sqrt{(v_t^2 + v_n^2)} \tag{21.26}$$

The direction of this force is given by the angle β such that

$$\cot \beta = F_n/F_t = \rho Z v_t \Gamma_b / \rho Z v_n \Gamma_b = v_t/v_n$$

but $v_t = \tfrac{1}{2}(v_{t_1} + v_{t_2})$, which may be expressed in terms of inlet and outlet angles α_1 and α_2 by using the trigonometric relationships

$$v_{t_1} = v_n \cot \alpha_1 \quad \text{and} \quad v_{t_2} = v_n \cot \alpha_2.$$

Therefore,

$$v_t = \tfrac{1}{2}v_n(\cot \alpha_1 + \cot \alpha_2)$$

and $\qquad \cot \beta = \tfrac{1}{2}(\cot \alpha_1 + \cot \alpha_2) \tag{21.27}$

It also follows from simple geometry that β is equal to β_∞, defined as the mean direction of flow and obtained by superposition of the inlet and outlet velocity triangles, as shown in Fig. 21.10. Thus the force F being perpendicular to the mean direction of flow is the lift on the blade.

It is now possible to consider an axial flow impeller by selecting an annular element δr at a radius r (Fig. 21.10) and developing it into a moving cascade having velocity $u = \omega r$.

In this configuration the velocity with which the fluid approaches the

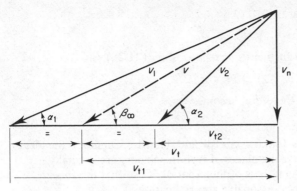

Figure 21.10   Combined inlet and outlet velocity triangles for a straight cascade

cascade is the relative velocity v_{r1} and takes the place of v_1 in the stationary cascade. Similarly

$$\left.\begin{aligned}
v_2 \text{ is replaced by } v_{r2}, \\
v_n \text{ is replaced by } v_f, \\
v_{t1} \text{ is replaced by } u, \\
v_{t2} \text{ is replaced by } (u - v_{w2}).
\end{aligned}\right\} \tag{21.28}$$

The full velocity triangles for the usual case of $v_{w1} = 0$ and, therefore, $v_1 = v_f$ are shown in Fig. 21.11. Note that the aerofoil blade is set at an

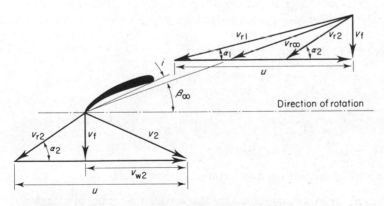

Figure 21.11   Velocity triangles for an axial flow impeller

angle of incidence i with respect to the mean flow direction. Thus the angle it makes with the direction of rotation, known as the blade *stagger* (or blade pitch) angle is $(i + \beta_\infty)$.

We can now obtain an expression for lift, which will then enable us to determine the energy transfer realized by a cascade formed by such blade elements. This may be obtained by first eliminating from equation (21.26) the term under the square root:

$$v_t^2 + v_n^2 = \{(v_{t1} + v_{t2})/2\}^2 + v_n^2$$

which may be modified further using relationships (21.28):

$$v_t^2 + v_n^2 = (u + v_{w2}/2)^2 + v_f^2 = v_{r\infty}^2.$$

The expression for lift, therefore, becomes

$$L = \rho Z \Gamma_b v_{r\infty}. \tag{21.29}$$

It is analogous to Kutta–Joukowski's law and, likewise, applies to the ideal (no losses) flow. However the blades of turbomachinery are generally efficient and the lift/drag ratio is of the order of 50, so that the above equation may be used as a reasonable approximation.

The lift given by the above expression is in terms of the blade circulation, which may be related to cascade data by equating equation (21.29) to equation (10.5),

$$\rho Z \Gamma_b v_r = \tfrac{1}{2} C_L \rho A v_{r\infty}^2,$$

but $A = cZ$, so that

$$\Gamma_b = \tfrac{1}{2} C_L c v_{r\infty}. \tag{21.30}$$

It is now possible to relate the cascade lift coefficient C_L to the blade angle at inlet α_1, at outlet α_2 and the mean angle β_∞. From equation (21.17),

$$\Gamma_b = s(v_{w2} - v_{w1}),$$

but $v_{w1} = 0$ and, from the outlet velocity triangle (Fig. 21.11),

$$v_{w2} = u - v_f \cot \alpha_2,$$

so that $\Gamma_b = s(u - v_f \cot \alpha_2).$

Equating this expression to equation (21.30),

$$s(u - v_f \cot \alpha_2) = \tfrac{1}{2} C_L c v_{r\infty};$$

dividing by v_f,

$$s(u/v_f - \cot \alpha_2) = \tfrac{1}{2} C_L c \, v_{r\infty}/v_f.$$

But $u/v_f = \cot \alpha_1$ and $v_{r\infty}/v_f = 1/\sin \beta_\infty$, therefore,

$$s(\cot \alpha_1 - \cot \alpha_2) = C_L c/2 \sin \beta_\infty,$$

so that $C_L = 2(s/c)(\cot \alpha_1 - \cot \alpha_2) \sin \beta_\infty.$ \hfill (21.31)

It is important to remember that the above equations apply to cascades and not to isolated aerofoils, because the theory is based on change of momentum of the fluid stream due to change of its direction (deflection) which only cascades achieve. Cascade data are obtained experimentally and a typical set may be as shown in Fig. 21.12.

The alternative approach is to use the isolated aerofoil data but corrected by the use of a *cascade coefficient K*, defined as

$$K = \frac{\text{Cascade lift coefficient}}{\text{Aerofoil lift coefficient}}. \tag{21.32}$$

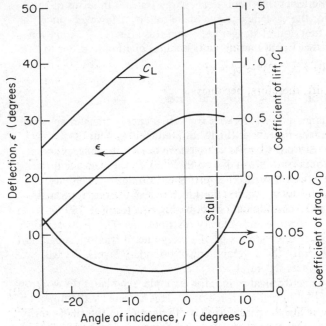

Figure 21.12   Typical cascade data

This approach is particularly appropriate for axial flow machines in which solidity varies considerably between hub and tip, but the values of K must be determined experimentally. Generally, K depends on the extent of overlap between adjacent blades, which is a function of solidity and the stagger angle.

The energy transfer occurring within a moving cascade may be obtained by considering the work done per second in the direction of motion, which, for a rotating impeller, is the plane of rotation.

$$\dot{W}_d = L \sin \beta_\infty u.$$

However, the lift for a radial element δr is

$$L = \tfrac{1}{2} C_L \rho c v_{r\infty}^2 \delta r,$$

so that $\dot{W}_d = \tfrac{1}{2} C_L \rho c u v_{r\infty}^2 \sin \beta_\infty \delta r.$

Now, the weight of fluid flowing per second through the elemental depth δr is

$$\rho g s v_f \, \delta r$$

and, hence, the energy transfer is

$$E = \tfrac{1}{2} C_L \rho c u v_{r\infty}^2 \sin \beta_\infty \delta r / \rho g s v_f \, \delta r$$

$$= \frac{1}{2g} C_L \frac{c}{s} \frac{u}{v_f} v_{r\infty}^2 \sin \beta_\infty,$$

but $c/s = \sigma$ and $\sin \beta_\infty = v_f/v_{r\infty}$ and, hence,

$$E = C_L \sigma \, v_{r\infty} u/2g. \tag{21.33}$$

The above expression gives the theoretical energy transfer in terms of the coefficient of lift for the cascade and the blade solidity. However, since the lift coefficient is a function of stagger angle and this affects $v_{r\infty}$, equation (21.33) has to be used in conjunction with Euler's equation in order to obtain a design solution.

21.4. Departures from Euler's theory and losses

There are two fundamental reasons why the actual energy transfer achieved by an hydraulic machine is smaller than that predicted by Euler's equation. The first reason is that the velocities in the blade passages and at the impeller outlet are not uniform due to the presence of blades and the real flow being three-dimensional. This results in a diminished velocity of whirl component and, hence, reduces the Euler's head. This effect is not caused by friction and, therefore, does not represent a loss but follows from ideal flow analysis of pressure and velocity distributions. The second reason is that in a real impeller there are losses of energy due to friction, separation and wakes associated with the development of boundary layers. We will consider these two effects separately.

In an impeller of a centrifugal pump, for example, the blades do work on the fluid by exerting an 'impelling' force on it. This is done by the upper or forward surface of the blade. It follows that the pressure in the fluid on this side of the blade is greater than that on the other side, as indicated in Fig. 21.13. Hence, the velocity near the back side of the blade is greater than that near the forward side. This difference of velocity on the two sides of the blade gives rise to the blade circulation Γ_b associated with the lift. But the

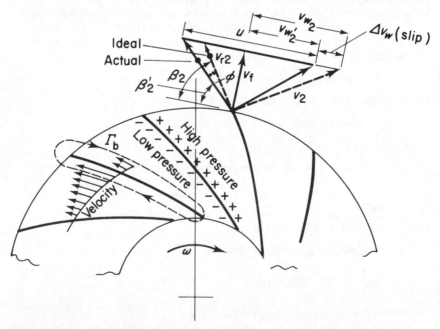

Figure 21.13   Effect of velocity distribution on the outlet velocity triangle

non-uniform velocity distribution is responsible for the mean direction of flow leaving the impeller being $\beta_2' = (\beta_2 - \phi)$ and not β_2, as assumed in the ideal flow situation. This effect is responsible for *deviation* and results in the reduction of the all-important tangential velocity component (velocity of whirl) from v_{w2} to v_{w2}', the reduction being Δv_w and called *slip*. Thus, the *slip factor* may be defined as:

$$S_F = v_{w2}'/v_{w2} \tag{21.34}$$

and the real velocity triangle is as shown by the full lines in Fig. 21.13, whereas that corresponding to the ideal case is shown by dashed lines.

There have been many attempts to predict the amount of slip and a number of theories have been formulated. The earliest, due to Stodola, stipulates the existence of a 'relative eddy' (shown in Fig. 21.14) which

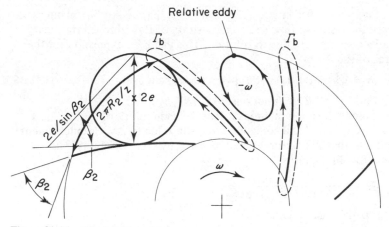

Figure 21.14   The 'relative eddy'

occurs between the adjacent blades. Since a frictionless fluid passes through the impeller without rotation, it must also leave the impeller without rotation. However, the impeller rotates at an angular velocity ω, which means that the fluid must have a rotation relative to the impeller of $(-\omega)$, which is the relative eddy. Stodola assumed that if the radius of a circle which may be inscribed between the two adjacent blades at outlet is e, then the slip may be considered to be the product of the relative eddy ω and e. Thus,

$$\Delta v_w = \omega e. \tag{21.35}$$

But the impeller circumference at outlet is $2\pi R_2$ and, hence, for z blades e may be obtained approximately from

$$2e/\sin \beta_2 \cong 2\pi R_2/z,$$

$$e = (\pi R_2/z) \sin \beta_2.$$

Substituting into equation (21.35) and taking $\omega = u_2/R_2$,

$$\Delta v_w = \frac{u_2}{R_2} \times \frac{\pi R_2 \sin \beta_2}{z} = \frac{u_2 \pi}{z} \sin \beta_2.$$

Also, since

$$v_{w2} = u_2 - v_f \cot \beta_2,$$

it follows that the slip factor may be expressed as

$$S_F = \frac{v_{w2} - \Delta v_w}{v_{w2}} = 1 - \frac{\Delta v_{w2}}{v_{w2}} = 1 - \frac{u_2 \pi \sin \beta_2}{z(u_2 - v_f \cot \beta_2)},$$

which becomes

$$S_F = 1 - \frac{\pi \sin \beta_2}{z[1 - (v_f/u_2) \cot \beta_2]}. \tag{21.36}$$

It is interesting to note that for an impeller with radial blades at the tip ($\beta_2 = 90°$) Stodola's slip factor reduces to

$$S_F = 1 - \pi/z. \tag{21.37}$$

The best known of the more exact theories is that due to Busemann, who considered the resultant flow as a superposition of flow through stationary vanes and a displacement due to rotation. This yields an expression for the slip factor of the form

$$S_F = [A - B(v_f/u_2) \cot \beta_2] [1 - (v_f/u_2) \cot \beta_2], \tag{21.38}$$

in which both A and B are functions of R_2/R_1, β_2 and z.

Stanitz used relaxation methods (blade-to-blade solution) for impellers having $45° < \beta_2 < 90°$ and concluded that slip velocity Δv_w is independent of β_2 and also that slip factor was not affected by compressibility. His expression for slip factor is

$$S_F = 1 - \frac{0 \cdot 63\pi}{z[1 - (v_f/u_2) \cot \beta_2]} \tag{21.39}$$

which, for radial blades, reduces to

$$S_F = 1 - 0 \cdot 63\pi/z. \tag{21.40}$$

On the whole, for pumps, the best agreement with experimental results is obtained in the following ranges of β_2:

$20° < \beta_2 < 30°$, Stodola's correction;

$30° < \beta_2 < 80°$, Busemann's correction;

$80° < \beta_2 < 90°$, Stanitz' correction.

The effects of losses due to friction, separation, wakes, etc., are a completely different issue, although of equal importance. In a cascade these losses manifest themselves as a drop in pressure downstream of the cascade. Assuming that the velocity remains unchanged, the actual pressure difference across the cascade becomes

$$p_2 - p_1 = (\rho/2)(v_1^2 - v_2^2) - \Delta p, \tag{21.41}$$

where Δp is the pressure loss in the cascade. This increased pressure difference affects the normal force component on the cascade and, hence, the resultant force, as shown in Fig. 21.15, which refers to a pump. In this case, assuming the same inlet pressure p_1 and the same inlet and outlet velocities, it is only the outlet pressure which is affected. Without losses, the

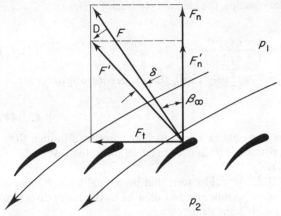

Figure 21.15 Effect of pressure losses on forces on a cascade

pump would generate greater pressure, so that the ideal pressure at outlet p_2' would be greater than p_2 actually achieved. Thus,

$$p_2 = p_2' - \Delta p.$$

It follows, therefore, that the actual force normal to the cascade is

$$F_n' = sZ(p_2 - p_1) = sZ(p_2' - p_1 - \Delta p),$$

whereas the theoretical or ideal force would be

$$F_n = sZ(p_2' - p_1).$$

These forces are shown in Fig. 21.15. The actual resultant force F' makes an angle $(\delta + \beta_\infty)$ with the normal. It is, therefore, no longer perpendicular to the mean direction of flow and, hence, is not equal to the lift. The angle δ is characteristic of the cascade efficiency, which may be defined as

$$\eta_c = \frac{p_2' - p_1 - \Delta p}{p_2' - p_1} = \frac{F_n'}{F_n}, \tag{21.42}$$

but $F_n' = F_t \cot(\beta_\infty + \delta)$ and $F_n = F_t \cot \beta_\infty$ so that

$$\eta_c = \frac{\cot(\beta_\infty + \delta)}{\cot \beta_\infty} = \frac{\cot \beta_\infty \cot \delta - 1}{\cot \beta_\infty (\cot \beta_\infty + \cot \delta)}.$$

This can be simplified to

$$\eta_c = \frac{1 - \tan \beta_\infty \tan \delta}{1 + \tan \delta \cot \beta_\infty}.$$

Now, let $\tan \delta = \epsilon$,

$$\tan \beta_\infty = \frac{v_f}{u - v_{w2}/2} = \frac{v_f}{u^*}$$

where $u^* = u - v_{w2}/2$ and

$$\cot \beta_\infty = \frac{u - v_{w2}/2}{v_f} = \frac{u^*}{v_f}.$$

Substituting the above relationships, the cascade (or blade) efficiency becomes

$$\eta_c = \frac{1 - \epsilon \tan \beta_\infty}{1 + \epsilon \cot \beta_\infty} = \frac{1 - \epsilon v_f/u^*}{1 + \epsilon u^*/v_f}.$$ (21.43)

Since, in practice, ϵ and v_f/u^* are fairly small, this expression approximates to

$$\eta_c = 1 - \epsilon u^*/v_f.$$ (21.44)

An identical expression may be obtained for a turbine cascade, although the initial reasoning regarding pressure differences must relate to the reversed direction of energy transfer.

Referring back to Fig. 21.15, it will be seen that the actual force F' may be resolved into components perpendicular and parallel to the mean direction of flow, thus giving the actual lift and drag:

$$L = F' \sin \delta,$$

$$D = F' \cos \delta.$$

It is, thus, apparent that the losses in a cascade are related to the drag.

Example 21.2

The impeller of a centrifugal pump has a diameter of 0·1 m and axial width at outlet of 15 mm. There are 16 blades swept backwards and inclined at $25°$ to the tangent to the periphery. The flow rate through the impeller is

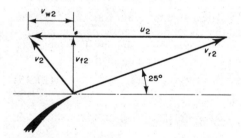

Figure 21.16

8·5 m³ h⁻¹ when it rotates at 750 rev min⁻¹. Calculate the head developed by the pump when handling water and assuming (a) one-dimensional ideal flow theory, (b) allowing for the relative eddy between the blades.

Solution

(a) Area at outlet, $A = \pi Dt = \pi \times 0·1 \times 15 \times 10^{-3} = 4·71 \times 10^{-3}$ m².

Velocity of flow at outlet, $v_{f2} = \dfrac{Q}{A} = \dfrac{8·5 \times 10^3}{3600 \times 4·71} = 0·501$ m s⁻¹.

Blade velocity at outlet, $u_2 = \dfrac{\pi ND}{60} = \dfrac{\pi \times 750 \times 0·1}{60} = 3·97$ m s⁻¹.

But, from the outlet velocity triangle,

$$\tan 25° = v_{f_2}/(u_2 - v_{w2})$$

and so $v_{w2} = u_2 - v_{f_2} \cot \beta_2 = 3.97 - 0.501 \cot 25°$

$$= 2.9 \text{ m s}^{-1}.$$

The theoretical head is given by Euler's equation:

$$H_{th} = E_{th} = \frac{u_2 v_{w2}}{g} = \frac{3.97 \times 2.9}{9.81} = 1.17 \text{ m of water.}$$

(b) Since $\beta_2 = 25°$, Stodola's slip factor may be used to allow for the relative eddy between the blades:

$$S_F = 1 - \frac{\pi \sin \beta_2}{z[1 - (v_{f_2}/u_2) \cot \beta_2]} = 0.886.$$

The head developed is

$$H = S_F H_{th} = 0.886 \times 1.17 = 1.04 \text{ m of water.}$$

21.5. Compressible flow through rotodynamic machines

Compressible flow involves appreciable changes of fluid density and, therefore, occurs most frequently in gases. Since, for gases, changes of density are related to changes of pressure and temperature by the equation of state, it is useful to make use of some of the fundamental thermodynamic concepts and relationships.

The first law of thermodynamics leads to the establishment of the steady-flow energy equation:

$$\dot{Q} - \dot{W}_d = \dot{m}(H_2 - H_1) + \tfrac{1}{2}(v_2^2 - v_1^2) + g(z_2 - z_1), \tag{21.45}$$

where $\dot{Q}$ = heat transfer per second, $\dot{W}_d$ = work done per second, $\dot{m}$ = mass flow per second and H is the specific enthalpy. Hence, assuming constant specific heat c_p,

$$H_2 - H_1 = c_p(T_2 - T_1). \tag{21.46}$$

For gases the term $g(z_2 - z_1)$ is small and, therefore, may be ignored. Rearranging equation (21.45) we obtain

$$\dot{Q} - \dot{W}_d = \dot{m}[(H_2 + \tfrac{1}{2}v_2^2) - (H_1 + \tfrac{1}{2}v_1^2)].$$

Now, we define stagnation enthalpy by

$$H_T = H + \tfrac{1}{2}v^2, \tag{21.47}$$

where

$$H_{T_2} = H_2 + \tfrac{1}{2}v_2^2, \quad H_{T_1} = H_1 + \tfrac{1}{2}v_1^2 \quad \text{and} \quad H_{T_2} - H_{T_1} = c_p(T_{T_2} - T_{T_1}) \tag{21.48}$$

in which T_T is the stagnation (or total) temperature and c_p is assumed to be constant. However, for most turbomachinery the flow processes are very nearly adiabatic and, therefore, it is justifiable to write $Q = 0$. As a result,

$$\dot{W}_d = \dot{m}(H_{T_1} - H_{T_2}) \text{ for turbines} \tag{21.49}$$

and $$\dot{W}_d = \dot{m}(H_{T_2} - H_{T_1}) \text{ for compressors.} \tag{21.50}$$

These processes may be most usefully represented in an enthalpy/entropy diagram (as shown in Fig. 21.17). For a compressor producing total pressure

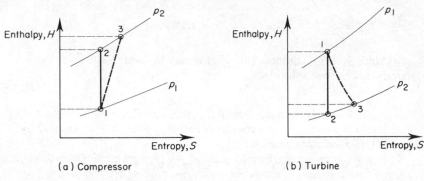

(a) Compressor   (b) Turbine

Figure 21.17   Enthalpy changes in turbomachinery

rise from p_1 to p_2, the increase of enthalpy corresponding to the assumption of no losses and, hence, of a reversible or isentropic process is represented by a straight line $1 - 2$ (S = constant). The actual process is not, however, isentropic because of losses due to friction and, therefore, the true increase of enthalpy is along line $1 - 3$ ($S_3 > S_1$). Thus, it is possible to define isentropic efficiency as

$$\eta_i = \frac{\text{Isentropic work}}{\text{Actual work}} = \frac{H_{T_2} - H_{T_1}}{H_{T_3} - H_{T_1}}. \tag{21.51}$$

Similar reasoning applied to a turbine (Fig. 21.17(b)) leads to the isentropic efficiency being expressed as

$$\eta_i = \frac{H_{T_1} - H_{T_3}}{H_{T_1} - H_{T_2}}. \tag{21.52}$$

It is now possible to relate the change of enthalpy in the impeller to the velocities of the fluid flowing through it by making use of Euler's equation. From equation (21.50), the work done per unit mass flow is

$$\dot{W}_d/\dot{m} = H_{T_2} - H_{T_1} \text{ for a compressor.}$$

But, from Euler's equation, the energy transfer per unit weight of fluid flowing is

$$E = (1/g)(u_2 v_{w2} - u_1 v_{w1})$$

and, therefore, the work done per unit mass of fluid flowing or specific energy is:

$$y = \dot{W}_d/\dot{m} = Eg = (u_2 v_{w2} - u_1 v_{w1}).$$

Thus, equating the two expressions, we obtain

$$H_{T_2} - H_{T_1} = u_2 v_{w_2} - u_1 v_{w_1}.$$

Using the usual assumption of no whirl at inlet ($v_{w_1} = 0$), the above equation reduces to:

$$H_{T_2} - H_{T_1} = u_2 v_{w_2}. \tag{21.53}$$

This relationship does not take into account any losses due to friction. If, however, the actual work is equated to Euler's equation instead of the isentropic work, we obtain

$$H_{T_3} - H_{T_1} = u_2 v_{w_2}.$$

Hence, by using equation (21.51), we get

$$u_2 v_{w_2} = (H_{T_2} - H_{T_1})/\eta_i. \tag{21.54}$$

This equation allows for the fact that the given impeller speed and fluid whirl velocity produce greater change of enthalpy than that anticipated by the ideal, frictionless conditions.

The problem now arises how to account for changes in velocities through the impeller or around the impeller blades and in their lift and drag properties caused by compressibility. It is possible to use as an approximation the Prandtl–Glauert similarity rules, which apply to an isolated aerofoil at small angles of incidence. They state that the aerodynamic performance in compressible flow may be related to that in the incompressible flow by the factor:

$$\lambda = (1 - Ma_0^2)^{-1/2}, \tag{21.55}$$

where Ma_0 is the Mach number of the free stream. The factor λ is obtained by a transformation of a linearized compressible-flow second-order differential equation in velocity potential to Laplace's equation for incompressible, two-dimensional, steady-potential flow and is beyond the scope of this book.

Thus, for the same aerofoil, if u is the velocity at any point on its profile in an incompressible flow and the velocity at the same point in the compressible flow is u', then

$$u'/u = (1 - Ma_0^2)^{-1/2}$$

or $\quad u' = \lambda u.$ $\hspace{4cm}$ (21.56)

Similarly,

$$C_L'/C_L = (1 - Ma_0^2)^{-1/2}$$

or $\quad C_L' \doteq \lambda C_L.$ $\hspace{4cm}$ (21.57)

It is also possible to consider two aerofoils, one in an incompressible flow (suffix i) and the other in the compressible flow (suffix c), such that their geometries are related by

$\hspace{2cm}$ (Camber)$_c = \lambda$(Camber)$_i$ $\hspace{3cm}$ (21.58)

and $\hspace{1.2cm}$ (Maximum thickness)$_c = \lambda$ (Maximum thickness)$_i$. $\hspace{1cm}$ (21.59)

For such related aerofoils, similarity rules due to Goethert are

$$\alpha_c = \lambda \alpha_i \text{ for angles of incidence,} \qquad (21.60)$$

$$(C_L)_c = \lambda^2 (C_L)_i \text{ for lift coefficients.} \qquad (21.61)$$

The Prandtl–Glauert and Goethert rules apply only to two-dimensional flow and may be used for $Ma < 0.8$. This limitation corresponds approximately to a likelihood of the Mach number becoming equal to unity at some point on the aerofoil. This will occur at a point of minimum pressure and, hence, maximum velocity. When the Mach number exceeds unity at such a point, a shock will be formed and, hence, the assumption of potential flow on which the Prandtl–Glauert similarity is based ceases to be valid.

Formation of shock causes immediate increase in losses because it constitutes an adverse pressure gradient and, therefore, induces boundary layer separation.

It is also possible for the velocity at some cross-section of the interblade passage to become equal to that of sound. When that happens, the flow becomes choked and it is not possible to increase the flow rate beyond the critical value corresponding to the choked flow.

Example 21.3

A two-dimensional aerofoil has a ratio of maximum thickness to chord equal to 0·035 and the camber to chord ratio of 0·015. It is tested in a low-speed wind tunnel and gives a lift coefficient of 0·6 at an angle of incidence of 3°. What would be the lift coefficient of this aerofoil at $Ma = 0.6$? What would be the geometric characteristics and the lift coefficient of a related aerofoil at $Ma = 0.6$?

Solution
Assume that the conditions at the low-speed wind tunnel correspond to incompressible flow. For the same aerofoil, equation (21.57) holds and, therefore,

$$C_L' = \lambda C_L.$$
$$\lambda = (1 - Ma^2)^{-1/2} = (1 - (0.6)^2)^{-1/2} = 1.25,$$

therefore,

$$C_L' = 1.25 \times 0.6 = \mathbf{0.75}.$$

A similar aerofoil would have a thickness/chord ratio of

$$0.035 \times (1 - (0.6)^2)^{-1/2} = \mathbf{0.043\ 75}$$

and a camber/chord ratio of

$$0.015 \times (1 - (0.6)^2)^{-1/2} = \mathbf{0.018\ 75}.$$

Its coefficient of lift would be

$$(C_L)_c = 0.6/(1 - Ma^2) = 0.6/0.64 = \mathbf{0.9375}$$

at an angle of incidence

$$\alpha_c = 3/(1 - Ma^2)^{1/2} = \mathbf{3.75°}.$$

21.6. Further reading

Betz, A. (1966). *Introduction to the Theory of Flow Machines*. Pergamon Press, New York.

Dixon, S. L. (1966). *Fluid Mechanics, Thermodynamics of Machinery*. Pergamon Press.

Ferguson, T. B. (1963). *The Centrifugal Compressor Stage*. Butterworth, London.

Lazarkiewicz, S. and Troskolanski, A. F. (1965). *Impeller Pumps*. Pergamon Press.

Shepherd, D. G. (1956). *Principles of Turbomachinery* Macmillan, London.

EXERCISES 21

21.1 A centrifugal fan delivering 2 m³/s of air (density 1·2 kg/m³) runs at 960 rev/min. The impeller outer diameter is 70 cm and inner diameter is 48 cm. The impeller width at inlet is 16 cm and is designed for constant radial flow velocity. The blades are backward inclined making angles of 22·5° and 50° with the tangents at inlet and outlet respectively. Draw the inlet and outlet velocity triangles and determine the theoretical head produced by the impeller.
[91·1 m of air]

21.2 A centrifugal pump delivers 0·3 m³/s of water at 1400 rev/min. The total effective head is 20 m. The impeller is 30 cm in diameter and 32 mm wide at exit, and is designed for constant velocity of flow. Both the suction and delivery pipes have the same bore. Calculate the following vane angles:

(*a*) for the impeller vanes at exit
(*b*) for entry to the stationary guide vanes surrounding the impeller.
[37°, 47·7°]

21.3 A centrifugal fan supplies air at a rate of 4·5 m³/s and total head of 100 mm of water. The outer dia. of the impeller is 50 cm and outer width is 18 cm. The blades are backward inclined and of negligible thickness. If the fan runs at 1800 rev/min and assuming that the conversion of velocity head to pressure head in the volute is counter balanced by the friction losses there and in the runner determine the blade angle at outlet. Assume zero whirl at inlet and take air density as 1·23 kg/m³.
[27·8°]

21.4 Starting from Euler's equation for the theoretical total head developed by a centrifugal fan, show that if the whirl component of the absolute velocity at inlet is zero, then the relationship between the theoretical head H_{th}, fan discharge Q and the blade angle at outlet β_2 is of the form.

$$H_{th} = A - BQ \cot \beta_2$$

where A and B are constants for a given fan speed, impeller diameter and width of outlet.

A centrifugal fan housing a 76 cm diameter impeller and rotating 960 rev/min is to deliver 155 m³/min at 75 mm total water column. If the air enters the impeller radially and the width of the impeller at outlet is

12 cm, determine the required blade angle at outlet. Assume that 45% of the theoretical head is dissipated as impeller and casing losses and take the density of air as $1 \cdot 25$ kg/m^3.
[$41°12'$]

21.5   When working at its best efficiency point, the blading at the mean radius, equal to 300 mm, of an axial flow pump deflects a stream approaching it at a relative angle of 60° to the axis, through 15°, so that the water leaves it at a relative angle of 45°. Assuming the water approaches it axially, and that the velocity of flow remains constant, draw the inlet and outlet velocity triangles under these conditions for a rotational speed of 600 rev/min and calculate the theoretical total head rise through the impeller.
[$15 \cdot 25$ m]

21.6   An axial flow pump operates at 500 rev/min. The outer diameter of the impeller is 750 mm and the hub diameter is 400 mm. At the mean blade radius, the inlet blade angle is 12° and the outlet blade angle is 15° both measured with respect to the plane of impeller rotation. Sketch the corresponding velocity diagrams at inlet and outlet and estimate from them (*a*) the head generated by the pump, (*b*) the rate of flow through the pump, (*c*) the shaft power consumed by the pump. Assume hydraulic efficiency of 87% and overall efficiency of 70%.
[$4 \cdot 12$ m, $1 \cdot 01$ m^3/s, $58 \cdot 1$ kW]

21.7   Show that for a 'free vortex' flow through an axial flow impeller the circulation round the blade does not vary with radius. An axial flow fan delivers $20 \cdot 0$ m^3/s of air and its major parameters are: rotational speed $N = 720$ rev/min; impeller diameter $D_2 = 1 \cdot 00$ m; hub diameter $D_1 = 0 \cdot 45$ m; number of blades $Z = 10$. The blades are of aerofoil cross-section which, for the optimum angle of incidence $i = 5°$, have the lift coefficient $C_L = 0 \cdot 80$ and the chord at hub is $C_h = 70$ mm. Using the isolated blade theory and assuming 'free vortex' flow and constant velocity of flow, determine the total head rise across the impeller, the blade angle and chord length at tip and the blade angular twist between hub and tip.
[$12 \cdot 38$m of air, $48 \cdot 7°$, 51mm, $11 \cdot 7°$]

21.8   A $0 \cdot 9$ m dia. axial flow fan having a $0 \cdot 45$ m dia. hub rotating at 720 rev/min delivers $4 \cdot 7$ m^3/s of air. The pitch/cord ratio at the mean diameter is $1 \cdot 4$. At this section the chord makes an angle of 22° with the direction of rotation and the angle of incidence is 4°. Determine the total head developed if, at the existing incidence, the blade profile has a lift coefficient of $0 \cdot 9$ and a drag coefficient of $0 \cdot 015$. (Density of air = $1 \cdot 2$ kg/m^3.)
[$33 \cdot 4$ mm water]

21.9   A downstream guide vane axial flow pump, $0 \cdot 6$ m diameter and running at 950 rev/min, is to deliver $0 \cdot 75$ m^3/s at a total head of 16 m. If the hub ratio is $0 \cdot 6$ and the blade solidity at hub and tip is $1 \cdot 0$ and $0 \cdot 55$ respectively, determine the blade angles at hub and tip and the guide vane inlet angles. Use the following aerofoil data:

Angle of incidence:	1·0°	4·0°	7·0°	10°	11°
Coefficient of lift:	0·46	0·87	1·16	1·39	stall

[$12°$, $26 \cdot 7°$, $51 \cdot 6°$, $65 \cdot 2°$]

21.10   An axial flow water pump with fixed stator blades downstream from the rotor has the following details:

Outer diameter of rotor = 1·8 m
Inner diameter of rotor = 0·75 m
Rotor blade outlet angle (at mean dia.) = 30°
Fixed blade inlet angle (at mean dia.) = 40°

For a rotor speed of 250 rev/min, given that the whirl velocity upstream from the rotor is zero at all radii, calculate:

(a) the axial velocity for which the incidence angle for the stator blades is zero, for zero deviation at the rotor outlet
(b) the rotor torque, if the axial velocity has the value found in (a) at all radii, and the change of whirl is also independent of the radius
(c) the rotor blade inlet angles for zero incidence at the root and tip.

[5·72 m/s, 52·3 x 10³ mN, 13·6°, 30·2°]

21.11 The impeller of a centrifugal pump rotates at 1450 rev/min and is 0·25 m diameter and 20 mm width at outlet. The blades are inclined backwards at 30° to the tangent at outlet and the whirl slip factor is 0·77. If the volumetric flowrate is 0·028 m³/s and neglecting shock losses and whirl at inlet, find the theoretical head developed by the impeller. Also, using Stodola's model of relative eddy, find the number of blades on the impeller.
[23·7 m, 8 blades]

21.12 The inlet and outlet diameters of a centrifugal fan rotating at 1450 rev/min are 475 mm and 700 mm respectively. The corresponding impeller widths are 190 mm and 145 mm. The performance of the fan is controlled by a series of inlet vanes which are set to produce a whirl component in the direction of rotation such that the relative velocity of air at inlet is 31 m/s making an angle of 15° with the tangent to the blade inlet circle and causing a shock loss of 0·6 $V_1^2/2g$ where V_1 is the absolute velocity of air at inlet. The impeller blades are backward inclined and the inlet and outlet angles are 12° and 38° respectively measured from the tangent.

Assuming that due to slip the actual whirl component at outlet is 0·8 of the theoretical and that the impeller loss is 0·4 of the velocity head at impeller outlet, calculate the total head developed by the impeller and the air flowrate through the fan.
[138·8 m of air, 2·28 m³/s]

21.13 Show that for a centrifugal pump, neglecting losses, the condition for maximum efficiency is

$$u_2 = \frac{2V_{f2}}{\tan \beta_2}$$

where u_2 is the blade peripheral speed at outlet, V_{f2} is the outlet velocity of flow and β_2 is the blade angle at outlet measured with respect to the tangent.

A centrifugal pump with impeller diameter 10 cm and axial width of 1·5 cm has swept back blades inclined at 25° to the tangent to the periphery. If the impeller speed is 12·4 rev/s calculate the flowrate when the pump is operating at maximum efficiency. Assume zero swirl at inlet.
[0·0043 m³/s]

22 Performance of rotodynamic machines

22.1. The concept of performance characteristics

In the introduction to this Part of the book a distinction was made between hydraulic machines in which work is done on the fluid (pumps, fans) and those machines in which work is done by the fluid and, therefore, energy is subtracted from it (turbines, motors). It was further stated that the process of energy transfer may be accomplished by either a 'positive displacement' action or by 'rotodynamic' action. In the first case, a volume of fluid fixed by the dimensions of the machine enters and leaves it at a frequency determined by the speed of operation of the machine. In the second case the flow is continuous through an impeller whose torque is equal to the rate of change of angular momentum of the fluid, as was shown in Chapter 21.

The fluid quantities involved in all hydraulic machines are the flow rate (Q) and the head (H), whereas the mechanical quantities associated with the machine itself are the power (P), speed (N), size (D) and efficiency (η). Although they are all of equal importance, the emphasis placed on certain of these quantities is different for pumps and for turbines. The output of a pump running at a given speed is the flow rate delivered by it and the head developed. Thus, a plot of head against flow rate at constant speed forms the fundamental performance characteristic of a pump. In order to achieve this performance, a power input is required which involves efficiency of energy transfer. Thus, it is useful to plot also the power P and the efficiency η against Q. Such a complete set of performance characteristics of a roto-dynamic pump is shown in Fig. 22.1.

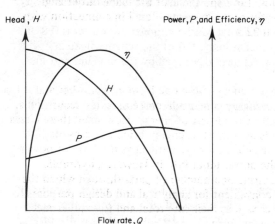

Figure 22.1 Typical pump characteristic

In the case of turbines, the output is the power developed at a given speed and, hence, the fundamental turbine characteristic consists of a plot of power against speed at constant head. The input in this case is the fluid flow rate and, therefore, this quantity as well as the efficiency are usually plotted against the speed to complete the set of turbine characteristics.

The performance characteristics, then, represent in a graphical form the relationships between variables relevant to an hydraulic machine. Each and every hydraulic machine has its own set of characteristics which represent its performance.

For rotodynamic machines the concept of a characteristic follows directly from Euler's equation. It was shown in Chapter 21 (equations (21.8) and (21.12)) that the theoretical energy transfer per unit weight of fluid flowing through the machine, or the fluid head, may be given by

$$H = (u_2/g)(u_2 - v_{f_2} \cot \beta_2).$$ (22.1)

But $v_{f_2} = Q/A_2$, where A_2 is the impeller outlet area. Substituting:

$$H = (u_2/g)[u_2 - (Q/A_2) \cot \beta_2]$$

or $\qquad H = u_2^2/g - (u_2/A_2 g)Q \cot \beta_2.$

This equation may, for a constant speed of rotation and given impeller diameter so that u_2 = constant and A_2 = constant, be rewritten in a general form:

$$H = K_1 - K_2 Q \cot \beta_2.$$ (22.2)

It is, thus, seen that there is a definite functional relationship between the head and flow rate of a rotodynamic machine. This relationship constitutes the performance characteristic and is determined experimentally by performance tests. Thus, a pump, for example, will generate a head dependent upon the quantity of fluid it is handling. Furthermore, since for machines of different design features and different sizes the values of K_1, K_2 and β_2 will be different, their characteristics will also be different.

Equation (22.2) also shows the importance of the blade outlet angle β_2. This particular point will be discussed in greater detail in connection with centrifugal pumps in Section 22.4.

22.2. Losses and efficiencies

All hydraulic machines convert energy from one form into another and it is a well-known fact that, in any energy conversion process, losses occur. Thus, hydraulic machines also suffer from losses of energy. How small these losses are or how good a machine is in converting energy is indicated by its efficiency. Efficiency of a machine is always defined as the ratio of power output of the machine to the power input into it. However, hydraulic machines are complex, consisting of a number of parts through which the fluid moves and, thus, it is convenient for analytical and design purposes to consider component losses as well as their sum total and to express each component loss in the efficiency form.

Let us now consider these component losses one by one. First, the actual energy transfer in a rotodynamic machine occurs in its impeller. Here the fluid passes through the blade passages and either receives energy from the moving blades or imparts energy to them. In any case, there are two major sources of energy loss within the impeller. The inevitable contact between the fluid moving over solid surfaces gives rise to boundary layer development and, hence, to frictional losses, whereas the need for the fluid to change direction often results in separation and, hence, leads to separation (or shock) losses. Both these losses may be augmented by secondary flows which may occur within the impeller due to pressure distribution across it and are usually prominent at off-design points of operation.

Thus if h_i is the head loss in the impeller and Q_i is the volumetric flow rate through the impeller, then the impeller power loss is

$$P_i = \rho g h_i Q_i. \tag{22.3}$$

Now, the flow rate through the impeller Q_i is usually not the same as that flowing through the machine, simply because some fluid passes through clearances between the impeller and the casing. In a pump, for example, of all the fluid passing through the impeller most flows into the discharge end but some passes through the inlet clearance and finds itself passing through the impeller again. Thus, the impeller always handles a greater volume than that discharged by the pump.

If we denote by q the volumetric flow rate leaking past the impeller and if H_i is the total head across the impeller, then the power loss due to the leakage may be expressed as

$$P_1 = \rho g H_i q. \tag{22.4}$$

In most machines the impeller is surrounded by a stationary casing so that the fluid passes through parts of the casing before it enters the impeller and after leaving it. Thus, losses due to friction (and possibly due to separation) occur in the casing as well. If the flow rate through the casing and, thus, through the machine is Q (greater or smaller than Q_i depending whether it is a turbine or a pump, the difference being q) and the loss of head in the casing is h_c, then the power loss in the casing is

$$P_c = \rho g h_c Q. \tag{22.5}$$

Finally, there are mechanical losses of energy such as in the bearings and sealing glands which must be accounted for. It is normal practice in hydraulic machines to include within this category losses due to disc friction, sometimes referred to as 'windage' loss. This is the power required to spin the impeller at the required velocity without any work being done by the impeller or on the impeller by the fluid. This would be possible only if the impeller did not have any blades. Thus, windage loss accounts for the friction between the outer surfaces of the impeller rotating in the fluid surrounding it within the casing.

It is now possible to consider the energy balance for the whole machine, but here we must begin to distinguish between pumps and turbines because what represents the output of one is the input of the other and vice versa. For a pump:

$$P = P_m + \rho g(h_i Q_i + H_i q + h_c Q + HQ). \tag{22.6}$$

Shaft power input	Mechanical loss	Impeller loss	Leakage loss	Casing loss	Useful fluid power
		Hydraulic losses			

For a turbine:

$$\underset{\substack{\text{Fluid} \\ \text{power} \\ \text{input}}}{\rho g H Q} = \underset{\substack{\text{Mechanical} \\ \text{loss}}}{P_m} + \underset{\substack{\text{Impeller} \\ \text{loss}}}{\underbrace{\rho g (h_i Q_i} + \underset{\substack{\text{Casing} \\ \text{loss}}}{h_c Q} + \underset{\substack{\text{Leakage} \\ \text{loss}}}{H_i q)}} + \underset{\substack{\text{Shaft} \\ \text{power} \\ \text{output}}}{P} \qquad (22.7)$$

$$\underbrace{\qquad\qquad\qquad\qquad\qquad\qquad}_{\text{Hydraulic losses}}$$

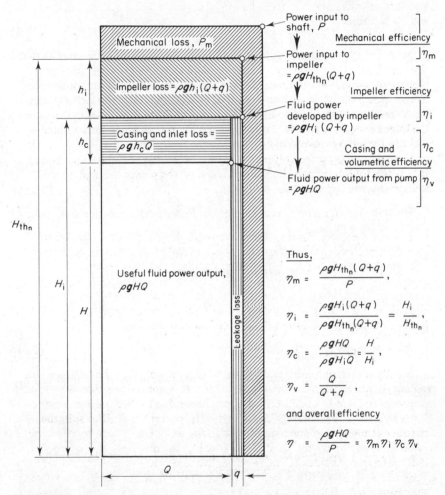

Figure 22.2   Energy balance for a pump and summary of efficiencies

Figure 22.2 represents graphically the energy balance for a pump. A similar diagram may be constructed for a turbine.

Having discussed all the component losses in an hydraulic machine and the complete energy balance, it is now possible to define efficiencies.

The most important is the *overall efficiency*. It refers to the machine as a whole and is, therefore, always plotted as one of the performance characteristics. It is defined as:

$$\eta = \frac{\text{Power output of the machine}}{\text{Power input to the machine}}.$$

Hence, for a pump,

$$\eta = \frac{\text{Fluid power output}}{\text{Power input to shaft}} = \frac{\rho g H Q}{P} \tag{22.8}$$

and for a turbine,

$$\eta = \frac{\text{Power output from shaft}}{\text{Fluid power input}} = \frac{P}{\rho g H Q} \tag{22.9}$$

where H is the actual total head difference between the inlet and outlet flanges of the machine. Thus, the overall efficiency relates the two extreme terms of equations (22.6) and (22.7).

If the mechanical power loss is P_m, then the power input to the impeller is $P - P_m$ and the *mechanical efficiency* may be defined as:

$$\eta_m = (P - P_m)/P \text{ for a pump} \tag{22.10}$$

and $\eta_m = P/(P + P_m)$ for a turbine. \hfill (22.11)

The *impeller efficiency* takes care of the losses in the impeller and, therefore, for a pump,

$$\eta_i = \frac{\text{Fluid power developed by impeller}}{\text{Mechanical power supplied to impeller}}$$

$$= \frac{\rho g H_i Q_i}{(P - P_m)}.$$

But, from equation (22.10), $P - P_m = \eta_m P$, so that

$$\eta_i = \rho g H_i Q_i / \eta_m P. \tag{22.12}$$

Alternatively, the denominator may be expressed in terms of the fluid loss in the impeller, giving

$$\eta_i = \frac{\rho g H_i Q_i}{\rho g H_i Q_i + \rho g h_i Q_i} = \frac{H_i}{H_i + h_i} = \frac{H_i}{H_{th_n}}, \tag{22.13}$$

where $H_i + h_i = H_{th_n}$ is sometimes used and denotes a theoretical total head deduced from the nett power input to the impeller from:

$$\eta_m P = \rho g Q_i (H_i + h_i) = \rho g Q_i H_{th_n}. \tag{22.14}$$

For a turbine,

$$\eta_i = \frac{\text{Mechanical power received by shaft}}{\text{Fluid power supplied to impeller}}$$

$$= (P + P_m)/\rho g H_i Q_i = P/\rho g H_i Q_i \eta_m = H_{th_n}/H_i. \tag{22.15}$$

The *volumetric efficiency* is not always appropriate, for example, in axial flow machines, but is of importance in the case of centrifugal pumps and especially fans. In general, for pumps,

$$Q = Q_i - q \tag{22.16a}$$

and, for turbines,

$$Q = Q_i + q, \tag{22.16b}$$

so that, for pumps, the volumetric efficiency is defined as

$$\eta_v = \frac{\text{Flow rate through machine}}{\text{Flow rate through impeller}}$$

$$= Q/Q_i = Q/(Q + q) \tag{22.17a}$$

and, for turbines,

$$\eta_v = \frac{\text{Flow rate through impeller}}{\text{Flow rate through machine}}$$

$$= Q_i/Q = (Q - q)/Q \tag{22.17b}$$

The *casing efficiency* accounts for the power loss in the casing. For a pump,

$$\eta_c = \frac{\text{Useful fluid power output}}{\text{Fluid power developed by impeller} - \text{Leakage loss}}$$

$$= \frac{\rho g H Q}{\rho g H_i Q_i - \rho g H_i q} = \frac{HQ}{H_i(Q_i - q)} = \frac{H}{H_i}. \tag{22.18a}$$

For a turbine,

$$\eta_c = \frac{\text{Fluid power supplied to impeller} + \text{Leakage loss}}{\text{Fluid power received by casing}}$$

$$= \frac{\rho g H_i Q_i + \rho g H_i q}{\rho g I I Q} = \frac{H_i(Q_i + q)}{HQ} = \frac{H_i}{H}. \tag{22.18b}$$

It is now possible to show that the overall efficiency is equal to the product of all the component efficiencies,

$$\eta = \eta_m \eta_i \eta_c \eta_v, \tag{22.19}$$

by substituting into the above the appropriate expressions as follows:

$$\eta = \frac{P - P_m}{P} \times \frac{\rho g H_i Q_i}{P - P_m} \times \frac{H}{H_i} \times \frac{Q}{Q + q},$$

which simplifies to

$$\eta = \frac{\rho g H Q_i}{P} \times \frac{Q}{Q + q}.$$

But, since $Q + q = Q_i$, we obtain

$$\eta = \rho g H Q / P,$$

which is the expression (22.8) for the overall efficiency.

The internal losses of the machine, i.e. those occurring in the impeller and in the casing due to friction and separation are sometimes called hydraulic losses, which gives rise to the *hydraulic efficiency*, defined as

$$\eta_h = \frac{\text{Actual head}}{\text{Theoretical head}} = \frac{H}{H_{th_n}} = \eta_i \eta_c, \tag{22.20}$$

where H_{th_n} is the theoretical head calculated from the nett power input to the impeller. Thus, equation (22.19) may be rewritten as

$$\eta = \eta_m \eta_h \eta_v. \tag{22.21}$$

The theoretical head calculated from Euler's equation (H_{th}) is not the same as the theoretical head calculated from the nett power input (H_{th_n}) used in equation (22.14). The difference is due to 'head slip' discussed in Chapter 21. Thus, if the slip factor $S_F = v'_{w2}/v_{w2}$ (equation 21.34) is introduced into Euler's equation,

$$H_{th} = u_2 v_{w2}/g$$

(equation (21.2) for $v_{w1} = 0$) then

$$H_{th_n} = u_2 v'_{w2}/g = H_{th} S_F \tag{22.22}$$

Thus, if the slip factor is known, the nett theoretical head H_{th_n} may be calculated from Euler's head.

It is now possible to relate the theoretical characteristic obtained from Euler's equation to the actual characteristic by accounting for various losses responsible for the difference. The theoretical characteristic for a given blade angle at outlet β_2 is a straight line determined by equation (22.2), as shown in Fig. 22.3.

The use of slip factor, which varies with flow rate, enables the H_{th_n}-curve to be obtained. This represents the nett head developed by the impeller, but as discussed earlier, does not account for losses. These, for the machine as a whole, may be considered separately under the following categories.

(i) *Shock losses*, which occur at the entry to the impeller, to the guide vanes, etc., especially at off-design operating conditions, may be simply expressed as

$$h_{sh} = k(Q - Q_N)^2, \tag{22.23}$$

where Q_N is the volumetric flow rate corresponding to the maximum efficiency point on the characteristic. Equation (22.23) assumes, therefore, that shock losses are zero at $Q = Q_N$. It is a parabola, which has a minimum at this point, as shown in Fig. 22.3.

(ii) *Friction losses*, which account for energy dissipation due to contact of the fluid with solid boundaries such as stationary vanes, impeller, casing, etc., are usually expressed in the form

$$h_f = k' Q^2, \tag{22.24}$$

where k' is a constant for a given machine. If it is assumed that when the machine operates at its maximum efficiency the shock losses are zero, then the hydraulic loss of head becomes equal to the friction losses at this point. Thus,

$$(H_{th_n} - H)_{h_{sh}=0} = (h_i + h_c)_{h_{sh}=0} = h_f.$$

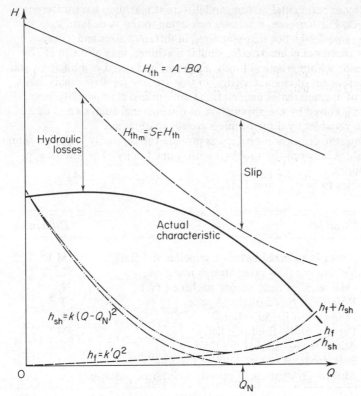

Figure 22.3   Losses and characteristics for a centrifugal fan

This relationship enables the value of k' to be established approximately. Equation (22.24) represents a parabola passing through the origin, also shown in Fig. 22.3.

The sum of the shock and friction losses when subtracted from the H_{th_n} curve gives the actual head/flow rate characteristic provided it is plotted against Q, the flow rate through the machine, and not $(Q + q)$, which represents the flow rate through the impeller. The disparity between these two quantities, defined by the volumetric efficiency, is usually dealt with during the determination of H_{th_n} curve.

(iii) The *mechanical losses*, which usually include 'disc friction' or 'windage' loss due to the rotation of 'bladeless' impeller, i.e. supporting discs or hub only, and bearing losses, do not affect the head/flow rate characteristic but only the power input and, hence, the overall efficiency.

Figure 22.3 shows the actual head/flow rate characteristic obtained by subtracting the hydraulic losses from the H_{th_n} curve.

22.3. Dimensionless coefficients and similarity laws

22.3.1. Dimensionless coefficients

The actual performance characteristics of rotodynamic machines have to be determined by experimental testing, and different machines have different characteristics. Furthermore, machines belonging to the same family, i.e. being of the same design but manufactured in different sizes and, thus, constituting a series of geometrically similar machines, may also run at different speeds within practical limits. Each size and speed combination will produce a unique set of characteristics, so that for one family of machines the number of characteristics needed to be determined is impossibly large. The problem is solved by the application of dimensional analysis and by replacing the variables by dimensionless groups so obtained.

In addition, the dimensionless groups provide the similarity laws governing the relationships between the variables within one family of geometrically similar machines.

The variables to be considered are:

Symbol	Variable	Dimensions
P	Power transferred between impeller and fluid	$M\,L^2\,T^{-3}$
Q	Volumetric flow rate through machine	$L^3\,T^{-1}$
H	Difference of head across machine $(=E)$	L
N	Rotational speed of the impeller	T^{-1}
D	Diameter of the impeller	L
ρ	Density of the fluid handled	$M\,L^{-3}$
μ	Absolute viscosity of the fluid	$M\,L^{-1}\,T^{-1}$
K	Bulk modulus of elasticity of the fluid	$M\,L^{-1}\,T^{-2}$
ϵ	Absolute roughness of machine's internal passages	L

Since the head H is the energy per unit weight of the fluid, it is appropriate to consider (gH) as a variable because it represents energy per unit mass, or specific energy y, which is more fundamental because it is independent of gravitational acceleration. Pumps, for example, develop the same specific energy (or work per unit mass of the fluid flowing) irrespective of the gravitational force.

Considering firstly the specific energy (head developed) as the dependent variable, the relationship between the variables may be written as:

$$gH = \phi(Q, N, D, \rho, \mu, K, \epsilon).$$

Using the indicial method, the power series reduces to:

$$gH = kQ^a N^b D^c \rho^d \mu^e K^f \epsilon^i$$

and substituting dimensions:

$$\frac{L^2}{T^2} = \left(\frac{L^3}{T}\right)^a \left(\frac{1}{T}\right)^b (L)^c \left(\frac{M}{L^3}\right)^d \left(\frac{M}{LT}\right)^e \left(\frac{M}{LT^2}\right)^f (L)^i.$$

Equating indices,

for [M]: $0 = d + e + f$, therefore, $d = -e - f$;

for [T]: $-2 = -a - b - e - 2f$, therefore, $b = 2 - a - e - 2f$;

for [L]: $2 = 3a + c - 3d - e - f - i$, therefore, $c = 2 - 3a - 2e - 2f - i$.

Substituting into the original equation,

$$gH = k\, Q^a N^{(2-a-e-2f)} D^{(2-3a-2e-2f-i)} \rho^{(-e-f)} \mu^e K^f \epsilon^i$$

$$= kN^2 D^2 \left(\frac{Q}{ND^3}\right)^a \left(\frac{\mu}{ND^2 \rho}\right)^e \left(\frac{K}{N^2 D^2 \rho}\right)^f \left(\frac{t}{D}\right)^i.$$

Therefore:

$$\frac{gH}{N^2 D^2} = \phi\left[\left(\frac{Q}{ND^3}\right); \left(\frac{\mu}{ND^2 \rho}\right); \left(\frac{K}{N^2 D^2 \rho}\right); \left(\frac{t}{D}\right)\right]. \qquad (22.25)$$

Now, $gH/N^2 D^2$ is the *head coefficient* K_H, and Q/ND^3 is the *flow coefficient, K_Q*. Also, since $ND \propto u$ it follows that:

$$\frac{\mu}{ND^2 \rho} \propto \frac{\mu}{uD\rho} \propto \frac{1}{Re}$$

where Re is the Reynolds number based on impeller diameter. Also, since $\sqrt{(K/\rho)} = c$ (equation (5.30)) it follows that:

$$\frac{K}{N^2 D^2 \rho} \propto \frac{c^2}{u^2} \propto \frac{1}{Ma}$$

Thus equation (22.25) may be rewritten as

$$K_H = \phi\,(K_Q, Re, Ma, \epsilon/D) \qquad (22.25a)$$

where ϵ/D is the relative roughness of the machine's internal passages.

Similarly, if the power is taken as the dependent variable the relationship between the variables may be written as:

$$P = \phi(Q, N, D, \rho, \mu, K, \epsilon)$$

and leads to:

$$\frac{P}{N^3 D^5 \rho} = \phi\left[\left(\frac{Q}{ND^3}\right); \left(\frac{\mu}{ND^2 \rho}\right); \left(\frac{K}{N^2 D^2 \rho}\right); \left(\frac{\epsilon}{D}\right)\right]. \qquad (22.26)$$

Now, $P/N^3 D^5 \rho$ is the *power coefficient, K_P*, and therefore the above equation may be restated as:

$$K_P = \phi(K_Q, Re, Ma, \epsilon/D) \qquad (22.26a)$$

The functional relationships between K_H, K_P and K_Q are determinable by experiment and constitute a set of performance characteristics, which are of the same shape as the H and P vs. Q characteristics, but represent the whole family of geometrically similar machines and are identical for all such machines if Re, Ma and relative roughness are the same.

22.3.2. Similarity laws

Since for all machines belonging to one family and operating under dynamically similar conditions the dimensionless coefficients are the same at corresponding points of their characteristics, it follows that the similarity laws governing the relationships between such corresponding points may be stated as follows:

since $K_Q = Q/ND^3$ = constant, $Q \propto ND^3$,   (22.27)

since $K_H = gH/N^2D^2$ = constant, $gH \propto N^2D^2$,   (22.28)

since $K_P = P/\rho N^3 D^5$ = constant, $P \propto \rho N^3 D^5$.   (22.29)

provided that Re, Ma, and ϵ/D are also the same. It will also be shown, using a particular example, which follows, that

$$\eta = \text{constant.} \qquad (22.30)$$

To illustrate the way in which the similarity laws are used in predicting the performance of a machine of a given size and running at a given speed from the known performance characteristics of a geometrically similar machine, consider a centrifugal pump whose characteristics when operating at a constant speed N_1 are as shown by the bold lines in Fig. 22.4. Let it be

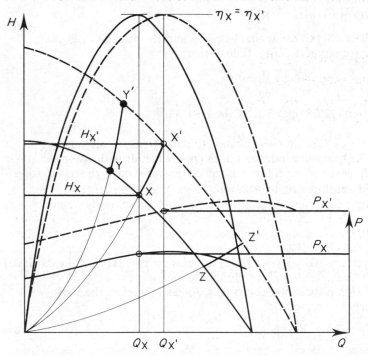

Figure 22.4

required to establish the performance characteristics of the same pump but running at a faster speed N_2.

If, at N_1, the pump is operating at point X such that it delivers Q_X, generates head H_X and consumes power P_X at efficiency η_X, the corresponding point at speed N_2, marked X', will be obtained by applying simultaneously the similarity laws as follows. From equation (22.27),

$$Q_X/N_1D^3 = Q_{X'}/N_2D^3,$$

but, since for the same pump the diameter D is the same,

$$Q_{X'} = Q_X N_2/N_1.$$

Similarly, from equation (22.28),

$$H_{X'} = H_X(N_2/N_1)^2.$$

Thus, plotting $H_{X'}$ against $Q_{X'}$, the point X' on the new characteristic is obtained. The power required at N_2 follows from equation (22.29),

$$P_{X'} = P_X(N_2/N_1),$$

which establishes a point on the power curve.

Now, the overall efficiency of the pump is defined as the ratio of the fluid power to the mechanical power supplied. The fluid power = $\rho g Q H$ (equation (6.30)), so that the efficiency,

$$\eta = \rho g H Q/P. \tag{22.31}$$

Let us now apply this expression to points X and X' corresponding to speeds N_1 and N_2. At N_1,

$$\eta_X = \rho g Q_X H_X/P_X.$$

At N_2, $\eta_{X'} = \rho g Q_{X'} H_{X'}/P_{X'}.$

Dividing one expression by the other,

$$\frac{\eta_X}{\eta_{X'}} = \frac{Q_X}{Q_{X'}} \times \frac{H_X}{H_{X'}} \times \frac{P_{X'}}{P_X} = \frac{N_1}{N_2} \left(\frac{N_1}{N_2}\right)^2 \left(\frac{N_2}{N_1}\right)^3 = 1.$$

Thus, $\eta_X = \eta_{X'}$, which proves equation (22.30). However, although $\eta_{X'}$ is the same as η_X it is now plotted against $Q_{X'}$ so that its position on the graph is changed.

The procedure as outlined above for point X may be applied to other points on the characteristic curves such as Y and Z resulting in points Y' and Z', as shown in Fig. 22.4. Thus, a new set of characteristics corresponding to N_2 may be drawn by joining the 'primed' points. These new characteristic curves are shown dashed in the figure.

A change in pump size and, therefore, impeller diameter, results also in a new set of characteristic curves, obtained by using equations (22.27) to (22.29), which yield the following relationships:

$$Q_{X''} = Q_X(D_2/D_1)^3, \qquad H_{X''} = H_X(D_2/D_1)^2,$$
$$P_{X''} = P_X(D_2/D_1)^5, \qquad \eta_{X''} = \eta_X.$$

Thus, the similarity laws enable us to obtain a set of characteristic curves for a machine from the known test data of a geometrically similar machine.

Example 22.1

A centrifugal pump, impeller diameter 0·50 m, when running at 750 rev min^{-1} gave on test the following performance characteristics:

Q (m^3 min^{-1})	0	7	14	21	28	35	42	49	56
H (m)		40·0	40·6	40·4	39·3	38·0	33·6	25·6	14·5	0
η (per cent)	0	41	60	74	83	83	74	51	0

Predict the performance of a geometrically similar pump of 0·35 m diameter and running at 1450 rev min^{-1}. Plot both sets of characteristics.

Solution
Let suffix 1 refer to the 0·5 m diameter pump and suffix 2 refer to the 0·35 m diameter pump. From equation (22.27),

$$Q_1/N_1 D_1^3 = Q_2/N_2 D_2^3.$$

Therefore,

$$Q_2 = Q_1 (N_2/N_1)(D_2/D_1)^3$$

$$= Q_1 (1450/750)(0·35/0·5)^3 = 0·663\, Q_1.$$

From equation (22.28),

$$H_1/N_1^2 D_1^2 = H_2/N_2^2 D_2^2.$$

Therefore,

$$H_2 = H_1 (N_2/N_1)^2 (D_2/D_1)^2$$

$$= H_1 (1450/750)^2 (0·35/0·50)^2 = 1·83\, H_1.$$

The values of Q_1 and H_1 are given by the table above. Therefore, by multiplying them by the multipliers calculated above, the values of Q_2 and H_2 may be tabulated. These, together with the same values of efficiency (equation (23.30)) constitute the predicted characteristic of pump 2 as follows:

Q_2 (m^3 min^{-1})	0	4·64	9·28	13·92	18·56	23·2	27·8	32·5	37·0
H_2 (m)		73·2	74·3	74·0	71·9	69·5	61·5	46·8	26·5	0
η (per cent)	0	41	60	74	83	83	74	51	0

The characteristics of both pumps are plotted in Fig. 22.5.

22.4. Computer program 'SIMPUMP'

The program calculates and displays the values of flow coefficient, head coefficient, power coefficient and efficiency for up to 15 points on a pump or fan characteristic curve. From these it calculates and tabulates the corresponding values of flow rate, head, power and efficiency for a specific machine of a given impeller diameter and running at a given rotational speed.

The required input is:

Density of fluid handled and
(a)   Up to 15 points on pump or fan characteristic in one of the following forms:
    (i)   flow rate, head, power;
    (ii)   flow rate, head, efficiency;
    (iii) flow coefficient, head coefficient, power coefficient;
    (iv)   flow coefficient, head coefficient, efficiency.
(b)   In all cases, (i) to (iv), the impeller diameter and rotational speed of the machine for which the specific performance characteristic is required.

Use is made of the following equations: (22.25), (22.27), (22.28), (22.29) and (22.31).

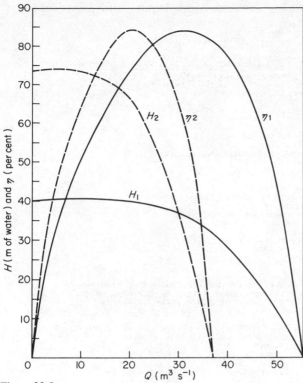

Figure 22.5

Input example: Fluid: water, density 1000 kg m^{-3}.

(a)

$Q(\text{m}^3/\text{s})$	0·05	0·10	0·15	0·20	0·25
H (m)	77·8	71·0	60·0	45·0	18·0
Efficiency (per cent)	66·0	79·0	78·0	60·0	12·0

diameter of pump impeller 0·4 m, rotational speed 1500 rev min^{-1}.

(b)   New pump impeller diameter 0·75 m, rotational speed 720 rev min^{-1}.

Output:

(a)

KQ	KH	KP	EFF
0·03	7·63	0·36	66·00
0·06	6·97	0·55	79·00
0·09	5·89	0·71	78·00
0·12	4·41	0·92	60·00
0·16	1·77	2·30	12·00

(b)

Q	H	P	EFF
0·16	63·02	148·19	66·00
0·32	57·51	225·96	79·00
0·47	48·60	290·10	78·00
0·63	36·45	377·13	60·00
0·79	14·58	942·82	12·00

List:

```
10 MODE 7: @%=&20209
20 REM: SIMPUMP
30 CLS: PRINT "PROGRAM: SIMPUMP"
40 DIM KQ(15),KH(15),KP(15),Q(15),H(15),P(15),E(15)
50 GOSUB 850: GOSUB 860
60 PRINT"HOW MANY POINTS ON THE CHARACTERISTIC"
70 INPUT"CURVES DO YOU WANT TO PROCESS";I
80 GOSUB850: INPUT"DENSITY OF FLUID HANDLED (KG/M^3)";RO
90 GOSUB850: PRINT "IS YOUR INPUT DATA IN TERMS OF"
100 INPUT"DIMENSIONLESS COEFFICIENTS (Y/N)";D$
110 IF D$="N" THEN 320
120 GOSUB 850: PRINT"ARE VALUES OF POWER"
130 INPUT"COEFFICIENT KNOWN (Y/N)";P$
140 IF P$="N" THEN 220
150 GOSUB 850: PRINT"WHAT ARE THE VALUES OF"
160 PRINT "THE COEFFICIENTS: KQ,KH, KP"
170 FOR X=1 TO I: INPUT"??";KQ(X),KH(X),KP(X)
180 E(X)=KQ(X)*KH(X)*100/KP(X)
190 NEXT
200 GOSUB 910
210 GOTO 560
220 GOSUB 840: INPUT"IS EFFICIENCY KNOWN (Y/N)";E$
230 IF E$="N" THEN 870
240 GOSUB 850: PRINT"WHAT ARE THE VALUES OF"
250 PRINT "COEFFICIENTS: KQ, KH AND"
260 PRINT "EFFICIENCY (%)"
270 FOR X=1TOI: INPUT"??";KQ(X),KH(X),E(X)
280 KP(X)=KQ(X)*KH(X)*100/E(X)
290 NEXT
300 GOSUB 910
310 GOTO 560
320 GOSUB 850: INPUT"WHAT IS THE IMPELLER DIAMETER, D1(M)";D
330 GOSUB 850: INPUT"ITS ROTATIONAL SPEED, N1(REV/MIN)";N
340 GOSUB 850: INPUT"ARE VALUES OF POWER KNOWN (Y/N)";P$
350 IF P$="N" THEN 460
360 GOSUB 850
370 FOR X=1 TO I
380 INPUT"INPUT: Q(M^3/S),H(M),P(KW)";Q(X),H(X),P(X)
390 M=N/60: KQ(X)=Q(X)/(M*D^3)
400 KH(X)=(9.81*H(X))/(M*D)^2
410 KP(X)=(1000*P(X))/(RO*D^5*M^3)
420 E(X)=KQ(X)*KH(X)*100/KP(X)
430 NEXT
440 GOSUB 910
450 GOTO 560
460 GOSUB 850: INPUT"IS EFFICIENCY KNOWN (Y/N)";E$
470 IF E$="N" THEN 870
480 CLS
490 FOR X=1 TO I
500 INPUT"Q(M^3/S),H(M),E(%)";Q(X),H(X),E(X)
510 M=N/60: KQ(X)=Q(X)/(M*D^3)
520 KH(X)=(9.81*H(X))/(M*D)^2
530 KP(X)=KQ(X)*KH(X)*100/E(X)
540 NEXT
550 GOSUB 910
560 GOSUB 860
570 GOSUB 850: PRINT"DO YOU WANT TO CALCULATE"
580 PRINT"PERFORMANCE FOR A SPECIFIC"
590 INPUT"MACHINE (Y/N)";A$
600 IF A$="N" THEN 770
610 GOSUB 850: INPUT"NEW DIAMETER: D2(M)";D2
620 GOSUB 850: INPUT"NEW SPEED N2(REV/MIN)";N2
630 CLS
640 PRINT TAB(7)"Q",TAB(15)"H",TAB(24)"P",TAB(32)"EFF"
650 STAR$=STRING$(40,"*"):PRINT STAR$
660 FOR X=1 TO I
670 Q2=KQ(X)*N2*D2^3/60
680 H2=KH(X)*(N2*D2)^2/9.81/(60)^2
690 P2=KP(X)*RO*(N2/60)^3*(D2)^5/1000
700 PRINT Q2,H2,P2,E(X)
710 NEXT
720 GOSUB 860
730 GOSUB 850: INPUT"WISH TO CHANGE DIAMETER (Y/N)";B$
740 IF B$="Y" THEN 610
750 GOSUB 840: INPUT"WISH TO CHANGE SPEED (Y/N)";C$
760 IF C$="Y" THEN 620
770 GOSUB 850: PRINT"DO YOU WANT TO RUN THE PROGRAM"
780 INPUT"AGAIN (Y/N)";D$
790 IF D$="Y" THEN CLS
800  IF D$="Y" THEN 50
810 IF D$="N" THEN 820
820 PRINT TAB(15,15)"THANK YOU"
830 END
```

```
840 PRINT: RETURN
850 PRINT: PRINT: RETURN
860 FOR X=1 TO 4000: NEXT: RETURN
870 CLS
880 PRINT TAB(5,10);CHR$(133);CHR$(136);"FOR THIS PROGRAM EITHER POWER"
890 PRINT TAB(5);CHR$(133);CHR$(136);"OR EFFICIENCY MUST BE KNOWN"
900 GOSUB860:GOSUB850: GOTO 770
910 CLS:   GOSUB 850
920 PRINT TAB(6)"KQ",TAB(16)"KH",TAB(24)"KP",TAB(32)"EFF"
930 STAR$=STRING$(40,"*"):PRINT STAR$
940 FOR X=1 TO I
950 PRINT KQ(X),KH(X),KP(X),E(X)
960 NEXT
970 RETURN
```

22.5. Scale effects

In the application of similarity laws as shown above, it was assumed that all criteria of dynamical similarity are satisfied, i.e. all the groups of equation (22.26) remain the same. This, however, is not true with regards to the dimensionless groups representing the Reynolds number, the Mach number and the relative roughness.

Consider first the Reynolds number, given by $Re = k(ND^2\rho/\mu)$, which shows that a change of either speed or diameter alters the value of Re. Thus, in practice, $Re \neq$ const. However, for water and air this effect is usually small because the values of Re are usually very high, the flow being fully turbulent.

A similar consideration of Mach number indicates that an increase of tip speed (by either an increase of N or of D) will make the Mach number higher. This not only means that one of the conditions of dynamical similarity is not satisfied but, in addition, may also mean that the compressibility effect (previously negligible) may now be of considerable importance. This second point must be watched carefully in the application of similarity laws to fans and compressors.

Consider now the effect of relative roughness. This again should be maintained constant, not only because it appears in equation (22.26) but also on account of geometrical similarity, which is the primary condition of any model laws to hold. Absolute roughness (ϵ) is the mean height of surface perturbances, which, therefore, remains the same for a given material and process used in the manufacture of the machine, irrespective of its size. Thus, any change of machine size involves a change of relative roughness (ϵ/D). On the whole, the larger the machine, the smaller the relative roughness will be. This will tend to make frictional losses relatively less important in larger machines.

In practice, it is also difficult to maintain geometrical similarity in clearances and some material thicknesses. The same gauge of sheet metal, for example, may be used for a range of sizes of fabricated impeller blades. Such deviations from geometrical similarity must obviously cause some departures from the idealized predictions based on the aforementioned similarity laws.

All such departures, which do occur in practice and which are due to the Reynolds number, Mach number, relative roughness or lack of strict geometrical similarity, are usually referred to as the *scale effect*. In general the scale effect tends to improve the performance of larger machines.

22.6. Type number

The performance of geometrically similar machines, i.e. machines belonging to one family, is governed by similarity laws and may be represented for the whole family by a single plot of dimensionless characteristics. Thus, the performance of machines belonging to different families may be compared by plotting their dimensionless characteristics on the same graph. Detailed comparison may then be achieved by analysing the various aspects of the sets of curves. This method of comparison is satisfactory and often needed, but it lacks the brevity required in machine classification. This is obtained by the use of the *type number*, also known as the *specific speed*.

Every machine is designed to meet a specific duty, usually referred to as the *design point*. For a pump, for example, this would be stated in terms of the flow rate and the head developed and, thus, represents a particular point on its basic performance characteristic. The design point is normally associated with the maximum efficiency of the machine.

It is, thus, informative to compare machines by quoting the values of K_Q, K_H and K_P corresponding to their design points. However, since for pumps K_Q and K_H are the two most important parameters, their ratio would indicate the suitability of a particular pump for large or small volumes relative to the head developed. Furthermore, if the ratio is obtained in such a way that the impeller diameter is eliminated from it, then the comparison becomes independent of machine size. This is achieved by raising K_Q to power $1/2$ and K_H to power $3/4$. The result is the *type number*:

$$n_s = \frac{(K_Q)^{1/2}}{(K_H)^{3/4}} = \left(\frac{Q}{ND^3}\right)^{1/2} \left(\frac{N^2 D^2}{gH}\right)^{3/4}$$

$$= N\frac{Q^{1/2}}{(gH)^{3/4}} = N\frac{Q^{1/2}}{y^{3/4}} \tag{22.32}$$

It must be realized that a value of type number can be calculated for any point on the characteristic curve. It will be equal to zero at point S in Fig. 22.6, because at that point the flow rate is zero, and will tend to infinity at point T because at large volumes head H tends to zero.

Such values are, however, of no practical interest and only the type number at the design point, usually referred to as *the* type number, is used for classification, comparison and design purposes.

The comparison of turbines is also achieved by the use of their type numbers. However, since for turbines the power developed is the most important variable, an alternative expression for type number in terms of power developed is obtained by eliminating D from the ratio of power and head coefficients. This is achieved by raising the power coefficient to the power of $1/2$ and the head coefficient to the power of $5/4$ and then taking their ratio:

$$n_s = \frac{(K_P)^{1/2}}{(K_H)^{5/4}} = \left(\frac{P}{\rho N^3 D^5}\right)^{1/2} \times \left(\frac{N^2 D^2}{gH}\right)^{5/4},$$

$$= N P^{1/2}/\rho^{1/2}(gH)^{5/4} = NP^{1/2}/(\rho^{1/2} y^{5/4}). \tag{22.33}$$

Equations (22.32) and (22.33) are fundamentally the same and are related

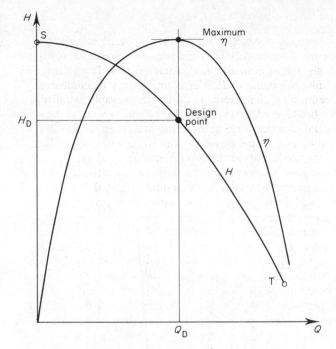

Figure 22.6 'Design point' on a pump characteristic

by equation (22.31), which, for a turbine, takes the form

$$P_{\text{(output)}} = \eta \rho g H Q.$$

Substituting into equation (22.33),

$$n_s = N(\eta \rho g H Q)^{1/2}/\rho^{1/2}(gH)^{5/4} = N\eta^{1/2}\,Q^{1/2}/(gH)^{3/4} \qquad (22.34)$$

The type number, since it is obtained from dimensionless coefficients, is also a dimensionless quantity provided a consistent system of units, such as SI, is used. Unfortunately, the units used in practice are often not consistent (e.g. revolutions per minute for N, litres per hour for Q and metres for H) and when this is the case a symbol N_s is used. It is then essential to state the units used for all the relevant quantities. The rotational speed may be either in rev s^{-1} or rad s^{-1}. Thus two forms of the type number are possible.

$$n_s = NQ^{1/2}/(gH)^{3/4} \quad \text{or} \quad \omega_s = \omega Q^{1/2}/(gH)^{3/4} \text{ for pumps} \qquad (22.35)$$

and

$$n_s = N\frac{(P/\rho)^{1/2}}{(gH)^{5/4}} \quad \text{or} \quad \omega_s = \omega\frac{(P/\rho)^{1/2}}{(gH)^{5/4}} \text{ for turbines} \qquad (22.36)$$

The relationship between the two forms is $\omega_s = 2\pi n_s$.

With the aid of type number the various types of pumps and turbines may be classified and compared, as shown in Fig. 22.7. Also, since the type number refers to the design point it is used as the most important design parameter.

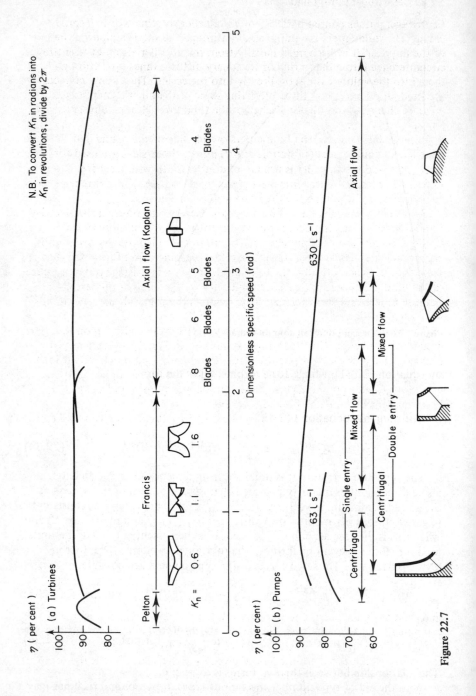

Figure 22.7

22.7. Centrifugal pumps and fans

Centrifugal pumps consist basically of an impeller rotating within a spiral casing. The fluid enters the pump axially through the suction pipe via the eye of the impeller; it is discharged radially from the impeller around the entire circumference into either a ring of stationary diffuser vanes (and through them into the volute casing) or directly into the casing. The casing 'collects' the fluid, decelerates it — thus converting some of the kinetic energy into pressure energy — and finally discharges the fluid through the delivery flange.

In single inlet pumps, the fluid enters on one side of the casing and impeller. In double inlet (or double entry) pumps, both sides are used for fluid entry and the impeller is usually of double width with a centre plate. It looks like two single entry impellers placed back to back. This arrangement has the effect of doubling the flow rate at the same head.

Single entry impellers may be arranged in series on a common shaft so that the fluid leaving one impeller is directed through a set of stationary vanes into the inlet of the next impeller. Such pumps are called multi-stage pumps, each impeller constituting a stage. The effect is an increase of head for the same volume. Theoretically, the head produced by a multistage pump is equal to the head produced by one stage multiplied by the number of stages. Because of losses in the interstage vane passages the overall head generated is somewhat smaller.

The blades of a centrifugal impeller vary in shape depending upon the design requirements. The blade angle at inlet (β_1 in Fig. 21.4) is chosen so that the relative velocity meets the blade tangentially ('no shock' condition) and since under design conditions $v_{w1} = 0$ it follows that β_1 depends upon u_1 and v_{f1} (hence Q) only. Thus, head considerations do not affect β_1. However β_2 is very much affected by them.

The theoretical head developed by the centrifugal pump is given by Euler's equation

$$H_{th} = v_{w2}u_2/g$$

(provided $v_{w1} = 0$) and, therefore, for given tip speed (u_2) depends entirely upon the outlet whirl component velocity v_{w2}.

Let us now examine the dependence of this component upon the blade angle at outlet β_2. Figure 22.8 shows three different blade angles, often referred to as inclined backwards ($\beta_2 < 90°$), inclined forward ($\beta_2 > 90°$) or radial ($\beta_2 = 90°$). The figure also shows a combined velocity diagram for the three types of blades.

The diagrams are drawn for the same u_2 and v_{f2} (and, hence, the same speed of rotation, diameter and discharge) for the three blade types. It is clear that as β_2 increases, the absolute velocity v_2 (and hence the whirl component v_{w2}) also increase. Therefore, the head developed depends upon β_2 and is larger for the forward inclined blades. However, it must be remembered that the theoretical head given by Euler's equation is the total head developed by the impeller and, hence, embraces both the static and velocity head terms. Reference to Fig. 22.8 will show that the large head developed by the forward inclined impeller blades includes a large proportion of velocity head since v_2 is very large. This presents practical

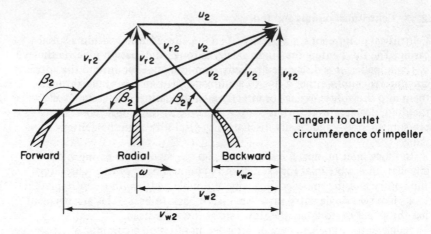

Figure 22.8   Effect of blade outlet angle on the outlet velocity triangle

difficulties in converting some of this kinetic energy into pressure energy. As shown in Section 8.8, the losses in a diffuser may be substantial, and are difficult to control.

The most common blade outlet angles for centrifugal pumps are from $15°$ to $90°$, but, for fans, the range extends into forward inclined blades (well-known multi-vane fans) with β_2 as large as $140°$.

The effect of outlet blade angle on the performance characteristics is shown in Fig. 22.9. It is seen that the forward-bladed impeller generates

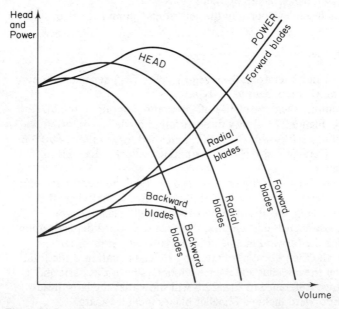

Figure 22.9   The effect of blade outlet angle on performance characteristics

greater head at a given volume, but it must be remembered that a substantial part of this total head is in fact the velocity head.

The power characteristics also show fundamental differences, which are of considerable practical importance. For the backward-bladed impeller, the maximum power occurs near the maximum efficiency point and any increase of flow rate beyond this point results in a decrease of power. Thus, an electric motor used to drive such a pump or fan may be safely rated at the maximum power. This type of power characteristic is called self-limiting.

This is not the case, however, for the radial- or forward-bladed impellers, for which the power is continuously rising. Choosing an appropriate motor, therefore, poses problems, because to have one rated for maximum power would mean over-rating and an unnecessary expenditure if the pump will operate only near the maximum efficiency point. On the other hand a smaller motor rated just for the operating point may be in danger of being overloaded should the pump be operated by mistake at a flow rate greater than the design value corresponding to the maximum efficiency point.

Centrifugal pumps and fans occupy the lower range of type numbers, up to approximately 1·8 as shown in Fig. 22.7. In general, the lower the type number of these machines the narrower is the impeller in relation to its diameter.

The overall efficiencies of centrifugal pumps are high, of the order of 90 per cent in the range of type numbers between 0·8 and 1·6. They tend to fall off rapidly at lower type numbers, mainly because of the increased frictional losses in the long interblade passages of these narrow impellers.

Also, the efficiencies depend upon the size of the machine and, hence, the capacity handled. The larger the machine, the higher is the efficiency (*see* Section 22.5).

For centrifugal fans, the highest efficiencies are realized by the 'aerofoil-bladed' fans. They are basically of the backward-bladed type, but the blades have an aerofoil profile rather than being of the same thickness. Their range of type number is from 0·5 to 1·6 and maximum efficiencies are of the order of 90 per cent.

22.8. Axial flow pumps and fans

Axial flow pumps and fans consist of an impeller rotating within a concentric cylindrical casing. Thus, the direction of flow through the machine is axial throughout. For this reason axial flow pumps and fans have the smallest transverse dimensions of all rotodynamic machines.

It is usual for a set of stationary guide vanes to be present in all but the cheapest and least efficient machines. The guide vanes ensure that the flow at the outlet from the pump or fan is without a tangential component, that is to say is axial and without a swirl. This result may be achieved by either a set of downstream or a set of upstream vanes. The choice affects the construction and to some extent the type number.

Axial flow pumps and fans extend over the far end of the type number spectrum, starting from about $\omega_s = 2\cdot8$ (*see* Fig. 22.7) to about 4·8. For fans, the range is somewhat wider, starting from about 1·4 to 2·4 for upstream guide vane fans, from 2·0 to 3·0, for downstream guide vane fans and from 2·5 to 4·5 for non-guide vane fans. Axial flow fans without a casing, called

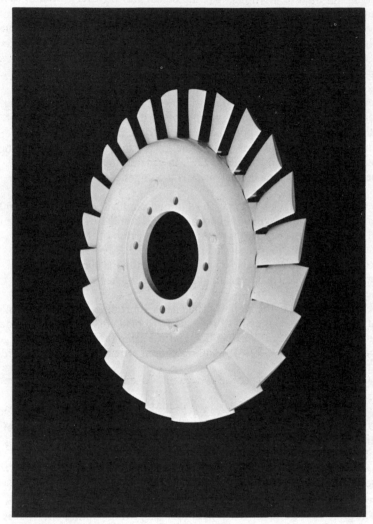

Figure 22.10   An impeller of an axial flow fan. (By courtesy of Airscrew-Howden Ltd)

propeller fans, cover the range from 3·5 to 5·0, whereas at the other end of the range are the contrarotating fans, for which the type numbers are from 1·2 to 1·6. In these fans, the aim of having axial flow at outlet is achieved not by the guide vanes, which are omitted, but by having two impellers rotating in opposing directions, so that the whirl component produced by one is cancelled by the whirl component of the other (which is equal in magnitude but opposite in direction).

The disadvantages of axial flow pumps and fans are that they develop a low head (up to 20 m per stage) and have steeply descending efficiency curves and, hence, are only economical if operated at discharges corresponding to or very near to the design point. Also, the pressure/volume characteristic on the

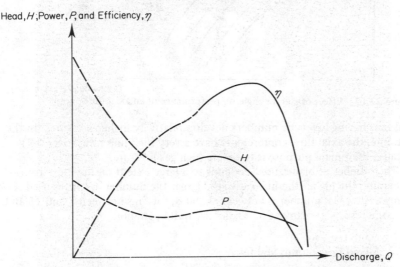

Head, H; Power, P, and Efficiency, η

Discharge, Q

Figure 22.11   Typical flow characteristics of an axial flow pump

left of the maximum efficiency has a region of instability, as shown dashed in Fig. 22.11.

In addition, axial flow pumps have a limited suction capacity and, thus, are prone to cavitation, which considerably restricts their selection for some applications.

The blades of an axial flow impeller are fixed to a hub, usually permanently. In some cases, however, for special applications, the blades are adjustable so that the stagger angle may be varied and, thus, the performance altered. The effect of changing the stagger angle on the characteristics is shown in Fig. 22.12.

The relationship between the blade radial length and the size of the hub is expressed by the hub ratio, defined as

$$\text{Hub ratio} = \frac{\text{Hub diameter}}{\text{Impeller diameter}} \qquad (22.37)$$

This parameter affects the type number of the machine and, for pumps, varies from 0·3 to 0·6. A low hub ratio (long blades) is associated with high type numbers, whereas a high ratio results in lower type numbers. Hence, the pumps

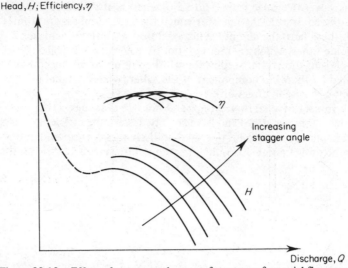

Figure 22.12   Effect of stagger angle on performance of an axial flow pump

and fans having low type numbers develop relatively higher pressures. In the extreme, the axial flow compressors have a very high hub ratio (over 0·9) because their main purpose is to create high gas pressure.

The number of blades used depends to a large extent on the hub ratio. Generally, the higher the hub ratio, the larger the number of blades used. For pumps, the usual number is between 2 and 8, for fans between 2 and 16 and for compressors as many as 32 blades are quite common.

22.9. Mixed flow pumps and fans

Mixed flow pumps occupy the position between the centrifugal and axial flow pumps. The impellers consist of a conical hub with blades attached in such a way that the flow into the impeller is axial, but through it the flow is partly axial and partly radial. On leaving the impeller, the fluid is usually diffused in the guide vanes, which lead into an axial outlet, as shown in Fig. 22.13. The alternative and less frequent solution is for the flow to be collected in by a volute casing and then discharged in a plane normal to the axis of the impeller rotation. Such an arrangement is shown in Fig. 22.14. Both configurations cover the type number range from about 1·0 to 2·2, the latter type being more common at the lower end of this type number range.

One of the advantages of mixed flow pumps with axial discharge is that while they offer large discharges, they may be easily arranged in multistage units, thus providing high pressure.

The efficiency of mixed flow pumps and fans is high, approaching the 90 per cent mark. Figure 21.2 shows a mixed flow impeller of a fan.

22.10. Water turbines

Turbines are subdivided into *impulse* and *reaction* machines. In the impulse turbines, the total head available is first converted into the kinetic energy.

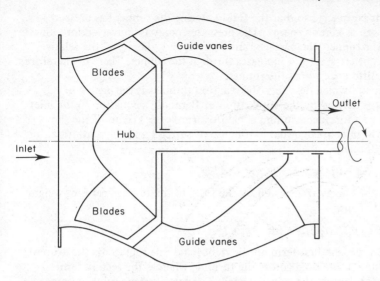

Figure 22.13   Mixed flow pump with axial discharge

This is usually accomplished in one or more nozzles. The jets issuing from the nozzles strike vanes attached to the periphery of a rotating wheel. Because of the rate of change of angular momentum and the motion of the vanes, work is done on the runner (impeller) by the fluid and, thus, energy is transferred. Since the fluid energy which is reduced on passing through the runner is entirely kinetic, it follows that the absolute velocity at outlet is smaller than the absolute velocity at inlet (jet velocity). Furthermore, the fluid pressure is atmospheric throughout and the relative velocity is constant except for slight reduction due to friction.

In the reaction turbines, the fluid passes first through a ring of stationary guide vanes in which only part of the available total head is converted into kinetic energy. The guide vanes discharge directly into the runner along the

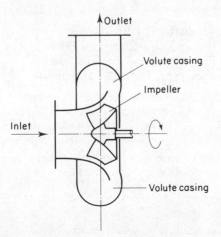

Figure 22.14   Mixed flow pump with volute casing

whole of its periphery, so that the fluid entering the runner has pressure energy as well as kinetic energy. The pressure energy is converted into kinetic energy in the runner (the passage running full) and, therefore, the relative velocity is not constant but increases through the runner. There is, therefore, a pressure difference across the runner.

A parameter which describes the reaction turbines is the *degree of reaction*. It is derived by the application of Bernoulli's equation to the inlet and outlet of a turbine, assuming ideal flow (no losses). Thus, if the conditions at inlet are denoted by the use of suffix 1 and those at outlet by the suffix 2, then,

$$p_1/\rho g + v_1^2/2g = E + p_2/\rho g + v_2^2/2g,$$

where E is the energy transferred by the fluid to the turbine per unit weight of the fluid. Thus,

$$E = (p_1 - p_2)/\rho g + (v_1^2 - v_2^2)/2g.$$

In this equation, the first term on the right-hand side represents the drop of static pressure in the fluid across the turbine, whereas the second term represents the drop in the velocity head. The two extreme solutions are obtained by making either of these two terms equal to zero. Thus, if the pressure is constant, so that $p_1 = p_2$, then $E = (v_1^2 - v_2^2)/2g$ and such a turbine is purely impulsive. If, on the other hand, $v_1 = v_2$, then $E = (p_1 - p_2)/\rho g$ and this represents pure reaction. The intermediate possibilities are described by the degree of reaction (R), defined as

$$R = \frac{\text{Static pressure drop}}{\text{Total energy transfer}}. \qquad (22.38)$$

But the static pressure drop is given by

$$(p_1 - p_2)/\rho g = E - (v_1^2 - v_2^2)/2g,$$

so that $R = \{E - (v_1^2 - v_2^2)/2g\}/E = 1 - (v_1^2 - v_2^2)/2gE.$

Substituting now from Euler's equation for $E = v_{w1} u_1/g$, we obtain

$$R = 1 - (v_1^2 - v_2^2)/2v_{w1} u_1 \qquad (22.39)$$

Water turbines are mainly used in power stations to drive electric generators. There are three well-known types which are used: the Pelton wheel, which is an impulse turbine, the Francis type and the axial flow (Kaplan) turbines, both being of the reaction type. Table 22.1 attempts to compare the three types.

	Pelton wheel	Francis	Kaplan
Type number ω_s range (rad)	0·05–0·4	0·4–2·2	1·8–4·6
Operating total head (m)	100–1700	80–500	up to 400
Maximum power output (MW)	55	40	30
Best efficiency (per cent)	93	94	94
Regulation mechanism	Spear nozzle and deflector plate	Guide vanes, surge tanks	Blade stagger

Table 22.1. Comparison of water turbines.

22.11. The Pelton wheel

The Pelton wheel is an impulse turbine in which vanes, sometimes called 'buckets', of elliptic shape are attached to the periphery of a rotating wheel, as shown in Fig. 22.15. One or two nozzles project a jet of water tangentially to the vane pitch circle. The vanes are of double outlet section, as shown in

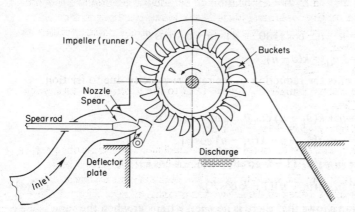

Figure 22.15 Diagrammatic arrangement of a Pelton wheel

Fig. 22.16, so that the jet is split and leaves symmetrically on both sides of the vane. In this way the end thrust on bearings and the shaft is eliminated.

The total head available at the nozzle is equal to the gross head less losses in the pipeline leading to the nozzle. If it is equal to H, then the velocity of jet issuing from the nozzle is:

$$v = C_v \sqrt{(2gH)} \qquad (22.40)$$

where C_v is the velocity coefficient and its value is between 0·97 and 0·99.

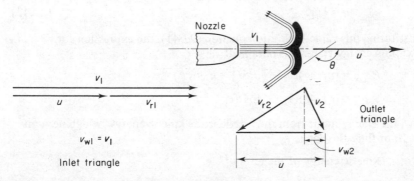

Figure 22.16 Velocity triangles for a Pelton wheel

The total energy transferred to the wheel is given by Euler's equation:

$$E = (v_{w1}u_1 - v_{w2}u_2)/g,$$

but, as shown by the velocity triangles in Fig. 22.16, the peripheral vane velocity at outlet is the same as that at inlet,

$$u_1 = u_2 = u,$$

so that $E = \dfrac{u}{g}(v_{w1} - v_{w2}).$

However,

$$v_{w2} = u - v_{r2} \cos(180° - \theta) = u + v_{r2} \cos \theta$$

and $v_{r2} = kv_{r1} = k(v - u),$

where k represents the reduction of the relative velocity due to friction. Thus,

$$v_{w2} = u + k(v_1 - u) \cos \theta \quad \text{and} \quad v_{w1} = v_1,$$

so that $E = (u/g)[v_1 - u - k(v_1 - u) \cos \theta]$

$$= (u/g)[v_1 (1 - k \cos \theta) - u(1 - k \cos \theta)]$$

$$= (u/g)(v_1 - u)(1 - k \cos \theta). \tag{22.41}$$

This equation shows that there is no energy transfer when the vane velocity is either zero or equal to the jet velocity. It is reasonable to expect, therefore, that the maximum energy transfer will occur at some intermediate value of the vane velocity. This may be obtained by differentiation as follows.

$$E = \{(1 - k \cos \theta)/g\} (v_1 u - u^2).$$

Therefore, for a maximum,

$$dE/du = \{(1 - k \cos \theta)/g\} (v_1 - 2u) = 0,$$

Hence,

$$v_1 - 2u = 0,$$

$$u = \tfrac{1}{2}v_1. \tag{22.42}$$

Substituting this value back into equation (22.41), the expression for maximum energy transfer is obtained:

$$E_{max} = (v_1/2g)(v_1 - \tfrac{1}{2}v_1)(1 - k \cos \theta)$$

$$= (v_1^2/4g)(1 - k \cos \theta).$$

Now, the energy input from the nozzle is the kinetic energy, which per unit weight of fluid flowing is

$$\text{Kinetic energy of the jet} = v_1^2/2g.$$

Thus, the maximum theoretical efficiency of the Pelton wheel becomes

$$\eta_{max} = E_{max}/\text{Kinetic energy of the jet}$$

$$= (v_1^2/4g)(1 - k \cos \theta) \div (v_1^2/2g)$$

$$= (1 - k \cos \theta)/2. \tag{22.43}$$

   In the ideal case, assuming no friction, there is no reduction of the relative velocity over the vane and, therefore, $k = 1$. Also, if $\theta = 180°$, the maximum efficiency becomes 100 per cent. In practice, however, friction exists and the value of k is in the region of 0·8 to 0·85. Also, the vane angle is usually 165°, to avoid the interference between the oncoming and outcoming jets. Thus, the ratio of the wheel velocity to the jet velocity becomes, in practice, somewhat smaller than the theoretical. Figure 22.17 shows the variation of the

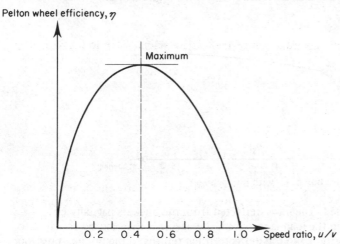

Figure 22.17   Pelton wheel efficiency as a function of speed ratio

Pelton wheel efficiency with the speed ratio. It will be seen that, for the maximum efficiency, this ratio is about 0·46.
   Since a Pelton wheel is usually employed to drive an electrical generator, it is required that its speed of rotation is constant, regardless of the load. Thus, u must be constant, but for maximum efficiency it is also important that the speed ratio is maintained constant as well. Since the jet velocity depends only upon the total head H, which for a given installation is also constant, the velocity ratio may be kept constant provided there is no reduction of head at the nozzle. This means that a throttling process using a valve in the penstock is not suitable, since a valve reduces flow by reducing head and dissipating energy. It follows, then, that any alteration of the load on the turbine must be accompanied by a corresponding alteration of the water power, but with u/v remaining constant. Since $P = \rho gQH$, it follows that this requirement can only be achieved by alteration in Q such that H is unchanged. But,

$$Q = Av = AC_v \sqrt{(2gH)}$$

and, therefore, to vary Q, the area of the jet must be changed. This is achieved by means of the needle (spear) shown in Fig. 22.15, which does not alter H.
   Small changes in efficiency result because the nozzle loss represented by the value of C_v will be changed and, also, jet windage as well as bearing losses

will change slightly. Figure 22.18 shows the typical variation of C_v with the jet opening.

The movement of the needle may be controlled automatically by a governor operated by a servo-motor or some similar arrangement.

The provision of a needle valve does not, however, cater satisfactorily for sudden load removal, because it is not possible to shut the needle rapidly without the serious risk of very high pressure build-up in the pipe system (*see* Chapter 20). Surge tanks are not suitable because of the large heads involved in the Pelton wheel installations. Therefore, a deflector plate (as shown in

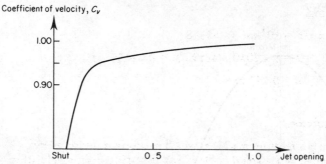

Figure 22.18   Variation of C_v with jet opening

Fig. 22.15) is used. The jet is deflected from the buckets partially or completely when the load is removed, the needle is then moved slowly into the required position and the deflector then returns to the original position.

Example 22.2

A Pelton wheel driven by two similar jets transmits 3750 kW to the shaft when running at 375 rev min^{-1}. The head from the reservoir level to the nozzles is 200 m and the efficiency of power transmission through the pipelines and nozzles is 90 per cent. The jets are tangential to a 1·45 m diameter circle. The relative velocity decreases by 10 per cent as the water traverses the buckets, which are so shaped that they would, if stationary, deflect the jet through 165°. Neglecting windage losses, find (a) the efficiency of the runner and (b) the diameter of each jet.

Solution

(a) Efficiency of runner,

$$\eta = (2u/v_j^2)(v_j - u)(1 - k \cos \theta)$$

If the efficiency of the pipeline and the nozzle is 90 per cent, the head at the base of the nozzle convertible to jet velocity is

$$h = 0·9 \times 200 = 180 \text{ m}$$

and the jet velocity is

$$v_j = \sqrt{(2gh)} = \sqrt{(2 \times 9·81 \times 180)} = 59·5 \text{ m s}^{-1}.$$

Now, the bucket speed is

$$u = \pi DN/60 = \pi \times 1·45 \times 375/60 = 28·5 \text{ m s}^{-1}$$

and, hence, the runner efficiency is

$\eta = \{2 \times 28 \cdot 5/(59 \cdot 5)^2\} \, (59 \cdot 5 - 28 \cdot 5) \times 1 \cdot 869 = \mathbf{0 \cdot 933 \ or \ 93 \cdot 3 \ per \ cent}$.

(b) If the runner is 93·3 per cent efficient, the total power of the jets is

$$P = 3750/0 \cdot 933 = 4021 \ kW$$

and power per jet is $4021/2 = 2010 \cdot 5$ kW. But, for a jet,

$$\text{Power} = \rho g A_j v_j (v_j^2/2g) = \rho (\pi d^2/8) v_j^3 \, .$$

Therefore, equating,

$$2010 \cdot 5 \times 10^3 = 10^3 \, \pi d^2 (59 \cdot 5)^3/8,$$

$$d^2 = 8 \times 2010 \cdot 5/\pi (59 \cdot 3)^3 = 0 \cdot 02455$$

$$d = \mathbf{0 \cdot 157 \ m}.$$

22.12. Francis turbines

A Francis turbine is a reaction machine, which means that during energy transfer in the runner (impeller) there is a drop in static pressure and a drop in velocity head. Only part of the total head presented to the machine is converted to velocity head before entering the runner. This is achieved in the stationary but adjustable guide vanes, shown in Fig. 22.19. It is important to

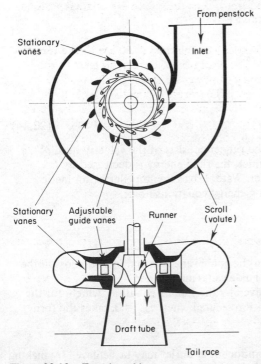

Figure 22.19   Francis turbine

realize that the machine is running full of water, which enters the impeller on its whole periphery. The guide vane ring may surround the runner on its outer periphery, in which case the flow of fluid is towards the runner centre. In such a case, the turbine is known as an *inward flow* type. The alternative arrangement is for the fluid to enter the guide vanes at the centre and to flow radially outwards into the runner which now surrounds the guide vanes. Such a turbine is known as the *outward flow* type.

Consider an inward flow Francis turbine, represented diagrammatically in Fig. 22.19. A section of the runner guide vane ring, showing the blades, vanes and velocity triangles, is given in Fig. 22.20.

The total head available to the machine is H and the water velocity on entering the guide vanes is v_0. The velocity leaving the guide vanes is v_1 and is related to v_0 by the continuity equation:

$$v_0 A_0 = v_{f_1} A_1 .$$

But $v_{f_1} = v_1 \sin \theta$, so that

$$v_0 A_0 = v_1 A_1 \sin \theta .$$

The direction of v_1 is governed by the guide vane angle θ. It is chosen in such a way that the relative velocity meets the runner blade tangentially, i.e. it makes an angle β_1 with the tangent at blade inlet. Thus,

$$\tan \theta = v_{f_1}/v_{w_1} \quad \text{and} \quad \tan \beta_1 = v_{f_1}/(u_1 - v_{w_1}).$$

Eliminating v_{w_1} from the two equations:

$$\tan \beta_1 = v_{f_1}/(u_1 - v_{f_1}/\tan \theta)$$

or $$\cot \beta_1 = \frac{u_1}{u_{f1}} - \cot \theta .$$

Therefore,

$$\frac{u_1}{v_{f_1}} = \cot \beta_1 + \cot \theta$$

or $$u_1 = v_{f_1}(\cot \beta_1 + \cot \theta) \tag{22.44}$$

The total energy at inlet to the runner consists of the velocity head $v_1^2/2g$ and pressure head H_1. In the runner, the fluid energy is decreased by E, which is transferred to the runner. Water leaves the impeller with kinetic energy $v_2^2/2g$. Thus, the following energy equations hold:

$$H = v_1^2/2g + H_1 + h_1'$$

and $$H = E + v_2^2/2g + h_1 ,$$

in which h_1' is the loss of head in the guide vane ring and h_1 is the loss in the whole turbine, including entry, guide vanes and runner.

The energy transferred E is given by the Euler's equation, which, for the maximum energy transfer condition secured when $v_{w_2} = 0$, takes the form

$$E = v_{w_1} u_1/g .$$

The condition of no whirl component at outlet may be achieved by making the outlet blade angle β_2, such that the absolute velocity at outlet v_2 is radial, as shown in Fig. 22.20. From the outlet velocity diagram, then, it follows that

$$\tan \beta_2 = v_2/u_2,$$

but, since $v_{w_2} = 0$, then $v_2 = v_{f_2}$ and, by the continuity equation,

$$A_1 v_{f_1} = A_2 v_{f_2},$$

so that β_2 can be determined.

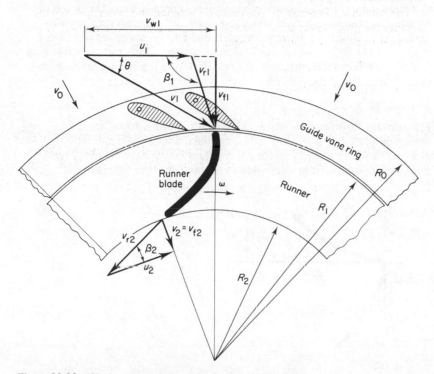

Figure 22.20   Section through part of a Francis turbine

If the condition of no whirl at outlet is satisfied, then the second energy equation takes the form

$$H = v_{w_1} u_1/g + v_2^2/2g + h_1.$$
(22.45)

The hydraulic efficiency is given by

$$\eta_h = E/H = v_{w_1} u_1/gH$$
(22.46)

and the overall efficiency by

$$\eta = P/\rho g Q H,$$
(22.47)

in which P is the power output of the machine, Q is the volumetric flow rate through it and H is the total head available at the turbine inlet.

The relationship between the runner speed and the spouting velocity, $\sqrt{(2gH)}$, for the Francis turbine is not so rigidly defined as for the Pelton wheel. In practice, the speed ratio $u_2/\sqrt{(2gH)}$ is contained within the limits 0·6 to 0·9.

Similarly to a Pelton wheel, a Francis turbine usually drives an alternator and, hence, its speed must be constant. Since the total head available is constant and dissipation of energy by throttling is undesirable, the regulation at part load is achieved by varying the guide vane angle θ, sometimes referred to as the gate. This is possible because there is no requirement for the speed ratio to remain constant. A change in θ results in a change in v_w and v_f. Thus, E is altered for given u. However, such changes mean a departure from the 'no shock' conditions at inlet and also give rise to the whirl component at outlet. As a result, the efficiency at part load falls off more rapidly than in the case of the Pelton wheel. Also, vortex motion in the draft tube resulting from the whirl component may cause cavitation in the centre. Sudden load changes are catered for either by a bypass valve or by a surge tank.

Example 22.3

In an inward-flow reaction turbine, the supply head is 12 m and the maximum discharge is 0·28 m³ s⁻¹. External diameter = 2 (internal diameter) and the velocity of flow is constant and equal to 0·15 $\sqrt{(2gH)}$. The runner vanes are radial at inlet and the runner rotates at 300 rev min⁻¹.

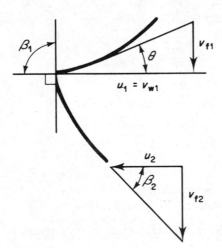

Figure 22.21

Determine (a) the guide vane angles, (b) the vane angle at exit for radial discharge, (c) widths of the runner at inlet and exit. The vanes occupy 10 per cent of the circumference and the hydraulic efficiency is 80 per cent.

Solution
  (a) Velocity of flow is given by

$$v_{f1} = v_{f2} = 0 \cdot 15 \sqrt{(2gH)} = 0 \cdot 15 \sqrt{(2 \times 9 \cdot 81 \times 12)} = 2 \cdot 3 \text{ m s}^{-1}.$$

Efficiency is given by equation (22.46) but $u_1 = v_{w1} [1 - (v_{f1}/v_{w1}) \tan \beta_1]$ and $v_{f1}/v_{w1} = \tan \theta$.

Therefore,

$$\eta = v_{w1}^2 (1 - \tan \theta / \tan \beta_1)/gH,$$

but, since $\beta_1 = 90°$ and $\eta = 80$ per cent, it follows that

$$v_{w1} = \sqrt{(0.8 \times 9.81 \times 12} = 9.7 \text{ m s}^{-1}.$$

Therefore, $u_2 = 9.7 \text{ m s}^{-1}$ and

$$\tan \theta = v_{f1}/u_1 = 2.3/9.7 = 0.237,$$

$$\theta = 13°20'.$$

(b) Since internal diameter $= \frac{1}{2} \times$ external diameter,

$$u_2 = \frac{1}{2} u_1 = 4.85 \text{ m s}^{-1}.$$

Therefore,

$$\tan \beta_2 = v_{f2}/u_2 = 2.3/4.85 = 0.475$$

$$\beta_2 = 25°20'.$$

(c) Now $u_1 = \omega r_1$, where $\omega = 300(2\pi/60) \text{ rad s}^{-1}$. Therefore,

$$r_1 = u_1/\omega = 9.7 \times 60/300 \times 2\pi = 0.31 \text{ m}.$$

Therefore, breadth at outlet is given by

$$b_1 = 0.28/2.3 \times 0.9 \times 2\pi \times 0.31 = 0.0696 \text{ m}$$

$$= 69.6 \text{ mm}.$$

Since velocity of flow is constant, and internal diameter $= \frac{1}{2}$ external diameter, breadth at outlet is given by

$$b_2 = 2 \times 69.6 = 139.2 \text{ mm}.$$

22.13. Axial flow turbines

The power developed by a turbine is proportional to a product of the total head available and the flow rate. Therefore, the power required from a turbine may, within limits, be obtained by a desired combination of these two quantities. For a Pelton wheel, in order to achieve high jet velocities, it is necessary that the total head is large and, consequently, the flow rate is usually small. However, the Pelton wheel becomes unsuitable if the head available is small, so that in order to achieve the desired power the quantity has to be greater. A Francis-type radial turbine is then used. Its proportions depend upon the flow rate which must pass through it. As in the case of pumps, for greater flow rate the size of the runner eye must be increased, the blade passages become shorter but wider, and a mixed-flow-type turbine results. If the process is carried further, an axial flow turbine is obtained because the maximum flow rate may be passed through when the flow is parallel to the axis.

Figure 22.22 shows that the arrangement of guide vanes for an axial flow turbine is similar to that for a Francis turbine. The guide vane ring is in a plane perpendicular to the shaft so that the flow through it is radial. The runner, however, is situated further downstream, so that between the guide vanes and the runner the fluid turns through a right angle into the axial direction. The purpose of the guide vanes is to impart whirl to the fluid so that when it approaches the runner it is essentially of a free vortex type, i.e. the tangential (whirl) velocity is inversely proportional to radius.

The runner blades must be long in order to accommodate the large flow rate and, consequently, considerations of strength required to transmit the tremendous torques involved impose the necessity for large blade chords. Thus, pitch/chord ratios of 1·0 to 1·5 are used and, hence, the number of blades is small, usually 4, 5 or 6.

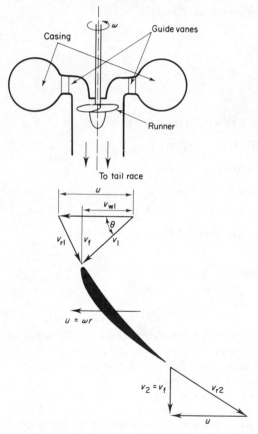

Figure 22.22   Axial flow turbines and velocity triangles

The velocity of the blades is directly proportional to the radius whereas, as stated earlier, the fluid whirl velocity is inversely proportional to it. To cater for this difference, the runner blades are twisted so that the angle they make with the axis is greater at the tip than at the hub.

The blades may be cast as integral parts of the runner or may be welded to the hub. In such cases, the blade angles are fixed, resulting in a rapid fall of efficiency under part-load conditions because then the reduction of flow rate through the machine results in a mismatch between the direction of the fluid velocity relative to the runner and the blade angle. To offset this difficulty, runners may have adjustable or variable pitch blades, whereby they may be turned about their own axes, thus altering the stagger angle to meet the fluid tangentially. By this arrangement a very wide band of high efficiencies may be achieved. Axial flow turbines with variable pitch blades are known as Kaplan turbines. The efficiencies of Kaplan turbines are between 90 and 93 per cent and powers developed are up to 85 MW.

The velocity triangles shown in Fig. 22.22 are similar to those for the axial flow pump. The velocity of flow is axial at inlet and outlet and, of course, remains the same. The whirl velocity is tangential. The blade velocity at inlet and outlet is the same, but varies along the blades with radius from hub to tip.

If the angular velocity of the runner is ω the blade velocity at radius r is given by $u = \omega r$; since at maximum efficiency $v_{w2} = 0$ and $v_2 = v_f$, it follows that

$$E = u v_{w1}/g,$$

in which $v_{w1} = v_f \cot \theta$. Since E should be the same at the blade tip and at the hub, but u is greater at the tip, it follows that v_{w1} must be reduced.

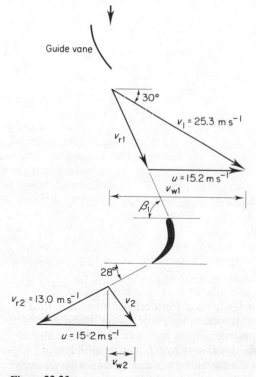

Figure 22.23

Similarly, the velocity of flow v_f should remain constant along the blade and, therefore, cot θ must be reduced towards the tip of the blade. Thus, θ has to be reduced and, consequently, the blade must be twisted so that it makes a greater angle with the axis at the tip than it does at the hub.

Example 22.4

Water is supplied to an axial flow turbine under a total head of 35 m. The mean diameter of the runner is 2 m and it rotates at 145 rev min^{-1}. Water leaves the guide vanes at 30° to the direction of runner rotation and at mean radius the angle of the runner blade at outlet is 28°. If 7 per cent of the total head is lost in the casing and guide vanes and the relative velocity is reduced by 8 per cent due to friction in the runner, determine the blade angle at inlet (at mean radius) and the hydraulic efficiency of the turbine.

Solution

$$H_{nett} = 0{\cdot}93 \times 35 = 32{\cdot}6 \text{ m},$$

$$v_1 = \sqrt{(2gH_{nett})} = \sqrt{(19{\cdot}62 \times 32{\cdot}6)} = 25{\cdot}3 \text{ m s}^{-1},$$

$$u = \pi ND/60 = \pi \times 145 \times 2/60 = 15{\cdot}2 \text{ m s}^{-1}.$$

Therefore, from the inlet velocity triangle (Fig. 22.23),

$$v_{r_1} = 14{\cdot}3 \text{ m s}^{-1}$$

and $\beta_1 = 62{\cdot}2°$.

Therefore,

$$v_{r_2} = 0{\cdot}92 \times 14{\cdot}3 = 13{\cdot}2 \text{ m s}^{-1}$$

and, from the velocity triangles,

$$v_{w_1} = 21{\cdot}9 \text{ m s}^{-1}, \qquad v_{w_2} = 4 \text{ m s}^{-1}.$$

Therefore,

$$E = (u/g)(v_{w_1} - v_{w_2})$$

$$= 15{\cdot}2 \times 17{\cdot}9/9{\cdot}81$$

$$= 27{\cdot}7 \text{ m}$$

and so $\eta_h = E/H = 27{\cdot}7/35 = 0{\cdot}792$

$$= 79{\cdot}2 \text{ per cent.}$$

22.14. Hydraulic transmissions

At the beginning of this Part, fluid machinery was primarily classified according to the direction in which energy is transferred. However, a class of machines exists in which the fluid is used as means of energy transfer. It receives energy from a moving mechanical part only to give it up to another moving mechanical part. If the fluid action is rotodynamic, this class of machines constitutes hydrodynamic transmissions and includes two distinctly different types: fluid (or hydraulic) couplings and torque converters.

Fundamentally, hydrodynamic transmissions consist of two elements: a pump, usually referred to as the primary, and a turbine, known as the secondary. The pump is driven by a prime mover, such as an electric motor or an internal combustion engine; it gives energy to the fluid, usually oil of low viscosity, which then enters the turbine and transmits its acquired energy to it. The turbine shaft provides the mechanical energy output. No solid contact exists between the primary and the secondary. In a fluid coupling, shown diagramatically in Fig. 22.24, two identical impellers are involved. They have radial blades within bowl-shaped shrouds. The space between the blades is full of oil. As the primary begins to rotate, the oil within its impeller moves towards the periphery and is discharged radially into the secondary at the outer radius. Within the secondary, it flows radially inwards towards the centre and is discharged back into the primary near the hub. For the flow to exist, the head produced by the primary must be

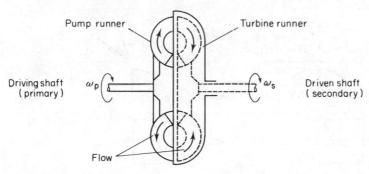

Figure 22.24   Fluid coupling

greater than the centrifugal head of the secondary resisting the flow. This is only possible if the speed of the primary is greater than that of the secondary. Thus, for torque transmission there must be a speed difference giving rise to the 'slip', defined as

$$s = (\omega_p - \omega_s)/\omega_p,$$ (22.48)

where ω_p = angular velocity of the primary, ω_s = angular velocity of the secondary.

The power input to the primary is

$$P_{in} = T_p\omega_p$$ (22.49)

and the power output from the secondary is

$$P_{out} = T_s\omega_s.$$ (22.50)

Therefore, the efficiency of the transmission is

$$\eta = P_{out}/P_{in} = T_s\omega_s/T_p\omega_p.$$ (22.51)

However, in a fluid coupling, the input torque T_p and the output torque T_s must be the same, because there is no other element between the two parts to provide a torque reaction. Hence, for a coupling,

$$\eta = \omega_s/\omega_p = 1 - s.$$ (22.52)

The efficiency of hydraulic couplings is high, usually in excess of 94 per cent. The losses are due to friction and turbulence created when the fluid enters each impeller because the blades are not shaped to meet the flow without shock.

The great advantage of fluid couplings over other types of transmission lies in applications involving unsteady operation, because then torsional vibrations in either of the two halves of the coupling are not transmitted to the other. Also, since the torque is proportional to the speed, the full torque is not developed until the full speed is reached. Thus, the starting load is low, which makes starting of prime movers considerably easier.

Figure 22.25 shows velocity triangles for the primary and secondary at their inlets and outlets. The suffixes used are as follows: 1, inlet to the primary at radius r_i; 2, outlet from primary at radius r_o; 3, inlet to the secondary at r_o; 4, outlet from secondary at r_i. The assumption made is that

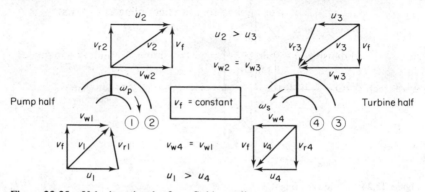

Figure 22.25   Velocity triangles for a fluid coupling

of 'zero whirl slip', which means that the fluid leaves one impeller and enters the other with the same tangential velocity component, e.g.

$$v_{w2} = v_{w3} \quad \text{and} \quad v_{w4} = v_{w1}.$$

Now, by Euler's equation the work done by the primary per unit weight of the fluid is

$$E_p = (v_{w2}u_2 - v_{w1}u_1)/g.$$

But $u_2 = \omega_p r_o$ and $u_1 = \omega_p r_i$. Also, from the velocity triangles,

$$v_{w2} = u_2 = \omega_p r_o \quad \text{and} \quad v_{w1} = v_{w4} = u_4 = \omega_s r_i.$$

Substituting into Euler's equation,

$$E_p = (\omega_p^2 r_o^2 - \omega_p \omega_s r_i^2)/g \tag{22.53}$$

Similarly, the work done on the secondary per unit weight of fluid flowing is:

$$E_s = (v_{w3}u_3 - v_{w4}u_4)/g.$$

But, $v_{w3} = v_{w2} = u_2 = \omega_p r_o,$ $u_3 = \omega_s r_o$ and $v_{w4} = u_4 = \omega_s r_i,$

so that $E_s = (\omega_p \omega_s r_o^2 - \omega_s^2 r_i^2)/g.$ \tag{22.54}

The energy dissipated may be obtained from the difference between equations (22.53) and (22.54):

$$\Delta E = E_p - E_s,$$

which, on substitution, gives

$$\Delta E = (\omega_p - \omega_s)(\omega_p r_0^2 - \omega_s r_i^2)/g. \tag{22.55}$$

Also, it is assumed that, in general,

$$\Delta E = kQ^2, \tag{22.56}$$

where Q is the flow rate through the coupling.

The above expressions are approximations to what actually happens in the coupling because of the difficulty in establishing radii r_0 and r_i and because the flow rate varies with the radius.

Dimensional analysis may be applied to a fluid coupling as follows:

$$T = f(\rho, D, \omega_p, \omega_s, \mu, V),$$

where ρ = fluid density, μ = fluid viscosity, V = volume of fluid, D = diameter of impellers, ω_p = angular velocity of the primary, ω_s = angular velocity of the secondary. It leads to the establishment of the 'torque coefficient',

$$T/\rho\omega_p^2 D^5 = \phi(s, \rho\omega_p D^2/\mu, V/D^3). \tag{22.57}$$

For a given coupling, V/D^3 is constant, the second term is proportional to Reynolds number and, since the flow is very turbulent and its effect is insignificant, the torque coefficient may be approximated to

$$T/\rho\omega_p^2 D^5 = \phi'(s). \tag{22.58}$$

Similarly, since power $P = \omega T$, a power coefficient may be obtained:

$$\frac{P}{\rho\omega_p^3 D^5} = \phi''(s). \tag{22.59}$$

The performance characteristics of an hydraulic coupling are shown in Fig. 22.26.

The difference between the fluid coupling and the torque converter is that while the former consists of only two runners, the latter also has a set of stationary vanes interposed between the two runners, as shown in Fig. 22.27.

Since the stationary vanes change the angular momentum of the fluid passing through them, they are subjected to a torque. Also, since they do not rotate, an equal and opposite torque must be exerted on them from the housing. Thus, the existence of this additional torque to the system means that the torque on the secondary runner is not the same as that on the primary. The relationship between the torques is now

$$T_s = T_p + T_v, \tag{22.60}$$

where T_s = torque on the secondary, T_p = torque on the primary, T_v = torque on the stationary vanes. The efficiency of the torque converter, therefore, is as given by equation (22.51), namely

$$\eta = T_s\omega_s/T_p\omega_p.$$

It is possible for T_v to be either positive or negative. If the vanes design is

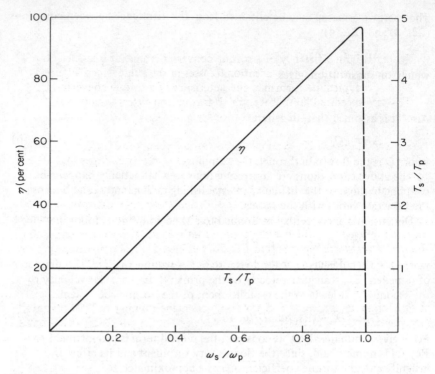

Figure 22.26   Typical characteristics of a fluid coupling

such that they receive torque from the fluid which is in the opposite direction to that exerted on the driven shaft, T_v is positive, signifying an increased output torque. It may in fact be as much as five times the input torque of the primary runner. If, on the other hand, the vanes by virtue of their design receive the torque which is in the same direction as that of the driven shaft, T_v is negative and the secondary torque is reduced.

The dimensional analysis for a torque converter yields the following

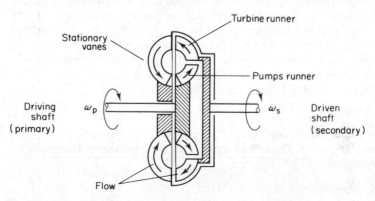

Figure 22.27   Torque converter

relationship after rejecting the Re and the V/D^3 terms,

$$T/\rho\omega_p^2 D^5 = \phi(s, T_s/T_p). \tag{22.61}$$

The maximum efficiency of a torque converter is smaller than that of a fluid coupling because of the additional losses in the guide vanes. Figure 22.28 shows typical performance characteristics of a torque converter.

If a large speed reduction is required, torque converters usually incorporate more than one set of stationary guide vanes and secondary runners.

Figures 22.26 and 22.28 show that when $\omega_s \to \omega_p$ (or $\omega_s/\omega_p \to 1$) the efficiency of a torque converter is low, while that of a coupling is high. It is advantageous, therefore, to design the guide vanes so that they are allowed to rotate. Under normal operation, when $\omega_p > \omega_s$, the stator is held in a fixed position by the torque on the stator vanes, but when the speed of the primary approaches that of the secondary, the vanes are allowed to freewheel. Thus, they cease to influence the torque transmitted, T_v becomes zero and the torque converter performs as a fluid coupling. Typical performance characteristics of such a torque converter are shown in Fig. 22.29.

Example 22.5

A fluid coupling transmits 185 kW from an engine running at 2250 rev min^{-1} to a gearbox. Oil of relative density 0·86 is used in the coupling, which has a slip of 3 per cent. The cross-sectional area of flow in the primary member is

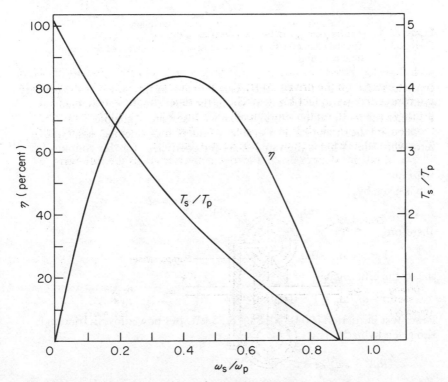

Figure 22.28   Typical characteristics of a torque converter

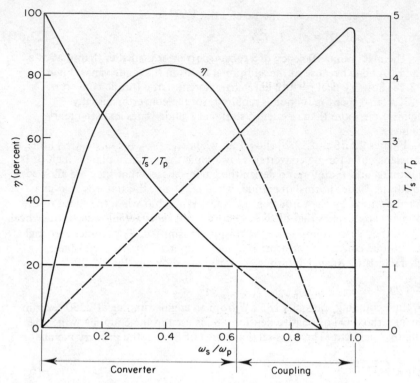

Figure 22.29   Typical characteristics of a torque converter capable of operation as a fluid coupling

constant at $2·8 \times 10^{-2}$ m^2 and the mean diameter at the outlet of the primary member is 460 mm. The frictional loss in the fluid circuit may be taken as 3·5 times the mean velocity head and shock losses are negligible.

Calculate the mean diameter at inlet to the primary member assuming zero 'whirl slip'. What is the value of torque coefficient for this coupling?

Solution
Slip is given by

$$s = (\omega_p - \omega_t)/\omega_p = 0·03.$$

Therefore,

$$1 - \omega_t/\omega_p = 0·03$$

and so the efficiency is

$$\eta = \omega_t/\omega_p = 1 - 0·03 = 0·97.$$

Power lost in friction is $0·03 \times 185 = 5·55$ kW. But power lost in friction is also given by $\rho g Q h_f$ and

$$h_f = 3·5 \, v^2/2g = 3·5(Q/A)^2/2g$$

$$= 3·5\{Q^2/(2·8 \times 10^{-2})^2\}/2g.$$

Therefore,

$$5 \cdot 55 \times 10^3 = 0 \cdot 86 \times 10^3 \ gQ \times 3 \cdot 5 \{Q^2/(2 \cdot 8 \times 10^{-2})^2 \} 2g$$

$$Q^3 = 5 \cdot 55 \times 2 \times (2 \cdot 8)^2 \times 10^{-4}/0 \cdot 86 \times 3 \cdot 5 = 29 \times 10^{-4}$$

$$Q = 0 \cdot 143 \ \text{m}^3 \ \text{s}^{-1}.$$

Energy transfer to the primary is

$$E = (u_o v_{wo} - u_i v_{wi})/g,$$

where the suffices o and i refer to the outlet and inlet, respectively. Therefore,

$$u_o = \omega_p r_o, \quad u_i = \omega_p r_i, \quad v_{wo} = \omega_p r_o, \quad v_{wi} = v_{w4} = v_{w1} = \omega_s r_i$$

and, hence,

$$E = (\omega_p^2 r_o^2 - \omega_p \omega_s r_i^2)/g = \omega_p^2 \{r_o^2 - (\omega_s/\omega_p) r_i^2\}/g.$$

The power transmitted is

$$P = \rho g Q \omega_p^2 \{r_o^2 - (\omega_s/\omega_p) r_i^2\}/g$$

$$= \rho Q \omega_p^2 \{r_o^2 - (\omega_s/\omega_p) r_i^2\}.$$

Now, $\omega_p = 2\pi N/60 = 2\pi \times 2250/60 = 236 \ \text{rad s}^{-1}$

and $\omega_s/\omega_p = 0 \cdot 97.$

Therefore,

$$185 = 0 \cdot 86 \times 0 \cdot 143 \times (236)^2 \{(0 \cdot 46/2)^2 - 0 \cdot 97 r_i^2\},$$

$$(0 \cdot 23)^2 - 0 \cdot 97 r_i^2 = 185/0 \cdot 86 \times 0 \cdot 143 \times (236)^2$$

$$0 \cdot 97 r_i^2 = 0 \cdot 53 - 0 \cdot 0272$$

$$r_i^2 = 0 \cdot 0276$$

$$r_i = 0 \cdot 167 \ \text{m} = 167 \ \text{mm}.$$

Hence, $D_i = 334 \ \text{mm}.$

Torque coefficient is calculated from

$$\text{Torque coefficient} = T/\rho \omega_p^2 D^5 = P/\rho \omega_p^3 D^5$$

$$= 185/0 \cdot 86 \times (236)^3 \times (0 \cdot 46)^5$$

$$= 0 \cdot 0008.$$

EXERCISES 22

22.1   The pressure rise Δp generated by a pump of given geometry depends on the impeller diameter D, its rotational speed N, the fluid density ρ, viscosity μ, and the rate of discharge Q. Show that the relationship between these variables may be expressed:

$$\frac{\Delta p}{\rho N^2 D^2} = f\left(\frac{Q}{ND^3}, \frac{\rho ND^2}{\mu}\right)$$

and give any consistent set of units for the quantities appearing in this equation.

A given pump rotates at a speed of 1000 rev/min and at its duty point it generates a head of 12·2 m when pumping water at a rate of 0·0151 m^3/s. Calculate the head generated by a similar pump of twice the size when operating under dynamically similar conditions and discharging 0·0453 m^3/s. Assume that the effects of viscosity are negligible.

Determine also the rotational speed of the larger pump.

[6·86 m, 375 rev/min]

22.2   A 0·4 m diameter fan running at 970 rev/min is tested when the air temperature is 10 °C and the barometric pressure is 772 mm Hg and the following data are observed: $Q = 0·7$ m^3/s; fan total pressure = 25 mm of water; shaft power = 250 W.

Find the corresponding volume flow, fan total pressure and shaft power of a geometrically similar fan 1 m diameter, running at 500 rev/min when the air temperature is 16 °C and the barometric pressure is 760 mm Hg. Assume that the fan efficiency is unchanged.

[5·65 m^3/s, 40 mm of water, 3·22 kW]

22.3   A centrifugal pump will operate at 300 rev/min delivering 6 m^3/s against 100 m head.

Laboratory facilities for a model are: maximum flow 0·28 m^3/s and maximum power available 225 kW. Using water and assuming that efficiencies of model and prototype are the same find the speed of the model and the scale ratio. Also calculate specific speed.

[1196 rev/min, 4· 4, 0·439 (rad)]

22.4   A centrifugal pump is to be designed to pump castor oil having density 944 kg/m^3 and viscosity 0·144 Ns/m^2. The design is to be tested by means of experiments carried out on a model of one-quarter scale in which air is to be used as the working fluid. The air density is 1·23 kg/m^3 and its viscosity is 1·82 x 10^{-5} Ns/m^2. It is important that viscous effects are accurately represented.

If the oil pump is to run at 105 rad/s calculate the rotational speed of the model air blower.

Determine also the ratio of the powers required for the two machines.

[163 rad/s, 210 x 10^3]

22.5   A centrifugal fan delivers 2·0 m^3/s when running at 960 rev/min. The impeller diameter is 70 cm and the diameter at blade inlet is 48 cm. The air enters the impeller with a small whirl component in the direction of impeller rotation, but the relative velocity meets the blade tangentially. The impeller width at inlet is 16 cm and at outlet 11·5 cm. The blades are backward inclined making angles of 22·5° and 50° with the tangents at inlet and outlet respectively.

Draw to scale the inlet and outlet velocity triangles and from them determine the theoretical total head produced by the impeller.

Assuming that the losses at inlet, in the impeller and in the casing amount to 70% of the velocity head at impeller outlet and that the velocity head at fan discharge is 0·1 of the velocity head at impeller outlet, calculate the fan

static pressure in mm of water column. Assume the air density to be $1 \cdot 2 \ kg/m^3$ and neglect the effect of blade thickness and interblade circulation.
[67·1 mm]

22.6 A centrifugal pump running at 2950 rev/min gave the following results at peak efficiency when pumping water during a laboratory test:

Effective head:	H = 75 m of water
Flowrate:	Q = 0·05 m^3/s
Overall efficiency:	η = 76%

(a) Calculate the specific speed of the pump in dimensionless terms and based on rotational speed in rev/s.

(b) A dynamically similar pump is to operate at a corresponding point of its characteristic when delivering 0·45 m^3/s of water against a total head of 117 m. Determine the rotational speed at which the pump should run to meet the duty and the ratio of its impeller diameter to that of the model pump tested in the laboratory, stating any assumptions made. What will be the power consumed by the pump?
[0·077; 1375 rev/min, 2·68, 679 kW]

22.7 A centrifugal pump delivers water at a rate of 0·022 m^3/s when running at 1470 rev/min. The gauge readings of suction and delivery pressure heads, taken at the same level, were −3 m and +12 m respectively and the power supplied to the pump was 4·8 kW. The cross-sectional area of the suction pipe is $14 \cdot 2 \times 10^{-3} \ m^2$ and the area of the delivery pipe is $10 \cdot 3 \times 10^{-3} \ m^2$. The pump has a 23 cm dia. impeller, which is 19 mm wide at outlet and has blades leaning backwards and making an angle of 60° with the radius at blade outlet.

Assuming that there is no whirl at impeller inlet and that due to interblade circulation the actual whirl component at outlet is equal to 2/3 of the theoretical, find the loss of head in the pump and its overall efficiency.
[2·89 m, 67·8%]

22.8 An axial flow pump handling oil of specific gravity 0·8 at a rate of 1·0 m^3/s is fitted with downstream guide vanes and its impeller rotates at 250 rev/min. The oil approaches the impeller axially, and the velocity of flow, which may be assumed constant from the hub to the tip, is 3·0 m/s. The pump consumes 60 kW, the overall efficiency being 77% and the hydraulic efficiency (including guide vanes) is 86%. If the impeller diameter is 0·8 m and the hub diameter is 0·4 m find the inlet and outlet blade angles and the guide vane inlet angles at the hub and at the tip. Assume that the total head generated at the hub and at the tip is the same.
[at hub: 29·8°, 158·5°, 13·2°; at tip: 16°, 36·5°, 25°]

22.9 In a Pelton wheel the diameter of the bucket circle is 2 m and the deflecting angle of the bucket is 162°. The jet is 165 mm diameter, the pressure behind the nozzle is 1000 kN/m^2 and the wheel rotates at 320 rev/min. Neglecting friction find the power developed by the wheel and the hydraulic efficiency.
[701 kW, 73·3%]

22.10 A Pelton wheel develops 8 MW under a net head of 130 m at a speed of 200 rev/min. Assuming the coefficient of velocity for the nozzle 0·98, hydraulic efficient 87%, speed ratio 0·46 and the jet diameter to wheel diameter ratio of 1/9 determine (a) the flow required, (b) the diameter of

the wheel, the diameter and the number of jets needed and the specific speed.
[7.21 m³/s, 2.17 m, 0.242 m, 3, 0.039]

22.11   A Pelton wheel nozzle, for which the coefficient of velocity is 0.97, is 400 m below the water surface of a lake. The jet diameter is 80 mm, the pipe diameter is 0.6 m, its length is 4 km and $f = 0.008$. The buckets deflect the jet through $165°$ and they run at 0.48 jet speed, bucket friction reducing the relative velocity at outlet by 15% of the relative velocity at inlet. The mechanical efficiency of the turbine is 90%. Find the flow rate and the shaft power developed by the turbine.
[0.42 m³/s, 1189 kW]

22.12   The velocity of the jet driving a Pelton wheel is 60 m/s, the bucket diameter 33 cm and the wheel speed N rev/min. The relative velocity at outlet is 0.85 of that at inlet and is deflected by the buckets by $160°$. Working from first principles deduce an expression for the hydraulic efficiency of the wheel and calculate it for $N = 400$ rev/min and $N = 800$ rev/min. What is the maximum efficiency?
[36.6%, 63.7%, 89.9%]

22.13   Three identical, double jet Pelton wheels operate under a gross head of 400 m. The nozzles are 75 mm dia. with coefficient of velocity 0.97. The pitch circle of the buckets is 1.2 m dia. and the bucket speed is 0.46 x (jet velocity). The buckets deflect the jet by $165°$ and due to friction the relative velocity is reduced by 18%. The mechanical efficiency is 96%. The water from the reservoir is supplied to the turbines by means of two parallel pipes, each of 0.5 m dia. and 450 m long, having friction factor $f = 0.0075$.

If the quantity of water supplied to each turbine is 0.65 m³/s calculate the shaft power developed by it and its rotational speed.
[1876 kW, 602 rev/min]

22.14   A vertical shaft Francis turbine has an overall efficiency of 90% and runs at 428 rev/min with a water discharge of 15.5 m³/s. The velocity at the inlet of the spiral casing is 9 m/s and the pressure head at this point is 260 m of water, the centre-line of the spiral casing inlet being 3.3 m above the tail-water level. The diameter of the runner at inlet is 2.4 m and the width at inlet is 0.3 m. The hydraulic efficiency is 93%. Determine the output power, the specific speed, the guide vane angle and the runner vane angle at inlet.
[36 MW, 0.073, $9°$, $41°$]

22.15   An axial flow turbine operates under a head of 21.8 m and develops 21 MW when running at 140 rev/min. The external runner diameter is 4.5 m and the hub diameter is 2.0 m. If the hydraulic efficiency is 94% and the overall efficiency is 88% determine the inlet and outlet blade angles at the mean radius.
[$30°$, $20°20'$]

22.16   An axial flow turbine, with fixed stator blades upstream of the rotor, running at 250 rev/min has an outer diameter of 1.8 m and an inner diameter of 0.75 m. At the mean diameter the outlet angle is $140°$ in the stator and the rotor blade angle at inlet is $30°$, both measured from the direction of blade velocity. Determine (a) the flowrate for which the angle of incidence for the rotor blades is zero assuming that the axial velocity is uniform, (b) the rotor blade angle at outlet if the whirl component there is zero, (c) the theoretical power output if the change of whirl is independent of the radius.
[12 m³/s, $18.9°$, 1360 kW]

22.17   A fluid coupling has a mean diameter at inlet equal to 0·6 of the mean diameter at outlet which is 0·38 m. The efficiency of the coupling is 96·5% and the specific gravity of the oil used is 0·85. Assuming that the cross-sectional area of the flow passage is constant and equal to 0.026 m² and that the loss round the fluid circuit is four times the mean velocity head, calculate the power transmitted by the coupling from an engine running at 2400 rev/min.
[170 kW]

22.18   The torque coefficient of a particular design of a hydraulic coupling is given approximately by:

$$\frac{T}{\rho\omega_p^2 D^5} = 0 \cdot 25 \left(1 - \frac{\omega_s}{\omega_p}\right)$$

A 250 mm diameter coupling using working fluid of specific gravity 0·85 is driven by a motor running at 1450 rev/min. What is the maximum torque which could be transmitted if the 'slip' is not to exceed 4%?

If the power output is to be increased by 20% to what value should input speed be increased in order to maintain the same slip?
[191 Nm, 1535 rev/min]

23 Positive displacement machines

23.1. Reciprocating pumps

A reciprocating pump consists essentially of a piston moving to and fro in a cylinder. The piston is driven by a crank powered by some prime mover such as an electric motor, I.C. engine or steam engine. Small portable reciprocating pumps may be hand-operated.

When a piston moves away from the valve end of the cylinder, that is to the right in Fig. 23.1, pressure is reduced in the cylinder. This enables the atmospheric pressure p_a acting on the free surface of the liquid in the lower reservoir to force the liquid up the suction pipe and into the cylinder. The suction valve (2) is a one-way valve and opens when the liquid is moving into the cylinder. Thus the outward motion of the piston constitutes a suction stroke. It is then followed by a delivery stroke during which the liquid in the cylinder is pushed out through the delivery valve (3) and into the upper reservoir. During the delivery stroke valve (2) is closed because of the fluid

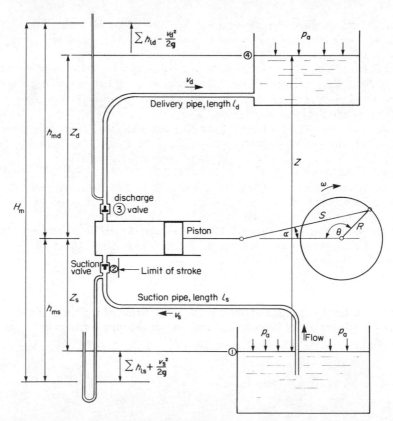

Figure 23.1 Reciprocating pump installation

pressure exerted on it. The whole cycle is then repeated at a frequency dependent upon the rotational speed of the crank ω. Each cycle may be represented by a plot of pressure in the cylinder against the volume of the liquid as shown in Fig. 23.2. During the suction stroke the pressure in the

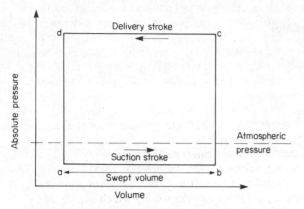

Figure 23.2 Basic pressure diagram for a reciprocating pump

cylinder is below atmospheric as represented by line ab in the diagram. On reversal of the direction of motion of the piston, that is at the end of suction stroke and beginning of the delivery stroke, the pressure rises abruptly along the line bc while the volume remains the same. The delivery stroke follows, during which the high delivery pressure is maintained. This is represented by the line cd. At the end of the delivery stroke the pressure falls along da and the cycle starts again.

However, the simplified analysis above does not take into account the effects which are actually present, namely that of the inertia of the liquid in the pipes which opposes any changes in velocity and that of the frictional losses in the pipes.

At the end of each stroke the liquid in the cylinder and in the relevant pipe must be brought to rest, i.e. decelerated. Immediately afterwards at the beginning of the following stroke, the fluid in the cylinder and in the associated pipe must be accelerated. These accelerations and decelerations result in additional pressures being involved. It was shown in Chapter 18 that the inertia pressure is given by:

$$p_i = \rho g h_i = \rho l \, \frac{\mathrm{d}v}{\mathrm{d}t},$$

where l is the length of pipe and $\mathrm{d}v/\mathrm{d}t$ is the acceleration of the fluid. If the cross-sectional area of the cylinder is A and that of the pipe is a then by continuity:

$$av = Au,$$

if u is the velocity of the piston, so that

$$\frac{\mathrm{d}v}{\mathrm{d}t} = \frac{A}{a} \, \frac{\mathrm{d}u}{\mathrm{d}t},$$

and

$$p_i = \rho l \, \frac{A}{a} \, \frac{du}{dt} \, .$$

(23.1)

Thus at the beginning of the suction stroke (point a) the liquid in the suction pipe must be accelerated so that the pressure in the cylinder must be lowered by an amount

$$(p_i)_{ae} = \rho l_s \, \frac{A}{a_s} \, \frac{du}{dt} \, ,$$

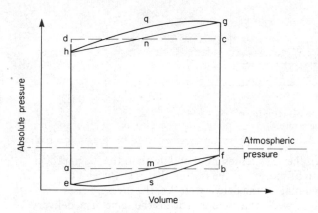

Figure 23.3   Theoretical pressure diagram for
a reciprocating pump

represented by the distance ae on the diagram in Fig. 23.3. At the end of the suction stroke the same liquid in the suction pipe is decelerated so that it exerts pressure on the cylinder by an amount $(p_i)_{fb}$ equal to $(p_i)_{ae}$.

Similarly the delivery stroke is affected at its beginning and its end by pressure changes:

$$(p_i)_{cg} = - (p_i)_{dh} = \rho l_d \, \frac{A}{a_d} \, \frac{du}{dt} \, .$$

Consequently the inertia effects modify the simple pressure diagram abcd so that it becomes emfgnh.

The frictional losses in the pipes are given by the Darcy equation:

$$h_f = \frac{4fl}{d} \, \frac{v^2}{2g} \, ,$$

and may be related to the piston velocity by substitution of v from the continuity equation:

$$v = \frac{A}{a} \, u,$$

so that for the delivery stroke during which the frictional losses in the delivery pipe become relevant:

$$h_{f_d} = \frac{4fl_d}{d_d} \left(\frac{A}{a_d} \right)^2 \frac{u^2}{2g} \tag{23.2}$$

Now, the piston velocity may be obtained from the displacement equation assuming simple harmonic motion, in terms of the crank radius r and crank angle θ and ultimately in terms of the angular velocity ω and time t giving:

$$u = \omega r \sin \theta = \omega r \sin \omega t.$$

The equation shows that when $\theta = 0$, that is at the beginning and at the end of each stroke, $u = 0$ and therefore the frictional effects are zero, but the acceleration (or deceleration) du/dt is a maximum and hence the inertia effects are maximum. When $\theta = 90°$ or $270°$, that is at the middle of each stroke, the velocity is maximum and hence the frictional effects reach a maximum, whereas the acceleration (or deceleration) is zero and therefore the inertia effects vanish. Also, it follows from the Darcy equation that $h_f \propto v^2$ so that curves esb and gqh representing frictional effects on the diagram are parabolae superimposed on lines emf and gnh. Thus the theoretical pressure diagram becomes esbgqh.

The work done by the piston may be obtained as follows. If the instantaneous pressure in the cylinder is p then the force exerted by the fluid on the piston is Ap. If now the piston moves a distance δx then the instantaneous work done is $Ap\delta x$. The total work done during the cycle is therefore

$$\text{W.D./cycle} = A \int p \, dx, \tag{23.3}$$

and is represented by the area of the pressure diagram.

The rate at which the liquid is delivered by the pump clearly depends upon the pump speed, since:

$$\text{Volume delivered in one stroke} = AS = V,$$

where S = piston stroke and V = swept volume.

Thus, if the pump speed is N (rev s^{-1}), then the theoretical volume delivered in 1 s is

$$Q_{th} = ASN,$$

and since $N = \omega/2\pi$,

$$Q_{th} = \frac{AS\omega}{2\pi} = \frac{V\omega}{2\pi}. \tag{23.4}$$

Thus the pump discharge is directly proportional to the rotational speed and is entirely independent of the pressure against which the pump is delivering.

Because of leakage of liquid through glands the actual pump discharge is smaller than the theoretical. If the leakage is q then the actual delivery,

$$Q = Q_{th} - q, \tag{23.5}$$

and the volumetric efficiency,

$$\eta_v = \frac{Q}{Q_{th}} = \frac{Q}{Q + q}. \tag{23.6}$$

Sometimes an expression known as *slip* is used:

$$\text{Slip} = \frac{Q_{th} - Q}{Q_{th}} = 1 - \eta_v \tag{23.7}$$

The pump pressure depends upon the system against which it is working, as shown in Fig. 23.1, and may be obtained as follows. Applying the energy equation to points (1) and (4) and using the liquid level in the lower reservoir as datum:

$$\begin{array}{c} \text{Total energy} \\ \text{at (1)} \end{array} + \text{W.D. by pump} = \begin{array}{c} \text{Total energy} \\ \text{at (4)} \end{array} + \begin{array}{c} \text{Losses in} \\ \text{the system,} \end{array}$$

but:

Total energy at (1) per unit mass of fluid flowing $= \dfrac{v_s^2}{2} + \dfrac{p_a}{\rho}$,

W.D. by the pump per unit mass of fluid flowing $= gH$ (where H = pump head)

Total energy at (4) per unit mass of fluid flowing $= gZ + \dfrac{p_a}{\rho}$,

Losses in the system $= g\Sigma h_{1s} + g\Sigma h_{1d} + \dfrac{v_d^2}{2}$.

Substituting,

$$\frac{p_a}{\rho} + \frac{v_s^2}{2} + gH = gZ + \frac{p_a}{\rho} + g\Sigma h_{1s} + g\Sigma h_{1d} + \frac{v_d^2}{2}.$$

Thus, rearranging, the pump head,

$$H = Z + \Sigma h_{1s} + \Sigma h_{1d} + \frac{v_d^2 - v_s^2}{2g}. \tag{23.8}$$

For reasons explained later it is sometimes useful to consider the pump suction side and delivery side separately and this is why the static lift Z (the difference between the water levels if the reservoirs are open to atmosphere) is considered to be the sum of the suction lift Z_s and delivery lift Z_d. If the difference in levels between the inlet and outlet to the pump is neglected (between (2) and (3) in the diagram) because it is usually very small compared with the rest of the system, the two sides of the pump may be analysed separately as follows.

Applying Bernoulli's equation to the lower reservoir (assuming constant water level) and to the pipe at pump inlet (2):

$$\frac{p_a}{\rho g} = \frac{p_s}{\rho g} + Z_s + \frac{v_s^2}{2g} + \Sigma h_{1s},$$

from which the static pressure at pump inlet,

$$\frac{p_s}{\rho g} = \frac{p_a}{\rho g} - \left(Z_s + \frac{v_s^2}{2g} + \Sigma h_{1s} \right). \tag{23.9}$$

The expression in brackets is known as the manometric suction head,

h_{ms}, because it represents the negative gauge pressure shown by a manometer attached to the pump inlet.

Similarly applying Bernoulli's equation to the delivery pipe at the pump outlet (3) and to the water level in the upper reservoir:

$$\frac{p_d}{\rho g} + \frac{v_d^2}{2g} = \frac{p_a}{\rho g} + Z_d + \Sigma h_{ld},$$

so that the static pressure at pump outlet,

$$\frac{p_d}{\rho g} = \frac{p_a}{\rho g} + (Z_d + \Sigma h_{ld} - \frac{v_d^2}{2g}). \tag{23.10}$$

Here the expression in the brackets is the manometric delivery head, h_{md}, because it represents the gauge pressure shown by a manometer connected to the pump outlet.

The pump manometric head H_m is defined by:

$$H_m = h_{md} + h_{ms}. \tag{23.11}$$

Therefore, substituting,

$$H_m = Z + \Sigma h_{ls} + \Sigma h_{ld} - \frac{v_d^2 - v_s^2}{2g} \tag{23.12}$$

Comparing equations (23.8) and (23.12):

$$H - H_m = \frac{v_d^2 - v_s^2}{2g}.$$

If the delivery and suction pipes are of the same diameter, then $v_d = v_s$ and $H = H_m$.

The internal head, H_i, generated by the pump is greater than the pump head, H, the difference accounting for the internal losses within the pump, h_{lp}. Thus,

$$H_i = H + h_{lp}. \tag{23.13}$$

The internal fluid power generated by the pump is:

$$P_i = \rho g H_i Q_{th}, \tag{23.14}$$

and the actual power output of the pump is:

$$P = \rho g H Q. \tag{23.15}$$

If the power input to the pump from the prime mover is P_o, then the overall pump efficiency is:

$$\eta = \frac{P}{P_o} = \frac{\rho g H Q}{P_o}. \tag{23.16}$$

Component efficiencies are also useful and they are as follows.

Hydraulic efficiency $\quad \eta_h = \frac{H_m}{H_i}.$ $\tag{23.17}$

Mechanical efficiency $\quad \eta_m = \frac{P_i}{P_o}.$ $\tag{23.18}$

It follows therefore that the overall pump efficiency,

$$\eta = \frac{P}{P_o} = \frac{P_i}{P_o} \times \frac{P}{P_i} = \frac{P_i}{P_o} \times \frac{\rho g H Q}{\rho g H_i Q_{th}} = \frac{P_i}{P_o} \times \frac{H}{H_i} \times \frac{Q}{Q_{th}},$$

so that

$$\eta = \eta_m \eta_h \eta_v. \tag{23.19}$$

When considering a pump installation an important limitation on the location of the pump, in relation to the level of the lower reservoir, is the manometric suction head. It represents the lowest pressure in the system (at pump inlet), in particular during the beginning of the suction stroke represented by point e on Fig. 23.3. If this pressure falls to the value of the liquid vapour pressure, cavitation will occur and delivery will cease. This phenomenon is discussed fully in Section 24.6.

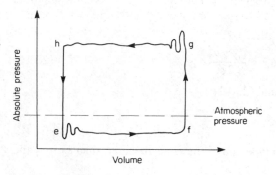

Figure 23.4   Indicator diagram for a reciprocating
             pump with air cylinders

One major disadvantage of reciprocating pumps is the fluctuating flow. It can be reduced by fitting air cylinders to either the suction pipe or the delivery pipe or both. Air cylinders are closed vessels which act similarly to surge tanks. The decelerating liquid moves into the cylinder compressing the enclosed air and thus storing energy in it. When the fluid is accelerated the energy in the air is released, thus augmenting the accelerating force. By this process the fluctuations in the flow are smoothed out to an extent dependent upon the size of the air vessels. Figure 23.4 shows an actual indicator diagram of a pump fitted with air cylinders. It shows that the effects of inertia have been largely eliminated.

The provision of an air chamber reduces the total friction loss to be overcome in the system by maintaining a steady flow from the pump. Without the air chamber the work done against friction would vary with piston position and, as the flow–friction loss relationship is parabolic, the mean frictional resistance is $\frac{2}{3}$ of the maximum, i.e. at mid discharge stroke.

$$\text{Mean friction loss} = \frac{2}{3} \times \frac{4fL}{2dg} \left(\frac{A\omega r}{A} \right)^2.$$

With an air chamber close to the pump, so that the inertia effects in the short connecting pipe may be ignored, then

Mean friction loss = Constant flow based loss

$$= \frac{4fL}{2dg} \left(\frac{A}{a} \frac{\omega r}{\pi} \right)^2.$$

Therefore,

$$\frac{\text{Work done with air vessel}}{\text{Work done without air vessel}} = \frac{1/\pi^2}{2/3} = \frac{3}{2\pi^2}.$$

Another method of dealing with the fluctuations in the flow is the use of multicylinder pumps. In these several cylinders act in parallel and out of phase as shown in Fig. 23.5.

A less effective solution is provided by a double-acting pump in which both sides of the piston are connected to the suction and delivery pipes and both

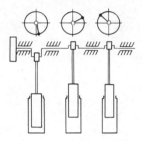

Figure 23.5 Three-cylinder, single acting ram pump

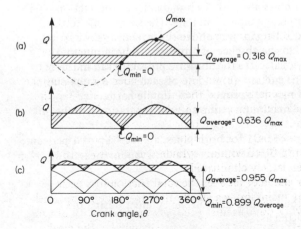

Figure 23.6 Variation of discharge with crank angle θ for
(a) single-cylinder, single acting pump
(b) single-cylinder, double acting pump
(c) two-cylinder, double acting pump

sides of the piston work on the fluid. When one side is on suction stroke the other side at the same time is on the delivery stroke.

Figure 23.6 compares the flowrate fluctuations and the mean deliveries of single acting, double acting and two-cylinder double acting pumps.

The theoretical plot of the pump head H against flowrate Q at a constant speed is a vertical straight line as shown in Fig. 23.7. However, as the head against which the pump is working is increased the flowrate is in practice slightly reduced because of internal leakage.

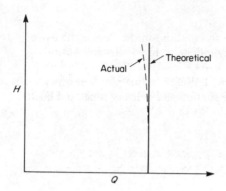

Figure 23.7 Characteristic of a reciprocating pump

Example 23.1

A single–acting single-cylinder positive displacement pump is used to drain an excavation. The pump has a bore of 150 mm and a stroke of 400 mm. The suction and discharge pipes are both 50 mm diameter, the suction pipe being 2 m long and the discharge pipe 15 m long. The suction lift to the pump is 1·5 m while the discharge is 6 m above the level of the water surface in the excavation. In the absence of any air chambers on either pump suction or discharge, calculate the absolute pressure head in the cylinder at the (a) start, (b) end and (c) middle of each stroke if the pump drive is at 0·2 rev s^{-1} and may be assumed to be simple harmonic.

Also, determine (d) the maximum pump speed if separation is to be avoided on the piston face.

Assume a friction factor of 0·01 for both pipes, a pump slip of 4 per cent, an atmospheric pressure of 10·3 m of water, and a fluid vapour pressure of 2·4 m.

Solution

A general expression for the absolute head in the cylinder during the suction stroke may be written as:

$$H = H_{at} \qquad - H_s \qquad - H_{si} \qquad - H_{sf}$$

        Atmosphere   Suction   Suction   Friction in suction
        pressure   lift   acceleration   pipe

While the suction lift term H_s remains a constant during the stroke the

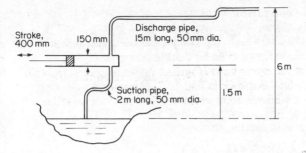

Figure 23.8.

values of the inertia head, H_{si}, associated with the acceleration of the
fluid along the suction pipe varies during the stroke depending on piston
speed. Similarly, as the inflow to the pump depends on piston speed, the
friction loss in the suction pipe depends on piston position. If simple harmonic motion is assumed then the piston speed is given by

$$V = \omega r \sin \omega t,$$

where ω is the rotational speed of the pump drive in rad s^{-1} and r is the
half stroke length, and the flow velocity in a pipe of area Ap is therefore

$$u = V \cdot \frac{A}{a}$$

where A is the cylinder cross-sectional area. At the beginning and end of the
stroke $V = 0$ but the acceleration or inertia pressure necessary to move fluid
is a maximum as du/dt is a maximum at these times

$$h_{si} = \pm \frac{L}{g} \frac{A}{a} \omega^2 r,$$

where L is the length of the pipe.

At mid-stroke time $du/dt = 0$ but the piston velocity is a maximum thus
giving a maximum friction loss along the pipe:

$$h_{sf} = \frac{4fL}{2dg} \left(\frac{VA}{a} \right)^2$$

where d is the diameter of the suction pipe.

Therefore the general equation is

$$H = H_{at} - H_s \pm \frac{L}{g} \frac{A\omega^2 r}{a} - \frac{4fL}{2dg} \left(\frac{VA}{a} \right)^2,$$

(a)   Cylinder head at start of suction stroke,
but $V = 0$, so $H_{sf} = 0$. Therefore

$$H = 10 \cdot 3 - 1 \cdot 5 - \frac{2}{9 \cdot 81} \left(\frac{150}{50} \right)^2 (0 \cdot 2 \times 2\pi)^2 \cdot \frac{200}{1000}$$

$$= 10 \cdot 3 - 1 \cdot 5 - 0 \cdot 58 = 8 \cdot 22 \text{ m of water}$$

$H = 8 \cdot 22$ m (Note inertia head is negative as flow is accelerated).

(b) Cylinder head at end suction stroke,

$$H = H_{at} - H_s + \frac{L}{g} \frac{A\omega^2}{a} r - \frac{4fL}{2dg} \left(\frac{VA}{a}\right)^2 .$$

Again $V = 0$; note inertia head is positive as flow is decelerated to rest.

$$H = 10 \cdot 3 - 1 \cdot 5 + 0 \cdot 58 = \mathbf{9 \cdot 38 \ m \ of \ water}$$

(c) Cylinder head at mid-suction stroke,

$$H = H_{at} - H_s - \frac{L}{g} \frac{dV}{dt} - \frac{4fL}{2dg} \left(\frac{VA}{a}\right)^2 ,$$

but $dV/dt = 0$ and $V = \omega r$.

$$H = 10 \cdot 3 - 1 \cdot 5 - \frac{4 \times 0 \cdot 01 \times 2 \times 1000}{2 \times 50 \times 9 \cdot 81} \left(0 \cdot 2 \times 2\pi \times \frac{200}{1000} \times \right.$$

$$\left. \times \left(\frac{150}{50}\right)^2 = \mathbf{8 \cdot 38 \ m.} \right)$$

During discharge the same form of the equation may be employed,

$$H = H_{at} + H_d \qquad + H_{di} \qquad\qquad + H_{df}$$
$$\quad\quad\quad \text{Delivery} \quad \text{Delivery} \qquad \text{Delivery}$$
$$\quad\quad\quad \text{lift} \qquad\quad \text{acceleration} \quad \text{pipe friction}$$

At the start and end of the stroke, $V = 0$, therefore $H_{df} = 0$ and at mid-stroke $dV/dt = 0$ so that $H_{di} = 0$.

(a) Cylinder head at start of discharge stroke,

$$H = 10 \cdot 3 + 4 \cdot 5 + \frac{L}{g} \frac{A}{a} \omega^2 r$$

$$= 14 \cdot 8 + \frac{15}{9 \cdot 81} \left(\frac{150}{50}\right)^2 (0 \cdot 2 \times 2\pi)^2 \cdot \frac{200}{1000}$$

$$H = \mathbf{19 \cdot 15 \ m \ of \ water}$$

(b) Cylinder head at end of discharge stroke,

$$H = 10 \cdot 3 + 4 \cdot 5 - \frac{L}{g} \frac{A}{a} \quad \omega^2 r$$

$$H = \mathbf{10 \cdot 45 \ m \ of \ water}$$

(c) Cylinder head at mid-discharge stroke,

$$H = 10 \cdot 3 + 4 \cdot 5 + . \frac{L}{g} \frac{dV}{dt} + \frac{4fL}{2dg} \left(\frac{VA}{a}\right)^2 .$$

Now $dV/dt = 0$, $V = \omega r$, and as slip - 4 per cent,

$$(VA)_{actual} = 0 \cdot 96 \ (VA)_{theory} .$$

Therefore,

$$H = 10\cdot3 + 4\cdot5 + \frac{4 \times 0\cdot01 \times 15 \times 1000}{2 \times 50 \times 9\cdot81}\left(0\cdot96 \times 0\cdot2 \times 2\pi\right)$$

$$\left(\frac{150}{50}\right)^2 \cdot \left(\frac{200}{1000}\right)^2$$

$H = 17\cdot68$ m of water

(d)   Separation may be defined as the production of vapour in the cylinder on pump suction (*see* Chapter 24.6) and this requires the absolute pump pressure to fall to $2\cdot4$ m absolute. The minimum pressure occurs at the start of the suction stroke, thus:

$$2\cdot4 = 10\cdot3 - 1\cdot5 - \frac{L}{g}\,\omega^2 r\,\frac{A}{a}\,,$$

$$-6\cdot4 = -\frac{L}{g}\,\frac{A}{a}\,\omega^2 r,$$

$$\omega^2 = 6\cdot4 \times \frac{9\cdot81}{2} \times \left(\frac{50}{150}\right)^2 \cdot \frac{1000}{200}\,,$$

$$\omega^2 = 17\cdot44$$

$$\omega = 4\cdot176 \text{ rad s}^{-1} = 4\cdot176/2\pi \text{ rev s}^{-1}$$

$$= 0\cdot655 \text{ rev s}^{-1}.$$

Drive speed for separation, and hence maximum pump rotation, is **40 rev min^{-1}**.

Example 23.2

A single–acting, single-cylinder positive displacement pump, driven at $0\cdot4$ rev s^{-1}, has a bore of 200 mm and a stroke of 500 mm. The suction and discharge pipes are both 100 mm in diameter. The suction lift is $0\cdot4$ m and the suction pipe is 3 m long. The water is discharged at a point 20 m above the pump level by means of a pipe 200 m long, fitted with a large air chamber 20 m from the pump. Calculate the absolute pump cylinder pressures at the (i) start, (ii) end and (iii) mid stroke times for both (a) suction and (b) discharge assuming no slip at the pump and a friction factor of $0\cdot008$ for both pipes.

Take atmospheric pressure as $10\cdot3$ m.

Further assess the effect of the introduction of the air chamber.

Solution

(a)   Suction stroke:

(i)   Cylinder head at start of suction stroke,

$$H = H_{at} - \underset{\substack{\text{lift}}}{H_{static}} - \underset{\substack{\text{acceleration}}}{H_{suction}} - \underset{\substack{\text{friction loss}}}{H_{suction}}$$

$$= 10\cdot3 - 0\cdot4 - \frac{L}{g}\,\omega^2 r\,\frac{A}{a}$$

i.e. as $V = 0$, dV/dt = maximum at start of stroke

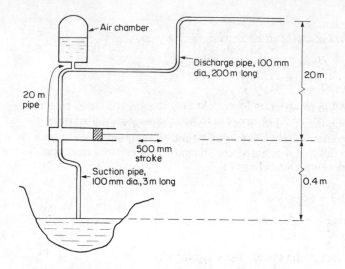

Figure 23.7.

$$H = 10 \cdot 3 - 0 \cdot 4 - \frac{3}{9 \cdot 81} \, (0 \cdot 4 \times 2\pi)^2 \cdot \left(\frac{200}{100}\right)^2 \cdot \frac{250}{1000}$$

$H = 9 \cdot 9 - 1 \cdot 93 = \textbf{7·97 m of water}$

(ii) Cylinder head at end of suction stroke,

$$H = H_{\text{at}} - H_{\substack{\text{static} \\ \text{lift}}} + H_{\text{suction deceleration}}$$

i.e. as $V = 0$ at end stroke, friction loss is zero.

$H = 10 \cdot 3 - 0 \cdot 4 + 1 \cdot 93 = \textbf{11·83 m of water}$

(iii) Cylinder head at mid suction stroke,

$$H = H_{\text{at}} - H_{\substack{\text{suction} \\ \text{lift}}} - H_{\text{friction loss}},$$

i.e. as $\mathrm{d}V/\mathrm{d}t = 0$ at mid-stroke, inertia head is zero.

$$H = 10 \cdot 3 - 0 \cdot 4 - \frac{4fL}{2dg} \, (\omega r)^2 \, \left(\frac{A}{a}\right)^2$$

$$= 10 \cdot 3 - 0 \cdot 4 - \frac{4 \times 0 \cdot 008 \times 3}{2 \times 100 \times 9 \cdot 81} \times 1000 \left(0 \cdot 4 \times 2\pi \times \right.$$

$$\left. \times \frac{250}{1000}\right)^2 \cdot \left(\frac{200}{100}\right)^4$$

$$= 10 \cdot 3 - 0 \cdot 4 - 0 \cdot 31$$

$H = \textbf{9·59 m}.$

(b) Discharge stroke:
   (i) Cylinder head at start of stroke,

$$H = H_{at} + H_{\substack{static \\ lift}} + H_{\substack{inertia \\ head}} + H_{friction}$$

As $V = 0$ at start of cycle, friction loss might at first sight appear to be zero; however, the action of the air chamber is to maintain a steady discharge from the system. Thus the discharge is a constant, at a value of:

$$Q = \text{piston area} \cdot \text{stroke} \cdot \text{rev s}^{-1}$$

$$= A \times 2r \times n$$

$$= 2r\,A\,\frac{\omega}{2\pi} = rA\,\omega/\pi.$$

Thus the friction loss in the system is a constant at:

$$h_f = 4f(L - l)\left(r\,\frac{A}{a}\,\frac{\omega}{\pi}\right)^2$$

where L is the total length of pipe and l is the length of the pipe from the pump to the air chamber, Therefore,

$$H = 10 \cdot 3 + 20 + \frac{l}{g}\,\omega^2 r\,\frac{A}{a} + \frac{4f}{2dg}\,(L - l)\left(r\,\frac{A}{a}\,\frac{\omega}{\pi}\right)^2$$

$$= 30 \cdot 3 + \frac{20}{9 \cdot 81}\,(0 \cdot 4 \times 2\pi)^2 \cdot \frac{250}{1000}\left(\frac{200}{100}\right)^2 +$$

$$+ \frac{4 \times 0 \cdot 008 \times 1000}{2 \times 100 \times 9 \cdot 81}\,(180)\left(r\,\frac{A}{a}\,\frac{\omega}{\pi}\right)^2,$$

where $r\,\dfrac{A}{a}\,\dfrac{\omega}{\pi} = \dfrac{250}{1000}\left(\dfrac{200}{100}\right)^2 \cdot \dfrac{0 \cdot 4 \times 2\pi}{\pi}$

$$= 0 \cdot 8.$$

$$H = 30 \cdot 3 + 12 \cdot 87 + 1 \cdot 88.$$

$$\boldsymbol{H = 45 \cdot 05 \text{ m of water}}$$

   (ii) Cylinder head at end of discharge stroke,

$$H = H_{at} + H_{\substack{static \\ lift}} - H_{\substack{deceleration \\ inertia\ head}} + H_{steady\ friction\ loss}$$

$$= 30 \cdot 3 - 12 \cdot 87 + 1 \cdot 93.$$

$$\boldsymbol{H = 19 \cdot 36 \text{ m of water}}$$

(iii) Cylinder head at mid discharge stroke,

$$H = H_{at} + H_{static} + H_{inertia} + H_{friction} + H_{steady\ friction}$$
$$\quad\quad\quad lift \quad\quad head = 0 \quad\quad in\ pipe\ 1$$

$$= 30\cdot3 + \frac{4fl}{2dg} \left(\omega r \, \frac{A}{a} \right)^2 + 1\cdot93$$

$$= 30\cdot3 + \frac{4}{2} \times \frac{0\cdot008}{100} \times \frac{20}{9\cdot81} \times 1000 \quad 0\cdot4 \times 2\pi \times \left(\frac{250}{1000}\right)^2 \left(\frac{200}{100}\right)^4$$

$$+ 1\cdot93.$$

$$H = 30\cdot3 + 2\cdot06 + 1\cdot93 = \textbf{34·3 m of water}$$

Example 23.3

For the reciprocating pump described in Example 23.1, calculate the increase in pump speed in rev min^{-1} if a large air chamber were fitted close to the pump suction valve.

Solution

The effect of introducing an air chamber would be to remove the inertia head effects calculated previously.

Therefore, as the minimum head occurs at the start of the suction cycle and to avoid separation it must be at least 2·4 m absolute, it follows that:

$$H = H_{at} - H_{suction} - H_{inertia} - H_{friction\ loss}$$
$$\quad\quad\quad lift \quad\quad head$$

$$= \textbf{Vapour head of 2·4 m of water}$$

Now $H_{inertia} = 0$, $H_{friction\ loss}$ = constant value based on the mean flow rate in the suction pipe, $\dfrac{A}{a} \omega \dfrac{r}{\pi}$:

$$H_{friction} = \frac{4fL}{2dg} \left(\frac{A}{a} \, \frac{\omega r}{\pi} \right)^2,$$

$$2\cdot4 = 10\cdot3 - 1\cdot5 - \frac{4fL}{2dg} \left(\frac{A}{a} \, \frac{\omega r}{\pi} \right)^2$$

$$6\cdot4 = \frac{4 \times 0\cdot01 \times 2 \times 1000}{2 \times 50 \times 9\cdot81} \left(\frac{150}{50}\right)^4 \left(\frac{\omega}{\pi^2} \times \frac{200}{1000}\right)^2$$

$$\omega = 15\cdot4\ rad\ s^{-1} = 140\ rev\ min^{-1}.$$

Increase in speed $= 140 - 40 = \textbf{100 rev min}^{-1}$.

23.2. Rotary pumps

In rotary pumps there is at least one rotating element which displaces a finite volume of fluid during each revolution. The most common are the gear pump, vane pump, screw pump and the rotating piston pump. All rotary pumps are distinguished by the following features.

(i) There is usually more than one rotating chamber so that the effects of inertia are minimized and the fluctuations in flowrate are small or effectively eliminated.

(ii) The chambers rotate so that in turn they come into direct contact with the inlet and outlet ports making valves unnecessary.

(iii) The rotational speed is usually quite high so that they can be coupled directly to high speed prime movers.

In general, rotary pumps, like all positive displacement pumps, are suitable for applications calling for large heads and small volumes, or specific speeds less than 0·2 (see Fig. 22.7). Rotary pumps are particularly used in oil hydraulics applications, not only because their 'positive' action is desirable, but also because of relatively large pressure and small volume requirements of such applications.

Rotary hydraulic pumps may be divided into two basic types:

(i) fixed capacity in which the swept volume per revolution is fixed;

(ii) variable capacity in which the swept volume per revolution may be varied within certain limits dependent upon the type of the pump.

23.3. Rotary gear pumps

Gear pumps belong to the fixed capacity category. A rotary gear pump consists essentially of two intermeshing spur gears which are identical and which are surrounded by a closely fitting casing as shown in Fig. 23.10. One of the pinions is driven directly by the prime mover while the other is allowed to rotate freely. The fluid enters the spaces between the teeth and the casing and moves with the teeth along the outer periphery until it reaches the outlet where it is expelled from the pump.

If the area enclosed by two adjacent teeth and the casing is a and the axial length of the pinion is l, then the volume of fluid enclosed between

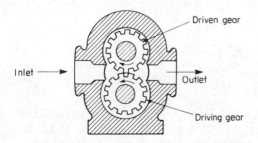

Figure 23.10 Gear pump

adjacent teeth is al and the total volume carried round by one pinion in one revolution is aln, where n is the number of teeth. If, further, the volumetric efficiency is η_v, then the pump discharge is given by:

$$Q = 2\eta_v alnN,$$ (23.20)

where N is the speed of rotation and a may be expressed in terms of the geometric parameters of the gears.

Gear pumps are used for flow rates up to about 400 m^3 h^{-1} working against pressures as high as 17 MN m^{-2}. The volumetric efficiency of gear pumps is in the order of 96 per cent at pressures about 4 MN m^{-2} but decreases as the pressure rises.

The overall efficiency, which takes into account the mechanical losses, is given by:

$$\eta_o = \frac{\rho gHQ}{P} = \frac{pQ}{P} = \eta_m\eta_v,$$ (23.21)

where η_m is the mechanical efficiency.

Figure 23.11 shows typical performance curves for a gear pump at different speeds.

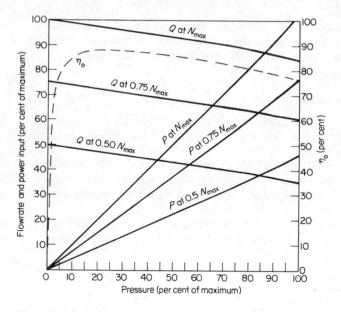

Figure 23.11   Typical performance curves of a gear pump

23.4. Rotary vane pumps

A vane pump consists of a rotor, which is fitted with vanes (blades) which are free to slide radially in the slots within the rotor, as shown in Fig. 23.12. The rotor is positioned eccentrically within a circular sleeve which is part of the casing. Thus as the rotor revolves the vanes subjected to a centrifugal force move radially outwards and their tips follow the contour of the casing sleeve.

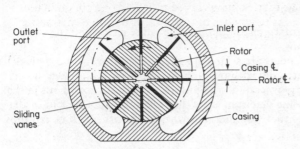

Figure 23.12 Vane pump

Fluid enters through the inlet port, is trapped between the moving vanes and pushed towards the outlet port through which it leaves the pump. Ten to twelve vanes are usually employed. The slots and hence the vanes may be slightly inclined backwards with respect to the direction of rotation to minimize friction and wear at the vanes' tips.

This type of pump can be either of the constant capacity type, in which case the rotor eccentricity is fixed, or the casing may be allowed to move in order to alter the eccentricity, in which case the pump is of the variable capacity type. Clearly the amount of eccentricity controls the volume swept by the vanes and hence the flowrate of the fluid handled.

The theoretical flowrate is given by:

$$Q_{\text{th}} = 2LenDN \sin\left(\frac{\pi}{n}\right),\tag{23.22}$$

(where L = vane width, e = eccentricity, n = number of vanes, D = casing inner diameter and N = rotor speed) and the actual flowrate is:

$$Q = Q_{\text{th}}\eta_v.\tag{23.23}$$

The eccentricity is usually about $0.3r$ but depends upon the number of vanes.

Figure 23.13 shows typical performance curves of a vane pump at constant speed.

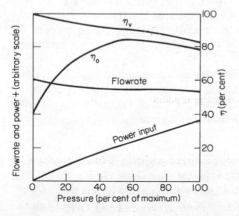

Figure 23.13 Typical characteristics of a vane pump at constant speed

23.5. Rotary piston pumps

There are two kinds of rotary piston pumps: the radial piston and the swash-plate (axial piston) pumps.

In the former the cylinders are arranged radially in a rotor (cylinder block) which is mounted eccentrically within a circular outer casing as shown in Fig. 23.14. As the rotor moves round the pistons reciprocate in and out in the cylinders, their stroke being determined by the eccentricity e. One half of the central stationary opening acts as fluid inlet whereas the other half constitutes the outlet.

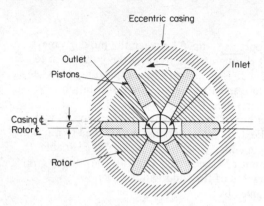

Figure 23.14 Rotary radial piston pump

Since the stroke is equal to $2e$ it follows that the theoretical discharge is given by

$$Q_{th} = \tfrac{1}{2} \ e\pi d^2 nN,\qquad\qquad\qquad (23.24)$$

where e = eccentricity, d = cylinder diameter, n = number of cylinders and N = rotational speed.

Such pumps are not suitable for high pressures because of excessive leakage and hence considerable drop in volumetric efficiency.

An alternative arrangement for a radial piston pump is to have a stationary cylinder block and the pistons operated by an eccentrically mounted shaft. Since in both cases it is possible to design these pumps in such a way that the eccentricity may be varied, they then become variable discharge pumps.

In a swash-plate type of pump, shown diagramatically in Fig. 23.15, the cylinders and pistons are arranged axially in a circle of a stationary casing. The driving shaft rotates an inclined swash plate which actuates the pistons.

For these pumps the theoretical discharge is given by:

$$Q_{th} = \tfrac{1}{2} \ \pi d^2 rnN \tan \theta,\qquad\qquad\qquad (23.25)$$

where d = piston diameter, r = cylinder centreline radius, n = number of cylinders, N = rotational speed and θ = swash-plate angle.

If the swash plate is of a variable angle type the pump discharge may be altered and therefore this type of pump may be designed as a variable delivery pump.

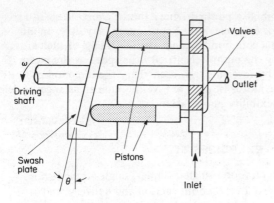

Figure 23.15   Rotary swash-plate pump

23.6. Hydraulic motors

Hydraulic and pneumatic motors are machines in which the fluid energy is converted into mechanical energy. The fluid energy is provided by either suitable oil supplied at high pressure or by compressed air. Most rotary positive displacement pumps, as described in the previous sections, can be used as motors.

For example, gear motors are very similar in construction to gear pumps and many may be used as either pumps or motors depending on whether the sealing arrangements are adequate for both modes of operation. Usually pumps tend to have plain bearings whereas motors are fitted with ball or roller bearings. This is to facilitate starting by reducing the initial torque required to overcome friction.

Similarly vane pumps may be run as motors, again subject to suitable sealing provisions.

Figure 23.16 shows typical performance curves of a swash-plate type hydraulic motor operating at constant pressure.

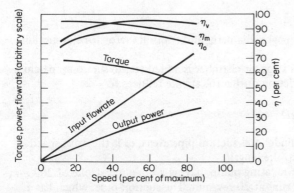

Figure 23.16   Typical characteristics of a swash-plate
type hydraulic motor at constant pressure

Hydraulic units which combine a pump driving a motor constitute self-sufficient hydraulic drives. For example, a vane type pump may drive an oil motor, both operating within a common housing. The advantage of such an arrangement is that by varying the pump eccentricity the speed of the motor may be varied. Furthermore, by reversing the eccentricity the direction of flow is reversed and hence the motor is put into reverse. Such an arrangement therefore offers great flexibility in operation.

EXERCISES 23

23.1 Sketch an indicator diagram for a single-cylinder, single-acting reciprocating pump, fitted with air chambers on both suction and delivery, and compare this to the corresponding diagram without air chambers.

A pump of this type has the following characteristic dimensions: piston diameter = 255 mm, piston stroke = 460 mm, delivery pipe diameter = 115 mm, delivery pipe length = 50 m, speed of pump drive = 20 rev/min, and the piston moves with s.h.m. If the friction factor applicable to the delivery pipe is 0·01, calculate the reduction in power required to overcome friction if a large air chamber is fitted to the pump discharge.
[0·2 kW]

23.2. Define % slip and separation with reference to positive displacement pumps.

Determine the maximum speeds at which a double-acting reciprocating pump can be run under the following conditions: (*a*) no air vessel on the suction side, (*b*) a large air vessel on the suction side close to the pump inlet. The suction lift is 3·65 m, the suction pipe length is 6 m of 100 mm diameter tubing. The cylinder diameter is 100 mm and the stroke is 460 mm. Assume simple harmonic motion for the piston movement and take the limit of pump operation at 2·4 m abs. at an atmospheric pressure of 10·3 m.
[35 rev/min, 169 rev/min]

23.3. In a reciprocating pump the velocity of the water in the suction pipe during the suction stroke varies between zero and V, the displacement of the piston being simple harmonic. Prove that the mean friction head during the stroke is

$$4fL V^2 /3gD$$

where L, D and f are the length, diameter and friction factor for the suction pipe.

Further show that, if a large air chamber is added to the suction pipe close to the pump, the mean friction during the stroke reduces to

$$4fL \left(\frac{A\,\omega r}{a\,\pi} \right) /2gD$$

where A and a are the cylinder and suction pipe areas, ω is the drive speed in rad/s and r is the half stroke length.

23.4 A single-acting reciprocating pump, having a bore and stroke of 200 mm and 400 mm respectively runs at 20 rev/min. The suction pipe, which has a diameter of 100 mm and a length of 9·1 m, has no air chamber fitted. The suction lift is 3·6 m. The discharge pipe is also 100 mm bore and is 470 m long, the discharge being 15·2 m above the pump level, and is fitted with an air chamber 15 m from the pump. Assuming simple harmonic motion for the piston motion and taking friction factor as 0·01 for all pipes, calculate the cylinder pressures at the start, middle and end of both suction and discharge strokes. Take atmospheric pressure as 10·34 m.

[3·4 m, 6·1 m, 9·97 m; 33·7 m, 31·2 m, 22·67 m]

23.5 A double-acting reciprocating pump, of 180 mm bore and 360 mm stroke, draws water from a level 3 m below the pump and delivers to a point 48 m above the water level. Both suction and delivery pipes are 100 mm diameter and of 6 and 76 m length respectively. The pump piston has simple harmonic motion and makes 40 double strokes per minute. Large air chambers are fitted to both pump suction and delivery, 1·5 m away on the suction side and 4·5 m away on the discharge side. The friction factor for both pipes may be taken as 0·008. Determine the pressure difference across the pump at the start of the stroke.

[57·5 m]

23.6 A double-acting positive displacement pump has a bore of 150 mm and a stroke of 300 mm. The suction pipe has a bore of 100 mm and is fitted with an air chamber. Calculate the rate of flow into or out of the air chamber when the crank driving the piston makes angles of 30°, 90° and 120° with inner dead centre.

Determine also the crank angles at which there is no flow to or from the air vessel. Assume drive speed to be 2 rev/s and that the piston has simple harmonic motion.

[17·5 litres/s, 35 litres/s, 30·3 litres/s, 39·5°, 140·5°]

23.7 A double-acting single-cylinder positive displacement pump of 190 mm bore and 380 mm stroke runs at 36 double strokes/min. The suction head is 3·65 m and the discharge static lift is 30·5 m. The suction pipe is 9 m long and the discharge pipe is 61 m long. Large air chambers are provided 3 m away on the suction side and 6 m away on the discharge side. Take the friction factor for both pipes as 0·008. Calculate at the beginning of the stroke the head at the two ends of the cylinder and the load on the piston rod, neglecting its size and assuming simple harmonic motion.

[3·36 m, 49·03 m, 12·8 kN]

23.8 A double-acting single-cylinder reciprocating pump has a cylinder of 90 mm bore and a stroke of 150 mm. The ratio of length of crank to connecting rod is 1 : 4. The suction lift is 2·43 m, the suction pipe being of 64 mm diameter and 4·25 m length. The separation losses at the suction side valves may be taken as the equivalent of an extra 2·4 m of suction pipe length. Calculate the speed at which separation occurs, assuming a vapour level of 1·2 m absolute and an atmospheric pressure of 10·34 m absolute, and the maximum friction loss at this speed.

[86·7 rev/min, 0·38 m]

23.9. An hydraulic motor provides a shaft power of 90 kW when running at 2500 rev/min. The actual capacity of the motor is $38 \times 10^{-6} m^3$/rad. What must be the pressure drop in the motor if the overall efficiency is 83%?

If the flow through the motor is to be provided by a pump with an actual capacity of $65 \times 10^{-6} m^3$/rad, at what speed should the pump operate?

[10 MN/m^2, 1470 rev/min]

24 Pipe-machine systems

Earlier chapters dealt with flow through pipes, assuming that the necessary energy required for the flow is available irrespective of its form. Other chapters dealt with the hydraulic machines, which either depend upon the availability of fluid energy supplied to them (turbines and motors) or convert mechanical energy to fluid energy (pumps and fans) which is then conveyed by a system of pipes or ducts to wherever it is required. Clearly, therefore, there is close interdependence between the performance of the machines and the losses incurred in the system of pipes connected to them. The following sections will deal with such problems.

24.1. Pump and the pipe system

In Section 12.4, it was shown how to relate the flow through a system of pipes in series to the available difference in potential energy due to a difference in levels between reservoirs. Section 12.7 introduced the concept of an equivalent pipe, which enables a complicated system of pipes in series and in parallel to be treated as a single pipe for which the relationship between the head loss h and the flow rate through the system is given by

$$h = KQ^2,$$ (24.1)

if, in equation (12.3), one takes $n = 2$, which is the most common practice. This equation represents what is commonly called the *system resistance*: h is the head loss in the pipeline due to flow rate Q. Thus, in order to maintain this flow rate, energy (per unit weight of fluid flowing) E must be supplied to the pipe system. In a two-reservoir system, if the difference in levels is ΔZ, it is this difference, which is in fact the difference in potential energy between the two reservoirs, that causes the flow. In such a case,

$$\Delta Z = E = h$$

and $\quad \Delta Z = KQ^2,$ (24.2)

which is a generalized expression for losses in Fig. 12.3, as the velocities may be replaced by flow rate, using the continuity equation, and all the resistance coefficients may be then added, their sum being K in equation (24.2). (23.2).

In such a case, there is no need for a pump because the energy required to maintain flow is provided by the difference in levels in the two reservoirs. Supposing, however, that the flow is to be in the opposite direction, i.e. from A to B, as shown in Fig. 24.1, where B is above A. Clearly a pump P will be needed to supply the energy loss in the pipe system and, in addition, to provide energy equal to the difference in levels, because work must be done on the fluid in order to raise it from A to B against the gravitational force. Thus, the total energy required for the flow from A to B to be maintained will be

$$E = Z_B - Z_A + KQ^2$$

or $\quad E = \Delta Z + KQ^2.$ (24.3)

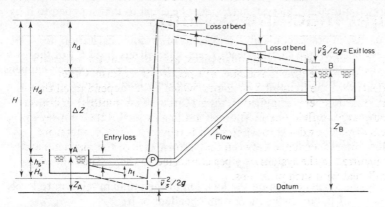

Figure 24.1 Pump and pipe system

This equation is known as the *system characteristic* and the term ΔZ is called the *static lift*.

The example of Fig. 24.1, which also shows the total energy line and the hydraulic gradient line for the system, clearly demonstrates that the loss term KQ^2 incorporates the following losses.

(i) On the suction side of the pump:

Entry loss + Friction loss in the suction pipe,

so that:

$$h_s = K_1 \frac{v_s^2}{2g} + \frac{4fl_s}{d_c} \frac{v_s^2}{2g}$$

(ii) On the delivery side of the pump:

Losses due to bends + Frictional losses + Exit loss,

so that:

$$h_d = K_2 \frac{v_d^2}{2g} + \frac{4fl_d}{d_d} \frac{v_d^2}{2g} + \frac{v_d^2}{2g}$$

and $KQ^2 = h_s + h_d$ ⠀⠀⠀⠀⠀⠀⠀⠀⠀⠀⠀⠀⠀⠀⠀⠀⠀⠀(24.4)

It must be understood that the losses enumerated above refer to the particular example being considered, but, in general, h_s includes all the separation and friction losses on the suction side of the pump whereas h_d includes all the separation and friction losses on the delivery side of the pump.

It will be seen from Fig. 24.1 that the total head rise in the pump H is, thus, equal to the sum of suction and delivery losses and the difference in levels between the reservoirs:

$$H = h_s + h_d + \Delta Z$$

or $H = \Delta Z + KQ^2,$ ⠀⠀⠀⠀⠀⠀⠀⠀⠀⠀⠀⠀⠀⠀⠀⠀⠀⠀(24.5)

which is the same as equation (24.3) because for the flow to be maintained the energy required by the system (E) must be equal to that supplied to it by the pump (H).

It is useful, in practice, mainly for reasons of avoiding cavitation, to distinguish between the suction head H_s and the delivery head H_d of the pump. The suction head includes losses h_s and that part of ΔZ which happens to be on the suction side of the pump. The delivery head includes losses h_d and that part of ΔZ which is on the delivery side (*see* Fig. 24.1). Thus,

$$H = H_s + H_d \tag{24.6}$$

and in our example (Fig. 24.1),

$$H_s = h_s \quad \text{and} \quad \dot{H}_d = h_d + \Delta Z.$$

Clearly H_d and H_s depend upon the location of the pump in relation to levels of points A and B, but their sum H is independent of it.

Returning to equation (24.5), the system resistance, that is the KQ^2 term is a parabola which when plotted offers a simple means of determination of head loss for any given value of Q. It must be remembered though that, if the pipe system is in any way modified or additional losses are introduced (such as partially closing a valve), a new and different parabola will result because the value of K will be changed.

The system characteristic takes into account the difference in elevation, ΔZ, in addition to the head loss. A typical system characteristic is shown in Fig. 24.2. In a system handling air, the Z term is usually negligible because of the low value of air density.

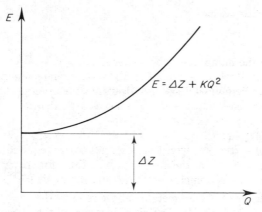

Figure 24.2   System characteristic

For a rotodynamic pump, the head generated is not constant but is a function of discharge (as discussed in Chapter 22), the relationship between the two being the pump characteristic:

$$H = f(Q).$$

Clearly, then, if a rotodynamic pump operates in conjunction with a pipe system, the two must handle the same volume and, at the same time, the head generated by the pump must be equal to the system energy requirement at that flow rate. Therefore,

$$H = E$$

or $$f(Q) = \Delta Z + KQ^2.$$ (24.7)

The solution of this equation is obtained graphically because, clearly, it is the intersection of the pump and system characteristics. This is shown in Fig. 24.3 and the point where the two characteristics cross is known as the

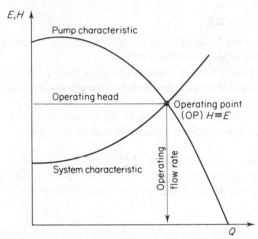

Figure 24.3   Pump and system characteristics

operating point. This is the point on the pump characteristic at which it operates and, at the same time, it is also the point on the system characteristic at which the system operates. 'Pump matching' usually means the process of selecting a pump to operate in conjunction with a given system so that it delivers the required flow rate, operating at its best efficiency, which corresponds to the pump's design point.

The point on the system characteristic which corresponds to the required flow rate through the system is known as the *duty required*. Thus, for correct matching, the operating point should coincide with the duty required. This is not always easy to achieve because the accuracy with which the system resistance is estimated, in practice, is rather poor. Figure 24.4 shows the effect on a fan application of the actual system resistance being different from that estimated. As a result, the flow rate delivered is greater than required and the fan operating efficiency is lower and, consequently, the power consumed would be unnecessarily in excess of that expected.

Figure 24.5 shows the negligible effect of system characteristic on the flow rate delivered by a positive displacement pump. Because of its almost vertical characteristic it is, for practical purposes, independent of the head requirement of the system. Since the efficiency is also almost constant, positive displacement pumps are selected on the flow rate basis only, subject to maximum pressure and power limitations for a given pump.

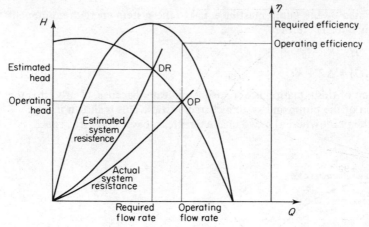

Figure 24.4   Fan operating against estimated and actual system characteristics

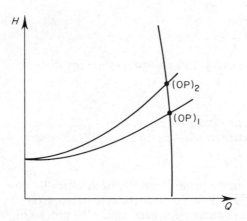

Figure 24.5   Positive displacement pump and system characteristic

Example 24.1

The characteristics of a centrifugal pump handling water are:

Q (m³ s⁻¹)	0·010	0·014	0·017	0·019	0·024
H (m)	9·5	8·7	7·4	6·1	0·9
η (per cent)	65	81	78	68	12

The system consists of 840 m of 0·15 m diameter pipes with absolute roughness 6×10^{-6} m joining two reservoirs; the difference between water levels being 3 m. The water is pumped from the lower to the upper reservoir. Neglecting all losses except friction determine the rate of flow between the two reservoirs and the power consumed by the pump. Take the dynamic viscosity of water as $\mu = 1·14 \times 10^{-3}$ N s m⁻².

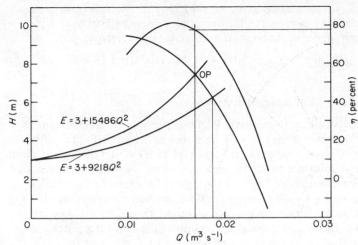

Figure 24.6

Solution
The pump characteristics are drawn in Fig. 24.6. In order to use equation (12.1),

$$h_f = flQ^2/3d^5,$$

to determine the frictional loss in the pipe, it is first necessary to establish the value of friction factor f. The relative roughness of the pipe is given by

$$e/d = 6 \times 10^{-6}/0.15 = 40 \times 10^{-6}.$$

Now, from Moody's chart (Fig. 8.7), since Re is not known, take as first approximation $f = 0.0025$. This gives

$$h_f = \{0.0025 \times 840/3 \times (0.15)^5\}Q^2 = 9218Q^2.$$

Therefore, the system characteristic becomes

$$E = 3 + 9218Q^2.$$

Plotting this, the intersection with the pump characteristic gives

$$Q = 0.0188 \text{ m}^3 \text{ s}^{-1}$$

and so the mean velocity in the pipe is

$$\bar{v} = 4Q/\pi d^2 = 4 \times 0.0188/\pi(0.15)^2 = 1.06 \text{ m s}^{-1}.$$

Hence, $Re = \rho\bar{v}d/\mu = 10^3 \times 1.06 \times 0.15/1.14 \times 10^{-3} = 0.14 \times 10^6.$

Referring to Moody's chart again, $f = 0.0042$, so that the system characteristic becomes

$$E' = 3 + 15486Q^2.$$

Plotting this, the intersection with the pump characteristic gives

$$Q = 0.017 \text{ m}^3 \text{ s}^{-1}.$$

Checking on Reynolds number gives Re = 0.13×10^6, which is close enough to the previous value to accept as the operating point: $Q = 0.017$ m^3 s^{-1}; $H = 7.45$ m; $\eta = 78$ per cent and so, power consumed is given by

$$P = \rho g H Q / \eta = 9.81 \times 10^3 \times 7.45 \times 0.017/0.78 = 1.59 \text{ kW}.$$

24.2. Parallel and series pump operation

It is sometimes necessary to use more than one pump in conjunction with a given system. The pumps may be used 'in series' or 'in parallel'. In the first case, the inlet of the second pump is connected to the outlet of the first pump so that the same flow rate passes through each pump, but the heads generated by the two pumps are added together for a given flow rate. In the parallel operation each pump handles part of the flow rate because the inlets of the pumps as well as the outlets are coupled together. Thus the total flow rate passing through the system is equal to the sum of the flow rates passing through the individual pumps at a given head, which is the same for each pump. Figure 24.7 shows the combined characteristics for two identical

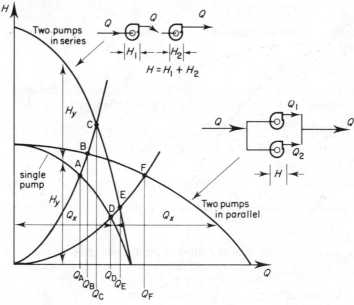

Figure 24.7 Parallel and series operation of two identical rotodynamic pumps

pumps operating in parallel and in series against two resistances R_1 and R_2. For clarity, a system characteristic for which $\Delta Z = 0$ has been used. From the single pump characteristic the combined characteristic of the pumps operating in parallel is obtained by the horizontal addition of values of Q_x for a given head to the characteristic of the single pump. Similarly the combined characteristic of the two pumps connected in series is obtained by vertical addition of values of H_y for every value of Q.

Having obtained the combined characteristics it will be seen that at the system resistance R_1, for example, the single pump will operate at point A, the two pumps connected in parallel will operate at B and when connected in series at point C. Similarly at R_2 the corresponding operating points will be D for a single pump, E for series operation and F for parallel operation.

It is also possible to use two or more dissimilar pumps either in series or in parallel. The procedure for obtaining the combined characteristics is the same as described above. Figure 24.8 shows a case of two dissimilar pumps and it

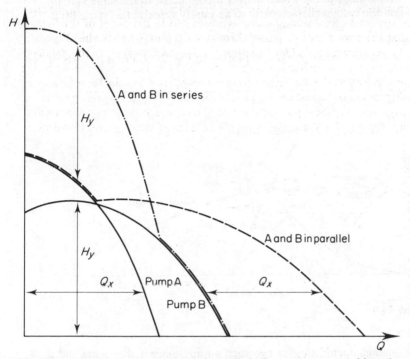

Figure 24.8 Combined characteristics of two dissimilar pumps connected in series and in parallel

should be noticed that certain parts of the combined characteristics are identical with that of the single pump and therefore no benefit whatsoever is achieved by the addition of the second pump if the system characteristic crosses the pump characteristic in these regions.

The cases represented by Figs. 24.7 and 24.8 are only general demonstrations that each case should be studied on its own, and the answer will obviously depend upon the shape of the pump characteristic and upon the system characteristic.

The shape of the pump characteristic is especially important in the case of parallel operation as it may, for certain system characteristics, lead to unstable operation. This occurs when the system characteristic crosses the combined 'pumps in parallel' characteristic at more than one point or is coincidental with it. Axial flow pumps and fans are particularly vulnerable to this danger because they have initially rising head/flow rate characteristics.

Example 24.2

The characteristics of an axial flow pump delivering water are as follows:

Q (m³ s⁻¹)	0		0·040	0·069	0·092	0·115	0·138	0·180
H (m)		5·6	4·20	4·35	4·03	3·38	2·42	0

When two such pumps are connected in parallel, the flow rate through the system is the same as when they are connected in series. Determine the flow rate that a single pump would deliver if connected to the same system. Assume the system characteristic to be purely resistive (no static lift).

Solution

The single pump characteristic is plotted in Fig. 24.9. From it, parts of the

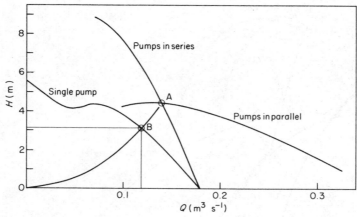

Figure 24.9

combined characteristics for two such pumps connected in series and in parallel are drawn. Since the volume delivered in each case is the same, the system resistance must pass through point A, which is the intersection of the 'parallel' and the 'series' characteristics. This is

$$Q = 0·14 \text{ m}^3 \text{ s}^{-1} \quad \text{and} \quad H = 4·45 \text{ m}.$$

Now, a parabola representing the system characteristic is drawn through this point and through the origin ($\Delta Z = 0$). It crosses the single pump characteristic at B, which would be the operating point for the single pump. It is

$$Q = 0·12 \text{ m}^3 \text{ s}^{-1} \quad \text{and} \quad H = 3·15 \text{ m}.$$

24.3. Change in the pump speed and the system

Pump characteristics, as shown in previous examples, refer to a given speed of pump operation. A change in pump speed will result in a different characteristic which may be predicted, subject to small scale effects, by the use of similarity laws, discussed in Section 22.3 and Chapter 26. These give the following relationships:

$$Q/ND^3 = \text{constant}, \quad gH/N^2D^2 = \text{constant}, \eta = \text{constant}.$$

For a given pump size, the impeller diameter is constant and, therefore, the above relationships reduce to:

$$\frac{Q}{N} = \text{constant}, \tag{24.8}$$

$$\frac{gH}{N^2} = \text{constant}. \tag{24.9}$$

Thus, if a pump characteristic at speed N_1 (Fig. 24.10) is known and it is

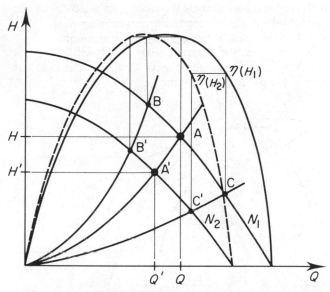

Figure 24.10 Effect of pump speed change on matching with the system without static lift

required to establish the pump's characteristic at speed N_2, the above relationships may be applied to any point on the 'N_1' characteristic — such as point A — giving values of Q' and H' which establish point A' on the 'N_2' characteristic. This point is the *corresponding* point to A and the pump efficiency at A' is the same as at A. Thus,

$$Q' = Q(N_2/N_1) \tag{24.10}$$

and $$H' = H(N_2/N_1)^2. \tag{24.11}$$

A similar procedure is applied to other, arbitrarily chosen, points on N_1 characteristic, such as points B or C, and the corresponding points B' and C' are established. Through them, the predicted characteristic at N_2 is drawn together with the efficiency curve, remembering that, since $\eta' = \eta$ at corresponding points, no calculations are required but merely a replot of efficiency values for the corresponding points at N_2.

It is interesting to note that the corresponding points lie on a parabola passing through the origin. This may be shown as follows. From equations (24.8) and (24.9),

$$Q = cN \quad \text{and} \quad H = kN^2;$$

eliminating N,

$$(Q/c)^2 = H/k \quad \text{or} \quad H = (k/c^2)Q^2,$$

which, in general, may be written as

$$K = CQ^2. \tag{24.12}$$

This equation is of the same form as equation (24.1) for the system resistance. Thus, if the system characteristic is purely resistive ($\Delta Z = 0$), the change in pump speed results in the corresponding points lying on the system characteristic. It means that, for any purely resistive system characteristic, the operating points at different pump speeds will be the corresponding points and, hence, the application of similarity laws does not necessitate replotting of the pump characteristic.

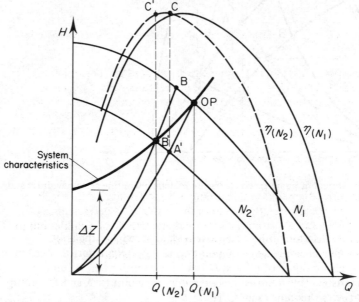

Figure 24.11   Effect of pump speed change on matching with system which contains static lift

This is not the case if $\Delta Z \neq 0$, as illustrated in Fig. 24.11. Here the operating point at N_1 has its corresponding point A' at N_2, but the system characteristic crosses the new pump characteristic (at N_2) at point B' and not at A'. Hence, the flow rate delivered will be $Q_{(N_2)}$ and not that corresponding to A′. Thus, the application of equation (24.10) to $Q_{(N_1)}$ will not give the correct result. Similarly, the efficiency will be that at C′ and not C. In such cases it is, therefore, necessary to replot part of the pump characteristic at the required new speed in order to establish the new operating point.

For positive displacement pumps, the flow rate is directly proportional to the pump speed and, hence, equation (24.8) holds. Since the pump characteristic is a vertical line (Fig. 24.12), a change in speed will result in a

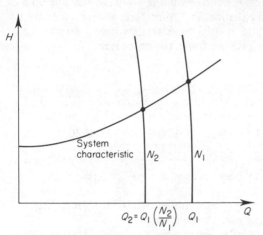

Figure 24.12   Matching of a positive displacement pump with a system characteristic

proportional change in flow rate and a change of head against which the pump will operate, this change depending entirely upon the system characteristic. However, this is of little consequence unless it exceeds the maximum for the system or the pump. The power consumed by the pump will, of course, be affected, but the efficiency may be assumed constant.

24.4. Change in pump size and the system

Geometrically similar pumps, i.e. of the same design, are made at different sizes. The impeller diameters are, consequently, different and it is again possible to apply the similarity laws in order to predict a performance of a pump having diameter D_2 from the known characteristic of a similar pump having diameter D_1, both pumps running at the same speed. Thus, if N = constant, the similarity laws give the following relationships:

$$Q/D^3 = \text{constant}, \tag{24.13}$$

$$gH/D^2 = \text{constant}. \tag{24.14}$$

Hence, selecting arbitrary points on the known characteristic (for D_1) it is possible to calculate the corresponding points on the required characteristic for D_2. If, for example, at point A the flow rate is Q and the head is H, then at A′,

$$Q' = Q(D_2/D_1)^3, \tag{24.15}$$

$$H' = H(D_2/D_1)^2. \tag{24.16}$$

It is important to observe, however, that these corresponding points do not lie on a square parabola such as the non-resistive system characteristic. This may be shown by eliminating D from equations (24.13) and (24.14):

$$D = Q^{1/3}/c = H^{1/2}/k,$$

so that: $H = CQ^{2/3}.$ (24.17)

Thus, for changes in diameter, the corresponding points ($\eta = \eta'$) lie on a curve represented by the above equation. It is, therefore, always necessary to plot part of the new characteristic in order to ascertain where it crosses the system characteristic. Example 24.3 illustrates the procedure.

Example 24.3

A centrifugal pump has an impeller diameter of 0·5 m and its characteristics are as follows:

Q (m³ s⁻¹)	0	0·10	0·15	0·20	0·25	0·30
H (m)		40	37·5	33·0	27·5	20·0	12·0
η (per cent)	0	73	82	81	71	48

Draw the characteristics of a geometrically similar pump having impeller diameter of 0·562 m and running at the same speed.

If the two pumps operate against a system which includes a static lift of 10 m and is such that the smaller pump delivers 0·22 m³ s⁻¹, establish the operating point of the larger pump and the operating efficiencies of both pumps. Show on your graph some lines connecting the corresponding points on the two characteristics.

Solution

The pump characteristic for D_1 = 0.5 m is drawn in Fig. 24.13. In order to obtain the pump characteristic for D_2 = 0·562 m, points A, B, C and D are selected and the similarity laws are applied as follows:

$$Q_2 = Q_1 (0·562/0·5)^3 = 1·42 Q_1,$$
$$H_2 = H_1 (0·562/0·5)^2 = 1·263 H_1.$$

Using these relationships, points A′, B′, C′ and D′ are obtained. Through them is plotted the new characteristic for D_2 = 0·562 m. The efficiency values corresponding to A, B, C and D are moved horizontally to correspond to points A′, B′, C′, D′, respectively.

The system resistance is obtained by drawing a parabola through the operating point and $\Delta Z = 10$ m. Its intersections with the two pump characteristics give

$$Q = 0·22 \text{ m}^3 \text{ s}^{-1}, H = 25 \text{ m}, \eta = 78 \text{ per cent, for } D_1 = 0·5 \text{ m};$$
$$Q = 0·28 \text{ m}^3 \text{ s}^{-1}, H = 35 \text{ m}, \eta = 82 \text{ per cent, for } D_2 = 0·562 \text{ m}.$$

The corresponding points are joined by lines AA′, BB′, CC′ and DD′, all of which pass through the origin.

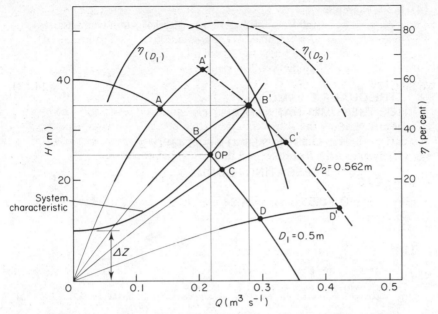

Figure 24.13

24.5. Computer Program 'MATCH'

(1) The program establishes the operating point at the intersection of pump or fan head-flowrate characteristic and a system characteristic. It then enables the user to alter the system parameters and/or change the pump/fan speed or impeller diameter and obtains the new operating point. Flow and pressure (head) units are not required provided consistency is maintained.

(2) The required input is:

(a) the pump or fan head-flowrate characteristic in any of the following forms:

   (i) tabular, of up to 11 points of head and flowrate;

   (ii) tabular, of up to 11 points of head coefficient and flow coefficient together with pump diameter and rotational speed;

   (iii) quadratic approximation of the form $H = C - BQ - AQ^2$ requiring A, B, C input;

   (iv) quadratic approximation of the form $K_H = C' - B'(K_Q) - A'(K_Q)^2$ requiring A', B', C' input together with pump diameter and rotational speed.

(b) the system characteristic in the form $h = z + kQ^2$ requiring z and k input.

(3) The concept of matching is based on the simultaneous solution of polynomials based on equations (22.26) with (24.3) so that the operating point is determined by successive calculation of $(H - h)$ for flowrates increasing in steps which are diminished every time $(H - h)$ changes sign. For dimensionless coefficients and similarity calculations use is made of equations (22.27), (22.28), (24.10), (24.11), (24.15) and (24.16).

(4) Input example:

$$K_Q = \quad 0.04 \quad 0.08 \quad 0.10 \quad 0.14$$
$$K_H = \quad 7.5 \quad 6.4 \quad 5.6 \quad 3.5$$
$$D = 0.862\text{m}; N = 720 \text{ rev min}^{-1};$$
$$z = 20 \text{ m}; k = 208 \text{ s}^2\,\text{m}^{-5}.$$

(5) Output:

THE QUADRATIC MODEL
OF THE PUMP CHARACTERISTIC
CURVE IS:

$$H = 86.53 + (-3.55)\, Q + (-38.46)\, Q^2$$

THE OPERATING POINT IS:

$$Q = 0.51 \qquad H = 74.64$$

(6)

List:

```
10 MODE 7
20 DIM Q(11),H(11),F1(11),U1(11),V1(11),P(11),F2(11),U2(11),V2(11),KQ(11),KH(
11)
30 CLS: PRINT "PROGRAM: MATCH": GOSUB 1560
40 GOSUB 1540: PRINT"IS THE PUMP PERFORMANCE INPUT DATA"
50 PRINT"IN A TABULAR FORM (T), OR IS THE"
60 PRINT"QUADRATIC APPROXIMATION (Q) KNOWN?"
70 INPUT "(T/Q)";A$
80 IF A$="Q" THEN 740
90 GOSUB 1540: PRINT "IS THE INPUT DATA FOR A SPECIFIC"
100 PRINT "PUMP OF KNOWN DIAMETER AND SPEED"
110 PRINT "(S) OR IS IT IN TERMS OF DIMENSIONLESS"
120 INPUT "COEFFICIENTS (C)";B$
130 IF B$="C" THEN 580
140 GOSUB 1540: PRINT"HOW MANY POINTS ARE YOU GOING"
150 GOSUB 1540: INPUT"TO INPUT, I=";I
160 GOSUB 1540: PRINT"WHAT ARE THEY?"
170 Q=0:V0=0
180 FOR X=1 TO I
190 INPUT"Q, H";Q(X),H(X)
200 Q=Q+Q(X)
210 V0=V0+H(X)
220 NEXT
230 A0=V0/I
240 QM=Q/I
250 V1=0:U1=0:P=0
260 FOR X=1 TO I
270 F1(X)=Q(X)-QM
280 U1(X)=F1(X)^2
290 V1(X)=H(X)*F1(X)
300 P(X)=Q(X)*U1(X)
310 U1=U1+U1(X)
320 V1=V1+V1(X)
330 P=P+P(X)
340 NEXT
350 A1=V1/U1
360 P1=P/U1
370 K=U1/I
380 V2=0: U2=0
390 FOR X=1 TO I
400 F2(X)=(Q(X)-P1)*F1(X)-K
410 U2(X)=F2(X)^2
420 V2(X)=H(X)*F2(X)
430 U2=U2+U2(X)
440 V2=V2+V2(X)
450 NEXT
460 A2=V2/U2
470 B1=A1-A2*(QM+P1)
480 C1=QM*(A2*P1-A1)+A0-K*A2
490 A3=INT(A2*100+.5)/100
500 B3=INT(B1*100+.5)/100
510 C3=INT(C1*100+.5)/100
520 CLS: PRINT"THE QUADRATIC MODEL"
530 PRINT"OF THE PUMP CHARACTERISTIC"
540 PRINT"CURVE IS:"
```

```
550 GOSUB 1530
560 PRINT TAB(8)"H=";SPC(1)C3"+(";SPC(1)B3")Q+("A3")Q^2"
570 A=-A2: B=-B1: C=C1: GOTO 990
580 GOSUB 1540: PRINT "HOW MANY POINTS ON THE CHARACTERISTIC"
590 INPUT "CURVE DO YOU WANT TO PROCESS";I
600 GOSUB 1540: PRINT " WHAT IS THE IMPELLER DIAMETER,"
610 INPUT " D1 (M)";D1
620 GOSUB 1540: PRINT "ITS ROTATIONAL SPEED,"
630 INPUT " N1 (REV/MIN)";N1
640 M=N1/60
650 GOSUB1540: PRINT "WHAT ARE THE VALUES OF"
660 PRINT "THE COEFFICIENTS KQ AND KH"
670 FOR X=1 TO I:INPUT "KQ, KH";KQ(X),KH(X)
680 M=N1/60
690 Q(X)=KQ(X)*M*D1^3
700 H(X)=KH(X)*(M*D1)^2/9.81
710 NEXT
720 Q=0: V0=0
730 FOR X=1 TO I: GOTO 200
740 GOSUB 1540: PRINT "IS THE QUADRATIC APPROXIMATION IN TERMS"
750 PRINT "OF FLOWRATE (F) OR IS IT IN TERMS OF"
760 PRINT "THE FLOW COEFFICIENT (C) ?"
770 INPUT "(F/C)";F$
780 IF F$="C" THEN 850
790 GOSUB 1540:CLS: PRINT"FOR A QUADRATIC MODEL OF A PUMP"
800 PRINT "CHARACTERISTIC OF THE FORM:"
810 GOSUB 1540: PRINT TAB(10)"H=C-BQ-AQ^2"
820 GOSUB 1540: PRINT"WHAT ARE THE VALUES OF COEFFICIENTS A, B AND C"
830 GOSUB 1540: INPUT "A,B,C";A,B,C
840 A1=A: B1=B: C1=C: GOTO 990
850 GOSUB 1540: PRINT "WHAT IS THE PUMP DIAMETER,"
860 INPUT "D1 (M)";D1
870 GOSUB1540: PRINT "WHAT IS ITS SPEED,"
880 INPUT "N1 IN (REV/MIN)";N1
890 M=N1/60
900 GOSUB 1540: PRINT "FOR A QUADRATIC MODEL OF THE PUMP"
910 PRINT "CHARACTERISTIC OF THE FORM:"
920 GOSUB 1540:   PRINT TAB(10)"KH=C-B(KQ)-A(KQ)^2"
930 GOSUB 1540:   PRINT "WHAT ARE THE VALUES OF COEFFICIENTS A, B AND C ?"
940 INPUT "A, B, C ";A,B,C
950 A2=-A/(9.81*D1^4)
960 B1=-B*M/(9.81*D1)
970 C1=C*(M*D1)^2/9.81
980 GOTO 490
990 GOSUB 1540: PRINT"FOR THE SYSTEM CHARACTERISTIC"
1000 PRINT"OF THE FORM:"
1010 GOSUB 1530: PRINT TAB(10)"H = Z + K*Q^2"
1020 GOSUB 1530:   INPUT* Z, K";Z,K
1030 IF (C-Z)<=0 THEN 1210
1040 QM=(SQR(B^2+4*A*C)-B)/(2*A)
1050 I=QM/3: Q=0
1060 Q=Q+I: GOSUB 1190
1070 IF D>0 THEN 1060
1080 IF ABS(D)<(5*F)/1000 THEN 1100
1090 Q=Q-I: I=I/3: GOTO 1060
1100 QO=INT(Q*100+.5)/100
1110 HO=(H+F)/2
1120 H1=INT(HO*100+.5)/100
1130 GOSUB 1540: PRINT TAB(7)"THE OPERATING POINT IS:"
1140 GOSUB 1540: PRINT TAB(9)"Q=";QO;:PRINT TAB(20)"H=";H1
1150 GOSUB 1540: GOSUB 1560
1160 GOSUB 1540: INPUT"OTHER SYSTEM? (Y/N)";P$
1170 IF P$="Y" THEN 1010
1180 IF P$="N" THEN 1260
1190 F=C-B*Q-A*Q^2: H=Z+K*Q^2
1200 D=F-H: RETURN
1210 CLS: PRINT TAB(8)"NO SOLUTION POSSIBLE"
1220 GOSUB 1550: GOSUB 1560: GOSUB 1530
1230 INPUT"WANT TO CARRY ON/ (Y/N)";W$
1240 IF W$="N" THEN 1490
1250 GOTO 1010
1260 GOSUB 1540: PRINT"DO YOU WANT TO CHANGE"
1270 INPUT"PUMP DIAMETER? (Y/N)";D$
1280 IF D$="N" THEN 1330
1290 GOSUB 1540: PRINT"WHAT ARE THE OLD AND NEW"
1300 PRINT"PUMP DIAMETERS: D1 AND D2 ?"
1310 INPUT "D1, D2";D1,D2
1320 D=D2/D1: GOTO 1340
1330 D=1
1340 GOSUB 1540: PRINT"DO YOU WANT TO CHANGE"
1350 INPUT"PUMP SPEED? (Y/N)";S$
1360 IF S$="N" THEN 1410
1370 GOSUB 1540: PRINT"WHAT ARE THE OLD AND NEW"
1380 PRINT"PUMP SPEEDS: N1 AND N2 ?"
```

```
1390 INPUT"N1, N2";N1,N2
1400 N=N2/N1: GOTO 1430
1410 N=1
1420 IF D=1 THEN 1470
1430 C=C*N*D
1440 B=B*N/D
1450 A=A/D^4
1460 GOTO 1030
1470 GOSUB 1540: INPUT"OTHER PUMP? (Y/N)";P$
1480 IF P$="Y" THEN 40
1490 CLS:,GOSUB 1540
1500 FOR X=1 TO 10:PRINT:PRINT:NEXT
1510 PRINT TAB(15)"THANK YOU"
1520 END
1530 PRINT:RETURN
1540 PRINT: PRINT: RETURN
1550 STAR$=STRING$(40,"*"):PRINT STAR$: RETURN
1560 FOR X=1 TO 6000: NEXT: RETURN
```

24.6. Cavitation in pumps and turbines

Cavitation is the name given to a phenomenon which consists, basically, of local vaporization of a liquid. When the absolute pressure falls to a value equal to or lower than the vapour pressure of the liquid at the given temperature, small bubbles of vapour are formed and boiling occurs. Since liquids normally have air dissolved in them, the lowering of pressure to a value near to the vapour pressure releases this air first. The combination of air release and vaporization is known as cavitation.

In practice, cavitation starts at pressures somewhat higher than the vapour pressure of the liquid. The actual mechanism of cavitation inception is not yet known, but it appears to be associated with the existence of microscopic gas nuclei which cause cavitation. One theory suggests that these nuclei are present in the pores of the solid material at the fluid boundary. It is because of the presence of these nuclei that a fluid cannot withstand tension. It is estimated that, in their absence, tension of 10 000 atm could be transmitted by water.

The nuclei give rise to the formation of bubbles during cavitation inception. These bubbles grow and collapse, producing pressure waves of high intensity, only to be followed by formation of successive bubbles. Each cycle lasts only a few milliseconds, but the local pressures are enormous (may be up to 4000 atm). Similarly, local temperatures may increase by as much as 800 °C.

The occurrence of cavitation is accompanied by a crackling noise and weak emission of light.

In a flowing system, the liquid may be subjected to changes of velocity and, consequently, changes in pressure. When the velocity increases, the pressure falls and, if it falls to a sufficiently low level, cavitation may occur. The bubbles may subsequently flow with the fluid into the region of higher pressure, where they collapse. Thus, cavitation may occur not only at pump inlets or draft tubes of turbines but also on hydrofoils, propellers, in venturi meters or syphons. In general, the effects of cavitation are noise, erosion of metal surfaces and the vibration of the system.

The most general and very useful cavitation parameter is the *cavitation coefficient*, σ, defined as

$$\sigma = (p_1 - p_c)/\tfrac{1}{2}\rho \bar{v}_1^2 \tag{24.18}$$

where p_1 = upstream or ambient static pressure, p_c = critical pressure at which cavitation occurs (usually taken as vapour pressure), $\bar{v}_1$ = mean upstream fluid velocity.

The value of σ at which cavitation starts is called critical cavitation coefficient σ_{crit} and is referred to as the *inception point*. Theoretically, cavitation starts when the pressure falls to the value of the vapour pressure of the liquid, but the latter is a function of temperature. Thus, a system which will operate satisfactorily without cavitation during winter may give cavitation trouble in the summer when the temperature is higher and,

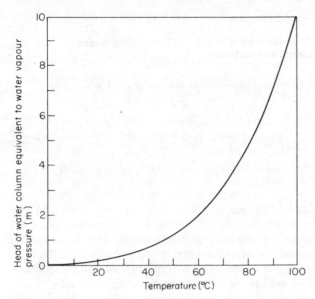

Figure 24.14   Variation of water vapour pressure with temperature

therefore, the vapour pressure of the liquid is also higher. Figure 24.14 gives values of vapour pressure of water as a function of temperature.

In rotodynamic pumps, the cavitation occurs at pump inlet, where the pressure is lowest, as demonstrated in Fig. 24.15, which shows the hydraulic gradient for a simplified pump system.

If the absolute static pressure at pump inlet is p_i, then,

$$p_i = p_{atm} - \rho g H_s, \tag{24.19}$$

where p_{atm} is the atmospheric pressure and H_s is the suction head which includes not only the suction lift Z_s but also the sum of the losses in the inlet pipe h_s and the velocity head, so that:

$$H_s = Z_s + h_s + \bar{v}^2/2g. \tag{24.20}$$

Now, if the vapour pressure is p_{vap} then theoretically cavitation starts when

$$p_i = p_{vap}$$

and the difference

$$H = (p_i - p_{vap})/\rho g$$

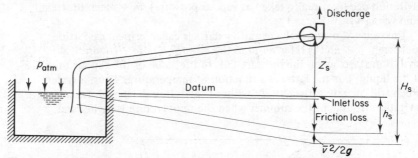

Figure 24.15 Total energy gradient for the suction side of a pump system

is a measure of the absolute head available at the pump inlet above the vapour pressure (above cavitation inception) and is known as the *nett positive suction head* or simply NPSH. Thus,

$$\text{NPSH} = (p_i - p_{\text{vap}})/\rho g = \frac{p_{\text{atm}}}{\rho g} - H_s - \frac{p_{\text{vap}}}{\rho g} \qquad (24.21)$$

In SI, the NPSH is replaced by the *nett positive suction energy* (NPSE), defined as

$$\text{NPSE} = g\,\text{NPSH} = \frac{p_{\text{atm}}}{\rho} - gH_s - \frac{p_{\text{vap}}}{\rho} \qquad (24.22)$$

Thus, NPSE or NPSH represents the difference between the total energy at the inlet flange of the pump and the vapour energy: it is the energy available to the pump on its suction side. If the pump total head is H, no more than the NPSH part of it should be used at the suction side if cavitation is to be avoided.

It was suggested by Thoma that NPSH is proportional to the pump total head H and he defined a cavitation coefficient,

$$\sigma_{\text{Th}} = \text{NPSH}/H = \text{NPSE}/gH. \qquad (24.23)$$

Another useful parameter is the *suction specific speed*, analogous to the type number or specific speed,

$$K_s = \omega Q^{1/2}/(\text{NPSE})^{3/4} = \omega Q^{1/2}/(g\text{NPSH})^{3/4}, \qquad (24.24)$$

where ω = pump rotational speed in radians per second, Q = pump flow rate in cubic metres per second, NPSE is in joules per kilogramme, NPSH is in metres.

The relationship between the Thoma cavitation coefficient and the suction specific speed may be obtained by dividing the type number of the pump by the suction specific speed:

$$\frac{\omega_s}{K_s} = \frac{\omega Q^{1/2}}{(gH)^{3/4}} \div \frac{\omega Q^{1/2}}{(g\text{NPSH})^{3/4}} = \frac{(\text{NPSH})^{3/4}}{H^{3/4}} = \sigma_{\text{Th}}^{3/4};$$

Thus, $\sigma_{\text{Th}} = (\omega_s/K_s)^{4/3}.$ \qquad (24.25)

For geometrically similar pumps the scaling laws may be obtained from:

$$\frac{\text{NPSH}_1}{\text{NPSH}_2} = \left(\frac{N_1}{N_2}\right)^2 \left(\frac{D_1}{D_2}\right)^2.$$ (24.26)

The main effect of cavitation on pumps besides erosion and vibration is the possibility of performance failure. Figure 24.16 shows the characteristic

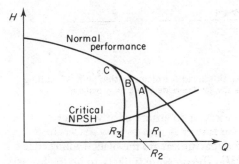

Figure 24.16   Cavitation effects on pump characteristic

curve of a centrifugal pump with performance failure due to the cavitation indicated.

If the pump on test is throttled at discharge, the normal characteristic is obtained. If, however, the inlet valve is partially closed, so that the inlet pressure is lowered by a resistance R_1, and then the pump is tested by opening the discharge valve (starting from shut-off), it will perform normally as far as point A, but a further opening of the discharge valve will no longer produce any increase in flow. Repeating the test with a greater inlet resistance, say R_2, will cause the falling off of performance earlier, namely at some point such as B. By testing a pump in the manner described, it is possible to determine the absolute pressure at inlet at which cavitation occurs (performance failure) and, hence, to calculate the corresponding NPSH (or NPSE). If, now, the critical NPSH is defined as the point at which the head falls by an arbitrary percentage, usually 2 or 3 per cent, below the normal non-cavitating performance, then the critical NPSH may be plotted against the flow rate, as shown in Fig. 24.16, alongside the other pump characteristics. It then shows the minimum NPSH required by the pump in order to avoid cavitation.

In positive displacement pumps, the manometric suction head must be always greater than the vapour pressure. The vapour pressure increases with temperature, as shown in Fig. 24.14. For example, at 90°C it is equivalent to a column of 7·14 m of water. Supposing the atmospheric pressure is equivalent to 740 mm of mercury, which is 0·74 x 13·6 = 10·06 m of water, then the available difference for the manometric suction head is 10·06 − 7·14 = 2·92 m of water. However, allowing for the drop of pressure due to inertia at the beginning of suction stroke and losses in the valve, the manometric suction head must be considerably smaller than that figure. For pumps handling cold water, the maximum manometric head is, in practice, between 6 and 6·5 metres. Figure 24.17 gives the relationship between the manometric suction head and temperature for reciprocating pumps handling water, from which it is seen that, if the water temperature is above 70 °C,

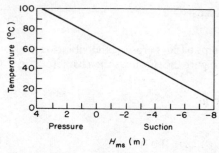

$$H_{ms} \text{ (m)}$$

Figure 24.17   Maximum manometric suction head for reciprocating pumps handling water

the pump must be below the lower reservoir to ensure positive water pressure.

In turbines, the areas susceptible to cavitation are the blade trailing edge and the draft tube, since in these places the pressure is likely to be the lowest. It is possible to avoid cavitation in turbines altogether by submerging them to a low level, but this usually means excavation work, which in view of the large size of average water turbines tends to be very costly. Therefore, cavitation is often accepted and provisions are made for periodic repair of damage caused by it. Because cavitation occurs on the downstream side of a turbine it has very little, if any, effect on its performance.

Turbine cavitation is usually defined by the Thoma coefficient:

$$\sigma_{Th} = (H_{atm} - Z - h_{vap})/H, \tag{24.27}$$

where H_{atm} = atmospheric head, Z = height of centre line of turbine above tailrace, h_{vap} = vapour pressure head, H = nett head across turbine.

As with pumps, a suction specific speed K_s is also used. An empirical relationship between $(K_s)_{crit}$ and ω_s, suggested by Noskievic, is of the form

$$(K_s)_{crit} = a/\sqrt{\omega_s} \tag{24.28}$$

where the value of a, a constant, is between 4·5 and 5·8.

Example 24.4

A centrifugal pump having dimensionless specific speed, based on rotational speed in radians per second equal to 0·45, is to pump 0·85 m³ s⁻¹ at a total head of 152 m. The pump will take water with a vapour pressure of 350 N m⁻² from a storage basin at sea level. For the pump speed consistent with the above requirements, calculate the elevation of the pump inlet relative to the water level based on an acceptable value of suction specific speed equal to 3·2.

Solution

$$\omega_s = \omega \, Q^{1/2}/(gH)^{3/4} = 0·45,$$

$$K_s = \omega \, Q^{1/2}/(gH_{t_1})^{3/4} = 3·2,$$

where H_{t_1} = NPSH = $H_{atm} - z - h - H_{vap}$. Therefore,

$$\omega_s/K_s = (H_{t_1}/H)^{3/4} = 0·45/3·2,$$

$$H_{t_1} = H(0.45/3.2)^{4/3} = H/13.6$$
$$= 152/13.6 = 11.2 \text{ m}.$$

But, $H_{t_1} = H_{atm} - Z - h - H_{vap}$

and, assuming $h = 0$, $H_{atm} = 10.3$ m,

$$H_{vap} = 350/9.81 \times 10^3 = 0.036 \text{ m}.$$

Therefore,

$$Z = H_{atm} - H_{t_1} - H_{vap} = 10.3 - 11.2 - 0.036$$
$$= -0.936 \text{ m}$$

and so the pump must be submerged below the water level.

24.7. Pump selection

Usually, the pump selection starts with the required flow rate and head being specified. These are the two essential quantities which the pump, when operating in conjunction with the given system, must deliver. In addition some constraints on the pump selected may be present. These can be:

(i) pump speed (which may be specified if the prime mover and its speed are known);

(ii) minimum operating efficiency (which may be specified and must be guaranteed by the manufacturer);

(iii) static lift (which may affect the location of the pump with regard to cavitation hazards);

(iv) type of pump (i.e. centrifugal or axial), because of ease of installation in a particular system;

(v) space available for the pump and the type of drive;

(vi) type of fluid to be handled (not only its density and viscosity, but also whether it carries abrasive particles, solid matter, acids, etc.);

(vii) minimum noise level (in the case of fans);

(viii) non-overloading power characteristic (which may be essential).

All these aspects will influence the type of pump or fan chosen. However, in the absence of all the above restrictions, in theory any pump or fan type may be selected for any duty. The difference will be in the size of the machine and its speed of operation. Also, some machine types are more efficient than the others. Generally speaking, if a low type number pump or fan is selected, it will be large and its speed of operation will be low. Conversely, if a high type number pump or fan is chosen, it will be smaller but it will have to run faster. This is well illustrated in Example 24.5.

If the only specification given is the flow rate and the operating head, usually referred to as the *duty required* (DR), the first step is to decide the pump speed. Since it is desirable for the pump to operate at its best efficiency and therefore, at its design point (DP) the pump type number is used because, by its definition, it describes uniquely the design point. Since,

$$n_s = N Q^{1/2}/(gH)^{3/4}$$

and Q as well as H are given, it is possible to establish the numerical relationship:

$$n_s = AN. \tag{24.29}$$

This equation demonstrates clearly that the decision with regards to the pump speed (N) immediately specifies the pump type number, and, hence, the pump type which has to be used.

The most common pump and fan drives are a.c. electric motors, which run at speeds governed by the a.c. frequency and the number of poles. Thus, the synchronous speed of an a.c. motor is given by,

$$N_{\text{synch}} = (2f/n)60 \text{ rev min}^{-1} \tag{24.30}$$

where f = frequency, n = number of poles. However motors run at speeds less than synchronous because of the required slip. The table below gives the synchronous and nominal motor speeds for 50 Hz supply.

No. of poles	N_{synch}	N_{nominal}
2	3000	2900
4	1500	1450
6	1000	960
8	750	720
10	600	575
12	500	480
14	430	410
16	375	360

If, therefore, an electric motor is to be used, the different values of nominal speeds may be substituted into equation (24.29), thus giving a choice of pump types that may be used. Some may have to be rejected because of any of the restrictions listed before. The final choice, therefore, will be based on the balance between the initial and the running costs of the machine chosen. A high type number machine will be small and, hence, initially cheaper. It may however, be less efficient and, hence, its running costs may be higher. Generally, for small pumps and fans requiring small powers, the capital expenditure is usually considered more important. It is certainly not the case for large pumps and fans consuming substantial amounts of power. In such cases it is essential to select a machine which will have the highest possible efficiency, thus reducing its running costs to a minimum.

Having established the pump speed and, hence, the type, it is next necessary to determine the pump size, i.e. the diameter of its impeller. For selection and comparison purposes, pump and fan characteristics are usually drawn in terms of dimensionless coefficients, namely the flow coefficient,

$$K_Q = Q/ND^3 \tag{24.31}$$

and the head coefficient,

$$K_H = gH/N^2D^2, \tag{24.32}$$

which were discussed in Chapter 22.

The value of either of the above two coefficients corresponding to the design point (hence n_s chosen) may be used to calculate the diameter required for the speed chosen. Alternatively, for each pump type it is useful to have, in addition to the value of the type number, the corresponding value of the *specific diameter*, defined as

$$D_s = D(gH)^{1/4}/Q^{1/2} = K_H^{1/4}/K_Q^{1/2}. \qquad (24.33)$$

If this is known for the particular pump selected, its diameter may be calculated immediately.

A common difficulty arises when, for the speed chosen, the calculated type number does not correspond to any of the type numbers of the pumps available. The ideal answer would be to design a new pump to suit the specification exactly. This very costly procedure may only be justified in the rare cases of very large pumps and/or if a substantial number of them is required.

In all other cases, it is necessary to chose a pump whose type number is closest to and greater than that required. It will mean that the pump will not operate at its design point but somewhat to the right of it. If the pump type number is less than that required, it will operate somewhere to the left of the design point, which for some types may lead to an unstable operation, depending upon the shape of the head/flow rate characteristic. In any case the operating efficiency will be less than maximum.

A procedure for pump selection is illustrated by Example 24.5.

The selection of positive displacement pumps presents less problems because their operation is (within limits) independent of the system characteristic. It is precisely this feature which makes them eminently suitable for applications where the system resistance is difficult to estimate or, more commonly, when it is subject to variation due to changes of the system. Hence, positive displacement rotary pumps are used in oil hydraulic applications, especially in the control systems in which the operating head depends upon the part of the circuit used at any time.

More generally, positive displacement pumps are used when the required head is high and the flow rate is small.

Example 24.5

It is required to pump water at a rate of 0.5 m^3 s^{-1} from a sump which is 7 m below the ground level to a reservoir whose water level is 20 m above the water level in the sump. The calculated losses due to friction and separation in the proposed pipeline amount to 52 m of water.

Select a suitable pump, which must be direct-driven by an a.c. synchronous electric motor, to meet the above duty. Pumps A, B and C are available for selection and their characteristics are given in Fig. 24.18. For the selected pump, specify its size, speed, efficiency, power consumed and any requirements regarding its location with respect to the water level in the sump. Assume that the cavitation characteristics given obey similarity laws and refer to saturated water vapour head of 0.2 m and barometric pressure of 750 mm of mercury, the conditions at which the pump will operate.

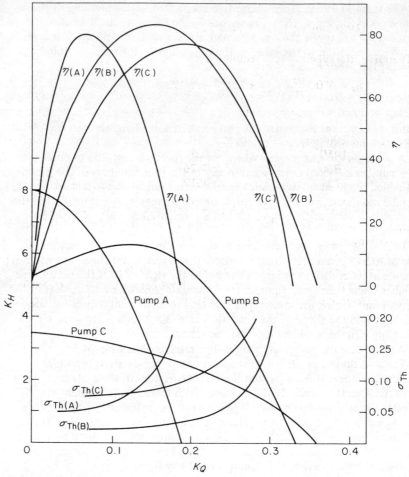

Figure 24.18

Solution
From the characteristics given, the values of K_Q and K_H corresponding to the maximum efficiency of each pump are established and the values of type number n_s calculated from:

$$n_s = Q^{1/2}/K_H^{3/4},$$

which gives:

	K_Q	K_H	n_s	σ_{Th}
Pump A	0·07	6·75	0·0631	0·055
Pump B	0·20	5·50	0·125	0·035
Pump C	0·16	2·80	0·185	0·085

Duty required is

$$Q = 0.5 \text{ m}^3 \text{ s}, \qquad H = 20 + 52 = 72 \text{ m}.$$

Therefore, the type number required is

$$n_s = N\, 0.5^{1/2}/(9.81 \times 72)^{3/4} = N/193.8.$$

Using nominal a.c. motor speeds:

for 2900 rev min^{-1} $n_s = 0.250$
     1450   0.125
     960   0.0826
     720   0.0619

Comparing these values of n_s with those corresponding to the pumps available, two possibilities emerge: (i) pump A at 720 rev min^{-1}; (ii) pump B at 1450 rev min^{-1}.

Consider first pump A.

The type numbers do not match exactly and, therefore, the pump would not operate at maximum efficiency but somewhere to the left of it, since the required n_s is less than the pump n_s. In order to establish the operating point, one possible method is to calculate the values of n_s' (not *the* type number n_s) at points on the pump characteristic to the left of K_N. Thus,

$$\text{at } K_Q = 0.060, K_H = 7.00 \quad \text{and} \quad n_s' = 0.057;$$

$$\text{at } K_Q = 0.065, K_H = 6.85 \quad \text{and} \quad n_s' = 0.062.$$

Therefore, the pump will operate at

$$K_Q = 0.065 \quad \text{and} \quad \eta = 79 \text{ per cent},$$

from which, since $K_Q = Q/ND^3$,

$$D = (Q/NK_Q)^{1/3} = (0.5 \times 60/720 \times 0.065)^{1/3} = 0.862 \text{ m}.$$

Power consumed is given by

$$P = 9.81 \times 0.5 \times 72/0.79 = 447 \text{ kW}.$$

Cavitation restrictions are

$$\sigma_{\text{Th}} = \text{NPSH}/H = 0.050$$

at the operating point, and so

$$\text{NPSH} = 72 \times 0.050 = 3.6 \text{ m}.$$

Now, $\text{NPSH} = p_{\text{atm}}/\rho g - H_s - p_{\text{vap}}/\rho g,$

$$p_{\text{atm}}/\rho g = 0.750 \times 13.6 = 10.2 \text{ m}.$$

Therefore,

$$H_s = 10.2 - 0.2 - 3.6 = 6.4 \text{ m}.$$

Thus, the maximum suction head including all pipe losses must not exceed (say) 6 m in order to avoid cavitation. Since the ground level is 7 m above the sump, the pump would have to be lowered below the ground level by at least 1 m, preferably more to account for pipe losses.

Consider now pump B.

The type number matches exactly and, therefore, the pump would operate at the maximum efficiency of 77 per cent.

$$D = (Q/NK_Q)^{1/3} = (0.5 \times 60/1450 \times 0.2)^{1/3} = 0.47 \text{ m}.$$

Power consumed is given by

$$P = 9.81 \times 0.5 \times 72/0.77 = 458.6 \text{ kW},$$

cavitation by

$$\text{NPSH} = \sigma_{\text{Th}}H = 0.035 \times 72 = 2.52 \text{ m}.$$

Therefore,

$$H_s = 10.2 - 0.2 - 2.52 = 7.48 \text{ m}.$$

Since, in this case, H_s is greater than the ground level above the sump, it will be possible to use this pump at ground level, provided it is close to the sump so that all pipe losses are on the delivery side of the pump. Also, pump B will be smaller than pump A. Conclusion: select pump B as follows:

$$D = 0.47 \text{ m}, N = 1450 \text{ rev min}^{-1}, \eta = 77 \text{ per cent}, P = 458.6 \text{ kW}.$$

EXERCISES 24

24.1   The characteristics of a fan are as follows:

Volume flow (m^3/h)	0	2000	4000	6000	8000	10 000	12 000
Fan total pressure (mm of water)	50	54·5	56	54·5	50	42·5	32
Fan input power (kW)		0·4	0·63	0·90	1·20	1·53	1·70	1·75

If the system resistance is 60 mm of total water column at 7000 m^3/h determine the fan operating point, the power consumed and fan total efficiency.

[950 W, 73%]

24.2   A centrifugal pump has the following characteristics:

Q (m^3/h)	0	23	46	69	92	115
h (m)	17	16	13·5	10·5	6·6	2
η (%)	0	49·5	61	63·5	53	10

The pump is used to pump water from a low reservoir to a high reservoir through a total length of 800 m of pipe 15 cm diameter. The difference between the water levels in the reservoirs is 8 m. Neglecting all losses except friction and assuming f = 0·004, find the rate of flow between the reservoirs. Also determine the power input to the pump.

[60 m^3/h, 3·04 kW]

24.3   Performance figures for a centrifugal fan are tabulated below. Plot these and superimpose a shaft power curve. From this determine the shaft power at the operating point if the system resistance is 100 mm of water at 40 m³/s and also the power if the output is reduced to 25 m³/s by damper regulation.

Q (m³/s)	0	10	20	30	40	50	60	70
h (mm of water)	85	92·5	95	90	80	65	47·5	25
η (%)	0	46	66	70	67	60	48	32

[44 kW, 33 kW]

24.4 The characteristics of two rotodynamic pumps at constant speed are as follows:

Q (m³/s)	0	0·006	0·012	0·018	0·024	0·030	0·036
Pump A H (m)	22·6	21·9	20·3	17·7	14·2	9·7	3·9
η(%)	0	32	74	86	85	66	28
Pump B H (m)	16·2	13·6	11·9	11·6	10·7	9·0	6·4
η(%)	0	14	34	60	80	80	60

One of the above pumps is required to lift water continuously through 3·2 m of vertical lift and the pipe to be used is 21 m long, 10 cm diameter and friction coefficient is 0·005.

Select the more suitable pump for this duty and justify your selection. What power input will be required by the selected pump?

[Pump B, 3·53 kW]

23.5 A centrifugal pump is used to circulate water in a closed loop experimental rig, consisting of: two vertical pipes, one 4 m long, and the other 3 m long, two horizontal pipes, each 1·3 m long, three 90° bends and a vertical 'working section' 1 m long. The pump is situated in one of the two low-level corners of the circuit. The pipes and the bends are 7·5 cm diameter and and the working section has a cross-sectional area of 125 cm². The friction factor for all pipes is 0·006 and the loss in each bend may be taken as 0·1 $v^2/2g$, where v is the mean velocity in m/s. The loss in the 'working section' may be taken as equivalent to a frictional loss in a 1 m long pipe 7·5 cm dia.

Determine the mean velocity in the 'working section' if the pump characteristic is as follows.

Q (m³/s)	0	0·006	0·012	0·018	0·024	0·027
H (m)	3·20	3·13	2·90	2·42	1·62	0·98

[1·32 m/s]

24.6 A 60 cm dia. impeller centrifugal pump has the following characteristic at 750 rev/min.

Q (m³/min)	H (m)	η(%)
0	40·0	0
7	40·6	41
14	40·4	60
21	39·3	74
28	38·0	83
35	33·6	83
42	25·6	74
49	14·5	51
56	0	0

(a) If the system resistance is purely frictional and is 40 m at 42 m³/min determine the pump operating point and power absorbed.

(b) The pump is used to pump water from one reservoir to another, the difference between levels being 13 m. The pipeline is 45 cm dia., 130 m long, $f = 0.005$ and contains 2 gate valves ($K = 0.2$) and ten 90° bends ($K = 0.35$). Obtain the volume delivered by the pump and power absorbed.

(c) If, for the system at (b) a geometrically similar pump but 50 cm dia. is used running at 900 rev/min, determine the volume delivered and power consumed.

[37 m³/min, 233 kW; 43·5 m³/min, 235 kW; 32·5 m³/min, 161 kW]

24.7   The characteristic of an axial flow pump running at 1450 rev/min is as follows:

Q (m³/s)	0	0·046	0·069	0·092	0·115	0·138	0·180
H (m)	5·6	4·2	4·35	4·03	3·38	2·42	0

When two such pumps are connected in parallel the flowrate through the system is the same as when they are connected in series. At what speed should a single pump run in order to deliver the same volume?

Assume the system characteristic to be purely resistive (no static lift).
[1691 rev/min]

24.8   The characteristics of a centrifugal pump at constant speed are as follows:

Q (m³/s)	0	0·012	0·018	0·024	0·030	0·036	0·042
H (m)	22·6	21·3	19·4	16·2	11·6	6·5	0·6
η(%)	0	74	86	85	70	46	8

The pump is used to lift water over a vertical distance of 6·5 m by means of a 10 cm dia. pipe, 65 m long, for which the friction coefficient $f = 0.005$.

(a) Determine the rate of flow and the power supplied to the pump.

(b) If it is required to increase the rate of flow, and this may be achieved only by an addition of a second, identical pump (running at the same speed), investigate whether it should be connected in series or in parallel with the original pump. Justify your answer by determining the increased rate of flow and power consumed by both pumps.

[0·0268 m³/s, 4·73 kW; parallel 7·9 kW, series 9·9 kW]

24.9   A centrifugal pump has the following characteristic:

H (m)	22·6	21·8	20	17·6	14·5	10·6	4·8
Q (m³/s)	0	0·009	0·018	0·027	0·036	0·045	0·054

The pump supplies water from a lake to a reservoir whose cross-sectional area is 40 m², via 65 m of 15 cm dia. pipe for which $f = 0.007$. The pump is switched on when the level in the reservoir is 5 m above the water level in the lake and is switched off when the level is 18 m.

By plotting the pump characteristic and the system resistance at say 30 min intervals (assuming constant discharge during the chosen time interval) obtain a graph showing a relationship between the pump discharge and time for one cycle of operation. How long does the cycle last?
[4 h 5 min]

24.10   The characteristic of a pump in terms of dimensionless coefficients may be approximated to: $K_H = 8 - 2K_Q + 210K_Q^2$. Such a pump having impeller diameter 0·4 m and running at 1450 rev/min operates against a system characteristic represented by: $h = 20 + 300Q^2$. Determine the flowrate delivered and the pump operating head. What would be the flowrate through the system if two such pumps were connected (a) in series and (b) in parallel. [0·217 m³/s, 43·12m, 0·255 m³/s, 0·326 m³/s]

24.11   A pump has the following characteristic when running at 1450 rev/min min:

Q (m³/s)	0	0·225	0·335	0·425	0·545	0·650	0·750	0·800
H (m)	20	17	15	13	10	7	3	0

A system is designed where the static lift is 5 m and the operating point is $H = 11·1$ m and $Q = 0·5$ m³/s using the pump as above. The system is redesigned, the static lift being 5 m as before but the frictional and other losses increase by 40%. Find the new pump speed such that the flowrate of 0·5 m³/s can be maintained.
[1551 rev/min]

24.12   Cavitation tests were performed on a pump giving the following results: $Q = 0·05$ m³/s; $H = 37$ m; barometric pressure 760 mm of mercury; ambient temperature 25 °C. Cavitation began when the total head at pump inlet was 4 m. Calculate the value of Thoma cavitation coefficient and the NPSH.

What could be the maximum height of this pump above water level if it is to operate at the same point on its characteristic in the ambient conditions of barometric pressure of 640 mm of mercury and temperature of 10 °C?
[0·165, 6·086 m, 2·5 m]

24.13   (a) Define Thoma coefficient and explain its use in conjunction with cavitation characteristics of rotodynamic pumps.

(b) Distinguish between available and required NPSH and state briefly how they are determined.

(c) A centrifugal pump of specific speed 0·683 (based on units of rev/s, m³/s, m) has a critical Thoma number equal to 0·2. The proposed installation of the pump requires its centre line to be 5·2 m above the sump water level. The pump when running at 1450 rev/min delivers 0·0637 m³/s. The losses in the suction pipe are estimated as 0·457 m of water. If the barometric pressure is 749 mm of Hg and the temperature of the water is 27 °C for which the vapour pressure is 26·2 mm of Hg, establish whether cavitation is likely to occur.
[Req. NPSH = 3·66 m; Av. NPSH = 4·18 m, therefore, no cavitation]

24.14   The critical Thoma number for a certain type of turbine varies in the following manner:

N_s (rev/min, kW, m units)	0	50	100	150	200	250
σ_{th}		0	0·04	0·1	0·18	0·28	0·41

A turbine runs at 300 rev/min under a net head of 50 m and produces 2 MW of power. The runner outlet velocity of fluid is 10·4 m/s and this point is 4·7 m above the tail race. The atmospheric pressure is equivalent to 10·3 m of water and the saturation pressure for water is 0·04 bar. Determine whether cavitation is likely to occur and find the head loss between runner outlet and tail race.

[No cavitation; 10·2 m]

24.15   Assuming that the volume rate of flow Q for a centrifugal pump depends on the coefficient of viscosity μ and mass density ρ of the fluid, the external diameter D and rotational speed N of the impeller, the effective head H and the acceleration g due to gravity, show that

$$\frac{Q}{ND^3} = \phi\left(\frac{\rho ND^2}{\mu} ; \frac{gH}{N^2 D^2}\right)$$

A single r stage centrifugal pump which has an impeller dia. 200 mm discharges at the rate of 12·3 litres against an effective head of 21 m when the impeller speed is 930 rev/min.

A multi-stage pump, required to run at 1430 rev/min, is built up from three similar impellers, each having a diameter of 250 mm. Assuming dynamically similar condition of operation for the two pumps calculate: (*a*) the maximum effective head against which the multi-stage pump will operate, (*b*) the rate of discharge.

[232·7 m, 37 l/s]

24.16   A centrifugal pump, having four stages in parallel, delivers 218 litres of liquid against a head of 26 m, the diameter of the impellers being 229 mm and the speed 1700 rev/min. A pump is to be made up with a number of identical stages in series of similar construction to those in the first pump to run at 1250 rev/min and to deliver 282 l/s against a head of 265 m. Find the diameter of the impellers and the number of stages required.

[439 mm, 5 stages]

24.17   Show that for geometrically similar rotodynamic pumps running at the same peripheral velocity the head developed is constant while volume and power are proportional to diameter squared.

A 50 cm dia. axial flow pump delivers 0·6 m³/s against a purely frictional resistance of 4·25 m when running at 650 rev/min and 80% overall efficiency. It is required to double the flow through the system using a number of identical pumps, geometrically similar to the 50 cm pump and running at the same peripheral velocity.

Determine the number of pumps required stating whether operating in parallel or in series, their diameter and speed and the total power consumed.

If the increased duty is met by the original pump what will be its peripheral velocity and the ratio of new to original peripheral velocity?

[4 in series, 70·7 cm, 459·6 rev/min, 250 kW, 68 m/s, 2]

PART VII
Dimensional analysis and similarity

The application of fluid mechanics in design, perhaps more than most engineering subjects, relies on the use of empirical results built up from an extensive body of experimental research. In many areas empirical data are supplied in the form of tables and charts that the designer may apply directly, an example being the values of friction factor for pipe flow and separation loss coefficients for duct and pipe fittings. However, even here, the tables and the underlying experimental work become too unwieldy and time consuming if no way can be found to replace the relationship between any two variables by generalized groupings. It is, therefore, in the organization of experimental work and the presentation of its results that dimensional analysis plays such an important role. This technique, which is dealt with first in this Part of the text, commences with a survey of all the likely variables affecting any phenomenon, and, to the experienced researcher, then suggests the formation of groupings of more than one variable. Experimental work may then be based on these groups rather than on individual variables, considerably reducing the testing programme and leading to simplified design guides, such as the Moody charts mentioned in Chapter 8.

The application of results from one test series, involving say a particular pipe flow situation to another case, depends on the full understanding of the principles of geometric and dynamic similarity which are covered in the second part of this section. Although similarity is inherent in the formation of relationships such as the Moody chart, it is more commonly associated with the use of models and model testing techniques. Examples of such applications as wind tunnel tests and river and harbour models are mentioned; however, the basic principles depend upon the equivalence of variable groupings formed initially by the use of dimensional analysis. Again, it will be appreciated that mathematics alone is not sufficient in the application of the similarity laws, in many cases it will be found that total equivalence of all the dimensionless groupings will be mutually impossible and here the experience of the researcher will be called upon, examples being found in the cases of ship model tests and pump or turbine modelling techniques utilizing gas in place of water.

Together, dimensional analysis, similarity and model testing techniques allow the design engineer to predict acccurately and economically the performance of the prototype system, whether it is an aircraft wing, ship hull, dam spillway or harbour construction. The basis of these interactive techniques are presented in this Part.

25 Dimensional analysis

25.1. Dimensional analysis

This is a useful technique for the investigation of problems in all branches of engineering and particularly in fluid mechanics. If it is possible to identify the factors involved in a physical situation, dimensional analysis can usually establish the form of the relationship between them. At first sight, the technique does not appear to be as precise as the usual algebraic analysis which seems to provide exact solutions, but these are usually obtained by making a series of simplifying assumptions which do not always correspond with the real facts. The qualitative solution obtained by dimensional analysis can usually be converted into a quantitative result, determining any unknown factors experimentally.

25.2. Dimensions

Any physical situation, whether it involves a single object or a complete system, can be described in terms of a number of recognizable properties which the object or system possesses. For example, a moving object could be described in terms of its mass, length, area or volume, velocity and acceleration. Its temperature or electrical properties might also be of interest, while other properties — such as the density and viscosity of the medium through which it moves — would also be of importance, since they would affect its motion. These measurable properties used to describe the physical state of the body or system are known as its *dimensions*.

25.3. Units

To complete the description of the physical situation, it is also necessary to know the magnitude of each dimension. It is not usually sufficient to know, for example, that a body has the dimension of length, we also need to know the magnitude of this length. For this purpose we use agreed *units* of measurement. A length would be measured in terms of a standardized unit of length, such as the metre. Similarly, other agreed units are used to measure other dimensions. There is more than one system of units in common use, but of course the system of units chosen does not affect the real size of the body or system — only the numerical value of its measurements. One foot is precisely the same length as 0·3048 metres. The distinction between units and dimensions is that dimensions are properties that can be measured and units are the standard elements in terms of which these dimensions can be described quantitatively and assigned numerical values.

25.4. Dimensional reasoning

In analysing any physical situation, it is necessary to decide what factors are involved and then to try to determine a quantitative relationship between them. The factors involved can often be assessed from observation, experiment or even, perhaps, intuition. However, it is sometimes difficult to establish precise quantitative relationships, because we cannot specify the conditions which exist exactly or the way in which the various factors interact. As a result, no suitable mathematical model can be constructed without making substantial assumptions to simplify the problem. A qualitative solution to the problem can sometimes be obtained by dimensional reasoning, and subsequent experimental investigation based on this analysis can frequently lead to a complete solution of the real problem.

In dimensional analysis, we are concerned only with the nature of the factors involved in the situation and not with their numerical values. The notation adopted to indicate this is to enclose the name or symbol of the quantity in square brackets, thus [length] means the dimension of length and not a particular length with a definite numerical value. For conciseness length is abbreviated to L and the dimension of length is written [L]. Similarly [M] is used for the dimension of mass, [T] for the dimension of time, [F] for the dimension of force, [Θ] for the dimension of temperature and so on. Whenever a property, in words or symbols, is enclosed in a square bracket it indicates that we are concerned with it only dimensionally and qualitatively, not quantitatively.

Dimensional reasoning is based on the proposition that, for an equation describing a physical situation to be true, the two sides must be equal both numerically and dimensionally. An equation must compare like with like. The simple equation 1 + 3 = 4 is numerically correct but, in physical terms, may be entirely untrue, depending upon the nature of each term. Thus, it would be *untrue* to say that

1 elephant + 3 aeroplanes = 4 days,

since elephants, aeroplanes and days are not the same sort of things; it would be *true* to say that

1 metre + 3 metres = 4 metres,

provided that our concern is restricted to the study of length, since each term has the dimensions of length.

An equation describing a physical situation will only be true if all the terms are of the same kind and have the same dimensions. The equation is then said to be *dimensionally homogeneous*, and is valid only in relation to these dimensions. If an equation does not compare like with like, it will be physically meaningless, even though it may balance numerically. In general any equation of the form

$$a_1^{m_1} b_1^{n_1} c_1^{p_1} + a_2^{m_2} b_2^{n_2} c_2^{p_2} + \ldots = X$$

will be physically true if, in addition to being numerically correct, the terms are dimensionally the same so that

$$[a_1^{m_1} b_1^{n_1} c_1^{p_1}] = [a_2^{m_2} b_2^{n_2} c_2^{p_2}] = \ldots = [X],$$

where $[a_1^{m_1} b_1^{n_1} c_1^{p_1}]$ means the dimensions of $a_1^{m_1} b_1^{n_1} c_1^{p_1}$.

25.5. Dimensionless quantities

In describing an object or system, we sometimes use quantities which are non-dimensional or abstract. For example, the shape of an ellipse is defined by the ratio of the major and minor axes. Other non-dimensional quantities are relative density, strain and angle measured in radians. Such quantities are ratios comparing one quantity with another of the same kind and their numerical values are independent of the system of units employed. For example, tensile strain is defined as extension divided by original length. Since both these quantities have the dimension [L],

$$[\text{Strain}] = \frac{[\text{Extension}]}{[\text{Original length}]} = \frac{[\text{L}]}{[\text{L}]} = [\text{L}^0] = [1],$$

indicating that the dimension of strain is a pure number and strain is, therefore, dimensionless.

25.6. Fundamental and derived units and dimensions

It would be possible to give independent units and dimensions to every physical property, but it would not help our understanding of their interrelationship if we did so. It is, therefore, desirable to select a number of fundamental dimensions and express other dimensions in terms of these. Length is, perhaps, an obvious choice for one such fundamental dimension. An area is defined as the product of two lengths. The area of a rectangle having sides of lengths a and b is given by the equation

Area of rectangle = Length a × Length b.

For this to be true, it must be dimensionally homogeneous, so that

[Area] = [Length] × [Length] = [L^2],

i.e. area has the dimensions of [L^2]. The corresponding unit of area will be the unit of length squared, for example m^2 in SI units.

Similarly, if a tank has sides of length a, b and c,

Volume of tank = Length a × Length b × Length c

and so   [Volume] = [L] × [L] × [L] = [L^3],

so that volume has the dimensions of [L^3].

In kinematics, the dimension of time [T] is required. The dimensions of other quantities can be expressed in terms of length and time using the established definitions and relationships, together with the principle of dimensional homogeneity. For linear motion,

Velocity = Distance/Time

and so   [Velocity] = [Distance]/[Time] = [L]/[T] = [LT^{-1}].

Similarly,

[Acceleration] = [Velocity]/[Time] = [LT^{-1}]/[T] = [LT^{-2}].

For angular motion, if we define the angle as being measured in radians,

Angle = Length of arc/Length of radius

[Angle] = [Length]/[Length] = [L]/[L] = [L^0],

which, dimensionally, is unity, indicating that angle is dimensionless.

$$[\text{Angular velocity}] = [\text{Angle}]/[\text{Time}] = 1/[T] = [T^{-1}]$$

and $[\text{Angular acceleration}] = [\text{Angular velocity}]/[\text{Time}]$
$$= [T^{-1}]/[T] = [T^{-2}].$$

In dynamics, an additional fundamental dimension is needed, since we are concerned with force and mass. Newton's second law provides the necessary relationship, which can be stated in the form

Force $\propto$ Mass $\times$ Acceleration.

In practice, for any given system of units, the constant of proportionality is made unity, so that

Force = Mass $\times$ Acceleration.

For this equation to be valid, it must be dimensionally homogeneous and, therefore,

$$[\text{Force}] = [\text{Mass}] \times [\text{Acceleration}].$$

Writing [F] for force, [M] for mass, [T] for time and [L] for length, then, since $[\text{Acceleration}] = [LT^{-2}]$,

$$[F] = [M] [LT^{-2}].$$

Thus, in dynamics, we can select [L] and [T] as fundamental dimensions together with either force [F] or mass [M], the remaining quantity being regarded as a derived dimension and expressed in terms of the three fundamental dimensions. In this book, we shall normally select the dimension of mass [M] as the third fundamental dimension and treat force as a derived dimension, having the dimensions $[MLT^{-2}]$. If we were to select [F], [L] and [T] as fundamental dimensions, mass would have the dimensions of $[FL^{-1} T^2]$.

The dimensions of other quantities can be expressed in terms of the fundamental dimensions using known relationships, definitions and equations and the principle of homogeneity of dimensions. For example,

Pressure = Force/Area,

$$[\text{Pressure}] = [F]/[L^2] \quad \text{or} \quad [MLT^{-2}]/[L^2]$$
$$= [FL^{-2}] \quad \text{or} \quad [ML^{-1}T^{-2}].$$

Similarly,

Mass density = Mass/Volume,

$$[\text{Mass density}] = [M]/[L^3] \quad \text{or} \quad [FL^{-1}T^2]/[L^3]$$
$$= [ML^{-3}] \quad \text{or} \quad [FL^{-4}T^2].$$

The dimensions of quantities commonly occurring in mechanics are given in Table 25.1.

The system of fundamental dimensions chosen for problems involving the thermodynamics of fluids will depend upon whether it is desirable to

Quantity	Defining equation	Dimensions, MLT system
Geometrical		
Angle	Arc/radius (a ratio)	$[M^0 L^0 T^0]$
Length	(Including all linear measurement)	$[L]$
Area	Length x Length	$[L^2]$
Volume	Area x Length	$[L^3]$
First moment of area	Area x Length	$[L^3]$
Second moment of area	Area x Length^2	$[L^4]$
Strain	Extension/Length	$[L^0]$
Kinematic		
Time	–	$[T]$
Velocity, linear	Distance/Time	$[LT^{-1}]$
Acceleration, linear	Linear velocity/Time	$[LT^{-2}]$
Velocity, angular	Angle/Time	$[T^{-1}]$
Acceleration, angular	Angular velocity/Time	$[T^{-2}]$
Volume rate of discharge	Volume/Time	$[L^3 T^{-1}]$
Dynamic		
Mass	Force/Acceleration	$[M]$
Force	Mass x Acceleration	$[MLT^{-2}]$
Weight	Force	$[MLT^{-2}]$
Mass density	Mass/Volume	$[ML^{-3}]$
Specific weight	Weight/Volume	$[ML^{-2} T^{-2}]$
Specific gravity	Density/Density of water	$[M^0 L^0 T^0]$
Pressure intensity	Force/Area	$[ML^{-1} T^{-2}]$
Stress	Force/Area	$[ML^{-1} T^{-2}]$
Elastic modulus	Stress/Strain	$[ML^{-1} T^{-2}]$
Impulse	Force x Time	$[MLT^{-1}]$
Mass moment of inertia	Mass x Length^2	$[ML^2]$
Momentum, linear	Mass x Linear velocity	$[MLT^{-1}]$
Momentum, angular	Moment of inertia x Angular velocity	$[ML^2 T^{-1}]$
Work, energy	Force x Distance	$[ML^2 T^{-2}]$
Power	Work/Time	$[ML^2 T^{-3}]$
Moment of a force	Force x Distance	$[ML^2 T^{-2}]$
Viscosity, dynamic	Shear stress/Velocity gradient	$[ML^{-1} T^{-1}]$
Viscosity, kinematic	Dynamic viscosity/Mass density	$[L^2 T^{-1}]$
Surface tension	Energy/Area	$[MT^{-2}]$

Table 25.1. Dimensions of quantities in mechanics (based on Newton's second law)

separate thermal quantities, such as heat and temperature, from other related properties. Since heat is a form of energy it could be expressed dimensionally in the same way as energy which, in terms of fundamental dimensions of mass length and time, is $[ML^2 T^{-2}]$. We also have the relationship

$$H = cm\theta,$$

where H is the quantity of heat required to raise a mass m of a substance of specific heat c through a temperature difference of θ. If c is treated as a ratio and, therefore, is dimensionless,

$$[\theta] = [Hm^{-1}] = [ML^2T^{-2}]\ [M^{-1}] = [L^2T^{-2}].$$

The dimensions of other thermal quantities can be derived in the same way are shown in Table 25.2 column 1.

If the thermal aspects of a problem are of particular interest, it is useful to introduce at least one additional fundamental dimension which is thermal in character, such as temperature Θ. In the resulting $MLT\Theta$ system, heat energy may be treated thermally as $[M\Theta]$, in which case the dimensions of other quantities are as shown in column 2 of Table 25.2, or heat energy can be expressed in mechanical terms as $[ML^2T^{-2}]$ with the results shown in column 3.

The quantity of heat H might also be treated as an additional fundamental quantity. The dimensions of other quantities using the $HLT\Theta$ system are shown in Table 25.2, column 4, and for the $HMLT\Theta$ system in column 5. The units used for each system would, of course, differ. Heat might be measured in calories, if considered as $[M\Theta]$, or in joules, if regarded as $[ML^2T^{-2}]$.

25.7. Dimensions of derivatives and integrals

The dimensions which should be assigned to partial or total derivatives can be determined easily, if it is remembered that, by definition, dy/dx is the limiting value of the ratio $\delta y/\delta x$ as δx tends to zero, where δy is the finite value of the increment of y which corresponds to a finite increment δx of x.

Clearly, the dimensions of δy are the same as those of y and the dimensions of δx are the same as those of x; thus, dimensionally,

$$\left[\frac{dy}{dx}\right] = \left[\frac{\delta y}{\delta x}\right] = \left[\frac{y}{x}\right].$$

Similarly,

$$\frac{d^2y}{dx^2} = \frac{d}{dx}\left(\frac{dy}{dx}\right) = \frac{\text{Increment of } dy/dx}{\text{Increment of } x},$$

so that, dimensionally,

$$\left[\frac{d^2y}{dx^2}\right] = \left[\frac{dy/dx}{x}\right] = \left[\frac{y/x}{x}\right] = \left[\frac{y}{x^2}\right]$$

or, in general,

$$\left[\frac{d^ny}{dx^n}\right] = \left[\frac{y}{x^n}\right].$$

The dimensions of integrals are found in the same way. The term

$$\int_b^a y\ dx$$

Quantity	Defining equation	Dimensions				
		MLT system	MLTΘ systems		HLTΘ system	HMLTΘ system
			Thermal	Dynamic		
Temperature, θ		$[L^2T^{-2}]$	$[\Theta]$	$[\Theta]$	$[\Theta]$	$[\Theta]$
Heat quantity, H		$[ML^2T^{-2}]$	$[M\Theta]$	$[ML^2T^{-2}]$	$[H]$	$[H]$
Enthalpy		$[ML^2T^{-2}]$	$[M\Theta]$	$[ML^2T^{-2}]$	$[H]$	$[H]$
Entropy, S	$dS = dH/\theta$	$[M]$	$[M]$	$[ML^2T^{-2}\theta^{-1}]$	$[H\Theta^{-1}]$	$[H\Theta^{-1}]$
Coeff. of thermal expansion	Change of length/unit length/degree	$[L^{-2}T^2]$	$[\Theta^{-1}]$	$[\Theta^{-1}]$	$[\Theta^{-1}]$	$[\Theta^{-1}]$
Thermal capacity	Heat required per degree temp. rise	$[M]$	$[M]$	$[ML^2T^{-2}\Theta^{-1}]$	$[H\Theta^{-1}]$	$[H\Theta^{-1}]$
Specific heat	Thermal capacity per unit mass	$[M^0L^0T^0]$	$[M^0L^0T^0\Theta^0]$	$[L^2T^{-2}\Theta^{-1}]$	—	$[HM^{-1}\Theta^{-1}]$
Specific heat ratio		$[M^0L^0T^0]$	$[M^0L^0T^0\Theta^0]$	$[M^0L^0T^0\Theta^0]$	$[H^0L^0T^0\Theta^0]$	$[H^0M^0L^0T^0\Theta^0]$
Thermal conductivity	Time rate of heat transmission per unit area and temp. gradient	$[ML^{-1}T^{-1}]$	$[ML^{-1}T^{-1}]$	$[MLT^{-3}\Theta^{-1}]$	$[HL^{-1}T^{-1}\Theta^{-1}]$	$[HL^{-1}T^{-1}\Theta^{-1}]$
Gas constant, R	Energy/Mass x Temp.	$[M^0L^0T^0]$	$[M^0L^0T^0\Theta^0]$	$[L^2T^{-2}\Theta^{-1}]$	—	$[L^2T^{-2}\Theta^{-1}]$
Coeff. of heat transfer		$[ML^{-2}T^{-1}]$	$[ML^{-2}T^{-1}]$	$[MT^{-3}\Theta^{-1}]$	$[HL^{-2}T^{-1}\Theta^{-1}]$	$[HL^{-2}T^{-1}\Theta^{-1}]$
Mechanical equivalent of heat		$[M^0L^0T^0]$	$[L^2T^{-2}\Theta^{-1}]$	$[M^0L^0T^0]$	—	$[H^{-1}ML^2T^{-2}]$

Table 25.2. Dimensions of common quantities in thermodynamics

means the limit of the sum of all the products of $y\delta x$ between $x = a$ and $x = b$. Thus, the dimensions of

$$\int_b^a y \, dx$$

will be the same as those of $y\delta x$ and, since $[\delta x] = [x]$,

$$\left[\int_a^b y \, dx \right] = [yx].$$

Similarly, a double integral $\iint a \, dz_1 \, dz_2$ means the limit sum of the products $a \, \delta z_1 \delta z_2$. Since $[\delta z_1] = [z_1]$ and $[\delta z_2] = [z_2]$,

$$[\iint a \, dz_1 \, dz_2] = [az_1z_2].$$

The dimensions of any multiple integral are found in the same way.

25.8. Use of dimensional reasoning to check calculations

The principle of dimensional homogeneity requires that any equation which fully and correctly represents a possible physical situation must be dimensionally homogeneous, so that each term has the same dimensional formula. It is, therefore, possible to make a rapid check of any algebraic analysis of such a situation by substituting the dimensions of the quantities in each term and checking that, in the resultant dimensional equation, each fundamental dimension appears to the same power in each term. So long as symbols are not replaced by numerical values, each line of working can be checked in this way, and it is, therefore, often advantageous to delay the insertion of numerical values until the last possible moment.

Many engineering formulae in day-to-day use are commited to memory and, sometimes, there may be a doubt as to whether they have been recalled correctly. A quick dimensional check will establish the correct form immediately, although, of course, it will not indicate whether the values of any numerical constants are correct.

25.9. Units of derived quantities

The units in which a derived quantity can be measured may be determined directly from its dimensional formula by substituting the appropriate unit for each fundamental quantity. Thus, the SI unit of dynamic viscosity is obtained by substituting the kilogramme for mass, the metre for length and the second for time in the dimensional formula $ML^{-1}T^{-1}$ and will be the kilogramme/metre-second.

The SI system is a rationalized system of metric units, in which the units for all physical quantities can be derived from six basic, arbitrarily-defined

Quantity	Unit	Symbol
BASIC UNITS		
Length	metre	m
Mass	kilogramme	kg
Time	second	s
Electric current	ampere	A
Absolute temperature	kelvin	K
Luminous intensity	candela	cd
Geometry		
Angle, plane	radian	rad
Angle, solid	steradian	sr
Area	square metre	m^2
Volume	metre cubed	m^3
First moment of area	metre cubed	m^3
Second moment of area	metre to fourth power	m^4
DERIVED UNITS		
Mechanics		
Frequency	hertz	Hz
Velocity,		
linear	metre per second	$m\ s^{-1}$
angular	radian per second	$rad\ s^{-1}$
Acceleration,		
linear	metre per second squared	$m\ s^{-2}$
angular	radian per second squared	$rad\ s^{-2}$
Force	newton (= kilogramme-metre per second squared)	$N(=kg\ m\ s^{-2})$
Density, mass	kilogramme per metre cubed	$kg\ m^{-3}$
Specific weight	newton per metre cubed	$N\ m^{-3}$
Momentum,		
linear	kilogramme-metre per second	$kg\ m\ s^{-1}$
angular	kilogramme-metre squared per second	$kg\ m^2\ s^{-1}$
Moment of inertia	kilogramme-metre squared	$kg\ m^2$
Moment of force	newton-metre	N m
Pressure or stress (intensity)	pascal (= newton per metre squared	$N\ m^{-2}$ (Pa)
Viscosity, dynamic	newton-second per metre squared (= 10 poise)	$N\ s\ m^{-2}$
kinematic	metre squared per second	$m^2\ s^{-1}$
Surface tension	newton per metre	$N\ m^{-1}$
Energy, work	joule (= newton-metre)	J (= N m)
Power	watt (= joule per second)	$W (=J\ s^{-1})$
Heat		
Temperature interval	degree Celsius	K
Linear expansion coefficient	expansion per unit length per degree Celsius	K^{-1}
Heat quantity	joule	J
Heat flow rate	watt	W
Entropy	joule per degree Kelvin	$J\ K^{-1}$
Thermal capacity	joule per degree Celsius	$J\ K^{-1}$
Thermal conductivity	watt per metre per degree Celsius	$W\ m^{-1}\ K^{-1}$
Coefficient of heat transfer	watt per metre squared per degree Celsuis	$W\ m^{-2}\ K^{-1}$

Table 25.3 SI units

units, which are:

Length	metre;
Mass	kilogramme;
Time	second;
Electric current	ampere;
Absolute temperature	kelvin;
Luminous intensity	candela.

The product or quotient of any two units is the unit of the resultant quantity. Certain of these derived units have been given special names. For example, force has the dimensions MLT^{-2} and is measured, therefore, in kg m s^{-2}. For convenience, this unit is referred to as the newton and abbreviated to N. Similarly, pressure is defined as force per unit area and measured in N m^{-2}, but this unit is sometimes called the pascal. Details of the basic and derived SI units are given in Table 25.3.

25.10. Conversion from one system of units to another

It is intended that SI units should (at the earliest reasonable date) be used everywhere, but other systems are currently in use in many parts of the world. These can be divided into the so called absolute systems — based on mass, length and time — and technical systems — based on force, length and time.

There are a number of units in common use which are not part of coherent systems. In many English-speaking countries, speeds on the roads are measured in miles per hour, weights are sometimes measured in tons, areas by the acre, volume by the Imperial gallon or the U.S. gallon and lengths by the chain or fathom, to name but a few. Similar local units are used in other countries and it seems probable that it will be a very long time before a single international system of coherent units will replace all others. In the meantime, the problem of converting from one system to another will remain important and dimensional reasoning can be of help in this process.

Clearly the true physical value of any quantity must be independent of the system of units by which it is measured. If a quantity Q is found to have a numerical value n_1 when measured in units of size u_1,

Physical value of $Q = n_1 u_1$.

Similarly, if the same quantity Q when measured in units of size u_2 has a numerical value of n_2,

Physical value of $Q = n_2 u_2$.

Since the value of Q must remain unchanged,

$Q = n_1 u_1 = n_2 u_2$.

Thus, the numerical value of a quantity is inversely proportional to the size of the units in which it is measured.

If the quantity Q has a dimensional formula $[M^a L^b T^c]$, the unit of measurement will be a derived unit. Suppose that the units of mass, length and time are m_1, l_1 and t_1 in the first system and m_2, l_2 and t_2 in the second system, then the derived units in these systems, based on the dimensional formulae, are

$$u_1 = m_1^a l_1^b t_1^c \quad \text{and} \quad u_2 = m_2^a l_2^b t_2^c.$$

The fundamental units in the two systems will also be related: the size of the unit of mass m_1 will be equal to $k_m m_2$, that of l_1 is $k_l l_2$ and that of t_1 is $k_t t_2$, where k_m, k_l and k_t are numerical constants.

$$Q = N_1 u_1 = N_1 m_1^a l_1^b t_1^c,$$

and so, replacing m_1 by $k_m m_2$, l_1 by $k_l l_2$ and t_1 by $k_t t_2$,

$$N_1 u_1 = N_1 (k_m^a m_2^a)(k_l^b l_2^b)(k_t^c t_2^c)$$
$$= N_1 k_m^a k_l^b k_t^c \times m_2^a l_2^b t_2^c,$$

But $N_1 u_1 = N_2 u_2 = Q$ and $m_2^a l_2^b t_2^c = u_2$, therefore,

$$N_2 u_2 = N_1 (k_m^a k_l^b k_t^c) u_2,$$
$$N_2 = N_1 k_m^a k_l^b k_t^c,$$

where k_m, k_l and k_t are the numbers of units in the second system required to make the corresponding fundamental units in the first system.

Example 25.1

An engine produces 57 horsepower. What is the corresponding value in kilowatts and what is the conversion factor?

Solution

The horsepower and the kilowatt are multiples of the basic units for power in the British Technical system and the SI system, respectively. Thus, if n_1 is the horsepower and N_1 is the corresponding number of British technical units (ft lbf/sec)

$$N_1 = 550 \, n_1$$

Similarly, since the basic unit of power is the watt in the SI system,

$$N_2 = 1000 \, n_2.$$

The dimensional formula for power is $[ML^2 T^{-3}]$. Using suffix 1 for British technical units and suffix 2 for SI units.

$$N_2 = N_2 k_m k_l^2 k_t^{-3},$$

where k_m, k_l and k_t are the ratios of the units of mass, length and time in system 1 to the corresponding units in system 2, as tabulated below

Quantity	System 1 (FPS technical)	System 2 (SI)	Ratio (1)/(2)
Mass	slug	kilogramme	14·6
Length	foot	metre	0·3048
Time	second	second	1

Therefore,

$$N_2 = N_1 \times 14.6 \times 0.3048^2 \times 1 = 1.356\, N_1$$

or $1000\, n_2 = 550\, n_1 \times 1.356$

so that $n_2 = 0.746\, n_1.$

Therefore,

Value in kilowatts = $0.746 \times$ Value in horsepower.

Putting $n_1 = 57$ horsepower,

Output = $0.746 \times 57 = 42.5$ kW,

Conversion factor = $n_2/n_1 = 0.746.$

25.11. Conversion of dimensional constants

Many equations commonly used by engineers in practice contain numerical constants. In some cases, these are pure numbers and are, therefore, unaffected by the system of units employed. In other cases, these numbers are not dimensionless and their numerical values will depend on the system of units being used. Conversion is carried out in the same way as has been explained in Section 25.10.

25.12. Construction of relationships by dimensional analysis using the indicial method

If the factors involved in any real physical situation can all be identified, the form of the equation relating them can be largely determined by dimensional reasoning. This is the case because the requirement that the resulting equations must be dimensionally homogeneous means that the variables can only be combined in a very limited number of ways, and in some cases this may lead to a unique solution. Since pure numbers are dimensionless, a complete numerical solution of the type provided by rigorous algebraic reasoning cannot be obtained by dimensional reasoning. However, the values of the missing numbers can usually be established with comparative ease by experiment. It should also be remembered that it is often necessary, when making an algebraic analysis, to simplify the situation by making a number of assumptions, some of which are not strictly correct. The choice is, therefore, between a complete answer to an idealized situation or the partial answer to the real problem provided by dimensional analysis.

Example 25.2

The thrust F of a screw propeller is known to depend upon the diameter d, speed of advance v, fluid density ρ; revolutions per second N, and the coefficient of viscosity μ of the fluid. Find an expression for F in terms of these quantities.

Solution

The general relationship must be $F = \phi(d, v, \rho, N, \mu)$, which can be expanded as the sum of an infinite series of terms giving

$$F = A(d^m v^p \rho^q N^r \mu^s) + B(d^{m'} v^{p'} \rho^{q'} N^{r'} \mu^{s'}) + \ldots,$$

where A, B, etc. are numerical constants and m, p, q, r, s are unknown powers. Since, for dimensional homogeneity, all terms must be dimensionally the same, this can be reduced to

$$F = Kd^m v^p \rho^q N^r \mu^s, \tag{I}$$

where K is a numerical constant.

The dimensions of the dependent variable F and the independent variables d, v, ρ, N and μ are

$$[F] = [\text{Force}] = [MLT^{-2}],$$

$$[d] = [\text{Diameter}] = [L],$$

$$[v] = [\text{Velocity}] = [LT^{-1}]$$

$$[\rho] = [\text{Mass density}] = [ML^{-3}],$$

$$[N] = [\text{Rotational speed}] = [T^{-1}],$$

$$[\mu] = [\text{Dynamic viscosity}] = [ML^{-1}T^{-1}].$$

For convenience, these can be set out in the form of a table or dimensional matrix, in which a column is provided for each variable and the power of each fundamental dimension in its dimensional formula is inserted in the corresponding row:

	F	d	v	ρ	N	μ
M	1	0	0	1	0	1
L	1	1	1	−3	0	−1
T	−2	0	−1	0	−1	−1

Substituting the dimensions for the variables in (I),

$$[MLT^{-2}] = [L]^m [LT^{-1}]^p [ML^{-3}]^q [T^{-1}]^r [ML^{-1}T^{-1}]^s$$

Equating powers of [M], [L] and [T]:

$$[M], 1 = q + s; \tag{II}$$

$$[L], 1 = m + p - 3q - s; \tag{III}$$

$$[T], -2 = -p - r - s. \tag{IV}$$

Since there are five unknown powers and only three equations, it is impossible to obtain a complete solution, but three unknowns can be determined in terms of the remaining two. If we solve for m, p and q, we get

$$q = 1 - s \text{ from (II)}$$

$$p = 2 - r - s \text{ from (II)}$$

$$m = 1 - p + 3q + s = 2 + r - s \text{ from (III)}.$$

Substituting these values in (I),

$$F = K d^{2+r-s} v^{2-r-s} \rho^{1-s} N^r \mu^s.$$

Regrouping the powers,

$$F = K\rho v^2 d^2 \, (\rho v d/\mu)^{-s}(dN/v)^r.$$

Since s and r are unknown this can be written

$$F = \rho v^2 d^2 \phi\{\rho v d/\mu, \, dN/v\} \qquad (V)$$

where ϕ means 'a function of'. At first sight, this appears to be a rather unsatisfactory solution, but (V) indicates that

$$F = C\rho v^2 d^2, \qquad (VI)$$

where C is a constant to be determined experimentally and the value of which is dependent on the values of $\rho v d/\mu$ and dN/v. (VI) could also be used to calculate the thrust of a full-size propeller from experiments on a model, providing that the values of $\rho v d/\mu$ and dN/v were made the same for both and, therefore, C would have the same value in the two cases. Then we would be able to state that

$$F_{\text{model}}/F_{\text{full size}} = (\rho v^2 d^2)_{\text{model}}/(\rho v^2 d^2)_{\text{full size}}$$

Equation V could have been written

$$F/\rho v^2 d^2 = \phi\{\rho v d/\mu, \, dN/v\}$$

or $$\phi\{F/\rho v^2 d^2, \, \rho v d/\mu, \, dN/v\} = 0. \qquad (VII)$$

Each of the terms in the bracket forms a dimensionless group since

$$F/\rho v^2 d^2 = [MLT^{-2}]/[ML^{-3}] \, [L^2 T^{-2}] \, [L^2] = [1]$$

$$\rho v d/\mu = [ML^{-3}] \, [LT^{-1}] \, [L]/[ML^{-1} T^{-1}] = [1]$$

$$dN/v = [L] \, [T^{-1}]/[LT^{-1}] = [1]$$

25.13. Dimensional analysis by the group method

The indicial method is rather lengthy if there are a large number of variables. The resulting relationship consists of a set of dimensionless groups and rules can be set out to determine the number of groups and the way in which the groups should be formed. In Example 25.2, it can be seen from the dimensional matrix that there were six variables expressed in terms of three fundamental dimensions M, L and T. The relationship obtained in equation (VII) from Example 25.2 consists of three dimensionless groups. This result is an example of Buckingham's Π theorem. Using the symbol Π to indicate an independent dimensionless group, Buckingham showed that if there are n variables, including the dependent variable, in a physical relationship and these variables contain m fundamental dimensions, for example M, L and T, the equation relating the variables will contain $n - m$ independent dimensionless groups and be of the form

$$\phi\{\Pi_1, \Pi_2, \Pi_3, \ldots, \Pi_{n-m}\} = 0.$$

Thus, in the above example, there are six variables F, d, v, ρ, N, μ and three fundamental dimensions M, L and T. The solution therefore contains

$6 - 3 = 3$ dimensionless groups which are

$$II_1 = F/\rho v^2 d^2, \quad II_2 = \rho v d/\mu, \quad II_3 = dN/v.$$

Independent dimensionless groups are defined as those which can be formed from any particular number of quantities, but are independent of each other in the sense that none of them can be formed by any combination of the others.

   An alternative version of this rule states that the number of dimensionless groups in the equation relating n variables will be $n - k$, where k is the largest number of variables which cannot be formed into a dimensionless group. Usually k is equal to the number of primary dimensions m, but it can be smaller (though never greater).

   In any particular problem, having determined the number of dimensionless groups as described above, the next step is to combine the variables to form these dimensionless groups using the following rules.

   (i) From the independent variables select certain variables to use as repeating variables which will appear in more than one group. The repeating variables should contain all the dimensions used in the problem and be quantities which are likely to have a substantial effect on the dependent variable.

   (ii) Combine the repeating variables with the remaining variables to form the required number of independent dimensionless groups, choosing well known groups, such as Reynolds number, if appropriate.

   (iii) The dependent variable should appear in one group only.

   (iv) A variable that is expected to have a minor influence should appear in one group only.

Example 25.3

The variables controlling the motion of a floating vessel through water are the drag force F, the speed v, the length l, the density ρ and dynamic viscosity μ of the water and the gravitational acceleration g. Derive an expression for F by dimensional analysis.

Solution

The resistance to motion will be partly due to skin friction which depends on viscosity μ and partly due to wave resistance which depends on the gravitational acceleration g. The relationship will be of the form

$$F = \phi \{v, l, \rho, \mu, g\}$$

The dimensions of the variables are

	F	v	l	ρ	μ	g
M	1	0	0	1	1	0
L	1	1	1	-3	-1	1
T	-2	-1	0	0	-1	-2

The dependent variable is F. The repeating variables could be v and l, both likely to be major factors. Since the dimensions of v and l do not include [M], a further repeating variable which contains M is needed, say ρ.

Total number of variables, $n = 6$,

Number of fundamental dimensions, $m = 3$,

Number of dimensionless groups to be formed $= n - m = 3$.

The required solution will be

$$II_1 = \phi\{II_2, II_3\}.$$

To find II_1. This group is formed so that it includes the dependent variable F. Since F contains all three fundamental dimensions M, L and T, all three of the repeating variables will be required to form a dimensionless group. We can write

$$II_1 = Fv^a l^b \rho^c.$$

Replacing the variables by their dimensions and remembering that the dimensional formula of a dimensionless number is $M^0 L^0 T^0$,

$$[M^0 L^0 T^0] = [MLT^{-2}][LT^{-1}]^a [L]^b [ML^{-3}]^c.$$

Equating powers of M, L and T,

$0 = 1 + c$ for M,

$0 = 1 + a + b - 3c$ for L,

$0 = -2 - a$ for T.

From which $a = -2$, $c = -1$, $b = -2$ and $II_1 = F/\rho v^2 l^2$.

To find II_2. This group is formed by combining one of the non-repeating independent variables μ with the required number of repeating variables. Since the dimensional formula of μ is $ML^{-1}T^{-1}$ all three of the repeating variables will be needed to form a dimensionless group.

$$II_2 = \mu v^d l^e \rho^f.$$

For dimensional homogeneity,

$$[M^0 L^0 T^0] = [ML^{-1}T^{-1}][LT^{-1}]^d [L]^e [ML^{-3}]^f.$$

Equating powers of M, L and T,

$0 = 1 + f$ for M,

$0 = -1 + d + e - 3f$ for L,

$0 = -1 - d$ for T.

From which $d = -1$, $f = -1$, $e = -1$ and $II_2 = \mu/\rho v l = 1/\mathrm{Re}$, where Re = Reynolds number.

To find II_3. The remaining independent variable g contains the dimensions [L] and [T] only and so must be combined to form a dimensionless group with the repeating variables which do not contain M, namely v and l:

$$II_3 = gv^p l^q.$$

For dimensional homogeneity,

$$[M^0 L^0 T^0] = [LT^{-2}][LT^{-1}]^p[L]^q.$$

Equating powers of L and T,

$0 = 1 + p + q$ for L,

$0 = -2 - p$ for T.

From which $p = -2$, $q = 1$ and $II_3 = lg/v^2 = 1/Fr^2$, where Fr = Froude number.

The required relationship is therefore

$$F/\rho v^2 l^2 = \phi\{1/Re, 1/Fr^2\}$$

or, since ϕ is an unknown function, we can write

$$F/\rho v^2 l^2 = \phi'(Re, Fr).$$

Alternative solution
Instead of forming the dimensionless groups by applying the indicial method, they can be formed by expressing the fundamental dimensions M, L and T in terms of the repeating variables v, l and ρ:

$$[v] = [LT^{-1}], \qquad [l] = [L], \qquad [\rho] = [ML^{-3}],$$

giving $\quad [L] = [l], \qquad [M] = [\rho l^3], \qquad [T] = [l/v].$

Now, select each of the remaining variables in turn, starting with the dependent variable, and write down their dimensional formulae first in terms of M, L and T and then in terms of v, l and ρ.

To find II_1. Selecting the dependent variable F,

$$[F] = [MLT^{-2}]$$
$$= [\rho l^3][l][v^2/l^2]$$

giving $\quad F = II_1 \rho l^2 v^2 \quad$ and $\quad II_1 = F/\rho v^2 l^2.$

To find II_2. Select the independent variable μ,

$$[\mu] = [ML^{-1}T^{-1}]$$
$$= [\rho l^3][l^{-1}][v/l] = [\rho v l],$$

giving $\quad \mu = II_2 \rho v l \quad$ and $\quad II_2 = \mu/\rho v l.$

To find II_3. Select the remaining independent variable g

$$[g] = [LT^{-2}]$$
$$= [L][v^2/l^2] = [v^2/l],$$

giving $\quad g = II_3 v^2/l \quad$ and $\quad II_3 = lg/v^2.$

In many cases the dimensionless groups can be determined by inspection from experience without formal calculation.

25.14. The significance of dimensionless groups

By definition, a dimensionless group is the ratio of two similar physical quantities. In Example 25.3, we obtain a dimensionless group $F/\rho v^2 l^2$. The numerator F is the force required to overcome the drag on the vessel and, therefore, the denominator should also represent a force existing in the system. Examining the term $\rho v^2 l^2$, it can be written $\rho l^3 \times v^2/l$. Taking l as some typical dimension, ρl^3 is the product of density and a typical volume and so represents the mass of a typical element of fluid. Similarly, the velocity v can be expressed as l/t, where t is the time required to traverse a typical distance l. Thus, $v^2/l \propto l^2/lt^2 \propto l/t^2$, which is a measure of acceleration. Thus, we have

$$\rho v^2 l^2 = \rho l^3 \times l/t^2$$

$$= \text{Mass} \times \text{Acceleration} = \text{Inertial force}.$$

The dimensionless group $F/\rho v^2 l^2$ is therefore the ratio Drag force/Inertial force.

In the same way, the Reynolds number $\rho v l/\mu$ can also be expressed as the ratio of two forces, since it can be written $\rho v^2 l^2/\mu(v/l)l^2$ and we have already seen that $\rho v^2 l^2$ represents an inertial force. By definition, the coefficient of dynamic viscosity μ is the shear force per unit area produced by unit velocity gradient in the fluid. The denominator is therefore equivalent to a viscous force since v/l is the velocity gradient and l^2 represents a typical area. The dimensionless group is, therefore, the ratio Inertial force/Viscous force.

The final group v^2/lg can also be expressed as a ratio of forces, since it can be written $\rho v^2 l^2/\rho l^3 g$. The denominator is the inertial force, as before, and, since ρl^3 is a typical mass and g the gravitational acceleration, $\rho l^3 g$ is a gravitational force. The dimensionless group v^2/lg is the ratio Inertial force/Gravitational force.

Other dimensionless groups can be shown to be related to other physical properties of the system. For example, the Mach number $\bar{v}/\sqrt{(k/\rho)}$ is the ratio of the velocity of flow to the velocity of propagation of a pressure wave in the fluid, and is of importance in the study of compressible flow.

These and other dimensionless groups are useful means of defining the conditions which exist in a physical system and indicating which properties are of importance. As shown above, the Reynolds number is the ratio of inertial force to viscous force and is a measure of the importance of viscous resistance in controlling the flow of a fluid. A low Reynolds number indicates that viscosity is a dominant factor; a high Reynolds number indicates that its effect is small. For example, the flow in a circular pipe will be laminar for values of Reynolds number below 2000 and will be turbulent for higher values, so that, by stating the Reynolds number, we can specify the type of flow. As different laws apply to the two types of flow, it is a matter of practical importance to be able to distinguish between them.

25.15. The use of dimensionless groups in experimental investigation

Dimensional analysis can be of assistance in experimental investigation by reducing the number of variables in the problem. The result of the analysis is to replace an unknown relation between n variables by a relationship between a smaller number, $n - m$, of dimensionless groups. Any reduction in the number of variables greatly reduces the labour of experimental investigation. A moment's consideration will show how great this saving can be. A function of one variable can be plotted as a single curve constructed from a relatively small number of experimental observations, perhaps six if the relation is simple, or the results can be presented as a single table which might require just one page.

A function of two variables will require a chart consisting of a family of curves, one for each value of the second variable, or, alternatively, the information can be presented in the form of a book of tables. A function of three variables will require a set of charts or a shelf-full of books of tables.

As the number of variables increases, the number of observations to be taken grows so rapidly that the situation soon becomes impossible. Any reduction in the number of variables is, therefore, extremely important.

It is also, sometimes, simpler to alter the value of a dimensionless group than to alter the value of a single quantity. For example, the value of the Reynolds number $\rho \bar{v} d / \mu$ can be changed by altering any one or more of the quantities $\rho, \bar{v}, d$ and μ. While the properties of individual fluids and the availability of pipes of a given diameter would considerably restrict the range of values available in each individual quantity, alteration of several variables making up the Reynolds number would provide a much wider range of values for the group. Thus, a low value of Reynolds number could be obtained by using a small diameter pipe or by reducing the velocity in a given pipe or by changing the fluid passing through the given pipe.

Considering, as an example, the resistance to flow through pipes, the shear stress or resistance R per unit area at the pipe wall when fluid of mass density ρ and dynamic viscosity μ flows in a smooth pipe can be assumed to depend on the velocity of flow $\bar{v}$ and the pipe diameter d. Selecting a number of different pipes and several different fluids, we could obtain a set of curves relating frictional resistance (measured as $R/\rho \bar{v}^2$) to velocity, as shown in Fig. 25.1.

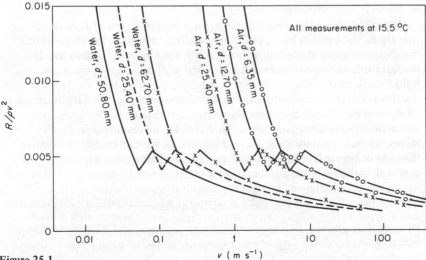

Figure 25.1

Such a set of curves would be of limited value both for use and for obtaining a proper understanding of the problem. However, it can be shown by dimensional analysis that the relationship can be reduced to-the form

$$R/\rho\bar{v}^2 = \phi(\rho\bar{v}d/\mu)$$

or, using the Darcy resistance coefficient $f = R/\tfrac{1}{2}\rho\bar{v}^2$,

$$f = \phi(\rho\bar{v}d/\mu).$$

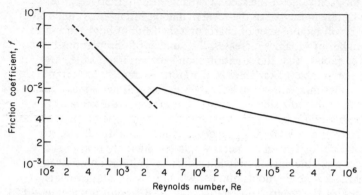

Figure 25.2

If the experimental points in Fig. 25.1 are used to construct a new graph of $\log f$ against $\log (\rho\bar{v}d/\mu)$ the separate sets of experimental data combine to give a single curve as shown in Fig. 25.2. For low values of Reynolds number, when flow is laminar, the slope of this graph is -1 and $f = 16/\text{Re}$, while for turbulent flow at higher values of Reynolds' number, $f = 0.0791\,(\text{Re})^{-1/4}$.

If the roughness of the pipe is taken into account, dimensional analysis shows that

$$f = \phi(\rho\bar{v}d/\mu, k/d),$$

where k is the roughness height. Since f is now a function of two variables, the result will be a set of curves, but experimental results for all types of fluids flowing in pipes of all sizes and roughnesses can be plotted together to form one such set of curves, which is the well-known Moody chart (Fig. 25.3).

In the laminar flow region, there is a single curve regardless of roughness, indicating that roughness is not significant and that the relationship remains $f = \phi(\rho\bar{v}d/\mu)$ in this region. At higher Reynolds numbers, when the flow is turbulent, we obtain a family of curves, one for each value of the relative roughness k/d, indicating that the relationship is of the form $f = \phi(\rho\bar{v}d/\mu, k/d)$. At high Reynolds numbers, when the flow is fully turbulent the curves become a set of straight lines parallel to the Reynolds number axis, one for each value of k/d, showing that f is no longer a function of Reynolds number but is given by $f = \phi(k/d)$. From this example, it can be seen that while a dimensional analysis may suggest that certain groups affect the relationship, this effect may not necessarily be significant and must be verified experimentally.

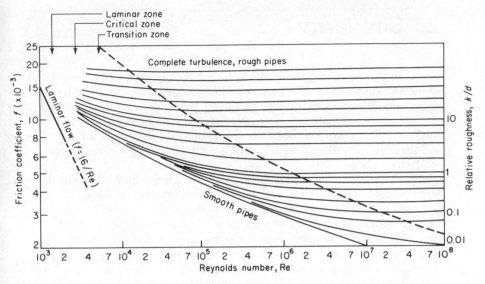

Figure 25.3

One further advantage of the dimensionless presentation of experimental data is that it is independent of the units employed and should, therefore, be internationally intelligible and convenient to use. Care should, however, be taken when results are given in terms of named dimensionless groups as the names and meanings of dimensionless groups have not been internationally agreed in all cases. It is desirable to ascertain the exact definition of the groups concerned and the precise quantities used in calculating them.

25.16. Further reading

Douglas, J. F. *An Introduction to Dimensional Analysis for Engineers* (1969). Pitman, London.

26 Similarity

Whenever the design engineer needs to take decisions at the design stage of a project, it will probably be necessary to initiate some form of model test programme. The basis of any such test series depends on the accurate use of instrumentation systems and the correct application of the theories of similarity. This, in turn, involves the application of dimensional analysis and the utilization of dimensionless groups such as the Reynolds, Froude or Mach numbers.

Model testing occurs in all areas of engineering based on fluid mechanics. Wind tunnel tests on aircraft, cars and trains, towing tests on ships and submarines, river/harbour tests carried out using models of high levels of intricacy to simulate tidal flow: all serve to illustrate such use of models. A recent application has arisen from the problems of airflow around buildings, which may be studied in wind tunnels and by examining smoke generation and propagation through building models. Model tests, then, depend on two basic types of similarity which may be considered separately: geometric and dynamic similarity.

26.1. Geometric similarity

The first requirement for model testing is a strict adherence to the principle of geometric similarity, i.e. the model be an exact geometric replica of the prototype. Thus, for an aerofoil model of 1/10 scale, say, both the span and chord must be exactly 1/10 of the full-scale dimensions. This principle is, however, not fully applied to river models, where distortion of the vertical scale is necessary to obtain meaningful results because it is necessary to keep the relationship between wave properties and depth correct. Generally, it may be assumed that geometric similarity is achieved in model testing.

26.2. Dynamic similarity

The definition of dynamic similarity is that the forces which act on corresponding masses in the model and prototype shall be in the same ratio throughout the area of flow modelled. If this similarity is achieved, then it follows that the flow pattern will be identical for both the model and the prototype flow fields. Before moving on to the consideration of particular flow situations, it is worthwhile to restate the derivation of the most common dimensionless groups whose respective values govern model testing. Consider a general, hypothetical flow situation where the pressure change Δp between two points is dependent on mean velocity $\bar{v}$, length l, density ρ and viscosity μ, bulk modulus K, surface tension σ and gravitational acceleration g:

$$\Delta p = f(\bar{v}, l, \rho, \mu, K, \sigma, g).$$

With eight variables, five dimensionless groups may be expected, and they may be recognized as

$$\frac{\Delta p}{\frac{1}{2}\rho v^2} = f_1\left(\frac{\rho \bar{v} l}{\mu}, \frac{\bar{v}}{\sqrt{K/\rho}}, \frac{\rho \bar{v}^2}{\sigma}, \frac{\bar{v}}{\sqrt{lg}}\right)$$

or   Pressure coefficient, $C_p = f_1$ (Re, Ma, We, Fr).   (26.1)

Equation (26.1) indicates that the pressure coefficient is dependent upon the other dimensionless groups and is defined if the other groups are defined.

The general condition for dynamic similarity is probably best understood by consideration of a particular flow situation. Consider the forces acting on the air stream passing over an aerofoil prototype and model (Fig. 26.1). The

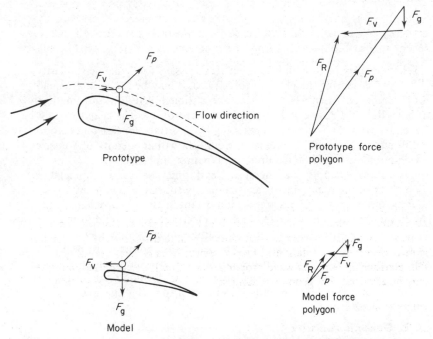

Figure 26.1   Forces acting on a fluid element passing over an aerofoil shape

forces acting are gravitational F_g, which may be disregarded but is included here as a general case, F_p the pressure force and F_v the viscous resistance force. The resultant force F_R will act on the particle and accelerate it in accordance with Newton's second law. As the force polygons in the model and prototype are similar, the magnitudes of the forces in the prototype and model will be in the same ratio as the magnitude of the mass × acceleration, ma, thus:

$$(ma)_p/(ma)_m = (F_g)_p/(F_g)_m,$$

where suffix m and p refer to the model and prototype, respectively.

Now, mass $m \propto \rho l^3$, acceleration $a \propto \bar{v}/t$, and $F_g \propto \rho g l^3$. Thus,

$$(\bar{v}/gt)_m = (\bar{v}/gt)_p;$$

but it can be arranged that

$$t_m/t_p \propto (l_m/\bar{v}_m)/(l_p/\bar{v}_p),$$

so that $(v^2/gl)_m = (v^2/gl)_p$ (26.2)

or, for similarity, the Froude numbers must be equal.

Considering the viscous forces, then, if $F_v \propto \mu \bar{v} l$, it follows that

$$(ma)_m/(ma)_p = (F_v)_m/(F_p)_p$$

or $(\rho l^3 \bar{v}/t)_m/(\rho l^3 \bar{v}/t)_p = (\mu \bar{v} l)_m/(\mu \bar{v} l)_p$ (26.3)

$$(\rho l^2/\mu t)_m = (\rho l^2/\mu t)_p$$

and, by substituting $t = l/\bar{v}$,

$$(\rho \bar{v} l/\mu)_m = (\rho \bar{v} l/\mu)_p$$ (26.4)

or $Re_m = Re_p$,

so that equality of Reynolds numbers is necessary for dynamic similarity. Finally, considering the pressure force ratio, it may be shown that equality of pressure coefficients is also necessary, although it follows automatically if both Re and Fr are equal for model and prototype. The analysis could be extended to include the forces resulting from surface tension and compressibility effects and such an extension would show that equality of Weber and Mach numbers would be required for dynamic similarity.

The above example is a general case and it may be readily appreciated that some of the requirements are practically superfluous in many cases. For example, gravity forces are not important in air flow problems, and Mach number may be neglected in most incompressible liquid flow situations.

However, a general definition of dynamic similarity may now be stated as a requirement that the significant dimensionless groups must be equal for model and prototype to ensure similarity of flow between model and prototype. The remainder of the chapter will be devoted to an analysis of a number of particular flow situations in order to define the governing dimensionless groups.

26.3. Model studies for flows without a free surface

Free surface effects are absent in bounded flow, e.g. pipes or ducts flowing full, or in the flow around submerged bodies, e.g. aircraft, submarines, cars and buildings. Under low flow velocity conditions, which apply in the majority of the above cases, the compressibility effects defined by flow Mach number may be ignored and the predominant factor that must be kept constant between model and prototype is the Reynolds number.

Example 26.1

A submarine-launched missile, 1 m diameter by 5 m long, is to be studied in a water tunnel to determine the loads acting on it during its underwater launch. The maximum speed during this initial part of the missile's flight is 10 m s^{-1}. Calculate the mean water tunnel flow velocity if a 1/20 scale model is to be employed and dynamic similarity is to be achieved.

Solution
For dynamic similarity, the Reynolds number must be constant for the model and the prototype:

$$Re_m = Re_p,$$

$$\bar{v}_m l_m \rho_m / \mu_m = \bar{v}_p l_p \rho_p / \mu_p.$$

The model flow velocity is given by

$$\bar{v}_m = \bar{v}_p (l_p/l_m)(\rho_p/\rho_m)(\mu_m/\mu_p),$$

but $\rho_p = \rho_m$ and $\mu_p = \mu_m$. Therefore,

$$\bar{v}_m = 10 \times 20 \times 1 \times 1 = \mathbf{200 \ m \ s^{-1}}.$$

This is a high flow velocity and illustrates the reason why few model tests are made with completely equal Reynolds numbers. At high Re values, however, the divergences become of lesser importance.

Example 26.2

An airship of 3 m diameter and 20 m length is to be studied in a wind tunnel. The airship speed range to be investigated is at the docking end of its range, a maximum of 2 m s^{-1}. Calculate the mean model wind tunnel speed if the model is made to 1/10 scale. Assume the same air pressure and temperature for model and prototype.

Solution
Equivalence of Reynolds numbers is required, i.e.

$$(\bar{v} l \rho / \mu)_m = (\bar{v} l \rho / \mu)_p.$$

Hence, $\bar{v}_m = \bar{v}_p (l_p/l_m)(\rho_p/\rho_m)(\mu_m/\mu_p),$

and, therefore,

$$\bar{v}_m = 2 \times 10 \times 1 \times 1 = \mathbf{20 \ m \ s^{-1}}.$$

Referring back to equation (26.1), it will be seen that the pressure coefficient $C_p = \Delta p / \frac{1}{2}\rho \bar{v}^2$ was defined by the dimensionless groups (Re, Ma, We and Fr). Therefore, it follows that, if dynamic similarity is achieved by equating these groups for the model and prototype, then the pressure coefficient for model and prototype will also be equal.

Thus, in Example 26.1, if the pressure difference between two points on the surface of the missile had been 5·0 N m^{-2}, then the pressure difference on the model would be given by

$$(C_p)_m = (C_p)_p,$$

$$(\Delta p)_m / \frac{1}{2}\rho_m \bar{v}_m^2 = (\Delta p)_p / \frac{1}{2}\rho_p \bar{v}_p^2,$$

$$(\Delta p)_m = (\Delta p)_p (\rho_m/\rho_p)(\bar{v}_m/\bar{v}_p)^2$$

$$= 5 \times 1 \times (200/10)^2$$

$$= 5 \times 400$$

$$= 2 \ kN \ m^{-2}$$

The importance of pressure coefficient may also be appreciated if the case of pipe flow modelling is considered.

Example 26.3

Flow through a heat exchanger tube is to be studied by means of a 1/10 scale model. If the heat exchanger normally carries water, determine the ratio of pressure losses between the model and the prototype if (a) water is used in the model and (b) air at normal temperature and pressure is used in the model.

Solution

For dynamic similarity, the Reynolds number must be constant:

$$\text{Re}_m = \text{Re}_p,$$

$$\bar{v}_m/\bar{v}_p = (l_p/l_m)(\rho_p/\rho_m)(\mu_m/\mu_p).$$

If the Reynolds numbers are equal, then so must be the pressure coefficients. Thus,

$$(C_p)_m = (C_p)_p,$$

$$(\Delta p)_m/(\Delta p)_p = (\rho_m/\rho_p)(\bar{v}_m/\bar{v}_p)^2,$$

$$(\Delta p)_m/(\Delta p)_p = (l_p/l_m)^2 (\mu_m/\mu_p)^2 (\rho_p/\rho_m).$$

(a) In the water model case,

$$(\Delta p)_m/(\Delta p)_p = 10^2 \times 1^2 \times 1,$$

$$(\Delta p)_m = 100(\Delta p)_p.$$

(b) If air is used, then

$$\rho_p/\rho_m = 1000/1 \cdot 23 = 10^3/1 \cdot 23,$$

$$\mu_m/\mu_p = 1 \cdot 8 \times 10^{-5}/1 \times 10^{-3} = 1 \cdot 8 \times 10^{-2},$$

$$(\Delta p)_m/(\Delta p)_p = 10^2 \times (1 \cdot 8 \times 10^{-2})^2 \times 10^3/1 \cdot 23,$$

$$(\Delta p)_m = 26 \cdot 34 (\Delta p)_p.$$

26.4. Zone of dependence of Reynolds and Mach numbers

The examples given, together with some realization of the way in which the model flow rates increase rapidly to maintain absolute dynamic similarity based on Reynolds number, suggest that, at high prototype Reynolds numbers, the model tests will become impractical. However, reference to, for example, the effect of Reynolds number variation on pressure losses along a duct will show that at high Reynolds numbers the effect becomes minimal. The use of air to simulate water on model tests of ducts or hydraulic machines is well known and is a convenient method of keeping the absolute flow rates down.

If some method such as this is not adopted, then two obvious consequences follow: either the flow velocities required over the model become too great to be practical, or, if the flow rates are achieved, there is a danger that compressibility effects may become important, resulting in erroneous results from the model tests.

Mach number becomes a significant parameter in flow situations where the ratio of flow velocity to sonic velocity exceeds about 0·25 to 0·3. It is normally difficult to satisfy both Reynolds number and Mach number equality simultaneously, and so it is important that testing decisions are made based on experience of the type of flow to be investigated. For example, if the viscous motion of a fluid close to a boundary in supersonic flow were the phenomenon under consideration, then Reynolds number would be the criterion, whereas, if the flow through the shock wave pattern around a body were to be investigated, then equivalence of Mach number would be the overriding criterion.

26.5. Model studies in cases involving free surface flow

In free surface model studies the effect of gravity becomes important and the governing parameter is Froude number. Generally the prototypes, i.e. large spillways, have Reynolds numbers large enough to be operating out of the range of dependence on Re; however, the model may be of such a size that, when Froude number equivalence is set up, the model Reynolds number is small enough to produce viscous effects not representative of the prototype. For this reason, the model must be large enough to place its Reynolds number above the viscous loss dependence level. One problem with free surface flow cases is that, generally, the same fluid is used for the model as for the prototype, so that the convenient expedient of substituting air for water in internal flows cannot be copied.

Example 26.4

A 1:50 scale model of a proposed power station tailrace is to be used to predict prototype flow. If the design load rejection bypass flow is 1200 $m^3 s^{-1}$, what water flow rate should be used on the model?

Solution
Equating Froude numbers,

$$Fr_m = Fr_p,$$

where $Fr = \bar{v}/\sqrt{(lg)}$. Therefore,

$$\bar{v}_m/\bar{v}_p = \sqrt{(l_m/l_p)}.$$

Flow rate may be determined by introducing the area ratio $A_m/A_p = 1/2500$ = (Scale)^2. Hence,

$$\frac{Q_m}{Q_p} = \frac{A_m \bar{v}_m}{A_p \bar{v}_p} = \frac{l_m^2}{l_p^2} \sqrt{\left(\frac{l_m}{l_p}\right)},$$

$$Q_m = Q_p (l_m/l_p)^{2\cdot5} = 1200 \times (\tfrac{1}{50})^{2\cdot5} \, m^3 \, s^{-1}$$

$$= 0\cdot067 \, m^3 \, s^{-1}.$$

This relatively simple approach is complicated for the case of ship resistance testing, as the phenomenon is made up of two factors, namely the surface resistance of the hull, dependent on Reynolds number and the wave resistance, which is Froude number dependent (*see* Section 10.4).

Consider the case of a model to be towed at a speed such that the Froude number is satisfied:

$$Fr_m = Fr_p,$$

$$v_m/\sqrt{(l_m g)} = v_p/\sqrt{(l_p g)},$$

$$v_m/v_p = \sqrt{(l_m/l_p)}.$$

Now consider the same model and equate Reynolds numbers:

$$Re_m = Re_p,$$

$$(\rho v l/\mu)_m = (\rho v l/\mu)_p,$$

$$v_m/v_p = (l_p/l_m)(\rho_p/\rho_m)(\mu_m/\mu_p) = l_p/l_m,$$

if $\rho_m = \rho_p$ and $\mu_m = \mu_p$. Obviously, then, the two criteria cannot be satisfied simultaneously and the approach followed is to equate Froude number to model wave resistance forces as these are the more difficult to analyse. Viscous hull resistance is then calculated by analytical techniques and added to the wave resistance measured.

26.6. Similarity applied to rotodynamic machines

Application of the techniques of dimensional analysis to fans and pumps yields relationships of the form

$$P/N^3 D^5 \rho = f(Q/ND^3, \mu/\rho ND^2, k/D, a/D, b/D, c/D),$$

$$P_s/\rho N^2 D^2 = f(Q/ND^3), \ \mu/\rho ND^2, k/D, a/D, b/D, c/D,$$

where P is shaft power, Q is volume flow rate, P_s is the pressure rise across the unit rotating at speed N, and of diameter D. The fluid type is defined by density ρ and viscosity μ while the detail dimensions of the machine are a, b, c with surface roughness k. For geometrically similar machines operating at high Reynolds numbers, so that the term $\mu/ND^2 \rho = Re$ becomes irrelevant, the expressions reduce to

$$P/N^3 D^5 \rho = f_2(Q/ND^3) \quad \text{and} \quad P_s/\rho N^2 D^2 = f_1(Q/ND^3)$$

Thus, for model testing to be valid, each of these groups should have identical values for the model and the prototype.

Model testing is of particular value in the design and manufacture of the larger scale fans, pumps and turbines, to which these relationships also apply, except that the power terms relate to power generated rather than power supplied.

Generally, the model scale is arranged so that the impeller diameters are less than 0·5 m and the same fluid is usually employed in the model tests as for the prototype. However, due to the lack of effect of Reynolds number, provided the flow is well into the fully turbulent region, it is possible to employ air or pressurized gas in order to obtain more manageable flow rates or machine scales.

Denoting the model by m and the full size machine by p, the following relations can be proved:

(i) Flow,

$$Q_m/Q_p = (ND^3)_m/(ND^3)_p \qquad\qquad (26.5)$$

(ii) Pressure rise (pumps) and pressure drop (turbines),

$$(P_s)_m/(P_s)_p = (\rho N^2 D^2)_m/(\rho N^2 D^2)_p, \qquad\qquad (26.6)$$

where the density ratio may vary from unity;

(iii) Power supplied (pumps) and power generated (turbines),

$$P_m/P_p = (N^3 D^5 \rho)_m/(N^3 D^5 \rho)_p. \qquad\qquad (26.7)$$

It will be appreciated that these relations are all independent of operating pressure, so that, in theory, any convenient operating test rig head may be employed. In practice, this is not entirely true as cavitation onset is dependent on absolute pressures and, for pumps and turbines, its occurrence is of major importance, so that pressure levels should be as close as possible to the full scale installation values.

Theoretically, the efficiency of model and prototype should be the same. However, there will be some excess inefficiency in the model due to scale effects relating to leakage flow, roughness variations and manufacturing constraints (*see* Section 22.5).

26.7. River and harbour models

River and harbour engineering projects are costly undertakings and, as analytical techniques only provide approximate predictions of the likely effects of any river widening or harbour improvement, the use of models at an early stage in the design has many advantages. The problems arise when a suitable scale is to be chosen for the model; the adoption of a scale that will give reasonable channel depths will usually result in a model too large to be practical, while choosing a scale on area–cost criteria yields channel depths that are very small.

The problems of shallow model channels are: (i) accuracy in level and level change measurement becomes impossible to achieve; (ii) the surface roughness of the channel beds would be impractically small and there is even a probability that channel flow would be laminar rather than turbulent, as normally found in practice.

In order to provide a solution to these problems, distorted scaling is adopted, vertical scales of 1/100 and larger being typical while horizontal scales vary from 1/200 to 1/500. Distortion of this sort is suitable if the overall discharge characteristics of a long length of river are being studied. However, it should be appreciated that the micro-situation is not well modelled, and situations such as the effects of breakwater positioning should be studied on as large and as undistorted a scale as possible.

In models of this type, strict geometric similarity is not achieved. However, if the mean flow velocity, $\bar{v}$, and depth, Z, are arranged so that there is an equivalence of Froude number $(\bar{v}/\sqrt{gZ})$, between model and river then it will be expected that the flow type, i.e. fast or slow, will be the same at corresponding points on the model and river. Thus,

$$\bar{v}_m^2/Z_m = \bar{v}_r^2/Z_r,$$

and, hence,

$$\bar{v}_m/\bar{v}_r = \sqrt{(Z_m/Z_r)}, \tag{26.8}$$

where m denotes model and r denotes full scale river. The discharge Q through the model must depend on both vertical and horizontal scales, thus $Q = \bar{v}lZ$ and

$$Q_m/Q_r = (Z_m/Z_r)^{3/2}(l_m/l_r), \tag{26.9}$$

where l_m/l_r is the horizontal scale $1:x$ and Z_m/Z_r is the vertical scale, $1:y$.

In order to manufacture models, it is necessary to have information on the effect of surface roughness, particularly the effects of model roughness size, which is often dictated by the manufacturing process chosen. From the Manning coefficient of surface roughness, n, and the Manning equation applied to both model and river,

$$\frac{\bar{v}_m}{\bar{v}_r} = \left(\frac{n_r}{n_m}\right) \left(\frac{Z_m}{Z_r}\right)^{2/3} \left(\frac{Z_m/l_m}{Z_r/l_r}\right)^{\frac{1}{2}},$$

$$n_m = n_r x^{1/2}/y^{2/3}. \tag{26.10}$$

Full scale river beds have values of surface roughness defined by $n > 0.03$ and as the normal model surface finish, obtained by use of cement mortar for model surfaces, is of the order of $n \simeq 0.012$, artificial roughening is normally necessary. The maximum value of n obtainable by artificial means, i.e. adding wire mesh, gravel and even small rods to the bed, is of the order of 0.04. As mentioned, the micro-flow situation, i.e. eddies of local currents, will differ between model and river, so adjustment by surface roughness only is not a reasonable course of action. Figure 26.2 illustrates a typical model and the artificial channel bed roughening used.

Before it is used to predict the effects of any modifications to the river channels, the model should first be checked for discharge and depth accuracy. Measurements of full scale depths and discharges should be checked against model discharge and depth by use of equation (26.9), relating total flow rates, and the vertical scale. Model discharge rates are produced using recirculating pump circuits and orifice plate or notch flow measuring instrumentation. If the model values are acceptable, then testing can continue. If the depth and flow rates are in poor agreement with the full scale results, then the model surface roughness should be adjusted or the scales altered.

It should be noted that the above analysis is based on the premise that the Reynolds number of both model and full scale river is such that the flow is totally turbulent and that variations in Reynolds number are not important. It is good practice, however, to check the values of model and channel Reynolds number to ensure that this simplification is valid, i.e. Re > 2000.

Estuary models may be constructed using the same general principles as outlined for river models; distorted scales are typically 1/50 to 1/150 vertical

Figure 26.2a *See* page 724

Figure 26.2b *See* page 724

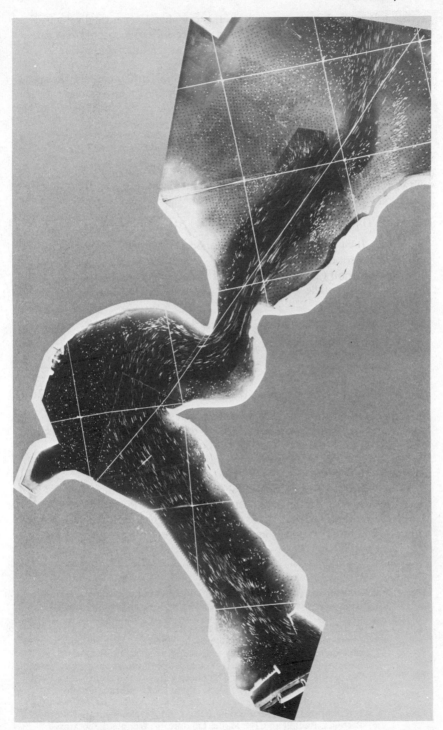

Figure 26.2c *See* page 724

and 1/300 and 1/2500 horizontal. As the available data on tidal velocity distributions are likely to be sketchy, it is necessary to incorporate in the mode the whole tidal channel system, as well as a substantial portion of adjustment coastline.

Although surface roughness is not so critical in estuary models, the speed of propagation of the tide becomes an important design criterion as the tidal period governs the time available for result recording. The tidal period of the model is, thus,

$$\frac{\text{Time in model}}{\text{Time in estuary}} = \frac{(\text{Distance}/\text{Velocity})_{\text{model}}}{(\text{Distance}/\text{Velocity})_{\text{estuary}}}$$

Now, as the Froude numbers are equal (as for river models) it follows from equation (26.8) that

$$(\text{Time})_m/(\text{Time})_e = (l_m/l_e)(\bar{v}_e/\bar{v}_m) = (l_m/l_e)\sqrt{(Z_e/Z_m)}. \qquad (26.11)$$

Thus, for a tidal period of 12·4 h and a model with 1/50, 1/500 scales, the model tidal period is 10·5 min.

While equation (26.11) refers to horizontal fluid velocity, the scale factor for vertical silt particle velocity is also of interest and may be derived in the same manner:

$$\frac{\text{Rate of fall in model}}{\text{Rate of fall in estuary}} = \frac{\text{Depth scale}}{\text{Time scale}}$$

$$= (Z_m/Z_e)\sqrt{\{(Z_m/Z_e)(l_e/l_m)\}}$$

$$= (Z_m/Z_e)^{3/2}(l_e/l_m). \qquad (26.12)$$

A further complication in estuary model studies, particularly of silting phenomena, is stratification effects due to density variations between salt and fresh water. If the use of saline solutions in the model testing is undesirable for corrosion reasons, then a stable clay solution may be employed.

Estuary models are now commonly used to investigate the effects of discharge of power station or industrial cooling water flows and, here, density variations may again have to be modelled, based on temperature. Other uses of estuary models include silting and erosion studies and the

---

Figure 26.2   (a) Full and (b) empty scale model of the entrance channel to Dar-es-Salaam harbour. Note channel roughening in the view of the empty model.
(c) A bird's eye view of the Dar-es-Salaam model showing flow patterns at ebb tide. The effect is obtained by scattering confetti on the water and then photographing it with a time exposure. Where the water is moving fastest, the confetti shows as a streak — the longer the streak, the faster the current. By studying results from a series of photographs representing various states of the tide, one is able to calibrate the model so that water flows in the model correspond closely to those recorded on site.
(Courtesy of UN Development Program, East African Harbours Corp., International Bank for Reconstruction and Development, Bertlin and Partners, Redhill and The Model Laboratory, Wimpey Laboratory, Middlesex.)

spread and deposit of effluent discharged into the sea. Generally, the role of the estuary model should be seen more as a method of comparing the attributes of various design solutions, rather than as an accurate method of predicting the effects of one design.

Harbour and coastal models require the inclusion of wave effects, and these are reproducible by means of mechanical wave-making devices. However, the type of wave motion encountered in coastal engineering is dependent on both water depth and wavelength for its propagation velocity, so that the degree of scale distortion acceptable in river and estuary models can no longer be applied. The best model studies are carried out with equal scaling; however, distortion of the vertical scale up to two or three times the vertical has been used.

Example 26.5

It is proposed to construct a model of 18 km length of river, for which the first 8 km are tidal. The normal discharge of the river is known to be in the region of 300 $m^3\,s^{-1}$, the average width and depth of the channel being 3 m and 65 m, respectively. Given a laboratory of 30 m length propose suitable scales and calculate the tidal period.

Solution

(i) The largest scale possible would be $30/18 \times 1000 = 1/600$ for the horizontal distances.

(ii) As the river is tidal, scale distortions of around 6 to 10 are acceptable, so a vertical scale of $1/60$ could be employed.

(iii) The model will be constructed to conform with the scales above. However, in doing so, the effect of Reynolds number is assumed negligible. It is good practice to check the Reynolds number.

Average river velocity = 300 $m^3\,s^{-1}/3 \times 65\ m^2 = 1.54\ m\,s^{-1}$.

From equation (26.8),

$$\bar{v}_m = \bar{v}_r \sqrt{(Z_m/Z_r)} = 1.54 \times \sqrt{(1/60)}$$

$$= 0.199\ m\,s^{-1}.$$

$$Re_m = \rho v m/\mu,$$

where m = Hydraulic mean depth = Area/Perimeter flow cross-section

$$= (3 \times \tfrac{1}{60} \times 65 \times \tfrac{1}{600})/(\tfrac{65}{600} + \tfrac{6}{60})$$

$$= 5.4 \times 10^{-3}/0.208 = 0.026\ m.$$

Thus, $Re_m = 1000 \times 0.199 \times 0.026/1.14 \times 10^{-3} = 4532$

which is sufficiently turbulent to allow Re effects to be ignored.

(iv) The tidal period can be calculated from the time scale (equation (26.11)):

$$\frac{(Time)_m}{(Time)_r} = \frac{l_m}{l_r} \sqrt{\left(\frac{Z_r}{Z_m}\right)} = \frac{1}{600} \times \sqrt{(60)} = 0.0129$$

Therefore,

Tidal period of model = 12.4 × 60 × 0.0129 = **9.6 min.**

726   Dimensional analysis and similarity

26.8. Further reading

Allen, J. *Scale Models in Hydraulic Engineering* (1952). Longman, London.
Bain, D. C. *et al.* (1971). *Wind Tunnels, An Aid to Engineering Structure Design*. British Hydraulics Research Association, Cranfield.
Bradshaw, P. (1964). *Experimental Fluid Mechanics*. Pergamon, Oxford.
B.S. 848 Part I 1963: *Fan Testing for General Purposes.*
Langhaar, H. L. (1951). *Dimensional Analysis and Theory of Models*. Wiley, New York.
Kline, S. J. (1965). *Similitude and Approximation Theory*. McGraw-Hill, New York.
Novak, P. and Cabelka, J. (1981). *Models in Hydraulic Engineering*. Pitman.
Sedov, L. I. (1959). *Similitude and Dimensional Analysis in Mechanics*. Academic Press, New York.
Streeter, V. L. (ed.) (1961). *Handbook of Fluid Dynamics*. McGraw-Hill, New York.
Yallin, M. S. (1971). *Theory of Hydraulic Models*. Macmillan, London.

EXERCISES 26

26.1   Water at 20 °C flows at 4 m/s in a 200 mm smooth pipe. Calculate air velocity in a 100 mm pipe at 40 °C if the two flows are dynamically similar.
[0·135 m/s]

26.2   A spherical balloon to be used in air at 20 °C is tested by towing a 1/3 model submerged in a water tank. If the model is 1 m in diameter and the drag force is measured as 200 newtons at a model speed of 1.2 m/s, what would be the expected prototype drag if the water temp. was 15 °C and dynamic similarity is assumed.
[42·2 N]

26.3   Determine the relationship between model and prototype kinematic viscosity if both Reynolds Number and Froude Number are to be satisfied.
[Linear scale to power 3/2]

26.4   A large venturi meter is calibrated by means of a 1/10 scale model using the same fluid as the prototype. Calculate the discharge ratio between model and prototype for dynamic similarity.
[1/10]

26.5   The velocity at a point in a model spillway for a dam is 1 m/s. For a scale of 1/10 calculate the corresponding velocity in the prototype.
[3·16 m/s]

26.6   If the scale ratio between a model spillway and its prototype is 1:25 what velocity and discharge ratio should apply between model and prototype? If the prototype discharge is 3000 m³/s what is the model discharge.
[1/5, 1/3125, 0·96 m³/s]

26.7   A 1/5 scale model of a missile has a drag coefficient of 3 at Mach No. 2. What would the model/prototype drag ratio be in air at the same temperature and one third the density for the prototype at the same Mach No
[1/25]

26.8   A ship model, scale 1:50, has a wave resistance of 30 N at its design speed. Calculate the prototype wave resistance.
[3750 kN]

26.9   A ship having a length of 200 m is to be propelled at 25 km/h. Calculate the prototype Froude Number and the scale of a model to be towed at 1·25 m/s.
[0·156, 1/30·6]

26.10   A fan running at 8 revs/s delivers 2·66 m³/s at a fan total pressure of 418 N/m², the air having a temperature of 0 °C and 101·325 kN/m² pressure. Given that the fan efficiency is 69%, calculate the air quantity delivered, the fan total pressure and the fan power when the air temperature is increased to 60 °C and the barometric pressure falls to 95 kN/m².
[2·66 m³/s, 311 N/m², 1·24 kW]

26.11   An axial flow water pump is to deliver 15 m³/s against a head of 20 m water. Calculate the air flow delivery rate and pressure rise for a 1/3 scale model using air at 1·3 kg/m³ density if the model and prototype are driven at the same speed.
[0·55 m³/s, 2·88 mm H₂O]

26.12   A model turbine employs 2 m³/s water flow when simulating a full scale prototype designed to be served by a 15 m³/s flow. If the scale is 1/5 calculate the speed ratio and the shaft delivered power ratio.
[16·66, 1·48]

26.13   In a test on a centrifugal fan it was found that the discharge was 2·75 m³/s and the total pressure 63·5 mm water column. The shaft power was 1·7 kW. If a geometrically similar fan having dimensions 25% smaller but having twice the rotational speed was used calculate the output, pressure generated and shaft power required. The air conditions are the same in both cases.
[2·32 m³/s, 142·8 mm H₂O, 3·2 kW]

26.14   A 1 : 5 scale model of a water pumping station piping system is to be tested to determine overall pressure losses. Air at 27°C, 100 kN/m² absolute is available. For a prototype velocity of 0·45 m/s in a 4·25 m diameter duct section with water at 15 °C determine the air velocity and quantity needed to model the situation.
[33·8 m/s, 31·4 m³/s]

26.15   The torque delivered by a water turbine depends upon discharge Q, head H, density ρ, angular velocity ω, and efficiency η. Determine the form of the equation for torque.

$$\left[\rho g H \cdot f \left(\omega \frac{H^3}{Q}, \eta \right) \right]$$

26.16   A 20 km length of river is to be modelled in a laboratory having only 12·5 m of available length. The river discharge is known to be in the range 400–500 m³/s and the average length and width are 3·5 and 55 m respectively. Propose suitable scales.
[1/625, 1/60]

26.17   If the model in Exercise 26.16 is tidal calculate the tidal period on the model.
[9·2 min]

Appendix 1   Some properties of common fluids

Temperature	Density, ρ	Viscosity, μ	Kinematic viscosity, ν	Surface tension, σ	Vapour-pressure head, $p_v/\rho g$	Bulk modulus of elasticity, K
(°C)	(kg m⁻³)	(kg m⁻¹ s⁻¹)	(m² s⁻¹)	(N m⁻¹)	(m)	(MN m⁻²)
0	999·9	1·792 (×10⁻³)	1·792 (×10⁻⁶)	7·62 (×10⁻²)	0·06	2040
5	1000·0	1·519	1·519	7·54	0·09	2060
10	999·7	1·308	1·308	7·48	0·12	2110
15	999·1	1·140	1·141	7·41	0·17	2140
20	998·2	1·005	1·007	7·36	0·25	2200
25	997·1	0·894	0·897	7·26	0·33	2220
30	995·7	0·801	0·804	7·18	0·44	2230
35	994·1	0·723	0·727	7·10	0·58	2240
40	992·2	0·656	0·661	7·01	0·76	2270
45	990·2	0·599	0·605	6·92	0·98	2290
50	988·1	0·549	0·556	6·82	1·26	2300
55	985·7	0·506	0·513	6·74	1·61	2310
60	983·2	0·469	0·477	6·68	2·03	2280
65	980·6	0·436	0·444	6·58	2·56	2260
70	977·8	0·406	0·415	6·50	3·20	2250
75	974·9	0·380	0·390	6·40	3·96	2230
80	971·8	0·357	0·367	6·30	4·86	2210
85	968·6	0·336	0·347	6·20	5·93	2170
90	965·3	0·317	0·328	6·12	7·18	2160
95	961·9	0·299	0·311	6·02	8·62	2110
100	958·4	0·284	0·296	5·94	10·33	2070

A.1.1. Variation of Some Properties of Water with Temperature

Pressure (bar)	Bulk modulus, K (MN m^{-2})			
	0 °C	20 °C	49 °C	93 °C
1	2040	2200	2289	2130
100	2068	2275	2358	2199
300	2186	2399	2496	2330
1000	2620	2827	2937	2792

A.1.2. Variation of Bulk Modulus of Elasticity of Water with Temperature and Pressure

Temperature, T	Density, ρ	Dynamic viscosity, μ	Kinematic viscosity, ν
(°C)	(kg m^{-3})	(kg m^{-1} s^{-1})	(m^2 s^{-1})
−40	1·52	14·94 (×10^{-6})	9·83 (×10^{-6})
−20	1·40	15·92	11·37
0	1·29	17·05	13·22
20	1·20	18·15	15·13
40	1·12	19·05	17·01
60	1·06	19·82	18·70
80	0·99	20·65	20·86
100	0·94	21·85	23·24
120	0·90	23·20	25·78

A.1.3. Variation of Some Properties of Air with Temperature at Atmospheric Pressure

Liquid	Density, ρ	Surface tension, σ	Dynamic viscosity, μ	Bulk modulus, K
	(kg m^{-3})	(N m^{-1})	(kg m^{-1} s^{-1})	(GN m^{-2})
Temperature, 20 °C				
Water, fresh	998	72·7(×10^{-3})	1·00(×10^{-3})	2·05
sea	1025			
Alcohol, ethyl	789	22·3	1·197	1·32
Benzene	879	28·9	0·647	1·10
Carbon tetrachloride	1632	26·8	0·972	1·12
Glycerol	1262	63	620	4·03
Mercury	13 546	472	1·552	26·2
Paraffin oil	800	26	1·9	1·62
Temperature, 38 °C				
Oil, s.a.e. 10	880–950	30	29	
s.a.e. 30	880–950	30	96	

A.1.4. Some Properties of Common Liquids

| Gas | Molecular weight | Density, ρ (kg m^{-3}) | Gas constant, R (J kg^{-1} K^{-1}) | Specific heats | | Specific heat ratio, γ (= c_p/c_v) | Dynamic viscosity, μ (kg m^{-1} s^{-1} ×10^{-6}) |
				c_p (J kg^{-1} K^{-1})	c_v (J kg^{-1} K^{-1})		
Air		1·293	287	993	708	1·402	17·05
Carbon monoxide	28·0	1·250	297	1050	748	1·404	16·6
Carbon dioxide	44·0	1·977	189	834	640	1·304	14
Helium	4·0	0·179	2077	5240	3157	1·66	18·6
Hydrogen	2·02	0·090	4121	14 300	10 140	1·41	8·35
Methane	16·04	0·717		2200	1676	1·313	10·3
Nitrogen	28·0	1·250	297	1040	741	1·404	16·7
Oxygen	32·0	1·429	260	913	652	1·40	19·2
Water vapour	18·0	0·800	462	2020 (373 K)	1519	1·33	8·7

A.1.5. Some Properties of Common Gases (at p = 1 atm, T = 273 K)

Altitude above sea level (m)	Absolute pressure (bar)	Absolute temperature (K)	Mass density (kg m^{-3})	Kinematic viscosity (m^2 s^{-1})	Velocity of sound (m s^{-1})
0	1·0132	288·15	1·2250	1·461 ($\times 10^{-5}$)	340·3
1000	0·8988	281·7	1·1117	1·581	336·4
2000	0·7950	275·2	1·0066	1·715	332·5
4000	0·6166	262·2	0·8194	2·028	324·6
6000	0·4722	249·2	0·6602	2·416	316·5
8000	0·3565	236·2	0·5258	2·904	308·1
10 000	0·2650	223·3	0·4134	3·525	299·5
11 500	0·2098	216·7	0·3375	4·213	295·1
14 000	0·1417	216·7	0·2279	6·239	295·1
16 000	0·1035	216·7	0·1665	8·540	295·1
18 000	0·075 65	216·7	0·1216	11·686	295·1
20 000	0·055 29	216·7	0·088 92	15·989	295·1
22 000	0·040 47	218·6	0·064 51	22·201	296·4
24 000	0·029 72	220·6	0·046 94	30·743	297·7
26 000	0·021 88	222·5	0·034 26	42·439	299·1
28 000	0·016 16	224·5	0·025 08	58·405	300·4
30 000	0·011 97	226·5	0·018 41	80·134	301·7
32 000	0·008 89	228·5	0·013 56	109·62	303·0

A.1.6. International Standard Atmosphere

Temperature (°C)	Volume of air dissolved* (m^3)
0	0·029
10	0·023
30	0·016
70	0·012
100	0·011

* Measured at 0 °C and 1 atm pressure per unit volume of water under a pressure of 1 atm.

A.1.7. Solubility of Air in Pure Water at Various Temperatures

Fluid	Absolute viscosity $\times 10^6$ (kg m⁻¹ s⁻¹)							
	−20 °C	0 °C	20 °C	40 °C	60 °C	80 °C	100 °C	120 °C
Carbon dioxide	13·4	14·6	15·8	16·8	17·7	18·7	19·4	20·1
Helium	18·7	19·4	20·5	21·6	22·5	23·4	23·9	24·4
Hydrogen	8·5	8·9	9·6	9·9	10·2	10·5	10·6	10·7
Alcohol, ethyl	2530	1720	1200	770	570	470	–	–
Castor oil	–	–	–	249 000	81 000	34 000	17 000	–
Crude oil (relative density 0·86)	–	17 000	8100	5300	3800	3000	2400	2000
Glycerine	–	–	–	163 000	43 000	21 000	–	–
Mercury	1820	1670	1580	1440	1380	1330	1260	1150
Paraffin	–	3160	1910	1290	–	–	–	–
Petrol	–	390	300	240	210	–	–	–

A.1.8. Absolute Viscosity of Some Common Fluids

Appendix 2 Values of drag coefficient C_D for various body shapes

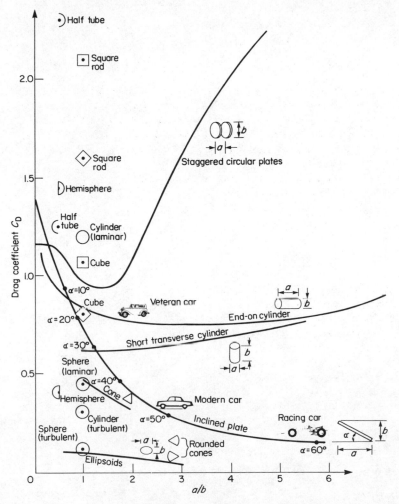

Appendix 2. Values of Drag Coefficient C_D for Various Body Shapes at Re $\simeq 10^5$, based on frontal area, except for the inclined plate, where $A = L \times W$.

$$C_D = 2(\text{Drag})/(\rho\, v^2 A)$$

Flow is from left to right with respect to the body shape indicated. Where not shown, a is the body dimension in the direction of flow and b is perpendicular to it.

Index